Guided Wave Optical Components and Devices

Guided Wave Optical Components and Devices

Basics, Technology, and Applications

Edited by

Bishnu P. Pal
INDIAN INSTITUTE OF TECHNOLOGY DELHI

ELSEVIER
ACADEMIC
PRESS

AMSTERDAM • BOSTON • HEIDELBERG • LONDON • NEW YORK
OXFORD • PARIS • SAN DIEGO • SAN FRANCISCO • SINGAPORE
SYDNEY • TOKYO

Original negative of cover image available from Prof. Wayne Knox,
Director of the Institute of Optics, Univ. of Rochester, (585) 273-5220; E-mail: wknox@optics.rochester.edu

Elsevier Academic Press
30 Corporate Drive, Suite 400, Burlington, MA 01803, USA
525 B Street, Suite 1900, San Diego, California 92101-4495, USA
84 Theobald's Road, London WC1X 8RR, UK

∞ This book is printed on acid-free paper.

Copyright © 2006, Elsevier Inc. All rights reserved.

No part of this publication may be reproduced or transmitted in any form or by any
means, electronic or mechanical, including photocopy, recording, or any information
storage and retrieval system, without permission in writing from the publisher.

Permissions may be sought directly from Elsevier's Science & Technology Rights
Department in Oxford, UK: phone: (+44) 1865 843830, fax: (+44) 1865 853333,
e-mail: permissions@elsevier.com. You may also complete your request on-line
via the Elsevier homepage (http://elsevier.com), by selecting
"Customer Support" and then "Obtaining Permissions."

Library of Congress Cataloging-in-Publication Data
Pal, Bishnu P., 1948-
Guided wave optical components and devices / Bishnu Pal.
    p. cm.
  Includes bibliographical references and index.
  ISBN 0-12-088481-X (alk. paper)
 1. Optoelectronic devices. 2. Integrated optics. 3. Optical wave guides. I. Title.
  TK8304.P35 2005   621.36'92–dc22
                            2005012672

British Library Cataloguing in Publication Data
A catalogue record for this book is available from the British Library

ISBN 13: 978-0-12-088481-0
ISBN 10: 0-12-088481-X

For information on all Academic Press publications
visit our Web site at *www.academicpress.com*

Printed in the United States of America
05 06 07 08 09   9 8 7 6 5 4 3 2 1

**Working together to grow
libraries in developing countries**

www.elsevier.com | www.bookaid.org | www.sabre.org

ELSEVIER   BOOK AID International   Sabre Foundation

Dedicated to

All my teachers, students, professional colleagues, friends, and close relatives for their encouragement and inspiration towards my academic pursuits and above all for their continued affection and support

Bishnu Pal
Editor

Contents

Preface xiii
Contributors xvii

CHAPTER 1
Optical Fibers for Broadband Lightwave Communication: Evolutionary Trends in Designs
B. P. Pal
1

1. INTRODUCTION   1
2. OPTICAL TRANSPARENCY   2
   2.1. Loss Spectrum   2
   2.2. Dispersion Spectrum   3
   2.3. Dispersion Shifted Fibers   8
3. EMERGENCE OF FIBER AMPLIFIERS AND DWDM SYSTEMS   9
   3.1. EDFAs   9
   3.2. DWDM   10
   3.3. Fibers for DWDM Transmission   10
   3.4. Dispersion Compensating Fibers   12
   3.5. Reverse/Inverse Dispersion Fibers   17
4. FIBERS FOR METRO NETWORKS   18
5. COARSE WAVELENGTH DIVISION MULTIPLEXING   21
6. COMBATING PMD IN A FIBER   21
7. CONCLUSION   22
8. ACKNOWLEDGMENTS   22
9. REFERENCES   22

CHAPTER 2
Recent Development of a Polymer Optical Fiber and its Applications
P. L. Chu
27

1. INTRODUCTION   27
2. TYPES OF POFs   27
   2.1. PMMA Fiber   27
   2.2. Deuterated PMMA POF   27
   2.3. Perfluorinated POF   27
3. MANUFACTURE OF POFS   28
   3.1. Preform and Drawing Method   28
   3.2. Extrusion Method   30
4. COMPARISON BETWEEN SILICA FIBER AND POLYMER FIBER   30
   4.1. Difference in Diameters   30
   4.2. Minimum Bend Radius   31
   4.3. Numerical Aperture   31
   4.4. Fiber Bandwidth   31
5. APPLICATIONS OF POFs   31
   5.1. Communication   32
   5.2. Illumination   36
6. POLYMER FIBER GRATINGS   36
7. SEGMENTED CLADDING POF   37
8. DYE-DOPED POLYMER FIBER AMPLIFIER   39
9. CONCLUSIONS   39
10. REFERENCES   40

CHAPTER 3
Microstructured Optical Fibers
T. M. Monro
41

1. FIBERS WITH MICRON-SCALE STRUCTURE   41
2. OVERVIEW OF OPTICAL PROPERTIES   43
   2.1. Introduction   43
   2.2. Nonlinearity Tailoring   44
   2.3. Dispersion   45
   2.4. Polarization   45
   2.5. Air–Light Overlap   46
3. FABRICATION APPROACHES   46
   3.1. Preform Fabrication   46
   3.2. Fiber Drawing   48
   3.3. State-of-the-Art   48
4. FIBER DESIGN METHODOLOGIES   49
   4.1. Effective Index Methods   49
   4.2. Structural Methods   49
   4.3. Predicting Confinement Loss   51
   4.4. Summary   51
5. SILICA HFS   51

5.1. Small-Core Fibers for Nonlinear
       Devices   51
  5.2. Large-Mode Area Fibers for High Power
       Applications   54
  5.3. Active Fibers   55
6. SOFT GLASS FIBERS   58
  6.1. Background   58
  6.2. Extreme Nonlinearity   58
  6.3. New Transmission Fibers   59
  6.4. Solid Microstructured Fibers   60
7. PBGFs   61
8. CONCLUSION AND THE
   FUTURE   63
9. ACKNOWLEDGMENTS   64
10. REFERENCES   64

CHAPTER 4

Photonic Bandgap-Guided Bragg Fibers
S. Dasgupta, B. P. Pal and M. R. Shenoy
71

1. INTRODUCTION   71
2. BRAGG FIBERS   72
  2.1. Bandgap in One-Dimensional Periodic
       Medium   72
  2.2. Light Propagation in Bragg Fibers   75
  2.3. Modal Characteristics   76
3. DISPERSION COMPENSATING BRAGG
   FIBER   78
4. BRAGG FIBERS FOR METRO
   NETWORKS   79
5. FABRICATION   80
6. CONCLUSION   81
7. REFERENCES   81

CHAPTER 5

Radial Effective Index Method for the Analysis of Microstructured Fibers
K. S. Chiang and V. Rastogi
83

1. INTRODUCTION   83
2. THE REIM   84
  2.1. Formulation of the Method   84
  2.2. Determination of the Effective Index
       Profile   84
3. SEGMENTED CLADDING
   FIBER   85
4. HOLEY FIBER   87
5. CONCLUSION   88
6. ACKNOWLEDGMENT   89
7. REFERENCES   89

CHAPTER 6

Some Important Nonlinear Effects in Optical Fibers
K. Thyagarajan and A. Ghatak
91

1. INTRODUCTION   91
2. NONLINEAR POLARIZATION   91
3. THIRD-ORDER NONLINEAR
   EFFECTS   91
  3.1. SPM   92
  3.2. Propagation of a Pulse   93
  3.3. Spectral Broadening due to SPM   94
  3.4. XPM   95
  3.5. FWM   98
4. CONCLUSIONS   100
5. REFERENCES   100

CHAPTER 7

Fiber Optic Parametric Amplifiers for Lightwave Systems
F. Yaman, Q. Lin and G. P. Agrawal
101

1. INTRODUCTION   101
2. THEORY OF FWM   101
3. SINGLE-PUMP PARAMETRIC
   AMPLIFIERS   103
4. DUAL PUMP PARAMETRIC
   AMPLIFIERS   107
5. FLUCTUATIONS OF ZDWL   109
6. EFFECT OF RESIDUAL FIBER
   BIREFRINGENCE   111
7. SUMMARY   114
8. ACKNOWLEDGMENTS   114
9. REFERENCES   114

CHAPTER 8

Erbium-doped Fiber Amplifiers
K. Thyagarajan
119

1. INTRODUCTION   119
2. EDFA   119
3. POPULATION INVERSION AND
   OPTICAL AMPLIFICATION   120
4. OPTICAL AMPLIFICATION IN
   EDFAs   121
5. GAIN FLATTENING OF EDFAs   124
  5.1. Gain Flattening Using External
       Filters   124
  5.2. Intrinsically Flat Gain Spectrum   125
6. NOISE IN AMPLIFIERS   126
7. EDFAs for the S-Band   128

8. CONCLUSIONS   129
9. ACKNOWLEDGMENTS   129
10. REFERENCES   129

CHAPTER 9

Fiber Optic Raman Amplifiers
G. P. Agrawal
131

1. INTRODUCTION   131
2. FUNDAMENTAL CONCEPTS   131
   2.1. Raman Gain Spectrum   132
   2.2. Simple Theory   134
   2.3. Gain Saturation   135
3. MODERN RAMAN AMPLIFIERS   136
   3.1. Broadband Raman Amplifiers   136
   3.2. Design of Raman Amplifiers   137
4. PERFORMANCE LIMITING FACTORS   138
   4.1. Spontaneous Raman Scattering   138
   4.2. Effective Noise Figure   140
   4.3. Rayleigh Backscattering   142
   4.4. Pump-Noise Transfer   143
   4.5. Effects of PMD   145
5. AMPLIFICATION OF OPTICAL PULSES   147
   5.1. Pulse-Propagation Equations   147
   5.2. Effects of Group-Velocity Mismatch   148
   5.3. Anomalous Dispersion Regime   149
   5.4. Normal Dispersion Regime   150
6. REFERENCES   151

CHAPTER 10

Application of Numerical Analysis Techniques for the Optimization of Wideband Amplifier Performances
N. Park, P. Kim, H. Lee and J. Park
155

1. FOREWORD   155
2. POWER EFFICIENCY: L-BAND EDFA   155
   2.1. Introduction   155
   2.2. Pump Wavelength Detuning   156
   2.3. Fiber Structural Detuning   157
   2.4. Conclusion   160
3. GAIN ENGINEERING: RAMAN AMPLIFIER   160
   3.1. Introduction   160
   3.2. Implementation of the Closed Form Raman Equation   160
   3.3. Application Example 1: Gain Prediction   162
   3.4. Application Example 2: Raman Gain Engineering—The Inverse Scattering Problem   163
   3.5. Application Example 3: Channel Reconfiguration   166
   3.6. Application Example 4: Analytic Solution for the Gain Clamping Problem   167
   3.7. Conclusion   169
4. TRANSIENT THULIUM-DOPED FIBER AMPLIFIER   169
   4.1. Introduction   169
   4.2. Average Inversion Analysis of TDFA Transient: Comparison with Experiment   169
   4.3. Conclusion   171
5. CONCLUSION   171
6. REFERENCES   171

CHAPTER 11

Analog/Digital Transmission with High-Power Fiber Amplifiers
P. Dua, K. Lu, N. K. Dutta and J. Jaques
173

1. INTRODUCTION   173
2. EXPERIMENT   173
   2.1. Analog Transmission   173
   2.2. Hybrid Digital/Analog Transmission   176
   2.3. Gain Tilt Measurement of the Er/Yb Co-Doped DCFA   177
3. RESULTS   178
4. REFERENCES   180

CHAPTER 12

Erbium-doped Fiber Amplifiers for Dynamic Optical Networks
A. Srivastava and Y. Sun
181

1. INTRODUCTION   181
2. EDFAS FOR HIGH CAPACITY NETWORKS   181
   2.1. Basic Characteristics of EDFAs   182
   2.2. System Issues   185
   2.3. Dynamic Network Related Issues   186
3. EDFAS FOR DYNAMIC NETWORKS   187
   3.1. Gain Dynamics of Single EDFA   187
   3.2. Fast Power Transients in EDFA Chains   190
   3.3. System Impairments due to Transients   192
   3.4. Channel Protection Schemes   195
4. ACKNOWLEDGMENTS   200
5. REFERENCES   201

CHAPTER 13

Fused Fiber Couplers: Fabrication, Modeling, and Applications
B. P. Pal, P. R. Chaudhuri, M.R. Shenoy and N. Kumar
205

1. INTRODUCTION 205
2. FABRICATION 205
3. MODELING 207
   3.1. Mode Analysis Algorithm 208
   3.2. Modeling the Propagation in the Coupling Region 210
   3.3. Supermodes and Beating 211
   3.4. Polarization Characteristics 212
   3.5. Results and Discussions 212
4. APPLICATIONS: FFC-BASED ALL-FIBER COMPONENTS 214
   4.1. Beam Splitter/Combiner 214
   4.2. WDM Coupler 215
   4.3. Principle of Operation of Classical WDM 216
   4.4. Wavelength Interleaver 216
   4.5. Fiber Loop Reflector 218
5. SUMMARY 221
6. ACKNOWLEDGMENTS 222
7. REFERENCES 222

CHAPTER 14

Side-Polished Evanescently Coupled Optical Fiber Overlay Devices: A Review
W. Johnstone
225

1. INTRODUCTION 225
2. PRINCIPLES OF OPERATION 226
3. DEVICES 227
4. APPLICATIONS 229
5. CONCLUSIONS 230
6. REFERENCES 231

CHAPTER 15

Optical Fiber Gratings
K. Thyagarajan
233

1. INTRODUCTION 233
2. FIBER BRAGG GRATINGS 233
   2.1. Coupled-Mode Theory for FBG 235
   2.2. Phase Matched Interaction 236
   2.3. Nonphase Matched Interaction 236
3. SOME APPLICATIONS OF FBGS 237
   3.1. Add/Drop Multiplexers 237
   3.2. Dispersion Compensation 238
4. LONG PERIOD GRATINGS 239
   4.1. Coupled-Mode Theory for LPG 239
5. SOME APPLICATIONS OF LPGs 240
   5.1. WDM Filter 240
   5.2. Broadband LPGs 241
   5.3. Gain Flattening of EDFAs 241
6. CONCLUSIONS 241
7. REFERENCES 242

CHAPTER 16

Enhancing Photosensitivity in Optical Fibers
N. K. Viswanathan
243

1. INTRODUCTION 243
2. UV SENSITIZATION OF FIBERS 243
   2.1. Effect of Ge Concentration 244
   2.2. Thermal Stability of Bragg Gratings 245
   2.3. Indication of Recirculating Catalyst 247
3. DILUTE H_2 SENSITIZATION OF FIBERS 247
   3.1. Comparison with Standard H_2-Loaded Fibers 248
   3.2. Effect of Diluent Gas 249
4. CONCLUSION 249
5. ACKNOWLEDGMENTS 250
6. REFERENCES 250

CHAPTER 17

Solitons in Fiber Bragg Grating
K. Porsezian and K. Senthilnathan
251

1. INTRODUCTION 251
   1.1. Optical Fiber Communications 251
   1.2. Soliton-Based OFC 251
2. LINEAR EFFECTS 252
   2.1. Optical Loss 252
   2.2. Chromatic Dispersion 252
3. NONLINEAR EFFECTS 253
   3.1 Kerr Nonlinearity 253
   3.2. Self-Steepening 254
   3.3 Stimulated Inelastic Scattering 254
   3.4. Effect of Birefringence 255
4. OPTICAL SOLITONS IN PURE SILICA FIBER 255
   4.1. Why Solitons in FBG? 257
5. FUNDAMENTALS OF FBG 257
   5.1. Introduction 257
   5.2. Types of Grating 258
   5.3. Properties of FBG 258
6. SOLITONS IN FBG 266
7. MAP SOLITONS 267
   7.1. Gap Solitons in Kerr Media 267
   7.2. Gap Solitons in Quadratic Media 271
8. BLOCH WAVE ANALYSIS 272

8.1. Bragg Grating Solitons   272
9. RESULTS AND DISCUSSION   278
   9.1. Experimental Considerations   278
10. ACKNOWLEDGMENTS   279
11. REFERENCES   279

CHAPTER 18

Advances in Dense Wavelength Division Multiplexing/Demultiplexing Technologies

V. Bhatia

281

1. INTRODUCTION   281
2. KEY PERFORMANCE CHARACTERISTICS   281
   2.1. Thin Film Filters   282
   2.2. Arrayed Waveguide Gratings   284
   2.3. Fiber Bragg Gratings   286
3. OPTICAL INTERLEAVERS   287
4. DISCUSSION   288
5. CONCLUSION   289
6. REFERENCES   289

CHAPTER 19

Dispersion-Tailored Higher Order Mode Fibers for In-Fiber Photonic Devices

S. Ramachandran

291

1. INTRODUCTION   291
2. DISPERSIVE PROPERTIES OF FEW-MODE FIBERS   293
3. MODE CONVERSION WITH LPGS: DEVICE PRINCIPLES   294
4. LPGS IN DISPERSION-TAILORED FEW-MODE FIBERS   295
   4.1. Broadband Mode Converters   295
   4.2. Bandwidth Control of TAP-LPGs   298
   4.3. Spectrally Flat Coupling for VOA   299
5. STATIC DEVICES USING TAP-LPGs   300
   5.1. HOM-DCM   300
   5.2. Dispersionless Bandpass Filtering   301
6. TUNABLE LPGS IN HOM FIBERS   304
   6.1. Amplitude Modulation: Novel Detuning Effects in TAP-LPGs   304
   6.2. Switching and Routing   305
7. TUNABLE TAP-LPG DEVICES   306
   7.1. Tunable/Adjustable HOM-DCMs   306
   7.2. Polarizers and PDL Controllers   308
8. CONCLUSION   309
9. ACKNOWLEDGMENTS   309
10. REFERENCES   309

CHAPTER 20

Acousto-Optic Interaction in Few-Mode Optical Fibers

H. E. Engan and K. Bløtekjær

311

1. INTRODUCTION   311
2. OPTICAL PROPERTIES   311
3. ACOUSTIC PROPERTIES   313
4. ACOUSTO-OPTIC INTERACTION   315
   4.1. Principles   315
   4.2. Experimental Setup   316
   4.3. Frequency Shifter   318
   4.4. Wavelength Dependence and Tunable Filters   318
   4.5. Fiber Nonuniformity   319
   4.6. Scanning Heterodyne Interferometer   320
5. PRACTICAL CONSIDERATIONS   321
6. INDUSTRIALIZATION   322
7. CONCLUSIONS   322
8. ACKNOWLEDGMENTS   323
9. REFERENCES   324

CHAPTER 21

Basic Theory and Design Procedures for Arrayed Waveguide Structures

C. R. Doerr

325

1. INTRODUCTION   325
2. ARRAYED WAVEGUIDE LENS   325
3. THE STAR COUPLER   325
4. WAVEGUIDE GRATING ROUTER   327
5. MUTUAL COUPLING-INDUCED ABERRATIONS   330
6. EXAMPLE DESIGN: DEMULTIPLEXER   331
7. EXAMPLE DESIGN: BAND DEMULTIPLEXER   332
8. CONCLUSION   333
9. ACKNOWLEDGMENTS   333
10. REFERENCES   334

CHAPTER 22

Photobleached Gratings in Electro-Optic Waveguide Polymers

V. N. P. Shivashankar, E. M. McKenna and A. R. Mickelson

335

1. INTRODUCTION   335
2. INTEGRATED OPTICS IN EO POLYMERS   335

2.1. The EO Effect   335
2.2. The Glass Transition Temperature   336
2.3. Poling Lifetime Issues   337
2.4. High Temperature Polymers   338
3. PHOTOBLEACHING DYE-DOPED POLYMERS   338
   3.1. Early Experiments   339
   3.2. A Theory of Radiation–Induced Chemical Reactions   339
   3.3. A Photobleaching Model for Dye-Doped Polymers   340
   3.4. Mechanical Effects of Photobleaching   341
4. DIFFRACTION GRATINGS IN DYE-DOPED WAVEGUIDE POLYMERS   343
   4.1. Diffraction Efficiency of Thin Sinusoidal Gratings   343
   4.2. Writing Gratings in Waveguide Polymers   344
   4.3. Some Study of Transient Gratings   345
   4.4. Describing the Grating Formation Process   346
   4.5. Irreversible Gratings in Waveguide Polymers   347
5. CONCLUSION   348
6. REFERENCES   349

CHAPTER 23

Optical MEMS using Commercial Foundries
D. Uttamchandani
353

1. INTRODUCTION   353
2. OPTICAL CHOPPERS FOR FIBER OPTIC APPLICATIONS   354
   2.1. Fabrication of MEMS Chopper   354
   2.2. Mechanical Design Considerations   355
   2.3. Optical Design Considerations   356
   2.4. Experimental Evaluation   357
3. THREE-DIMENSIONAL VARIABLE OPTICAL ATTENUATOR   360
   3.1. Self-Assembly Applied to a MEMS VOA   362
   3.2. Design Parameters   362
   3.3. Device Fabrication   363
   3.4. VOA Performance   363
4. HYBRID TUNABLE FILTER   365
   4.1. Component Characteristics and Fabrication   366
   4.2. Experiments and Results   366
5. CONCLUSIONS   368
6. ACKNOWLEDGMENTS   369
7. REFERENCES   369

CHAPTER 24

Principles of Fiber Optic Sensors
B. Culshaw
371

1. INTRODUCTION   371
2. FIBER OPTIC SENSORS: THE BASIC PRINCIPLE   371
3. FIBER OPTICS IN PHYSICAL SENSING   374
4. CHEMICAL SENSORS   376
5. CHEMICAL SENSORS: SOME APPLICATION CASE STUDIES   379
   5.1. Multiplexed Fiber Optic Spectroscopy   379
   5.2. Olive Oil   380
   5.3. Distributed Chemical Sensing   383
6. CONCLUSIONS   386
7. REFERENCES   387

CHAPTER 25

Structural Strain and Temperature Measurements Using Fiber Bragg Grating Sensors
W. Jin, T. K. Y. Lee, S. L. Ho, H. L. Ho, K. T. Lao, L. M. Zhou and Y. Zhou
389

1. INTRODUCTION   389
2. STRAIN MEASUREMENT IN A COMPOSITE-STRENGTHENED CONCRETE BAR   390
3. DYNAMIC STRAIN MEASUREMENT OF A COMPOSITE SAMPLE USING AN EMBEDDED FBG SENSOR   391
4. TEMPERATURE MEASUREMENT ON A HEATED CYLINDER IN A CROSS-FLOW   392
5. MULTI-POINT STRAIN MEASUREMENT OF KOWLOON CANTON RAILWAY TRAIN BODY SHELL   396
6. SUMMARY   399
7. ACKNOWLEDGMENTS   399
8. REFERENCES   399

CHAPTER 26

Principles and Status of Actively Researched Optical Fiber Sensors
B. Lee, Y. W. Lee and M. Song
401

1. INTRODUCTION   401
2. FIBER GRATING SENSORS   401
3. FIBER OPTIC GYROSCOPES   409
4. FIBER OPTIC CURRENT SENSORS   414

5. OTHER SENSORS 416
6. CONCLUSION 418
7. REFERENCES 418

Author Biography 425
Index 437

Preface

Optical fibers, from the point of view of telecommunication, is now taken for granted in view of its wide-ranging application as the most suitable singular transmission medium for voice, video, and data signals. In the last two decades, the world has witnessed an enormous growth in Lightwave communication and associated technologies. Initial revolution in this field centered on achieving optical *transparency* in terms of exploiting the low-loss and low-dispersion transmission wavelength windows of high-silica optical fibers. The development of *broadband* optical fiber *amplifiers* in the form of Erbium Doped Fiber Amplifiers (EDFA), which led to the birth of the era of *dense wavelength* division *multiplexing* (DWDM) technology in the mid-1990s, ushered in the next revolution in fiber optics. Two other competing amplifying devices have been Raman fiber amplifiers and semiconductor amplifiers. The DWDM technology has revolutionized the backbone networks by exceeding twice its capacity each year. Fueled by the Internet, this technology has already penetrated the metro network, where it can provide a flexible and scalable support to a broad range of services for enterprise, access and business-to-business applications. These relatively recent developments in optical transmission technology have given rise to demand for newer fiber designs. Chapter 1 gives an exposition on the evolution of dispersion-tailored single-mode fiber designs for broadband telecommunication applications in the long haul and metro networks. Recent years have also witnessed remarkable progress on polymer/plastic fibers. Plastic fibers offer great potential for niche applications in connector-intensive optical networks, e.g. enterprise/business, automobile, back plane interconnection as well as an alternative signage technology in place of neon signs. Chapter 2 describes recent developments in plastic fiber technology from various emerging applications point of view. New generation microstructured optical fibers like holey, Photonic Crystal Fibers (PCF) and Bragg fibers offer new potentials to realize a variety of optical components, not necessarily for telecommunications alone. Chapters 3-5 are devoted to describing evolution and techniques to model these emerging new generation optical fibers and their potential applications. As is often true with any new technology, DWDM technology has also posed certain problems for system designers. Due to multi-wavelength transmission, it supports greater throughput of optical power, which under certain circumstances could induce nonlinear effects such as four wave mixing or cross phase modulation thereby leading to signal impairments. In chapter 6 basics of such nonlinear effects in fibers are described. Under certain circumstance these nonlinearity-induced effects could be exploited to configure potentially interesting devices as is spelled out in chapter 7, which discusses fiber-optic parametric amplifiers for Lightwave systems.

As mentioned above EDFA, has been indeed the driver for DWDM revolution. Chapters 8-12 are devoted to fiber amplifiers – chapter 8 on the basic functional principle of an EDFA, chapter 9 on the fundamentals of Raman fiber amplifiers (RFA), chapter 10 on numerically modeling EDFAs and RFAs, chapter 11 on use of high power EDFAs for CATV applications, and chapter 12 on dynamic control of EDFAs to overcome transient effects encountered during add-drop of signal wavelength channels.

In addition to optical fibers and amplifiers, today's optical networks require a host of *branching components* that distribute optical signals in a network both in a pre-determined as well as dynamic fashion. Several of these components are passive, data-format transparent, and are able to combine/split optical power or multiplex wavelengths, regardless of the information content of the signal. Three generic technology platforms are exploited to realize many such components in an *all-fiber* form namely, fused bi-conical tapered fiber couplers, side-polished fiber half-couplers, and in-fiber gratings. Chapters 13-16 are presented to introduce and describe to readers these technologies and realizable branching components thereof. In an all-fiber form, such devices could be introduced in a communication link through splicing to the transmission fiber with relatively low insertion loss. Nonlinear effects in Bragg gratings e.g. solitons propagation are discussed in chapter 17. Chapter 18 makes a comparative study on advances in different DWDM multiplexing technologies including the very popular one based on multilayer thin film filters while chapter 19 introduces readers to the intense emerging interest on dispersion tailored higher order mode (HOM) or two-mode in-fiber devices. Acousto-optic interaction in such few mode fibers, which have been exploited to realize several fiber-based optical components e.g. frequency shifters, tunable filters, and scanning heterodyne interferometers including a commercial device for gain flattening of an EDFA's non-uniform gain spectrum, is described in chapter 20.

Remarkable progress in the field of integrated optics in recent years has also led to development and commercialization of various guided wave optical components – one of which finds extensive applications in DWDM optical networks as wavelength multiplexers, routers, etc is known as arrayed waveguide (AWG) structures based around silica-on-silicon technology platform. Chapter 21 presents a semi-tutorial introduction to AWG structures. Polymer-based integrated optics has generated great deal of interest in recent years. Chapter 22 that follows describes the fabrication and underlying physics of photo bleached gratings in electro-optic polymer waveguides. This process of combining optical wave guidance with holographic processing of light is a relatively recent development, which holds promises to yield several signal processing functional components.

Chapter 23 forms a tutorial on commercial foundry-based MEMS (micro-electro-mechanical systems) technology, which could yield a host of micro-sensors and actuators, as well as certain Microsystems consisting of assemblies of moving parts that could range in size from a few to a few hundred microns. It became evident since the mid-1990s that one could derive great benefit from the huge investment and subsequent progress that has been made in Photonics to develop the field of optical-MEMS (O-MEMS), which has already yielded components like variable attenuators, filters, switches, optical cross-connects, add-drop multiplexers and so on

Besides telecommunication, optical fibers also find extensive applications in a variety of sensing environments. In fact, basic ideas on optical fiber-sensing first emerged nearly 40 years ago – predating the first realization that optical fibers could be used for optical signal transmission. Optical fiber sensors have already assumed great importance vis-Á-vis other sensor technologies especially from the point of view of distributed measurements and immunity from electromagnetic interference. Chapter 24 is devoted to a review of the scientific and technological principles that underlie optical fiber sensor technology and it also explores the potential applications of more recent research results in this area. Health monitoring of civil and aerospace structures through in-fiber grating-based sensor technology is described in chapter 25. Finally chapter 26 deals with details of relatively more actively researched specific fiber sensor technologies namely, fiber Bragg gratings, fiber gyroscopes, and fiber optic current sensors besides brief description of few other fiber sensors like the ones based on measurements through Optical Time Domain Reflectometer (OTDR).

Each chapter, written by leaders in the field, is aimed to provide a semi-tutorial introduction to state-of-the-art and exposition towards the emerging applications of guided wave optical components and devices. It is hoped that the book should benefit both graduate students and young researchers interested in learning about guided wave optical components as an introductory material containing basics and applications as well as an advanced source book for reference purposes and class room use.

Acknowledgments

The book was conceived on one evening at the suggestion of Prof. Ajoy Ghatak, who has been one of my Ph.D. thesis advisors and currently a close colleague while I was having dinner with him in a restaurant at a city known as Vadodara in the western part of India in December 2003. I am indebted to Professor Ghatak for this and also for his constructive criticisms. The final die for the book was cast within the next few days during a short visit to our Institute by Prof. Govind Agrawal, who is one of the editors of the Elsevier Science and Technology series on Optics and Photonics. I am indeed grateful to Govind for his keen interest on this project; Govind and I were graduate students at

IIT Delhi at the same time almost 34 years ago! Thanks are due to Prof. Wayne Knox of Institute of Optics at the University of Rochester for taking the cover photograph of super-continuum generation in his laboratory from a tapered holey fiber (detailed in F. Lu, Y. Deng, and W.H. Knox, *Opt. Letts.* pp. 1566–1568, 2005).

I gratefully acknowledge academic interactions with several of my colleagues in our Fiber Optics group and with my graduate students over the last two decades. In addition to Professor Ghatak, I wish to mention K. Thyagarajan and graduate student Ms. Kamma Pande for their constructive criticism and help with drawing some of the figures and dispersion calculation. My colleague Dr. R.K. Varshney is thanked for his help with the figure on group delay vs wavelength. I also wish to thank Dr. R.S. Vodhanel of the Research Division of Corning Inc. and Dr. D. Bayart of the Research and Innovations Division of Alcatel and his colleague Dr. P. Sansonetti at Draka Comteq (Marcousse) for their help with the description of metrocentric fibers.

I thank my wife Subrata for her tremendous patience and encouragement without which it would not have been possible to complete the project in such a short time. Finally I thank my daughter Parama, who is pursuing graduate studies at the Institute of Optics at Rochester (NY) for her help in editing certain portions of the book.

Bishnu Pal
Editor

Contributors

Govind P. Agrawal,[*] (101, 131)
F. Yaman, Q. Lin (101)
Institute Of Optics
Rochester, New York, USA
E-mail: gpa@optics.rochester.edu

Vikram Bhatia (281)
Avanex Corporation
Painted Post, New York, USA
E-mail: vikram_bhatia@avanex.com

Pak L. Chu (27)
Optoelectronics Research Center
Department of Electronic Engineering
City University of Hong Kong
Kowloon, Hong Kong
E-mail: eepchu@cityu.edu.hk

Brian Culshaw (371)
Electronic & Electrical Engineering Department
University of Strathclyde
Glasgow, Scotland
E-mail: b.culshaw@eee.strath.ac.uk

Kin S. Chiang[*] (83)
Department of Electronic Engineering
City University of Hong Kong
Kowloon, Hong Kong
E-mail: eeksc@cityu.edu.hk

V. Rastogi (83)
Physics Department
IIT Roorkee
Uttaranchal, India
E-mail: vipulfph@iitr.ernet.in

Bishnu P. Pal[*] (1, 71, 205)
Sonali Dasgupta, M.R. Shenoy (71)
P. Roy Chaudhuri, M.R. Shenoy, and Naveen Kumar (205)
Physics Department
IIT Delhi, Hauz Khas
New Delhi, India
E-mail: bppal@physics.iitd.ernet.in

Christopher R. Doerr (325)
Lucent Technologies
Bell Laboratories
Holmdel, New Jersey, USA
E-mail: crdoerr@lucent.com

Puneit Dua, Kunzhong Lu, Niloy K. Dutta[*] (173)
Department of Physics
University of Connecticut
Storrs, Connecticut, USA
E-mail: niloy@engr.uconn.edu

James Jaques (173)
Lucent Technologies
Government Communications Lab
Murray Hill, New Jersey, USA

Helge E. Engan,[*] **Kjell Bløtekjær** (311)
Department of Electronics & Telecommunications
Norwegian University of Science and Technology O.S.
Trondheim, Norway
E-mail: helge.engan@iet.ntnu.no

Wei Jin,[*] **T.K.Y. Lee, S.L. Ho, H.L. Ho** (389)
Department of Electrical Engineering
The Hong Kong Polytechnic University
Hong Kong
E-mail: eewjin@polyu.edu.hk

K.T. Lau, L.M. Zhou and Y. Zhou (389)
Department of Mechanical Engineering
The Hong Kong Polytechnic University
Hong Kong

Walter Johnstone (225)
Department of Electronic and Electrical Engineering
University of Strathclyde
Glasgow, Scotland
E-mail: w.johnstone@eee.strath.ac.uk

Byoungho Lee,* **Yong Wook Lee** (401)
School of Electrical Engineering
Seoul National University
Seoul, Korea
E-mail: byoungho@snu.ac.kr

Minho Song (401)
Division of Electronics and Information Engineering
Chonbuk National University
Jeonju, Korea

Tanya M. Monro (41)
School of Chemistry & Physics
University of Adelaide SA
Adelaide, Australia
E-mail: tanya.monro@adelaide.edu.au

Namkyoo M. Park,* **P. Kim, H. Lee and J. Park** (155)
Optical Communication Systems Laboratory
School of Electrical Engineering
Seoul National University
Seoul, South Korea
E-mail: nkpark@plaza.snu.ac.kr

K. Porsezian (251)
C/O J. Herrmann
Max Born Institute for Nonlinear Optics & Short Pulse Spectroscopy
Berlin, Germany

K. Senthilnathan (251)
Department of Physics
Anna University
Chennai, India
E-mail: ponzsol@yahoo.com

Siddharth Ramachandran (291)
OFS Laboratories
Murray Hill, New Jersey, USA
E-mail: sidr@ofsoptics.com

V. N. P. Sivashankar, E. M. McKenna, Alan R. Mickelson* (335)
Department of Electrical and Computer Engineering
University of Colorado at Boulder
Boulder, Colorado, USA
E-mail: mickel@schof.colorado.edu

Atul Srivastava (181)
Bookham Technology
Morganville, New Jersey, USA
E-mail: asrivastava@bookham.com

Yan Sun (181)
Onetta Inc.
Piscataway, New Jersey, USA

K. Thyagarajan (91, 119, 233)
Ajoy K. Ghatak (91)
Physics Department
IIT Delhi
New Delhi, India
E-mail: ktrajan@physics.iitd.ac.in

Deepak Uttamchandani (353)
Deptartment of E&EE
University of Strathclyde
Glasgow, Scotland
E-mail: du@eee.strath.ac.uk

Nirmal K. Viswanathan (243)
Photonics Division
IRDE
Dehradun, India
E-mail: nirmal@irde.res.in

1
CHAPTER

Optical Fibers for Broadband Lightwave Communication: Evolutionary Trends in Designs

Bishnu P. Pal*
Physics Department
Indian Institute of Technology Delhi
New Delhi, India

1. INTRODUCTION

Development of optical fiber technology is considered to be a major driver behind the information technology revolution and the tremendous progress on global telecommunications that has been witnessed in recent years. Fiber optics, from the point of view of telecommunication, is now almost taken for granted in view of its wide-ranging application as the most suitable singular transmission medium for voice, video, and data signals. Indeed, optical fibers have now penetrated virtually all segments of telecommunication networks, whether transoceanic, transcontinental, intercity, metro, access, campus, or on-premise. The first fiber optic telecom link went public in 1977. Since that time, growth in the lightwave communication industry until about 2000 has been indeed mind boggling. According to a Lucent technology report [1], in the late 1990s optical fibers were deployed at approximately 4800 km/hr, implying a total fiber length of almost three times around the globe each day until it slowed down when the information technology bubble burst!

The Internet revolution and deregulation of the telecommunication sector from government controls, which took place almost globally in the recent past, have substantially contributed to this unprecedented growth within such a short time, which was rarely seen in any other technology. Initial research and development (R&D) in this field had centered on achieving optical *transparency* in terms of exploitation of the *low-loss* and *low-dispersion* transmission wavelength windows of high-silica optical fibers. Though the low-loss fiber with a loss under 20 dB/km that was reported for the first time was a single-mode fiber (SMF) at the He–Ne laser wavelength [2], the earliest fiber optic lightwave systems exploited the first low-loss wavelength window centered on 820 nm with graded index multimode fibers forming the transmission media. However, primarily due to the unpredictable nature of the bandwidth of jointed multimode fiber links, since the early 1980s the system focus shifted to SMFs by exploiting the zero material dispersion characteristic of silica fibers, which occurs at a wavelength of 1280 nm [3] in close proximity to its second low-loss wavelength window centered at 1310 nm [4].

The next revolution in lightwave communication took place when *broadband* optical fiber *amplifiers* in the form of erbium-doped fiber amplifiers (EDFA) were developed in 1987 [5], whose operating wavelengths fortuitously coincided with the lowest-loss transmission wavelength window of silica fibers centered at 1550 nm [6] and heralded the emergence of the era of *dense wavelength division multiplexing* (DWDM) technology in the mid-1990s [7]. By definition, DWDM technology implies simultaneous optical transmission through one SMF of at least four wavelengths within the gain bandwidth of an EDFA (Fig. 1.1). Recent development of the so-called AllWave™ and SMF-28e™ fibers devoid of the characteristic OH^- loss peak (centered at 1380 nm) extended the low-loss wavelength window in high-silica fibers from 1280 nm (235 THz) to 1650 nm (182 THz), thereby offering, in principle, an enormously broad 53 THz of optical transmission bandwidth to be potentially tapped through the DWDM technique! These fibers are usually referred to as enhanced SMF (G.652.C) and are characterized with an additional low-loss window in the E-band (1360–1460 nm), which is about 30% more than the two low-loss windows centered about 1310 and 1550 nm in legacy SMFs. The emergence of DWDM technology has also driven development of various

*E-mail: bppal@physics.iitd.ernet.in

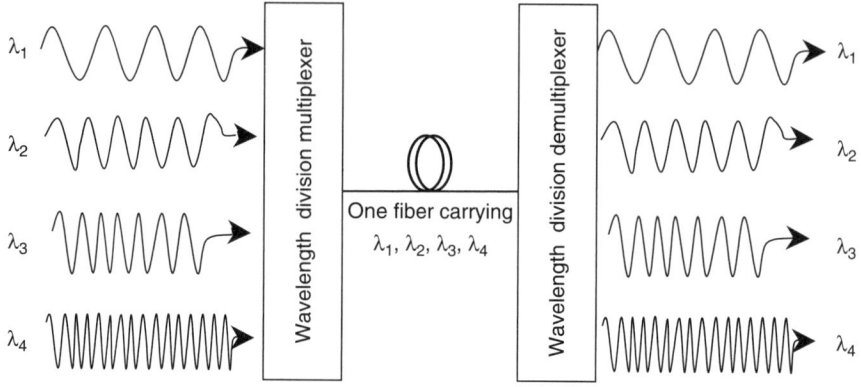

FIGURE 1.1 Schematic representing DWDM optical transmission with a minimum of four wavelengths in the EDFA band as a sample.

specialty fibers and all-fiber components for seamless growth of the lightwave communication technology. These fibers were required to address new features like nonlinearity-induced potential impairments in optical transmission due to large optical throughput, broadband dispersion compensation, bend-loss sensitivity to variation in signal wavelengths, and so on.

In this chapter we attempt to present evolutionary trends in the design of single-mode optical transmission fibers, in particular for lightwave communication seen in the last 30 years or so. Multimode fibers in the form of plastic fibers, which hold promise for on-premise and other applications, are discussed in Chapter 2 authored by P. L. Chu.

2. OPTICAL TRANSPARENCY

2.1. Loss Spectrum

SMFs constitute an integral component of any DWDM link meant to transport high volumes of signals and data. Its characteristics greatly influence the choice of auxiliary components that go into a network and also the overall cost and eventual performance of the communication system. *Loss* and *dispersion spectra* are the two most important propagation characteristics of a single-mode optical fiber. Figure 1.2 gives an example of the loss spectrum of a state-of-the-art, commercially available, conventional G.652 type of SMF. Except for a portion of the loss spectrum around 1380 nm at which a peak appears due to absorption by minute traces (in parts per billion) of OH^- present in the fiber, the rest of the spectrum in a G.652 fiber more or less varies, with wavelength as $A\lambda^{-4}$, meaning that signal loss in a state-of-the-art SMF is essentially caused by Rayleigh scattering. Rayleigh scattering loss coefficient A in a fiber may be approximately modeled through the relation [8]

$$A = A_1 + A_2\Delta^n \quad (1)$$

where $A_1 = 0.7\,dB\text{-}\mu m^4/km$ in fused silica and $A_2 \approx 0.4 - 1\,dB\text{-}\mu m^4/km$, whereas Δ is the relative core-cladding index difference expressed in % and exponent $n = 0.7 - 1$, whose value depends on the dopant. With GeO_2 as the dopant and $\Delta \approx 0.37\%$, estimated Rayleigh scattering loss in a high-silica fiber is about 0.18–0.2 dB/km at 1550 nm. Superimposed on this curve over the wavelength range of 1360–1460 nm (often referred to as the E-band) is a dotted curve, which is devoid of any peak but otherwise overlaps with rest of the loss spectrum; this modified spectrum corresponds to that of a low water peak fiber (LWPF), classified by the International

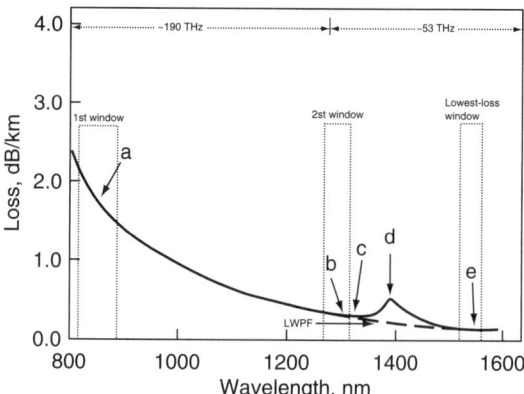

FIGURE 1.2 A sample loss spectrum (full curve) of a state-of-the-art G.652 type single-mode fiber, e.g. SMF-28 (adapted from Corning product catalog © Corning Inc.): (a) 1.81 dB/km at 850 nm, (b) 0.35 dB/km at 1300 nm, (c) 0.34 dB/km at 1310 nm, (d) 0.55 dB/km at 1380 nm, (e) 0.19 dB/km at 1550 nm. The dashed portion of the curve corresponds to that of an LWPF due to reduction of the OH^- peak in enhanced SMF; available theoretical transmission bandwidths at different low loss spectral windows are also shown.

Telecommunications Union (ITU) as G.652.C fibers, examples of which are AllWave™/SMF-28e™ fibers. LWPFs, also referred to as enhanced SMFs, opened up an additional 30% transmission bandwidth over and above standard SMFs of the G.652 type.

In real-world systems, however, there are other sources of loss, which are required to be budgeted on a case-to-case basis and which could add up to more than the inherent loss. Examples of these are cabling-induced losses due to microbending, bend-induced losses along a fiber route, and losses due to splices and connectors, including those involving insertion of various components in a fiber link. Several of these also depend to a certain extent on the refractive index profile of the fiber in question. Detailed studies have indicated that these extraneous sources of loss could be addressed by optimizing mode field radius (W_P, known as Petermann spot size) and effective cutoff wavelength (λ_{ce}) [9]. The parameter W_P effectively determines transverse offset-induced loss at a fiber splice and sensitivity to microbend-induced losses, and λ_{ce} essentially determines sensitivity to bend-induced loss in transmission. Both of these are well-known important characteristic parameters of a SMF [10]. For operating at 1310 nm, optimum values of these parameters turned out to be $4.5 < W_P(\mu m) < 5.5$ and $1100 < \lambda_{ce}$ (nm) < 1280 [11]. An indirect way to test that λ_{ce} indeed falls within this range is to determine whether the measured excess loss of 100 turns of the fiber loosely wound around a cylindrical mandrel of diameter 7.5 cm falls below 0.1 dB at 1310 nm and below 1.0 dB at 1550 nm [8].

In the early 1980s it was observed that many of the installed (multimode) fibers exhibited an increase in transmission loss at $\lambda > 1200$ nm after aging as little as 3 years. Detailed studies indicated that this phenomenon was attributable to formation of certain chemical bonds between silicon and various dopants with hydroxyl ions in the form of Si—OH, Ge—OH, and P—OH because of trapping of hydrogen at the defect centers formed in the Si—O network due to incorporation of various dopants. Of the two most often used dopants, namely GeO_2 and P_2O_5, the latter was found to be more troublesome because defect densities formed with it were much higher in number because of a difference in valency between Si and P [11]. The phenomenon of hydrogen-induced loss increase with aging of a fiber could be effectively eliminated (at least over the life of a system assumed to be about 25 years) by avoiding phosphorus as a refractive index modifier of silica and through complete polymerization of the fiber coating materials, for which ultraviolet-curable epoxy acrylates were found to be the best choice; in fact, these are now universally chosen as the optimum fiber coating material before a fiber is sent to a cabling plant.

2.2. Dispersion Spectrum

Chromatic dispersion is another important transmission characteristic (along with loss) of an SMF, and its magnitude is a measure of its information transmission capacity. In a fiber optical network, signals are normally transmitted in the form of temporal optical pulses (typically in a high-speed fiber optical link, individual pulse width is about 30 ps [12]). For example, under the most extensively used pulse code modulation technique for digitized signal transmission, a telephone voice analog signal is at first converted into 64,000 electrical pulses. Likewise, several other telephone signals are converted to 64,000 electrical pulses, and all of these are electronically time division multiplexed in the electric domain at the telephone exchange before being fed to the electronic circuit driving a laser diode to produce equivalent optical pulses. Thereby, these optical pulses replicate the time division multiplexed signals in the optical domain, which are then launched into an SMF for transmission to a distant receiver. The above-mentioned process steps are schematically depicted in Fig. 1.3. Due to the phenomenon of *chromatic dispersion*, these signal pulses in general broaden in time with propagation through the fiber. Chromatic dispersion arises because of the dispersive nature of an optical fiber due to which the group velocity of a propagating signal pulse becomes a function of frequency (usually referred to as *group-velocity dispersion* [GVD] in the literature), and this phenomenon of GVD induces frequency chirp to a propagating pulse, meaning thereby that the *leading edge* of the propagating pulse differs in frequency from the *trailing edge* of the pulse. The resultant frequency chirp [i.e. $\omega(t)$] leads to intersymbol interference, in the presence of which the receiver fails to resolve the digital signals as individual pulses when the pulses are transmitted too close to each other [13, 14]. Thus these pulses, though started as individually distinguishable pulses at the transmitter, may become indistinguishable at the receiver depending on the amount of chromatic dispersion–induced broadening introduced by the fiber (Fig. 1.4). In fact, the phenomenon of GVD limits the number of pulses that can be sent through the fiber per unit time. For self-consistency of our discussion on pulse dispersion, in the following we

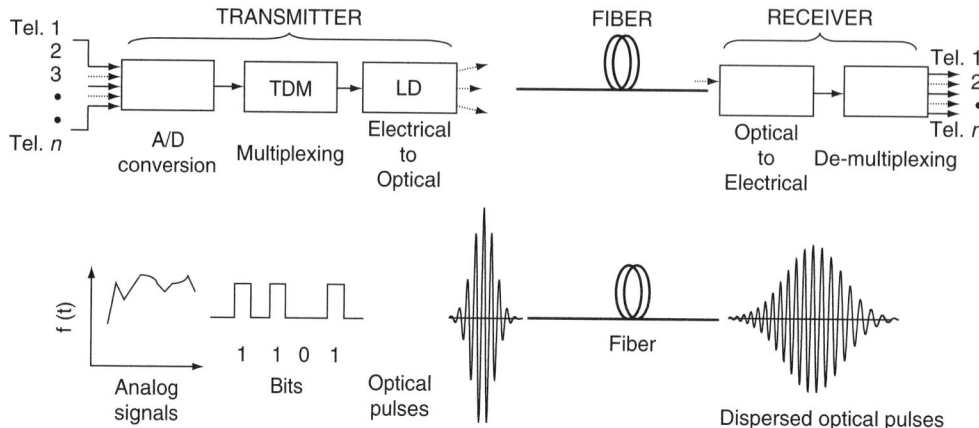

FIGURE 1.3 Block diagram depicting major components of an optical communication system; lower part of the figure depicts conversion of analog to digital (A/D) form (in electrical domain) and subsequent transmission of same in the form of equivalent optical pulses through the fiber where these get chirped and broadened due to chromatic dispersion. LD, laser diode; TDM, time division multiplexing.

outline basic principles that underlay pulse dispersion in an SMF [13, 14].

The *phase velocity* of a plane wave propagating in a medium of refractive index n having propagation constant k, which determines the velocity of propagation of its phase front, is given by

$$v_p = \frac{\omega}{k} = \frac{c}{n(\omega)} \quad (2)$$

On the other hand, in case of a propagating optical pulse in a *dispersive medium* (for which k is not a linear function of ω), the characteristic velocity at which pulse energy propagates through the medium is decided by its *group velocity* v_g given by

$$\frac{1}{v_g} = \frac{1}{c}\left[n(\omega) + \omega\frac{dn}{d\omega}\right] = \frac{1}{c}\left[n(\lambda_0) - \lambda_0\frac{dn}{d\lambda_0}\right] \quad (3)$$

As an optical pulse propagates through a SMF, it propagates via the LP_{01} guided mode of the fiber, and in that case the group velocity of the pulse is decided by the following equation in place of Eq. (3)

$$\frac{1}{v_g} = \frac{1}{c}\left[n_{\text{eff}}(\lambda_0) - \lambda_0\frac{dn_{\text{eff}}}{d\lambda_0}\right] = \frac{N_g}{c} \quad (4)$$

where $n_{\text{eff}}(\omega)(=\beta/k_0;\beta$ being propagation constant of the guided mode, and k_0 is k in free space) and N_g respectively represent *effective index* and *group index* of the LP_{01} mode. Likewise, for determining phase velocity of the propagating pulse in an optical fiber, $n(\omega)$ in Eq. (2) is required to be replaced by $n_{\text{eff}}(\omega)$. We now consider a *Fourier transform limited* Gaussian-shaped pulse, which is launched into an SMF such that at $z = 0$

$$f(x, y, z = 0, t) = E_0(x, y)e^{-t^2/2\tau_0^2}\exp(i\omega_c t) \quad (5)$$

The quantity $E_0(x,y)$ represents the LP_{01} modal field of the fiber, ω_c represents the frequency of the optical carrier, which is being modulated by the pulse, and τ_0 corresponds to the characteristic width of the Gaussian pulse, and $\omega_c t$ represents the phase of the pulse. By obtaining Fourier transform of this pulse in the frequency domain and incorporating Taylor's series expansion (retaining only terms up to second order) of the phase constant β around $\omega = \omega_c$ (in view of the fact that Fourier transform of such a Fourier transform limited pulse is sharply peaked around a narrow frequency domain $\Delta\omega$ around $\omega = \omega_c$), we would get

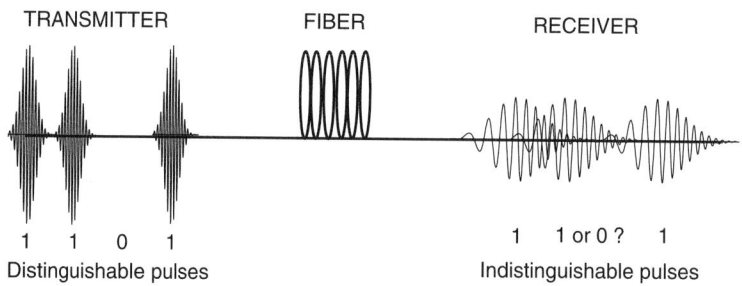

FIGURE 1.4 Dispersion-induced broadening of optical pulses and the resultant intersymbol interference at the receiver.

$$f(x, y, z = L, t) = E_0(x, y) \exp[i(\omega_c t - \beta|_{\omega=\omega_c})] \cdot \Psi(z = L, t) \qquad (6)$$

where

$$\Psi(z = L, t) = \frac{1}{2\pi} \int_{-\infty}^{+\infty} \left\{ F(z = 0, \Omega) \right. \\ \left. \exp\left[-i\left(\frac{\Omega}{v_g} + \frac{1}{2}\beta''\Omega^2\right)L\right] \right\} e^{i\Omega t} d\Omega \qquad (7)$$

represents the envelope of the dispersed pulse at a fiber length L, and $F(\Omega)$ is Fourier transform of the pulse in the frequency domain with $\Omega = \omega - \omega_c$; the ratio L/v_g simply represents the delay, that is, the time taken by the pulse to propagate through a fiber length L. In Eq. (7), the quantity $\beta'' = (d^2\beta/d\omega^2)|_{\omega=\omega_c}$ is known as the GVD parameter. We may rewrite the Fourier transform [cf. Eq. (7)] in a new time frame, which includes effect of delay due to propagation through $z = L$, that is, we replace t by $\left(t - \frac{L}{v_g}\right)$ and obtain

$$\text{F.T. of } \left\{ \Psi\left(z = L, t - \frac{L}{v_g}\right) \right\} \\ = F(z = 0, \Omega) \exp\left(-i\frac{1}{2}\beta''\Omega^2 L\right) \qquad (8)$$

Thus, the right-hand side of Eq. (8) may be taken as the transfer function of an SMF of length L. By carrying out the integration in Eq. (7), we would obtain from Eq. (6)

$$f(x, y, L, t) = \frac{E_0(x, y)}{(\tau(L)/\tau_0)^{1/2}} \exp\left[-\frac{\left(t - \frac{L}{v_g}\right)^2}{2\tau^2(L)}\right] \\ \exp[i(\Phi(L, t) - \beta_c L)] \qquad (9)$$

where

$$\Phi(L, t) = \omega_c t + \kappa\left(t - L/v_g\right)^2 - \frac{1}{2}\tan^{-1}(\bar{\alpha}) \qquad (10)$$

$$\kappa = \frac{1}{2\tau_0^2} \frac{\bar{\alpha}}{(1 + \bar{\alpha}^2)} \qquad (11)$$

$$\bar{\alpha} = \frac{\beta'' L}{\tau_0^2} \qquad (12)$$

and

$$\tau^2(L) = \tau_0^2(1 + \bar{\alpha}^2) \qquad (13)$$

Thus, the Gaussian temporal pulse remains Gaussian in shape with propagation but its characteristic width increases to $\tau(L)$ [given by Eq. (13)] as it propagates through a fiber of length L. After propagating through a distance of $L_{\text{DISP}} = (\tau_0^2/\beta'')$, the pulse assumes a width of $\sqrt{2}\tau_0$. This quantity L_{DISP} is a characteristic (dispersion-related) parameter of an SMF and is referred to as the *dispersion length*, which is often used to describe dispersive effects of a medium. The smaller the dispersion, the longer the L_{DISP}. Further, it can be seen from Eq. (13) through differentiation that for a given length of a fiber there is an optimum input pulse width $\tau_0^{\text{opt}} = \sqrt{\beta'' L}$, for which the output pulse width is minimum given by $\tau_{\min} = \sqrt{2\beta'' L}$. The quantity

$$\Delta\tau = \left[\tau^2(L) - \tau_0^2\right]^{1/2} = \frac{L}{\tau_0}\beta'' \qquad (14)$$

may be defined as the *chromatic dispersion–induced* increase in the width of a Fourier transform limited temporal pulse due to propagation through the fiber. The broadening of a Gaussian temporal pulse with propagation is shown in Fig. 1.5.

The GVD parameter β'' may be alternatively expressed as

$$\beta'' = \frac{d^2\beta}{d\omega^2}\bigg|_{\omega=\omega_c} = \frac{\lambda_0^3}{2\pi c^2}\frac{d^2 n_{\text{eff}}}{d\lambda_0^2} = \frac{\lambda_0^2}{2\pi c} D \qquad (15)$$

where D, known as the dispersion coefficient, is given by

$$D = \frac{1}{L}\frac{d\tau}{d\lambda_0} = -\frac{\lambda_0}{c}\frac{d^2 n_{\text{eff}}}{d\lambda_0^2} \qquad (16)$$

expressed in units of ps/nm·km. This parameter D is very extensively used in the literature to describe pulse dispersion in an optical fiber. The mode effective index n_{eff} is often expressed through [14]

$$n_{\text{eff}} \approx n_{\text{cl}}(1 + b\Delta) \qquad (17)$$

where $0 < b < 1$ represents the normalized propagation constant of the LP$_{01}$ mode

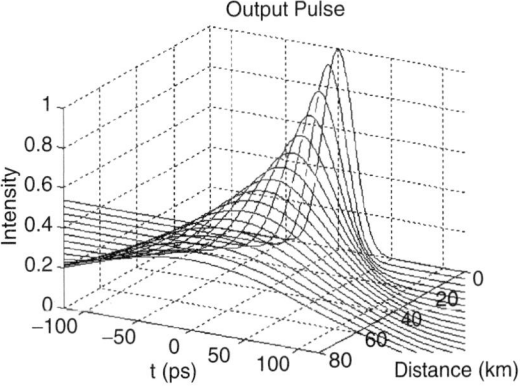

FIGURE 1.5 Broadening of a Gaussian temporal pulse with propagation through a single-mode fiber.

$$b = \frac{n_{\text{eff}}^2 - n_{\text{cl}}^2}{n_{\text{c}}^2 - n_{\text{cl}}^2} \qquad (18)$$

and $\Delta \approx ((n_c - n_{cl})/n_{cl})$ stands for relative core-cladding index difference in an SMF. Thus, n_{eff} is composed of two terms: The first is purely material related (n_{cl}) and the second arises due to the waveguide effect (b), and hence total dispersion in an SMF could be attributed to two types of dispersion, namely material dispersion and waveguide dispersion. Indeed, it can be shown that the total dispersion coefficient (D_T) is given by the following algebraic sum to a very good accuracy [14,15]:

$$D_T \approx D_M + D_{WG} \qquad (19)$$

where D_M and D_{WG} correspond to material and waveguide contributions to D, respectively. The material contribution can be obtained from Eq. (16) by replacing n_{eff} with n_{cl} and the waveguide contribution is

$$D_{WG} = -\frac{n_2 \Delta}{c\lambda_0}\left(V \frac{d^2(bV)}{dV^2}\right) \qquad (20)$$

where V is the well-known normalized waveguide parameter defined through

$$V = \frac{2\pi}{\lambda_0} a n_c \sqrt{2\Delta} \qquad (21)$$

In Eq. (21), a represents the core radius of the fiber. A plot of a typical dispersion spectrum of an SMF (i.e., D vs. λ) along with its components D_M and D_{WG} are shown in Fig. 1.6. It is apparent from Fig. 1.6 that D_{WG} is negative and D_M changes from negative to positive (going through zero at a wavelength of ~ 1280 nm [16]) as the wavelength increases; as a result, the two components cancel each other at a wavelength of about 1300 nm, which is referred to as the *zero dispersion wavelength* (λ_{ZD}), a very important design parameter of SMFs. Realization of this fact led system operators to choose the operating wavelength of first generation SMFs as 1310 nm. These fibers optimized for transmission at 1310 nm are now referred to as G.652 fibers per ITU standards; millions of kilometers of these fibers are laid underground all over the world. Though it appears that if operated at λ_{ZD} one might get infinite transmission bandwidth, in reality zero dispersion is only an approximation (albeit a very good approximation) because it signifies that only the second-order dispersive effects would be absent. In fact, as per ITU recommendations, for G.652 fibers to qualify for deployment as telecommunication media provided at the 1310-nm wavelength the D_T is < 3.5 ps/nm·km. At a wavelength around λ_{ZD}, higher order dispersions, namely third-order dispersions characterized by $d^3\beta/d\omega^3$, would determine the net dispersion of a pulse. Thus, in the absence of second-order dispersions, pulse dispersion is quantitatively determined by the dispersion slope S_0 at $\lambda = \lambda_{ZD}$ through

$$\Delta\tau = \frac{L(\Delta\lambda_0)^2}{2} S_0 \qquad (22)$$

where $S_0 = (dD/d\lambda_0)|_{\lambda_0 = \lambda_{ZD}}$, which is measured in units of ps/nm²·km; S_0 in G. 652 fibers at 1310 nm is ≤ 0.09 ps/nm·km. If third-order dispersion is the sole determining factor of pulse dispersion, the output pulse does not remain symmetric and acquires some oscillation at the tails [17]. A knowledge of D and S_0 enables the determination of the dispersion (D) at any arbitrary wavelength within a transmission window, for example, the EDFA band in which D in G.652 fibers varies approximately linearly with λ. This feature often finds applications in component designs, and $D(\lambda)$ is usually explicitly stated in commercial fiber data sheets as

$$D(\lambda_0) = \frac{S_0}{4}\left[\lambda_0 - \frac{\lambda_{ZD}^4}{\lambda_0^3}\right] \qquad (23)$$

The genesis of this relation lies in the following three-term polynomial equation often used as a fit to measured data for delay (τ) versus λ in an SMF:

$$\tau(\lambda_0) = \frac{A}{\lambda_0^2} + B + C\lambda_0^2 \qquad (24)$$

where coefficients A, B, and C are determined through a least-square fit to the data for measured $\tau(\lambda_0)$. A typical plot for τ versus λ for a G.652 fiber is

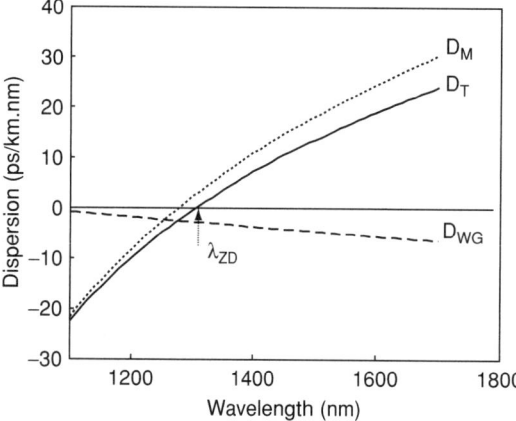

FIGURE 1.6 Material (D_M), waveguide (D_{WG}), and total (D_T) dispersion coefficients of a high-silica matched clad SMF.

Optical Fibers for Lightwave Communication

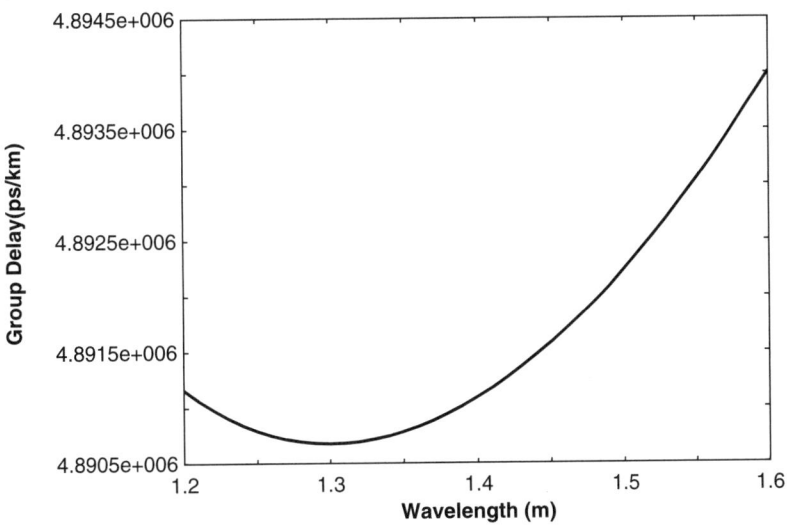

FIGURE 1.7 Typical wavelength dependence of group delay of a standard G.652 type single-mode fiber.

shown in Fig. 1.7. Equation (24) yields the following for $D(\lambda_0)$:

$$D(\lambda_0) = \frac{d\tau}{d\lambda_0} = 2C\lambda_0 - \frac{2A}{\lambda_0^2} \quad (25)$$

Setting $D(\lambda_0 = \lambda_{ZD})$ to zero yields $\lambda_{ZD} = (A/C)^{1/4}$. Thus,

$$S_0 = 8C \quad (26)$$

In addition to pulse broadening, because the energy in the pulse gets reduced within its time slot, the corresponding signal-to-noise ratio decreases, which could be compensated by increasing the power in the input pulses. This additional power requirement is termed the *dispersion power penalty*. For estimating the dispersion power penalty, a more general definition of pulse width in terms of root mean square (rms) width becomes more useful, especially for pulse shapes, which may not be of a well-defined shape like a Gaussian pulse. The rms width of a temporal pulse is defined through

$$\tau_{rms} = \sqrt{\langle t^2 \rangle - \langle t \rangle^2} \quad (27)$$

where

$$\langle t^n \rangle = \frac{\int\limits_{-\infty}^{\infty} t^n |f(z,t)|^2 dt}{\int\limits_{-\infty}^{\infty} |f(z,t)|^2 dt} \quad (28)$$

For a Gaussian output pulse, its τ_{rms} is related to the input Gaussian pulse width τ_0 through $\tau_{rms} = \tau_0/\sqrt{2}$. If a system is operating at a bit rate of B bits per second, the bit period is $1/B$ s. To keep the interference between neighboring bits at the output below a specified level, τ_{rms} of the dispersed pulse must be kept below a certain fraction ε of the bit period B. Accordingly, for maximum allowed rms pulse width (assuming it is Gaussian in shape) of the output pulse, τ_{rms} of the dispersed pulse should satisfy

$$\tau_{rms} = \frac{\tau_0}{\sqrt{2}} < \frac{\varepsilon}{B} \quad (29)$$

which implies

$$B\tau_0 < \sqrt{2}\varepsilon \quad (30)$$

On substitution of the optimum value for output pulse width, that is, $\tau_{min} = \sqrt{2\beta''L}$ in Eq. (30) and making use of Eq. (15), we get

$$B^2 DL < \frac{2\pi c}{\lambda_0^2} \varepsilon^2 \quad (31)$$

For a 2-dB power penalty, it is known that $\varepsilon = 0.491$, whereas for a 1-dB dispersion power penalty it is 0.306; these values for ε were specified by Bellcore standards (Document no. TR-NWT-000253). For a 1-dB dispersion power penalty at the wavelength of 1550 nm, we can write the inequality [Eq. (31)] approximately as

$$B^2 DL \leq 10^5 \text{Gb}^2 \cdot \text{ps/nm} \quad (32)$$

where B is measured in Gbits, D in ps/nm·km, and L in km. Based on Eq. (32), Table 1.1 lists the maximum allowed dispersion for different standard bit rates assuming a dispersion power penalty of 1 dB.

TABLE 1.1 Maximum allowed dispersions for different standard bit rates and for 1-dB dispersion power penalty.

Data rate (B)	Maximum allowed dispersion ($D.L.$)
2.5 Gb/s (OC-48)	~16,000 ps/nm
10 Gb/s (OC-192)	~1,000 ps/nm
40 Gb/s (OC-768)	~60 ps/nm

OC: optical channels.

By the mid-1980s, proliferation of optical-trunk networks and a steady increase in bit transmission rates gave rise to the requirement for a standard transmission format of the digital optical signals so that they could be shared without any need for an interface (this is the so-called mid-fiber meet) between the American, European, and Japanese telephone networks, each of which until then were following a different digital hierarchy [18]. This lack of standards led to the development of synchronous digital hierarchy (SDH), which has now become the global industry standard (in the United States this is known as SONET, for Synchronous Optical NETworks) for transmission of digital *optical channels* (OCs) [19]. An SDH network has a master clock that controls the timing of events all through the network. The base rate (OC-1) in SDH is taken as 51.84 Mb/s; thus, OC-48 rate corresponds to a transmission rate of 2.488 ($\approx$ 2.5 Gb/s), which became the standard signal transmission rate at the 1310-nm wavelength window over standard G.652 SMFs. Although in principle a very high bit transmission rate was achievable at 1310 nm, maximum link length/repeater spacing was limited by transmission loss ($\sim$ 0.34 dB/km at this wavelength) to ($\sim$ 40 km in these systems. Thus, it became evident that it would be an advantage to shift the operating wavelength to the 1550-nm window, where the loss is lower (cf. Fig. 1.2), to overcome transmission loss-induced distance limitation of the 1310-nm window.

2.3. Dispersion Shifted Fibers

As early as 1979 the Nippon Telephone and Telegraph (NTT) team from Japan had succeeded in achieving a single-mode silica fiber that exhibited a loss of 0.2 dB/km at 1550 nm, which was close to the fundamental limit determined by Rayleigh scattering alone [6]. This figure for fiber loss is 40% less than its value at 1310 nm! Thus, it was only natural for systems engineers to exploit this lowest-loss window by shifting the operating wavelength from 1310 to 1550 nm. However, commercial introduction of next generation systems based on the exploitation of the 1550-nm lowest-loss window had to wait almost another 10 years. This was mainly because conventional G.652 type fibers optimized for transmission at the 1310-nm window exhibited excessive chromatic dispersion of approximately +17 ps/nm·km at 1550 nm (Fig. 1.6). For these fibers, which are characterized by a nominal step refractive index profile, typically, the chosen Δ and core diameter ($2a$) were $\leq 0.36\%$ and ≥ 8.2 μm, respectively, which led to a resultant mode field diameter of 9.2 $\pm$ 0.4 μm at 1310 nm (10.4 $\pm$ 0.8 μm at 1550 nm) and a cutoff wavelength (λ_c) ≤ 1260 nm. Due to a D of approximately +17 ps/nm·km, repeater spacing at 1550 nm with such fiber-based links were limited by chromatic dispersion. By the mid-1980s, it was realized that repeater spacing of 1550-nm-based systems could be pushed to a much longer distance if the fiber designs could be tailored to shift λ_{ZD} to coincide with this wavelength so as to realize dispersion shifted fibers (DSFs), which are given the generic classification as G.653 fibers by the ITU. Comparative plots of dispersion spectra for G.652 and G.653 types of SMFs are shown in Fig. 1.8.

On the laser transmitter front, the most common types of laser diodes available for operation at 1550 nm were *Fabry-Perot* (FP) and *distributed feedback* (DFB) lasers, which were based on InGaAsP semiconductors. FP lasers are characterized by spectral widths (full width at half maximum = 2–5 nm), which are broader than the more expensive DFB lasers (continuous wave (cw) spectral width $\ll 1$ nm). However, because of chirping and mode hopping, DFB lasers exhibit broader dynamic spectral widths ($\sim$ 1–2 nm) at high modulation rates (≥ 2 Gb/s). In view of these factors, FP lasers in conjunction with DSFs seemed to be the best *near-term option* for optimum exploitation of the 1550-nm wavelength window [20]. Typically, D_T is

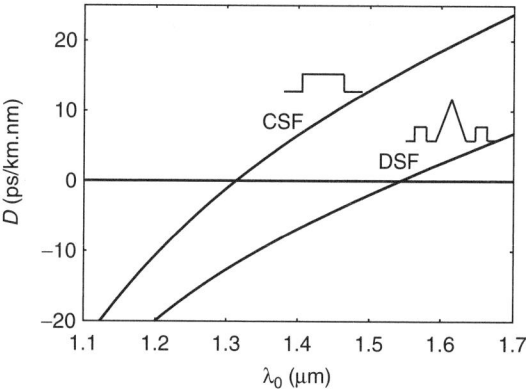

FIGURE 1.8 Dispersion spectra of a conventional single-mode fiber (CSF) of the type G.652 and a G.653 type DSF along with typical refractive index profiles as their labels. For CSF, λ_{ZD} falls at 1310 nm, and for DSF it occurs at 1550 nm.

≤ 2.7 ps/nm·km with $S_0 \leq 0.058$ ps/nm²·km at the 1550-nm wavelength in a DSF while maintaining inherent scattering loss low (~ 0.21 dB/km at 1550 nm). One serious problem with these fibers is their sensitivity to microbend and bend-induced losses at a $\lambda_0 > 1550$ nm due to their low effective cutoff wavelength of about 1100 nm [11]. A theoretical design was suggested in 1992 to overcome this issue of bend-loss sensitivity by which the theoretical cutoff wavelength was moved up close to about 1550 nm through a spot size optimization scheme [21]. These new generation systems operating at 2.5 Gb/s became commercial in 1990, with the potential to work at bit rates in excess of 10 Gb/s with careful design of sources and receivers and use of DSFs. DSFs were extensively deployed in the backbone trunk network in Japan in the early 1990s [11]. This trend persisted for a while before EDFAs emerged on the scene and led to a dramatic change in technology trends.

3. EMERGENCE OF FIBER AMPLIFIERS AND DWDM SYSTEMS

3.1. EDFAs

In the late 1980s typical state-of-the-art, repeaterless, transmission distances were about 40–50 km at the 560 Mb/s transmission rate. Because maximum launched optical power was below 100 μW, it was difficult to improve system lengths beyond this specification, and the use of electronic repeaters became inevitable. At a repeater, the so-called 3R-regeneration functions (*reamplification, retiming,* and *reshaping*) are performed in the electric domain on the incoming attenuated as well as distorted (due to dispersion) signals after detection by a photodetector and before the revamped signal is fed to a laser diode drive circuit, from which these cleaned optical pulses are launched to the next section of the fiber link. However, these complex functions are expensive and unit replacement is required when network capacity is to be upgraded to higher bit transmission rates because electronic components are bit rate sensitive. Because these units are required to convert photons to electrons and back to photons, often at modulation rates approaching the limits of the then available electronic switching technology, a bottleneck was encountered in the late 1980s. *What was needed was an optical amplifier to bypass this electronic bottleneck.* In 1986, the research group at Southampton University in England reported success in incorporating rare earth trivalent erbium ions into host silica glass during fiber fabrication [22]. Erbium is a well-known lasing species characterized by strong fluorescence at 1550 nm. Subsequently, the same group demonstrated that excellent noise and gain performance is feasible in a large part of the 1550-nm window with erbium-doped standard silica fibers [5].

The concept of optical amplification in fiber is almost as old as the laser itself. Today, EDFAs seem like an outstanding breakthrough, but they are really an old idea. In 1964, Koester and Snitzer [23] demonstrated a gain of 40 dB at 1.06 μm in a 1-meter-long Nd-doped fiber side pumped with flash lamps. The motivation at that time was to find optical sources for communication, but the impressive development of semiconductor lasers that took place in subsequent years pushed fiber lasers to the background. The operation of an EDFA is very straightforward [24]. The electrons in the $4f$ shell of the erbium ions are excited to higher energy states by absorption of energy from a pump. Absorption bands most suitable as pumps for obtaining amplification of 1550-nm signals are the 980- and 1480-nm wavelengths (Fig. 1.9a); Fig. 1.9b shows a schematic of an EDFA. When pumped at either of these wavelengths, an erbium-doped fiber was found to amplify signals over a band of almost 30–35 nm at the 1550-nm wavelength region (see Fig. 8.9 in Chapter 8).

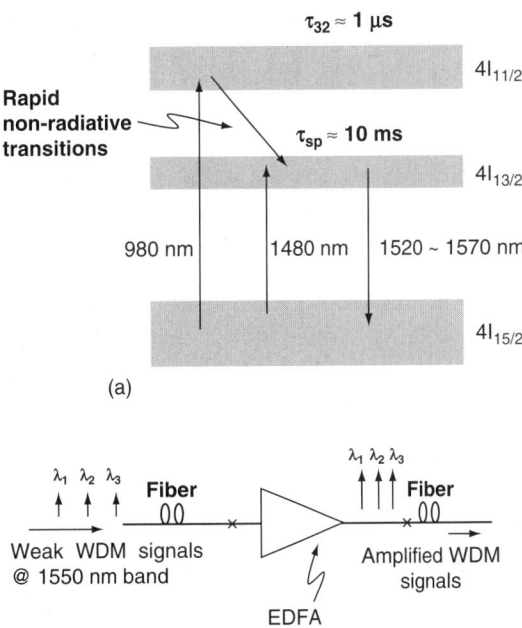

FIGURE 1.9 (a) Energy level diagram of an Er^{+3} ion. (b) Schematic layout of an EDFA (for details see Chapter 8).

Typical pump powers required for operating an EDFA as an amplifier range from 20 to 100 mW. Absorption of pump energy by the erbium ions leads to *population inversion* of these ions from the ground state ($4I_{15/2}$) to either $4I_{11/2}$ (980 nm) or $4I_{13/2}$ (1480 nm) excited states: the $4I_{13/2}$ level effectively acts as a storage of pump power from which the incoming weak signals may stimulate emission and experience amplification [13]. Stimulated events are extremely fast, and hence the amplified signal slavishly follows the amplitude modulation of the input signal. EDFAs are accordingly *bit rate transparent*. EDFAs are new tools that system planners now almost routinely use for designing networks. EDFAs can be incorporated in a fiber link with an insertion loss of ~ 0.1 dB, and almost full population inversion is achievable at the 980-nm pump band. Practical EDFAs with an output power of around 100 mW (20 dBm), a 30-dB small signal gain, and a noise figure of < 5 dB are now available commercially. In a 10-Gb/s transmission experiment, a record receiver sensitivity of 102 photons per bit was attained in a two-stage composite EDFA having a noise figure of 3.1 dB [25].

3.2. DWDM

A very important and attractive feature of EDFAs is that they exhibit a fairly smooth gain versus wavelength curve (especially when the fiber is doped with Al) almost 30–35 nm wide (see Chapter 8). Thus, multichannel operation via WDM within this gain spectrum became feasible, each wavelength channel being simultaneously amplified by the same EDFA. A relatively long lifetime of the excited state (~ 10 ms) leads to slow gain dynamics and therefore minimal cross-talk between WDM channels. In view of this system, designers were blessed with a great degree of freedom to boost the capacity of a system as and when required due to the flexibility to make the network responsive to future demands simply by introducing additional wavelengths through the same fiber, each multiplexed at, for example, 2.5 or 10 Gb/s or even higher. With the development of L-band EDFA characterized by a gain spectrum that extends from 1570 to 1620 nm [26], the potential for large-scale increase in capacity of already installed links through DWDM became enormous. To avoid haphazard growth of multiple wavelength transmitting fiber links, the ITU introduced certain wavelength standards that are now referred to as ITU wavelength grids for DWDM operation. As per ITU standards, the reference wavelength is chosen to be 1552.52 nm corresponding to the Krypton line, which is equivalent to 193.1 THz (f_0) in the frequency domain; the chosen channel spacing away from f_0 in terms of frequency (Δf) is supposed to follow the relation $\Delta f = 0.1 I$ THz, with $I = $ positive/negative integers [27].

Recommended channel spacings are 200 GHz ($\equiv 1.6$ nm), 100 GHz ($\equiv 0.8$ nm), and 50 GHz ($\equiv 0.4$ nm); the quantities within parentheses have been calculated by assuming a central wavelength as 1550 nm.

Today's DFB lasers can be tuned to exact ITU wavelength grids. All the terabit transmission experiments that were reported in recent years took the route of DWDM.

The possibility of introducing broadband services in the 1990s led to a boom in the communication industry, blurring the distinction between voice, video, and data service providers, and the demand for more and more bandwidth began to strain the capacity of installed communications links [28]. DSF in combination with EDFA appeared to be the ideal solution to meet the demand for high data rate and long-haul applications. However, it was soon realized that large optical power throughput that is encountered by a fiber in a DWDM system due to simultaneous amplification of multiple wavelength channels poses problems because of the onset of nonlinear propagation effects, which may completely offset the attractiveness of multichannel transmission with DSF. It turned out that due to negligible temporal dispersion that is characteristic of a DSF at the 1550-nm band, these are highly susceptible to nonlinear optical effects, which may induce severe degradation to the propagating signals [29] (see Chapter 6).

3.3. Fibers for DWDM Transmission

By late 1990s the demand for bandwidth had been steadily increasing, and it became evident that new fibers were required to handle transmission of large number of wavelengths, each of which was expected to carry a large amount of data especially for undersea applications and long-haul terrestrial links. A very useful figure of merit (FOM) for a DWDM link is known as *spectral efficiency*, which is defined as the *ratio* of bit rate to channel spacing [11]. Because bit rate cannot be increased arbitrarily due to the constraints of availability of electronic components relevant to bit rates > 40 Gb/s, a decrease in channel spacing appeared to be the best near-term option to tap the huge gain bandwidth of an EDFA for DWDM applications. However, it was soon realized that with a decrease in channel spacing, the fiber became more strongly sensitive to detrimental nonlinear effects such as four-wave mixing (FWM), which could be relaxed by allowing the propagating signals to experience a finite dispersion in the fiber (Fig. 1.10). If the number of wavelength channels in a DWDM stream is N, the FWM effect, if present, leads to the generation of $N^2(N-1)$ sidebands! These sidebands would naturally draw power from the propagating signals and hence could result in serious cross-talk [29]. Therefore, for DWDM applications, fiber designers came up with new designs for the signal fiber for low-loss and dedicated DWDM signal transmission at the 1550-nm band, which were generically named nonzero (NZ) DSF. These fibers were designed to meet the requirement of low sensitivity to nonlinear effects especially to counter FWM, which could be substantially suppressed if each wavelength channel is allowed to experience a finite amount of dispersion during propagation [30] (see Chapter 6). This requirement is precisely counter to the dispersion characteristics of a DSF, hence the name of NZ-DSF [31–33].

ITU has christened such fibers as G.655 fibers, which are generally characterized with refractive index profiles, which are relatively more complex than a simple matched clad step index design of G.652 fibers; a schematic of a specimen is shown in Fig. 1.11. The need to deploy such advanced fibers became so imminent that the ITU-T Committee had to evolve new standards for G.655 fibers (ITU-T G.655). As per ITU recommendations, G.655 fibers should exhibit finite dispersion $2 \leq D$ ps/nm·km ≤ 6 in the 1550-nm band to detune the phase matching condition required for detrimental nonlinear

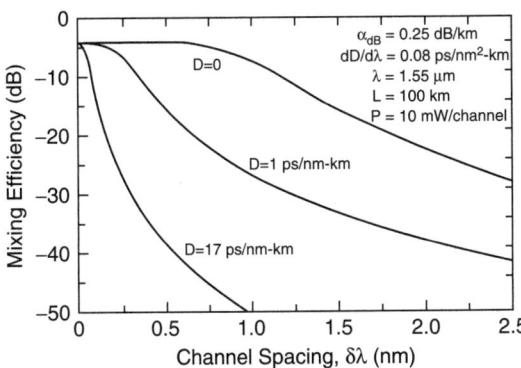

FIGURE 1.10 Nonlinearity-induced FWM efficiency in a DWDM system as a function of channel separation in fibers having different values for D, which are shown as labeling parameters. Other important fiber parameters are shown in the inset. (Adapted from [29]; © IEEE 1995.)

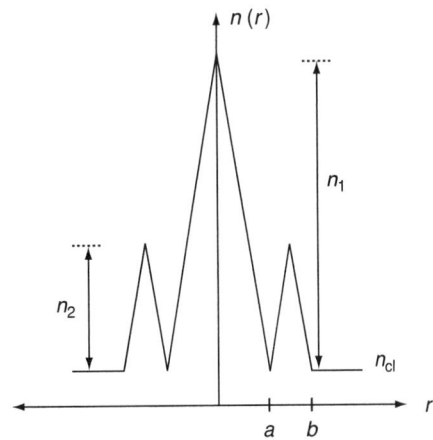

FIGURE 1.11 A sample schematic refractive index profile of an NZ-DSF.

propagation effects like FWM and cross-phase modulation to take place during multichannel transmission. ITU arrived at the above range for D by assuming a channel spacing of 100 GHz or more. For smaller channel spacing like 50 or 25 GHz, it turned out that the above range for D is insufficient to suppress potential nonlinear effects unless (1) the power per channel is reduced substantially and (2) the number of amplifiers is limited. Unfortunately, these steps would amount to a decrease in the repeater spacing, which would be counterproductive because this would defeat the very purpose for which G.655 fibers were proposed [34].

To overcome this disadvantage of an NZ-DSF, a more advanced version of NZ-DSF as a transmission fiber was proposed for super-DWDM, which was christened by ITU as advanced NZ-DSF/G.655b fibers. As per the ITU recommendation, the upper limit on D for an advanced NZ-DSF should be 10 ps/nm·km at the longer wavelength edge of the C-band, that is, at 1565 nm. For an advanced NZ-DSF, its D falls in the range $6.8 \leq D$ ps/nm·km ≤ 8.9 (C-band) and $9.1 \leq D$ ps/nm·km ≤ 12.0 (L-band) [35]. In view of its reduced sensitivity to detrimental nonlinear effects, it can accommodate transmission of as many as 160 channels of 10 Gb/s at 25 GHz channel spacing within the C-band alone and also offers ease in deployment of 40-Gb/s systems. Another attractive feature of advanced NZ-DSFs is that it can be used to transmit signals in the S-band as well, for which gain-shifted thulium-doped fiber amplifiers and Raman fiber amplifiers have emerged as attractive options. Figure 1.12 shows typical dispersion spectra for G.652, G.653, and G.655 types of transmission fibers. There are variations in G.655 fibers, for example, large effective area fibers (LEAF®) [32],

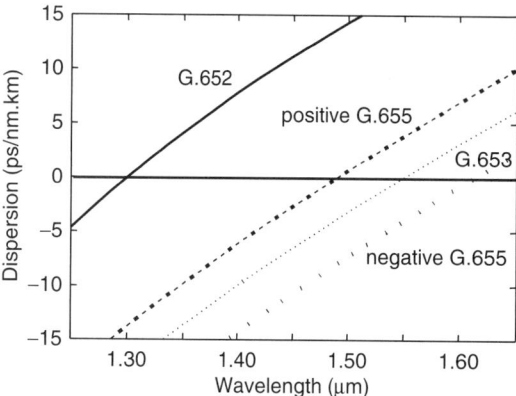

FIGURE 1.12 Typical dispersion spectra of ITU standard G.652, G.653, and G.655 (both positive and negative versions) fibers.

reduced slope (TrueWave RS®) [33], and Tera-Light® [36], each of which are proprietary products of well-known fiber manufacturing giants. Table 1.2 depicts typical characteristics of some of these fibers at 1550 nm along with those of G.652 fibers

3.4. Dispersion Compensating Fibers

In G.655 fibers for DWDM transmission, because a finite (albeit low) D is deliberately kept at the design stage itself to counter potentially detrimental nonlinear propagation effects, one would accumulate dispersion between EDFA sites! Assuming a D of 2 ps/nm·km, though a fiber length of about 500 km could be acceptable at 10 Gb/s before requiring correction for dispersion, at 40 Gb/s a corresponding unrepeated span would hardly extend to 50 km [see Eq. (32)]. The problem is more severe in G.652 fibers, when at 2.5 Gb/s a link length of about 1000 km would be feasible at the 1550-nm window; if the bit rate is increased to 10 Gb/s, a tolerable D in this case over 1000 km would be hardly about 1 ps/nm·km! Thus, repeater spacing of 1550-nm links based on either of these fiber types would be chromatic dispersion limited for a given data rate.

Around the mid-1990s before the emergence of G.655 fibers, it was believed that for network upgrades it would be prudent to exploit the EDFA technology to "mine" the available bandwidth of the already embedded millions of kilometers of G.652 fibers all over the world by switching over transmission through these at the 1550-nm window. This route appeared to be economically very attractive for attaining network upgrade(s) because it would not require any new fibers to be installed—one could make use of the already installed huge base of G.652 fibers and simply replace the 1310-nm transmitter with a corresponding transmitter operating at 1550 nm. Though transmission of 1550-nm signals through G.652 fibers would imply an accumulation of unacceptably high chromatic dispersion at EDFA sites (ideally spaced approximately every 80–120 km) due to the fiber's D being +17 ps/nm·km, though such a large D could substantially reduce the possibility of potentially harmful nonlinear effects. Hence, to reap the benefit of the availability of EDFAs, system upgrade through the DWDM technique in the 1550-nm band of the already installed G.652 fiber links would necessarily require some dispersion compensation scheme. Realization of this immediately triggered a great deal of R&D efforts to develop some dispersion compensating modules, which could be integrated to an SMF optic link so that net dispersion of the link could be maintained/managed within desirable limits. Three major, state-of-the-art, fiber-based, optical technologies available as options for dispersion management are *dispersion compensating fibers* (DCFs) [37–42], *chirped fiber Bragg gratings* [43–46] (see Chapter 15), and *high-order-mode fibers* [35, 47–49] (see Chapter 19).

To understand the logic behind dispersion compensation techniques, we notice from Eqs. (9) and (10) that the instantaneous frequency of the output pulse is given by

$$\omega(t) = \frac{d\Phi}{dt} = \omega_c + 2\kappa\left(t - \frac{L}{v_g}\right) \quad (33)$$

TABLE 1.2 Characteristics of fibers.

ITU standard fiber type	Dispersion coefficient D (ps/nm·km) at 1550 nm	Dispersion slope S (ps/nm²·km)	Mode effective area A_{eff} (μm²) at 1550 nm
G.652	~17	~0.058	~80
G.655			
LEAF®	~4.2	~0.085	~65–80
TrueWave RS®	~4.5	~0.045	~55
TeraLight®	~8	~0.058	—

A_{eff}: mode effective area [see Eq. (15) in Chapter 6].

The center of the pulse corresponds to $t = L/v_g$. Accordingly, the leading and trailing edges of the pulse correspond to $t < L/v_g$ and $t > L/v_g$, respectively. In the normal dispersion regime where $d^2\beta/d\lambda_0^2$ is positive [implying that the parameter κ defined in Eq. (11) is positive], the leading edge of the pulse is down-shifted, that is, *red-shifted*, in frequency whereas the trailing edge is up-shifted, that is, *blue-shifted*, in frequency with respect to the center frequency ω_c. Thus, red spectral components (i.e., longer wavelength components) of the signal pulse would travel faster than their blue spectral components (thereby meaning shorter wavelength components), that is, the group delay would increase with a decrease in wavelength. The converse would be true if the signal pulse wavelength corresponds to the anomalous dispersion region where $d^2\beta/d\lambda_0^2$ is negative. Hence, as the pulse broadens with propagation due to this variation in its group velocity with wavelength in a dispersive medium like SMF, it also gets chirped, meaning thereby that its instantaneous frequency within the duration of the pulse itself changes with time. Chirping of a temporal pulse with propagation both in the normal and anomalous dispersion regimes in a fiber is schematically shown in Fig. 1.13. If we consider propagation of signal pulses through a G.652 fiber at the 1550-nm wavelength band at which its D is positive (i.e., its group delay increases with increase in wavelength), it would exhibit *anomalous* dispersion. If this broadened temporal pulse were transmitted through a DCF, which exhibits *normal* dispersion (i.e., its dispersion coefficient D is negative) at this wavelength band, then the broadened pulse would get compressed with propagation through the DCF. This could be understood with the help of the transfer function of a fiber given by Eq. (8) and studying evolution of the pulse as it propagates through different segments of the fiber link shown in Fig. 1.14. At stage (1), let the F.T. of the input pulse $f_1(t)$ be $F_1(\Omega)$, which transforms in view of Eq. (8) at subsequent stages to [12]

$$stage(2): F_2(\Omega) = F_1(\Omega)\exp(-i\beta_T'' L_T \Omega^2)$$
$$stage(3): F_3(\Omega) = F_2(\Omega)\exp(-i\beta_D'' L_D \Omega^2)$$
$$= F_1(\Omega)\exp[-i(\beta_T'' L_T + \beta_D'' L_D)\Omega^2]$$

where β_T'' and β_D'' represent GVD parameters of the transmission fiber and the DCF, respectively, and $L_{T,D}$ refer to corresponding fiber lengths traversed by the signal pulse. It is apparent from the above that if the following condition is satisfied,

$$\beta_T'' L_T + \beta_D'' L_D = 0 \qquad (34)$$

the dispersed pulse would recover its original shape [i.e., evolve back to $f_1(t)$] at stage (3). This is more explicitly illustrated in Fig. 1.15. Thus, GVD parameters (and hence coefficients $D_{T,D}$) of the transmission fiber and the DCF should be opposite in sign. Consequently, if a G.652 fiber as the transmission fiber is operated at the EDFA band, corresponding DCF must exhibit negative dispersion at this wavelength band [50]. Figure 1.16 illustrates as an example the concept of dispersion compensation by a DCF for the dispersion suffered by 10 Gb/s externally modulated signal pulses over a 600-km span of a G.652 fiber; the solid line represents delay versus length in the transmission fiber and dashed line corresponds to delay versus length for the DCF [51]. Figure 1.16 shows three possible routes for introducing the DCF: (1) at the beginning of the span (dashed line), (2) at the end of the span (solid

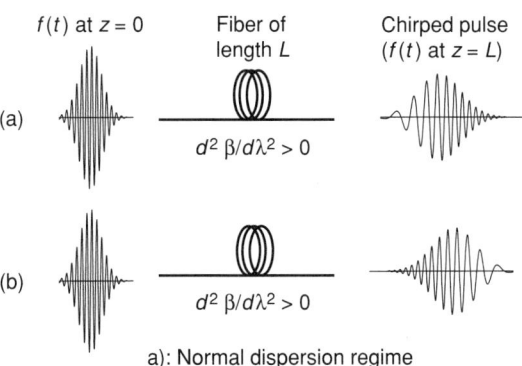

FIGURE 1.13 Propagation of a Gaussian pulse at a wavelength (a) in the normal dispersion regime and (b) in the anomalous dispersion regime and resulting chirping of the propagated pulse.

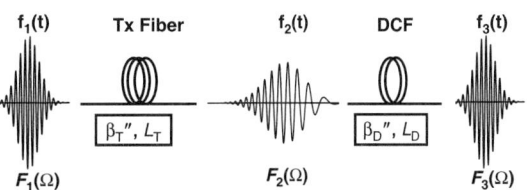

FIGURE 1.14 Schematic illustration of dispersion-induced pulse broadening with propagation in terms of fiber transfer function through Fourier transform pairs and compression of the dispersed pulse through propagation in a DCF.

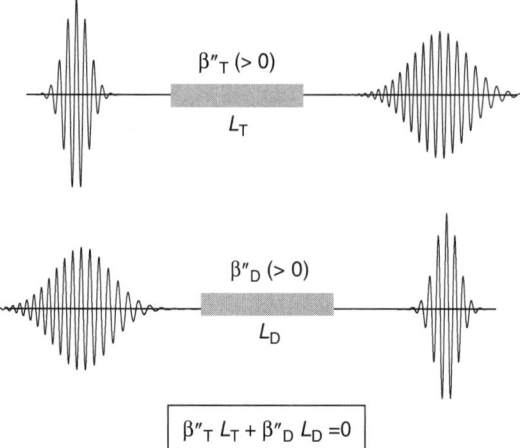

FIGURE 1.15 Illustration of the role of a DCF to neutralize dispersion-induced broadening of a pulse due to propagation through a fiber having GVD parameter > 0 at the operating wavelength.

line), or (3) every 100 km (dotted line), say, at EDFA sites. Further, the larger the magnitude of D_D, the smaller the length of the required DCF. This is achievable if D_{WG} of the DCF far exceeds its D_M in absolute magnitude. Large negative D_{WG} is achievable through an appropriate choice of the refractive index profile of the fiber so that at the wavelengths of interest a large fraction of its modal power rapidly spreads into the cladding region for a small change in the propagating wavelength. In other words, the rate of increase in mode size is large for a small increase in λ. Accordingly, the modal power distribution is strongly influenced by the cladding, where the mode travels effectively faster due to lower refractive index [51].

The *first-generation* DCFs relied on narrow core and high core cladding refractive index contrast (Δ) fibers to fulfill this task; typically, in these the fiber refractive index profiles were similar to those of the matched clad SMFs with the difference that Δ of the DCF were much larger ($\geq 2\%$) (Fig. 1.17). These DCFs were targeted to compensate for dispersion in G.652 fibers at a single wavelength and were characterized with a D of approximately -50 to -100 ps/nm·km and a positive dispersion slope. A schematic for D_D versus λ for a first-generation DCF relative to D_T for the transmission fiber is shown in Fig. 1.18. Because of high Δ and poor mode confinement as explained above, a DCF necessarily involved large insertion loss (typically attenuation in DCFs could vary in the range of 0.5–0.7 dB/km [11]) and sensitivity to bend-induced loss. To simultaneously achieve a high negative dispersion coefficient and low attenuation coefficient, α_D, DCF designers have ascribed an FOM to a DCF, which is defined through

$$\text{FOM} = \frac{-D_D}{\alpha_D} \quad (35)$$

expressed in ps/nm·dB. Total attenuation and dispersion in dispersion-compensated links is given by

$$\alpha = \alpha_T L_T + \alpha_D L_D \quad (36)$$
$$D = D_T L_T + D_D L_D \quad (37)$$

It could be shown from Eqs. (35) to (37) that for $D = 0$ [42]

$$\alpha = \left(\alpha_T + \frac{D_T}{\text{FOM}}\right) L_T \quad (38)$$

which shows that any increase in total attenuation in dispersion-compensated links would be solely through FOM of the DCF; thus, the larger the

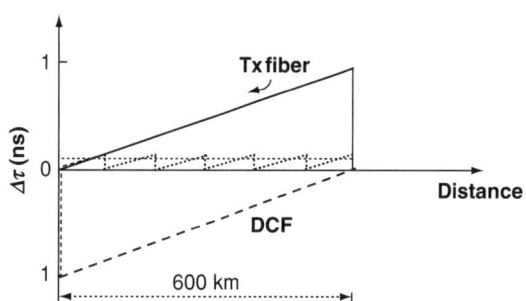

FIGURE 1.16 Schematic illustrating the concept of dispersion compensation of 10-Gb/s pulses transmitted through 600 km of a G.652 fiber. The full line represents variation of dispersion with length for a G.652 fiber; dashed lines represent corresponding variations for a compatible DCF when placed at the beginning or at the end (full line) or at every 100 km (dotted line) along the link. (Adapted from [51]; © IEEE.)

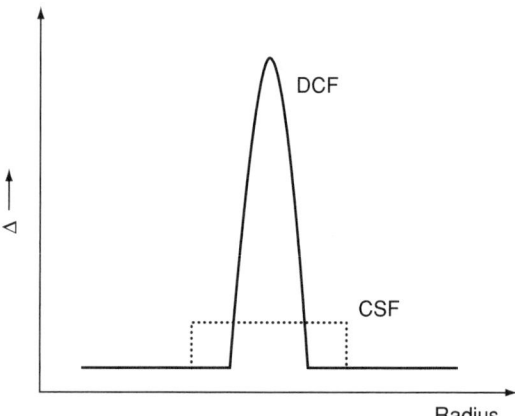

FIGURE 1.17 Typical refractive index profile of a first-generation DCF relative to that of a conventional single-mode fiber (CSF).

Optical Fibers for Lightwave Communication

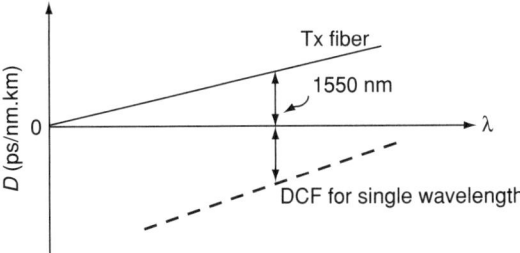

FIGURE 1.18 Dispersion spectra for a single wavelength DCF relative to that of a CSF at the 1550-nm band.

FOM, the smaller the incremental attenuation in the link due to insertion of the DCF. Dispersion slopes for the first-generation DCF were of the same sign as the transmission fiber (cf. Fig. 1.18) and hence perfect dispersion compensation was realizable only at a single wavelength in this class of DCFs.

Because DWDM links involve multichannel transmission, ideally one would require a broadband DCF so that dispersion could be compensated for at all the wavelength channels simultaneously. The key to realize a broadband DCF lies in designing a DCF in which not only that D versus λ is negative at all those wavelengths in a DWDM stream, but also that the dispersion slope is negative. The broadband dispersion compensation ability of a DCF is quantifiable through a parameter known as relative dispersion slope (RDS) expressed in units of nm^{-1}, which is defined through

$$\text{RDS} = \frac{S_D}{D_D} \quad (39)$$

A related parameter, referred to as κ (in units of nm), also referred to in the literature, is simply the inverse of RDS [52]. Values of RDS for LEAF®, TrueWave RS®, and Teralight® are 0.026, 0.01, and 0.0073, respectively. Thus, if a DCF is so designed that its RDS (or κ) matches that of the transmission fiber, then that DCF would ensure perfect compensation for all wavelengths. Such DCFs are known as dispersion slope compensating fibers (DSCFs). A schematic for the dispersion spectrum of such a DSCF along with single-wavelength DCF is shown in Fig. 1.19. Any differences in the value of RDS (or κ) between the transmission fiber and the DCF would result in under- or over-compensation of dispersion, leading to increased *bit error rates* (BERs) at those channels. In practice, a fiber designer targets to match values of the RDS for the Tx fiber and the DCF at the median wavelength of a particular amplification band, that is, C- or L-band. In principle, this is sufficient to achieve dispersion slope compensation across that particular amplifier band

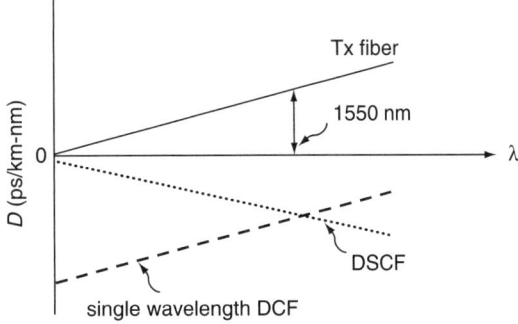

FIGURE 1.19 Dispersion spectra for a broadband DSCF relative to that of a conventional single-mode fiber (Tx) and a single wavelength DCF at the 1550 nm band.

because the dispersion spectrum of the transmission fiber is approximately linear within a specific amplification band. However, other propagation issues, such as bend loss sensitivity and countenance of nonlinear optical effects through large-mode effective area, often demand a compromise between 100% dispersion slope compensation and the largest achievable mode effective area. In such situations, the dispersion slope S may not precisely match those wavelength channels, which fall at the edges of a particular amplification band. Table 1.3 shows the impact of this mismatch (ΔS) in S at different bit rates [11].

A typical value of RDS for G.652 type of SMF at 1550 nm is about 0.00335 nm^{-1} (or κ as 298 nm), whereas in LEAF®, TrueWave®, and Teralight® fibers, its values are about 0.026 nm^{-1}, 0.01 nm^{-1} and 0.0073 nm^{-1}, respectively. The insertion loss of an approximately matching commercial DCF for an 80-km span is $\sim$ 6–7 dB, which includes the loss due to splices. Ideally, a DCF design should be so optimized that it has a low insertion loss, a low sensitivity to bend loss, a large negative D with an appropriate negative slope for broadband compensation, a large mode effective area (A_{eff}) for reduced sensitivity to nonlinear effects, and a low polarization mode dispersion (PMD).

A few proprietary designs of DCF index profiles are shown in Fig. 1.20. It is apparent from these index profiles that multiple claddings were introduced to achieve better control on the mode expansion–induced changes in the guided mode's effective index. One recent design (Fig. 1.21), which yielded

TABLE 1.3 Tolerable mismatch (ΔS) in S at different bit rates.

Bit rate	$\Delta S/S$ (%)
10 Gb/s	$\sim$20
40 Gb/s	$\sim$1

After [11].

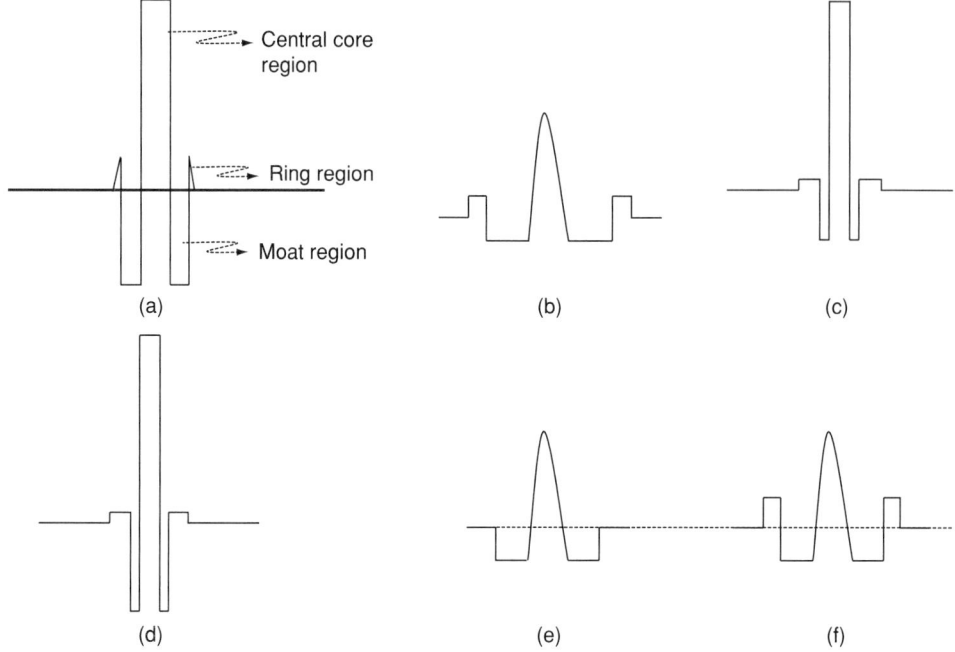

FIGURE 1.20 Schematic refractive index profiles of DCFs of (a) Corning Inc.; (b), (c), and (d) Lucent Technology; (e) and (f) Furukawa as available in the literature.

the record for largest negative D (-1800 ps/nm·km at 1558 nm), was based on a coaxial dual-core refractive index profile [53]; Fig. 1.22 [54] shows the fabricated DCF's actual index profile and its measured chromatic dispersion spectrum. The two cores are interconnected through a matched cladding, in contrast to most of the other designs in which there is usually a depressed index clad or a moat region in the cladding. In the fabricated preform, the central core was more of a triangular shape (Fig. 1.21, dashed line) than a step distribution. The two cores essentially function like a directional coupler. Because these two concentric fibers are not significantly identical, through adjustments of index profile parameters their mode effective indices could be made equal at some desired wavelength (called phase matching wavelength, λ_p),

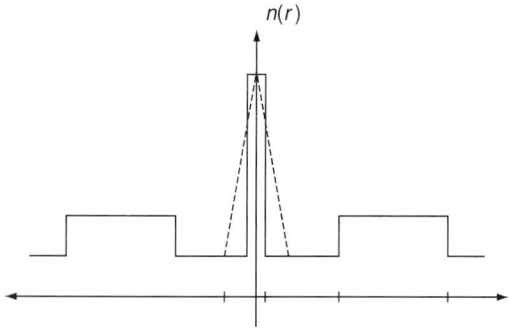

FIGURE 1.21 Refractive index profile of the coaxial dual-core DCF.

in which case the effective indices as well as modal field distributions of the normal modes of this dual-core fiber would exhibit rapid variations with λ around λ_p [53].

Typical modal power distributions for three different wavelengths around the phase matching wavelength (λp) are shown in Fig. 1.23. These rapid variations in modal distributions with wavelengths could be exploited not only to tailor dispersion characteristics of the DCF, for example, to achieve very large negative dispersion or broadband dispersion compensation characteristics, but also to achieve inherent gain flattened EDFAs (see Chapter 8) or broadband Raman fiber amplifiers [55]. Because the index profile involves multitudes of profile parameters, one could optimize the design through an appropriate choice of these parameters to achieve control of its propagation characteristics. Further research in this direction indeed has led to designs of coaxial dual-core DSCFs for broadband dispersion compensation in G.652 as well as G.655 fibers within various amplifier bands such as the S-, C- and L-bands [56–59]. Mode effective areas of various DCF designs based on Fig. 1.20 typically range between 15 and 25 μm^2, which make these susceptible to nonlinear effects unless care is taken to reduce launched power (≤ 1 dBm per channel) into the so designed DCFs. In contrast, the designs based on dual-core DSCFs could be designed to

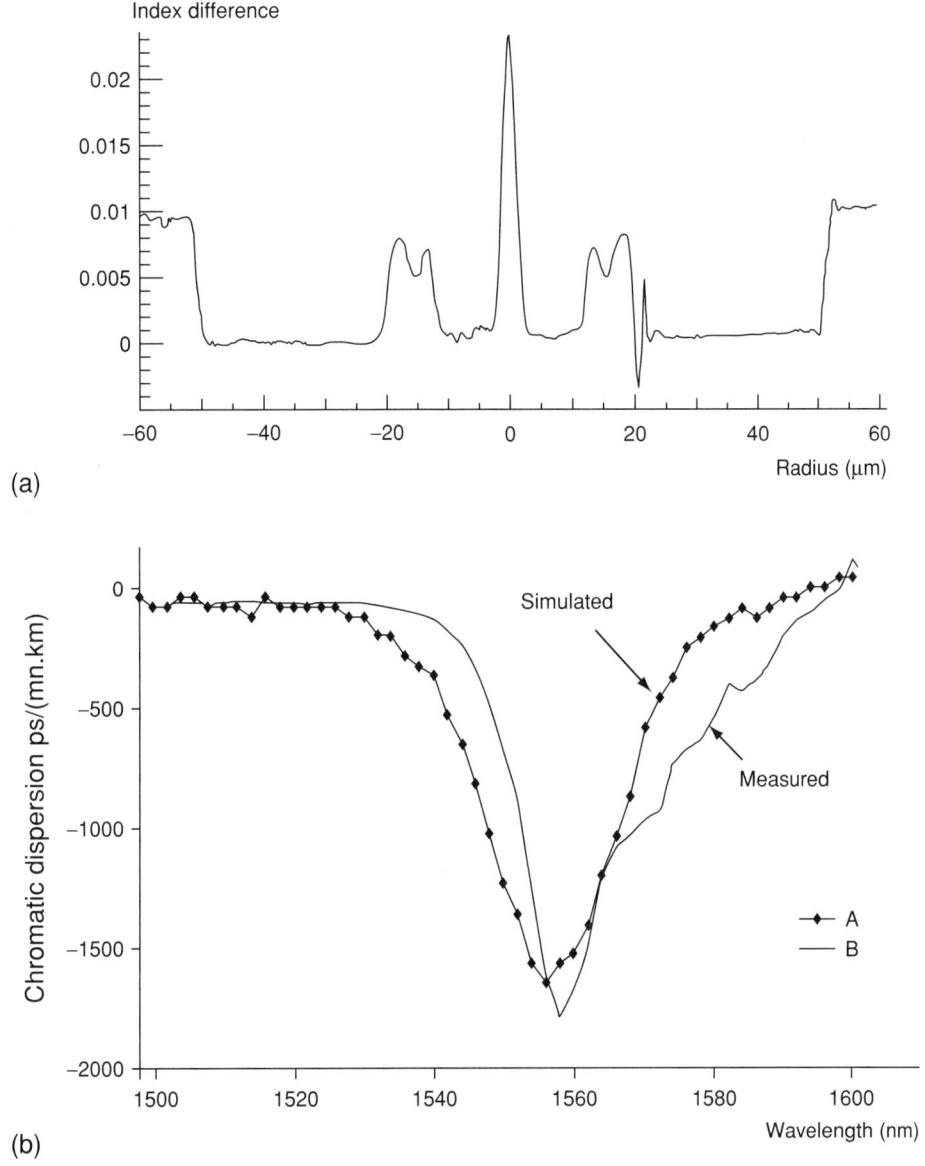

FIGURE 1.22 (a) Refractive index profile of the fabricated dual-core DCF preform. (b) Measured dispersion spectrum of the fabricated DCF (Adapted from [54]; © Copyright IEE).

attain A_{eff}, which are comparable with that of the G.652 fiber ($\sim$70–80 μm^2), as shown in Fig. 1.24. The net residual dispersion spectra of a 100-km-long G.652 fiber link along with so designed DSCFs (in a ratio of approximately 10:1) at each of the amplifier bands are shown in Fig. 1.25. It can be seen that residual average dispersion is well within ± 1 ps/nm·km within all three bands. Other estimated performance parameters of the designed DSCFs are shown in Table 1.4. Results for residual dispersion for standard G.655 fibers jointed with the designed dual-core DSCFs for various amplifier bands are shown in Fig. 1.26 [58]. Table 1.5 displays performance parameters similar to those shown in Table 1.4 of these coaxial dual-core design-based DSCFs for standard G.655 fibers.

3.5. Reverse/Inverse Dispersion Fibers

The above-mentioned DCFs are normally introduced in a link in the form of a spool as a stand-alone module at an EDFA site. An alternative scheme involves the use of the compensating fiber within the cable itself after having been jointed with the transmission fiber as part of the overall fiber link. Such DCFs are referred as *reverse/inverse* dispersion fibers (RDF/IDF) [60–62]. Reverse

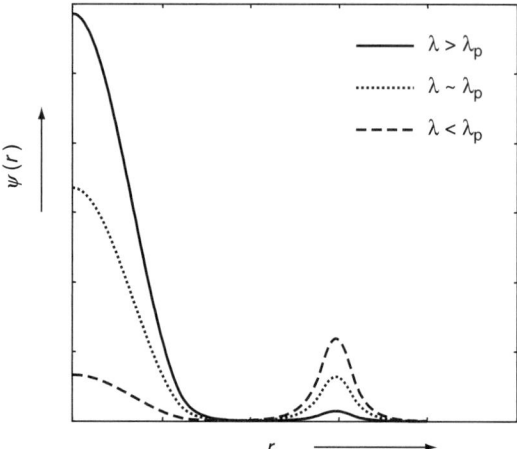

FIGURE 1.23 Typical modal power distributions for three different operating wavelengths relative to phase matching wavelength in a coaxial dual-core DCF.

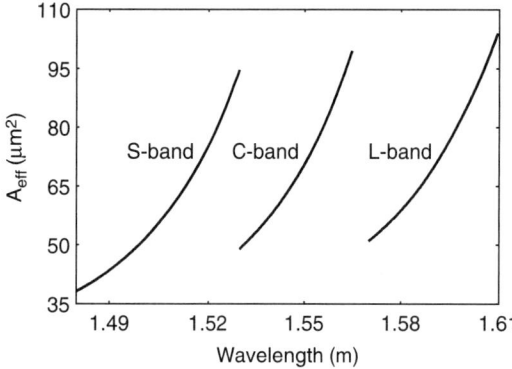

FIGURE 1.24 Wavelength dependence of mode effective area of the various coaxial dual-core DSCFs designed (see Table 1.4) for S-, C-, and L-band fiber amplifiers [57].

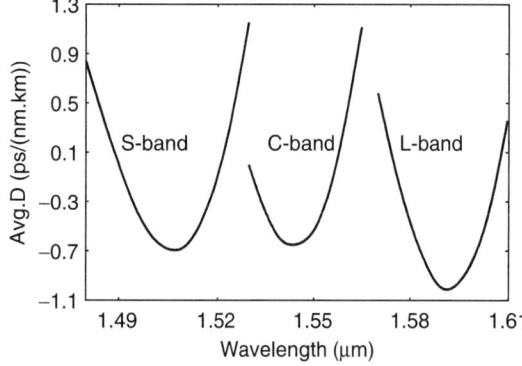

FIGURE 1.25 Net residual dispersion spectra at different amplifier bands of a dispersion compensated link consisting of 100 km of G.652 fiber jointed with about 10 km of the designed coaxial dual-core DSCFs [57].

dispersion fibers are used with the transmission fiber almost in the ratio of 1:1, meaning their characteristic dispersion parameters D and S are almost the same as that of the transmission fiber except for the signs. Thus, for the reverse dispersion fiber used in [60], D and S were -16 ps/nm·km and -0.050 ps/nm^2·km, respectively. Dispersion ramped fiber links, which involve use of RDFs, alternate fiber segments with the link and consist of positive and negative dispersion fibers, so that the overall dispersion is low, and D versus length curve mimics a sawtooth curve. One major attribute of these fibers was that loss in such fibers was low (in contrast to high negative D DCFs), ~ 0.25 dB/km. Terabit transmission experiments through a repeater span of over 125 km were reported in such dispersion managed fiber links involving RDFs in conjunction with G.652 fibers over a bandwidth of 50 nm [63]. Alternate fiber segments in these links consist of positive and negative D (in 1:1 ratio) fibers such that overall dispersion is low and D versus length curve mimics a sawtooth curve. RDFs were also used with NZ-DSFs (in a ratio of 1:2) [64].

4. FIBERS FOR METRO NETWORKS

In recent years, metro optical networks have attracted a great deal of attention from lightwave communication engineers due to potentials for high growth. A metro network provides generalized telecommunication services transporting any kind of signal from point to point in a metro. First-generation metropolitan optical networks based on SDH/SONET relied on rings laced with nodes at which information is electronically exchanged. *Access* that aggregates a wide variety of traffic from business and residential end users is required to port this traffic directly to the transport product for distribution throughout an optical network. In *transport*, DWDM is the key enabling technology to expand the capacity of existing and new fiber cables without optical-to-electrical-to-optical conversions. Metro networks with distances of less than 200 km, which bridges the gap between local/access networks and long distance telecommunication networks, were originally designed to essentially link local switching offices, from which telephone lines branch out to individual customers in the access portion of the network. In larger cities, it may consist of two layers, *metro access or edge loops* (20–50 km), that connect groups of switching offices, and these loops, in turn, link to a *metro core* or *backbone* (50–200 km) network (having several nodes) that serves the whole metropolitan area [65].

TABLE 1.4 Important performance parameters of the designed DSCFs for G.652 fibers [57].

Amplifier band (central λ) (nm)	D (ps/nm·km)	RDS (nm⁻¹)	A_{eff} (μm²)	Mode field diameter (μm)	Estimated FOM (ps/nm·dB)	Bend loss* (dB)
S (1500)	−182	0.0056	51	7.15	771	0.0221
C (1550)	−191	0.0027	70	7.97	941	0.095
L (1590)	−162	0.0034	70	8.12	837	0.0014

*For a single-turn bend of diameter 32 mm.

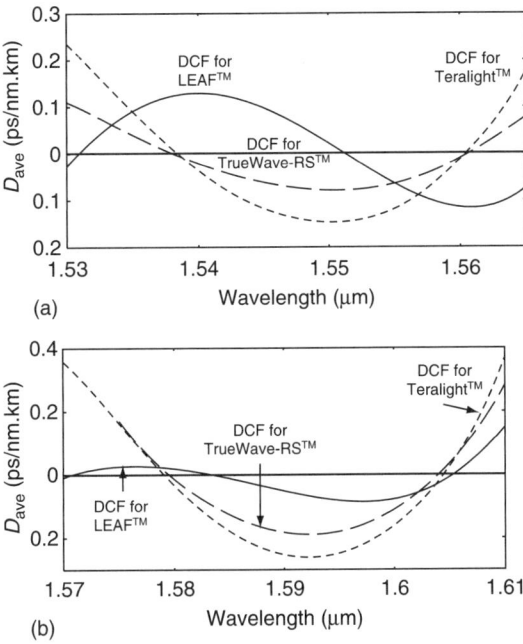

FIGURE 1.26 (a) Net residual dispersion spectra at the C-band of standard NZ-DSF fiber links jointed with designed dual-core DSCFs. (b) Net residual dispersion spectra at the L-band of standard NZ-DSF fiber links jointed with designed dual-core DSCFs.

has been to move toward *transparent* rings, in which wavelength channels are routed past or dropped off at the nodes [66]. The futuristic transparent metro optical networks must have flexibility to route/drop off signals at any node in the network so that the signal(s) may travel along the entire ring in the network. New services are now fed optically at the network edge, for example, *ESCON* (enterprise system connection, which is a 200-Mb/s protocol meant for providing connectivity of a main frame to other mainframe storage devices), *Gigabit Ethernet* (which is a simple protocol that uses inexpensive electronics and interfaces for sharing a local area network), *FDDI* (a local area network standard covering transmission at 100 Mb/s, which can support up to 500 nodes on a dual ring network with an internode spacing up to 2 km), and digital video. Gigabit Ethernet is fast evolving as a universal protocol for optical packet switching. For example, most of the data on the Internet start as Ethernet packets generated by system servers. Thus, in addition to voice, video, and data, a metro network should be able to support various protocols like Ethernet, fast Ethernet, and 10-Gb/s Ethernet.

A modern metro network is a *distribution system* and not just an information transport pipeline connecting a pair of points. It may be described as an updated version of a regional network designed to function as a high capacity system, which provides connectivity between many sites. For example, telephone switching offices, Internet service provider

However, with the growth in Internet usage and increasing demand in data services, a need arose to migrate from conventional network infrastructure to an *intelligent* and *data-centric* metro network. Accordingly, the network trend in the metro sector

TABLE 1.5 Important performance parameters of the designed DSCFs for 6.655 fibers (After [58]; copyright Optical Society of America).

Designed dual-core DSCF for	D (ps/nm·km)	RDS (nm⁻¹)	A_{eff} (μm²)	Mode field diameter (μm)	FOM (ps/nm·dB)	Bend loss* (dB)
LEAF® (C-band)	−264	0.033	41.8	3.34	1248	2.54 × 1.0e−02
LEAF® (L-band)	−172	0.017	43.63	3.47	885	4.28 × 1.0e−03
TrueWave RS® (C-band)	−173	0.0099	49.92	3.68	760	2.35 × 1.0e−06
TrueWave RS® (L-band)	−173	0.0075	61.4	4.02	851	1.89 × 1.0e−06
TeraLight® (S-band)	−201	0.01	45.97	3.42	844	5.2 × 1.0e−02
TeraLight® (C-band)	−187	0.0084	49.99	3.55	873	6.51 × 1.0e−03
TeraLight® (L-band)	−150	0.006	57.77	3.92	771	1.4 × 1.0e−02

Values correspond to central λ in each band as in Table 1.4.
*For a single-turn bend of diameter 32 mm.

servers, corporate, and university campuses are often structured as loops with add/drops at nodes of similar importance.

To offer service providers the flexibility to increase network capacity in response to customer demand, a metro network is required to address unique features like low first cost, high degree of scalability to efficiently accommodate unpredicted traffic growth, flexibility to add/drop individual signals at any central office in the network, interoperability to carry a variety of signals, and provide connectivity to variety of equipment such as cell phones, SONET/SDH, legacy equipment, asynchronous transfer mode (ATM), and Internet protocol (IP) [65]. Legacy metro networks relied heavily on low-cost G.652 fibers having near-zero dispersion at 1310 nm. With G.652 and G.652.C fibers, dispersion limited distances at 10 Gb/s transmission rate would be about 70 km beyond which a conventional DCF with a negative D is required to achieve signal reach in longer metro rings [66]. However, this would add to overall cost, more so because EDFA(s) may be required to offset the extra insertion loss introduced by a DCF.

Directly modulated, distributed feedback (DBF)-lasers in contrast to externally modulated DFB lasers, are an economic advantage in a metro environment. However, directly modulated lasers are usually accompanied with a positive chirp, which could introduce severe pulse dispersion in the EDFA band if the transmission fiber is characterized with a positive D. The positive chirp-induced pulse broadening can be countered with a transmission fiber if it is so designed that it exhibits normal dispersion (i.e., negative D) at the EDFA band. This is precisely the design philosophy followed by certain fiber manufacturers for deployment in a metro network such as MetroCor™ fiber [67]. As a typical example, a schematic of the dispersion spectrum of one such negative dispersion metro fiber relative to that of a standard SMF is shown in Fig. 1.27a. Before the introduction of metro-specific fibers, capacity upgrades were based on lighting of new fibers. In a metro network design, technological adaptability is extremely important to outweigh obsolescence because outside plant fixed costs such as trenching and ducts are almost twice that of long-haul networks [67]. DWDM in a metro environment is attractive in this regard for improved speed in provisioning due to possibility of allocating dynamic bandwidth to customers on demand and for better cost efficiency in futuristic transparent networks running up to 200 km or more. For such distances, a DCF in the form of a standard SMF with positive D could be used to compensate for dispersion in a MetroCor™

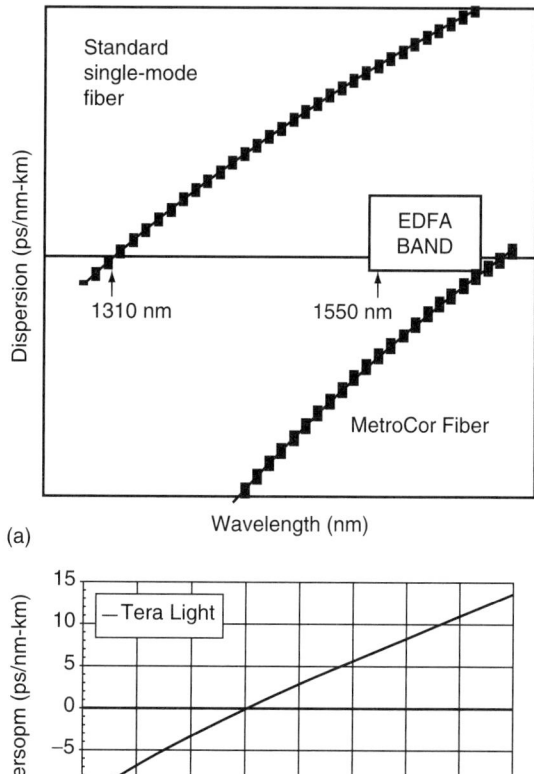

FIGURE 1.27 (a) Dispersion spectrum of a negative dispersion MetroCor™ fiber relative to a standard G.652 single-mode fiber. (Adapted from [67]; © Corning Inc.) (b) Dispersion spectrum of a positive dispersion NZ-DSF TeraLight™ brand metro fiber. (Adapted from [66]; © Alcatel Telecommunications Review 2002) (© Alcatel Telecommunications Review)

kind of fiber. However, due to a relatively low magnitude of D in an SSMF, long lengths of standard SMF would be required, which might require careful balancing of overall loss budget and system cost.

An alternative type of metro fiber has also been proposed and realized, which exhibits positive D ~ 8 ps/nm·km at 1550 nm as shown in Fig. 1.27b [66]. The advantage cited by the manufacturer of such positive dispersion metro fibers is that dispersion compensation could be achieved with efficient already available conventional DCFs of much shorter length(s) as compared with standard SMFs required for negative dispersion equivalent metro fibers. Nevertheless a careful cost analysis for the network as a whole on a case-by-case basis would decide whether to deploy positive or negative NZ-DSFs in a metro network.

5. COARSE WAVELENGTH DIVISION MULTIPLEXING

With the availability of LWPF SMFs (Fig. 1.2) christened as G.652C fibers by ITU, examples of which are AllWave™ and SMF-28e™, a new market has emerged for configuring short distance communication (up to ~ 20 km, e.g., for campus backbone) networks with these fibers. Because of the low-loss wavelength window being wider by 30% (in the E-band) offered by these fibers in comparison with G.652 legacy fibers, one could use these to transmit widely spaced wavelengths as carriers with low-cost, uncooled, FP lasers as the light sources. Recently introduced ITU standards allow channel spacing as wide as 20 nm under a coarse wavelength division multiplexing scheme due to which laser cooling becomes unnecessary because environmental temperature–induced wavelength wander would not be an issue. This would bring down system complexity and overall network cost because demands on wavelength tolerance of other ancillary components like filters would also be much less stringent as compared with those required for DWDM systems.

6. COMBATING PMD IN A FIBER

With growth in transmission capacity to 10 Gb/s and beyond, systems engineers began to encounter a problem in about 20–30% of legacy fibers, which were deployed before 1990, due to what is known as polarization mode dispersion (PMD). PMD arises because of a certain in-built geometric birefringence, as a result of slight noncircularity/asymmetry in fiber cross-section, which could be attributed to the process of fiber manufacture. Because of asymmetry, however small it may be, and the resultant birefringence, an SMF in general supports two orthogonal polarization modes. The birefringence may also be independently caused by mechanical stress due to diurnal or seasonal heating and cooling of a deployed fiber. It could also arise due to, for example, vibrations from a moving train, if deployed in the neighborhood of railway tracks (sometimes laid to overcome the problem of right of way) [68]. Even if deployed as an aerial cable, it be subjected to stress due to swaying caused by wind.

Because of birefringence, whatever the cause, the input pulse power gets distributed among the two *eigen polarization* modes of the fiber as shown in Fig. 1.28. The eigen modes could be understood by considering the example of an elliptic core fiber of dimension $2a$ for the major axis (say, along x) × $2b$ for the minor axis (say, along y). A light polarized

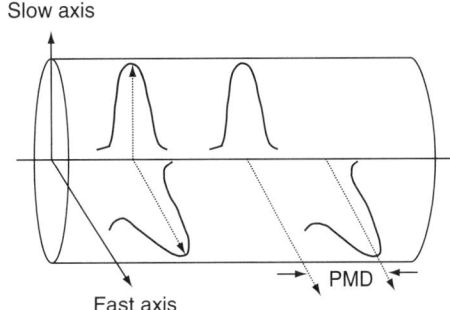

FIGURE 1.28 Schematic demonstrating the build-up of PMD with propagation in a single-mode fiber. (After [70].)

parallel to the major axis of the fiber would preferentially launch light into the natural mode of the fiber (of, say, effective index n^x_{eff}), which is polarized along the major axis because light guidance would be strongly dictated by the V-number, which would involve a as the core dimension. Likewise, a y-polarized incident light would preferentially excite a y-polarized natural mode (of, say, effective index n^y_{eff}) of the fiber. As a result, the pulse energy via one of these modes would travel at a speed slower or faster than the other, giving rise to a differential group delay (DGD), which may be sufficient to cause enough pulse spread to make one pulse overlap with the other and cause intersymbol interference. This group delay difference is formally known as the PMD, and the PMD coefficient (D_{PMD}) is defined for a short length of the fiber as $\Delta N_g/c$, where ΔN_g represents a difference in the group indices [cf. Eq. (4)] of the two eigen polarization modes of the fiber and c is the velocity of light in a vacuum [69]. It is expressed in ps/km; the product D_{PMD} × fiber length yields short-fiber length PMD in ps. It can create problems at high bit rates, such as for 10 and 40 Gb/s systems [14]. It would be an alarming effect if the short-length PMDs were extrapolated linearly to long fiber lengths!

However fortuitously, usually cross-coupling of power between the eigen modes takes place along the fiber length, which eventually populates both the polarization states equally [69]. The instantaneous value of PMD varies randomly with time, temperature, and wavelength. Accordingly, it is described in the form of an average of the Maxwellian density function whose mean value increases with square root of fiber length [70]. The maximum instantaneous DGD is approximately 3.2 times its average value [68]. The PMD of a fiber along with other optical components in a network should be less than $140/B$; B is measured in Gb/s [71]. At 40 and

10 Gb/s, the tolerable total PMD across a link has been estimated to be 3.5 and 14 ps, respectively. For a given bit rate and D_{PMD}, and assuming all mode mixing between the eigen modes, the possible link length L (in km) is given by [72]

$$L = \left[\frac{10^3 f}{B(\text{Gb/s}) \times D_{PMD}(\text{ps}/\sqrt{\text{km}})}\right]^2 \quad (40)$$

The parameter f representing *allowed bit period fraction* is given by [72]

$$f = \text{PMD}(\text{ps}) \times B(\text{Gb/s}) \quad (41)$$

with PMD as the $\text{PMD}_{\text{total}}$, that is, it includes PMD of all the components that exhibit PMD and that populate the link between the transmitter and the receiver

$$\text{PMD}_{\text{total}} = (\text{PMD}_1^2 + \text{PMD}_2^2 + \text{PMD}_3^2 + \ldots)^{1/2} \quad (42)$$

where $\text{PMD} = D_{PMD} \times \sqrt{L}$. Equation (40) is a generalization of a result that assumed $f = 0.1$ [69,72]. It is generally recommended that PMD on a cabled fiber for high-capacity systems should be less than $0.5\,\text{ps}/\sqrt{\text{km}}$ [71]. In some of the legacy fibers installed before the early 1990s, D_{PMD} was $1-2\,\text{ps}\sqrt{\text{km}}$. Such fibers would introduce intolerable DGD even over distances below 500 km at 10 Gb/s. However, since the early 1990s through certain innovations introduced during the fiber drawing step, for example, controlled spinning [73,74], state-of-the-art transmission fibers may exhibit PMD below $0.1\,\text{ps}/\sqrt{\text{km}}$, which is sufficient to maintain DGD below 3.5 ps at 40 Gb/s.

7. CONCLUSION

In this chapter, evolving R&D trends in SMF designs since the early 1980s as well as certain specialty fibers like NZ-DSFs, DCFs, LWPFs, and fibers for metro networks were described. Issues such as PMD, which emerged in more recent years, were also discussed. It should serve as a state-of-the-art review of SMFs as well as an introduction to other chapters on specialty fibers that follow in this book.

8. ACKNOWLEDGMENTS

I gratefully acknowledge academic interactions with several of my colleagues in our Fiber Optics group and with my graduate students over the last two decades. In particular, I acknowledge Professors Ajoy Ghatak and K. Thyagarajan and graduate student Ms. Kamna Pande for their constructive criticism and help in drawing some of the figures on dispersion calculation.

My colleague Dr. Ravi Varshney is thanked for his help in drawing the figure on group delay versus wavelength. I also wish to thank Dr. R. S. Vodhanel of Corning Inc.'s Research Department, Dr. P. Sansonetti of Drake/Alcatel, and Dr. D. Bayert of Alcatel's Research and Innovations Department (Marcoussis) for their help in obtaining published materials related to metro network-centric NZ-DSFs.

9. REFERENCES

1. A.M. Glass, D.J. DiGiovanni, T.A. Straser, A.J. Stenz, R.E. Slusher, A.E. White, A. Refik Kortan, and B. Eggleton, Advances in Fiber Optics, Bell Labs Tech. J., **Jan–Mar.**, pp. 163–187 (2000).
2. F.P. Kapron, D.B. Keck, and R.D. Maurer, Radiation losses in glass optical waveguides, *App. Phys. Lett.* **17**, pp. 423–425 (1970).
3. D.N. Payne and W.A. Gambling, Zero material dispersion in optical fibers, *Electron. Lett.* **11**, pp. 176–178 (1975).
4. J.I. Yamada, S. Machida, and T. Kimura, 2 Gb/s optical transmission experiments at 1.3 µm with 44 km single-mode fiber, *Electron. Lett.* **17**, pp. 479–480 (1981).
5. R.J. Mears, L. Reekie, I.M. Jauncy, and D.N. Payne, Low-noise erbium-doped fiber amplifier operating at 1.54 µm, *Electron. Lett.* **23**, pp. 1026–1027 (1987).
6. T. Miya, Y. Terunume, T. Hosaka, and T. Miyashita, An ultimate low-loss single-mode fiber at 1.55 µm, *Electron. Lett.* **15**, pp. 106–108 (1979).
7. S.K. Kartalopoulos, *Introduction to DWDM Technology*, SPIE Press, Bellingham, Washington and IEEE Press, Piscataway, NJ (2000).
8. F.P. Kapron, Chapter on "Transmission properties of optical fibers" in *Optoelectronic Technology and Lightwave Communication Systems*, Chinlon Lin (Ed.), Van Nostrand, New York (1989).
9. K. Kitayama, Y. Kato, M. Ohashi, Y. Ishida, and N. Uchida, Design considerations for the structural optimization of a single-mode fiber, IEEE J. Lightwave Tech. **LT-1**, pp. 363–369 (1983).
10. B.P. Pal, Chapter on "Transmission characteristics of telecommunication optical fibers" in *Fundamentals of Fiber Optics in Telecommunication and Sensor Systems*, B.P. Pal (Ed.), John Wiley, New York (1992).
11. N. Uchida, Development and future prospects of optical fiber technologies, IEICE Trans. Electron. **E85 C**, pp. 868–880 (2002).
12. A. Yariv, *Optical Electronics in Modern Communication*, Oxford University Press, New York (1997).
13. A. Ghatak and K. Thyagarajan, *Introduction to Fiber Optics*, Cambridge University Press, Cambridge (1998).

14. B.P. Pal, Chapter on "Optical Transmission" in *Perspective in Optoelectronics*, S.S. Jha (Ed.), World Scientific, Singapore (1995).
15. D. Marcuse, Interdependence of waveguide and material dispersion, *App. Opt.* **18**, pp. 2930–2932 (1979).
16. D.N. Payne and W.A. Gambling, Zero material dispersion in optical fibers, *Electron. Lett.* **11**, pp. 176–178 (1975).
17. G.P. Agrawal, *Lightwave Technology: Components and Devices*, John Wiley and Sons, New Jersey (2004).
18. J. Powers, *An Introduction to Fiber Optic Systems*, 2nd ed., Irwin, Chicago (1997).
19. R.J. Boehm, Y.C. Ching, C.G. Griffith, and F.A. Saal, Standardized fiber optic transmission system—a synchronous optical network view, *IEEE J. Selected Areas in Commn.* **SAC-4**, pp. 1424–1431 (1986).
20. B.P. Pal, Chapter on "Optical fibers for lightwave communication: design issues," in *Fiber Optics and Applications*, A.K. Ghatak, B. Culshaw, V. Nagarajan and B.D. Khurana (Eds.), VIVA publishers, New Delhi (1995).
21. R. Tewari, B.P. Pal, and U.K. Das, Dispersion-shifted dual-shape-core fibers: optimization based on spot size definitions, *IEEE J. Lightwave Tech.* **10**, pp. 1–5 (1992).
22. R.J. Mears, L. Reekie, S.B. Poole, and D.N. Payne, Low-threshold tunable cw and Q-switched fiber laser operating at 1.55 μm, *Electron. Lett.* **22**, pp. 159–160 (1986).
23. C.J. Koester and E. Snitzer, Amplification in a fiber laser, *App. Opt.* **3**, pp. 1182–1184 (1964).
24. E. Desurvire, *Erbium Doped Fiber Amplifier: Principle And Applications*, Wiley-Interscience, New York (1994).
25. R.I. Laming, A.H. Gnauck, C.R. Giles, M.N. Zervas, and D.N. Payne, "High sensitivity optical preamplifier at 10 Gbit/s employing a low noise composite EDFA with 46 dB gain," in Post-deadline session, Opt. Amp. and their Applications, OSA Topical Meeting, Washington, DC, Post-deadline Paper PD13, (1992).
26. A.K. Srivastava, Y. Sun, J.W. Sulhoff, C. Wolf, M. Zirngibl, R. Monnard, A.R. Chraplyvy, A.A. Abramov, R.P. Espindola, T.A. Strasser, J.R. Pedrazzani, A.M. Vengsarkar, J.L. Zyskind, J. Zhou, D.A. Ferrand, P.F. Wysocki, J.B. Judkins, S.W. Granlund, and Y.P. Li, "1 Tb/s Transmission of 100 WDM 10 Gb/s Channels over 400 km of Truewave® Fiber," Technical Digest *Optical Fiber Communication Conference* 1998, Opt. Soc. Am., Washington, DC, Post-deadline paper PD10 (1998).
27. ITU-T Recommendation G.692: *Optical Interfaces for Multichannel Systems With Optical Amplifiers*, International Telecommunications Union, Geneva, Switzerland.
28. K.M. Abe, Optical fiber designs evolve, *Lightwave* (February 1998).
29. T. Li, The impact of optical amplifiers on long-distance lightwave telecommunications, *Proc. IEEE*, **81**, pp. 1568–1579 (1995).
30. T. Li and C.R. Giles, "Optical amplifiers in lightwave telecommunications," Chapter 9, in *Perspective in Optoelectronics*, S.S. Jha (Ed.), World Scientific, Singapore, 1996.
31. A.R. Chraplyvy, R.W. Tkach, and K.L. Walker, "Optical fiber for wavelength division multiplexing," U.S. Patent 5,327,516 (issued July 5, 1994).
32. D.W. Peckham, A.F. Judy, and R.B. Kummer, "Reduced dispersion slope, non-zero dispersion fiber," in *Proceedings of the 24th European Conference on Optical Communication ECOC'98* (Madrid, 1998), pp. 139–140 (1998).
33. Y. Liu, W.B. Mattingly, D.K. Smith, C.E. Lacy, J.A. Cline, and E.M. De Liso, "Design and fabrication of locally dispersion-flattened large effective area fibers," in *Proceedings of the 24th European Conference on Optical Communication ECOC'98* (Madrid, 1998), pp. 37–38 (1998).
34. J. Ryan, Special report: ITU G.655 adopts higher dispersion for DWDM, *Lightwave*, **18**, no. 10 (2001).
35. Y. Danziger and D. Askegard, High-order-mode fiber an innovative approach to chromatic dispersion management that enables optical networking in long-haul high-speed transmission systems, *Opt. Networks Mag.* **2**, pp. 40–50 (2001).
36. Y. Frignac and S. Bigo, "Numerical optimization of residual dispersion in dispersion managed systems at 40 Gb/s," in *Proceedings of the Optical Fiber Communications Conference OFC 2000* (Baltimore, MD, 2000), pp. 48–50 (2000).
37. H. Izadpanah, C. Lin, H. Gimlett, H. Johnson, W. Way, and P. Kaiser, "Dispersion compensation for upgrading interoffice networks built with 1310 nm optimized SMFs using an equalizer fiber, EDFAs, and 1310/1550 nm WDM," Tech Digest Optical Fiber Communication Conference, Post-deadline paper PD15, pp. 371–373 (1992).
38. Y. Akasaka, R. Suguzaki, and T. Kamiya, "Dispersion Compensating Technique of 1300 nm Zero-dispersion SM Fiber to get Flat Dispersion at 1550 nm Range," in *Digest of European Conference on Optical Communication*, paper We.B.2.4 (1995).
39. A.J. Antos and D.K. Smith, Design and characterization of dispersion compensating fiber based on LP_{01} mode, *IEEE J. Lightwave Tech.* **LT-12**, pp. 1739–1745 (1994).

40. D.W. Hawtoff, G.E. Berkey, and A.J. Antos, "High Figure of Merit Dispersion Compensating Fiber," in *Technical Digest Optical Fiber Communication Conference* (San Jose, CA), Post-deadline paper PD6 (1996).
41. A.M. Vengsarkar, A.E. Miller, and W.A. Reed, "Highly efficient single-mode fiber for broadband dispersion compensation," in *Proceedings of Optical Fiber Communications Conference OFC'93* (San Jose, CA), pp. 56–59 (1993).
42. L. Grüner-Nielsen, S.N. Knudsen, B. Edvold, T. Veng, B. Edvold, T. Magnussen, C.C. Larsen, and H. Daamsgard, Dispersion Compensating Fibers, *Opt. Fib. Tech.* **6**, pp. 164–180 (2000).
43. F. Ouellette, Dispersion cancellation using linearly chirped Bragg grating filters in optical waveguides, *Opt. Lett.* **12**, pp. 847–849 (1987).
44. H.G. Winful, Pulse compression in optical fibers, *Appl. Phys. Lett.* **46**, pp. 527–529 (1985).
45. B.J. Eggleton, T. Stephens, P.A. Krug, G. Dhosi, Z. Brodzeli, and F. Ouellette, "Dispersion compensation over 100 km at 10 Gb/s using a fiber grating in transmission," in *Technical Digest Optical Fiber Communication Conference* (San Jose, CA), Post-deadline paper PD5 (1996).
46. B.P. Pal, "All-fiber guided wave components," in *Electromagnetic Fields in Unconventional Structures and Materials*, O.N. Singh and A. Lakhtakia (Eds.), John Wiley, New York, pp. 359–432 (2000).
47. C.D. Poole, J.M. Weisenfeld, and D.J. Giovanni, Elliptical-core dual-mode fiber dispersion compensator, *IEEE Photon. Tech. Lett.* **5**, pp. 194–197 (1993).
48. S. Ramachandran, B. Mikkelsen, L.C. Cowsar, M.F. Yan, G. Raybon, L. Boivin, M. Fishteyn, W.A. Reed, P. Wisk, D. Brownlow, R.G. Huff, and L. Gruner-Nielsen, All-fiber, grating-based, higher-order-mode dispersion compensator for broadband compensation and 1000-km transmission at 40 Gb/s, *IEEE Photon. Tech. Lett.* **13**, pp. 632–634 (2001).
49. S. Ramachandran, S. Ghalmi, S. Chandrasekhar, I. Ryazansky, M. Yan, F. Dimarcello, W.A. Reed, and P. Wisk, Wavelength-continuous broadband adjustable dispersion compensator using higher order mode fibers and switchable fiber-gratings, *IEEE Photon. Tech. Lett.* **15**, pp. 727–729 (2003).
50. M. Onishi, Y. Koyana, M. Shigematsu, H. Kanamori, and M. Nishimura, Dispersion compensating fiber with a high figure of merit of 250 ps/nm·dB, *Electron. Letts.* **30**, pp. 161–163 (1994).
51. B. Jopson and A. Gnauck, Dispersion compensation for optical fiber systems, *IEEE Comm. Mag.* **June**, pp. 96–102 (1995).
52. V. Srikant, "Broadband dispersion and dispersion slope compensation in high bit rate and ultra long haul systems," in *Technical Digest Optical Fiber Communication Conference*, Anaheim, CA, paper TuH1 (2001).
53. K. Thyagarajan, R.K. Varshney, P. Palai, A.K. Ghatak, and I.C. Goyal, A novel design of a dispersion compensating fiber, *IEEE Photon. Tech. Letts.* **8**, pp. 1510–1512 (1994).
54. J.L. Auguste, R. Jindal, J.M. Blondy, M. Clapeau, J. Marcou, B. Dussardier, G. Monnom, D.B. Ostrowsky, B.P. Pal, and K. Thyagarajan, −1800 ps/km-nm chromatic dispersion at 1.55 μm in a dual concentric core fiber, *Electron. Lett.* **36**, pp. 1689–1691 (2000).
55. K. Thyagarajan and C. Kakkar, Fiber design for broadband, gain flattened Raman fiber amplifier, *IEEE Photon. Tech. Letts.* 15, pp. 1701–1703 (2003).
56. P. Palai, R.K. Varshney, and K. Thyagarajan, A dispersion flattening dispersion compensating fiber design for broadband dispersion compensation, *Fib. Int. Opt.* **20**, pp. 21–27 (2001).
57. B.P. Pal and K. Pande, Optimization of a dual-core dispersion slope compensating fiber for DWDM transmission in the 1480–1610 nm band through G.652 single-mode fibers, *Opt. Comm.* **201**, pp. 335–344 (2002).
58. K. Pande and B.P. Pal, Design optimization of a dual-core dispersion compensating fiber with high figure of merit and a large effective area for dense wavelength division multiplexed transmission through standard G.655 fibers, *Appl. Opt.* **42**, pp. 3785–3791 (2003).
59. I.C. Goyal, R.K. Varshney, B.P. Pal, and A.K. Ghatak, "Design of a small residual dispersion fiber along with a matching DCF," U.S. Patent 6,650,812 B2 (2003).
60. K. Mukasa, Y. Akasaka, Y. Suzuki, and T. Kamiya, "Novel network fiber to manage dispersion at 1.55 μm with combination of 1.3 μm zero dispersion single-mode fiber," in *Proceedings of the 23rd European Conf on Opt. Commn. (ECOC)*, Edinburgh, Session MO3C, pp. 127–130 (1997).
61. K. Mukasa and T. Yagi, "Dispersion flat and low nonlinear optical link with new type of reverse dispersion fiber (RDF-60), in *Tech. Dig. Opt. Fib. Comm. Conference (OFC'2001)*, Anaheim, CA, Paper TuH7 (2001).
62. T. Grüner-Nielsen, S.N. Knudsen, B. Edvold, P. Kristensen, T. Veng, and D. Magnussen, Dispersion compensating fibers and perspectives for future developments," in *Proceedings of the 26th European Conference Opt. Comm. (ECOC'200)*, pp. 91–94 (2000).
63. Y. Miyamoto, K. Yonenaga, S. Kuwahara, M. Tomizawa, A. Hirano, H. Toba, K. Murata, Y. Tada, Y. Umeda, and H. Miyazawa, "1.2 Tbit/s (30×42.7 Gb/s ETDM optical channel) WDM transmission over 376 km with 125-km spacing using forward error

correction and carrier suppressed RZ format," in *Tech. Dig. Opt. Fib. Comm. Conference (OFC'2000)*, Baltimore, MD, Post-deadline paper, PD26 (2000).

64. L.-A. de Montmorrilon, F. Beamont, M. Gorlier, P. Nouchi, L. Fleury, P. Sillard, V. Salles, T. Sauzeau, C. Labatut, J.P. Meress, B. Dany, and O. Leclere, "Optimized Teralight®/reverse Teralight© dispersion-managed link for 40 Gbit/s dense WDM ultra-long haul transmission systems," in *Proceedings of the 27th Europ. Conf. Opt. Comm. (ECOC'2001)*, Amsterdam, pp. 464–465 (2001).

65. J. Hecht, *Understanding Fiber Optics*, 4th ed., Prentice Hall, Upper Saddle River, NJ (2002).

66. J. Ryan, Fiber considerations for metropolitan networks, *Alcatel Telecom. Review* **1**, pp. 52–56 (2002).

67. D. Culverhouse, A. Kruse, C-C. Wang, K. Ennser, and R. Vodhanel, "Corning® MetroCore™ fiber and its applications in metropolitan networks," Corning Inc. White Paper WP5078 (2000).

68. ProForum Tutorial on Polarization mode dispersion: http://www.iec.org

69. F. Kapron A. Dori, J. Peters, and H. Knehler, "Polarization-mode dispersion: should you be concerned?" in *Proceedings of the Nat. Fiber Optic Eng. Conf. (NFOEC'96)*, pp. 756–758 (1997).

70. J.J. Refi, Optical fibers for optical networking, Bell Labs. Tech. J., **January-March**, pp. 246–261 (1999).

71. C.D. Poole and J. Nagel, "Polarization effects in lightwave systems," in *Optical Fiber Telecommunications*, Vol. IIIA, I.P. Kaminow and T.L. Koch (Eds.), Academic Press, San Diego, CA, pp. 114–161 (1997).

72. F.P. Kapron, "System considerations for polarization mode dispersion," in *Proceedings of the Nat. Fiber Optic Eng. Conf. (NFOEC'97)*, San Diego, CA, pp. 433–444 (1997).

73. A.C. Hart, R.G. Huff, and K.L. Walker, "Method of making a fiber having low polarization mode dispersion due to a permanent spin," U.S. Patent 5,298,047 (1994).

74. A.F. Judy, "Improved PMD stability in optical fibers and cables," in *Proceedings of the 43rd Intl. Wire and Cable Symposium*, Atlanta, GA, Nov. 14–17 (1994).

2

CHAPTER

Recent Development of a Polymer Optical Fiber and its Applications

Pak L. Chu*
Optoelectronics Research Centre
Department of Electronic Engineering
City University of Hong Kong
Kowloon, Hong Kong

1. INTRODUCTION

Glass optical fiber has received intense research and development over the past 30 years, forming the backbone of optical communication and sensing. More recently, this fiber has found applications in medical diagnosis and surgery. New types of glass optical fiber have also been developed, notably the photonic crystal fiber or holey fiber [1]. On the other hand, polymer optical fiber (POF) was developed more than 20 years ago by DuPont with poly-methal-methacrylate (PMMA) as the constitutent material. However, its attenuation was very large (>200 dB/km) compared with that of glass optical fiber. Hence, it found applications only in systems involving the use of short lengths of fiber, such as back plane interconnection and decorations. In the past 10 years, however, POF has received renewed attention mainly because of its applications in motor vehicles and home networks and also because of the development of low-loss POF. The purpose of this article is to give a brief review of the latest development of POF and its various applications.

2. TYPES OF POFs

The most important parameter of an optical fiber for applications in communication is its attenuation. Whereas silica optical fiber has attenuation less than 0.2 dB/km at 1550 nm, polymer fiber has a much larger figure, making it suitable only for short distance application. From the attenuation point of view, there are three types of POFs that receive popular attention: PMMA fiber, deuterated PMMA fiber, and perfluorinated POF.

2.1. PMMA Fiber

The most popular POF is made of PMMA material, whose chemical composition is shown in Fig. 2.1. The molecules in the PMMA chain oscillate with a resonant absorption wavelength between 3.3 and 3.5 μm. Its harmonics extend into the visible range of light so that the theoretical minimum attenuation of this fiber is 100 dB/km at 0.65 μm. However, the practical figure is about 200 dB/km. Thus, this fiber is used for distances of over several hundred meters. The bandwidth of the fiber depends on its refractive index profile, and multimode fiber is often used. The glass transition temperature is about 110° C.

2.2. Deuterated PMMA POF

The large attenuation in the PMMA POF occurs because the resonant wavelength of the carbon–hydrogen bond is too close to the visible wavelength of light. Hence, one way to reduce the attenuation is to somehow remove the resonant wavelength to a larger value. This has been done by replacing the hydrogen atom in the molecular chain with a heavy hydrogen atom, that is, one with two protons in the nucleus instead of one as in the ordinary hydrogen atom. The resultant resonant wavelength then appears at 4.5 μm, and its minimum attenuation can be as low as 25 dB/km at 650 μm. However, this fiber tends to absorb water easily. In doing so, the attenuation rapidly increases. This fiber therefore is not widely used.

2.3. Perfluorinated POF

To reduce the loss further, it is necessary to remove the hydrogen atom completely from the polymer and replace it by a heavy atom. This has been achieved by means of introducing fluorine atoms into the polymer; the fiber thus fabricated is called

---

*E-mail: eepchu@cityu.edu.hk

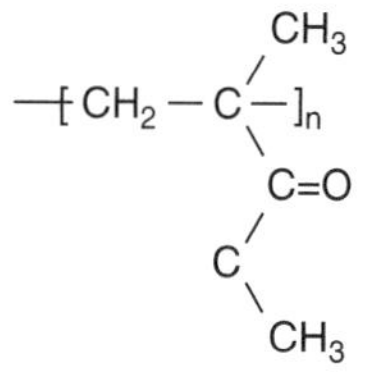

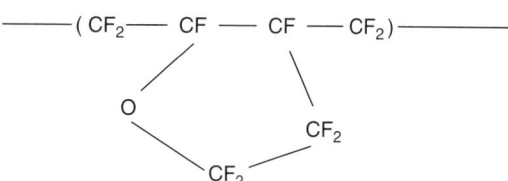

FIGURE 2.1 Chemical composition of PMMA.

FIGURE 2.2 Chemical composition of perfluorinated POF.

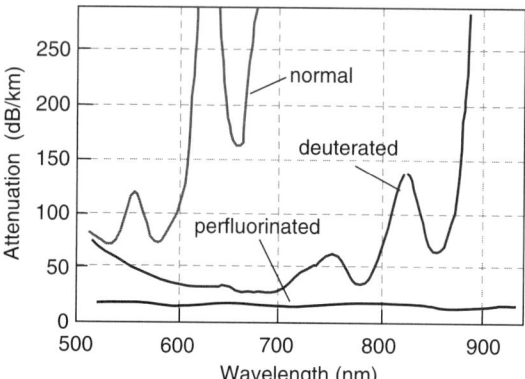

FIGURE 2.3 Attenuation spectra of POFs.

perfluorinated POF [2]. Its chemical composition is shown in Fig. 2.2.

The resonant absorption wavelength of this fiber extends from 7.7 μm to 10 μm. Because this is very far from the visible range, its attenuation is correspondingly small. It has been shown [3] that its theoretical value at 1300 nm is 0.5 dB/km and at 1550 nm is 0.3 dB/km. These figures happen to coincide with those of silica fiber. The best perfluorinated POF made so far has an attenuation figure of 10 dB/km. This is a great improvement over the PMMA fiber. There are other attractive features of this fiber that are explained in the following sections.

Figure 2.3 shows the attenuation spectra of the three types of POF. It is of interest to note that the perfluorinated POF has a nearly flat and low attenuation over a wide range of wavelengths.

3. MANUFACTURE OF POFS

We concentrate our discussion on the manufacture of PMMA POF. There are two methods of fabricating POF: the preform and drawing method and the extrusion method. In the extrusion method, the starting material is PMMA in powder or in pebble form that can be purchased directly from polymer companies. In the preform and drawing method, the starting material is the MMA monomer. We discuss this method first [4].

3.1. Preform and Drawing Method

3.1.1. Material Preparation

The MMA monomer can be bought from a commercial company. Some chemicals have been added to inhibit polymerization taking place during storage and transport. Therefore, the first step in preparing the material is to remove these inhibitors. The common inhibitors are hindered phenols such as butylated hydroxy toluene. The inhibitor stops polymerization by reacting with free radicals in the monomer, as shown in Fig. 2.4. The inhibitor can be removed by distillation, extraction, or chromatography. Alternatively, one can simply use extra initiator to overwhelm the inhibitor.

The monomer is then degassed to remove all the air bubbles. It is also necessary to get rid of the impurities in the monomer to avoid excessive scattering loss of the resultant fiber. To start the polymerization process, we need to add to the MMA monomer with an initiator such as lauryl peroxide or AIBN (Azobisisobutyronitrile), the chemical composition of which is given in Fig. 2.5. In addition,

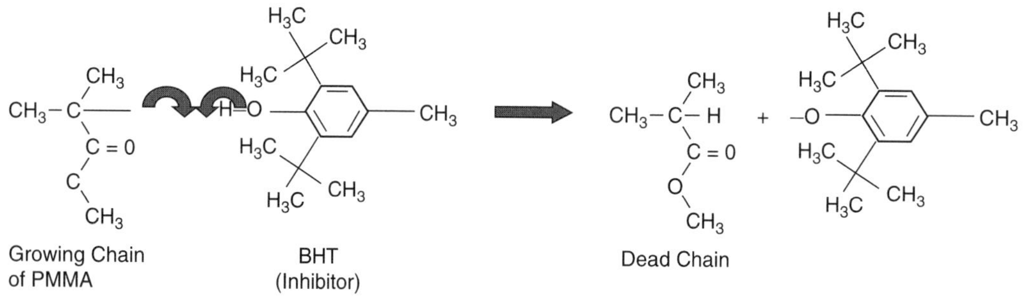

FIGURE 2.4 Function of inhibitor. BHT, butylated hydroxy toluene.

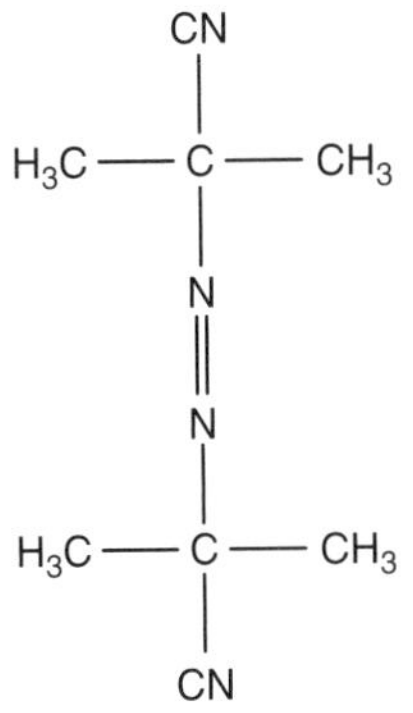

FIGURE 2.5 Chemical Composition file, initiator, AIBN.

we need to put in a chain transfer agent to adjust the molecular weight of the polymer. A chain transfer is a kind of termination reaction in that the growing chain radical reacts with the agent that has an even number of electrons. The growing chain is terminated, but another radical is formed. That new radical may or may not reinitiate polymerization, depending on its structure. Various compounds can act as chain transfer agents, but the most common ones are structures with an S—H bond known as mercaptans, the chemical composition of which is shown in Fig. 2.6, or thiols.

The growing polymer chain end radical abstracts the hydrogen atom from the mercaptan along with one of the two electrons in the S—H bond of the mercaptan. This terminates the growing chain with a hydrogen atom. One electron from the former S—H bond remains behind, creating a new radical on sulfur. This new radical can act as an initiator, reacting with additional monomer to grow a new chain. In this way, the deliberately added chain transfer agent terminates the growing polymer chain sooner than would be the case without the agent. This reduces the molecular weight.

The MMA monomer has a refractive index of about 1.402. This is used as the cladding material of the fiber. However, for the core material, we need to raise its refractive index. This is achieved by adding an appropriate amount of dopant. The index increase is linearly proportional to the amount of dopant. There is a variety of dopants that can be used, such as benzyl methAcrylate, bromobenzen, benzyl butyl phthalate, benzyl benzoate, or diphenyl phthalate. Similarly, to reduce the index instead, one has to simply add trifluro ethyl methacrylate.

$H\text{-}S\text{-}C_4H_9$

FIGURE 2.6 Chemical composition of mercaptan.

3.1.2. Polymerization

We can summarize the polymerization process by the following chemical equation:

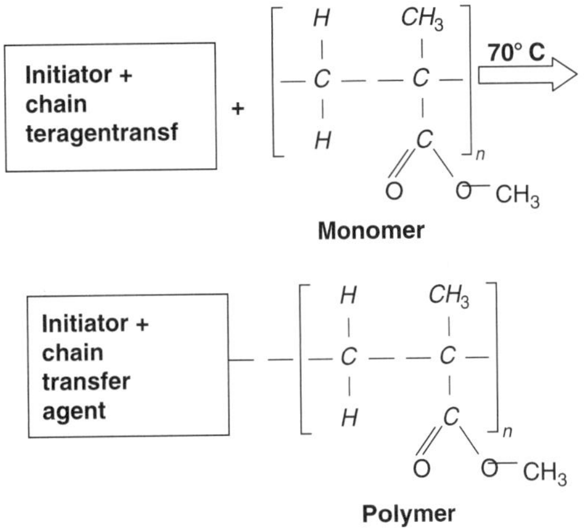

We start to polymerize the cladding of the preform first. The MMA monomer together with the initiator and transfer agent is poured into a glass tube that is rotated at high speed (3000 rpm) in an oven set at a temperature of 70° C. The high-speed rotation is needed to create an axial space for the core monomer to be introduced after the cladding is polymerized. The polymerization takes 1 or 2 days.

The next step after the cladding is polymerized is to start the polymerization of the core by pouring index raising dopant, initiator, and chain transfer agent into the space with prepared monomer material, that is, MMA monomer. The filled tube is again placed in an oven at 95° C and rotated at 50 rpm for 24 h. The whole preform is then polymerized and is ready for drawing into fiber. Figure 2.7 shows a sample of the preform.

To fabricate a graded index preform, we rely on the interfacial gel effect [5] in which the polymerization begins from the core–cladding boundary. We also assume that the dopant molecules are larger than the MMA molecules so that the latter molecules can diffuse toward the boundary and become polymerized first. In this way, the dopant concentration gets larger toward the center. Thus, a graded index preform is created. The exact profile is much more difficult to achieve than a silica preform by chemical vapor deposition.

3.1.3. Fiber Drawing

The drawability of a POF preform is very sensitive to its glass transition temperature, which is about 110° C, and its average molecular weight, which

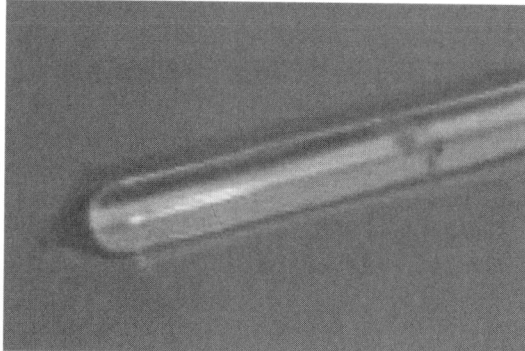

FIGURE 2.7 POF preform.

should be less than 80,000. In the drawing process, the tip of the preform is slowly fed into the furnace under computer control at a temperature between 280 and 290° C, as shown in Fig. 2.8. The preform method allows us to manufacture a variety of POFs, for example, single mode and multimode fiber, twin-core fiber, dye-doped fiber, inorganic-doped fiber (e.g., erbium), and electrooptic fiber. The disadvantage of the preform method is that only a finite length of the preform is made each time, which means that a limited amount of fiber is produced. This is not commercially attractive.

3.2. Extrusion Method

With this method, the starting material is not monomer but polymer in the form of powder or pebbles that can be bought directly from commercial suppliers. Figure 2.9 shows the extrusion setup [6,7]. The core powder is introduced through feeder 1 while the cladding powder is fed through feeder 2. Both powders are pushed toward the output die at the exit end of the diffusion zone by a set of feed

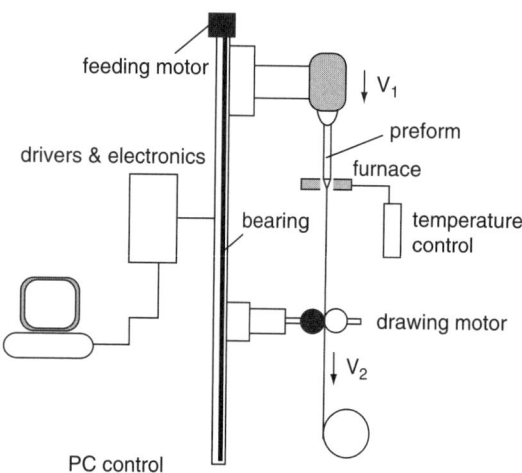

FIGURE 2.8 POF drawing machine.

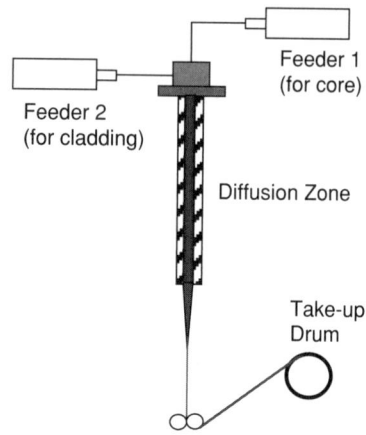

FIGURE 2.9 Extrusion of POF.

screws. There is a temperature gradient throughout the diffusion zone, reaching a melt temperature of about 280° C. This serves to help the creation of graded-index fiber. The extrusion rate varies between 93 and 245 g/hr. The length of the diffusion zone is about 6.5 cm. The die has two concentric nostrils serving to create the core and cladding of the fiber. The diameters of the nostrils determine the dimensions of the fiber core and cladding.

The advantage of the extrusion method is its ability to produce a very long length of fiber, which is commercially attractive. However, the fiber purity cannot be controlled but depends on the purity of the starting material.

4. COMPARISON BETWEEN SILICA FIBER AND POLYMER FIBER

4.1. Difference in Diameters

The standard multimode silica fiber for communication has a core diameter of 62.5 μm and an overall fiber diameter of 125 μm. In the case of polymer fiber, the standard core diameter is 900 μm and the overall fiber diameter, 1000 μm. Thus, the cross-section of polymer fiber is much larger than that of silica fiber. Consequently, it is easier to join two polymer fibers with low loss than the silica fiber. This is significant because it leads to more economical installation of optical fiber systems.

The recent low-loss POF development [8] leads to a smaller core diameter in the order of 120 μm and a cladding diameter of 250 μm. The reason for such a reduction is the cost of the polymer, especially the perfluorinated polymer. However, such a reduction removes the advantage of the ease of joining. To keep the advantage, the fiber is jacketed with a sleeve so that the overall diameter is still about

1 mm. The multilayer structure of the fiber may introduce an undesirable effect such as the nonconcentricity of the core, leading to excess joint loss [9].

4.2. Minimum Bend Radius

Polymer fiber has a much smaller Young's modulus than silica, about 30 times less. It is therefore easier to negotiate bends without breakage. Hence, the minimum bend radius of a polymer fiber is smaller than that of silica fiber for the same diameter. For example, a 125 μm overall diameter polymer fiber has a minimum bend radius of 1 cm compared with 3 cm for silica fiber. On the other hand, a 1 mm diameter polymer fiber has a minimum bend radius of 5 cm compared with 22 cm for silica fiber of the same diameter. It is noted that the bend radius increases linearly with the overall fiber diameter.

The ability to negotiate a sharp corner finds applications in very confined spaces, such as inside an automobile or a cramped compartment. This is the main reason polymer fiber is being increasingly used in automobiles [10].

4.3. Numerical Aperture

The numerical aperture of an optical fiber is determined by the relative magnitudes of the refractive indices in the core and in the cladding. In the case of silica fiber, the difference between these indices cannot be too large, otherwise the residual thermal stresses in these regions will be so large the fiber will fracture. It is rare to find silica fiber with a numerical aperture larger than 0.3. However, in the case of polymer fiber, the stress build-up is less severe because of the small Young's modulus. In fact, the fiber does not fracture. It simply yields gradually if the strain is too great. Hence, it is common to find these fibers with a numerical aperture as large as 0.7 or more.

Because the optical power carrying capacity of the fiber is linearly proportional to the square of its numerical aperture, it is desirable to have a fiber with a large numerical aperture for applications in local area networks, especially in a bus configuration in which computers hang down from the fiber and drain some parts of the light.

4.4. Fiber Bandwidth

Figure 2.10 shows the comparison of bandwidths of the silica multimode fiber and polymer multimode fiber. Both fibers have a parabolic index profile. It can be seen that the polymer fiber has a larger bandwidth. The reason is that its material dispersion

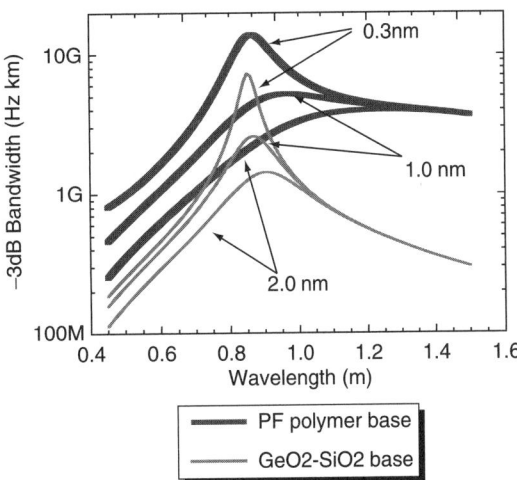

FIGURE 2.10 Bandwidth of graded-index multimode fibers.

is less than that of silica fiber, as shown in Figure 2.11 [11].

Figure 2.11 shows that the material dispersions of all different kinds of polymer fiber approach the horizontal axis asymptotically, whereas the silica fiber crosses the axis at a wavelength near 1.3 μm and continues to increase nearly linearly, so much so that at 1.55 μm it reaches a value of 17 ps/nm·km. This shows that polymer fiber basically provides more bandwidth than silica fiber.

5. APPLICATIONS OF POFs

POF finds applications in situations where short length of fiber is required. Broadly, it can be divided into (1) short distance communication, (2) sensors, and (3) illumination.

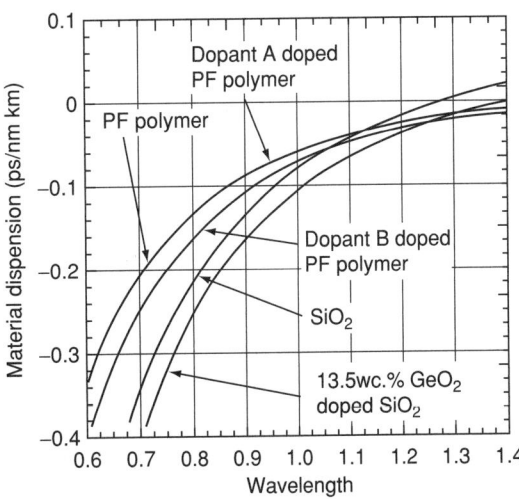

FIGURE 2.11 Fiber material dispersion.

5.1. Communication

PMMA fiber has an attenuation figure of about 100 dB/km. Its useful length would therefore be less than 100 m. For multimode step index fiber, its bandwidth is on the order of 10 Mb/s·km. Hence, this fiber would be useful for communication systems less than 100 m with a bandwidth 100 Mb/s. It finds applications in automobile optical networks and back plane data communication. However, the low-loss perfluorinated fiber can be used for distances up to 1 km because its attenuation is now reduced to 10 dB/km. Furthermore, its bandwidth can reach 2.5 Gb/s. Hence, this fiber is useful for office networks and home networks.

5.1.1. Automobile Optical Networks

Modern motor vehicles are equipped with many communication devices, as shown in Fig. 2.12. If these are wired together by copper wires, there will be severe interference problems. Some of the European automobile manufacturers have come together to develop a standard communication protocol for motor cars, called MOST (media-oriented system transport) [12]. It separates the communication devices given in Fig. 2.12 into different networks in accordance with the bandwidth required as shown in Fig. 2.13.

In-Car Communication Domains

— Power Train
— Body Electronics
  - Instrument Cluster
  - Air-Conditioning
  - Anti-Theft
  - Window Control
  - Mirror Adjust
  - Seat Adjust
— Chassis Systems (x-by-wire)
  - Steer-by-wire
  - Brake-by-wire
— Surround Sensing
  - Access to Diagnosis Data
  - Electronic Logbook
  - Electronic User Manual
— Infotainment
  - Navigation
  - Telephone
  - Audio/Video
  - Internet
  - Multimedia

FIGURE 2.12 Different communication devices inside a modern car.

Figure 2.14 shows the network configurations for different groups of devices. Each network must satisfy the ISO/OSI seven-layer model. The total length of fiber used in the car is less than 30 m.

Figure 2.15 shows the locations of different networks within the automobile.

5.1.2. Back Plane Interconnect

There two types of back plane interconnections: intrasystem and intersystem. In the first system, it is the connection by POF from one subsystem to another within a large system. In this case, because

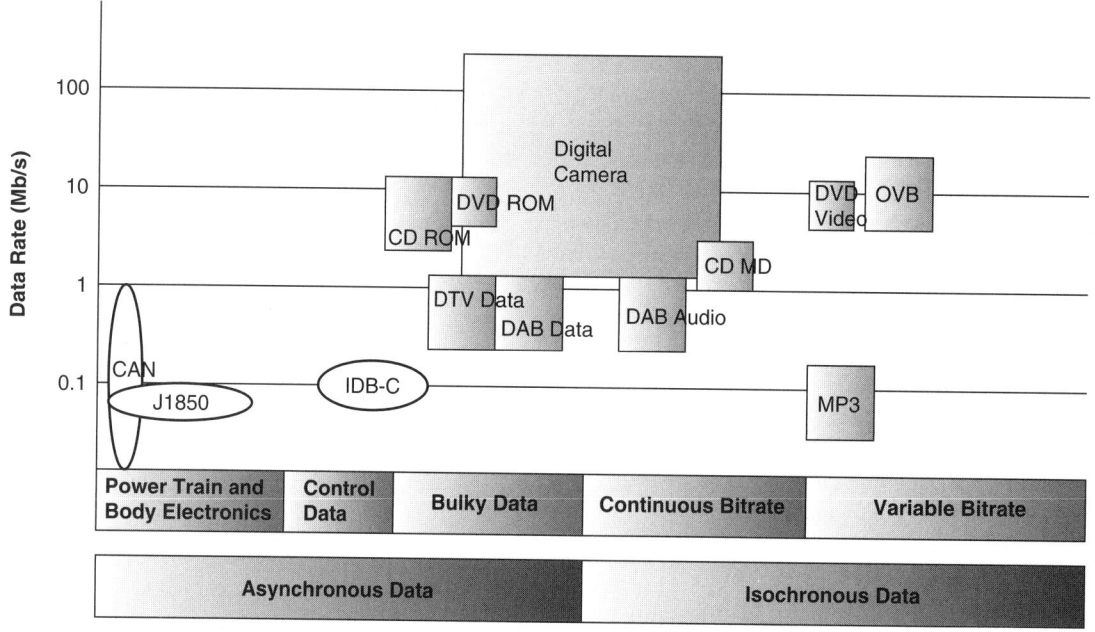

FIGURE 2.13 Data type and data rate of automobile networks.

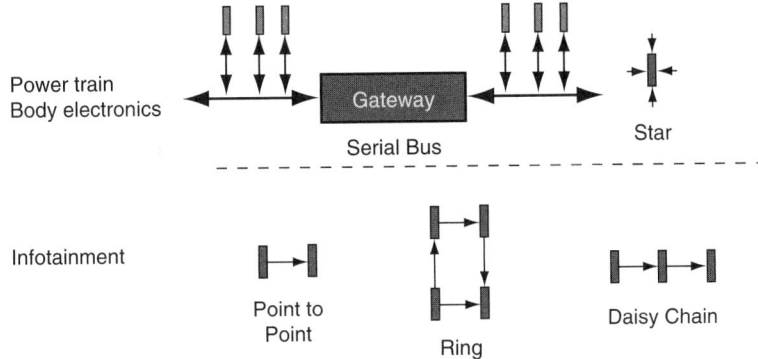

FIGURE 2.14 Network topologies for automobiles.

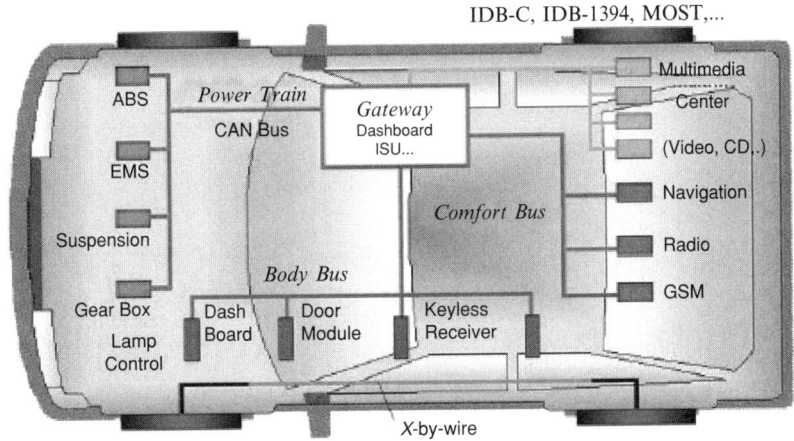

FIGURE 2.15 POF network within the automobile.

the system is self-contained, standards for interconnection are not needed. The fiber usually replaces the massive copper wires at the back plane as shown in Fig. 2.16. The fiber length is usually less than 30 m. The bit rate required at present varies from 155 to 622 Mb/s, but in the near future this will upgrade to more than 2.5 G/s.

In the intersystem interconnection, systems from the same vendor or different vendors are connected together. The length of fiber may be as long as

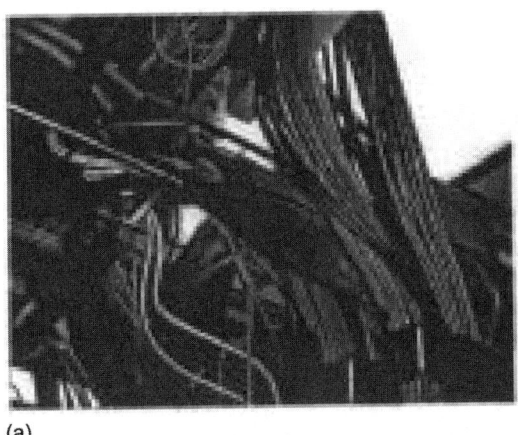
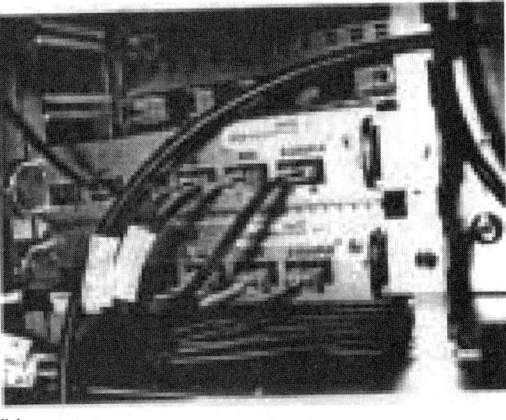

(a) (b)

FIGURE 2.16 (a) UTP back plane. (b) POF back plane.

300 m. It is important for the vendors to follow international standards, and the bit rate may reach 10 Gb/s with a bit error rate (BER) between 10^{-9} and 10^{-12}. A typical example of an intersystem connection is in the storage access network (SAN), as shown in Fig. 2.17.

5.1.3. Office Network

The penetration of POF into office network faces challenges from UTP (untwisted pair) Cat5e and Cat6 cables and from multimode silica fiber. Cat5e can have a bandwidth of 100 Mb/s over a distance of 100 m, and Cat6 has a bandwidth of 600 Mb/s over the same distance. They are extensively used at present in most office networks. Hence, it is unlikely that POF will replace UTP for network distances less than 100 m. In fact, the POF to be used here must be graded to provide for sufficient bandwidth. The disadvantage of a UTP is that it is affected by electromagnetic interference.

Future office networks will run at a bit rate greater than 2 Gb/s. UTP will not be able to compete in this area. However, silica multimode fiber may be a strong competitor. It certainly has an advantage over POF in fiber attenuation and hence the distance used. POF still has some other advantages over silica fiber, such as its ease of connection and ability to negotiate a small bend. Finally, silica fiber is more brittle and its end may fracture during installation, posing a danger to the installer in that the fractured glass pieces may get into the fingers.

Figure 2.18 shows the wiring of UTP, silica fiber, or POF within a building leading to the office. The signal from outside the building arrives at the main distribution frame located at the basement through the access network. This signal is usually optical and, if UTP is still used inside the building, it has to be converted into an electrical signal within the main

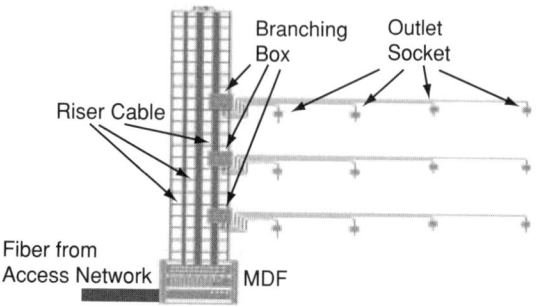

FIGURE 2.18 Office network.

distribution frame (MDF) before it is carried upward to different floors by the riser coaxial cable. However, if the optical fiber is already installed in the office (FTTO), the MDF does not have to do any conversion but it does have to distribute the optical signal to different riser fibers for different floors. The branching box on each floor then distributes the signal to different floor fibers which are eventually terminated at some wall socket outlets.

5.1.4. Home Network

The structure of the home network is similar to the office network. The MDF located at the outside wall of the house receives signals from a variety of media such as Internet, cable television, and satellite Internet and distributes them to different rooms inside the house via coaxial cables or optical fibers (silica or polymer) terminating at wall outlets as shown in Fig. 2.19.

The devices to be connected within the home can be summarized as shown in Fig. 2.20. The residential gate is equivalent to the MDF. There are two types of networks used to connect these devices depending on the quality of service required. There are some services that require guaranteed connection and transmission without loss of packets. For example, videophone, high-density television transmission, and indispensable sensors monitoring patients and infants. These services require high bit rate transmission. There are other services that do not required guaranteed service but the best of effort will do. In this category, we include voice over IP telephone, home electric appliances, sensors with redundancy, that is, more than one sensor to monitor a single event and conventional personal computers. These services require low bit rate transmission. Hence, there are two kinds of home networks: the wavelength division multiplexing (WDM)-based network which guarantees the required quality of service, and the token-based network, which offers best of effort service. Figure 2.21 shows their bandwidth requirements.

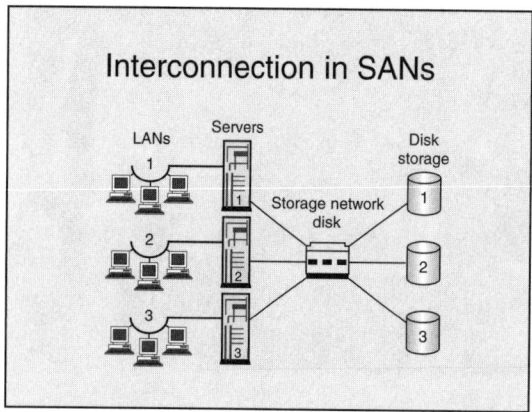

FIGURE 2.17 Back plane interconnection in SAN.

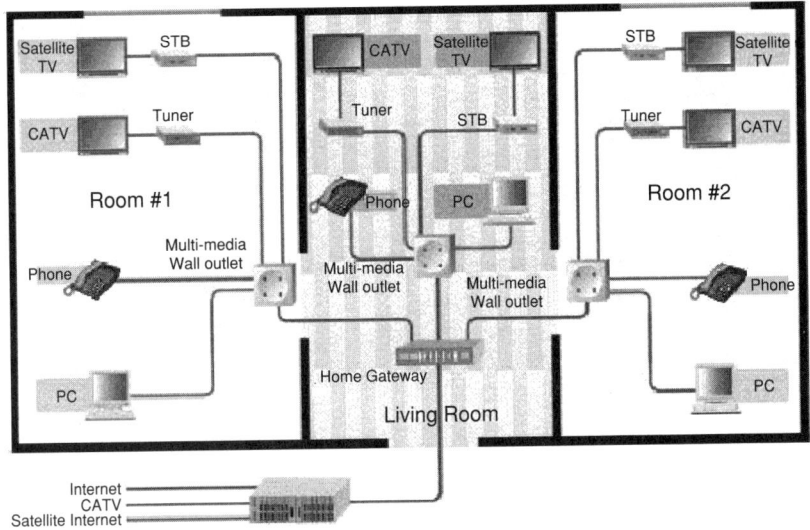

FIGURE 2.19 Home network.

For WDM networks, the coarse WDM will do because the maximum bandwidth required would not exceed a few Gb/s. It has been estimated that a total of eight wavelength channels is sufficient. However, because multimode fibers are used within the home, it is now necessary to develop CWDM over these fibers, which is a relatively new technology. On the other hand, the token-based network needs to transmit at low bit rates. The well-developed OTDM will be sufficient. The important thing to take into account is the intermodal dispersion of the fiber.

Fiber to the home is not yet popular simply because of the cost of the optical system and the need for a very large bandwidth. Although optical fiber (either silica or polymer) is as cheap or even cheaper than copper cable, there is an extra cost of the transceivers at both ends of the fiber. Unless this cost is reduced to a very insignificant level, fiber to the home is unlikely to be realized. The existing copper pair joining our plain old telephone is still useful even for high bandwidth transmission if it is configured in asymmetric digital subscriber line (ADSL) or its associated

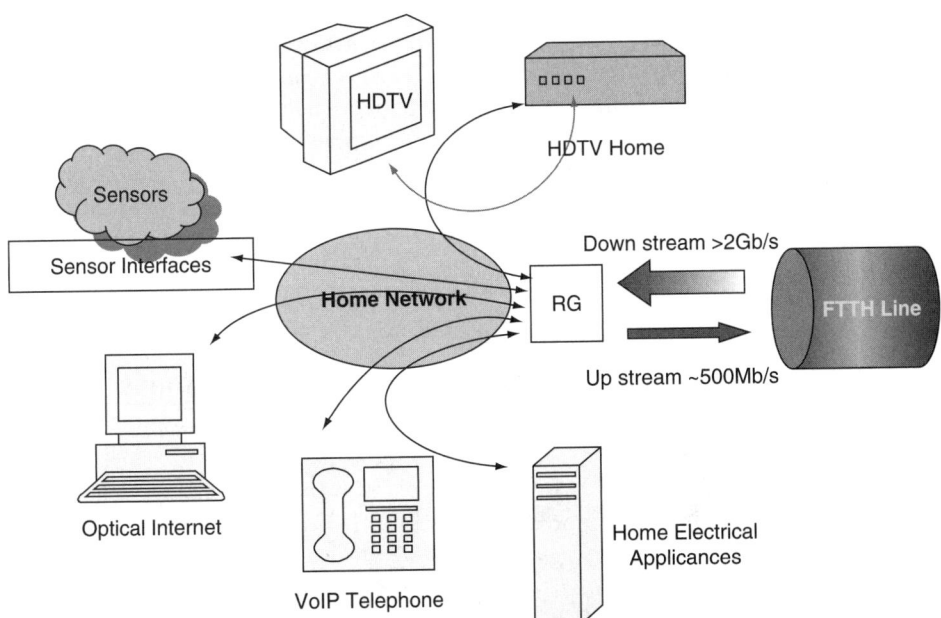

FIGURE 2.20 Home network applications. FTTH, fiber to the home; HDTV, high-density television; RG, residential gate; VoIP, voice over IP.

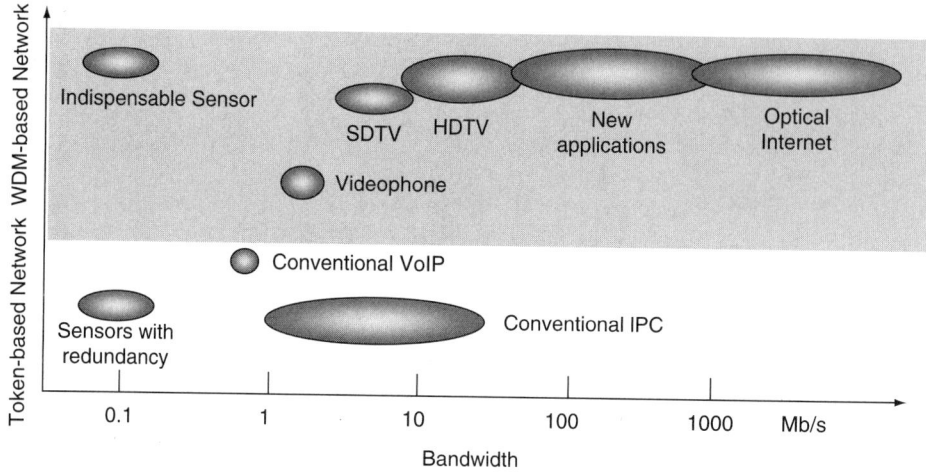

FIGURE 2.21 Bandwidth requirements in home network. HDTV, high-density television; IPC, IP communication; SDTV, standard definition television; VoIP, voice over IP.

format. However, it is noted that at the point of writing this chapter, a company in Germany had just produced a pair of transceivers at a market value of U.S.$5.00. This is certainly a significant step toward putting fiber into the home. In this case, polymer fiber may be a preferred candidate instead of silica fiber for the reasons mentioned above. Another competitor is wireless communication, but it is unlikely to compete strongly.

5.2. Illumination

A new application of POF is in the illumination and signage areas. Instead of making the fiber to deliver light from one end to the other for transmission of information, it is possible to design the fiber so that it leaks light uniformly from the side throughout the whole length of the fiber. This can be achieved by inserting a thin inner cladding with a refractive index in between those of core and outer cladding, as shown in Fig. 2.22. Fibers for illumination normally have larger diameters than communication fiber.

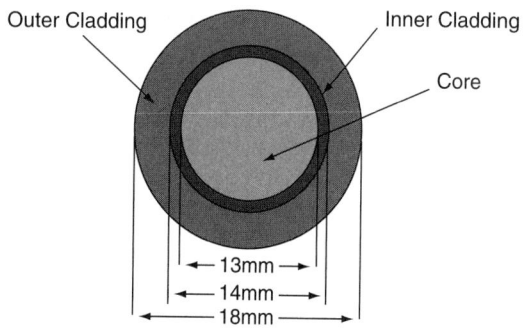

FIGURE 2.22 Side emitting POF for illumination.

For the fiber illustrated in Fig. 2.22, the core refractive index is 1.522, for inner cladding 1.343, and for outer cladding 1.484. In the core, some insoluble particles are introduced so that light traveling in it is scattered. Those with scattering angles larger than the critical angle defined by the core and the inner cladding indices leave the core and eventually leave the fiber sideways. It is obvious that the inner cladding is used to control the rate of leakage of light. The difficulty is in ensuring a uniform leakage throughout the fiber because the intensity of the light within the core necessarily decays along its length unless the concentration of the scattering particles increases with fiber length. This obviously is difficult to achieve. However, the uniformity of light leakage may be improved if it is launched from both ends simultaneously.

The unique feature of polymer fiber is its elasticity compared with glass fiber, and this enables POF to be used for signage. Figure 2.23 shows such a sign. The three letters are formed by bending a single length of POF and powered by one halogen lamp of 150 W. It is conceivable that POF will one day replace neon signs.

6. POLYMER FIBER GRATINGS

Polymer is a very photosensitive material. Figure 2.24 shows its change in refractive index when it is irradiated by an ultraviolet (UV) light of 488 nm at 1.84 mW [13]. It is therefore possible to create Bragg gratings in a single-mode POF.

To write a grating into the core of a single-mode POF, we use the setup shown in Fig. 2.25 [14]. A UV beam at a wavelength of 355 nm irradiates the POF through a phase mask. The zero-order diffraction

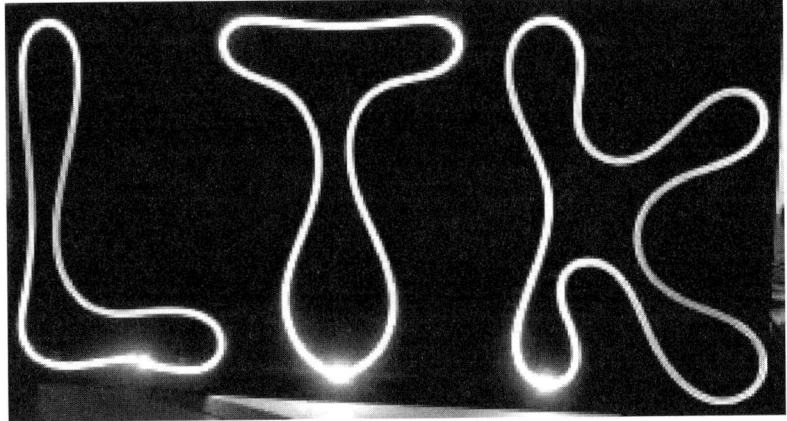

FIGURE 2.23 POF for signage.

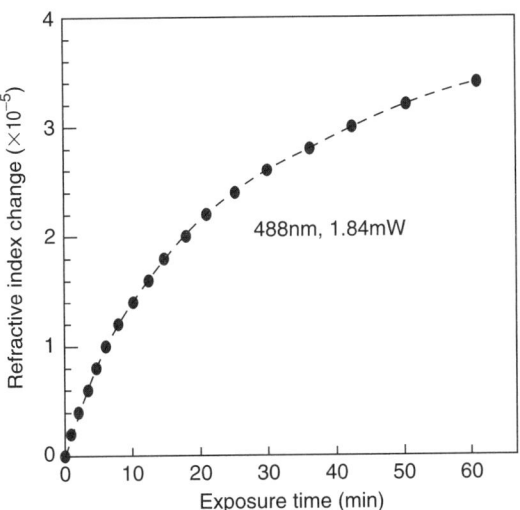

FIGURE 2.24 Photosensitivity of PMMA fiber.

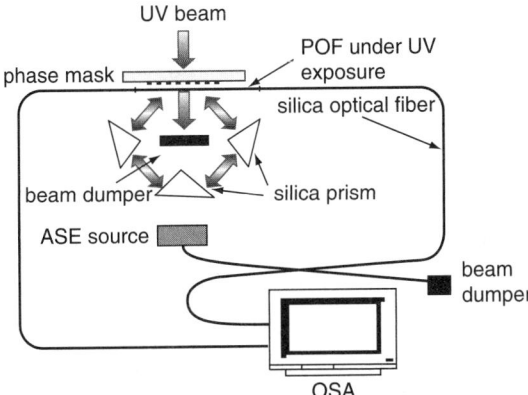

FIGURE 2.25 Writing of Bragg gratings in single-mode POF. ASE, amplified spontaneous emission.

beam from the phase mask is blocked by the beam dumper, and the + and − first-order beams are allowed to interfere at the POF core through the ring arrangement of the prisms. The photosensitivity of the polymer creates a grating in the core. A broadband light source (amplified spontaneous emission; ASE) is connected to the POF via a coupler and a length of silica single mode fiber. The build-up of the grating during the irradiation is observed through the optical spectrum analyzer.

Figure 2.26 shows that the index modulation of the core increases linearly with time until it reaches 62 min. This is called type 1 grating. After that, the modulation increases with a steep slope until 87 min. This is called type 2 grating. The grating then remains unchanged for a longer irradiation. The energy of the irradiation beam is about 6 mJ. The spectrum of the final grating is shown in Figure 2.27. The stop band is less than 0.5 nm, and the extinction ratio is 28 dB.

The unique feature of the POF grating is its large thermal and mechanical tuning ranges. Figure 2.28a shows that a change of temperature over 50° C shifts the Bragg wavelength by about 8 nm, and Fig. 2.28b shows a mechanical strain can tune the wavelength by 10 nm [15]. More importantly, these tunings are linear, and no hysteresis is observed. These features can be utilized for constructing very sensitive sensing devices, such as current sensors and acoustic sensors.

7. SEGMENTED CLADDING POF

Photonic crystal fibers (PCFs; also called holey fibers) have received intense study recently (see Chapter 3). These fibers have been made in both silica and polymer form [16]. The common feature of these fibers is that longitudinal holes are distributed either randomly or regularly over the cross-section of the fiber. One of

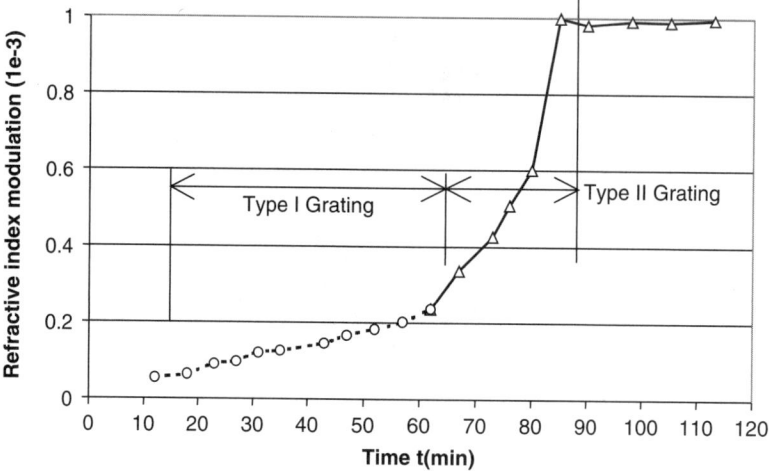

FIGURE 2.26 Build-up of Bragg gratings in single-mode POF.

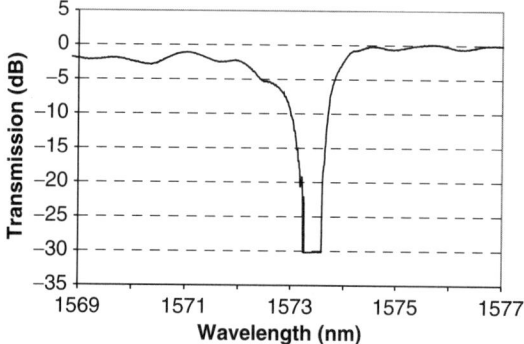

FIGURE 2.27 Spectrum of Bragg grating in single-mode POF.

the distinguishing characteristics of holey fiber is its large wavelength range for single-mode operation. This is based on the principle of differential leakage of the higher order modes in the fiber. As long as the effective index of the cladding is less than the mode index of the fundamental mode and larger than the mode indices of all higher order modes, all the higher order modes will eventually leak away at different rates provided the fiber is long enough, leaving the single mode in the core. Based on this principle, it is possible to design new fibers without holes that also possess large wavelength range single-mode operation. An interesting design is the segmented cladding fiber shown in Fig. 2.29 [17]. In this fiber, the core material fans out into the cladding in the form of spokes, thus segmenting the cladding. A polymer version of this fiber has been fabricated with four segments, as shown in Fig. 2.30 [18]. Its core has a diameter of 20 μm, and the overall fiber diameter is 200 μm.

With such a large core diameter, a conventional fiber would be very much multimodal even at a wavelength of 1.55 μm. However, Fig. 2.31 shows that this fiber is in single-mode operation at this wavelength when its length is 36.5 cm. When the signal wavelength is reduced to 0.633 μm, the fiber is few-moded as shown in Fig. 2.32.

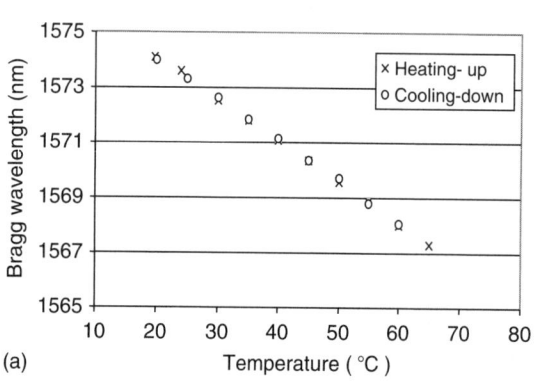
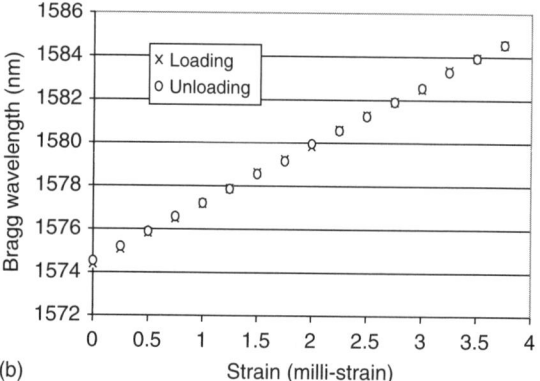

FIGURE 2.28 (a) Thermal tuning of grating. (b) Mechanical tuning.

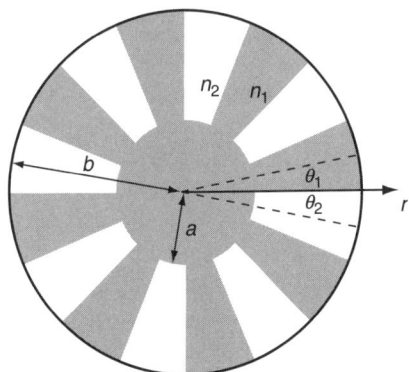

FIGURE 2.29 Segmented cladding fiber.

FIGURE 2.30 Four-segmented cladding fiber.

8. DYE-DOPED POLYMER FIBER AMPLIFIER

Chemical dyes such as rhodamine B can be easily introduced into the MMA monomer during preform fabrication. This enables us to manufacture dye-doped polymer fibers that can function as an optical amplifier [19]. Figure 2.33 shows the gain characteristic of a rhodamine B–doped polymer fiber amplifier. It can be seen that a signal gain more than 20 dB is obtained for a fiber length of 90 cm at a wavelength of about 620 nm. In this experiment, both the pump power and signal are in pulse form with a pulse width of 5 ns and a repetition rate of 10 Hz. The peak pump power is 920 W. This means that the average pump power is only 0.046 mW.

9. CONCLUSIONS

POF is gaining increasing attention from both industrialists and researchers because it has unique characteristics not available in silica fiber. In this chapter, we described how polymer fibers are manufactured and their applications in short distance communication and illumination for signage. It is expected that polymer fiber will capture the markets in these two areas. The wide tuning range of polymer fiber grating is an advantage that silica fiber grating cannot match. Hence, many tunable optical devices will be constructed from polymer fiber grating. The segmented cladding polymer fiber is a new addition to the host of holey fibers, with the advantage of the absence of holes. More study is required to ascertain unique applications of this fiber. Finally, polymer fiber amplifier opens up a new field of research, especially because dye-doped fiber can be easily fabricated. Different dyes will lead to fiber amplifiers operating at different signal wavelengths. Again, more study is needed to make these amplifiers useful in practical communication and sensing systems.

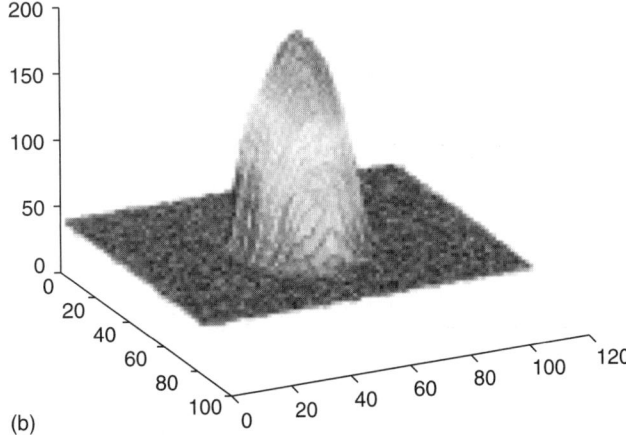

FIGURE 2.31 (a) Far-field spot of SCF. (b) Intensity distribution.

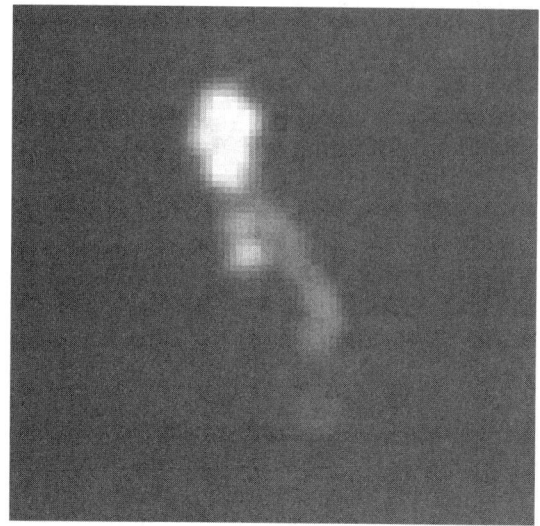

FIGURE 2.32 Four-segmented cladding polymer fiber becomes few-moded.

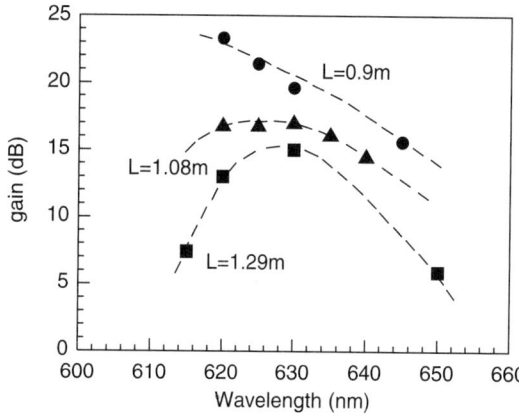

FIGURE 2.33 Rhodamine B–doped polymer fiber amplifier.

10. REFERENCES

1. J.C. Knight, T.A. Birks, P. St. J. Russell and D.M. Atkin, "All-silica single-mode optical fibre with photonic crystal cladding," Optics Lett., **21**, No. 19, pp. 1457–1459 (1996).
2. Y. Koike, Proc. ECOC'96, paper MoB.3.1, 1.41, (1996).
3. N. Tanio and Y. Koike: "What is the most transparent polymer?," Proc. POF Conf.'97, pp. 33–34, September 1997, Hawaii.
4. G.D. Peng, P.L. Chu, L. Xia and R.A. Chaplin, "Fabrication and characterization of polymer optical fibres," J. Electr. Electron. Eng. Australia, **15**, No. 3, pp. 289–296, 1995.
5. Y. Koike, T. Ishigure and E. Nihei, "High-bandwidth graded-index polymer optical fibre," IEEE J. Lightw. Technol., **13**, No. 7, pp. 1475–1489, 1995.
6. B.C. Ho, J.H. Chen, W.-C. Chen, Y.-H. Chang, S.Y. YaTseng, "Gradient-index polymer fibres prepared by extrusion," Polymer J. **27**, No. 3, pp. 310–313, 1995.
7. I.S. Sohn, K. Yoon, T. Kang, O. Kwon and C.W. Park, "Fabrication of GI-POF by a co-extrusion method with enhanced diffusion," 11th Intern. Conf. POF, pp. 61–64, September 2002, Tokyo, Japan.
8. R. Ratnagiri, M. Park, W.R. White and L.L. Blyler Jr., "Control of properties of extruded perfluorinated graded index polymer optical fibres," Proc. 12th Intern. Conf. POF, pp. 212–214, September 2003, Seattle, WA, USA.
9. P.L. Chu, V. Yau and W.A. Gambling, "Effect of core non-concentricity on joint loss in polymer optical fibre," Proc. 12th Intern. Conf. POF, pp. 195–198, September 2003, Seattle, WA, USA.
10. O. Ziemann, L. Giehmann, PE. Zamzow, H. Steinberg and D. Tu, "Potential of PMMA based SI-POF for Gbps transmission in automotive applications," Proc. Intern. Conf. POF, pp. 44–48, September 2000, Cambridge, MA, USA.
11. Y. Koiko, "Progress in GI-POF: status of high speed plastic optical fibre and its future prospect," Proc. Intern. Conf. POF, pp. 1–5, September 2000, Cambridge, MA, USA.
12. W. Baierl, "Evolution of automotive networks," 10th Intern. Conf. POF, pp. 161–168, September 2001, Amsterdam.
13. G.D. Peng, Z. Xiong and P.L. Chu, "Photosensitivity and gratings in dye-doped polymer optical fibres," Optical Fibre Techn., **5**, pp. 242–251, 1999.
14. H.Y. Liu, G.D. Peng and P.L. Chu, "Polymer fiber bragg gratings with 28dB transmission rejection," IEEE Photon. Technol. Let., **14**, No. 7, pp. 935–937, 2002.
15. H.B. Liu, H.Y. Liu, G.D. Peng and P.L. Chu, "Strain and temperature sensor using a combination of polymer and silica fibre Bragg gratings," Optics Commun. **219**, pp. 139–142, 2003.
16. G. Barton, M. van Eijkelenborg, G. Henry, N. Issa, K. F. Klein, M. Large, S. Manos, W. Padden, W. Pok and L. Poladian, "Characteristics of multimode microstructured POF performance," Proc. 12th Intern. Conf. POF, pp. 81–84, September 2003, Seattle, WA, USA.
17. V. Rastogi and K.S. Chiang, "Propagation characteristics of a segmented cladding fibre," Optics Lett., **26**, No. 8, pp. 491–493, 2001.
18. A. Yeung, K.S. Chiang and P.L. Chu, "Polymer segmented cladding optical fibre," 12th Intern. Conf. POF, pp. 77–80, September 2003, Seattle, WA, USA.
19. G.D. Peng, P.L. Chu, Z. Xiong, T. Whitbread and R.P. Chaplin, "Broadband tunable optical amplification in rhodamine B-doped step index polymer optical fibre," Optics Commun., **129**, pp. 353–357, 1996.

CHAPTER 3

Microstructured Optical Fibers

Tanya M. Monro*
School of Chemistry & Physics
University of Adelaide
Adelaide, Australia

1. FIBERS WITH MICRON-SCALE STRUCTURE

The development of core-clad silica glass optical fibers has revolutionized communications systems over the past 30 years. These "conventional" optical fibers have also made a significant impact in areas as diverse as sensing, medical imaging, laser welding, and machining and have allowed the realization of new classes of lasers and amplifiers. All these advances have been enabled by one key factor: reduction of the fiber loss. Reducing loss was a topic of intensive research and development for two decades, and dramatic improvements in the transmission of silica-based fibers in the 1.5-μm telecommunications window were achieved as a result. The widely used Corning SMF-28 fiber has a loss of less than 0.2 dB/km at 1550 nm.

In the early 1970s, when the fabrication processes for the manufacture of core-clad preforms had not yet reached maturity, Kaiser et al. proposed an alternative route to achieving low fiber losses. Kaiser et al.'s concept was to confine light within a pure (undoped) silica core by surrounding it with air [1, 2]. The core was supported by a sub-wavelength strand of silica glass and then jacketed in a silica cladding for strength. Although this new class of fibers showed promise, the fabrication methods used to produce these early single-material fibers were limited. Therefore, this new technology was quickly overtaken by improvements in the modified chemical vapor deposition process, which allowed the definition of high-quality preforms for the production of low-loss core-clad silica fibers.

In the late 1980s, work by Yablonovitch [3] on the development of three-dimensional photonic crystals identified micron-scale structuring to be a powerful means of modifying the optical characteristics of a material. The earliest photonic crystal samples were formed by drilling centimeter-scale holes to produce photonic bandgaps within which light propagation was forbidden. These samples were confirmed to have photonic bandgaps located at microwave wavelengths. In the 1990s, a number of groups worked to extend this concept to infrared and visible wavelengths by scaling down the dimensions of the photonic crystal structure to micron-scale feature sizes. The technique used most extensively for defining two-dimensional photonic crystals is electron beam lithography [4]. However, this technique is not well suited for defining structures that are truly extended in the third dimension to avoid nonuniformities in this direction modifying the properties of the photonic bandgaps. Fabricating two-dimensional photonic crystals is an engineering challenge, and although a number of approaches exist, there is a continued drive to develop cheap and flexible techniques for the large-scale production of high-quality photonic crystals.

In 1995, Birks et al. [5] proposed a novel technique for producing two-dimensionally structured silica/air photonic crystal structures by taking advantage of optical fiber manufacturing techniques. The fabrication concept was to stack macroscopic silica capillary tubes together into a hexagonal lattice to form a preform with millimeter-scale features and then to pull this preform to a fiber with micron-scale features on a drawing tower. Thus, the scale reduction and longitudinal uniformity inherent to the fiber drawing process could be utilized to produce the first photonic crystals that could truly be considered infinite in the third dimension.

Although the index contrast between air and silica is not sufficient to form photonic bandgaps for all polarizations of light propagating within the transverse plane, by considering out-of-plane propagation, it is possible to form full photonic bandgaps in such structures [5]. Hence, it was proposed that by introducing a "defect" to the otherwise periodic

---

*E-mail: tanya.monro@adelaide.edu.au

transverse structure, light could be localized with the bandgap and thus guided within the defect, which acts as the fiber core. In addition, introducing an air defect raises the possibility of guiding light within an air core, something that cannot be achieved in conventional optical fibers, which guide light due to total internal reflection in a high index core material.

In 1996 the first silica/air microstructured fiber was made by this stack-and-draw technique [6]. This fiber had a hexagonal arrangement of small air holes and a central solid core, which was formed by replacing one of the capillaries within the stack with a solid rod. Although the fabrication of this fiber represented a significant breakthrough, it also raised some interesting questions. Calculations indicated that the holes within the cross-section of the fabricated fiber were too small to lead to the formation of photonic bandgaps, and thus light could not be guided within this fiber via photonic bandgap effects. Despite this, Knight and colleagues [7] demonstrated that light could be guided within the solid core of this fiber, and a number of interesting optical characteristics began to emerge within this new fiber type, most notably that they can be "endlessly single mode," guiding just a single mode at all wavelengths.

In this way, an important new class of optical fibers was discovered during the drive to produce photonic bandgap fibers (PBGFs). These "index-guiding" microstructured fibers guide light due to a modified variant of the law of total internal reflection. The arrangement of air holes acts to lower the effective refractive index in the cladding region, and so light is confined to the solid core, which has a relatively higher index. This class of fibers has attracted significant attention from both university and industrial research groups in recent years largely because they can exhibit many novel optical properties that cannot be achieved in conventional optical fibers. Note that it is not essential to use a periodic arrangement of air holes in this class of fibers [8]. The optical properties of this class of fibers are reviewed in Section 2.

A number of names have been given to these fibers, including photonic crystal fibers [6], holey fibers (HFs) [9], and microstructured optical fibers [10]. Within this chapter the term holey fiber is used to differentiate index-guiding fibers from those that guide via photonic bandgap effects, which are described as PBGFs; and the term microstructured optical fiber is used to describe all types of fibers with micron-scale transverse structural features.

In 1998, Broeng et al. [11] reported that the use of a honeycomb air hole lattice led to the formation of larger bandgaps than the triangular/hexagonal lattice both for in-plane and out-of-plane propagation. These honeycomb fibers were the first PBGFs to be fabricated [12]. The light guided by these fibers is guided within a ring-shaped mode located in the silica surrounding a central air defect in the honeycomb lattice structure. The first air-guiding PBGFs were realized in 1999 [13], 3 years after the first air/silica microstructured optical fibers were made by the stack-and-draw fabrication technique. These fibers have air holes arranged on a hexagonal lattice and a very large air-filling fraction (Fig. 3.1c). Research in this field is reviewed in Section 7.

In the single-material microstructured fibers described thus far, light is solely confined by the holes in the cladding. Hybrid microstructured fibers are another class of fiber that combines a doped core with a holey cladding. At one extreme, in "hole-assisted" fibers light is guided by the relatively higher index of the doped core, and the air holes located in the cladding of a conventional solid fiber act to modify properties such as dispersion [14] or bend loss. In air-clad fibers, an outer cladding with a high air-filling fraction creates a high numerical aperture (NA) inner cladding that allows the realization of cladding-pumped high power lasers [15]. Alternatively, dopants can be added to the core of an HF to create novel HF-based amplifiers and lasers (see, e.g., [16]). Work performed to date on active microstructured fibers is described in Section 5.3.

Note that the parameters Λ and d/Λ are widely used to label the feature sizes in fibers that have hexagonal hole arrangements, where Λ is the hole-to-hole spacing and d is the hole diameter. Figure 3.1 presents a gallery of scanning electron microscope (SEM) images of a representative selection of the microstructured optical fibers realized to date. The images presented are all of fibers made at the Optoelectronics Research Centre (ORC) at the University of Southampton, United Kingdom. Figure 3.1, a–d, are SEM images of silica microstructured fibers. The fabrication of silica microstructured fibers has now reached a level of maturity that allows a broad range of high-quality fiber profiles to be defined. More detailed information about fiber fabrication is given in Section 3, and work in silica fibers is described further in Section 5. There are now a number of groups (both university groups and companies) worldwide capable of producing silica fiber structures of similar quality. Figure 3.1, e and f, are images of two new classes of soft glass microstructured

Microstructured Optical Fibers

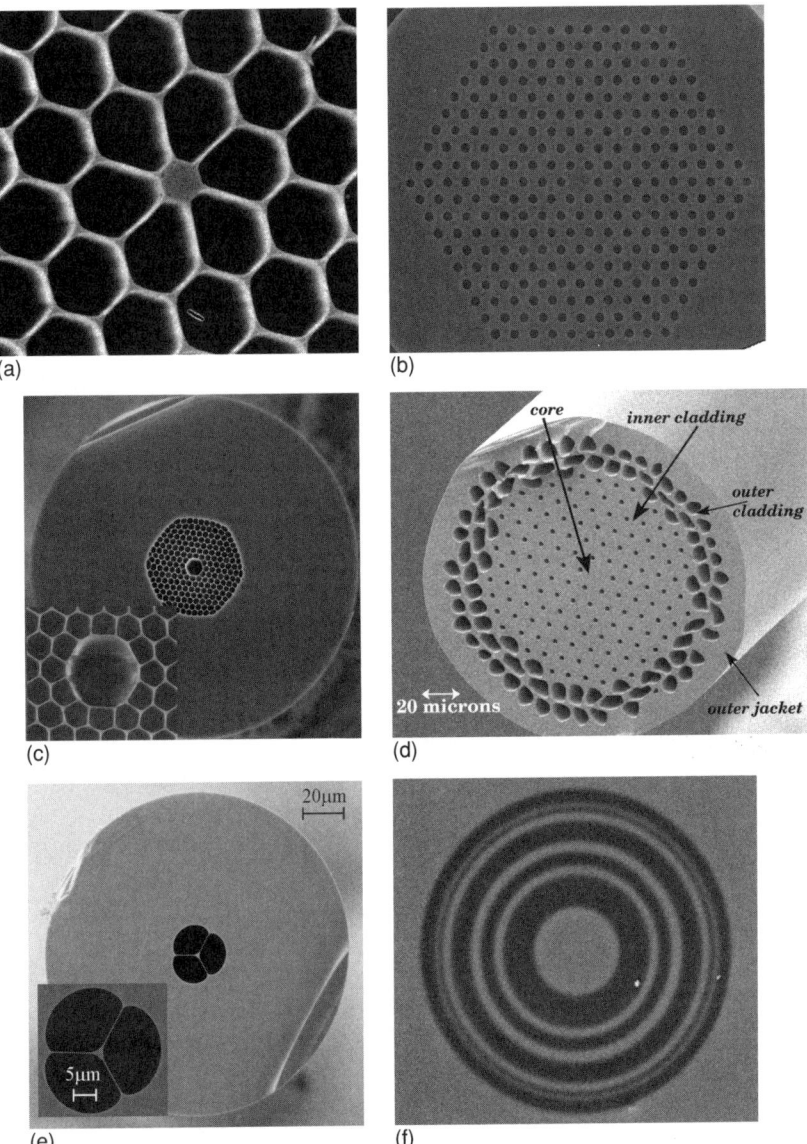

FIGURE 3.1 A representative selection of microstructured optical fibers. The images shown are all fibers fabricated at the ORC (Southampton, UK). (a) Small-core nonlinear pure silica holey fiber. (b) Large-mode area pure silica holey fiber. (c) Air-core photonic bandgap fiber. (d) Double-clad Yb^{3+}-doped large-mode area fiber. (e) Extruded highly nonlinear bismuth holey fiber. (f) One-dimensional layered soft glass microstructured fiber.

fibers. An overview of soft glass microstructured fibers is presented in Section 6.

2. OVERVIEW OF OPTICAL PROPERTIES

2.1. Introduction

This section broadly reviews the optical characteristics that have been identified thus far in both classes of microstructured optical fibers. As described in Section 1, HFs guide light due to the effective refractive index difference between the solid core and the arrangement of air holes that forms the cladding region. The effect of the holes on the fiber properties depends on the hole distribution and size(s) relative to the wavelength of light guided in the fiber. Hence, the effective refractive index of the structured cladding region (and thus the fiber's NA) can be a strong function of wavelength in these fibers. For this reason, it is possible to design fibers with spectrally unique properties not possible in conventional core-clad optical fibers. In addition, the optical properties of microstructured fibers are determined by the spatial configuration of air holes used to form the cladding, and many different arrangements can be envisaged within this flexible fiber type.

PBGFs guide light for use with a fundamentally different mechanism than conventional fibers, and so it is not surprising that the optical properties of modes guided with these fibers can be radically different from conventional fibers. It is worth noting that research in the area of PBGFs is less mature than in index-guiding fibers, and so the full range of optical properties possible in PBGFs is not yet known.

2.2. Nonlinearity Tailoring

Microstructured fibers provide a powerful means of tailoring the effective nonlinearity that can be achieved in a fiber form over at least five orders of magnitude, a much wider range than can be achieved in conventional core-clad fiber designs. The effective nonlinearity (γ) of a fiber is defined as in [17] to be

$$\gamma = \frac{2\pi}{\lambda} \frac{n_2}{A_{\text{eff}}}$$

where A_{eff} is the effective mode area of the fundamental guided mode (defined in [17]), λ is the wavelength of light, and n_2 is the effective nonlinear refractive index of the material. For example, the standard Corning SMF-28 fiber has $\gamma \approx 1/(\text{W}\cdot\text{km})$. Observe that the effective nonlinearity of a fiber can be tailored by either modifying the mode area of the fiber or using different host materials, or both.

Fibers with high γ values are attractive for a broad range of applications in nonlinear fiber devices, and some examples are presented in Section 5.1. By modifying conventional fiber designs, values of γ as large as $20/(\text{W}\cdot\text{km})$ have been achieved [18]. This is done by reducing the diameter of the fiber core and using high germanium concentrations within the core, which both increases the NA and enhances the intrinsic nonlinearity (n_2) of the material. Both modifications act to confine light more tightly within the fiber core and thus increase the nonlinearity γ by reducing the mode area A_{eff}. However, the NA that can be achieved limits the nonlinearity of conventional fiber designs.

When a large air-filling fraction is used in the fiber cladding (Fig. 3.1a), index-guiding HFs can offer a very high effective core-cladding refractive index contrast (and hence a high NA) relative to that which can be achieved in conventional fiber designs. Combining this with small-scale core dimensions allows this class of fibers to offer tight mode confinement. One useful way of estimating the minimum mode area (and thus maximum γ) that can be achieved in any given material system is to consider the theoretical limit of a rod of glass suspended in air. For an air-suspended silica rod, the minimum value of A_{eff} is $\sim 1.5\,\mu\text{m}^2$ at 1550 nm [19]. This is illustrated in Fig. 3.2, which shows the effective mode area at 1550 nm for both an air-suspended rod of silica and a range of small-core silica HF designs. As Fig. 3.2 shows, once the core size becomes significantly smaller than the optical wavelength, the rod becomes too small to confine the light well and the mode broadens again. The fiber with the largest air-filling fraction ($d/\Lambda = 0.9$) in Fig. 3.2 has a minimum effective mode area of $1.7\,\mu\text{m}^2$, only slightly larger than the air-suspended silica rod ($1.5\,\mu\text{m}^2$).

The value of n_2 in undoped silica glass is $\sim 2.2 \times 10^{-20}\,\text{m}^2/\text{W}$. Hence, the ultimate nonlinearity limit in silica air/glass fibers is $\gamma \approx 70/(\text{W}\cdot\text{km})$, and fibers with effective nonlinearities close to this value have been manufactured [20]. Note that glasses with a higher linear refractive index can provide better confinement and thus smaller mode areas. Nonlinearity tailoring can also be achieved via choice of the glass host material, and significantly higher values of n_2 can be achieved using high index glass. For example, fiber nonlinearities as high as $1100/(\text{W}\cdot\text{km})$ have been reported in index-guiding HFs made from bismuth oxide glass [21]. Progress in this area is reviewed in Section 6.2.

At the opposite extreme, low values of γ (and large effective mode areas) are attractive for avoiding both nonlinear effects and fiber damage for a

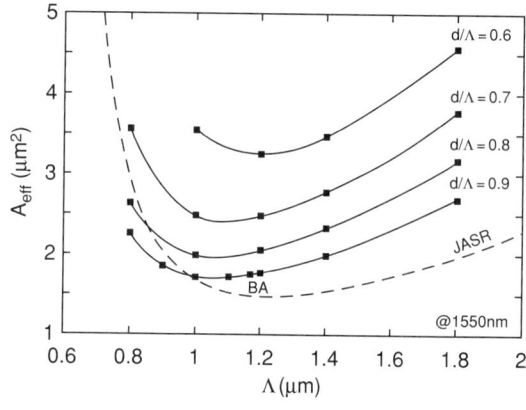

FIGURE 3.2 Effective mode area (A_{eff}) as a function of the hole-to-hole Λ spacing for a range of small-core silica holey fiber designs with air holes arranged on a hexagonal lattice. JASR corresponds to the case of an air-suspended core of diameter Λ. (From ORC, Southampton, UK.)

range of high power applications. Although large-mode area fibers can be produced using conventional core-clad designs, microstructured fiber designs are attractive for a number of reasons. First, pure silica index-guiding HFs offer a number of useful features for high power delivery applications, including broadband single-mode guidance and the option of using a pure silica glass core (which offers better power handing than a doped core). Single-mode silica HFs with mode areas as large as 680 μm^2 have been achieved at 1550 nm (corresponding to $\gamma \approx 0.1$) [22]. Large-mode area silica fibers are discussed in more detail in Section 5.2. PBGFs can have less than 1% of the guided mode located in the silica within the cladding, and typical mode areas for fibers produced to date are $A_{\text{eff}} \approx 20\ \mu m^2$ at 1.06 μm [23]. Hence, values of γ as low as an order of 0.01 are attainable in PBGFs. Thus, these fibers are very attractive for a broad range of high power applications. Note that for these fibers, the value of γ is a sensitive function of the air–mode overlap, and so nonlinearity measurements are a useful way of measuring this overlap. The air–mode overlap in microstructured fibers is discussed further in Section 2.5.

2.3. Dispersion

A broad range of novel dispersive properties of microstructured fibers has been predicted and observed within silica microstructured fibers. In index-guiding silica HFs, the dispersion can be particularly strongly influenced by the cladding configuration, particularly when the features within the cladding are on the scale of the wavelength of light and the core is small. In such fibers, the waveguide contribution to the total dispersion can dominate the material dispersion of silica, and so new regimes are possible, including anomalous dispersion at wavelengths down to 550 nm [24]. Such fibers have applications in new soliton-based devices (see the example in Section 5.1) and for the generation of broadband supercontinuum [25]. In both applications, the required dispersive properties are achieved using small-core designs, which offer the additional advantage of allowing nonlinear effects at low powers in short fiber lengths. Silica fibers can also be defined in which the waveguide contribution to the dispersion is finely tuned to compensate for the material dispersion over a broad wavelength range, resulting in broadband flat near-zero dispersion [26].

The impact of fiber dispersion on the nonlinear applications of these fibers is discussed in more detail in Section 5.1. Although a vast range of dispersion profiles can be achieved, it is worth noting that it is not always possible to combine all the properties that might be desired for any particular device application within the one fiber. One good example of this is the concept of development HFs for dispersion compensating devices [27]. HF designs can have very large values of normal dispersion (as large as −2000 ps/nm·km at 1550 nm), which at first appears attractive for compensating the dispersion of conventional communications fibers. However, to achieve such values, it is necessary to use fiber designs with very small effective mode areas, which results in poor integration with conventional systems, significantly higher losses, and degradation of the linear polarization of the fiber mode.

The dispersion properties of nonsilica index-guiding HFs are not yet as well mapped out as is the case for silica fibers. However, it is clear that due to the generally steeper material dispersion characteristics (as a function of wavelength) and the higher index contrast in these fibers, in general it will be more challenging to target specific values of dispersion–dispersion slope than in silica fibers.

The highly wavelength-selective nature of the guidance mechanism in PBGFs naturally gives rise to extraordinary dispersion properties in this fiber type. In other words, waveguide dispersion dominates in this class of fibers. In particular, zero dispersion can be realized within the bandgap, and in general the dispersion is anomalous across most of the bandgap [28], and the dispersion values at the edges of the bandgap can be extremely large [29, 30].

Note that the accurate prediction of dispersion properties in microstructured fibers is challenging and is in general a good test of the accuracy of any numerical method. In addition, the dispersive properties of microstructured fibers are more sensitive than most of the other fiber properties (such as mode area, number of guided modes, and birefringence) to subtle variations of or imperfections in the fiber profile.

2.4. Polarization

Microstructured optical fibers with air holes arranged on a hexagonal lattice support a degenerate pair of fundamental modes and thus are not birefringent. More generally, any fiber structure with a greater than 2-fold symmetry is not birefringent, if a perfect cross-section profile is assumed. This is proved in [31] on the basis of arguments based on group theory and symmetry classes.

However, in reality, no real fabricated fibers are perfectly symmetric. When any structural imperfections are combined with the small feature sizes and large index contrast that is possible in microstructured fiber designs, small asymmetries can translate into significant birefringence effects. For example, the fiber described in 16 has a somewhat elliptical-shaped core and has a measured birefringent beat length at 1550 nm of 0.3 mm, the record for the shortest beat length at the time it was reported. Unsurprisingly, this effect is more noticeable in highly nonlinear fibers, and typical beat lengths for such fibers are on the order of a few millimeters at 1550 nm, even for relatively symmetric structures.

Highly birefringent microstructured fibers can also be made by deliberately introducing a 2-fold symmetry to the fiber profile, and this symmetry can either be introduced in the core shape or into the cladding configuration. This can be done for either index-guiding [32] or photonic bandgap [33, 34] guidance mechanisms.

One tested way of reducing unwanted polarization effects in fibers is the process of fiber spinning, in which the fiber preform is rotated during the fiber drawing process. Recently, an order of magnitude decrease in the polarization mode dispersion of HFs was reported in spun HFs [35].

2.5. Air–Light Overlap

The holes in the cladding of a microstructured fiber open up new opportunities for exploiting the interaction of light with gases and liquids via evanescent field effects. For example, the concentration of pollutants in a gas could be determined by measuring the absorption that occurs as light propagates through the gas for a range of wavelengths. One attraction of the fiber geometry is that it can naturally provide extremely long optical path lengths in a compact fashion and does not require large gas volumes. Two approaches can be envisaged: either to exploit PBGFs, which guide light in air, or to use the overlap between the guided mode of an index-guiding HF and the cladding air holes.

For index-guiding HFs to be useful as evanescent field devices, a significant fraction of the modal field must be located within the holes. However, calculations indicate that there is typically only a very small overlap between the guided mode and the holes [36]. Indeed, for many HF designs, less than 1% of the guided mode's power is located in the air holes. The field distribution for the fundamental mode depends strongly on the size of the features in the HF relative to the wavelength, and so the overlap between the guided mode and air can be tailored. This leads to the requirement $\lambda < 2.2\Lambda$ for this overlap to be significant. For example, the air–mode overlap is 30% at 1.5 μm for an HF with hexagonally arranged holes with a relative hole size $d/\Lambda = 0.8$ and spacing Λ of 0.75 μm. Note that if Λ is made too small, it becomes difficult to fill the holes with gas in a reasonable time. One alternative fiber type that offers the potential for easier filling is a fiber structure like the one shown in Fig. 3.1e, which has a small-scale core suspended on a number of thin spokelike membranes. In either case, the fiber designs required to allow significant air–mode overlap have small mode areas, which makes them more challenging to integrate with conventional systems.

PBGFs are the most promising route for evanescent field applications, because these fibers can offer nearly perfect overlap between the guided mode and the material that fills the holes. For example, Bouwmans et al. [37] reported that more than 99% of the fundamental mode is guided within the air core. This was exploited in [38], which demonstrated the measurement of acetylene and hydrogen cyanide absorptions within a PBGF.

The excellent overlap of the guided mode with air within a PBGF also makes these fibers attractive for high power delivery applications [39]. The light intensities within the silica material within the fiber cladding are low relative to the mode peak intensities, and so such fibers promise higher damage thresholds than fibers that guide light within glass. However, one important consideration with the use of these fibers for high power applications is damage of the fragile cladding structure caused by beam wander or contamination. Further work needs to be done to ascertain the relative merit of this fiber type for extending the power handling capabilities beyond those possible in more conventional fibers.

3. FABRICATION APPROACHES

3.1. Preform Fabrication

The first stage in the fabrication of any microstructured optical fiber is the production of a preform, which is a macroscopic version of the structure that is to be defined in the final fiber. For most microstructured fibers, the preform fabrication stage is the most challenging step of the fiber fabrication process.

The vast majority of microstructured fibers produced to date have been made from pure silica glass, and the preforms for these fibers are usually fabricated using capillary stacking techniques. Capillary tubes are stacked in a hexagonal configuration. If the fiber to be produced is an index-guiding HF, the central capillary is typically replaced with a solid glass rod, which ultimately forms the fiber core. To produce a bandgap fiber, either 7 or 19 of the capillaries are removed to produce a large central air core. The stacking procedure is flexible: For example, active fibers can be made using rare-earth–doped core rods, and off-center or multiple core fibers can be readily made by replacing noncentral or multiple capillaries, respectively. The reproducibility that can be achieved in the transverse cross-section of the preform is determined by the length and uniformity of the capillaries that are used to form the stack.

Stacking is a flexible technique and has been used successfully by a number of groups worldwide. Note that the preforms for the fibers in Fig. 3.1, a–d, were all produced via capillary stacking. One significant drawback of the stacking approach is that the preform fabrication is labor intensive and the quality of the final fiber depends significantly on the craft of the fabricator in forming the preform.

Another widely applicable and flexible technique for the definition of structured preforms is drilling of bulk glass samples. Drilling has been used to produce both polymer [40] and soft glass microstructured fibers [41]. Drilling allows the definition of a broad range of hole configurations and can be applied to a very broad range of optical materials. Some drawbacks of drilling include the length of time required to produce a complex preform structure containing many holes and the reduction in preform yield with increasing complexity.

Recent attention has been focused on the fabrication of microstructured fibers in a range of nonsilica glasses (see Section 6). In general, such compound or "soft" glasses have low softening temperatures relative to silica glass, which allows new techniques to be used for the fabrication of structured preforms, such as rotational casting and extrusion. The technique used for most nonsilica glass microstructured fibers made thus far is extrusion [21, 42–49]. Casting techniques have also been used for the production of low-loss microstructured fibers in tellurite glass [50].

Extrusion techniques allow the fabrication of structured fiber preforms with millimeter-scale features directly from bulk glass billets. In this process, a glass billet is forced through a die at elevated temperatures near the softening point, whereby the die orifice determines the preform geometry. Once the optimum die geometry and process parameters have been established, the preform fabrication process can be automated. In this way good reproducibility in the preform geometry can be achieved. One advantage of the extrusion technique is that the preform for the microstructured part of a fiber can be produced in one step. In addition, extrusion allows access to a more diverse range of cladding structures, because the holes are not restricted to hexagonal arrangements.

As an example, consider the structured preform shown in Fig. 3.3a. In this geometry, the core (center) is attached to three long, fine, supporting struts. The outer diameter of this preform is approximately 16 mm. The preform was reduced in scale on a fiber drawing tower to a cane of approximately 1.7 mm in diameter (Fig. 3.3b). In the last step, the cane was inserted within an extruded jacket tube, and this assembly was drawn to the final fiber (Fig. 3.3c). The illustrative examples shown in Fig. 3.3 are made from the lead silicate glass SF57 and are described in more detail in [47]. This procedure has also been used to make nonlinear fibers from other materials, including bismuth glass [48]. A similar technique has been used to produce small-core tellurite

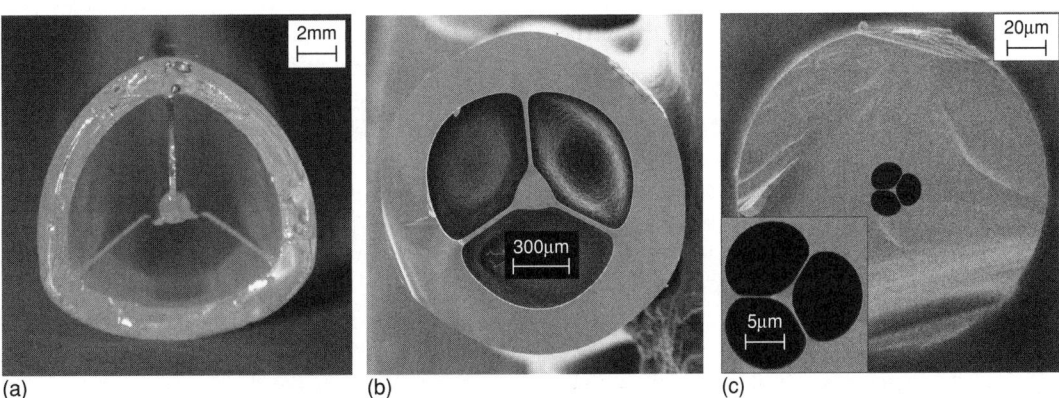

FIGURE 3.3 (a) Photograph of a cross-section through a typical extruded SF57 glass preform. FWHM, full width at half-maximum. (b) SEM image of cane cross-section. (c) SEM image of resulting holey fiber cross-section. (Pictures from the ORC, Southampton, UK.)

microstructured fibers in which the core is supported by six long fine struts [44]. In [44], the preform extrusion is performed directly on a fiber drawing tower, which allows the preform to be reduced to an ~1-mm-diameter cane directly at the time of extrusion.

3.2. Fiber Drawing

To produce fibers that have relatively large-scale features, such as large-mode area fibers (Fig. 3.1b), which typically have hole-to-hole spacing (Λ) in the range of 7–15 µm, fibers are drawn directly in a single step from the preform. To achieve microscale hole-to-hole spacings (Λ), it is in generally necessary to first reduce the preform to a cane of 1–2 mm in diameter, and in a second step this cane is inserted in a jacketing tube and then drawn to the final fiber.

One of the most challenging aspects of air/glass microstructured fiber fabrication is to prevent collapse of the holes and to achieve the target hole size and shape during caning and fiber drawing. The microstructured profile can be affected by the pressure inside the holes, surface tension of the glass, and temperature gradient in the preform. Hence, it is useful to have predictive tools for evaluating the best choice of fiber drawing conditions for producing any predetermined target fiber structure. Work on modeling the drawing of silica capillary tubes has led to some progress in this direction [51]. This model has produced a range of useful rules of thumb and asymptotic limits that aid the selection of appropriate drawing conditions.

A recent highlight in this area is the prediction that the air holes in microstructured fibers can survive the spinning of the preform during fiber drawing and indeed that spinning can be used as a means of controlling the geometry of the drawn fiber [52]. Indeed, it has been demonstrated in practice that the holes within the fiber can survive spinning during the fiber drawing process [35].

Note that compared with silica glass, most non-silica glasses have significantly steeper viscosity curves, which leads to greater demands on the process control during fiber drawing. Nevertheless, a high degree of reproducibility for the HF geometry has already been demonstrated for both lead silicate glass [47] and bismuth glass [48]. The core diameter can be adjusted during fiber drawing by an appropriate choice of the external fiber diameter. Small-core dimensions are chosen to provide tight mode confinement and thus high effective nonlinearity. For fiber designs of the type shown in Fig. 3.3, core diameters in the range of 1.7–2.3 µm correspond to struts that are typically >5 µm long and <250 nm thick. These long thin struts act to isolate the core optically from the external environment and thus ensure that confinement loss is negligible (see Section 4.3). Excellent structural reproducibility has been demonstrated using this fabrication technique.

3.3. State-of-the-Art

A diverse range of high-quality transverse fiber cross-sections based on a hexagonal lattice configuration can now be fabricated in silica glass via the capillary stacking fabrication technique. Continued improvements in the fabrication procedures have reduced the losses of microstructured fibers dramatically in recent years. At the time of this writing, transmission losses of 0.28 dB/km for silica index-guiding HFs [53] and 1.7 dB/km for silica PBGFs [54] were reported at 1550 nm. It is worth noting that losses are typically larger for small-core (highly nonlinear) fibers of the type described in Section 2.2. This is principally caused by nanometer-scale surface roughness at the air–glass boundaries near the core [55]. However, this is not a significant limitation, because most applications of highly nonlinear fibers require the use of short fiber lengths.

As the numbers quoted above attest, within a decade the loss of microstructured fibers has been reduced to values close to those achieved in conventional solid transmission fibers. The prospect of microstructured silica fibers with a lower transmission loss than conventional fibers represents a tantalizing possibility with the potential to revolutionize telecommunications. Index-guiding fibers are one potential route to lower fiber losses, given that the core can be made from pure silica, which potentially offers lower loss than a doped core. Gas core bandgap fibers are another attractive potential route to lower fiber losses, given that they can have as little as 1% of the light propogating in the silica within the cladding.

The fabrication of soft glass microstructured fibers is currently less mature. The first good quality single-mode fibers were reported in 2002 [42], and it is now possible to produce fibers in a broad range of glasses with losses on the order of 2 dB/m at 1550 nm. It is anticipated that further fabrication work should reduce the losses of these fibers to below 1 dB/m in a number of soft glass materials. Although this loss value is high relative to silica glass, it still allows the development of soft glass–based nonlinear fiber devices with better figures of merit than is possible using existing technologies (see Section 6.2).

4. FIBER DESIGN METHODOLOGIES

The presence of wavelength-scale holes in microstructured fibers leads to challenges in the accurate modeling of their optical characteristics. A wide variety of fiber design techniques can be used, ranging from effective step-index fiber models to approaches that incorporate the full complexity of the transverse cross-section. Here these methods are reviewed and assessed in terms of their suitability for modeling optical properties of both index-guiding HFs and PBGFs. Some of the issues associated with designing and modeling practical fibers are highlighted. Note that when dispersion predictions are required, many of the approaches described below allow the dispersion of the material to be included ab initio through the usual Sellmeier formula.

4.1. Effective Index Methods

The complex nature of the cladding structure of the microstructured optical fiber does not generally allow for the direct use of analytical methods from traditional fiber theory. However, for index-guiding HFs, a scalar model based on an effective index of the cladding has proven to give a good qualitative description of their operation [7].

The fundamental idea behind this approach is to first evaluate the properties of the periodically repeated air hole lattice that forms the cladding. By solving the scalar wave equation in a hexagonal cell centered on a single air hole, the propagation constant of the lowest order mode that could propagate in the infinite cladding material is determined. In this work, the hexagonal unit cell is approximated by a circular one to simplify the analysis. This procedure allows the effective cladding index of the fundamental cladding mode, sometimes called the fundamental space-filling mode (FSM), to be determined as a function of the wavelength ($n_{FSM}(\lambda)$).

The next step of the method is then to model the fiber as a standard step-index fiber, using n_{FSM} as the cladding index. The core of this equivalent fiber is assumed to have the refractive index of pure silica with a core radius typically taken to be 0.62Λ. (This assumes that the core is created by the omission of one cladding air hole.)

Despite ignoring the spatial distribution of the refractive index profiles within HFs, the effective index method can provide some useful insight into their operation. For example, it correctly predicts the endlessly single-mode guidance regime for fibers with small d/Λ. This method has also been used as a basis for the approximate dispersion and bending analysis presented in [56]. However, this reduced model cannot accurately predict modal properties such as dispersion, birefringence, or other polarization properties that depend critically on the spatial configuration of holes within the cladding.

One difficulty that arises when using this approach is the question of how to define the properties of the equivalent step-index fiber. One method for making this choice was described above. In this work, the core radius was taken to be 0.62Λ, and the results obtained using the effective index method were made to agree well with full simulations for a particular fiber structure via the appropriate choice of this constant of proportionality. However, for different structures or wavelengths, different choices can become necessary. This restricts the usefulness of this approach, because it is typically necessary to determine the best choice of equivalent structure by referring to results from a more complete numerical model. Riishede and coworkers [57] explored the possibility of choosing the step-index fiber parameters in a more general fashion by allowing a wavelength-dependent core or cladding index. However, to date no entirely satisfactory method for ascribing parameters to the equivalent fiber has been found.

It is noteworthy that recent developments in the fabrication of structures with relatively large air holes have made it relevant to approximate the fiber by an isolated strand of glass surrounded by air (Fig. 3.1, a and e), and this case has been labeled the air suspended rod limit in the literature. In this limit, the step-index fiber analogy can provide useful information about the fiber properties (see, e.g., [48]).

4.2. Structural Methods

To accurately model index-guiding HFs, it is typically necessary to account for the complex spatial distribution of air holes that define the cladding. A number of these techniques have also been successfully applied to PBGFs. Note that when the effective index contrast between core and cladding regions is large, the weak-guidance (scalar) approximation breaks down, leading to inaccurate results, and it is often necessary to adopt a vectorial method that includes polarization effects. This is typically necessary when the air-filling fraction of the cladding is large. Vectorial methods are also appropriate for asymmetric structures. Most of the techniques described in this section can be implemented in both scalar and vector forms.

As noted in Section 1, much of the early research in microstructured fibers was driven by the desire to fabricate a fiber operating by the photonic bandgap effect, which may be obtained in periodically structured material. It is not possible to analyze PBGFs using simple scale modeling—or effective index approaches—because the full vectorial nature of the electromagnetic waves must be accounted for to paint an accurate picture of the resulting photonic bandgaps.

In 1990, the first method for finding photonic bandgaps in photonic crystals was described [58]. The method was closely related to methods used for calculating electronic bandgaps in semiconductor crystals in that it described the magnetic field as a plane wave multiplied by a Bloch function with the two-dimensional periodicity of the photonic crystal. From Maxwell's equations, an eigenvalue equation can be formulated that is well suited for calculating the bandgaps of a periodic dielectric structure, because it describes both the field and the structure as a Bloch function. To include a core, one has to impose an artificial periodicity that is handled numerically by creating a supercell with periodically repeated core defects. This approach yields reasonable solutions if the supercell is much larger than the guided mode area [59]. Such plane-wave techniques have been used to make a broad range of useful predictions both for photonic bandgap guiding fibers and index-guiding fibers.

Beam propagation methods (BPMs) can also be used to calculate the modal properties of HFs. For example, Eggleton et al. [60] used a commercial BPM package to investigate a modified conventional germanium-doped fiber in which six large holes were added around the doped core region. A Bragg grating written in the core of the fiber was used to investigate the cladding modes of the structure, and good agreement with the BPM predictions was found. However, because BPMs calculate the modes of a fiber indirectly by propagating a light distribution along a fiber, they are relatively computationally intensive.

An alternative approach was initially suggested by Mogilevtsev et al. [61], which described the modal fields using localized functions. This technique takes advantage of mode localization and so is more efficient than the plane-wave methods; however, it cannot be accurate unless the refractive index is also represented well.

A hybrid approach that combines some of the best features of the localized function and plane-wave techniques described above was developed [9, 62–63], and some extensions to this approach are outlined briefly here. The air hole lattice was described using a plane-wave decomposition, as in the plane-wave techniques described above, and the solid core and the modal fields were described using localized functions [9, 62]. This allows for an efficient description, particularly for idealized periodic structures, because only symmetric terms need to be used in the expansions. To model HFs with asymmetric profiles or to obtain accurate predictions for higher order modes, it is necessary to extend this approach to use a complete basis set [63]. When more complex fiber profiles are considered, the advantages of describing the localized core separately from air holes is diminished, and the best combination of efficiency and accuracy is obtained by describing the entire refractive index distribution using a plane-wave expansion while using localized functions only for the modal fields. The general implementation of this hybrid approach can be used to explore the full range of both index guiding and photonic bandgap structures and modes and can predict the properties of real fibers by use of SEM photographs to define the refractive index profile (see, e.g., [64]). This allows the deviations in optical properties that are caused by the subtle changes in structure to be explored.

In this implementation, the entire transverse refractive index profile is described using a plane-wave expansion, and the Fourier coefficients are evaluated by performing overlap integrals, which only need to be calculated once for any given structure. The modal electric field is expanded into orthonormal Hermite-Gaussian functions (both even and odd functions are included). These decompositions can be used to convert the vector wave equation into a simple eigenvalue problem (as in the plane-wave method) that can be solved for the modal propagation constants and fields. To solve the system, a number of overlap integrals between the various basis functions need to be evaluated. For the choice of decompositions made here, these overlaps can be performed analytically, which is a significant advantage of this approach.

Most of the modeling done to date has considered ideal hexagonal arrangements of air holes. As discussed in Section 2.4, group theory arguments can be used to show that all symmetric structures with higher than 2-fold symmetry are not birefringent [28]. However, as the techniques described thus far perform calculations based on a Cartesian grid, they typically predict a small degree of birefringence that can be reduced (but not eliminated) by using a finer grid. However, when modeling asymmetric structures have a form birefringence that is significantly larger than this false birefringence, it is possible to make reasonably accurate predictions for fiber birefringence.

4.3. Predicting Confinement Loss

To design practical single-material fibers, it is important to have a means of predicting confinement loss, because all guided modes are intrinsically leaky modes. Even in index-guiding HFs with high index dopants in the core (which can have true bound modes), it can be important to understand the leakage characteristics of any cladding modes. PBGFs can also suffer significantly from confinement loss (see Section 7).

The confinement loss associated with a given mode can be extracted from the imaginary part of the modal propagation constant. Of the techniques described in Section 4.2, only the BPM can currently calculate complex propagation constants.

The finite element method is being used increasingly for analyzing microstructured fibers and is capable of calculating complex propagation constants [65, 66]. With this method, the classical Maxwell differential equations are solved for a large set of properly chosen subspaces, taking into account the continuity of the electromagnetic fields. More specifically, the modeled waveguide is split into distinct homogeneous subspaces of triangular and quadrilateral shapes. Maxwell's equations are discretized for each element, and the resulting set of elementary matrices is combined to create a global matrix system for the entire structure. This method has been shown to lead to fast and accurate numerical solutions for both classes of microstructured fibers.

Another technique that is well suited to this problem is the multipole approach [67]. This approach is suitable for studying effects caused by the finite cladding region, because it does not make use of periodic boundary conditions. Another advantage of this method is that it calculates the modal fields using decompositions that are based in each of the cladding air holes, and so it avoids the false birefringence problems associated with using a Cartesian coordinate system described above. For this reason, this method is also particularly well suited to exploring the symmetry properties of new fiber geometries. However, it cannot be used to investigate fibers with arbitrary cladding configurations and is limited to circular or at most elliptical hole shapes.

Another technique that can predict confinement loss is based on representing the refractive index distribution as a series of annular segments [68]. The algorithm uses a polar coordinate Fourier decomposition method with adjustable boundary conditions to model the outward radiating fields. The use of annular segments allows the overlap integrals between the structure and the field components to be performed directly, and so this method can be efficient. It is possible to represent arbitrary fiber profiles in this way.

4.4. Summary

A number of techniques have been adapted or developed to model microstructured optical fibers, and a range of novel guidance regimes has been identified in these fibers that promise to lead to a new generation of optical devices with tailor-made optical properties. Many of the techniques described herein complement one another and can often be used in conjunction with each other to paint a complete picture of the optical characteristics of any given microstructured fiber. The extremes that are possible in these fibers have highlighted a number of challenges in accurate modeling of their properties, and it seems likely that as the technology for fabricating these structures matures, further challenges will emerge.

Almost all the work performed to date designing and modeling microstructured optical fibers has relied on a combination of guesswork and experience to establish the fiber cross-section. By applying one or more numerical design tools, the optical characteristics of the fiber are determined, and this is often computationally intensive. In practice, it would be more useful to know what (if any) fiber cross-section(s) allows any desired optical properties to be realized. Hence, this field has reached a point where there is a great need for the application of inverse solution methodologies, and work in this area has begun to be reported. For example, genetic algorithm techniques were used to define fiber structures with optimized optical properties [69].

5. SILICA HFS

5.1. Small-Core Fibers for Nonlinear Devices

5.1.1. Background

One of the most promising practical applications of index-guiding HF technology is the opportunity to develop fibers with a high optical nonlinearity per unit length [70]. Figure 3.1a shows a typical highly nonlinear silica HF made at the ORC (Southampton, UK). A small-core diameter combines with a large air-filling fraction in the cladding to result in a fiber that confines light tightly within the solid central core region. In this case, $\Lambda \approx 1.5\,\mu m$ and $d/\Lambda \approx 0.95$.

Even though silica is not intrinsically a highly nonlinear material, its nonlinear properties can be used if high light intensities are guided within the core, as described in Section 2.2. The breakthrough that first demonstrated the promise of nonlinear applications of microstructured fibers came with the experimental demonstration of the supercontinuum generation in microstructured silica fibers reported by Ranka et al. in 2000 [25].

The high NA that can be achieved in HFs allows the realization of nonlinearities more than 50 times higher than in standard telecommunications fiber and two times higher than the large NA conventional designs. The key principle is that modest optical powers can induce significant nonlinear effects within these fibers, and so they offer an attractive new route toward efficient, compact, fiber-based, nonlinear devices. Nonlinear effects can be used for a wide range of optical processing applications in telecommunications and beyond, and examples include optical data regeneration, wavelength conversion, optical demultiplexing, and Raman amplification. Here we review the optical properties of these small-core fiber designs and present an overview of some of the emerging device applications of this new class of fibers.

5.1.2. Design Considerations

Small-core HFs can also exhibit a range of novel dispersive properties of relevance for nonlinear applications (see Section 2.3). By modifying the fiber profile, it is possible to tailor both the magnitude and the sign of the dispersion to suit a range of device applications. They can exhibit anomalous dispersion down to 550 nm, which has made soliton generation in the near-infrared (IR) and visible spectrum possible for the first time. An application of this regime was reported [16] in which the soliton self-frequency shift (SSFS) in an ytterbium-doped HF amplifier was used as the basis for a femtosecond pulse source tunable from 1.06 to 1.33 µm. Shifting the zero dispersion wavelength to regimes where there are convenient sources also allows the development of efficient supercontinuum sources [25], which are attractive for dense wavelength division multiplexing transmitters, pulse compression, and the definition of precise frequency standards. It is also possible to design nonlinear HFs with normal dispersion at 1550 nm [27]. Fibers with low values of normal dispersion are advantageous for optical thresholding devices, because normal dispersion reduces the impact of coherence degradation [71] in a nonlinear fiber device.

Highly nonlinear fibers with zero dispersion at 1.55 µm have long been pursued because these fibers are very attractive for a range of telecom applications, such as 2R regeneration [72], multiple clock recovery [73], optical parametric amplifiers [74], pulse compression [75], wavelength conversion [76], all-optical switching [77], supercontinuum-based wavelength division multiplexing telecom sources [78], and demultiplexing [79].

It is worth noting that small-core HFs pose a number of challenges for effective modeling. The high index contrast inherent in these fibers necessitates the use of a full-vectorial method. In addition, any asymmetries or imperfections in the fiber profile, when combined with this large contrast and the small structure–scale, can lead to significant form birefringence. In general, even small asymmetries can lead to noticeable birefringence for these small-core fibers. Hence, it is often necessary to use the detailed fiber profile to make accurate predictions.

When the core diameter is reduced to scales comparable with (or less than) the wavelength of light guided within the fiber, confinement loss arising from the leaky nature of the modes can contribute significantly to the overall fiber loss. Confinement loss is described in more detail in Section 4.3. Indeed, the small-core HFs fabricated to date are typically more lossy than their larger core counterparts. Here, we briefly outline some general design rules for designing low-loss high-nonlinearity HFs as described [19].

The range of effective mode areas that can be achieved using silica glass at 1550 nm is shown in Fig. 3.2. As Fig. 3.2 shows, the hole-to-hole spacing (Λ) can be chosen to minimize the value of the effective mode area, and this is true regardless of the air-filling fraction (d/Λ). However, it is not always desirable to use the structures with the smallest effective mode area, because they typically exhibit higher confinement losses [19]. In other words, in the limit of core dimensions that are much smaller than the wavelength guided by the fiber, many rings (more than six) of air holes are required to ensure low-loss operation, which increases the complexity of the fabrication process. In this small-core regime, unless many rings of holes are used, the mode can *see over* the finite cladding region. A relatively modest increase in the structure scale in this small-core regime can lead to dramatic improvements in the confinement of the mode without compromising the achievable effective nonlinearity significantly.

With careful design, it is possible to envisage practical HFs with small-core areas ($<2 \mu m^2$) and low confinement loss (<0.2 dB/km). Note that although fiber loss limits the effective length of any nonlinear device, for highly nonlinear fibers short lengths (<10 m) are typically required, and so loss values of orders of 1 dB/km can be readily tolerated. In addition, note that reducing the core diameter to dimensions comparable with the wavelength of light generally increases the fiber loss for another reason: In relatively small-core fibers, light interacts more with the air–glass boundaries near the core, and so the effect of surface roughness can be significant [54].

5.1.3. Device Demonstrations

A range of device demonstrations have now been performed using highly nonlinear silica HFs. The first such demonstration was 2R data regeneration, a function that is a crucial element in any optical network, because it allows a noisy stream of data to be regenerated optically. The first demonstration of regeneration used a silica HF with a mode area A_{eff} of just $2.8\,\mu\text{m}^2[\gamma = 35/(\text{W}\cdot\text{km})]$ at 1550 nm [72]. Typically, devices based on conventional fibers are ~1 km long, whereas in these early experiments just 3.3 m of HF was needed for an operating power of 15 W. Subsequent experiments used an 8.7-m-long variant of this switch for data regeneration within an optical code division multiple access system [80]. Significant improvements in system performance were obtained in this way.

The 2R regeneration scheme is reviewed briefly here as an example. A schematic of the HF-based data regenerator is shown in Fig. 3.4. Pulses of light propagating in a highly nonlinear fiber broaden spectrally due to self-phase modulation, and Fig. 3.4b shows the spectrum of 2.5-ps soliton pulses before and after propagation through the HF. Figure 3.4c shows the pulse power that is transmitted through a 1.0-nm narrowband filter (offset by +2.5 nm relative to the incident pulses) as a function of incident pulse peak power. The S-shaped characteristic is suitable for thresholding because at low powers the pulses do not broaden, and so transmission through the filter is negligible. This corresponds to a "0" in the data stream. For higher powers (~2 W here), substantial self-phase modulation occurs, and so transmission through the filter becomes appreciable. This corresponds to a "1." This device acts to remove noise from an incoming data stream by nullifying all noisy "0" bits and by equalizing all noisy "1" bits.

Fibers with a high effective nonlinearity also offer length/power advantages for devices based on other processes such as Brilluoin and Raman effects. The demand for increased optical bandwidth in telecommunications systems has generated enormous interest in the S- and L-bands, outside the gain band (C-band) of conventional erbium-doped fiber amplifiers (see Chapter 1). Fiber amplifiers based on

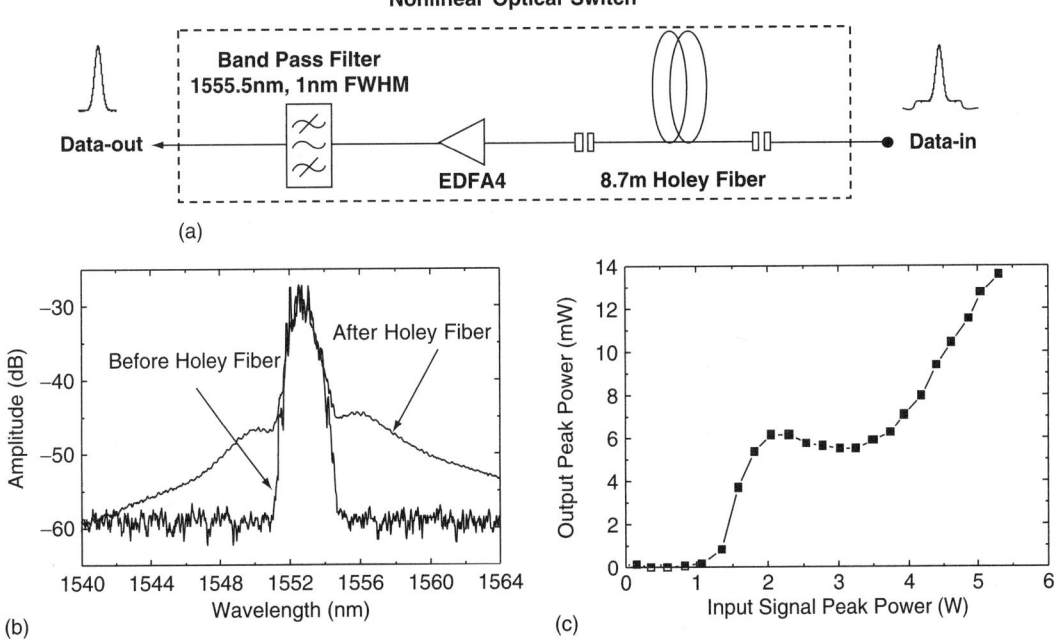

FIGURE 3.4 (a) Schematic of thresholder. (b) Pulse spectra before and after HF. (c) Power transmitted (including offset narrowband filter). (From ORC, Southampton, UK.)

Raman effects offer an attractive route to extending the range of accessible amplification bands. In addition, the fast response time (<10 fs) of the Raman effect can also be used for all-optical ultrafast signal processing applications. Despite these attractions, there is one significant drawback to Raman devices based on conventional fibers: Long lengths (~10 km) are generally required, and so Rayleigh scattering ultimately limits their performance. High nonlinearity fibers offer a method for obtaining sufficient Raman gain in a short fiber length, which eliminates this problem. For example, Yusoff et al. [81] demonstrated a 70-m fiber laser-pumped Raman amplifier. The amplifier was pumped using a pulsed fiber laser and provided gains of up to 43 dB in the L+-band for peak powers of ~7 W.

Other nonlinear device applications of HFs that have been demonstrated include a continuous-wave Raman laser [82], a wavelength division multiplexing wavelength converter [83], and pulse compression down to 20 fs [84]. Note the continuous-wave power density at the facet ($2 W/\mu m^2$) in the continuous-wave Raman device mentioned above demonstrates that HFs can exhibit a good resilience to damage.

One of the applications of highly nonlinear silica HFs to be investigated most intensively is the generation of supercontinuum (see, e.g., [25, 85–88]). The continua have been used in applications including optical coherence tomography [89], spectroscopy, and frequency metrology [90]. Supercontinua covering several octaves as well as multiwatt output have been demonstrated [91]. The absence of dopants in the core of a pure silica HF has also allowed the generation and propagation of ultraviolet wavelengths in fiber [92]. Considerable effort has been made to develop a better appreciation of the complex interplay of nonlinear processes behind supercontinuum generation, and today many of the basic mechanisms (e.g., soliton fission [93, 94], self-phase modulation [95], four-wave-mixing, and stimulated Raman scattering [85]) are understood.

5.2. Large-Mode Area Fibers for High Power Applications

The development of large-mode area fibers is important for a wide range of practical applications, most notably those requiring the delivery of high power optical beams. For many of these applications, spatial mode quality is a critical issue, and such fibers should preferably support just a single transverse mode. Large-moded single-mode fibers can be made using conventional fiber doping techniques such as modified chemical vapor deposition by reducing the NA of the fiber and increasing the fiber core size.

HFs are an attractive route toward such fibers [96], and single-mode HFs with effective areas as large as $1000 \mu m^2$ have been reported [97]. Large-mode HFs (LMHFs) can be produced by designing fibers with a large hole-to-hole spacing ($\Lambda > 5 \mu m$) and/or small air holes ($d\Lambda < 0.3$). A typical pure silica LMHF is shown in Fig. 3.1b. In addition to offering large-mode areas, LMHFs offer other unique and valuable properties; most notably they can be single-moded at all wavelengths.

The models described in Section 4 can be applied to model the optical properties of these fibers, although typically extra care is needed because of the wide range of spatial scales present. Polarization effects are typically less important in this class of fibers, and it is often sufficient to use a scalar model.

Macroscopic bend loss ultimately limits the practicality of such large-mode fibers, and so understanding bend loss is important in the design of this class of fiber. Note that single-material fibers exhibit a short wavelength bend loss edge in addition to the long wavelength loss edge found in conventional fiber designs. Generally, larger holes result in broader operational windows, whereas the hole-to-hole distance roughly determines the center position of the window (as a first approximation, the minimum bend loss occurs at a wavelength around $\Lambda/2$) [98]. Hence, standard telecommunications wavelengths fall on the short wavelength loss edge. Therefore, macrobending losses effectively limit the operational wavelength range of this class of fibers. Despite this, LMHFs with hexagonally arranged holes possess comparable bending losses to similarly sized conventional fibers at 1550 nm [99]. Note that the effective index model described in Section 4.1 is capable of predicting accurately the spectral location of the short wavelength bend loss edge [96].

Two distinct bend loss mechanisms have been identified in conventional fibers: transition loss and pure bend loss [100]. Pure bend loss occurs continually along any curved section of fiber: At some radial distance, the tails of the mode need to travel faster than the speed of light to negotiate the bend and are thus lost. The loss associated with the transition from straight to bent fiber sections is called the transition loss and is typically negligible for macrobends [99].

In any fiber, bend loss increases for decreasing values of NA and with increasing effective mode area (A_{eff}). When evaluating the relative bending losses of different fiber designs, it is thus essential to consider fibers that are equivalent in terms A_{eff} and NA at the wavelength of interest to ensure that the loss predictions for different designs can be compared. Microstructured fiber fabrication techniques permit a high level of flexibility, and bending loss can be reduced by using modified cladding hole configurations. For example, the effective mode area can be enlarged, without increasing the bending losses, by using three adjacent rods to form the fiber core instead of a single central rod [101]. This "tri-core" HF is described in more detail below.

The degree of modal distortion and the associated attenuation both increase with the severity of the bend. Baggett et al. [99] described a method for modeling the bending losses of HFs. This model uses an orthogonal function method together with a conformal transformation to obtain the distorted modal fields of the bent fiber. The bend loss is extracted by estimating the fraction of the modal field lost to radiation. This bend loss model has very few restrictions on the refractive index profile that can be considered and has been experimentally validated for LMHFs. Note that this approach does not approximate the fiber profile as a step-index fiber, as previous work does, which allows the effect of the angular orientation of the fiber relative to the bend to be explored for the first time. The results presented below show that the bend loss of HFs cannot be evaluated with effective index methods alone and that the detailed nature of the cladding configuration, which is reflected in the shape of the bent mode, is an essential consideration.

The bent mode of a standard LMHF design with $\Lambda = 12.7\,\mu\text{m}$ and $d/\Lambda = 0.45$ is shown in Fig. 3.5a for a bend of radius 3.4 cm. This fiber has an effective mode area of $\sim 190\,\mu\text{m}^2$ at 1064 nm when straight. Figure 3.5b shows a bent mode in a tri-core LMHF chosen to have a similar A_{eff} and NA to the design shown in Fig. 3.5a. The tri-core design has $\Lambda = 7.4\,\mu\text{m}$ and $d/\Lambda = 0.2$. Figure 3.5 clearly demonstrates that the tri-core structure is better able to confine the bent mode to the core than the single-rod HF for this bend radius even though the effective indices of the two (straight) fibers are similar. The predicted critical bend radius for the tri-core fiber is ~ 3.0 cm, approximately 20% smaller than for traditional single-rod designs. This level of improvement is in excellent agreement with observations and corresponds to an ability to increase A_{eff} by 15% without increasing bending losses. Preliminary results for a tri-core fiber fabricated at the ORC are in excellent agreement with these predictions.

5.3. Active Fibers

5.3.1. Background

The capillary-stacking techniques that are generally used to make single-material silica microstructured fibers can be readily adapted to allow the production of doped fibers simply by replacing the solid silica core rod with a doped rod in the preform stack. The doped core rod is typically formed by extracting the core region from a conventional doped fiber. In the example given in [103], the starting point was an aluminosilicate ytterbium–doped rod with an NA

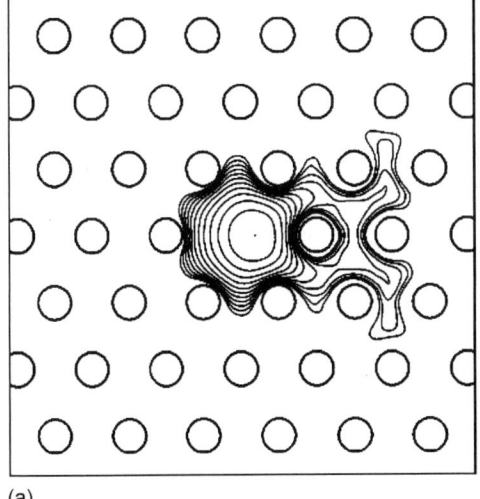 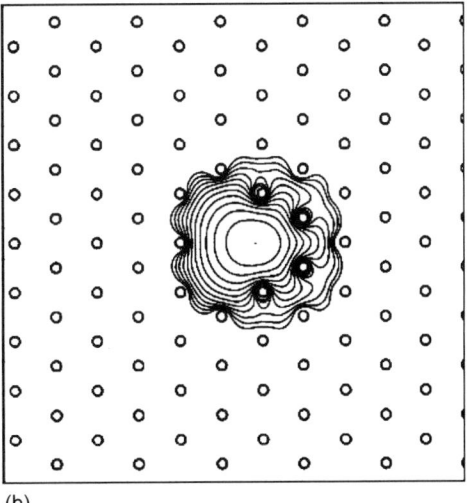

(a) (b)

FIGURE 3.5 Contour plots of the modal intensity and refractive index profile for (a) a standard large-mode area silica fiber and (b) a tri-core holey fiber. Contour lines are spaced by 2 dB.

of 0.05 produced using conventional modified chemical vapor deposition process techniques.

Many properties of the index-guiding HFs have benefits for active devices. HFs incorporating Er^{3+} [102, 104] and Yb^{3+} have been fabricated, and continuous-wave [105] and mode-locked [106] Yb^{3+} HF lasers have been realized experimentally. As is the case for the passive applications of HFs, two regimes of particular practical relevance are low NA large-core active fibers and high NA small-core active fibers. Large-mode areas offer the potential for high power generation without nonlinear effects or fiber damage and in addition allow broadband single-mode laser output. At the other extreme, nonlinear fibers allow the realization of very high gain efficiency (>8.5 dB/mW was reported in [104]).

Below one active device of each type is reviewed: a tunable fs-soliton source based on highly nonlinear fiber and a cladding pumped laser based on double-clad large-mode area fiber.

5.3.2. Tunable fs-Soliton Source

Fibers with anomalous dispersion extending down through the visible regions of the spectrum allow the prospect of extending soliton laser techniques that have been developed for 1550 nm to Yb^{3+}-doped and Nd^{3+}-doped fiber lasers. Wavelength tuneable femtosecond (fs) pulse sources have applications in areas such as ultrafast spectroscopy, materials processing, and nonlinear optics. Traditional fs pulse sources based on bulk crystals offer a limited wavelength range, particularly above 1.1 μm. The SSFS in fibers [107] has opened up the possibility of wavelength tuneable fs-soliton fiber sources. To use the SSFS effect, the frequency shifting fiber must exhibit anomalous dispersion at the seed wavelength and across the required tuning range. Although conventional fibers can have anomalous dispersion beyond 1.3 μm, as discussed in Section 2.3, small-core, large, air-filling fraction HFs extend this range to 550 nm. Such fibers also have a high effective nonlinearity, which allows soliton formation with just pJ pulse energies in meter long fiber lengths [108].

Here we outline a continuously tuneable soliton source operating from 1.06 to 1.33 μm based on HF technology. A fuller exposition of this work can be found in [16]. This source is seeded by a diode-pumped 1.06-μm Yb^{3+}-doped silica fiber laser and relies on SSFS effects in a Yb^{3+}-doped holey fiber amplifier. The fiber used in these experiments was single mode for all wavelengths considered here and has an effective mode area of just $A_{\text{eff}} \approx 2.5 \ \mu m^2$ at 1550 nm (an SEM image of this fiber is shown in the inset to Fig. 3.6).

The mode-locked seed laser produces pulses at 1.06 μm with a positive linear chirp, which are launched into Yb^{3+}-doped HF amplifiers, together with a pump beam from a diode laser to control the gain, as shown by Fig. 3.6. Because of the amplification and nonlinear evolution of the pulses as they pass through the amplifier, Raman solitons form and are continuously wavelength shifted via the SSFS. Because the nonlinear evolution of the pulses depends on the pulse peak power, the wavelength of the Raman solitons at the amplifier output can be tuned by varying the gain in the amplifier. Using ~5 m of amplifier fiber, monocolor soliton output pulses have been wavelength tuned from 1.06 to 1.33 μm, as shown at the bottom of Fig. 3.6. At higher pump powers, the change in gain distribution causes the Raman solitons to form earlier within the amplifier, thereby leaving them a greater length of fiber within which to walk-off to longer wavelengths through the SSFS. The final central wavelength of the pulses varies in an almost linear fashion with the level of incident pump power. The maximum wavelength shift of the Raman soliton increases with the length of fiber and is ultimately only limited by the absorption of silica near 2.3 μm.

5.3.3. Cladding Pumped Fiber Laser

Capillary stacking techniques can be extended to allow the production of all-glass double-clad fibers, and an example is shown in Fig. 3.1d. This approach is attractive for high power active fiber devices, because it allows the use of cladding pumping, and for such applications large-mode area fiber designs are of particular interest (see Section 5.2 for a review of passive LMHFs). However, the use of rare-earth dopants in LMHFs is challenging. The presence of dopants (and associated co-dopants such as germanium, aluminium, and boron that are required to incorporate the rare-earth ions at reasonable concentrations and to maintain laser efficiency) modifies the refractive index of the host glass. This affects the NA of the fiber and can lead to the loss of some of the most attractive LMHF features, such as broadband single-mode guidance, unless care is taken in the fiber designs.

Here we review laser development based on double-clad HFs [103]. The ultimate advantages of these fibers relative to polymer-coated dual-clad fibers are that they allow for all-glass structures, with inner cladding NAs in excess of 0.5, and good pump/

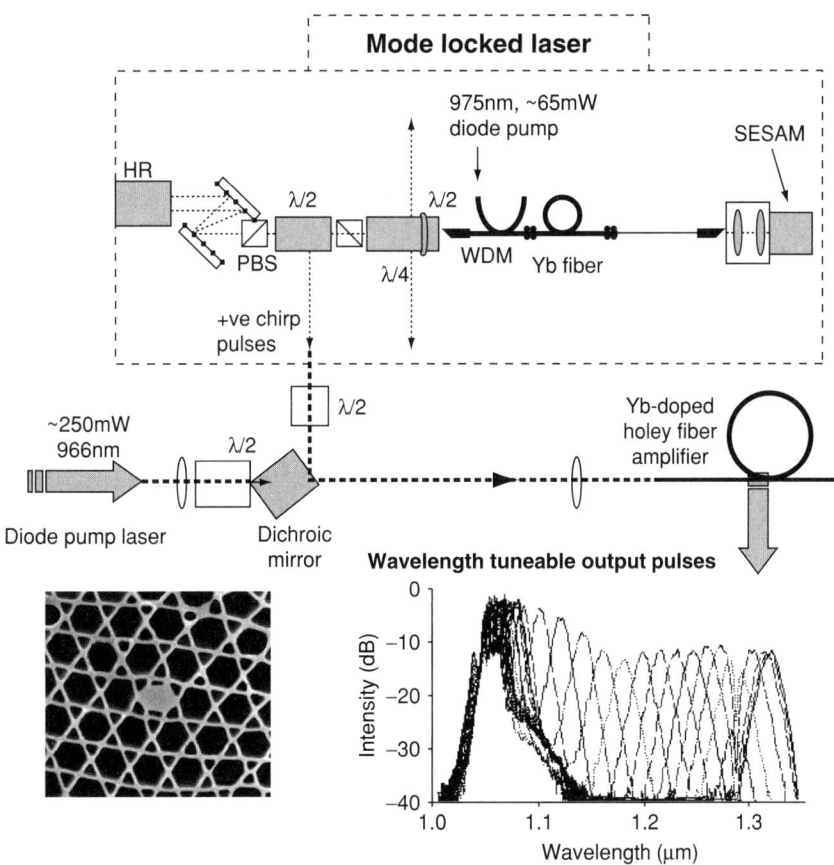

FIGURE 3.6 Experimental setup of the mode-locked Yb^{3+} fiber seed laser (diode-pumped), the launch arrangement for seeding the pulses and pump laser to the Yb^{3+}-doped holey fiber amplifier, and the wavelength tunable output pulses. The Yb^{3+}-doped holey fiber used here is shown in the inset. PBS, polarizing beam splitter; SESAM, semiconductor saturable absorber mirror; WDM, wavelength division multiplexing.

mode mixing. In addition, they offer the combination of single-mode guidance and large-core dimensions. In terms of device, these features translate to the possibility of higher coupled diode powers, shorter device lengths, and extended tuning ranges. Consequently, HF technology represents a most interesting proposition for application in the future power scaling of both continuous-wave and pulsed fiber laser systems.

The fiber in Fig. 3.1d has a Yb^{3+}-doped core surrounded by an inner cladding consisting of five rings of small ($d \approx 2.7$ μm) holes separated by $\Lambda \approx 9.7$ μm. The inner cladding NA was measured to be 0.3 in a ~10-cm fiber length. Two rings of larger holes define the outer cladding. Note that the core is offset from the center of the fiber, which breaks the cladding symmetry and hence enhances the pump absorption.

Slope efficiencies as high as 82% were recorded in a 4-m fiber length, comparable with the best conventional ytterbium fiber lasers. As expected, the output beam was robustly single mode. Because of the low NA of the inner core/cladding structure, a significant fraction of the pump was launched into the inner cladding, and so this laser acts as a hybrid core/cladding pumped fiber laser. A cladding pumped laser was realized using a low brightness 915-nm fiber-coupled laser diode. In this first demonstration, average powers in excess of 1 W were achieved at a 7.5-m fiber length, with a measured slope efficiency of 70% [103]. Both Q-switched and mode-locked operations were demonstrated in this cladding-pumped HF laser. In the mode-locking experiments, fundamental mode locking was obtained over a wavelength tuning range in excess of 60 nm. The pulse duration was estimated to ~100 ps. An output power of more than 500 mW was achieved for a pump power of 1.33 W.

Further progress in this class of fibers was reported [109], which presents an HF laser generating up to 80 W of output power with a slope efficiency of 78%. Single transverse mode operation is achieved using a fiber with a mode-field area of 350 μm². No thermo-optical limitations are observed at the extracted power level of 35 W/m, which implies that such fibers allow scaling to even higher powers. Limpert et al. [110] explored the issue of thermal management of air-clad microstructured fibers in comparison with conventional double-clad fibers. This work revealed that the temperature in the fiber core is determined primarily by heat transport through the outer surface of the fiber. If the dimensions of the air-cladding region are properly designed, the temperature profile can be comparable with a conventional double-clad fiber. Hence, the air-clad region is not necessarily a limitation to the power scaling capabilities of microstructured fiber lasers, and therefore air-clad microstructure fibers are likely to be scalable to power levels of several kilowatts.

6. SOFT GLASS FIBERS

6.1. Background

The combination of the microstructured fiber concept with nonsilica glasses is an emerging field that promises the development of a host of new fibers and operational regimes not achievable with existing fiber technology. For example, recently, compound or "soft" glasses with high intrinsic optical nonlinearity (such as lead silicate, tellurite, and bismuth oxide) have been used to produce HFs with extreme values of effective fiber nonlinearity [21, 42–49]. Work on high nonlinearity soft glass HFs is reviewed within this section. A number of nonsilica glasses transmit at wavelengths substantially beyond silica into the mid-IR (e.g., chalcogenide glasses), and microstructured fibers made from such materials promise a new means of providing power delivery at these wavelengths. Other classes of microstructured fiber have also begun to emerge, such as microstructured fibers with solid cladding designs, and these fibers are also described in this section.

The fabrication of soft glass microstructured fibers was reviewed briefly in Section 3. Note that in contrast to conventional fibers, microstructured fibers can be made from a single material, which eliminates the problems induced by the requirement that the core/cladding glasses are thermally and chemically matched. This opens up the prospect of using microstructured fiber technology as a tool for realizing optical fibers from an extremely broad range of optical materials.

6.2. Extreme Nonlinearity

As described in Section 2.2, the combination of a highly nonlinear glass composition and small-core/high NA HF geometry allows a dramatic increase of the fiber nonlinearity relative to conventional fibers. Soft glasses typically exhibit a higher linear refractive index than silica and thus can provide better confinement and smaller mode areas, in addition to higher material nonlinearity. For example, an air-suspended rod of bismuth oxide glass (which has a refractive index of 2.02 [111]) has a minimum $A_{\text{eff}} \approx 0.6\,\mu\text{m}^2$ at 1550 nm. When this is combined with the nonlinear refractive index of bismuth oxide glass, which is $3.2 \times 10^{-19}\,\text{m}^2/\text{W}$ [111], the maximum γ that could be achieved in an HF made from this material is $\sim 2200/(\text{W·km})$, a factor of 2 higher than the record nonlinearity achieved thus far for a chalcogenide fiber with conventional solid cladding [112].

Lead silicate glasses are one promising family of glasses for use in highly nonlinear index-guiding HFs. Although their intrinsic material nonlinearity (and indeed linear refractive index) is lower than chalcogenide and heavy metal oxide glasses [113], they offer higher thermal and crystallization stability and less steep viscosity–temperature curves while exhibiting low softening temperatures [114]. Among commercially available lead silicate glasses, the Schott glass SF57 exhibits the highest nonlinearity. The softening temperature of SF57 is 520° C [115], the nonlinear refractive index of this glass was measured to be $4.1 \times 10^{-19}\,\text{m}^2/\text{W}$ at 1060 nm [116], and the linear refractive index of SF57 is ~ 1.8 at 1550 nm [115]. The zero dispersion wavelength for this glass is 1970 nm, and the material dispersion is strongly normal at 1550 nm.

Some key results that have been achieved in microstructured lead silicate fibers include effective nonlinearity coefficients (γ) as high as $640/(\text{W·km})$ with a loss ~ 2.6 dB/m at 1550 nm, anomalous dispersion at 1550 nm, and Raman soliton generation [46]. Observations of the SSFS and pulse compression have also been made in SF57 HF [46]. Supercontinuum generation has been observed in SF6 glass HF [43].

Bismuth oxide–based glasses are also attractive materials for nonlinear devices. It shows high nonlinearity but without containing toxic elements such as Pb, As, Se, and Te [117]. Moreover, the bismuth-based glass exhibits good mechanical, chemical, and thermal stability, which allows easy fiber fabrication process. A nonlinear fiber [118] and a short Er-doped fiber amplifier with broadband emission [119] have been developed from this glass. In addition, bismuth oxide–based fibers can be fusion spliced to a silica fiber [120], which offers easy integration to silica-based networks. Because of the high bismuth content, the glass exhibits a high linear and nonlinear refractive index of $n = 2.02$ and $n_2 = 3.2 \times 10^{-19} \text{m}^2/\text{W}$ at 1550 nm, respectively [118], and has a softening temperature of 550° C.

Bismuth oxide–based glass HFs with effective fiber nonlinearities as high as $\gamma = 1100/(\text{W·km})$ have been realized [21]. In addition, the splicing of small-core bismuth HFs to conventional fibers has been reported [49]. To reduce the overall mode-mismatch loss, two intermediate buffer stages were used. The splices were mechanically strong with respect to strain in the axial direction. Although the total splicing losses achieved to date are still quite high (5.8 dB), they can largely be accounted for by mode mismatches at the buffer fiber interfaces. The introduction of an additional buffer stage should help to reduce the mode mismatch. Splicing of bismuth glass HF to silica fiber has resulted in two benefits in performance. One is the reduction of coupling losses by 0.9 dB relative to butt coupling. The other is the achievement of single-mode guidance in the bismuth HF at 1550 nm, although the fiber can support more than one mode in the case of free-space coupling.

Tellurite glasses, like lead silicate and bismuth glasses, offer high refractive index and high optical nonlinearity ($n_2 = 2.5 \times 10^{-19} \text{m}^2/\text{W}$) [44]. In addition, tellurite glass has good infrared transmittance and has a low phonon energy relative to other oxide glasses [121]. Furthermore, tellurite glasses are more stable than fluoride glasses, have higher rare-earth solubilities than chalcogenide glasses [121], and have an order of magnitude larger Raman gain peak than fused silica [122]. Tellurite glasses have a low softening temperature around 350° C [121].

Tellurite HFs with $\gamma = 48/(\text{W·km})$ and a loss of 5 dB/m at 1550 nm were reported [44]. First- and second-order Stokes stimulated Raman scattering were observed in 1 m of this tellurite HF. More recently, low-loss index-guiding tellurite fibers with a low of just 0.18 dB/m at 1550 nm and $\gamma \approx 675/(\text{W·km})$ have been reported [50]. This is the lowest low reported to date in a nonsilica microstructured fiber. In addition, this fiber also has a zero-dispersion wavelength that has been shifted to near 1550 nm.

Both the effective fiber nonlinearity (γ) and effective fiber length (L_{eff}) determine the performance of a nonlinear device. The effective length of a fiber depends on the fiber's propagation loss α via $L_{\text{eff}} = [1 - \exp(-\alpha L)]/\alpha$. Small-core high-NA HFs based on bismuth oxide and lead silicate glasses have clearly higher fiber nonlinearity but also higher propagation loss compared with silica HFs or conventional nonlinear fiber. However, for short devices using ≤ 1 m fiber length, fiber losses of ≤ 2 dB/m can generally be tolerated. When the propagation loss is less than 2 dB/m, the effective fiber length is more than 80% of the real fiber length of ≤ 1 m, whereas the nonlinearity of nonsilica fibers can be up to 10 times higher than that of silica HFs. In other words, in compact devices using ≤ 1 m fiber length with ≤ 2 dB/m loss, the increase in fiber nonlinearity obtained by using nonlinear glass compositions clearly outweighs the decrease of the effective fiber length due to higher propagation losses. Thus, provided that relatively low-loss fibers can be produced, highly nonlinear compound glass HFs provide a route to better nonlinear performance than existing silica fibers (in terms of lower power consumption and/or shorter fiber length).

Most soft glasses have a high normal material dispersion at 1550 nm, which tends to dominate the overall dispersion of fibers with a conventional solid cladding structure, and for most glass compositions, near-zero dispersion at a wavelength of about 1550 nm cannot readily be achieved using conventional technologies. However, for many nonlinear device applications, anomalous or near-zero dispersion is required. Fortunately, the cladding geometry of a microstructured fiber can result in a large enough waveguide dispersion to allow the highly normal material dispersion to be overcome. (Indeed, as mentioned above, for example, anomalously dispersive lead silicate HFs at 1550 nm have been demonstrated [47].) The fact that fiber dispersion is anomalous also enables us to exploit soliton effects [46].

6.3. New Transmission Fibers

A range of nonsilica glasses can exhibit properties such as transparency in the mid-IR region and high solubility of rare-earth ions that are not available in silica glass. Conventional fiber approaches have thus far made limited progress in developing low-loss single-mode soft glass fibers due to difficulties in finding compatible core/cladding materials. For this reason, single material nonsilica index-guiding HFs are of particular promise for applications in mid-IR region and active devices.

One challenge associated with the use of soft glass–based fibers for high power applications is the onset of intensity-dependent nonlinear effects and optical damage. The use of single-mode LMHF designs [96] is one means of minimizing nonlinear effects. Because the nonlinear refractive index (n_2)

of nonsilica glass is typically in the range of $(1-50) \times 10^{-19} m^2/W$, higher than that of silica glass by 1–2 orders of magnitude, large-mode area designs ($\gg 100\,\mu m^2$) are required to reduce the effective nonlinearity (γ) even to the 60/(W·km) level. In contrast to the nonlinear fiber designs described in Section 6.2, an LMHF cladding needs to contain a high feature count to provide low NA guidance (i.e., by containing a large number of relatively small air holes).

An HF with a mode area of $40\,\mu m^2$ at 800 nm fabricated from the Schott lead silicate glass SF6 using the conventional capillary-stacking technique was reported [123]. This fiber had a four-ring microstructured cladding with a hole-to-hole spacing Λ of $4.3\,\mu m$. Robust single-mode guidance at 800 nm was observed. Although, as described above, this value of mode area is still not large enough to avoid intensity-dependent nonlinear effects, this work demonstrates that the capillary-stacking technique could be exploited to fabricate soft glass HFs with a complex holey cladding. It should be feasible to fabricate single-mode fibers with larger-mode areas in high-index glasses using this technique in the future.

6.4. Solid Microstructured Fibers

Research to date on silica microstructured fibers has shown that the combination of wavelength-scale features and a large refractive index contrast is a powerful means of obtaining fibers with a broad range of useful optical properties. However, there are some practical drawbacks associated with the use of air/glass fibers. When compared with solid fibers, air/glass microstructured fibers are challenging to splice, polish, and taper, and when the cross-section is largely composed of air, they can be fragile. In addition, it can be challenging to fabricate kilometer-scale HFs with identical and controllable cladding configurations. This is because the transverse profile of a drawn microstructured fiber is sensitive to the effect of pressure inside the holes, surface tension at the air–glass boundaries, and temperature gradients present during the fiber drawing process. To prevent the collapse of the holey microstructure during drawing, the holes within the preform are often sealed, and consequently the air pressure inside the structure changes during fiber drawing. Hence, the precise details of the final fiber microstructure are typically time dependent as well as dimension dependent. Note that the optical characteristics of microstructured fibers can be sensitive to the cladding configuration, and even minor changes in the microstructure can cause noticeable deviations in properties such as dispersion.

The development of microstructured fibers that have a solid microstructured cladding promises to eliminate these drawbacks. These fibers combine the practicality of a solid cladding with the design flexibility provided by the transverse microstructure. Such fibers have the potential to advance the development of both index-guiding and photonic bandgap-guiding microstructured fibers. Two different geometric implementations of this solid microstructured fiber concept have recently been realized.

The first approach is to replace the air holes in the transverse structure of an air/glass microstructured preform with low index glass regions to produce an index-guiding solid variant of an HF [41]. Another approach is to form a fiber in which the cladding is defined by a series of thin-nested concentric layers of two or more different glasses [124]. Such fibers are often referred to as Bragg fibers (see Chapter 4). In both approaches, the basic requirement is the identification of materials that are thermally and chemically matched so that they can be drawn into optical fiber and provide a sufficient refractive index contrast to allow light to be confined either by index-guiding or photonic bandgap effects without requiring unfeasibly large numbers of cladding features. Note that although all-solid microstructured fibers may have some practical advantages, as suggested above, this approach clearly restricts the range of materials that can be used relative to single-material fiber designs. Here an example of each of the four possible classes of solid fiber is briefly reviewed.

Feng et al. [41] described the development of all-solid index-guiding HF based on two thermally matched silicate glasses with a large index contrast. A borosilicate glass with a refractive index of $n = 1.76$ at 1550 nm was selected as the background material for this fiber. A glass with $n = 1.53$ at 1550 nm was used as the material to fill the holes. These glasses were selected because of their mechanical, rheological, thermodynamic, and chemical compatibility. Calculations indicate that it is possible to design fibers from these two materials with negligible confinement loss provided that the d/Λ ratio is sufficiently large, as is the case for the fibers described below [41].

Rods of the low index glass were inserted into the tubes made from the higher index glass, both of which were drilled from bulk glass samples. These rod–tube structures were caned on a fiber drawing tower and stacked around a core rod of the high index glass within a high index jacket tube. This preform was then drawn using a two-stage procedure to produce two fibers: one with an outer diameter of 440 μm and the other with an outer diameter of 220 μm. Figure 3.7 compares the cladding configurations of the 1-mm cane, 440-μm fiber, and 220-μm fiber. Note that all the low index (black) regions retain their circularity, regardless of the draw-down ratio. In addition, the value of the d/Λ ratio is ~0.81 regardless of the fiber outer diameter. These fibers thus provide a practical way of avoiding structural deformations during fiber drawing.

Even though the index contrast between silica/air leads to a similar minimum mode area as this combination of materials, the significantly larger material nonlinearity (n_2) of these soft glass materials results in a dramatic improvement in the nonlinearity. Hence, although the maximum nonlinearity that can be achieved in a silica/air HF is ~60/(W·km), more than 500/(W·km) is possible in using this combination of high index glasses. The fiber in Fig. 3.7c has a measured nonlinearity of $\gamma \approx 230$/(W·km).

Using the commercial Schott glasses LLF1 and SF6, Luan et al. [125] reported the fabrication of an all-solid PBGF. The periodic cladding in this fiber was formed by stacking LLF1/SF6 tubes in a hexagonal pattern, as was done in the index-guiding fiber described above. In this case, light was guided via photonic bandgap effects within the relatively lower index core region made from LLF1 glass. One advantage of this approach relative to air/silica PBGFs is that it appears to avoid the problem of surface modes with the bandgaps, which can restrict the practically useful wavelength range in these fibers.

Index-guiding fibers with a one-dimensionally multilayered structured cladding and a solid core have also been realized [126], and an example is shown in Fig. 3.1f (dark regions represent the low index regions). In this case, the structured preform was formed by extruding an alternating stack of thin films into a preform with a macroscopic ring structure. Single-mode guidance and high effective fiber nonlinearity [$\gamma \approx 260$/(W·km)] have been observed in this fiber.

The first reported example of a fiber with a high index-contrast solid multilayered cladding guided light in an air core via photonic bandgap effects [124]. To achieve high index contrast in the layered portion of the fiber, a chalcogenide glass with a refractive index of 2.8 (arsenic triselenide) was combined with a high glass-transition temperature thermoplastic polymer with a refractive index of 1.55, poly(ether sulfone). The glass layers were thermally evaporated onto a polymer film and the coated film rolled to form a hollow multilayered tubular preform, which was drawn into a fiber with submicron layer thicknesses. Note that the polymer material is not transparent at the operating wavelength, and fiber losses below 1 dB/m were made possible by the short penetration depths of electromagnetic waves in the high refractive index contrast photonic crystal structure. Using a fiber with a fundamental photonic bandgap centered near 10.6 μm and a hollow core with a diameter of 700 μm, CO_2 laser emission was successfully transmitted. No damage to the fibers was reported when a laser power density of ~300 W/cm^2 was coupled into the hollow fiber core.

7. PBGFs

The index-guiding HFs discussed in Sections 5 and 6 can to a first approximation be regarded as a variant of the traditional step-index fiber with a larger and wavelength dependent index contrast between core and cladding. In contrast, with PBGFs, light is

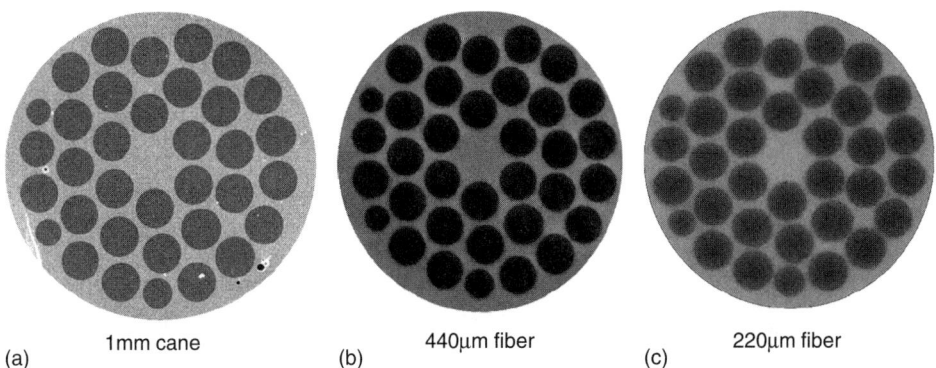

FIGURE 3.7 SEM imags of microstructured cladding in (a) 1-mm cane, (b) 440-μm outer diameter (OD) fiber with $\Lambda = 4$ μm, and (c) 220-μm OD fiber with $\Lambda = 2$ μm.

localized at a defect placed in a photonic bandgap material. The periodically structured material acts acts to suppress the transverse propagation of light at the frequency of the guided mode. Here progress in the development of PBGFs is reviewed for the two fiber types that have been realized experimentally: *honeycomb fibers*, whose properties in many respects resemble those of index-guiding microstructured fibers, and *air-guiding fibers*, which can only be realized through the photonic bandgap effect.

For the photonic bandgap-guiding mechanism to operate, the bandgap must extend over the whole plane perpendicular to the direction of propagation. In silica/air structures such a gap can only be achieved for a finite (non-zero) longitudinal propagation constant, the minimum value required being dependent on the fiber structure. The simple hexagonal arrangement of cladding holes commonly used for making index-guiding microstructured fibers only provides a complete bandgap for air-filling fractions of 30% or higher. In contrast, arranging the holes on a honeycomb lattice makes it possible to open up complete gaps at much lower air-filling fractions (<1%) [127]. For this reason, the first PBGF experimentally fabricated was based on the honeycomb lattice. The perfect honeycomb lattice can be thought of as an array of silica rods, separated by rings of air holes. The core region of the waveguide is created by introducing an extra air hole in the central silica rod, thereby lowering the effective index of this region.

It is a general feature of PBGFs that the central defect is typically created by lowering the effective index of the core region (whereas for index-guiding fibers the effective core index is raised to trap the guided mode). The magnitude and position of the bandgaps are controlled by the radius of the cladding holes. Depending on the size of the central core hole, the fundamental defect mode may be pushed into either the first or a higher order bandgap. In principle, any of the bandgaps may support a number of guided modes. By proper tuning of the core and cladding hole configurations, one can create fibers that are either single or multimode. Note that this class of fibers allows the possibility of guiding higher order modes without guiding the fundamental mode. For example, the honeycomb PBGF reported in [128] guided third-order mode only. These provided the first proof-of-principle of the photonic bandgap waveguiding mechanism for honeycomb fiber designs.

Somewhat surprisingly, modeling results imply that the honeycomb PBGFs have a similar potential for dispersion engineering as do index-guiding fibers [129]. For example, chromatic group-velocity dispersion curves may be flattened over large wavelength intervals, zero dispersion may be obtained at short wavelengths, and fibers can have large anomalous group-velocity dispersion coefficients (several hundred ps/nm·km). As is the case for index-guiding fibers, by breaking the 6-fold symmetry of the structure, it is possible to produce polarization-maintaining PBGFs [130]. This can be done either by introducing an asymmetric core defect or an asymmetric cladding configuration. One unique possibility birefringent PBGFs offer is a means of realizing fibers that support a single polarization (i.e., one of the polarization states can be pushed out of the photonic bandgap).

One of the greatest attractions of PBGF technology is that it can allow light to be guided in an air core. This occurs when modes with an effective index of ~1 fall within the bandgap. In other words, for light to be guided in air, the effective index of the cladding mode at the lower boundary of the gap falls above the light line. As mentioned above, the hexagonal lattice structure has this property for sufficiently large air-filling fractions. As an example, Fig. 3.8 shows the bandgap diagram for a structure with a filling fraction of 70% [131]. It can be seen that several bandgaps cross the air line. To confine an air-guided mode within these bandgaps, a core defect consisting of a rather large air hole must be introduced. In the inset of Fig. 3.8, a structure is shown in which the core defect has been obtained by replacing seven holes by a single air hole. This structure was found to support both a fundamental and a second-order guided mode, though not at the same frequency. The traces of the two modes are shown in the upper inset of Fig. 3.8. Inside the bandgap the modes are bound, whereas outside they appear as leaky resonances within the bands of cladding modes. For both modes, guidance only takes place within a rather narrow frequency interval. Furthermore, the guidance properties of the structure depend strongly on the radius of the core defect.

The fabrication of air-guiding PBGFs was first reported in [13]. In these first fibers, structures with $\Lambda \approx 5\,\mu m$ and an air-filling fraction of 30–50% resulted in several air-guiding transmission bands over a fiber length of a few centimeters, in good agreement with predictions. Since this first report, considerable progress has been made to reduce the loss of PBGFs, and losses of 1.7 dB/km can now be achieved at 1550 nm [54]. As noted in Section 3.1, to produce a bandgap fiber preform, the fiber core can

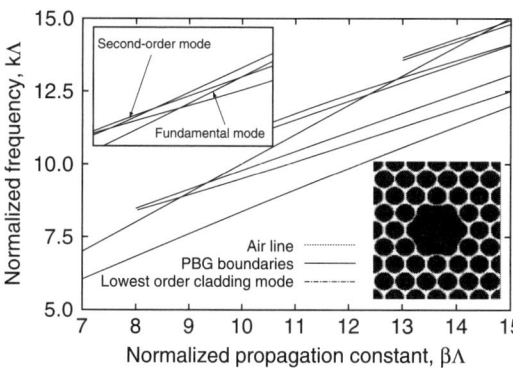

FIGURE 3.8 Photonic bandgap (PBG) boundaries and defect mode traces (upper inset) for the air-guiding PBGF discussed in the text. The defect modes are confined when they fall inside the PBG (and leaky elsewhere). The fiber microstructure is shown in the lower inset. (From [131].)

be formed in a number of ways, and the two cases explored in the most detail thus far are 7-cell and 19-cell air cores. The typical air-core PBGF, seen in Fig. 3.1c, has a 7-cell core.

In addition to using microstructured optical fibers as light pipes, they can be used to manipulate light propagating across the fiber [132]. Fundamental and higher order bandgaps have been observed experimentally via the spectral features that result from the periodic nature of the fiber microstructure in the transverse direction.

The dispersion properties of PBGFs were reviewed briefly in Section 2.3. It is important to note that the dispersion of PBGFs is strongly wavelength dependent. Typically, negative group-velocity dispersion values (normal dispersion) are found at the short wavelength edge of the bandgap, passing through zero dispersion near the center of the bandgap with positive values (anomalous dispersion) at longer wavelengths. The dispersion slope increases rapidly near the upper band edge and decreases rapidly near the lower band edge [133, 134], and values of more than ± 2000 ps/nm·km are typical. In effect, the dispersion of the fiber can be tuned by tuning the operating wavelength with respect to the location of the bandgap (or vice versa). The dispersion slope can be significantly modified by changing the geometry of the core region [133].

The loss, dispersion, and leakage properties of the fundamental mode of a PBGF can all be affected by the presence of surface modes within the bandgap [135]. This occurs because the air-guided modes can couple to surface modes within the bandgap. Surface modes can be supported within the glass region surrounding the air core, and the quantity and configuration of glass surrounding the core determines the properties of these states. Some fiber designs do not have surface states within the bandgap, and this is desirable because surface modes can reduce the effective bandgap width and increase the fiber loss. On a more positive note, surface modes can lead to narrowband regions of strong anomalous dispersion (via avoided crossings) that may find use in novel device applications [136, 137].

Now that high-quality, low-loss, air-guiding fibers can be made, many new applications can be envisaged. These include low-loss/high-power transmission at wavelengths where the absorption in silica is high, in gas-sensing applications, and for spectral filtering exploiting the narrow-band nature of the guided-mode transmission bands. Applications for laser-assisted atom transport or even as particle accelerators have also been proposed [138].

One recently demonstrated example of an application of air-guiding PBGF is all-fiber chirped pulse amplification [137]. Pulses from a wavelength- and duration-tunable femtosecond/picosecond source at 10 GHz were dispersed in 100 m of dispersion compensating fiber before being amplified in an erbium-doped fiber amplifier and subsequently recompressed in 10 m of the anomalously dispersive PBGF. Pulses as short as 1.1 ps were obtained. The advantage of air-core fibers for this application is that they present negligible nonlinearity and thus can potentially be used to obtain ultra-high pulse peak powers. The novel dispersion properties of air-core PBGFs have also been used for dispersion compensation within a fiber laser [139].

Hollow-core PBGFs are also ideally suited to deliver high power laser beams. Recent work in this area has concluded that 7-unit cell cores are currently most suitable for transmission of femtosecond and sub-picosecond pulses, whereas larger cores (e.g., 19-cell cores) are better for delivering nanosecond pulses and continuous-wave beams [140].

8. CONCLUSION AND THE FUTURE

Microstructured optical fibers have now developed to the point where they are not only of interest from a research perspective, but are also becoming available commercially. As this chapter demonstrated, good quality index-guiding HFs and PBGFs based on a cladding structure with a triangular lattice of air holes embedded in pure silica glass can now routinely be fabricated over a wide parameter range. More complicated structures, such as double-clad fibers and active fibers, have also been demonstrated.

The unique properties and design flexibility of these fibers opens up a wide range of possible applications as functional components in fiber communication networks (including devices for amplification or dispersion compensation), in novel broadband sources, or for high-power transmission, to name a few. One area in which significant progress has been made recently is the field of highly nonlinear index-guiding HFs. The generation of broad supercontinuum spectra and all-optical data regeneration are examples of the significant advances that have resulted from this silica-based HF technology. Moreover, nonsilica index-guiding HFs with extremely high nonlinearity can now be routinely fabricated [21], and such fibers promise to offer nonlinear fiber devices at unprecedentedly low operating powers (1–10 mW) and short device lengths (0.1–1 m).

By filling the holes of these fibers with a range of materials, it should be possible to significantly extend their functionality. One example is the filling of index-guiding HFs with a high index liquid to form tunable PBGFs [141].

The synergy of novel nonsilica glass materials and microstructured fiber technology promises a broad range of new and potentially useful optical fibers. The large index contrast possible in nonsilica microstructured fibers is promising for the development of fibers based on photonic bandgap effects. However, it has been demonstrated that increasing the refractive index contrast beyond that available in air/silica does not necessarily broaden the photonic bandgaps that are available [142]. One particularly promising application of this new fiber type will be the development of new air-guiding fibers for broadband high power IR transmission.

It was demonstrated that solid microstructured fibers can be produced using a relatively small number of cladding features. It is expected that the use of a solid fiber structure may lead to a number of practical advantages relative to air/glass fibers. For example, edge polishing, angle polishing, and splicing should all be more straightforward in solid microstructured fibers. By combining a solid cladding with an air core, it is possible to demonstrate that low fiber attenuation can be achieved through structural design rather than high-transparency material selection [124]. As the fabrication techniques used to produce structured preforms and to draw high quality fibers continue to improve, it is anticipated that more novel optical properties and promising applications will continue to emerge.

9. ACKNOWLEDGMENTS

I thank a number of colleagues at the ORC, University of Southampton who have made many important contributions to the research described within this chapter. In particular, warm thanks to Joanne Baggett, Heike Ebendorff-Heidepriem, Xian Feng, Vittoria Finazzi, Kentaro Furusawa, John Hayes, Ju Han Lee, Periklis Petropoulous, Jonathan Price, and David Richardson. I also acknowledge the support of a Royal Society University Research Fellowship.

10. REFERENCES

1. P. Kasier, E.A.J. Marcatili and S.E. Miller, "A new optical fiber," Bell Sys. Tech. J. **52**, 265–269 (1973).
2. P. Kaiser and H.W Astle, "Low-loss single-material fibers made from pure fused silica," Bell Sys. Tech. J. **54**, 1021–1039 (1974).
3. E. Yablonovitch, "Inhibited spontaneous emission in solid-state physics and electronics," Physical Rev. Lett. **58**, 2059–2062 (1987).
4. M.A. McCord and R. Michael J., "Electron beam lithography," in P.R. Coudry (Ed.), *Handbook of Microlithography, Micromachining and Microfabrication*, **1**, SPIE Press, Bellingham, WA (1997).
5. T.A. Birks, P.J. Roberts, P.St.J. Russell, D.M. Atkin and T.J. Shepherd, "Full 2-D photonic bandgaps in silica/air structures," Electron. Lett. **31**, 1941–1943 (1995).
6. J.C. Knight, T.A. Birks, P.St.J. Russell and D.M. Atkins, "All-silica single-mode optical fiber with photonic crystal cladding," Optics Lett. **21**, 1547–1549 (1996).
7. T.A. Birks, J.C. Knight, P.St.J. Russell, "Endlessly single-mode photonic crystal fiber," Opt. Lett. **22**, 961–963 (1997).
8. T.M Monro, P.J. Bennett, N.G.R. Broderick and D.J. Richardson, "Holey fibers with random cladding distributions," Opt. Lett. **25**, 206–208 (2000).
9. T.M. Monro, D.J. Richardson, N.G.R. Broderick and P.J. Bennett, "Holey optical fibers: an efficient modal model," J. Lightwave Technol. **17**, 1093–1102 (1999).
10. B.J. Eggleton, C. Kerbage, P. Westbrook, R.S. Windeler and A. Hale, "Microstructured optical fiber devices," Opt. Express **9**, 698–713 (2001).
11. J. Broeng, S.E. Barkou, A Bjarklev, J.C. Knight, T.A. Birks and P.St.J. Russell, "Highly increased photonic bandgaps in silica/air structures," Opt. Commun. **156**, 240–244 (1998).
12. J.C Knight, J. Broeng, T.A. Birks and P.St.J. Russell, "Photonic bandgap guidance in optical fibers," Science **282**, 1476–1478 (1998).

13. R.F. Cregan, B.J. Mangan, J.C. Knight, T.A. Birks, P.St.J. Russell, P.J. Roberts and D.C. Allan, "Single-mode photonic bandgap guidance of light in air," Science **285**, 1537–1549 (1999).
14. T. Hasegawa, E. Sasaoka, M. Onishi, M. Nishimura, Y. Tsuji and M. Koshiba, "Hole-assisted lightguide fiber for large anomalous dispersion and low optical loss," Opt. Express **9**, pp. 681–686 (2001).
15. J.K. Sahu, C.C. Renaud, K. Furusawa, R. Selvas, J.A. Alvarez-Chavez, D.J. Richardson and J. Nilsson, "Jacketed air-clad cladding pumped ytterbium-doped fibre laser with wide tuning range," Electron. Lett. **37**, pp. 1116–1117 (2001).
16. J.H.V. Price, K. Furusawa, T.M. Monro, L. Lefort and D.J. Richardson, "Tunable, femtosecond pulse source operating in the range 1.06–1.33µm based on an Yb^{3+}-doped holey fiber amplifier," J. Opt. Soc. Amer. B **19**, pp. 1286–1294 (2002).
17. G. P. Agrawal, *Nonlinear Fiber Optics,* Academic Press, Boston (2001).
18. T. Okuno, M. Onishi, T. Kashiwada, S. Ishikawa and M. Nishimura, "Silica-based functional fibers with enhanced nonlinearity and their applications," IEEE J. Sel. Top. Quant. **5**, 1385–1391 (1999).
19. V. Finazzi, T.M. Monro and D.J. Richardson, "Small core silica holey fibers: nonlinearity and confinement loss trade-offs," J. Opt. Soc. Amer. B **20**, 1427–1436 (2003).
20. W. Belardi, J.H. Lee, K. Furusawa, Z. Yusoff, P. Petropoulos, M. Ibsen, T.M. Monro and D.J. Richardson, "A 10Gbit/s tuneable wavelength converter based on four-wave mixing in highly non-linear holey fibre," in Proc. ECOC, Copenhagen, Denmark, September 2002. Postdeadline paper PD1.2.
21. H. Ebendorff-Heidepriem, P. Petropoulos, S. Asimakis, V. Finazzi, R. C. Moore, K. Frampton, F. Koizumi, D. J. Richardson and T. M. Monro, "Bismuth glass holey fibers with high nonlinearity," Opt. Express **12**, 5082–5087 (2004).
22. J.C. Baggett, T.M. Monro, K. Furusawa and D.J. Richardson, "Comparative study of large mode holey and conventional fibers," Opt. Lett. **26**, 1045–1047 (2001).
23. M. N. Petrovich, V. Finazzi, T. M. Monro, D. J. Richardson, E. J. O'Driscoll, M. A. Watson, T. Delmark, "Photonic bandgap fibers for infrared applications," 2nd EMRS DTC Conference, Edinburgh, UK (2005).
24. J.C. Knight, J. Arriaga, T.A. Birks, A. Ortigosa-Blanch, J.W. Wadsworth and P.St.J. Russell, "Anomalous dispersion in photonic crystal fiber," IEEE Photon. Technol. Lett., **12**, 807–809 (2000).
25. J.K. Ranka, R. S. Windeler and A.J. Stentz, "Visible continuum generation in air silica microstructure optical fibers with anomalous dispersion at 800 nm," Opt. Lett. **25**, 25–27 (2000).
26. K. Saitoh, M. Koshiba, T. Hasegawa and E. Sasaoka, "Chromatic dispersion control in photonic crystal fibers: application to ultra-flattened dispersion," Opt. Express **11**, 843–852 (2003).
27. T.A. Birks, D. Mogilevtsev, J.C. Knight and P.St.J. Russell, "Dispersion compensation using single material fibers," IEEE Photon. Technol. Lett. **11**, 674–676 (1999).
28. D.G. Ouzounov, F.R. Ahmed, D. Muller, N. Venkataraman, M.T. Gallagher, M.G. Thomas, J. Silcox, K.W. Koch and A.L. Gaeta, "Generation of megawatt optical solitons in hollow core photonic band-gap fibers," Science **301,** 1702–1704 (2003).
29. K. Saitoh and M. Koshiba, "Leakage loss and group velocity dispersion in air-core photonic bandgap fibers," Opt. Express **11**, 3100–3109 (2003).
30. H. Lim and F. W. Wise, "Control of dispersion in a femtosecond ytterbium laser by use of hollow-core photonic bandgap fiber," Opt. Express **12**, 2231–2235 (2004).
31. M.J. Steel, T.P. White, C.M. de Sterke, R.C. McPhedran and L. C. Botten, "Symmetry and degeneracy in microstructured optical fibers," Opt. Lett. **26**, 488–490 (2001).
32. T. P. Hansen, J. Broeng, E. B. Libori, E. Knudsen, A. Bjarklev, J. R. Jensen and H. Simonsen, "Highly birefringent index-guiding photonic crystal fibers," IEEE Photon. Technol. Lett. **13**, 588–590 (2001).
33. A. Bjarklev, J. Broeng, S. E. Barkou, E. Knudsen, T. Søndergaard, T. W. Berg and M. G. Dyndgaard, "Polarization properties of honeycomb-structured photonic bandgap fibres," J. Opt. A **2**, 584–588 (2000).
34. X. Chen, M. Li, N. Venkataraman, M. T. Gallagher, W. A. Wood, A. M. Crowley, J. P. Carberry, L. A. Zenteno and K. W. Koch, "Highly birefringent hollow-core photonic bandgap fiber," Opt. Express **12**, 3888–3893 (2004).
35. M. Fuochi, J.R. Hayes, K. Furusawa, W. Belardi, J.C. Baggett, T.M. Monro and D.J. Richardson, "Polarization mode dispersion reduction in spun large mode area silica holey fibres," Opt. Express **12**, 1972–1977 (2004).
36. T.M. Monro, D. J. Richardson and P. J. Bennett, "Developing holey fibres for evanescent field devices," Electron. Lett. **35**, 1188–1189 (1999).
37. G. Bouwmans, F. Luan, J. C. Knight, P. St. J. Russell, L. Farr, B. J. Mangan and H. Sabert, "Properties of a hollow-core photonic bandgap fiber at 850nm wavelength," Opt. Express **11**, 1613–1620 (2003).

38. T. Ritari, J. Tuominen, H. Ludvigsen, J. C. Petersen, T. Sørensen, T. P. Hansen and H. R. Simonsen, "Gas sensing using air-guiding photonic bandgap fibers," Opt. Express **12**, 4080–4087 (2004).
39. G. Humbert, J. C. Knight, G. Bouwmans, P. S. J. Russell, D. P. Williams, P. J. Roberts and B. J. Mangan, "Hollow core photonic crystal fibers for beam delivery," Opt. Express **12**, 1477–1484 (2004).
40. M. van Eijkelenborg, M. Large, A. Argyros, J. Zagari, S. Manos, N.A. Issa, I.M. Bassett, S.C. Fleming, R.C. McPhedran, C.M. deSterke and N.A.P. Nicorovici, "Microstructured polymer optical fibre," Opt. Express **9**, 319 (2001).
41. X. Feng, T.M. Monro, P. Petropoulos, V. Finazzi and D.W. Hewak, "Solid microstructured optical fiber," Opt. Express **11**, pp. 2225–2230 (2003).
42. K.M. Kiang, K. Frampton, T.M. Monro, R. Moore, J. Trucknott, D.W. Hewak and D.J. Richardson, "Extruded singlemode non-silica glass holey optical fibres," Electron. Lett. **38**, 546–547 (2002).
43. V.V.R.K. Kumar, A.K. George, W.H. Reeves, J.C. Knight and P.S.J. Russell, "Extruded soft glass photonic crystal fiber for ultrabroadband supercontinuum generation," Opt. Express **10**, 1520–1525 (2002).
44. V.V.R.K. Kumar, A.K. George, J.C. Knight and P.S.J. Russell, "Tellurite photonic crystal fiber," Opt. Express **11**, pp. 2641–2645 (2003).
45. T.M. Monro, K.M. Kiang, J.H. Lee, K. Frampton, Z. Yusoff, R. Moore, J. Trucknott, D.W. Hewak, H.N. Rutt and D.J. Richardson, "High nonlinearity extruded single-mode holey optical fibers," Proc. Optical Fiber Communications Conference (OFC 2002), Anaheim, California, postdeadline paper FA1-1 (2002).
46. P. Petropoulos, T.M. Monro, H. Ebendorff-Heidepriem, K. Frampton, R.C. Moore, H.N. Rutt, D.J. Richardson, "Soliton-self-frequency-shift effects and pulse compression in anomalously dispersive high nonlinearity lead silicate holey fiber," Proc. Optical Fiber Communications Conference (OFC), Atlanta, Georgia, postdeadline paper PD03 (2003).
47. P. Petropoulos, H. Ebendorff-Heidepriem, V. Finazzi, R.C. Moore, K. Frampton, D.J. Richardson and T.M. Monro, "Highly nonlinear and anomalously dispersive lead silicate glass holey fibers," Opt. Express **11**, 3568–3573 (2003).
48. H. Ebendorff-Heidepriem, P. Petropoulos, V. Finazzi, K. Frampton, R.C. Moore, D.J. Richardson and T.M. Monro, "Highly nonlinear bismuth-oxide-based glass holey fiber," Proc. Optical Fiber Communications Conference (OFC), Los Angeles, California, paper ThA4 (2004).
49. P. Petropoulos, H. Ebendorff-Heidepriem, T. Kogure, K. Furusawa, V. Finazzi, T.M. Monro and D.J. Richardson, "A spliced and connectorized highly non-linear and anomalously dispersive bismuth-oxide glass holey fiber," Proc. Conference on Lasers and Electro-Optics (CLEO), San Francisco, California, paper CTuD (2004).
50. A. Mori, K. Shikano, K. Enbutsu, K. Oikawa, K. Naganuma, M. Kato and S. Aozasa, "1.5 μm band zero-dispersion shifted tellurite photonic crystal fibre with a nonlinear coefficient γ of 675 W^{-1}km^{-1}," Proc. European Conference on Optical Communications (ECOC) (2004).
51. A. D. Fitt, K. Furusawa, T. M. Monro, C. P. Please and D. J. Richardson, "The mathematical modelling of capillary drawing for holey fibre manufacture," J. Eng. Math. **43**, 201–227 (2002).
52. C. J. Voyce, A. D. Fitt and T. M. Monro, "Mathematical model of the spinning of microstructured fibres," Opt. Express 12, 5810–5820 (2004).
53. K. Tajima, J. Zhou, K. Kurokawa and K. Nakajima, "Low water peak photonic crystal fibres," Proc. European Conference on Optical Communication (ECOC), Rimini, Italy, postdeadline paper Th4.1.6 (2003).
54. B. Mangan, L. Farr, A. Langford, P.J. Roberts, D.P. Williams, F. Couny, M. Lawman, M. Mason, S. Coupland, R. Flea, H. Sabert, T.A. Birks, J.C. Knight and P.St.J. Russell, "Low loss (1.7 dB/km) hollow core photonic bandgap fiber," Proc. Optical Fiber Communications (OFC), Anaheim, paper PDP24 (2004).
55. L. Farr, J.C. Knight, B.J. Mangan and P.J. Roberts, "Low loss photonic crystal fibre," in Proc ECOC, paper PD1.3, Copenhagen, Denmark (2002).
56. A. Bjarklev, J. Broeng, S.E. Barkou and K. Dridi., "Dispersion properties of photonic crystal fibres," 24th European Conference on Optical Communication (ECOC), Madrid 1, 135–136 (1998).
57. J. Riishede, S.B. Libori, A. Bjarklev, J. Broeng and E. Knudsen, "Photonic crystal fibers and effective index approaches," Proceedings of the 27th European Conference on Optical Communications (ECOC), Th.A.1.5 (2001).
58. K.M. Ho, C.T. Chan and C.M. Soukoulis, "Existence of a photonic gap in periodic dielectric structures," Phys. Rev. Lett. **65**, 3152–3155 (1990).
59. R.D. Meade, A.M. Rappe. K.D. Brommer, J.D. Joannopoulos and O.L. Alerhand, "Accurate theoretical analysis of photonic band-gap materials," Phys. Rev. B **48**, 8434–8437 (1993).
60. B.J. Eggleton, P.S. Westbrook, R.S. Windeler, S. Spalter and T.A Strasser, "Grating resonances in

air/silica microstructured optical fibers," Opt. Lett. **24**, 1460–1462 (1999).
61. D. Mogilevtsev, T.A. Birks and P.St.J. Russell, "Group-velocity dispersion in photonic crystal fibers," Opt. Lett. **23**, 1662–1664 (1998).
62. T.M. Monro, D.J. Richardson, N.G.R. Broderick and P.J. Bennett, "Modelling large air fraction holey optical fibers," J. Lightwave Technol. **18**, 50–56 (2000).
63. T.M. Monro, N.G.R. Broderick and D.J. Richardson, "Exploring the optical properties of holey fibres," NATO Summer School on Nanoscale Linear and Nonlinear Optics, Erice, Sicily, July (2000).
64. P.J. Bennett, T.M. Monro and D.J. Richardson, "Towards practical holey fibre technology: fabrication splicing modeling and characterization," Opt. Lett. **24**, 1203–1205 (1999).
65. F. Brechet, J. Marcou, D. Pagnoux and P. Roy, "Complete analysis of the characteristics of propagation into photonic crystal fibers, by the finite element method," Opt. Fiber Technol. **6**, 181–191 (2000).
66. H.P. Uranus and H. Hoekstra, "Modelling of microstructured waveguides using a finite-element-based vectorial mode solver with transparent boundary conditions," Opt. Express **12**, 2795–2809 (2004).
67. T.P. White, R.C. McPhedran, C.M. deSterke, L.C. Botten and M.J. Steel, "Confinement losses in microstructured optical fibers," Opt. Lett. **26**, 1660–1662 (2001).
68. L. Poladian, N.A. Issa and T.M. Monro, "Fourier decomposition algorithm for leaky modes of fibres with arbitrary geometry," Opt. Express **10**, 449–454 (2002).
69. E. Kerrinckx, L. Bigot, M. Douay and Y. Quiquempois, "Photonic crystal fiber design by means of a genetic algorithm," Opt. Express **12**, 1990–1995 (2004).
70. N.G.R. Broderick, T.M. Monro, P.J. Bennett and D.J. Richardson, "Nonlinearity in holey optical fibers: measurement and future opportunities," Opt. Lett. **24**, 1395–1397 (1999).
71. N. Nakazawa, H. Kubota and K. Tamura, "Random evolution and coherence degradation of a high-order optical soliton train in the presence of noise," Opt. Lett. **24**, 318–320 (1999).
72. P. Petropoulos, T.M. Monro, W. Belardi, K. Furusawa, J.H. Lee and D.J. Richardson, "2R-regenerative all-optical switch based on a highly nonlinear holey fiber," Opt. Lett. **26**, 1233–1235 (2001).
73. F. Futami, S. Watanabe and T. Chikama, "Simultaneous recovery of 20×20 GHz WDM optical clock using supercontinuum in a nonlinear fiber," in Proc. ECOC, (2000).
74. J. Hansryd and P.A. Andrekson, "Broad-band continuous-wave-pumped fiber optical parametric amplifier with 49-dB gain and wavelength-conversion efficiency," Photon. Technol. Lett. **13**, 194–196 (2001).
75. F. Druon, N. Sanner, G. Lucas-Leclin, P. Georges, R. Gaumé, B. Viana, K.P. Hansen and A. Petersson, "Self-compression of 1-um femtosecond pulses in a photonic crystal fiber," in Proc. CLEO (2002).
76. J.H. Lee, Z. Yusoff, W. Belardi, M. Ibsen, T.M. Monro, B. Thomsen and D.J. Richardson, "A holey fiber based WDM wavelength converter incorporating an apodized fiber Bragg grating filter," in Proc. CLEO (2002).
77. J.E. Sharping, M. Fiorentino, P. Kumar and R.S. Windeler, "All optical switching based on cross-phase modulation in microstructure fiber," Photon. Technol. Lett. **14**, 77–79 (2002).
78. H. Takara, T. Ohara, K. Mori, K. Sato, E. Yamada, K. Jinguji, Y. Inoue, T. Shibata, T. Morioka and K.-I. Sato, "Over 1000 channel optical frequency chain generation from a single supercontinuum source with 12.5 GHz channel spacing for DWDM and frequency standards," in Proc. ECOC (2000).
79. K.P. Hansen, J.R. Jensen, C. Jacobsen, H.R. Simonsen, J. Broeng, P.M.W. Skovgaard, A. Petersson and A. Bjarklev, "Highly nonlinear photonic crystal fiber with zero-dispersion at 1.55 μm," OFC '02 postdeadline paper (2002).
80. J.H. Lee, P.C. Teh, Z. Yusoff, M. Ibsen, W. Belardi, T.M. Monro and D.J. Richardson, "A holey fiber-based nonlinear thresholding device for optical CDMA receiver performance enhancement," IEEE Photon. Technol. Lett. **14**, 876–878 (2002).
81. Z. Yusoff, J.H. Lee, W. Belardi, T.M. Monro, P.C. the and D.J. Richardson, "Raman effects in a highly nonlinear holey fiber: amplification and modulation," Opt. Lett. **27**, 424–426 (2002).
82. J. Nilsson, R. Selvas, W. Belardi, J.H. Lee, Z. Yusoff, T.M. Monro, D.J. Richardson, K.D. Park, P.H. Kim and N. Park, "Continuous-wave pumped holey fiber Raman laser," in Proc. OFC, OSA Technical Digest, Anaheim, California, (2002).
83. J.H. Lee, Z. Yusoff, W. Belardi, M. Ibsen, T.M. Monro, B. Thomsen and D.J. Richardson, "A holey fiber based WDM wavelength converter incorporating an apodized fiber Bragg grating filter," CLEO/QELS 2002, Long Beach, California, (2002).
84. F. Druon and P. Georges, "Pulse-compression down to 20 fs using a photonic crystal fiber seeded by a diode-pumped Yb:SYS laser at 1070 nm," Opt. Express **12**, 3383–3396 (2004).
85. S. Coen, A.H.L. Chau, R. Leonhardt and J.D. Harvey, "Supercontinuum generation via stimulated Raman scattering and parametric four-wave mixing in photonic crystal fibers," J. Opt. Soc. Amer. B (2002).

86. K.P. Hansen, J. J. Larsen, J.R. Jensen, S. Keiding, J. Broeng, H.R. Simonsen and A. Bjarklev, "Super continuum generation at 800 nm in highly nonlinear photonic crystal fibers with normal dispersion," in Proc. LEOS (2001).

87. K. M. Hilligsøe, T. V. Andersen, H. N. Paulsen, C. K. Nielsen, K. Mølmer, S. Keiding, R. Kristiansen, K. P. Hansen and J. J. Larsen, "Supercontinuum generation in a photonic crystal fiber with two zero dispersion wavelengths," Opt. Express **12**, 1045–1054 (2004).

88. S.G. Leon-Saval, T.A. Birks, W.J. Wadsworth, P.St.J. Russell and M.W. Mason, "Supercontinuum generation in submicron fibre waveguides," Opt. Express **12**, 2864–2869 (2004).

89. I. Hartl, X.D. Li, C. Chudoba, R.K. Ghanta, T.H. Ko and J.G. Fujimoto, "Ultrahigh-resolution optical coherence tomography using continuum generation in an air-silica microstructure optical fiber," Opt. Lett. **26**, 608–610 (2001).

90. R.E. Drullinger, S.A. Diddams, K.R. Vogel, C.W. Oates, E.A. Curtis, W.D. Lee, W.M. Itano, L. Hollberg and J.C. Bergquist, "All-optical atomic clocks," International Frequency Control Symposium and PDA Exhibition, 69–75 (2001).

91. P.A. Champert, S.V. Popov and J.R. Taylor, "Generation of multiwatt, broadband continua in holey fibers," Opt. Lett. **27**, 122–124 (2002).

92. J.H.V. Price, T.M. Monro, K. Furusawa, W. Belardi, J.C. Baggett, S. Coyle, C. Netti, J.J. Baumberg, R. Paschotta and D.J. Richardson, "UV generation in a pure silica holey fiber," Appl. Phys. B **77**, 291–298 (2003).

93. A.V. Husakou and J. Herrmann, "Supercontinuum generation of higher-order solitons by fission in photonic crystal fibers," Phys. Rev. Lett. **87**, (2001).

94. J. Herrmann, U. Griebner, N. Zhavoronkov, A. Husakou, D. Nickel, J.C. Knight, W.J. Wadsworth, P.St.J. Russell, and G. Korn, "Experimental evidence for supercontinuum generation by fission of higher-order solitons in photonic fibers," Phys. Rev. Lett. **88**, 173–201 (2002).

95. K.P. Hansen, J.R. Jensen, D. Birkedal, J.M. Hvam and A. Bjarklev, "Pumping wavelength dependence of super continuum generation in photonic crystal fibers," in Proc. Conference on Optical Fiber Communication (2002).

96. J.C. Knight, T.A. Birks, R.F. Cregan, P.St.J. Russell and J.-P. De Sandro, "Large mode area photonic crystal fibre," Electron. Lett. **34**, 1347–1348 (1998).

97. J. Limpert, A. Liem, M. Reich, T. Schreiber, S. Nolte, H. Zellmer, A. Tunnermann, J. Broeng, A. Petersson and C. Jakobsen, "Low-nonlinearity single-transverse-mode ytterbium-doped photonic crystal fiber amplifier," Opt. Express **12**, 1313–1319 (2004).

98. T. Sørensen, J. Broeng, A. Bjarklev, E. Knudsen, S.E. Barkou, H.R. Simonsen and J. Riis Jensen, "Macrobending loss properties of photonic crystal fibres with different air filling fractions," in Proceedings of ECOC'2001, Amsterdam, The Netherlands (2001).

99. J.C. Baggett, T.M. Monro, K. Furusawa and D.J. Richardson, "Understanding bending losses in holey optical fibers," Opt. Commun. **227**, 317–335 (2003).

100. A. Gambling, H. Matsumura, C.M. Ragdale, R.A. Sammut, "Measurement of radiation loss in curved single-mode fibres," Microwaves Opt. Acoust. **2**, 134–140 (1978).

101. N.A. Mortensen, M.D. Nielsen, J.R. Folkenberg, A. Petersson and H.R. Simonsen, "Improved large-mode-area endlessly single-mode photonic crystal fibers," Opt. Lett. **28**, 393–395 (2003).

102. R.F. Cregan, J.C. Knight, P.St.J. Russell and P.J. Roberts, "Distribution of spontaneous emission from an Er^{3+}-doped photonic crystal fiber," IEEE J. Lightwave. Technol. **17**, 2138–2141 (1999).

103. K. Furusawa, A. Malinowski, J.H.V. Price, T.M. Monro, J.K. Sahu, J. Nilsson and D.J. Richardson, "A cladding pumped Ytterbium-doped fiber laser with holey inner and outer cladding," Opt. Express **9**, 714–720 (2001).

104. K. Furusawa, T. Kogure, T.M. Monro and D.J. Richardson, "High gain efficiency amplifier based on an erbium doped aluminosilicate holey fiber," Opt. Express **12**, 3452–3458 (2004).

105. W.J. Wadsworth, J.C. Knight, W.H. Reeves, P.St.J. Russell and J. Arriaga, "Yb^{3+}-doped photonic crystal fiber laser," Electron. Lett. **36**, 1452–1454 (2000).

106. K. Furusawa, T.M. Monro, P. Petropoulos and D.J. Richardson, "A mode-locked laser based on ytterbium doped holey fiber," Electron. Lett. **37**, 560–561 (2001).

107. F.M. Mitschke and L.F. Mollenauer, "Discovery of the soliton self-frequency shift," Opt. Lett. **11**, 659–661 (1986).

108. J.H.V Price, W. Belardi, L. Lefort, T.M. Monro and D.J. Richardson, "Nonlinear pulse compression, dispersion compensation, and soliton propagation in holey fiber at 1 micron," in Proc Nonlinear Guided Waves and Their Applications (NLGW), paper WB1-2 (2001).

109. J. Limpert, T. Schreiber, S. Nolte, H. Zellmer, T. Tunnermann, R. Iliew, F. Lederer, J. Broeng, G. Vienne, A. Petersson and C. Jakobsen, "High-power air-clad large-mode-area photonic crystal fiber laser," Opt. Express **11**, 818–823 (2003).

110. J. Limpert, T. Schreiber, A. Liem, S. Nolte, H. Zellmer, T. Peschel, V. Guyenot and A. Tünnermann, "Thermo-optical properties of air-clad photonic crystal fiber lasers in high power operation," Opt. Express **11**, 2982–2990 (2003).

111. K. Kikuchi, K. Taira and N. Sugimoto, "Highly nonlinear bismuth oxide-based glass fibers for all-optical signal processing," Electron. Lett. **38**, 166–167 (2002).

112. R.E. Slusher, J.S. Sanghera, L.B. Shaw and I.D. Aggarwal, "Nonlinear optical properties of As-Se fiber," in Proceedings of OSA Topical meeting on Nonlinear Guided Waves and their Applications, Stresa, Italy, 1–4 Sept. (2003).

113. E.M. Vogel, M.J. Weber and D.M. Krol, "Nonlinear optical phenomena in glass," Phys. Chem. Glasses **32**, 231–254 (1991).

114. S. Fujino, H. Ijiri, F. Shimizu and K. Morinaga, "Measurement of viscosity of multi-component glasses in the wide range for fiber drawing," J. Jpn. Inst. Met. **62**, 106–110 (1998).

115. Schott Glass Catalogue, 2003. Available from: www.us.schott.com/optics-devices/english/download

116. S.R. Fribergand and P.W. Smith, "Nonlinear optical-glasses for ultrafast optical switches," IEEE J. Quant. Electron. **23**, 2089–2094 (1987).

117. N. Sugimoto, H. Kanbara, S. Fujiwara, K. Tanaka, Y. Shimizugawa and K. Hirao, "Third-order optical nonlinearities and their ultrafast response in Bi_2O_3-B_2O_3-SiO_2 glasses," J. Opt. Soc. Am. B **16**, 1904–1908 (1999).

118. K. Kikuchi, K. Taira and N. Sugimoto, "Highly nonlinear bismuth oxide-based glass fibers for all-optical signal processing," Electron. Lett. **38**, 166–167 (2002).

119. N. Sugimoto, Y. Kuroiwa, K. Ochiai, S. Ohara, Y. Furusawa, S. Ito, S. Tanabe and T. Hanada, "Novel short-length EDF for C+L band amplification," in Proceedings of Optical Amplifiers and their Applications, Quebec City, Canada (2000).

120. Y. Kuroiwa, N. Sugimoto, K. Ochiai, S. Ohara, Y. Furusawa, S. Ito, S. Tanabe and T. Hanada, "Fusion spliceable and high efficient Bi2O3-based EDF for short length and broadband application pumped at 1480 nm," in Proc. Optical Fiber Communications (OFC), Anaheim, California, paper TuI5 (2001).

121. J.S. Wang, E.M. Vogel and E. Snitzer, "Tellurite glass: a new candidate for fiber devices," Opt. Mat. **3**, 187–203 (1994).

122. R. Stegeman, L. Jankovic, H. Kim, C. Rivero, G. Tegeman, K. Richardson, P. Delfyett, Y. Guo, A. Schulte and T. Cardinal, "Tellurite glasses with peak absolute Raman gain coefficients up to 30 times that of fused silica," Opt. Lett. **28**, 1126–1128 (2003).

123. X. Feng, A.K. Mairaj, D.W. Hewak and T.M. Monro, "Towards high-index glass based monomode holey fiber with large mode area," Electron. Lett. **40**, 167–169 (2004).

124. B. Temelkuran, S.D. Hart, G. Benoit, J.D. Joannopoulos and Y. Fink, "Wavelength-scalable hollow optical fibres with large photonic bandgaps for CO_2 laser transmission," Nature **420**, 650–653 (2002).

125. F. Luan, A.K. George, T.D. Hedley, G.J. Pearce, D.M. Bird, J.C. Knight and P.St.J. Russell, "All-solid photonic bandgap fiber," Opt. Lett. **29**, 2369–2371 (2004).

126. X. Feng, T.M. Monro, P. Petropoulos, V. Finazzi and D.J. Richardson, "Single-mode high-index-core one-dimensional microstructured fiber with high nonlinearity," Proc. Optical Fiber Communications (OFC), 2005.

127. S.E. Barkou, J. Broeng and A. Bjarklev, "Silica-air photonic crystal fiber design that permits waveguiding by a true photonic bandgap effect," Opt. Lett. **24**, 46–48 (1999).

128. J.C. Knight, J. Broeng, T.A. Birks and P.St.J. Russell, "Photonic bandgap guidance in optical fibers," Science **282**, 1476–1478 (1998).

129. S.E. Barkou, J. Broeng and A. Bjarklev, "Dispersion properties of photonic bandgap guiding fibers," in Proc. Optical Fiber Communications (OFC), paper FG5, pp. 117–119 (1999).

130. A. Bjarklev, J. Broeng, S.E. Barkou, E. Knudsen, T. Søndergaard, T.W. Berg and M.G. Dyndgaard, "Polarization properties of honeycomb-structured photonic bandgap fibres," J. Opt. A **2**, 584–588 (2000).

131. J. Broeng, S.E. Barkou, T. Søndergaard and A. Bjarklev, "Analysis of air-guiding photonic bandgap fibers," Opt. Lett. **25**, 96–98 (2000).

132. H.C. Nguyen, P. Domachuk, B.J. Eggleton, M.J. Steel, M. Straub, M. Gu and M. Sumetsky, "A new slant on photonic crystal fibers," Opt. Express **12**, 1528–1539 (2004).

133. K. Saitoh and M. Koshiba, "Leakage loss and group velocity dispersion in air-core photonic bandgap fibers," Opt. Express **11**, 3100–3109 (2003).

134. J.A. West, N. Venkataraman, C.M. Smith and M.T. Gallagher, "Photonic crystal fibers," in Proc. European Conference on Optical Communications (ECOC), Th.A.2.2 (2001).

135. K. Saitoh, N.A. Mortensen and M. Koshiba, "Air-core photonic band-gap fibers: the impact of surface modes," Opt. Express **12**, 394–400 (2004).

136. D.G. Ouzounov, F.R. Ahmad, D. Müller, N. Venkataraman, M.T. Gallagher, M.G. Thomas, J. Silcox, K.W. Koch and A.L. Gaeta, "Generation of

137. C.J.S. de Matos, J.R. Taylor, T.P. Hansen, K.P. Hansen and J. Broeng, "All-fiber chirped pulse amplification using highly-dispersive air-core photonic bandgap fiber," Opt. Express **11**, 2832–2837 (2003).
138. N. Venkataraman, M.T. Gallagher, C.M. Smith, D. Müller, J.A. West, K.W. Koch and J.C. Fajardo, "Low loss (13 dB/km) air core photonic band-gap fibre," in Proc. European Conference on Optical Communications (ECOC) Copenhagen, Denmark, postdeadline paper PD1.1 (2002).
139. H. Lim and F.W. Wise, "Control of dispersion in a femtosecond ytterbium laser by use of hollow-core photonic bandgap fiber," Opt. Express **12**, 2231–2235 (2004).
140. G. Humbert, J.C. Knight, G. Bouwmans, P.S.J. Russell, D.P. Williams, P.J. Roberts and B.J. Mangan, "Hollow core photonic crystal fibers for beam delivery," Opt. Express **12**, 1477–1484 (2004).
141. N.M. Litchinitser, S.C. Dunn, P.E. Steinvurzel, B.J. Eggleton, T.P. White, R.C. McPhedran and C.M. de Sterke, "Application of an ARROW model for designing tunable photonic devices," Opt. Express **12**, 1540–1550 (2004).
142. L.B. Shaw, J.S. Sanghera, I.D. Aggarwal and F.H. Hung, "As-S and As-Se based photonic bandgap fiber for IR laser transmission," Opt. Express **11**, 3455–3460 (2003).

4
CHAPTER

Photonic Bandgap–Guided Bragg Fibers

Sonali Dasgupta, Bishnu P. Pal,* and M. R. Shenoy

Department of Physics
Indian Institute of Technology Delhi
New Delhi, India

1. INTRODUCTION

Consequent to the mind-boggling progress in high-speed optical telecommunications witnessed in recent times, it appeared that it would only be a matter of time before the huge theoretical bandwidth of 53 THz offered by low-loss transmission windows (extending from 1280 nm [235 THz] to 1650 nm [182 THz]) in OH^--free high-silica optical fibers would be tapped for telecommunication through dense wavelength division multiplexing techniques! In spite of this possibility, there has been a considerable resurgence of interest among researchers to develop specialty fibers, that is, fibers in which transmission loss of the material would not be a limiting factor and in which nonlinearity or dispersion properties could be conveniently tailored to achieve transmission characteristics that are otherwise almost impossible to realize in conventional high-silica fibers. Research targeted at such fiber designs gave rise to a new class of fibers, known as *microstructured* optical fibers (see Chapter 3). One category of such microstructured fibers is known as *photonic bandgap* fibers (PBGFs). In a conventional optical fiber, light is guided by total internal reflection because of the refractive index contrast that exists between a finite-sized cylindrical core and the cladding of lower refractive index that surrounds it. On the other hand, in a PBGF, light of *certain frequencies* cannot propagate along directions perpendicular to the fiber axis but instead are free to propagate along its length confined to the fiber core. This phenomenon that forbids the propagation of photons (of certain frequencies) transverse to the axis of microstructured fibers, led to the christening of these specialty fibers as photonic bandgap-guided optical fibers, in analogy with the electronic bandgaps encountered by electrons in semiconductors. In contrast to the electronic bandgap, which is the consequence of periodic arrangement of atoms/molecules in a semiconductor crystal lattice, photonic bandgap arises due to a *periodic distribution of refractive index* in certain dielectric structures, generically referred to as *photonic crystals* (reminiscent of semiconductor crystals in solid-state physics) [1]. If the frequency of incident light happens to fall within the photonic bandgap, which is characteristic of the photonic crystal, then light propagation is forbidden in it. Depending on the number of dimensions in which periodicity in refractive index exists, photonic crystals are classified as one-, two-, or three-dimensional photonic crystals (Fig. 4.1) [2]. Opals are naturally occurring photonic crystals consisting of a three-dimensional lattice of dielectric spheres, and *opalescence* is a direct consequence of their photonic crystal nature. The iridescent wings of some butterflies and beetles are also essentially the result of naturally occurring photonic crystal effects that evolved on their surface with time. In 1987, Yablonovitch first proposed the possibility of controlling properties of light through the photonic bandgap effect in man-made photonic crystals.

PBGFs essentially consist of a core surrounded by a periodic cladding having a photonic crystal-like structure (i.e., a periodic refractive index distribution). As explained above, the inherent periodicity of the cladding results in a photonic bandgap, which forbids propagation of certain frequencies of light into the cladding. Consequently, if light of these frequencies is launched into the core of the microstructured fiber, it could be confined within this region and be guided along the fiber length. PBGFs can be broadly classified as either *Bragg fibers* [4] or *photonic crystal fibers* [5, 6]. The former consists of a one-dimensional cladding that is periodic along the radial direction, whereas the latter consists of a two-dimensional periodic cladding extending along the directions radial as well as

---

*E-mail: bppal@physics.iitd.ernet.in

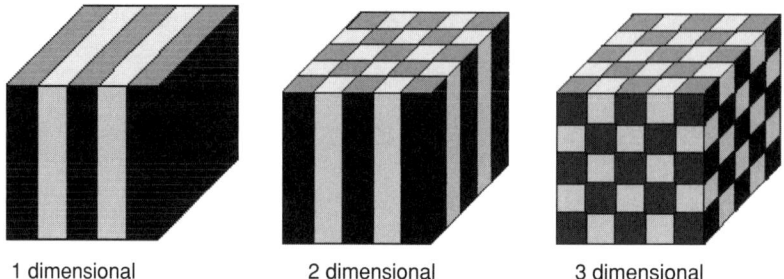

FIGURE 4.1 Schematic representations of photonic crystals in one, two, and three dimensions. (After [2].)

azimuthal. In contrast to conventional fibers, a PBGF can guide light through the low-index core (air) and hence offers several advantages as compared with light transmission through conventional fibers. In air-core PBGFs, because light is guided in air, two major benefits gained are much lower transmission loss and reduced sensitivity to nonlinear optical impairments such as four-wave mixing, cross-phase modulation, and so on. Following the same reasoning, the power-handling capability of these fibers would be much better and material dispersion effects would be much reduced. In addition, because of a multitude of physical parameters that can be altered independently, their propagation characteristics can be tuned with ease to control and maneuver light guidance for a variety of applications, ranging from telecommunications to sensors [7]. In this chapter our focus will remain only on Bragg fibers.

2. BRAGG FIBERS

Bragg fibers consist of a low-index core surrounded by periodic multilayer cladding (concentric layers) of alternate high and low refractive index materials (each of which has a refractive index higher than that of the core). As mentioned above, the inherent periodicity in the cladding forms a photonic bandgap does not allow light of certain frequencies from leaking through the cladding and thereby confines it within the core region. Figure 4.2 shows the cross-sectional view of a conventional optical fiber (a) and a Bragg fiber (b) (air core). The Bragg fiber was proposed as early as 1978 [4]. However, because of the lack of advanced fabrication techniques and doubts related to their applicability [8], there was hardly any further research in this field for almost a decade. Interest revived again in the late 1990s in the context of a flurry of research on photonic crystals, and several articles reported successful fabrication of broadband, low-loss, hollow-core fiber waveguides suitable for various ranges of wavelengths [9, 10].

2.1. Bandgap in One-Dimensional Periodic Medium

The multilayer cladding of the Bragg fiber is periodic along the radial direction and results in a one-dimensional photonic bandgap. It is functionally similar to a multilayer planar stack that consists of thin films of alternate refractive indices, n_1 and n_2, having thickness l_1 and l_2, respectively, such that $\Lambda = l_1 + l_2$ (Fig. 4.3). Accordingly, wave guidance in a Bragg fiber can be conveniently understood in terms of the physics that underlies the formation of the bandgap and the concept of decaying Bloch waves in a multilayer planar stack. Following the analysis presented in [11], the electric field of the wave propagating in the x–y plane in the qth unit cell of a planar stack (periodic in the x-direction, Fig. 4.3) may be written as the superposition of a forward and a backward propagating plane wave as follows:

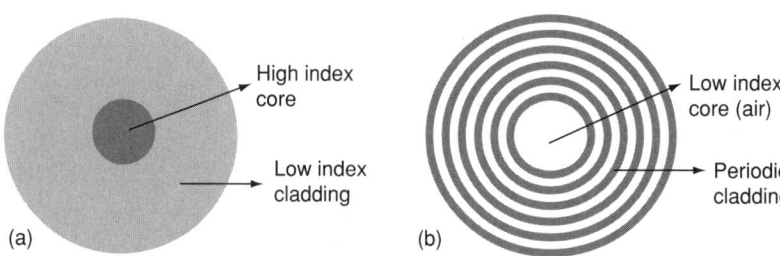

FIGURE 4.2 (a) Cross-sectional view of a conventional fiber. (b) Cross-sectional view of a Bragg fiber.

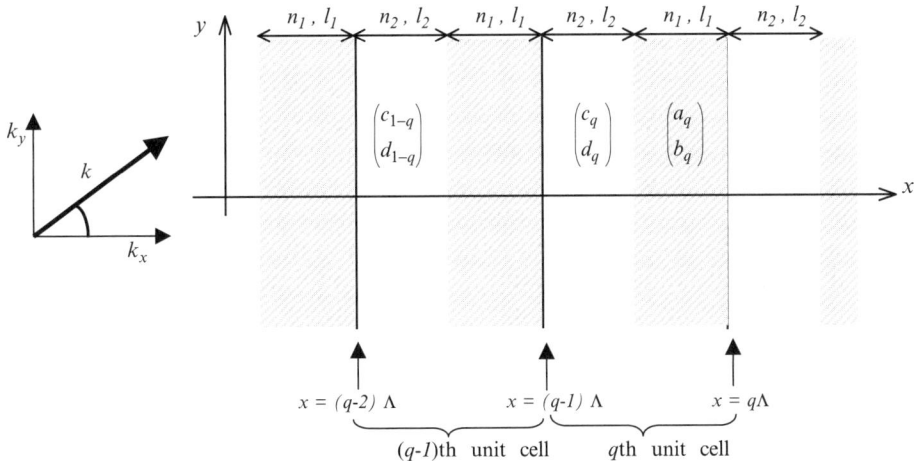

FIGURE 4.3 Planar Bragg structure (Adapted from (11); © John Woley & Sons, 1991).

$$E_z(x, y) = \left[a_q e^{-ik_{1x}(x-q\Lambda)} + b_q e^{ik_{1x}(x-q\Lambda)}\right] e^{-ik_y y}$$
$$E_z(x, y) = \left[c_q e^{-ik_{2x}(x-q\Lambda)} + d_q e^{ik_{2x}(x-q\Lambda)}\right] e^{-ik_y y} \quad (1)$$

where (a_q, b_q) and (c_q, d_q) are the coefficients of the forward and backward propagating fields in the layers having refractive indices n_1 and n_2, respectively; k_y is the y-component of the propagation vector k, and $k_{jx} = \left[\left(\frac{n_j \omega}{c}\right)^2 - k_y^2\right]^{1/2}$, $j = 1, 2$. In general, the Bloch theorem can be used to obtain the normal modes of such a periodic medium. According to this theorem, the normal modes of a one-dimensional periodic structure (having a periodicity Λ along the x-direction) can be written as the product of a periodic Bloch wave function and a plane wave function:

$$E_z(x, y) = \left[E_K(x)e^{-iKx}\right]e^{-ik_y y}$$
$$H_z(x, y) = \left[H_K(x)e^{-iKx}\right]e^{-ik_y y} \quad (2)$$

where K is the periodic Bloch wave vector; and $E_K(x)$ and $H_K(x)$ satisfy the following periodicity conditions:

$$E_K(x) = E_K(x + \Lambda)$$
$$H_K(x) = H_K(x + \Lambda) \quad (3)$$

Combining Eqs. (1) and (2) along with (3), the coefficients of electric field amplitudes in two consecutive layers can be related through

$$\begin{pmatrix} a_{q-1} \\ b_{q-1} \end{pmatrix} = e^{iK\Lambda} \begin{pmatrix} a_q \\ b_q \end{pmatrix} \quad (4)$$

Alternatively, the complex amplitudes of the plane waves in a layer characterized by a refractive index n_1 in a certain unit cell can be related to that of the equivalent layer in the next unit cell by the unit cell *translation matrix* [see Eq. (5)]. The continuity conditions for the transverse electric and magnetic field components E_z and H_y (for TE modes) and H_z and E_y (for TM modes) at the interfaces and $x = (q-1)\Lambda$ and $x = (q-1)\Lambda + l_2$ yield the translation matrix as

$$\begin{pmatrix} a_{q-1} \\ b_{q-1} \end{pmatrix} = \begin{pmatrix} A_{TE/TM} & B_{TE/TM} \\ C_{TE/TM} & D_{TE/TM} \end{pmatrix} \begin{pmatrix} a_q \\ b_q \end{pmatrix} \quad (5)$$

where

$$A_{TE} = e^{ik_{1x}l_1}\left[\cos(k_{2x}l_2) + \frac{i}{2}\left(\frac{k_{2x}}{k_{1x}} + \frac{k_{1x}}{k_{2x}}\right)\sin(k_{2x}l_2)\right]$$

$$B_{TE} = e^{-ik_{1x}l_1}\left[\frac{i}{2}\left(\frac{k_{2x}}{k_{1x}} - \frac{k_{1x}}{k_{2x}}\right)\sin(k_{2x}l_2)\right]$$

$$C_{TE} = e^{ik_{1x}l_1}\left[-\frac{i}{2}\left(\frac{k_{2x}}{k_{1x}} - \frac{k_{1x}}{k_{2x}}\right)\sin(k_{2x}l_2)\right]$$

$$D_{TE} = e^{-ik_{1x}l_1}\left[\cos(k_{2x}l_2) - \frac{i}{2}\left(\frac{k_{2x}}{k_{1x}} + \frac{k_{1x}}{k_{2x}}\right)\sin(k_{2x}l_2)\right] \quad (6)$$

$$A_{TM} = e^{ik_{1x}l_1}\left[\cos(k_{2x}l_2) + \frac{i}{2}\left(\frac{n_2^2 k_{1x}}{n_1^2 k_{2x}} + \frac{n_1^2 k_{2x}}{n_2^2 k_{1x}}\right)\sin(k_{2x}l_2)\right]$$

$$B_{TM} = e^{-ik_{1x}l_1}\left[\frac{i}{2}\left(\frac{n_2^2 k_{1x}}{n_1^2 k_{2x}} - \frac{n_1^2 k_{2x}}{n_2^2 k_{1x}}\right)\sin(k_{2x}l_2)\right]$$

$$C_{TM} = e^{ik_{1x}l_1}\left[-\frac{i}{2}\left(\frac{n_2^2 k_{1x}}{n_1^2 k_{2x}} - \frac{n_1^2 k_{2x}}{n_2^2 k_{1x}}\right)\sin(k_{2x}l_2)\right]$$

$$D_{TM} = e^{-ik_{1x}l_1}\left[\cos(k_{2x}l_2) - \frac{i}{2}\left(\frac{n_2^2 k_{1x}}{n_1^2 k_{2x}} + \frac{n_1^2 k_{2x}}{n_2^2 k_{1x}}\right)\sin(k_{2x}l_2)\right] \quad (7)$$

The eigenvalue equation that governs the photonic bandgap, is obtained by substituting Eq. (5) into Eq. (4) as

$$e^{iK\Lambda} = \frac{1}{2}(A+D) \pm \left[\left\{\frac{1}{2}(A+D)\right\}^2 - 1\right]^{1/2} \quad (8)$$

Subscripts TM and TE have been dropped for simplicity. Note that if $\left|\frac{1}{2}(A+D)\right| < 1$, it would imply that the Bloch wave vector K is real, which would correspond to a propagating Bloch wave. However, if $\left|\frac{1}{2}(A+D)\right| > 1$, K would be complex and this condition would correspond to an evanescent Bloch wave, whose amplitude decays exponentially along the direction of periodicity. Thus, light frequencies that satisfy the condition $\left|\frac{1}{2}(A+D)\right| > 1$ constitute the "forbidden" bands of the periodic medium.

Figure 4.4 shows the typical band structure of such a one-dimensional periodic medium. The white (blank) regions in the figure correspond to decaying Bloch waves, and light frequencies that fall within this region are forbidden to propagate in the medium. However, the one-dimensional bandgap is not *omnidirectional* (*omni* means *all* in Latin) because, in general, such a structure is unable to reflect light at arbitrary incident angles. An omnidirectional bandgap can confine light within a certain frequency range irrespective of the incident angle and state of polarization of light [12]. It occurs if the forbidden frequencies in the bandgaps overlap along all possible directions in space. Intuitively, it might seem that this would require a system that is periodic along the three orthogonal directions (three-dimensional photonic crystal). However, omnidirectional reflection can be achieved even from a one-dimensional photonic crystal (one-dimensional planar stack, in this case) if the wave incident from the ambient medium does not couple to the propagating states of the photonic crystal [12, 13].

To comprehend this feature, we first explain the concept of *light line*. Figure 4.6a shows the dispersion relation for light waves in free space. The solid line corresponds to $\omega = ck$, where ω is the frequency of light, k is the magnitude of the propagation wave vector, and c is the speed of light in a vacuum. We assume the x–y plane as the plane of propagation of the light wave. In general, the propagation vector k would have both x- and y-components, such that $\omega = c(k_y^2 + k_x^2)^{1/2}$ and $\omega \geq ck_y$. Again, referring to the planar stack in Fig. 4.3, $k_y = k \sin\theta$, where θ is the angle of incidence on the planar stack such that $-90° \leq \theta \leq 90°$. Hence, the range of all possible values of k_y would be $-k \leq k_y \leq k$. Figure 4.5b shows the variation of k_y with ω, wherein any horizontal line (e.g., the dashed line) would correspond to all possible values of the incident wave vector for a given frequency. The y-axis corresponds to $k_y = 0$ and the *light line* corresponds to $\omega = ck_y = ck$ ($k_x = 0$, $\theta = \pm 90°$). The entire region above the light line satisfies $\omega > ck_y$, and hence all allowed states of the ambient medium lie in this region. To obtain the condition of omnidirectionality, we assume Fig. 4.6b to be superimposed on the band structure of the planar stack (Fig. 4.4). If there now exists a range of "forbidden" frequencies within the region $\omega > ck_y$, the incident wave from the ambient medium at those frequencies would encounter the bandgap for all possible angles of incidence. Consequently, the incident

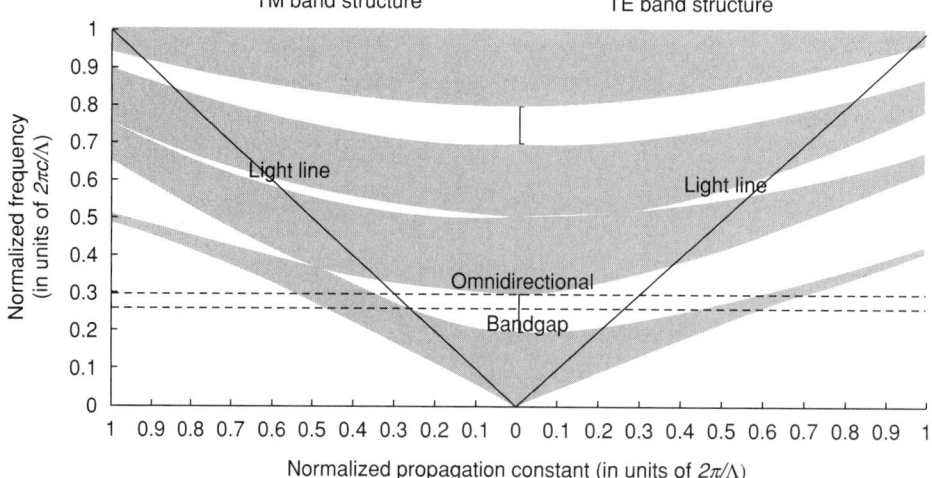

FIGURE 4.4 Photonic bandgap structure for TE and TM polarization states for a one-dimensional planar stack.

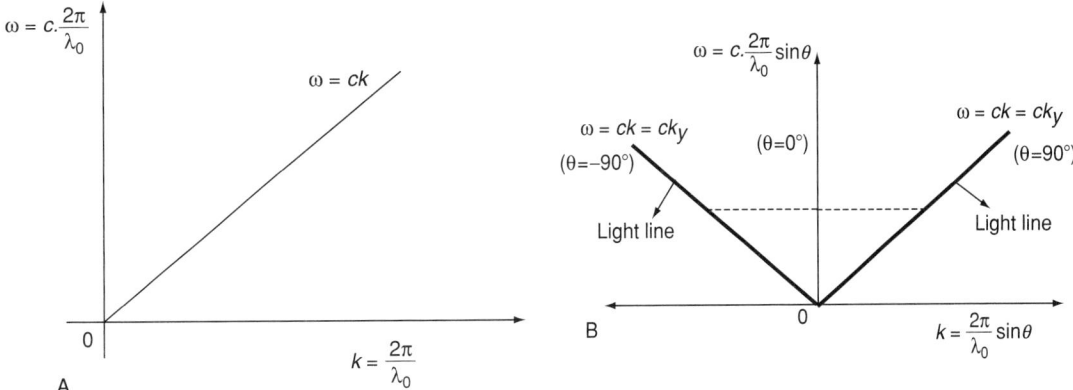

FIGURE 4.5 (a) Variation of ω vs. k (dispersion relation of light wave) in free space. (b) Variation of ω vs. k_y (dispersion relation of light wave) in free space; θ is the angle at which the wave vector is incident on the planar stack in Fig. 4.3.

wave would be unable to couple to the propagating states of the periodic structure, and hence omnidirectional reflection could be achieved [12]. In Fig. 4.4, the frequencies lying within the two horizontal dashed lines constitute the omnidirectional bandgap. For any further details on band structure and reflection characteristics of periodic planar stacks, we refer readers to [11].

2.2. Light Propagation in Bragg Fibers

Propagation of electromagnetic waves in Bragg fibers is governed by the well-known Maxwell's equations in charge-free space as

$$\nabla \cdot D = 0$$
$$\nabla \times E = -\frac{\partial B}{\partial t}$$
$$\nabla \cdot B = 0 \qquad (9)$$
$$\nabla \times H = \frac{\partial D}{\partial t}$$

The electric and magnetic fields associated with the guided modes propagating in the z-direction (in cylindrical coordinates) can be assumed to have the following form:

$$\Psi(r, \theta, z, t) = \psi(r, \theta) e^{-i(\Omega t - \beta z)} \qquad (10)$$

where $\psi(r, \theta)$ could be any of the field components: $E_z, E_r, E_\theta, H_z, H_r$, and H_θ. The field components E_z and H_z satisfy the following wave equation:

$$\left(\nabla_t^2 + (\omega^2 \mu \varepsilon - \beta^2)\right) \begin{pmatrix} E_z \\ H_z \end{pmatrix} = 0 \qquad (11)$$

where $\nabla_t^2 = \nabla^2 - \frac{\partial^2}{\partial z^2}$ is the transverse Laplacian operator. Bragg fibers possess cylindrical symmetry, and the general solutions of the field components E_z and H_z, that satisfy the wave equation, can be written as follows (see, e.g., [3]):

$$E_z = [A_j J_l(k_j r) + B_j Y_l(k_j r)] \cos(l\theta + \Phi_j) \quad j = 1, 2$$
$$H_z = [C_j J_l(k_j r) + D_j Y_l(k_j r)] \cos(l\theta + \chi_j) \quad j = 1, 2$$
(12)

where $k_j = \left[\left(\frac{n_j \omega}{c}\right)^2 - \beta^2\right]^{1/2}$ and $j = 1, 2$ corresponds to the cladding layers of refractive index n_1 and n_2 and thickness l_1 and l_2. The other transverse field components of the guided modes $(E_\theta, E_r, H_\theta, H_r)$ can be written in terms of E_z and H_z as [15]

$$E_r = \frac{i\beta}{\omega^2 \mu \varepsilon - \beta^2} \left(\frac{\partial}{\partial r} E_z + \frac{\omega \mu}{\beta} \frac{\partial}{r \partial \theta} H_z\right)$$
$$E_\theta = \frac{i\beta}{\omega^2 \mu \varepsilon - \beta^2} \left(\frac{\partial}{r \partial \theta} E_z - \frac{\omega \mu}{\beta} \frac{\partial}{\partial r} H_z\right)$$
$$H_r = \frac{i\beta}{\omega^2 \mu \varepsilon - \beta^2} \left(\frac{\partial}{\partial r} H_z - \frac{\omega \varepsilon}{\beta} \frac{\partial}{r \partial \theta} E_z\right) \qquad (13)$$
$$H_\theta = \frac{i\beta}{\omega^2 \mu \varepsilon - \beta^2} \left(\frac{\partial}{r \partial \theta} H_z + \frac{\omega \varepsilon}{\beta} \frac{\partial}{\partial r} E_z\right)$$

2.2.1. Minimization Procedure

In the case of conventional optical fibers, the effective index of a mode can be easily obtained by using the boundary condition that the transverse field decays exponentially to zero, outside the cladding–air interface. However, this is not applicable to Bragg fibers because they support oscillatory solutions even in the cladding layers. Moreover, the aperiodic nature of Bessel functions does not allow the Bloch theorem to be applied to the fields in the multilayer cladding. Hence, to obtain the effective index of a guided mode in such fibers, the outward flowing power flux has to be minimized by appropriately choosing the thickness of the cladding layers [4]. The radial component of the net Poynting vector is

$$S_r = \left(\frac{1}{2}\right) \text{Re}\left[E_\theta H_z^*\right] \qquad (14)$$

S_r is zero for any Bragg fiber structure with an infinite or finite number of cladding layers if an infinitely extended source is used to excite the modes. However, in a real-world situation, the net outward flow of power is finite and constitutes the propagation loss. Optimum confinement is achieved when the cladding bilayers are of "half-wave" thickness, implying that each layer satisfies the following *quarter-wave thickness* condition:

$$k_1 l_1 = k_2 l_2 = \pi/2 \quad (15)$$

Because the zeroes of Bessel functions are not exactly periodic, theoretically the cladding layers of a Bragg fiber that satisfy the quarter-wave thickness condition cannot be of equal thickness!

2.2.2. Asymptotic Matrix Theory

The minimization procedure, explained above, is relatively easy for pure TE and TM modes but becomes tedious and time-consuming for hybrid modes. However, Xu et al. proposed an *asymptotic matrix theory* that yielded an analytical equation for obtaining propagation constants of the guided modes in a Bragg fiber, and it required much less computation time [15]. It was based on the fact that in the asymptotic limit, Bessel functions essentially resemble the plane waves with an additional amplitude factor of $\sqrt{r}$ in its denominator. In this method, the field in the core of the Bragg fiber and up to a certain finite number of cladding layers (say N_1) are assumed to be exact solutions of Maxwell's equations (in the form of Bessel functions). In the subsequent cladding layers beyond the (N_1)th layer, the field is assumed to follow the asymptotic form of Bessel functions as follows [15]:

$$\left. \begin{array}{l} E_z = A_{j'} J_l(k_{j'} r) + B_{j'} Y_l(k_{j'} r) \\ H_z = C_{j'} J_l(k_{j'} r) + D_{j'} Y_l(k_{j'} r) \end{array} \right\} \quad j' = 1, 2, \ldots, N_1$$

where $k_{j'} = \left[\left(\frac{n_{j'} \omega}{c} \right)^2 - \beta^2 \right]^{1/2}$ and $n_{j'}$ is the refractive index in the (j')th layer.

$$\left. \begin{array}{l} E_z = \dfrac{a_j^q e^{ik_q(r-\rho_j^q)} + b_j^q e^{-ik_q(r-\rho_j)}}{\sqrt{k_q r}} \\ H_z = \dfrac{c_j^q e^{ik_q(r-\rho_j)} + d_j^q e^{-ik_q(r-\rho_j)}}{\sqrt{k_q r}} \end{array} \right\} \quad \begin{array}{l} \rho_j^q \leq r < \rho_j^q + l_q \\ j = 1, 2, \ldots N - N_1, q = 1, 2 \end{array}$$

$$(16)$$

where $k_q = \left[\left(\frac{n_q \omega}{c} \right)^2 - \beta^2 \right]^{1/2}$ and where $q = 1, 2$ corresponds to the cladding layers having refractive index n_1 and n_2, and N is the total number of cladding layers. The eigenvalue equation for the guided modes is obtained by applying the Bloch theorem to the field in the outer cladding layers and using the continuity conditions for the electric and magnetic fields (i.e., E_z, E_θ, H_z, H_θ) at the layer interfaces. The accuracy of the solutions could be improved by increasing the number of inner cladding layers for which the solutions were assumed to be (rigorously correct) Bessel functions. The asymptotic approximation is valid only if $k_j r \gg 1$ [16]. The condition that $k_j r \gg 1$ may not be true if the fiber has a very small core radius or if the refractive index of the cladding layers is not sufficiently large relative to the refractive index of air. Other numerical techniques, like the finite difference time domain method [17] and perturbation method [18, 19], are also widely used to obtain the propagation constants of guided modes in Bragg fibers.

2.3. Modal Characteristics

In general, bandgap structures for the TE and TM polarizations do not overlap. However, a large refractive index contrast in the cladding layers can result in an omnidirectional bandgap [20]. TE and TM modes are the fundamental modes of a Bragg fiber, for which the electric fields are purely parallel (TE) and normal (TM) to the layer interfaces (Fig. 4.6). Unlike the LP modes of a conventional fiber, the TE and TM modes of an air-core Bragg fiber are nondegenerate. Hence, Bragg fibers, which support a single TE or TM mode, are inherently free of polarization mode dispersion. Though omnidirectionality is not essential for a mode to be guided along the Bragg fiber, the TE_{01} modal properties of a dielectric waveguide with a large index contrast in the cladding closely resemble the modal structure of hollow metallic waveguides if the TE_{01} mode lies within the omnidirectional bandgap [21]. Similar to metallic waveguides, TE_{01} mode is the lowest-loss mode out of the several (including TM and hybrid) modes supported in Bragg fibers, all of which exhibit much larger differential loss relative to the TE_{01} mode. In view of this feature, single-mode propagation of TE_{01} mode could effectively be realized even in relatively large-core Bragg fibers [19]. Accordingly, the TE_{01} mode is relatively more tightly confined inside the core as compared with other modes (Fig. 4.7). In addition to the bandgap-guided modes (TE, TM, and hybrid), Bragg fibers also support other index-guided modes (which decay exponentially in the air regions but extend throughout the dielectric cladding) and surface modes (which decay exponentially in both air and dielectric layers).

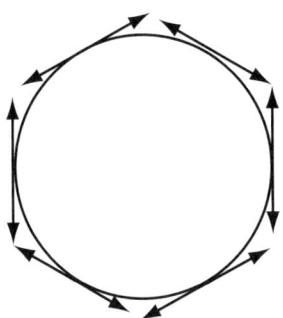 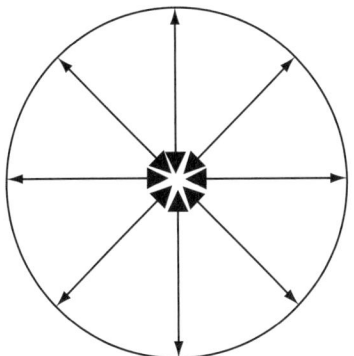

FIGURE 4.6 Schematic of electric field components of TE and TM modes in a Bragg fiber.

2.3.1. Propagation Loss

Propagation loss of Bragg fibers primarily arises from material absorption and scattering losses and waveguide-related radiation loss. The material-related losses are dictated by the choice of materials used for the core and the cladding layers, uniformity of the layer interfaces and penetration of the field into the cladding. However, because a large fraction of the TE_{01} modal energy essentially propagates through the hollow core in an air-core Bragg fiber, material absorption losses of this mode are not very significant [22]. The radiation loss of the propagating modes depends on the refractive index contrast and number of cladding layers. The guiding region (air core) of a Bragg fiber has a refractive index that is lower than that of all the cladding layers and it supports only leaky modes. Ideally, a Bragg fiber with infinitely extended cladding layers would have zero radiation loss. However, in practice, a fabricated Bragg fiber has only a finite number of cladding layers, and consequently, the modes supported by it would invariably suffer a finite radiation loss. Hence, the modes supported by a Bragg fiber are termed "quasi-modes." However, due to the low loss of the TE_{01} mode relative to higher order modes, these would leak out within a characteristic length (L_{SM}) of the fiber, beyond which the fiber becomes effectively single moded. Furthermore, because the guided mode (i.e., TE_{01}) itself is also leaky by nature, there exists a characteristic fiber length L_{max} beyond which the power of the guided mode reduces below useful limits [23]. The quarter-wave stack condition [Eq. (15)] is an optimization condition that optimizes the radiation loss of TE modes in

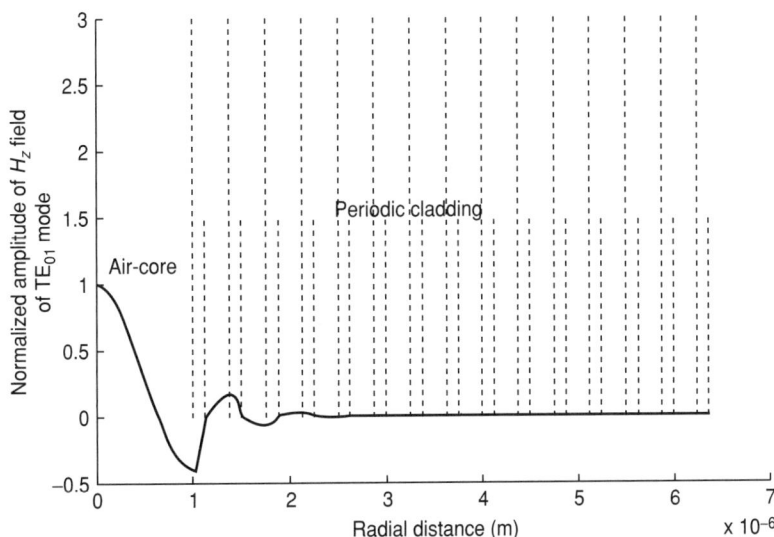

FIGURE 4.7 Typical variation of H_z field of the TE_{01} mode along the radial direction. Note that the field rapidly decays to zero within a few cladding layers. Vertical dashed lines correspond to the cladding layer interfaces; the length of the lines is proportional to the refractive index in the particular layer.

large-core Bragg fibers, and the radiation loss is proportional to the square of the field amplitude in the cladding. Modal loss also depends on the refractive index contrast in the cladding layers and core radius of the fiber. For TE_{01} modes, it is proportional to $1/r_{co}^3$, whereas for TM_{01} modes, the loss is proportional to $1/r_{co}$, where r_{co} is the core radius of the fiber [22]. Besides the refractive index contrast, modal radiation loss of the Bragg fiber also depends on the number of cladding layers, N, and scales as Δ^N, where Δ is a constant whose value depends on the mode being considered [22].

2.3.2. Dispersion

Because of the multitude of index profile parameters involved, Bragg fibers offer fiber designers wide latitude in the choice of parameters to tailor their dispersion characteristics for diverse functionality. Judicious choice of the core radius, cladding layer thickness, and refractive index contrast in the cladding layers can yield dispersion characteristics that are virtually unique to Bragg fibers and almost unachievable in conventional single-mode fibers. For example, single-mode light propagation at 1.06 μm with zero dispersion at a wavelength ≈ 1 μm through a silica core Bragg fiber was experimentally demonstrated in 2000 [9]. Subsequently, various designs have been proposed to achieve multiple zero dispersion [24] and high negative dispersion for dispersion compensation purposes [24, 25] and single polarization single-mode propagation [26].

3. DISPERSION COMPENSATING BRAGG FIBER

Dispersion compensators are an integral part of any long-haul fiber optic telecommunications link. Conventional high-silica single-mode fibers exhibit positive temporal dispersion beyond a wavelength of 1310 nm. Consequently, a signal carrying the LP_{01} mode acquires positive dispersion as it propagates through a conventional single-mode transmission fiber, which unless canceled through insertion of a component to combat it would limit signal transmission capacity of the fiber. This is achieved by inserting a dispersion compensating module, which in most cases is a dispersion compensating fiber (DCF; see Chapter 1). A DCF is normally designed to exhibit a large negative dispersion coefficient (D) across the signal wavelength band such that a relatively short length (l_C) of the DCF could cancel the dispersion that the signal acquires while propagating through much longer lengths of the transmission fiber (l_{Tx}). Equation (17) determines the length of the DCF required for compensating the dispersion accumulated in a transmission fiber.

$$D_C l_C + D_{Tx} l_{Tx} = 0 \quad (17)$$

where D_C and D_{Tx} are the dispersion coefficients of the DCF and the transmission fiber, respectively. Besides high negative dispersion, an efficient DCF must also exhibit low loss in the operating wavelength range. The overall efficiency of a DCF is measured in terms of an integral parameter known as figure of merit (FOM), which is defined as the ratio of dispersion to loss of the DCF:

$$\text{FOM} = D/\alpha \quad (18)$$

where the loss is measured in dB/km, so that FOM is expressed in units of ps/nm·dB. The larger the value of FOM, the more efficient is the DCF. One can achieve large negative dispersion in conventional silica fibers through suitable tailoring of its refractive index profile. The largest dispersion coefficient of −1800 ps/nm·km demonstrated to date was based on a dual-core DCF design [27, 28]. However, the material loss of silica limits the FOM that can be achieved, and typically FOM ranges from ~300 to 400 ps/nm·dB in commercially available high FOM DCFs. Moreover, typical refractive index profiles required to generate a large waveguide (negative) dispersion to substantially offset the positive material dispersion (within the gain spectrum of an erbium-doped fiber amplifier) are necessarily characterized with a relatively small mode effective area (A_{eff}) ≈ 15–20 μm², which makes these DCFs sensitive to detrimental nonlinear effects unless the launched signal power into the DCF is restricted.

Bragg fibers have evolved as an attractive alternative to conventional DCFs, and various designs have been proposed for realizing high negative dispersion fibers with a relatively large A_{eff}. One of the earliest designs reported an estimated D up to approximately −25,000 ps/nm·km through the hybrid HE_{11} mode of a Bragg fiber [25]. However, the hybrid nature and higher loss of HE_{11} mode (as compared with TE_{01} mode) and small core radius were major issues that limited the practical use of this otherwise attractive design. Subsequently, it was reported that high negative D could be achieved for the TE_{01} mode as well by intentionally incorporating a defect layer within the periodic cladding layers [24]. The fiber is so designed that the fundamental (TE_{01}) mode of the defect-free Bragg fiber and the defect mode supported by the defect layer are in weak resonance, analogous to phase matching of the supermodes in a fiber directional coupler at a particular wavelength. Near the resonance wavelength, the fiber exhibits very large negative dispersion. The thickness and localization of the defect layer were crucial in determining precise dispersion characteristics of such Bragg fibers. These parameters can be altered suitably to match the dispersion and dispersion-slope for achieving broadband dispersion compensation characteristics. A fundamental design rule for such Bragg fiber–based DCFs with an intentional defect layer(s) is that its modal field must penetrate sufficiently into the cladding for interactions with the defect layer so as to generate large negative dispersion. This procedure, however, inadvertently increases the material-related loss of the structure besides making the modal field sensitive to nonuniformities in cladding thickness. An alternate design was recently proposed to realize a DCF with a high FOM [29] in which the quarter-wave stack condition [Eq. (15)] was modified to

$$\frac{2\pi}{\lambda_0}n_1 l_1 = \frac{2\pi}{\lambda_0}n_2 l_2 = n\pi/2 \qquad (19)$$

where n is an odd integer. The multiple quarter-wave stack condition [Eq. (19) with $n = 3$] was found to confine the TE_{01} mode tightly within the core with a low radiation loss while at the same time exhibiting high negative dispersion. The dispersion spectrum of one such designed DCF is shown in Fig. 4.8. It can be seen from Fig. 4.8 that the so designed DCF's high negative D extends over a wavelength range of about 60 nm. The design was targeted to minimize the radiation loss (Fig. 4.9), thereby yielding a very high estimate for the FOM (~4000 ps/nm·dB at 1550 nm) for the designed fiber. Though the proposed design supports more than one mode in the operating wavelength band, the differential losses suffered by the higher order

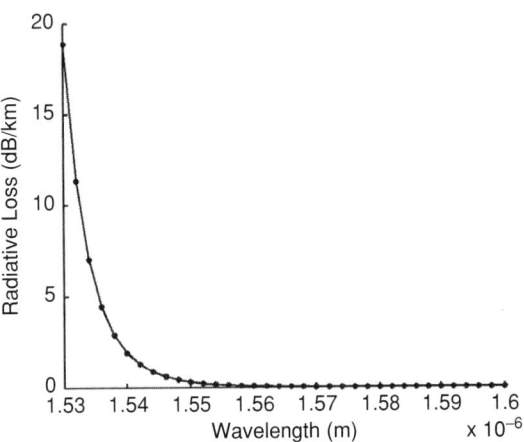

FIGURE 4.9 Radiation loss spectrum of TE_{01} mode of Bragg fiber based DCF.

modes with respect to the TE_{01} mode were found to be an order of magnitude higher, thus effectively making the fiber single moded.

4. BRAGG FIBERS FOR METRO NETWORKS

With the growth in Internet usage and increasing demand in data services and network efficiency, metro optical networks are increasingly evolving as *intelligent* and *data-centric* networks. Metropolitan networks bridge the gap between local access networks and long distance telecommunications networks. As a rule of thumb, metro networks are required to be suitably designed to address features like low installation cost, high degree of scalability, dynamism that is capable of accommodating unpredicted traffic growth besides flexibility to add/drop individual signals at any central office in the network, and interoperability to support protocols such as cell phones, synchronous optical networks/synchronous digital hierarchy, and legacy equipment. The low cost requirement implies that metro-specific fiber designs are aimed at minimizing use of components such as amplifiers, dispersion compensators, gain flattening filters, and so on. In any case, dispersion as well as loss spectra are the dictating factors in estimating maximum signal reach in a network.

Reported designs for metro fibers are based on their operation either as negative dispersion fibers [30, 31] or positive dispersion fibers across the C-band [32], so as to achieve a span length of ~100 km, without the need for a dispersion compensating device. However, these networks still need an amplifier after approximately every 80 km. Because Bragg fibers offer the potentiality of realizing ultra-low-loss transmission, deployment of Bragg fibers

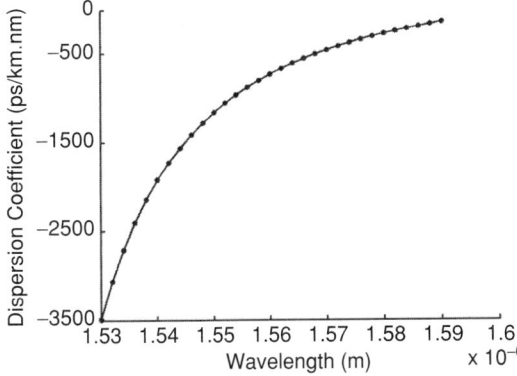

FIGURE 4.8 Dispersion spectrum of TE_{01} mode of Bragg fiber based DCF.

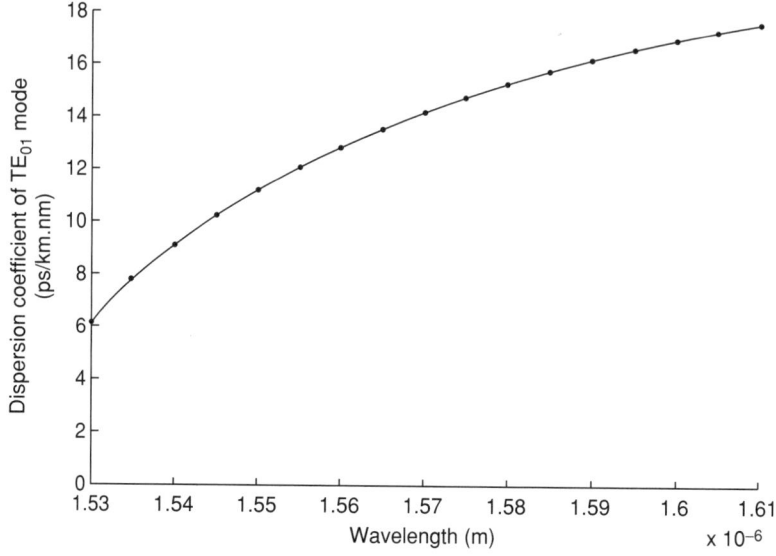

FIGURE 4.10 Dispersion spectrum of TE_{01} mode of metro-centric Bragg fiber.

could minimize investment cost on amplifier head and provide a useful platform for dispersion tailoring, both of which should be very attractive for a metro network. Dispersion spectrum of one such Bragg fiber designed in-house having an air-core radius of 10.0 μm with 15 cladding bilayers and with a refractive index contrast of 2.0 is shown in Fig. 4.10. Loss suffered by higher order modes of the fiber is much larger as compared with the TE_{01} mode, thus making the fiber effectively single moded. It can be seen from Fig. 4.10 that the average dispersion of the fiber across the C-band is ~9.9 ps/nm·km, with a dispersion slope of 0.23 ps/nm² · km. This should enable a dispersion-limited fiber length of ~100 km in the C-band at 10 Gb/s. The average dispersion across the L-band is ~16.0 ps/nm·km, with a dispersion slope of 0.07 ps/nm² · km at 1.59 μm. Because the TE_{01} mode is very well confined within the air core, the material absorption losses in the cladding could be ignored as compared with the radiation loss. The radiation loss of the TE_{01} mode across the C- and L-bands is tabulated in Table 4.1. To achieve sufficient coupling into the TE_{01} mode from a laser pigtailed source, the fiber may be tapered to attain significant modal overlap between the TE_{01} mode and the approximate Gaussian field of the single-mode fiber pigtail.

5. FABRICATION

In spite of their many appealing features, Bragg fibers did not gain much popularity until the late 1990s because of the unavailability of suitable fabrication techniques. Recently, development in fabrication technologies has enabled the realization of broadband, low-loss, omnidirectional Bragg fibers. In the year 2000, the classical modified chemical vapor deposition technique was used to manufacture the first "depressed core index photonic bandgap" fiber, which allowed single-mode propagation of light at 1060 nm [9]. The core of the depressed core index PBGF was made of slightly F-doped silica and the cladding layers were made of Ge-doped silica such that the refractive index contrast of consecutive layers was 0.009. In 1999, a combination of techniques that included dip coating on a capillary, serving as the core, and thermal evaporation and deposition was used to fabricate a broadband, low-loss, hollow waveguide in the 10 μm range [10]. The transmission through the waveguide around a 90° bend was also measured to demonstrate the low-loss features of such structures even in the presence of bends in the spectral band corresponding to the photonic bandgap. In 2003, a large effective area Bragg fiber with a silica core was fabricated using

TABLE 4.1 Variation of radiation loss of the metro-centric Bragg fiber with wavelength.

Wavelength (nm)	1530	1540	1550	1560	1570	1580	1590	1600	1610
Radiation loss ($\times 10^{-2}$ dB/km)	10.5	4.03	1.69	0.76	0.37	0.19	0.10	0.006	0.003

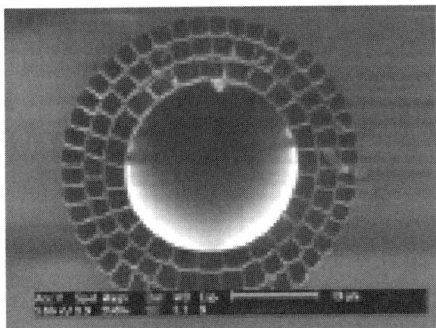

FIGURE 4.11 Air-silica Bragg fiber (reproduced from [34] © Optical Society of America).

the modified chemical vapor deposition process [33], which supported a single HE_{11} mode with a mode effective area of $\sim 526\,\mu m^2$ (core radius $\approx 17.5\,\mu m$) and exhibited losses of 0.4 dB/m at 1550 nm with only three cladding bilayers. Air-core Bragg fibers with four pairs of Si/Si_3N_4 cladding layers have also been fabricated, yielding propagation loss below 1 dB/cm [22]. The experimental loss figures are a consequence of additional loss in the fiber due to the formation of gas inlet ports in the fiber during the chemical vapor depositine fabrication process. Very recently, an air-silica Bragg fiber supporting the TE_{01} mode was demonstrated for the first time [34] (see Fig. 4.11). The air-core Bragg fiber consisted of silica rings separated by 2.3-μm-thick air rings and 45-nm-thick support bridges and exhibited a propagation loss of $\sim$1.5 dB/m at the transmission peak.

Though the transmission loss of Bragg fibers fabricated to date is orders of magnitude higher than conventional silica fibers ($\sim$0.2 dB/km), further improvement in fabrication techniques to achieve more uniform layer thickness and larger number of cladding layers should enable propagation loss of Bragg fibers to be brought down to well below the 0.2-dB/km figure. Additionally, the low loss of these fibers would not be restricted to the 1550-nm window. These distinct advantages in wavelength tunability for loss and dispersion should open up avenues for a wide range of potentially newer applications and devices.

6. CONCLUSION

In this chapter we attempted to describe relatively recent history, underlying physics, modeling of propagation characteristics, design issues, and fabrication of one-dimensional photonic bandgap-guided Bragg fibers. Bragg fibers have indeed emerged as a specialty fiber with the potential for several attractive applications.

7. REFERENCES

1. J.D. Joannopoulos, R.D. Meade, and J.N. Winn, *Photonic Crystals: Molding the Flow of Light*, Princeton University Press, Princeton, NJ (1995).
2. http://ab-initio.mit.edu/photons/tutorial/photonic-intro.pdf
3. E. Yablonovitch, "Inhibited spontaneous emission in solid-state physics and electronics," *Phys. Rev. Lett.*, **58**, 2059 (1987).
4. P. Yeh, A. Yariv, and E. Marom, "Theory of Bragg fiber," *J. Opt. Soc. Amer.*, **68**, 1196 (1978).
5. J.C. Knight, "Photonic crystal fibers," *Nature*, **424**, 847 (2003).
6. T.A. Birks, J.C. Knight, B.J. Mangan, and P.St.J. Russel, "Photonic crystal fibers: an endless variety," *IEICE Trans Electron.*, **E84-C**, 585 (2001).
7. B.J. Eggleton, C. Kerbage, P.S. Westbrook, R.S. Windeler, and A. Hale, "Microstructured optical fiber devices," *Opt. Exp.*, **9**, 698 (2001).
8. N.J. Doran and K.J. Blow, "Cylindrical Bragg fibers: a design and feasibility study for optical communications," *J. Lightwave Technol.*, **LT-1**, 588 (1983).
9. F. Brechet, P. Roy, J. Marcou, and D. Pagnoux, "Single-mode propagation into depressed-core-index photonic bandgap fiber designed for zero-dispersion propagation at short wavelengths," *Electron. Lett.*, **36**, 514 (2000).
10. Y. Fink, D.J. Ripin, S. Fan, C. Chen, J.D. Joannopoulos, and E.L. Thomas, "Guiding optical light in air using an all-dielectric structure," *J. Lightwave Technol.*, **17**, 2039 (1999).
11. P. Yeh, *Optical Waves in layered media*, John Wiley & Sons. Inc., Singapore (1991).
12. J.N. Winn, Y. Fink, S. Fan, and J.D. Joannopoulos, "Omnidirectional reflection from a one-dimensional photonic crystal," *Opt. Lett.*, **23**, 1573 (1998).
13. Y. Fink, J.N. Winn, S. Fan, C. Chen, J. Michel, J.D. Joannopoulos, and E.L. Thomas, "A dielectric omnidirectional reflector," *Science*, **282**, 1679 (1998).
14. J.A. Stratton, *Electromagnetic Theory*, McGraw-Hill, New York (1941).
15. Y. Xu, G.X. Ouyang, R.K. Lee, and A. Yariv, "Asymptotic matrix theory of Bragg fibers," *J. Lightwave Technol.*, **20**, 428 (2002).
16. S. Guo, S. Albin, and R.S. Rogowski, "Comparative analysis of Bragg fibers," *Opt. Exp.*, **12**, 198 (2004).
17. A. Taflove and S.C. Hagness, *Computational Electrodynamics: The Finite-Difference-Time-Domain Method*, Artech House, Boston (2000).

18. D.Q. Chowdhary and D.A. Nolan, "Perturbation model for computing optical fiber birefringence from a two-dimensional refractive-index profile," *Opt. Lett.,* **20**, 1973 (1995).
19. S.G. Johnson, M. Ibanescu, M. Skorobogatiy, O. Weisberg, T.D. Engeness, M. Soljačić, S.A. Jacobs, J.D. Joannopoulos, and Y. Fink, "Low-loss asymptotically single-mode propagation in large-core OmniGuide fibers," *Opt. Exp.,* **9**, 748 (2001).
20. S.D. Hart, G.R. Maskaly, B. Temelkuran, P.H. Prideaux, J.D. Joannopoulos, and Y. Fink, "External reflection from omnidirectional dielectric mirror fibers," *Science,* **296**, 510 (2002).
21. M. Ibanescu, S.G. Johnson, M. Soljačić, J.D. Joannopoulos, Y. Fink, O. Wiesberg, T.D. Engeness, S.A. Jacobs, and M. Skorobogatiy, "Analysis of mode structure in hollow dielectric waveguide fibers," *Phys. Rev. E,* **67**, e046608 (2003).
22. Y. Xu, A. Yariv, J.G. Fleming, and S.Y. Lin, "Asymptotic analysis of silicon based Bragg fibers," *Opt. Exp.,* **11**, 1039 (2003).
23. A. Argyros, "Guided modes and loss in Bragg fibres," *Opt. Exp.,* **10**, 1411 (2002).
24. T.D. Engeness, M. Ibanescu, S.G. Johnson, O. Weisberg, M. Skorobogatiy, S. Jacobs, and Y. Fink, "Dispersion tailoring and compensation by modal interactions in omniguide fibers," *Opt. Exp.,* **11**, 1175 (2003).
25. G. Ouyang, Y. Xu, and A. Yariv, "Theoretical study on dispersion compensation in air core Bragg fibers," *Opt. Exp.,* **10**, 899 (2002).
26. G.R. Hadley, J.G. Fleming, and S.Y. Lin, "Bragg fiber design for linear polarization," *Opt. Lett.,* **29**, 809 (2004).
27. K. Thyagarajan, R.K. Varshney, P. Palai, A.K. Ghatak, and I.C. Goyal, "A novel design of a dispersion compensating fiber," *IEEE Photon. Technol. Lett.,* **8**, 1510 (1996).
28. J.L. Auguste, R. Jindal, J.M. Blondy, J. Marcou, B. Dussardier, G. Monnom, D.B. Ostrowsky, B.P. Pal, and K. Thyagarajan, "−1800 (ps/nm)/km chromatic dispersion at 1.55 μm in dual concentric core fiber," *Electron. Lett.,* **36**, 1689 (2000).
29. S. Dasgupta, B.P. Pal, and M.R. Shenoy, "Design of a low loss bragg fiber with high negative dispersion for the TE01 mode," *2004 Frontiers in Optics*, Pres. no. FWH49, Rochester, New York, USA (2004).
30. I. Tomkos, B. Hallock, I. Roudas, R. Hesse, A. Boskovic, J. Nakano, and R. Vodhanel, "10-Gb/s transmission of 1.55-μm directly modulated signal over 100 km of negative dispersion fiber," *IEEE Photon. Technol. Lett.,* **13**, 735 (2001).
31. T. Okuno, H. Hatayama, K. Soma, T. Sasaki, M. Onishi, and M. Shigematsu, "Negative dispersion-flattened fiber suitable for 10 Gbit/s directly modulated signal transmission in whole telecommunication band," *Electron. Lett.,* **40**, 723 (2004).
32. D. Culverhouse, A. Kruse, C. Wang, K. Ennser, and R. Vodhanel, "Corning® MetroCor® fiber and its application in metropolitan networks," 2000 Corning Incorporated, White paper at http://www.corning.com/docs/opticalfiber/wp5078_7-00.pdf
33. S. Férvier, P. Viale, F. Gérôme, P. Leproux, P. Roy, J.M. Blondy, B. Dussardier, and G. Monnom, "Very large effective area singlemode photonic bandgap fibers," *Electron. Lett.,* **39**, 1240 (2003).
34. G. Vienne, Y. Xu, C. Jakobsen, H.J. Deyerl, T.P. Hansen, B.H. Larsen, J.B. Jensen, T. Sorensen, M. Terrel, Y. Huang, R. Lee, N.A. Mortensen, J. Broeng, H. Simonsen, A. Bjarklev, and A. Yariv, "First demonstration of air-silica Bragg fiber," OFC 2004 Post deadline paper, **PDP25,** Los Angeles, California, USA (2004).

5
CHAPTER

Radial Effective Index Method for the Analysis of Microstructured Fibers

Kin Seng Chiang*
Optoelectronics Research Centre
Department of Electronic Engineering
City University of Hong Kong
Kowloon, Hong Kong, China

Vipul Rastogi[†]
Department of Physics
Indian Institute of Technology Roorkee
Uttaranchal, India

1. INTRODUCTION

Modal analysis is of the most fundamental importance in the optical waveguide theory, yet modal solutions in exact analytical forms are available only for a limited number of special waveguide geometries. Over the years, there have been rapid advances in optical waveguide technology, and a large number of modal analysis methods have been developed to cope with the increasingly sophisticated waveguide structures (see, e.g., [1]). In recent years, because of the dramatic increase in computer power and decrease in computing cost, most waveguide analyses can now be performed with commercial software packages based on versatile numerical methods, such as the beam propagation method and the finite difference time domain method. Although such general black-box tools, if properly used, can provide accurate numerical results for the waveguide modes, which are certainly useful for the design of practical devices, they do not usually provide much physical insight into the waveguide structure under investigation. In many cases, a semianalytical approximate method, when used within its domain of applicability, can offer a more intuitive understanding of the characteristics of the waveguide. In this chapter, we describe such a method for the analysis of optical fibers with arbitrary shapes and refractive index profiles. The method is known as the radial effective index method (REIM) [2].

The original version of the effective index method (EIM) [3] was proposed more than three decades ago as an empirical approach to improve upon Marcatili's method [4] for the analysis of the fundamental mode of a simple rectangular-core dielectric waveguide. The idea of the EIM is to approximate a rectangular-core waveguide by a slab waveguide with an effective refractive index. The effective index concept is so attractive that the method has since been extended to the analysis of a wide range of composite rectangular structures (see, e.g., [5–8]) and nonrectangular structures, including optical fibers [9–12] and inhomogeneous waveguides [13], which has resulted in tremendous savings in computing effort. For some relatively simple rectangular structures, the approximate nature of the method has been established rigorously with the perturbation theory [8,14–16]. In fact, a perturbation analysis of the EIM has led to the development of an improved EIM that can provide a much higher level of accuracy without compromising the computing efficiency [17–19].

Naturally, the EIM considered in all the studies mentioned above is formulated in the Cartesian coordinate system, which is well suited for rectangular structures. The REIM considered in the present chapter, however, is an EIM formulated in the cylindrical coordinate system. The REIM was proposed originally in 1987 for the analysis of noncircular fibers [2]. The idea is to approximate a noncircular fiber by a circular fiber with an effective refractive index profile. Compared with its Cartesian counterpart, the REIM has not gained much popularity

*E-mail: eeksc@cityu.edu.hk
[†]E-mail: vipulfph@iitr.ernet.in

due to its relatively limited applications, because even common noncircular fibers, such as elliptical-core fibers and fused tapered fiber couplers, can be analyzed accurately with the Cartesian EIM [9–11]. Recently, however, the REIM has found new and important applications in the study of microstructured fibers, which today is a topical area of research. In particular, the REIM has been applied with advantages to the analysis of two special fibers: the segmented cladding fiber [20–23] and the holey fiber [23, 24], which cannot be treated accurately with the Cartesian EIM. In this chapter, the application of the REIM to the analysis of these fibers is demonstrated with examples.

2. THE REIM

2.1. Formulation of the Method

The REIM is formulated for solving the scalar wave equation expressed in cylindrical coordinates. The following treatment is more general than the original one given in [2], which deals with only the fundamental guided mode.

We consider an optical fiber whose cross-sectional shape and refractive index profile can be quite arbitrary. The transverse component of the electric field in the fiber is assumed to satisfy the scalar wave equation, which is given below in cylindrical coordinates:

$$\frac{\partial^2 \phi}{\partial r^2} + \frac{1}{r}\frac{\partial \phi}{\partial r} + \frac{1}{r^2}\frac{\partial^2 \phi}{\partial \theta^2} + k^2[n^2(r,\theta) - n_{\text{eff}}^2]\phi = 0 \quad (1)$$

where $\phi(r, \theta)$ is the field, $k = 2\pi/\lambda$ is the free-space wave number with λ as the wavelength, $n(r,\theta)$ is the refractive index profile, and $n_{\text{eff}} = \beta k$ is the mode index with β as the propagation constant. Without losing generality, the mode field can be expressed as

$$\phi(r, \theta) = \phi_r(r)\phi_{r\theta}(r, \theta) \quad (2)$$

Substituting Eq. (2) into Eq. (1) gives

$$\phi_r \frac{\partial^2 \phi_{r\theta}}{\partial r^2} + \phi_{r\theta}\frac{d^2 \phi_r}{dr^2} + 2\frac{d\phi_r}{dr}\frac{\partial \phi_{r\theta}}{\partial r} + \frac{1}{r}\left(\phi_r \frac{\partial \phi_{r\theta}}{\partial r} + \phi_{r\theta}\frac{d\phi_r}{dr}\right)$$
$$+ \frac{\phi_r}{r^2}\frac{\partial^2 \phi_{r\theta}}{\partial \theta^2} + k^2[n^2(r,\theta) - n_{\text{eff}}^2]\phi_r \phi_{r\theta} = 0 \quad (3)$$

Here we assume that $\phi_{r\theta}$ is a slowly varying function of r compared with ϕ_r, that is, ϕ_r accounts for most of the variation in the r direction, so that the terms associated with $\partial \phi_{r\theta}/\partial r$ and $\partial^2 \phi_{r\theta}/\partial r^2$ are neglected. With this assumption, Eq. (3) is simplified to

$$\phi_{r\theta}\frac{d^2 \phi_r}{dr^2} + \frac{\phi_{r\theta}}{r}\frac{d\phi_r}{dr} + \frac{\phi_r}{r^2}\frac{\partial^2 \phi_{r\theta}}{\partial \theta^2}$$
$$+ k^2[n^2(r,\theta) - n_{\text{eff}}^2]\phi_r \phi_{r\theta} = 0 \quad (4)$$

We then define a radially varying effective index profile $n_{\text{eff}r}(r)$ that satisfies

$$\frac{\partial^2 \phi_{r\theta}}{\partial \theta^2} + k^2[n^2(r,\theta) - n_{\text{eff}r}^2(r)]r^2\phi_{r\theta} = 0 \quad (5)$$

With the above definition, Eq. (4) is reduced to

$$\frac{d^2\phi_r}{dr^2} + \frac{1}{r}\frac{d\phi_r}{dr} + k^2\left[\tilde{n}_{\text{eff}r}^2(r) - \frac{l^2}{k^2 r^2} - n_{\text{eff}}^2\right]\phi_r = 0 \quad (6)$$

where the general effective index profile $\tilde{n}_{\text{eff}r}(r)$ is given by

$$\tilde{n}_{\text{eff}r}^2(r) = n_{\text{eff}r}^2(r) + \frac{l^2}{k^2 r^2}, \quad l = 0, 1, 2, \ldots \quad (7)$$

The REIM thus reduces the two-dimensional wave equation Eq. (1) into a one-dimensional wave equation Eq. (6), which is recognized as the wave equation for the LP$_{lm}$ mode of a circular fiber with a refractive index profile $\tilde{n}_{\text{eff}r}(r)$. In other words, with the REIM, the fiber is approximated by a circular fiber with an effective index profile $\tilde{n}_{\text{eff}r}(r)$. The central task is therefore to calculate the effective index profile $\tilde{n}_{\text{eff}r}(r)$ from Eqs. (5) and (7) for the fiber of concern. Once the effective index profile is known, we can use a numerical method to solve the resultant wave equation, Eq. (6).

2.2. Determination of the Effective Index Profile

In general, the effective index profile $n_{\text{eff}r}(r)$ can be obtained by solving Eq. (5) at each value of r (i.e., at $r = r_i$), namely,

$$\frac{\partial^2 \phi_{r\theta}(r_i, \theta)}{\partial \theta^2} + k^2[n^2(r_i, \theta) - n_{\text{eff}r}^2(r_i)]r_i^2 \phi_{r\theta}(r_i, \theta) = 0 \quad (8)$$

which is a one-dimensional wave equation with $n_{\text{eff}r}(r_i)$ as the eigenvalue. Here we consider two special refractive index profiles that are needed for the analysis of the special fibers considered in the present chapter.

The first profile is one that has no variation in the angular direction, that is, $n(r, \theta) = n(r)$. The effective index profile $n_{\text{eff}r}(r)$ for this profile can be solved explicitly from Eq. (5) as

$$n_{\text{eff}r}^2(r) = n^2(r) - \frac{l^2}{k^2 r^2} \quad (9)$$

which applies to all the LP$_{lm}$ modes.

The second profile is a periodic one as shown in Fig. 5.1, in which $2\theta_1$ and $2\theta_2$ are the angular widths of a period with $n(r,\theta) = n_1$ and $n(r,\theta) = n_2$, respectively ($n_1 > n_2$ are constants). Because $n(r,\theta)$ is a periodic function of θ, $\phi_{r\theta}(r,\theta)$ must satisfy the periodic boundary conditions $\phi_{r\theta}(r,0) = \phi_{r\theta}(r,2\pi)$ and $\frac{\partial \phi_{r\theta}}{\partial \theta}|_{\theta=0} = \frac{\partial \phi_{r\theta}}{\partial \theta}|_{\theta=2\pi}$. In addition, the field must undergo $2l$ zero crossings in $0 \le \theta \le 2\pi$ (e.g., two zero crossings for the LP$_{1m}$ mode). Applying these boundary conditions to the periodic profile, we solve Eq. (5) analytically and obtain the following characteristic equations:

$$\tilde{u} \tan \tilde{u} = \tilde{w} \tanh\left(\tilde{w} \frac{\theta_2}{\theta_1}\right) \qquad (10)$$

for the LP$_{0m}$ mode and

$$\cosh\left(2\tilde{w}\frac{\theta_2}{\theta_1}\right) \cos 2\tilde{u} + \frac{\tilde{w}^2 - \tilde{u}^2}{2\tilde{u}\tilde{w}} \sinh\left(2\tilde{w}\frac{\theta_2}{\theta_1}\right) \sin 2\tilde{u} = \cos\frac{2\pi}{N} \qquad (11)$$

for the LP$_{1m}$ mode, where $\tilde{u} = \theta_1 rk(n_1^2 - n_{\text{effr}}^2)^{1/2}$, $\tilde{w} = \theta_1 rk(n_{\text{effr}}^2 - n_2^2)^{1/2}$, and N is the number of periods. Characteristic equations for the LP$_{lm}$ mode with $l > 1$ could be derived, if necessary.

3. SEGMENTED CLADDING FIBER

The segmented cladding fiber (SCF) was first proposed [20] as an alternative design to the holey fiber for the provision of single-mode operation over an extended range of wavelengths. It was then shown that an SCF could be designed as an ultra-large-core single-mode fiber for optical communications [21], which could suppress effectively nonlinear optical effects because of its large core size and, at the same time, provide a much higher transmission capacity because of its potentially weak birefringence, compared with a holey fiber.

The cross-section of an SCF is shown in Fig. 5.2, where a uniform core region ($0 < r < a$) of refractive index n_1 is surrounded by a segmented cladding region ($a < r < b$). The cladding consists of alter-

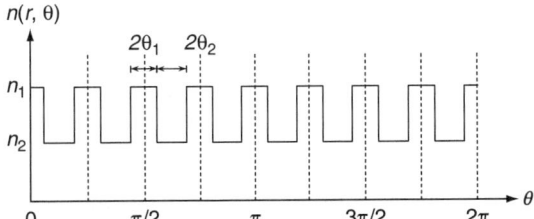

FIGURE 5.1 Periodic refractive index variation $n(r,\theta)$ in the angular direction at a given value of r.

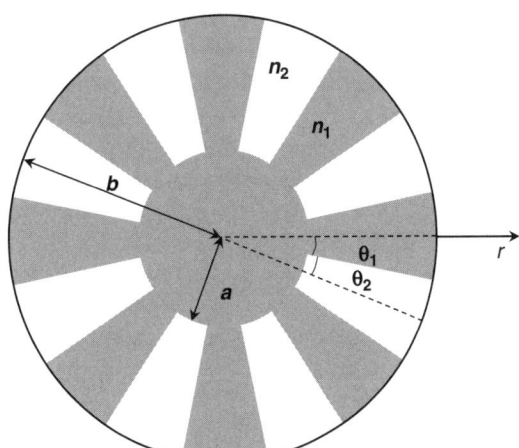

FIGURE 5.2 Cross-section of an SCF with core radius a and cladding radius b. n_1 and n_2 are the refractive indices of the segments, and $2\theta_1$ and $2\theta_2$ are the corresponding angular widths.

nate segments of high-index (n_1) medium of angular width $2\theta_1$ and low-index (n_2) medium of angular width $2\theta_2$, which are arranged periodically in the angular direction. The index difference between the two media is characterized by a relative index height, $\Delta = (n_1^2 - n_2^2)/2n_1^2$, which is assumed to be much smaller than one. The fiber shown in Fig. 5.2 contains eight segments of high and low refractive indices ($N = 8$). The period and the duty cycle of the segmentation are given by $\Lambda = 2\theta_1 + 2\theta_2$ and $\gamma = 2\theta_2/\Lambda$, respectively. The cladding has a radius b. For a bare fiber, the index in the region beyond $r = b$ is 1 (air), whereas for a coated fiber, a suitable surrounding index should be used.

In the light of the REIM, the SCF is replaced by a circular fiber with an effective index profile. In the core region, the effective index profile is simply given by Eq. (9) with $n(r) = n_1$, because the core index of the SCF is a constant. In the cladding region, the refractive index profile of the SCF is a periodic function of θ, which is just the same as that shown in Fig. 5.1. The effective index profile for this region can therefore be computed from Eqs. (10) and (11) for the LP$_{0m}$ and LP$_{1m}$ modes, respectively.

The propagation characteristics of a segmented cladding fiber have been discussed extensively elsewhere [20–23]. Here we use a large-core polymer SCF to demonstrate the application of the REIM. The fiber consists of eight periods of segmentation with $n_1 = 1.49$, $n_2 = 1.41$ ($\Delta = 0.052$), $a = 25\,\mu\text{m}$, $b = 62.5\,\mu\text{m}$, and duty cycle $\gamma = 0.5$. The effective index profile of the fiber, $\tilde{n}_{\text{effr}}(r)$, for the LP$_{01}$ mode is shown in Fig. 5.3 for the wavelength $1.35\,\mu\text{m}$. Clearly, the effective index is nonuniform in the cladding and, in fact, increases monotonically with r.

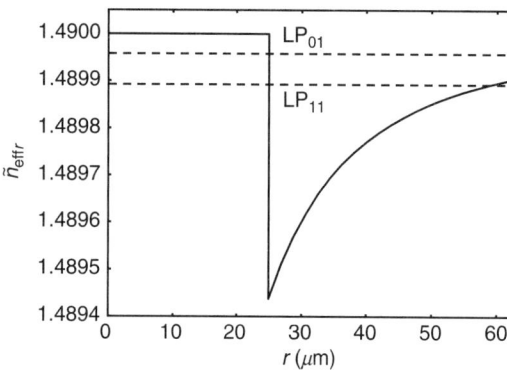

FIGURE 5.3 Effective index profile of the SCF with $n_1 = 1.49, n_2 = 1.41, a = 25\,\mu m, b = 62.5\,\mu m, N = 8$, and $\gamma = 0.5$ at wavelength $1.35\,\mu m$. The dashed horizontal lines mark the mode indices of the first two modes of the fiber.

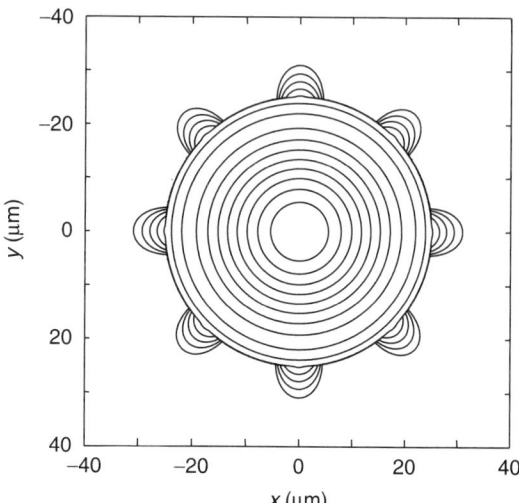

FIGURE 5.4 Intensity distribution of the fundamental mode of the SCF at wavelength $1.35\,\mu m$.

The effective index profile for the LP_{11} mode is almost the same as that for the LP_{01} mode and therefore not shown in the figure. To find the mode indices and the fields of the fiber, we use the well-established transverse matrix method (TMM) [25].

The mode indices calculated for the LP_{01} and LP_{11} modes are also shown in Fig. 5.3. It can be seen from Fig. 5.3 that the LP_{01} mode index is larger than the maximum cladding index, whereas the LP_{11} mode index is smaller than the maximum cladding index, which suggests that only the LP_{01} mode is well guided by the core and all the higher order modes are cladding modes. When all the higher order modes are stripped off by using a suitable index-matching coating, the fiber becomes effectively single moded. In fact, because the cladding index is highly dispersive, the LP_{01} mode can remain as the only core mode even at visible wavelengths [20–22]. Figure 5.4 shows the contour plot of the intensity distribution for the LP_{01} mode at $1.35\,\mu m$. As expected, the mode field is mainly confined in the core region of the fiber with some penetration into the cladding region through the high-index segments.

In practice, the fiber is truncated at a finite cladding radius b and coated with a high-index material, which can be a soft jacket or an additional uniform polymer ring of index n_1 covered with a soft jacket. The coated SCF becomes a leaky structure, and all the modes of the fiber suffer from finite leakage losses [22,23]. It is through the comparison of the leakage losses of the LP_{01} and LP_{11} modes that the single-mode operation of the fiber is established [22,23].

We replace the air region with a medium that has an index equal to the core index and calculate the leakage losses of the first two modes from the effective index profiles by the TMM. The dependence of the leakage losses of the first two modes on the wavelength is shown in Fig. 5.5. The leakage loss of the LP_{01} mode tells how lossy the fiber is, and the differential loss between the two modes shows how effectively the fiber is single moded. It can be seen from the loss curves in Fig. 5.5 that the leakage loss of the LP_{01} mode is nearly two orders of magnitude smaller than that of the LP_{11} mode in the wavelength range from 0.65 to $1.35\,\mu m$. The SCF thus shows single-mode behavior in the entire wavelength range. To be more specific, at $1.31\,\mu m$, the leakage losses of the LP_{01} and LP_{11} modes are 0.0138 and 0.590 dB/m, respectively, which means that a 35-m-long fiber is sufficient to strip off the LP_{11} mode with an extinction ratio of 20 dB. At

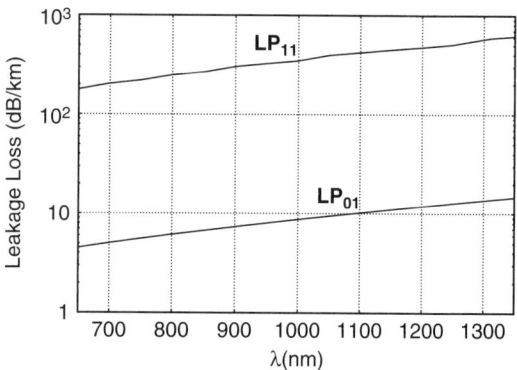

FIGURE 5.5 Dependence of the leakage losses of the fundamental mode and the first higher order mode of the coated SCF on the wavelength, showing an extended single-mode operation from 0.65 to $1.35\,\mu m$.

0.65 μm, the leakage losses of the modes decrease to 0.0045 and 0.178 dB/m, respectively, and the LP$_{11}$ mode can be stripped off effectively with a 115-m long fiber. As a comparison, a conventional step-index fiber with $\Delta = 0.052$ and $a = 25$ μm supports 1670 and 6770 modes at 1.31 and 0.65 μm, respectively. Therefore, an SCF can be designed to function as a single-mode fiber with an ultra-large effective core area [21]. The leakage losses, as well as the differential loss, can be varied over a wide range by changing the fiber parameters. A detailed study of the dependence of the leakage losses on the parameters of an SCF can be found in [22]. A polymer SCF could offer a bandwidth comparable with that of a conventional single-mode fiber and yet reduce significantly the connection cost associated with a conventional single-mode fiber. Recently, some large-core SCF samples have been fabricated with PMMA and a single-mode operation with a four-segment design has been demonstrated experimentally [26].

4. HOLEY FIBER

A holey fiber is characterized by a distribution of air holes in the cladding along the entire length of the fiber [27–29] (see also Chapter 3). Light in such a fiber can be confined by two different mechanisms: the photonic bandgap effect [27] and the average index effect [28, 29]. The photonic bandgap guidance can be achieved when large air holes are arranged in a strict periodic fashion, such as a honeycomb pattern [27]. In the average index model, where the cladding full of air holes is replaced by an average refractive index, light guidance is possible in the absence of a photonic bandgap [28, 29]. It is known that a holey fiber can provide endless single-mode operation, which can be explained by the strong wavelength dependence of the average cladding index [28, 29]. By means of the REIM, we have shown that a holey fiber is a leaky structure, and for the fiber to operate effectively, it is necessary to keep the leakage loss of the fundamental mode low and that of the higher order mode sufficiently high [24].

The cross-section of a holey fiber with a circular distribution of air holes [30, 31] is shown in Fig. 5.6, where rings of equally spaced holes of diameter d are distributed around the central core with a ring separation r_0. The number of holes on the ith ring is i times n, where n is the number of holes on the first ring. The maximum number of holes that can be accommodated on the ith ring depends on the sizes of the holes and the rings and is limited by an integer smaller than π [$2\sin^{-1}(d/4ir_0)$].

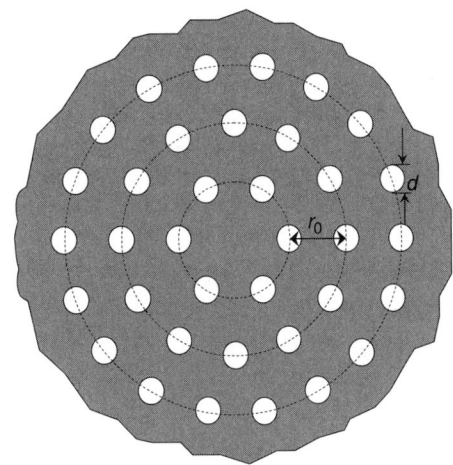

FIGURE 5.6 Cross-section of a holey fiber with a circular distribution of holes, where d is the hole diameter and r_0 is the ring separation.

Again, the REIM replaces the holey fiber by an effective circular fiber. In the core, the effective index profile is given by Eq. (9) with $n(r) = n_1$, where n_1 is the refractive index of the material. In the cladding region, the refractive index of the holey fiber varies periodically along the angular direction across the holes with a period depending on the number of holes and a high-index/low-index angular-width ratio depending on the distance r. Therefore, the effective index profile for the cladding region can be calculated from Eqs. (10) and (11) for the LP$_{0m}$ and LP$_{1m}$ modes, respectively.

As an example, we consider a polymer holey fiber with $n_1 = 1.49$, $r_0 = 10$ μm, $d = 6$ μm, $n = 6$, and three rings of holes. Figure 5.7 shows the effective index profiles $\tilde{n}_{\text{eff}r}(r)$ of the fiber at four different wavelengths, 1550, 1300, 800, and 633 nm. As shown

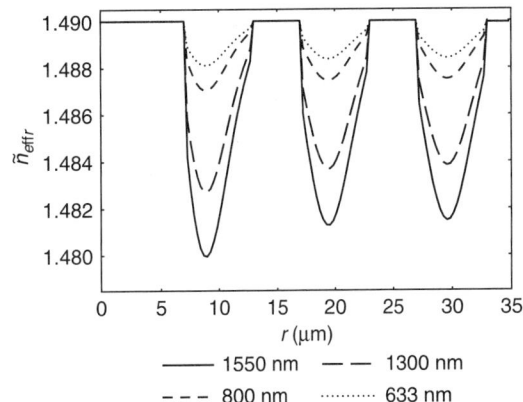

FIGURE 5.7 Effective index profiles of the holey fiber with $n_1 = 1.49$, $d = 6$ μm, $r_0 = 10$ μm, $n = 6$, and three rings of holes at wavelengths 1550, 1300, 800, and 633 nm.

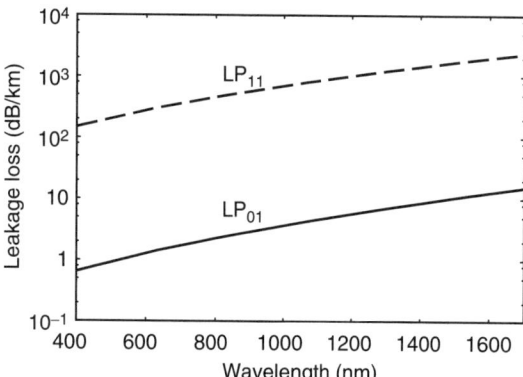

FIGURE 5.8 Dependence of the leakage losses of the first two modes of the holey fiber on the wavelength.

by the results in Fig. 5.7, the effective cladding profile is highly dispersive, that is, it changes significantly with wavelength. The difference between the effective index profiles calculated for the LP_{01} and LP_{11} modes is too small to show up in Fig. 5.7. Because the refractive index beyond the rings of holes is equal to the core index, all the modes suffer from finite leakage losses. As in the case of a coated SCF, it is through the comparison of the leakage losses of the LP_{01} and LP_{11} modes that the single-mode operation of the fiber is established. The dependence of the leakage losses of the LP_{01} and LP_{11} modes of the fiber on wavelength is shown in Fig. 5.8. As shown in Fig. 5.8, the leakage losses of the modes decrease with the wavelength, and the loss of the LP_{11} mode is more than two orders of magnitude higher than that of the LP_{01} mode in the wavelength range 400–1700 nm. Because of its high leakage loss, the LP_{11} mode can be stripped off quickly, and the fiber remains effectively single moded in the entire wavelength range. At wavelength 1550 nm, the present fiber incurs a leakage loss of 12 dB/km to the fundamental mode and 1700 dB/km to the LP_{11} mode, which means that a 12-m-long fiber is sufficient to strip off the LP_{11} mode with an extinction ratio of 20 dB. At the wavelength 400 nm, the leakage losses of the LP_{01} and LP_{11} modes drop to 0.65 and 130 dB/km respectively, and the LP_{11} mode can be stripped off effectively with a 155-m-long fiber.

The leakage losses of the modes depend critically on the number of rings of holes. As shown in Fig. 5.9, the leakage loss decreases rapidly with an increase in the number of rings; on the other hand, the differential loss between the first two modes increases with the number of rings. In our example with only four rings, the leakage loss of the LP_{01} mode can be kept well below 1 dB/km over the wavelength range 400–1700 nm, which is lower than the loss of a typical polymer fiber by two to three orders of magnitude. A detailed study of the dependence of the leakage losses on the hole size and the hole spacing is given in [24].

5. CONCLUSION

The REIM for the analysis of optical fibers with arbitrary shapes and refractive index profiles has been described and demonstrated with applications to a segmented cladding fiber and a holey fiber. A comparison between the results for the segmented cladding fiber and the holey fiber shows that the two fibers, regardless of their very different structures, operate on the same physical principle. Both fibers are leaky structures, and their highly dispersive claddings give rise to effective single-mode operation over a wide range of wavelengths. The REIM provides not only an efficient means for calculating the characteristics of these fibers but also an intuitive understanding of these characteristics. It is expected that the REIM can be applied to other types of microstructured fibers.

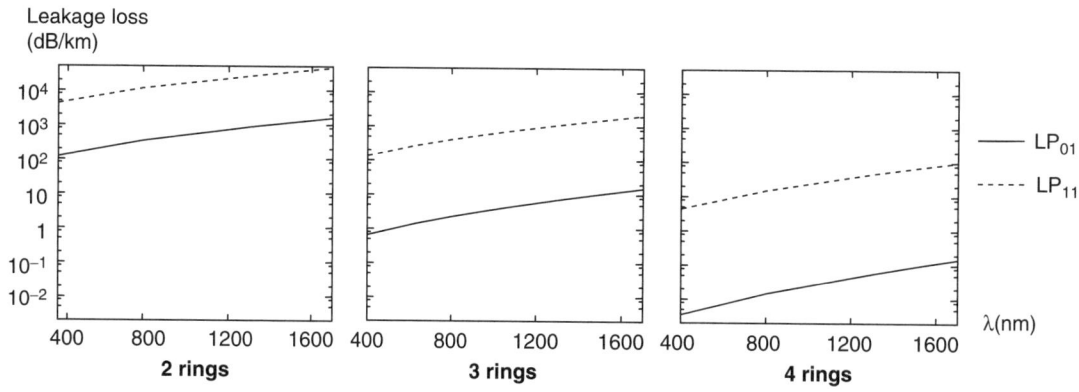

FIGURE 5.9 Dependence of the leakage losses of the first two modes of the holey fiber on the number of rings of holes.

6. ACKNOWLEDGMENT

This work was supported by a grant from the Research Grants Council of the Hong Kong Special Administrative Region, China (Project No. CityU 1034/02E).

7. REFERENCES

1. K. S. Chiang, "Review of numerical and approximate methods for the modal analysis of general optical dielectric waveguides," *Opt. Quantum Electron.*, **26**, S113–S134 (1994).
2. K. S. Chiang, "Radial effective-index method for the analysis of optical fibers," *Appl. Opt.*, **26**, 2969–2973 (1987).
3. R. M. Knox and P. P. Toulios, "Integrated circuits for the millimeter through optical frequency range," in *Proceedings, Symposium on Submillimeter Waves*, Polytechnic Press, Brooklyn, 497–516 (1970).
4. E. A. J. Marcatili, "Dielectric rectangular waveguide and directional coupler for integrated optics," *Bell Syst. Tech. J.*, **48**, 2071–2102 (1969).
5. W. V. McLevige, T. Itoh, and R. Mittra, "New waveguide structures for millimeter-wave and optical integrated circuits," *IEEE Trans. Microwave Theory Tech.*, **MTT-23**, 788–794 (1975).
6. A. A. Oliner, S. T. Peng, T. I. Hsu, and Sanchez, "Guidance and leakage properties of a class of open dielectric waveguides: part II—new physical effects," *IEEE Trans. Microwave Theory Tech.*, **MTT-29**, 855–870 (1981).
7. K. S. Chiang, "Effects of cores in fused tapered single-mode fiber couplers," *Opt. Lett.*, **12**, 431–433 (1987)
8. K. S. Chiang, "Effective-index method for the analysis of optical waveguide couplers and arrays: an asymptotic theory," *J. Lightwave Technol.*, **9**, 62–72 (1991).
9. K. S. Chiang, "Analysis of optical fibers by the effective-index method," *Appl. Opt.*, **25**, 348–354 (1986).
10. K. S. Chiang, "Geometric birefringence in a class of step-index fiber," *J. Lightwave Technol.*, **LT-5**, 737–744 (1987).
11. K. S. Chiang, "Analysis of fused couplers by the effective-index method," *Electron. Lett.*, **22**, 1221–1222 (1986).
12. K. S. Chiang, "Stress-induced birefringence fibers designed for single-polarization single-mode operation," *J. Lightwave Technol.*, **7**, 436–441 (1989).
13. K. Van de Velde, H. Thienpont, and R. Van Geen, "Extending the effective index method for arbitrarily shaped inhomogeneous optical waveguides," *J. Lightwave Technol.*, **6**, 1153–1159 (1988).
14. A. Kumar, D. F. Clark, and B. Culshaw, "Explanation of errors inherent in the effective-index method for analyzing rectangular-core waveguides," *Opt. Lett.*, **13**, 1129–1131 (1988).
15. K. S. Chiang, "Performance of the effective-index method for the analysis of dielectric waveguides," *Opt. Lett.*, **16**, 714–716 (1991).
16. K. S. Chiang, "Analysis of the effective-index method for vector modes of rectangular-core dielectric waveguides," *IEEE Trans. Microwave Theory Tech.*, **44**, 692–700 (1996).
17. K. S. Chiang, K. M. Lo, and K. S. Kwok, "Effective-index method with built-in perturbation correction for integrated optical waveguides," *J. Lightwave Technol.*, **14**, 223–228 (1996).
18. K. S. Chiang, C. H. Kwan, and K. M. Lo, "Effective-index method with built-in perturbation correction for the vector modes of rectangular-core optical waveguides," *J. Lightwave Technol.*, **17**, 716–722 (1999).
19. C. H. Kwan and K. S. Chiang, "Study of polarization-dependent coupling in optical waveguide directional couplers by the effective-index method with built-in perturbation correction," *J. Lightwave Technol.*, **20**, 1018–1026 (2002).
20. V. Rastogi and K. S. Chiang, "Propagation characteristics of a segmented cladding fiber," *Opt. Lett.*, **26**, 491–493 (2001).
21. K. S. Chiang and V. Rastogi, "Ultra-large-core single-mode fiber for optical communications: the segmented cladding fiber," in *Optical Fiber Communication Conference—OFC 2002*, paper ThGG6, 620–621 (2002).
22. V. Rastogi and K. S. Chiang, "Analysis of the segmented cladding fiber by the radial effective-index method," *J. Opt. Soc. Amer. B*, **21**, 258–265 (2004).
23. V. Rastogi and K. S. Chiang, "Leakage losses in segmented cladding fibers," in *Optical Fiber Communication Conference—OFC 2003*, paper FI4, 697–699 (2003).
24. V. Rastogi and K. S. Chiang, "Holey optical fiber with circularly distributed holes analyzed by the radial effective-index method," *Opt. Lett.*, **28**, 249–251 (2004).
25. K. Thyagarajan, S. Diggavi, A. Taneja, and A. K. Ghatak, "Simple numerical technique for the analysis of cylindrically symmetric refractive-index profile optical fibers," *Appl. Opt.*, **30**, 3877–3879 (1991).
26. A. Yeung, K. S. Chiang, V. Rastogi, P. L. Chu, and G. D. Peng, "Experimental demonstration of single-mode operation of large-core segmented cladding fiber" in *Optical Fiber Communication Conference—OFC 2004*, Paper ThI4 (2004).
27. J. C. Knight, J. Broeng, T. A. Birks, and P. St. J. Russell, "Photonic band gap guidance in optical fibers," *Science*, **282**, 1476–1478 (1998).
28. T. A. Birks, J. C. Knight, and P. St. J. Russell, "Endlessly single-mode photonic crystal fiber," *Opt. Lett.*, **22**, 961–963 (1997).

29. T. M. Monro, P. J. Bennett, N. G. R. Broderick, and D. J. Richardson, "Holey fibers with random cladding distributions," *Opt. Lett.*, **25**, 206–208 (2000).
30. J. Xu, J. Song, C. Li, and K. Ueda, "Cylindrically symmetrical hollow fiber," *Opt. Commun*, **182**, 343 (2000).
31. A. Argyros, I. M. Bassett, M. A. Van Eijkelenborg, M. C. J. Large, J. Zagari, N. A. P. Nicorovici, R. C. McPhedran, and C. M. de Sterke, "Ring structures in microstructured polymer optical fibers," *Opt. Express*, **9**, 813 (2001).

6
CHAPTER

Some Important Nonlinear Effects in Optical Fibers

K. Thyagarajan* and Ajoy Ghatak[†]

Physics Department
Indian Institute of Technology Delhi
New Delhi, India

1. INTRODUCTION

The development of low loss optical fibers, compact and efficient semiconductor lasers operating at room temperatures, optical detectors, and optical amplifiers has truly revolutionized the field of telecommunications. When information-carrying light pulses propagate through an optical fiber, they suffer from attenuation, temporal broadening, and even interact with each other through nonlinear effects in the fiber. These effects tend to distort the signals, resulting in loss of information or in cross-talk among different channels. Increased channel capacity is obtained either by increasing the bit rate of transmission or by using the technique of wavelength division multiplexing (WDM) wherein multiple wavelengths carrying independent channels are multiplexed and propagate simultaneously through the same fiber. Economics requires increased spacing between optical repeaters in the link, which in turn requires higher launched optical powers to achieve the required signal-to-noise ratio. With the increased launched optical powers, bit rates, and the number of wavelength channels, the total optical power propagating through the optical fiber increases and leads to nonlinear optical effects. Nonlinear effects become all the more important because of long propagation distances and small core areas of optical fibers, leading to very high intensities. These nonlinear effects include self-phase modulation (SPM), cross-phase modulation (XPM), four-wave mixing (FWM), stimulated Brillouin scattering, and stimulated Raman scattering (SRS) (see, e.g., [1]). Although most of these nonlinear effects lead to signal degradation and signal cross-talk, some of them can be used to an advantage, such as in the formation of dispersionless pulses, namely solitons, with the help of SPM: in the realization of low noise optical amplifiers using SRS, in all optical signal processing using XPM, or indeed in the realization of wavelength converters using FWM. In this chapter, we discuss mainly SPM, XPM, and FWM, which affect pulse propagation through optical fibers.

2. NONLINEAR POLARIZATION

In a linear medium, the electric polarization P is assumed to be a linear function of the electric field E:

$$P = \varepsilon_0 \chi E \quad (1)$$

where for simplicity a scalar relation has been written. The quantity χ is termed as linear dielectric susceptibility. At high optical intensities (which correspond to high electric fields), all media behave in a nonlinear fashion. Thus, Eq. (1) is modified to

$$P = \varepsilon_0(\chi E + \chi^{(2)} E^2 + \chi^{(3)} E^3 + \cdots) \quad (2)$$

where $\chi^{(2)}, \chi^{(3)}, \ldots$ are higher order susceptibilities giving rise to the nonlinear terms. The second term on the right-hand side is responsible for second harmonic generation, sum and difference frequency generation, and parametric interactions whereas the third term is responsible for third harmonic generation, intensity dependent refractive index, SPM, and FWM. For media possessing an inversion symmetry, $\chi^{(2)}$ is zero, and there is no second-order nonlinear effect. Thus, silica optical fibers, which form the heart of today's communication networks, do not posses second-order nonlinearity.

3. THIRD-ORDER NONLINEAR EFFECTS

SPM, XPM, and FWM represent some of the very important consequences of third-order nonlinearity. These effects have become all the more important as they play a significant and important role in WDM optical fiber communication systems.

---

*E-mail: Ktrajan@physics.iitd.ac.in
[†]E-mail: ajoyghatak@yahoo.com

3.1. SPM

Consider the propagation of a plane light wave at frequency ω through a medium having $\chi^{(3)}$ nonlinearity. The polarization generated in the medium is given by

$$P = \varepsilon_0 \chi E + \varepsilon_0 \chi^{(3)} E^3 \tag{3}$$

If we consider light at frequency ω propagating along the z-direction with propagation constant k, then its electric field is given by

$$E = E_0 \cos(\omega t - kz) \tag{4}$$

then

$$P = \varepsilon_0 \chi E_0 \cos(\omega t - kz) + \varepsilon_0 \chi^{(3)} E_0^3 \cos^3(\omega t - kz) \tag{5}$$

Expanding $\cos^3 \theta$ in terms of $\cos \theta$ and $\cos 3\theta$, we obtain the following expression for the polarization at frequency ω:

$$P = \varepsilon_0 \left(\chi + \frac{3}{4} \chi^{(3)} E_0^2 \right) E_0 \cos(\omega t - kz) \tag{6}$$

For a plane wave given by Eq. (4), the intensity is

$$I = \frac{1}{2} c \varepsilon_0 n_0 E_0^2 \tag{7}$$

where n_0 is the refractive index of the medium at low intensity. Then

$$P = \varepsilon_0 \left(\chi + \frac{3}{2} \frac{\chi^{(3)}}{c \varepsilon_0 n_0} I \right) E \tag{8}$$

The polarization P and electric field are related through the following equation

$$P = \varepsilon_0 (n^2 - 1) E \tag{9}$$

where n is the refractive index of the medium. Comparing Eqs. (8) and (9), we get

$$n^2 = n_0^2 + \frac{3}{2} \frac{\chi^{(3)}}{c \varepsilon_0 n_0} I \tag{10}$$

where

$$n_0^2 = 1 + \chi \tag{11}$$

Since the last term in Eq. (10) is usually very small, we get

$$n \approx n_0 + n_2 I \tag{12}$$

where

$$n_2 = \frac{3}{4} \frac{\chi^{(3)}}{c \varepsilon_0 n_0^2} \tag{13}$$

is the nonlinear coefficient. Thus, Eq. (12) shows that the refractive index of the medium becomes intensity dependent and the coefficient n_2 represents the strength of the intensity dependence.

For fused silica $n_0 \approx 1.47$, $n_2 \approx 3.2 \times 10^{-20} \, \text{m}^2/\text{W}$, and if we consider the power of 100 mW having a cross-sectional area of 100 μm², the resultant intensity is $10^9 \, \text{W/m}^2$ and the corresponding change in refractive index is

$$\Delta n = n_2 I = 3.2 \times 10^{-11}$$

Although this change in refractive index is extremely small, when the beam propagates over an optical fiber over long distances (a few hundred to a few thousand kilometers), the accumulated nonlinear effects can be significant.

In the case of an optical fiber, the light beam propagates as a mode having a specific transverse electric field distribution, and thus the intensity is not constant across the cross section. In such a case, it is convenient to express the nonlinear effect in terms of the power carried by the mode (rather than in terms of intensity). If the linear propagation constant of the mode is represented by β, then in the presence of nonlinearity the effective propagation constant is given by

$$\beta_{\text{NL}} = \beta + \frac{k_0 n_2}{A_{\text{eff}}} P \tag{14}$$

where $k_0 = 2\pi/\lambda_0$, and P is the power carried by the mode. The quantity A_{eff} represents the effective transverse cross-sectional area of the mode and is defined by

$$A_{\text{eff}} = \frac{\left[\int_0^\infty \int_0^{2\pi} \psi^2(r) r \, dr \, d\Phi \right]^2}{\int_0^\infty \int_0^{2\pi} \psi^4(r) r \, dr \, d\Phi} \tag{15}$$

where $\psi(r)$ represents the transverse mode field distribution of the mode. For example, under the Gaussian approximation (see, e.g., [2])

$$\psi(r) = \psi_0 e^{-r^2/w_0^2} \tag{16}$$

where ψ_0 is a constant and $2w_0$ represents the mode field diameter, we get

$$A_{\text{eff}} = \pi w_0^2 \tag{17}$$

It is usual to describe the nonlinear characteristic of an optical fiber by the coefficient γ given by

$$\gamma = \frac{k_0 n_2}{A_{\text{eff}}} \tag{18}$$

Thus, for the same input power and same wavelength, smaller values of A_{eff} lead to greater nonlinear effects in the fiber. Typically

$A_{\text{eff}} \approx 50\text{--}80\,\mu\text{m}^2$ and $\gamma \approx 2.4\text{--}1.5\,\text{W}^{-1}\text{km}^{-1}$

Values of the effective areas of different fiber types are given in Table 6.1. As seen from Table 6.1 dispersion compensating fibers have a very small effective area and thus lead to strong nonlinear effects.

When a light beam propagates through an optical fiber, the power decreases because of attenuation. Thus, the corresponding nonlinear effects are also reduced. Indeed, the phase change suffered by a beam in propagating from 0 to L is given by

$$\Phi = \int_0^L \beta_{\text{NL}}\, dz = \beta L + \gamma \int_0^L P\, dz \quad (19)$$

If α represents the attenuation coefficient, then

$$P(z) = P_0 e^{-\alpha z} \quad (20)$$

and we get

$$\Phi = \beta L + \gamma P_0 L_{\text{eff}} \quad (21)$$

where

$$L_{\text{eff}} = \frac{1 - e^{-\alpha L}}{\alpha} \quad (22)$$

is referred to as the effective length. For $\alpha L \gg 1$, $L_{\text{eff}} \approx 1/\alpha$, and for $\alpha L \ll 1$, $L_{\text{eff}} \approx L$.

The effective length represents the length of the fiber over which most of the nonlinear effects has accumulated. For a loss coefficient of 0.20 dB/km, $1/\alpha \approx 21$ km.

If we consider a fiber length much longer than L_{eff}, then to have a reduced impact of SPM, we must have

$$\gamma P_0 L_{\text{eff}} \ll 1$$

or

$$P_0 \ll \frac{1}{\gamma L_{\text{eff}}} \approx \frac{\alpha}{\gamma}$$

For $\alpha = 4.6 \times 10^{-2}\,\text{km}^{-1}$ (which corresponds to an attenuation of 0.2 dB/km) and $\gamma = 2.4\,\text{W}^{-1}\text{km}^{-1}$, we get

$$P_0 \ll 19\,\text{mW}$$

TABLE 6.1 A_{eff} at 1550 nm for different fiber types.

Fiber type	Effective area (μm^2)
Single-mode fiber (SMF) G652	≈ 85
Dispersion shifted fiber (DSF)	≈ 46
Nonzero dispersion shifted fiber (NZ-DSF)	$\approx 52(D > 0), 56(D < 0),$ and 73
Dispersion compensating fiber (DCF)	$\approx 23(D < 0)$

3.2. Propagation of a Pulse

When an optical pulse propagates through a medium, it suffers from attenuation, dispersion, and nonlinearity. Attenuation refers to the reduction in the pulse energy due to various mechanisms, such as scattering and absorption. Dispersion is caused by the fact that a light pulse consists of various frequency components and each frequency component travels at a different group velocity. Dispersion causes the temporal width of the pulse to change. In most cases it results in an increase in pulse width; however, in some cases the temporal width could also decrease. Dispersion is accompanied by chirping, the variation of the instantaneous frequency of the pulse within the pulse duration (see Chapter 1). Because both attenuation and dispersion cause a change in the temporal variation of the optical power, they closely interact with nonlinearity in deciding the pulse evolution as it propagates through the medium.

Let $E(x, y, z, t)$ represent the electric field variation of an optical pulse. It is usual to express E in the following way:

$$E(x, y, z, t) = \frac{1}{2}\left[A(z, t)\psi(x, y)e^{i(\omega_0 t - \beta_0 z)} + cc\right] \quad (23)$$

where $A(z, t)$ represents the slowly varying complex envelope of the pulse, $\psi(x, y)$ represents the transverse electric field distribution of the mode, ω_0 represents center frequency, and β_0 represents the propagation constant at ω_0.

In the presence of attenuation, second-order dispersion, and third-order nonlinearity, the complex envelope $A(z, t)$ can be shown to satisfy the following equation (see, e.g., [3]):

$$\frac{\partial A}{\partial z} = -\frac{\alpha}{2}A - \beta_1 \frac{\partial A}{\partial z} + i\frac{\beta_2}{2}\frac{\partial^2 A}{\partial t^2} - i\gamma|A|^2 A \quad (24)$$

Here

$$\beta_1 = \frac{d\beta}{d\omega}\bigg|_{\omega=\omega_0} = \frac{1}{v_g} \quad (25)$$

represents the inverse of the group velocity of the pulse, and

$$\beta_2 = \frac{d^2\beta}{d\omega^2}\bigg|_{\omega=\omega_0} = -\frac{\lambda_0^2}{2\pi c}D \quad (26)$$

where D represents the group velocity dispersion (measured in ps/nm·km).

The various terms on the right-hand side of Eq. (24) represent the following:

I term : attenuation
II term : group velocity term

III term : second-order dispersion
IV term : third-order nonlinearity

If we change to a moving frame defined by coordinates $T = t - \beta_1 z$, Eq. (24) becomes

$$\frac{\partial A}{\partial z} = -\frac{\alpha}{2} A + i \frac{\beta_2}{2} \frac{\partial^2 A}{\partial T^2} - i\gamma |A|^2 A \quad (27)$$

If we neglect the attenuation term, we obtain the following equation, which is also referred to as the nonlinear Schrödinger equation:

$$\frac{\partial A}{\partial z} = i \frac{\beta_2}{2} \frac{\partial^2 A}{\partial T^2} - i\gamma |A|^2 A \quad (28)$$

The above equation has a solution given by

$$A(z, t) = A_0 \,\text{sech}\, (\sigma T) e^{-igz} \quad (29)$$

with

$$A_0^2 = -\frac{\beta_2}{\gamma} \sigma^2, \quad g = -\frac{\sigma^2}{2} \beta_2 \quad (30)$$

In terms of pulse peak power $P_0 = A_0^2$, the soliton pulse can be written as

$$A(z, t) = \sqrt{P_0} \,\text{sech}\left(\sqrt{\frac{P_0 \gamma}{|\beta_2|}}\, T\right) e^{-i\gamma P_0 z/2}$$

Equation (29) represents an envelope soliton and has the property that it propagates undispersed through the medium. The full width at half-maximum of the pulse envelope is given by $\tau_f = 2\tau_0$ where

$$\text{sech}^2 \sigma \tau_0 = \frac{1}{2}$$

which gives the FWHM τ_f

$$\tau_f = 2\tau_0 = \frac{2}{\sigma} \ln(1 + \sqrt{2}) \approx \frac{1.7627}{\sigma} \quad (31)$$

The peak power of the pulse is:

$$P_0 = |A_0|^2 = \frac{|\beta_2|}{\gamma} \sigma^2 \quad (32)$$

Replacing σ by τ_f, we obtain

$$P_0 \tau_f^2 \approx \frac{\lambda_0^2}{2\pi c} D \quad (33)$$

where we used Eq. (26). The above equation gives the required peak power for a given τ_f for the formation of a soliton pulse. A heuristic derivation of the required power for the soliton formation can be found in [2]. As an example, we have $\tau_f = 10$ ps, $\gamma = 2.4\, \text{W}^{-1}\text{km}^{-1}$, $\lambda_0 = 1.55\, \mu\text{m}$, $D = 2$ ps/nm·km, and the required peak power is $P_0 = 33$ mW.

Soliton pulses are being extensively studied for application to ultra-long distance optical fiber communication. In actual systems, the pulses have to be optically amplified at regular intervals to compensate for the loss suffered by the pulses. The amplification could be carried out using erbium-doped fiber amplifiers *FI* or fiber Raman amplifiers. Figure 6.1 shows results of experiments on dispersionless soliton propagation at a bit rate of 10 Gb/s over several tens of millions of kilometers of a single-mode fiber.

3.3. Spectral Broadening due to SPM

In the presence of only nonlinearity, Eq. (27) becomes

$$\frac{dA}{dz} = -i\gamma |A|^2 A \quad (34)$$

whose solution is given by

$$A(z, t) = A(z = 0, t) e^{-i\gamma P z} \quad (35)$$

where $P = |A|^2$ is the power in the pulse. If P is a function of time, then the time-dependent phase term at $z = L$ becomes

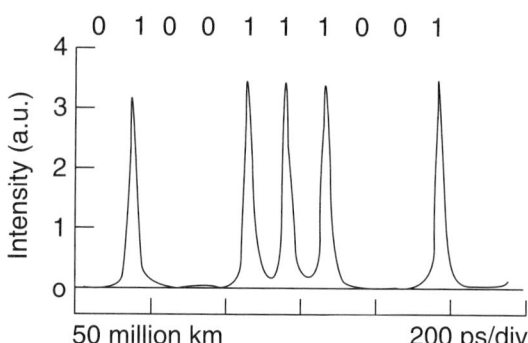

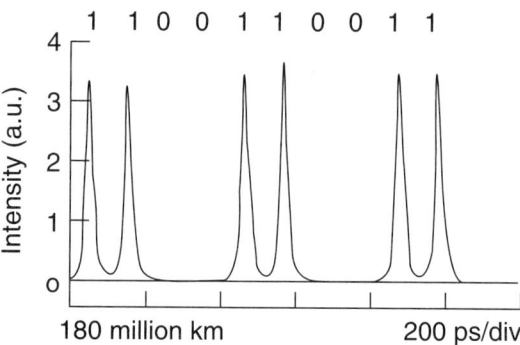

FIGURE 6.1 Distortionless propagation of soliton pulses at a bit rate of 10 Gb/s through 50 million and 180 million kilometers of single-mode fiber in a recirculating loop experiment. (After [4]; 2005 IEEE)

$$e^{i\Phi(t)} = e^{i[\omega_0 t - \gamma P(t)L]} \quad (36)$$

We can define an instantaneous frequency as

$$\omega(t) = \frac{d\Phi}{dt} = \omega_0 - \gamma L \frac{dP}{dt} \quad (37)$$

For a Gaussian pulse

$$P = P_0 e^{-2T^2/\tau_0^2} \quad (38)$$

giving

$$\omega(t) = \omega_0 + \frac{4\gamma L T P_0 e^{-2T^2}}{\tau_0^2} \quad (39)$$

Thus, the instantaneous frequency within the pulse changes with time, leading to chirping of the pulse (Fig. 6.2). Note that because the pulse width has not changed but the pulse is chirped, according to Fourier transform theory the frequency spectrum of the pulse

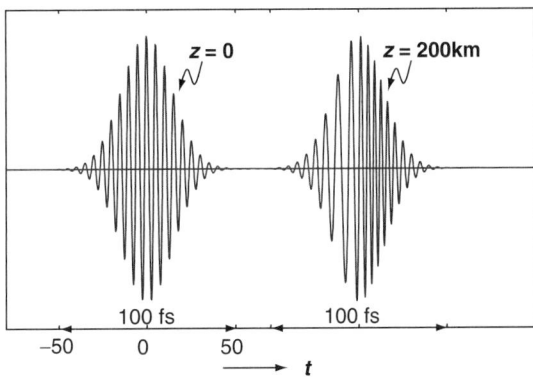

FIGURE 6.2 Due to self phase modulation, the instantaneous frequency within the pulse changes with time, leading to chirping of the pulse. Calculations correspond to $P_0 = 15\,\text{mW}$, $\lambda_0 = 1550\,\text{nm}$, $\tau_0 = 20\,\text{fs}$, $A_{\text{eff}} = 50\,\mu\text{m}^2$, and $\nu_g = 2 \times 10^8\,\text{m/s}$.

has increased. Thus, SPM leads to the generation of new frequencies. By Fourier transform theory, an increased spectral width implies that the pulse can now be compressed in the temporal domain by passing it through a medium with the proper sign of dispersion. This is indeed one of the standard techniques to realize ultrashort femtosecond optical pulses. Figure 6.3 shows results of simulation of the evolution of the pulse over a distance of 1000 km (neglecting attenuation) in the time and spectral domain due only to SPM showing clearly no change in temporal envelope but broadening in the spectral domain.

In the presence of dispersion and nonlinearity, the pulse evolution becomes more complex. Depending on the sign of the dispersion coefficient, the effective pulse broadening could increase or decrease in the presence of nonlinearity. Figure 6.4 shows simulation results of the broadening of the pulse as a function of distance in the normal ($D < 0$) and anamolous ($D > 0$) group-velocity dispersion regimes. In the normal group-velocity dispersion region, the pulse suffers additional dispersion due to SPM, whereas in the anamolous group-velocity dispersion region the dispersion is reduced. Thus, the amount of dispersion compensation of each span of a fiber optic link has to be appropriately chosen to take account of the nonlinear effects in the propagation.

3.4. XPM

If we consider light beams at two different frequencies propagating through an optical fiber, then the change in refractive index brought about by each of the frequencies will affect the propagation of the other frequency. This effect is termed cross-phase modulation (XPM). If the signals at both

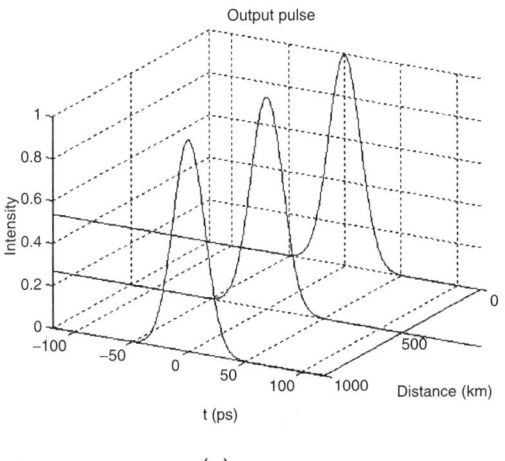

(a)

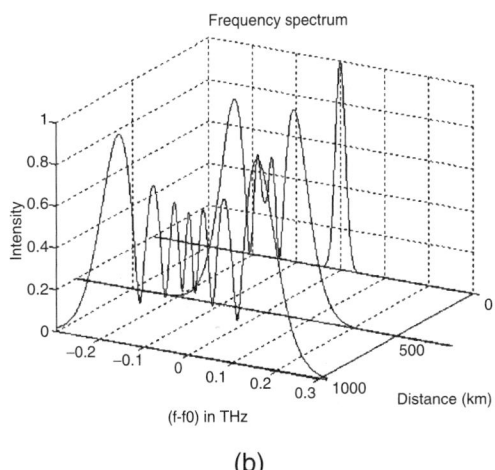

(b)

FIGURE 6.3 Temporal and spectral evolutions of an optical pulse in the presence of SPM only. (After [5].)

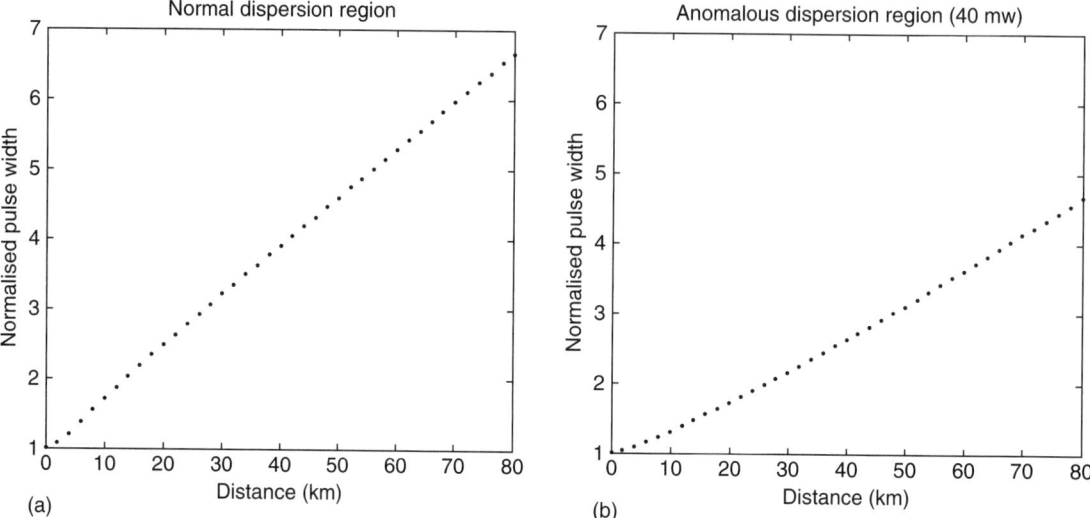

FIGURE 6.4 Variation of normalized pulse width (ratio of output pulse width to input pulse width) in the normal and anamolous group-velocity dispersion regimes in the presence of SPM. In the anamolous group-velocity dispersion region, the nonlinear effects result in reduced dispersion as compared with the case of normal dispersion. (After [6].)

frequencies are pulses, then because of difference in group velocities of the pulses, there is a walk-off between the two pulses, that is, if they start together, they separate as they propagate through the medium. Nonlinear interaction takes place as long as they physically overlap in the medium. The smaller the dispersion, the smaller the difference in group velocities (assuming closely spaced wavelengths) and the longer they will overlap. This would lead to stronger XPM effects. At the same time, if two pulses pass through each other, then because one pulse will interact with both the leading and the trailing edge of the other pulse, XPM effects will be nil provided there is no attenuation. In the presence of attenuation in the medium, the pulse will still get modified due to XPM.

To study XPM, we assume simultaneous propagation of two waves at two different frequencies through the medium. If ω_1 and ω_2 represent the two frequencies, then the variation of the amplitude A_1 of the frequency ω_1 can be obtained as

$$\frac{dA_1}{dz} = -i\gamma(\tilde{P}_1 + 2\tilde{P}_2)A_1 \quad (40)$$

where $\tilde{P}_1$ and $\tilde{P}_2$ represent the powers at frequencies ω_1 and ω_2, respectively. The first term in Eq. (40) represent SPM, whereas the second term corresponds to XPM. If the powers are assumed to attenuate at the same rate, that is,

$$\tilde{P}_1 = P_1 e^{-\alpha z}, \quad \tilde{P}_2 = P_2 e^{-\alpha z} \quad (41)$$

then the solution of Eq. (40) is

$$A_1(L) = A_1(0)e^{-i\gamma(P_1 + 2P_2)L_{\text{eff}}} \quad (42)$$

where, as before, L_{eff} represents the effective length of the medium. When we are studying the effect of power at ω_2 on the light beam at frequency ω_1, we refer to the wave at frequency ω_2 as the pump and the wave at frequency ω_1 as the probe or signal. From Eq. (42) it is apparent that the phase of signal at frequency ω_1 is modified by the power at another frequency. This is referred to as XPM. Note also that XPM is twice as effective as SPM.

Similar to the case of SPM, we can now write for the instantaneous frequency in the presence of XPM as [cf. Eq. (37)]

$$\omega(t) = \omega_0 - 2\gamma L_{\text{eff}} \frac{dP_2}{dt} \quad (43)$$

Hence, the part of the signal that is influenced by the leading edge of the pump is down-shifted in frequency (since in the leading edge $dP_2/dt > 0$) and the part overlapping the trailing edge is up-shifted in frequency (since $dP_2/dt < 0$). This leads to a frequency chirping of the signal pulse just as in the case of SPM.

If the probe and pump beams are pulses, then XPM can lead to induced frequency shifts depending on whether the probe pulse interacts only with the leading edge or trailing edge or both as the bulses propagate through the medium. Let us consider a case when the group velocity of pump pulse is greater than that of the probe pulse. Thus, if both pulses enter the medium together, then because the pump pulse travels faster, the probe pulse interacts only with the trailing edge of the pump. Because in this case dP_2/dt is negative, the probe pulse suffers a blue-induced frequency shift. Similarly, if the pulses enter at different instants but completely overlap at the end of the medium, then $dP_2/dt > 0$ and the probe pulse suffers a red-induced frequency shift. Indeed, if the two pulses start separately and walk through each other, then there is no induced shift due to cancellation of shifts induced by leading and trailing edges of the pump (neglecting attenuation). Figure 6.5 shows a simulation of the pump-induced frequency shift of the probe pulse (see [7]).

We can define a parameter termed *walk-off* length L_{wo} that is the length of the fiber required for the interacting pulses to walk off relative to each other. The walk-off length is given by

$$L_{wo} = \frac{\Delta \tau}{D \Delta \lambda} \quad (44)$$

where D represents the dispersion coefficient and $\Delta \lambda$ represents the wavelength separation between the interacting pulses. For return to zero pulses, $\Delta \tau$ represents the pulse duration, whereas for non-return to zero pulses, $\Delta \tau$ represents the rise term or fall time of the pulse. Closely spaced channels thus interact over longer fiber lengths, leading to greater XPM effects. Larger dispersion coefficients reduce L_{wo} and thus the effects of XPM. Because the medium is attenuating, the power carried by the pulses decreases as they propagate, thus leading to a reduced XPM effect. The characteristic length for attenuation is the effective length L_{eff} defined by Eq. (22). If $L_{wo} \ll L_{eff}$, then over the length of interaction of the pulses, the intensity levels do not change appreciably and the magnitude of the XPM-induced effects will be proportional to the wavelength spacing $\Delta \lambda$. For small $\Delta \lambda$, $L_{wo} \gg L_{eff}$ and the interaction length is now determined by the fiber losses (rather than by walk-off) and the XPM-induced effects become almost independent of $\Delta \lambda$. Indeed, if we consider XPM effects between a continuous-wave probe beam and a sinusoidally intensity modulated pump beam, then the amplitude of the XPM induced phase shift ($\Delta \Phi_p$) in the probe beam is given by [8]

$$\begin{aligned} \Delta \Phi_p &\approx 2\gamma P_{2m} L_{eff} \quad \text{for } L_{wo} \gg L_{eff} \\ \Delta \Phi_p &\approx 2\gamma P_{2m} L_{wo} \quad \text{for } L_{wo} \ll L_{eff} \end{aligned} \quad (45)$$

Here P_{2m} is the amplitude of the sinusoidal power modulation of the pump beam. XPM-induced intensity interference can be studied by simultaneously propagating an intensity modulated pump signal and a continuous-wave probe signal at a different wavelength. The intensity modulated signal induces phase modulation on the continuous-wave probe signal and the dispersion of the medium converts the phase modulation to intensity modulation of the probe. Thus, the magnitude of the intensity fluctuation of the probe signal serves as an estimate of the XPM-induced interference. Figure 6.6 shows the variation of the root mean square (RMS) value of probe intensity modulation with the wavelength separation between the intensity modulated signal and the probe. The experiment was performed over four amplified spans of 80 km of a standard SMF and a nonzero dispersion shifted fiber. The large dispersion in SMF has been compensated for using dispersion compensating chirped gratings. The probe modulation in the case of SMF decreases approximately linearly with $1/\Delta \lambda$ for all $\Delta \lambda$, the modulation

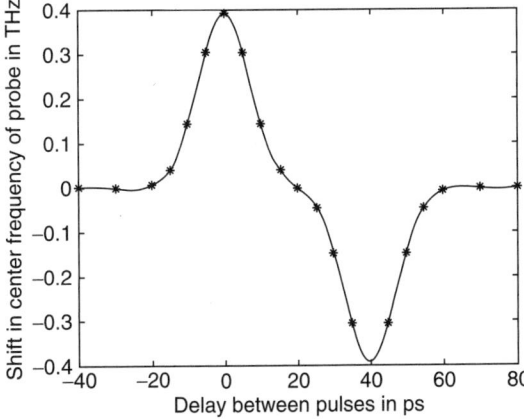

FIGURE 6.5 Simulation showing the shift in the frequency of the probe pulse with time delay between the pump and the probe pulse. The probe pulse corresponds to a wavelength of 1560.8 nm, whereas the pump pulse is at a wavelength of 1560 nm. (After [6].)

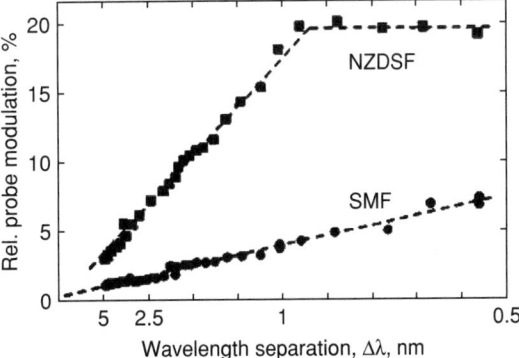

FIGURE 6.6 Variation of the RMS value of probe intensity modulation with the wavelength separation between the intensity modulated signal and the probe. NZDSF, nonzero dispersion shifted fiber; SMF, single-mode fiber. (After [9]; © 2005 IEEE)

is independent of $\Delta\lambda$. This is consistent with the earlier discussion in terms of L_{wo} and L_{eff}.

3.5. FWM

FWM is a nonlinear interaction that occurs in the presence of multiple wavelengths in a medium, leading to the generation of new frequencies. Thus, if light waves at three different frequencies ω_2, ω_3, and ω_4 are launched simultaneously into a medium, the same nonlinear polarization that led to the intensity dependent refractive index leads to the nonlinear polarization component at a frequency

$$\omega_1 = \omega_3 + \omega_4 - \omega_2 \qquad (46)$$

This nonlinear polarization, under certain conditions, leads to the generation of electromagnetic waves at ω_1. This process is referred to as FWM due to the interaction between the four different frequencies. In a WDM system carrying multiple channels, FWM can cause severe cross-talk. It is thus necessary that FWM effects be minimized in WDM systems.

During the FWM process, there are four different frequencies present at any point in the medium. If we write the electric field of the waves as

$$E_i = \frac{1}{2}\left[A_i(z)\psi_i(x,y)e^{i(\omega_i t - \beta_i z)} + cc\right], \quad i = 1, 2, 3, 4 \qquad (47)$$

where, as before, $A_i(z)$ represents the amplitude of the wave, $\psi_i(x, y)$ the transverse field distribution, and β_i the propagation constant of the wave. The total electric field is given by

$$E = E_1 + E_2 + E_3 + E_4 \qquad (48)$$

Substituting for the total electric field in the equation for nonlinear polarization, the term with frequency ω_1 comes out to be

$$P_{NL}(\omega_1) = \frac{1}{2}\left[P_{NL}^{(\omega_1)} e^{i(\omega_1 t - \beta_1 z)} + cc\right] \qquad (49)$$

where

$$P_{NL}^{(\omega_1)} = \frac{3\varepsilon_0}{2}\chi^{(3)} A_2^* A_3 A_4 \psi_2 \psi_3 \psi_4 e^{-i\Delta\beta z} \qquad (50)$$

and

$$\Delta\beta = \beta_3 + \beta_4 - \beta_2 - \beta_1 \qquad (51)$$

In writing Eq. (50), we only considered the FWM term, neglecting the SPM and XPM terms.

Substituting the expression for $P_{NL}^{(\omega_1)}$ in the wave equation for ω_1 and making the slowly varying approximation (in a manner similar to that used in the case of SPM and XPM), we obtain the following equation for $A_1(z)$:

$$\frac{dA_1}{dz} = -2i\gamma A_2^* A_3 A_4 e^{-i\Delta\beta z} \qquad (52)$$

where γ is defined by Eq. (18) with $k_0 = \omega/c$, ω represents the average frequency of the four interacting waves and A_{eff} the average effective area of the modes.

Assuming all waves have the same attenuation coefficient α and neglecting depletion of waves at frequencies ω_2, ω_3, and ω_4, due to nonlinear conversion we obtain the power in the frequency ω_1 as

$$P_1(L) = 4\gamma^2 P_2 P_3 P_4 L_{eff}^2 \eta e^{-\alpha L} \qquad (53)$$

where

$$\eta = \frac{\alpha^2}{\alpha^2 + \Delta\beta^2}\left[1 + \frac{4e^{-\alpha L}\sin^2\frac{\Delta\beta L}{2}}{(1 - e^{-\alpha L})^2}\right] \qquad (54)$$

and L_{eff} is the effective length [see Eq. (22)]. Maximum FWM takes place when $\Delta\beta = 0$, because in such a case $\eta = 1$. Now

$$\Delta\beta = \beta(\omega_3) + \beta(\omega_4) - \beta(\omega_2) - \beta(\omega_1) \qquad (55)$$

Because the frequencies are usually close to each other, we can make a Taylor series expansion about any frequency, say ω_2. In this case, we obtain

$$\Delta\beta = (\omega_3 - \omega_2)(\omega_3 - \omega_1)\left.\frac{d^2\beta}{d\omega^2}\right|_{\omega=\omega_2} \qquad (56)$$

In optical fiber communication systems, the channels are usually equally spaced. Thus, we assume the frequencies to be given by

$$\omega_4 = \omega_2 + \Delta\omega, \quad \omega_3 = \omega_2 - 2\Delta\omega \quad \text{and} \quad \omega_1 = \omega_2 - \Delta\omega$$

Using these frequencies and Eq. (26), Eq. (56) gives us

$$\Delta\beta = -\frac{4\pi D\lambda^2}{c}(\Delta\nu)^2 \qquad (57)$$

where $\Delta\omega = 2\pi\Delta\nu$. Thus, maximum FWM takes place when $D = 0$. This is the main problem in using WDM in dispersion shifted fibers characterized by zero dispersion at the operating wavelength of 1550 nm, because FWM will then lead to cross-talk among the various channels. FWM efficiency can be reduced by using fiber with nonzero dispersion. This has led to the development of nonzero dispersion shifted fiber that have a finite nonzero dispersion of about ± 2 ps/nm·km at the operating wavelength.

From Eq. (57) we notice that for a given dispersion coefficient D, FWM efficiency reduces as $\Delta\nu$ increases. Thus, for systems with small channel spacings, the dispersion coefficient required to overcome FWM effects is higher. In this context new fibers, such as Teralite fibers having a dispersion coefficient of 8 ps/nm·km, have been developed for use in smaller channel spacings. Indeed, the conventional fiber, which has zero dispersion in the 1310-nm band, has a significant dispersion (~17 ps/nm·km) in the 1550-nm band and thus poses no problem with regard to FWM even for small channel spacings.

To get a numerical appreciation, we consider the case with $D = 0$, that is, $\Delta\beta = 0$. For such a case $\eta = 1$. If all channels were launched with equal power P_{in}, then

$$P_1(L) = 4\gamma^2 P_{\text{in}}^3 L_{\text{eff}}^2 e^{-\alpha L} \tag{58}$$

Thus, the ratio of power generated at ω_1 due to FWM and that existing at the same frequency is

$$\frac{P_g}{P_{\text{out}}} = \frac{P_1(L)}{P_{\text{in}} e^{-\alpha L}} = 4\gamma^2 P_{\text{in}}^2 L_{\text{eff}}^2 \tag{59}$$

Typical values are $L_{\text{eff}} = 20\,\text{km}$, $\gamma = 2.4\,\text{W}^{-1}\text{km}^{-1}$. Thus

$$\frac{P_g}{P_{\text{out}}} \approx 0.01\, P_{\text{in}}^2 (\text{mW}^2)$$

Figure 6.7 shows the output spectrum measured at the output of a 25-km-long dispersion shifted fiber ($D = -0.2\,\text{ps/nm·km}$) when three 3-mW wavelengths are launched simultaneously. Notice the generation of many new frequencies by FWM. Figure 6.8 shows the ratio of generated power to the output as a function of channel spacing $\Delta\lambda$ for different

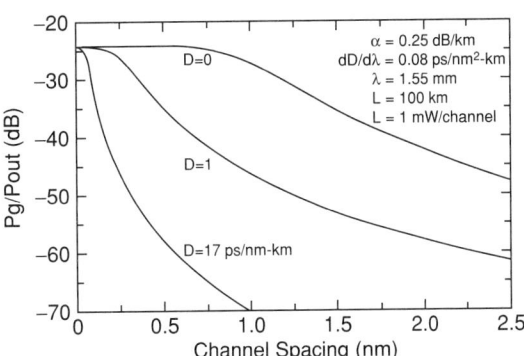

FIGURE 6.8 Ratio of generated power to the output as a function of channel spacing $\Delta\lambda$ for different dispersion coefficients. (After [10]; © 2004 IEEE)

dispersion coefficients. It can be seen that by choosing a nonzero value of dispersion, the FWM efficiency can be reduced. The larger the dispersion coefficient, the smaller the channel spacing for the same cross-talk.

Because dispersion leads to increased bit error rates in fiber optic communication systems, it is important to have low dispersion. On the other hand, lower dispersion leads to cross-talk due to FWM. This problem can be resolved by noting that FWM depends on the local dispersion value in the fiber, whereas the pulse spreading at the end of a link depends on the overall dispersion in the fiber link. If one chooses a link made up of positive and negative dispersion coefficients, then by an appropriate choice of the lengths of the positive and negative dispersion fibers, it would be possible to achieve a zero total link dispersion while at the same time maintaining a large local dispersion. This is referred to as dispersion management in fiber optic systems.

Although FWM leads to cross-talk among different wavelength channels in an optical fiber communication system, it can be used for various optical processing functions such as wavelength conversion, high-speed time division multiplexing, pulse compression, and so on. For such applications, there is a concerted worldwide effort to develop highly nonlinear fibers with much smaller mode areas and higher nonlinear coefficients. Some of the novel fibers developed recently include holey fibers, photonic bandgap fibers, or photonic crystal fibers, which are very interesting because they posses extremely small-mode effective areas (~2.5 μm² at 1550 nm) and can be designed to have zero dispersion even in the visible region of the spectrum (see Chapters 3 and 4). Photonic crystal fibers based on lead silicate glass have been fabricated with γ nearly equal to

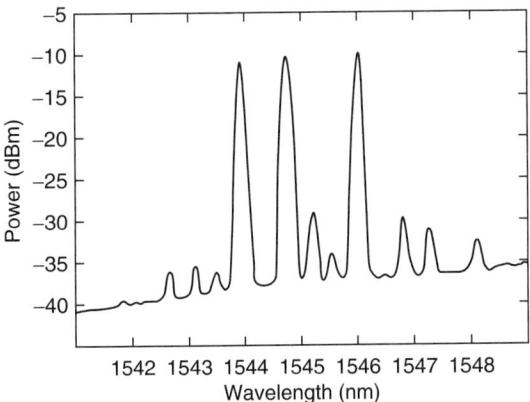

FIGURE 6.7 Generation of new frequencies because of FWM when waves at three frequencies are incident in the fiber. (After [10]; © 2005 IEEE)

1860 w^{-1} km^{-1} which is 775 times larger than a conventional SMF. Such fibers are expected to revolutionize nonlinear fiber optics by providing new geometries to achieve highly efficient nonlinear optical processing at lower powers. When light pulses from a laser propagate through nonlinear fibers having appropriate dispersion, the nonlinear effects broaden the spectrum of the input laser into a broad continuum. This is referred to as supercontinuum generation [11] (see, e.g., the cover of this book). Such light sources are finding wide application in meteorology, spectroscopy, telecommunications, etc.

4. CONCLUSIONS

The small cross sectional areas and long interaction lengths of optical fibers lead to significant nonlinear effects. These nonlinear effects ultimately limit the information carrying capacity of the fiber. At the same time, some of the nonlinear effects can be profitably used for the processing of optical signals.

5. REFERENCES

1. Chraplyvy, A. R. (1990) Limitations on lightwave communications imposed by optical fiber nonolinearities, *J. Lightwave Tech.*, **8**, 1548.
2. Ghatak, A. and Thyagarajan, K. (1998) *Introduction to Fiber Optics*, Cambridge University Press, Cambridge, UK.
3. Agrawal, G. P. (1989) *Nonlinear Fiber Optics*, Academic Press, Boston.
4. Nakazawa, M. (1994) Soliton transmission in telecommunication networks, *IEEE Comm. Magazine*, **March**, p. 34.
5. Kumar, G. (2002) Simulation of long haul optical fiber communication systems, M Tech. (Optoelectronics) Thesis, IIT Delhi.
6. Hemant, N. A. (2004) Nonlinear effects in multichannel lightwave systems, M Tech. (Optoelectronics) Thesis, IIT Delhi.
7. Baldeck, P. L., Alfano, R. R. and Agrawal, G. P. (1998) Induced frequency shift of copropagating ultrafast optical pulses, *App. Phys. Letts.* **52**, 1939–1941.
8. Chiang, T. K., Kagi, N., Marhic, M. E. and Kazovsky, L. G. (1996) Cross phase modulation in fiber links with multiple optical amplifiers and dispersion compensators, *J. Lightwave Tech.* **14**, 249–259.
9. Shtaif, M., Eiselt, M. and Garret, L. D. (2000) Cross phase modulation distortion measurements in multispan WDM systems, *IEEE. Photon. Tech. Lett.* **12**, 88–90.
10. Tkach, R. W., Chraplyvy, A. R., Forghiari, F., Gnanck, A. H. and Derosier, R. M. (1995) Four-photon mixing and high speed WDM systems, *J. Lightwave Tech.* **13**, 841–849.
11. Hilligsoe, K. M., Anderson, T. V., Paulsen, H. N., Nielsen, C. K., Molmer, K., Keiding, S., Kristiensen, R., Hansen, K. P. and Larsen, J. J. (2004) Supercontinuum generation in a photonic crystal fiber with two zero dispersion wavelengths, *Opt. Exp.* **12**, 1045–1054.

CHAPTER 7

Fiber Optic Parametric Amplifiers for Lightwave Systems

F. Yaman, Q. Lin and Govind P. Agrawal*
Institute of Optics
University of Rochester
Rochester, New York, USA

1. INTRODUCTION

Modern optical communication systems require not only signal amplification periodically but also devices that are capable of ultrafast all-optical signal processing. Fiber optic parametric amplifiers (FOPA), based on four-wave mixing (FWM) occurring inside optical fibers (see Chapter 6), are attracting considerable attention because they can provide broadband amplification and can thus replace erbium-doped fiber amplifiers used commonly for signal amplification. However, although not yet fully appreciated, FOPAs are also an ideal candidate for ultrafast all-optical signal processing because of an instantaneous electronic response of the silica nonlinearity responsible for FWM in optical fibers. Moreover, amplification provided by FOPAs is accompanied with relatively low noise, allowing operation close to the quantum limit [1].

Reasonably large values of optical gain with a relatively flat and wide gain spectrum can be realized when FOPAs are either pumped in the vicinity of zero dispersion wavelength (ZDWL) using a single pump laser [2, 3] or pumped using two lasers at two well-selected wavelengths located on each side of the ZDWL [4–6]. Such FOPAs have been used for applications such as broadband wavelength division multichannel amplification, wavelength conversion, optical sampling, and pulse compression [1–6]. A feature unique to FOPAs is that the idler field generated during signal amplification is phase conjugated. Such phase conjugation provides an efficient way for dispersion compensation [7], as already demonstrated experimentally [8]. It can also be used to reduce timing jitter [9] as well as phase jitter [10] in long-haul lightwave systems. Although FOPAs can be used to manipulate quantum noise with a proper control of phase differences among the four interacting waves [11–13], modern communications systems do not yet use phase-sensitive amplification.

In this review, we consider the recent progress that has been realized in designing FOPAs. Among many applications of such devices, we focus on signal amplification and wavelength conversion. The chapter is organized as follows. We first consider in Section 2 the theory behind the nonlinear phenomenon of FWM and use it in Section 3 to discuss the performance of single-pump FOPAs. Section 4 considers the more general case of dual-pump FOPAs and shows how the use of two pumps at suitable wavelengths can provide uniform gain over a wide bandwidth. Sections 5 and 6 discuss the impact of two phenomena that affect the performance of all FOPA-based devices. The ZDWL of a fiber can vary along its length in a random fashion due to core-diameter variations that occur invariably during fiber manufacturing; the effects of ZDWL fluctuations are discussed in Section 5. Section 6 then focuses on birefringence fluctuations that also occur in all practical fibers and lead to a phenomenon known as polarization mode dispersion (PMD). The main results are then summarized in the concluding section.

2. THEORY OF FWM

FWM process originates from the nonlinear response of bound electrons to intense optical waves inside a nonlinear medium such as silica fibers. When two intense pump waves at frequencies ω_1 and ω_2 copropagate inside a silica fiber, electrons, which only have a tiny mass, can be driven almost instantaneously at any frequency stemming from the mixing of these waves. Even though the potential provided by silica molecules confines electrons to their original atom, electrons respond to the applied electromagnetic field by emitting secondary waves not only at the original frequencies ω_1 and ω_2 (linear response), but also at two new frequency

*E-mail: gpa@optics.rochester.edu

components denoted as ω_3 and ω_4 (third-order nonlinear response). Physically, two photons at the original frequencies are scattered elastically into two new photons at frequencies ω_3 and ω_4. In the absence of absorption, the total energy and momentum is conserved during FWM. Noting that photon energy and momentum are $\hbar\omega$ and $\hbar\beta$, respectively, for an optical field of frequency ω propagating with the propagation constant β, the conservation relations take the form

$$\omega_1 + \omega_2 = \omega_3 + \omega_4 \qquad (1)$$

$$\beta_1 + \beta_2 = \beta_3 + \beta_4 \qquad (2)$$

where β_j is the propagation constant at frequency ω_j ($j = 1$–4). Only the scalar form of the wave vector is relevant for single-mode fibers because all four waves propagate along the fiber axis. Because β_j governs the phase shift experienced by the jth wave, Eq. (2) is also referred to as the phase-matching condition [14].

What determines the frequencies ω_3 and ω_4 during the FWM process is a question that must be answered. If only the pump beams are incident on an optical fiber, the new waves grow from noise and their frequencies are determined by the phase-matching condition through spontaneous FWM. In practice, the efficiency of the FWM process is enhanced by seeding it. Seeding is accomplished by launching a signal wave at the frequency ω_3. The probability of creating photons at the frequency ω_4 depends on how many photons at ω_3 already exist inside the fiber. As a result, the FWM process is stimulated, and new photons at ω_3 and ω_4 are created with an exponential growth rate provided the phase-matching condition is nearly satisfied. The fourth wave at the frequency ω_4 is commonly referred to as the *idler wave* following the terminology used in the microwave literature. It is not obligatory to launch two separate pump beams for FWM to occur. The same process can occur even when the two pump photons have the same frequency, and this is referred to as degenerate FWM. The more general case of two distinguishable pump beams is called nondegenerate FWM.

Mathematically, the description of FWM is relatively simple for optical fibers because all four waves propagate in the form of a fiber mode and maintain their spatial profile [1]. Because they also propagate along the same fiber axis (assumed to coincide with the z-axis), transverse effects can be completely ignored, and one can use a one-dimensional model. Moreover, if we assume perfect cylindrical symmetry for optical fibers (no residual birefringence),

all waves can be assumed to maintain their initial state of polarization, and we can use the scalar approximation. We relax this approximation in Section 6 because silica fibers do exhibit some residual birefringence that changes randomly along the fiber length.

The FWM analysis is simplified considerably if we assume that the signal and idler powers remain relatively small compared with pump powers throughout the fiber length. This amounts to assuming that pumps are not depleted during the FWM process. The evolution of the signal and idler waves is then governed by the following two coupled but linear equations [1]:

$$\frac{dB_3}{dz} = \frac{i}{2}\kappa B_3 + 2i\gamma B_1 B_2 B_4^* \qquad (3)$$

$$\frac{dB_4}{dz} = \frac{i}{2}\kappa B_4 + 2i\gamma B_1 B_2 B_3^* \qquad (4)$$

where $\kappa = \Delta\beta + \gamma(P_1 + P_2)$ describes the total phase mismatch, whereas $\Delta\beta = \beta_3 + \beta_4 - \beta_1 - \beta_2$ is the linear phase mismatch resulting from fiber dispersion. Moreover, P_1 and P_2 are the input powers of two pumps, and γ is the nonlinear parameter defined as $\gamma = 2\pi n_2/(\lambda_p A_{\text{eff}})$, where $n_2 \approx 2.6 \times 10^{-20}\,\text{m}^2/\text{W}$ is for silica fibers, λ_p is the average pump wavelength, and A_{eff} is the effective core area of the fiber. The optical field amplitudes for the signal and idler waves, A_3 and A_4, are related to B_3 and B_4 by a phase factor through $A_j = B_j \exp\{iz[-\kappa/2 + 2\gamma(P_1 + P_2)]\}$ [1]. In Eqs. (3) and (4), we included the contribution of self-phase modulation (SPM) and cross-phase modulation (XPM) induced by the two pumps, but these nonlinear effects are neglected for the signal and idler waves. Fiber losses are also neglected because only a relatively short length of fiber (~ 1 km) is generally used for making FOPAs.

From Eqs. (3) and (4), the signal and idler power $P_3 = |B_3|^2$ and $P_4 = |B_4|^2$ are found to satisfy the same equation:

$$\frac{dP_3}{dz} = \frac{dP_4}{dz} = 2\xi\sqrt{P_3 P_4}\sin\theta \qquad (5)$$

where $\xi = 2\gamma\sqrt{P_1 P_2}$ is a measure of the FWM efficiency in the nondegenerate case and $\theta = \phi_3(z) + \phi_4(z) - \phi_1(z) - \phi_2(z)$ describes the accumulated phase mismatch among the four waves. Here ϕ_j is the phase of the field B_j, that is, $B_j = \sqrt{P_j}\exp(i\phi_j)$. When the two pumps are assumed to remain undepleted, ϕ_1 and ϕ_2 remain fixed to their initial values, and the accumulated phase mismatch is governed by

$$\frac{d\theta}{dz} = \kappa + \xi \cos\theta \frac{(P_3 + P_4)}{\sqrt{P_3 P_4}} \quad (6)$$

Equation (5) shows clearly that the growth of the signal and idler waves inside a fiber is determined by the phase matching condition. When $\theta = \pi/2$, the signal and the idler extract energy from the two pumps. In contrast, when $\theta = -\pi/2$, energy can flow back to the two pumps from the signal and idler. If only the two pumps and the signal are launched into FOPA initially, the idler wave is automatically generated by the FWM process. This can be seen from Eq. (4). Even if $B_4 = 0$ at $z = 0$, its derivative is not zero as long as $B_3(0)$ is finite. If we integrate this equation over a short fiber section of length Δz, we obtain $\Delta B_4 \approx 2i\gamma B_1 B_2 B_3^*(0)\Delta z$. The factor of i provides an initial value of $\pi/2$ for θ and shows that the correct phase difference is automatically picked up by the FWM process [15]. If $\kappa = 0$ initially (perfect phase matching), Eq. (6) shows that θ remains frozen at its initial value of $\pi/2$. However, if $\kappa \neq 0$, θ changes along the fiber as dictated by Eq. (6), and energy flows back into the two pumps in a periodic fashion. Thus, phase matching is critical for signal amplification and idler generation.

3. SINGLE-PUMP PARAMETRIC AMPLIFIERS

In this section we focus on the simpler case in which a single intense pump is launched into a fiber together with the signal, and a single idler wave is generated through the degenerate FWM process. Equations (3)–(6) remain unchanged in the degenerate case provided we define ξ and κ as $\xi = \gamma P_1$ and $\kappa = \Delta\beta + 2\gamma P_1$. Integrating Eqs. (3) and (4) with the initial condition $B_4(0) = 0$, the signal power at the end of a fiber of length of L is found to be [1]

$$P_3(L) = P_3(0)[1 + (1 + \kappa^2/4g^2)\sinh^2(gL)] \quad (7)$$

where the parametric gain coefficient g and the phase mismatch κ are given by

$$g = \sqrt{(\gamma P_1)^2 - (\kappa/2)^2}, \quad \kappa = \Delta\beta + 2\gamma P_1 \quad (8)$$

Equation (8) shows that the parametric gain is reduced by phase mismatch κ and is maximum when $\kappa = 0$. Both the nonlinear (SPM and XPM) and the linear effects (fiber dispersion) contribute to κ. Although the nonlinear contribution is constant, the linear phase mismatch depends on the wavelengths of the three waves. To realize net amplification of the signal, parametric gain g should be real. Thus, tolerable values of the linear phase mismatch $\Delta\beta$ are limited to the range $-4\gamma P_1 \leq \Delta\beta \leq 0$. The FOPA gain is maximum when the phase mismatch κ approaches zero or when $\Delta\beta = -2\gamma P_1$. This relation indicates that optimal operation of FOPAs requires some amount of negative linear mismatch to compensate for the nonlinear phase mismatch. In fact, the bandwidth of the gain spectrum is determined by the pump power and the nonlinear parameter γ. Figure 7.1 shows this dependence clearly by plotting the parametric gain as a function of $\Delta\beta$ at three different power levels of a single pump [1].

The linear phase mismatch $\Delta\beta$ depends on the dispersion characteristics of the fiber. As the signal and idler frequencies are located symmetrically around the pump frequency ($\omega_4 = 2\omega_1 - \omega_3$), it is useful to expand $\Delta\beta$ in a Taylor series around the pump frequency as [16]

$$\Delta\beta = \beta_3 + \beta_4 - 2\beta_1 = 2 \sum_{m=1}^{\infty} \left(\frac{d^{2m}\beta}{d\omega^{2m}}\right)_{\omega=\omega_1} \frac{(\omega_3 - \omega_1)^{2m}}{(2m)!} \quad (9)$$

This equation shows that only even-order dispersion parameters evaluated at the pump frequency contribute to the linear phase mismatch. Clearly, the choice of the pump wavelength is very critical while designing a FOPA. The linear phase mismatch $\Delta\beta$ is dominated by the second-order dispersion parameter, $\beta_{2p} = (d^2\beta/d\omega^2)_{\omega=\omega_1}$, when the signal wavelength is close to the pump but by the fourth- and higher order dispersion parameters (β_{4p}, β_{6p}, etc.) while the signal deviates far from it. Thus, the ultimate FOPA bandwidth depends on the spectral range over which the linear phase mismatch is negative but large enough to balance the constant positive nonlinear phase mismatch of $2\gamma P_1$. This can be achieved

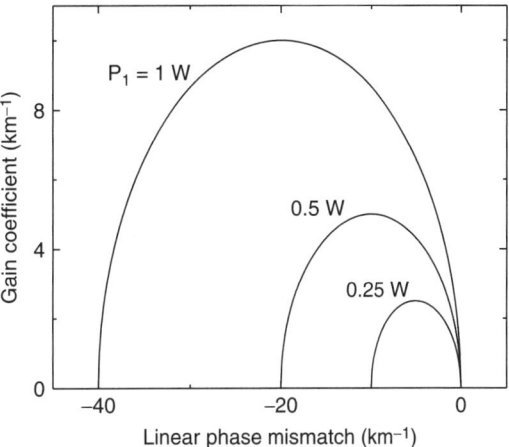

FIGURE 7.1 Parametric gain for a single-pump FOPA as a function of linear phase mismatch at three pump powers for a fiber with $\gamma = 10\,\text{W}^{-1}/\text{km}$.

by slightly displacing the pump wavelength from the ZDWL of the fiber such that β_{2p} is negative but β_{4p} is positive, or viceversa.

We should relate the parameters β_{2p} and β_{4p} to the fiber dispersion parameters, $\beta_m = (d^m\beta/d\omega^m)_{\omega=\omega_0}$, calculated at the ZDWL of the fiber. This can be accomplished by expanding $\beta(\omega)$ in a Taylor series around ω_0. If we keep terms up to fourth-order in this expansion, we obtain

$$\beta_{2p} \approx \beta_3(\omega_1 - \omega_0) + \frac{\beta_4}{2}(\omega_1 - \omega_0)^2, \quad \beta_{4p} \approx \beta_4 \quad (10)$$

Depending on the values of the fiber parameters β_3 and β_4, we can choose the pump frequency ω_1 such that β_{2p} and β_{4p} have opposite signs. More specifically, because both β_3 and β_4 are positive for most silica fibers, one should choose $\omega_1 < \omega_0$, that is, the pump wavelength should be longer than the ZDWL of the fiber.

Figure 7.2 shows the gain spectra at several different pump wavelengths in the vicinity of the ZDWL λ_0 (chosen to be 1550 nm) by changing the pump detuning $\Delta\lambda_p = \lambda_1 - \lambda_0$ in the range -0.1 to $+0.15$ nm. The dotted curve shows the case $\Delta\lambda_p = 0$ for which the pump wavelength coincides with the ZDWL exactly. The peak gain is about 8 dB, and the gain bandwidth is limited to below 40 nm. When the pump is tuned toward the shorter wavelength side, the bandwidth actually decreases. In contrast, both the peak gain and the bandwidth are enhanced by tuning the pump toward the longer wavelength side. The signal gain in the vicinity of pump is the same regardless of the pump wavelength. When the signal wavelength moves away from the pump, the linear phase mismatch $\Delta\beta$ strongly depends on the pump wavelength. If both the third- and fourth-order dispersion parameters are positive at ZDWL, according to Eq. (10) the second-order dispersion at the pump is negative when $\Delta\lambda_p$ is slightly positive and thus can compensate for the nonlinear phase mismatch. This is the reason why gain peak is located at a wavelength far from the pump when $\lambda_1 > \lambda_0$. When phase matching is perfect ($\kappa = 0$), FOPA gain grows exponentially with the fiber length L as $G = 1 + \exp(2\gamma P_1 L)/4$. For the parameters used for Fig. 7.2, the best case occurs when $\Delta\lambda_p = 0.106$ nm. However, when $\Delta\lambda_p < 0$, both the second- and fourth-order dispersion parameters for the pump are positive. As a result, the linear phase mismatch adds up with the nonlinear one, making κ relatively large. As a result, the FOPA bandwidth is reduced.

From a practical standpoint, one wants to maximize both the peak gain and the gain bandwidth at a given pump power P_1. Because peak gain is approximately given by $G_p \approx \exp(2\gamma P_1 L)/4$, its value increases exponentially with $\gamma P_1 L$ and can be increased by increasing fiber length. However, the gain bandwidth scales inversely with L because phase mismatch increases for longer fibers. The obvious solution is to use as short a fiber as possible. However, as the available amount of gain is a function of $\gamma P_1 L$, shortening of fiber length must be accompanied with an increase in the value of γP_1 to maintain the same amount of gain. This behavior is illustrated in Fig. 7.3 where the gain bandwidth is shown to increase significantly when large values of γP_1 are combined with shorter lengths of fiber. The solid curve obtained for the 250-m-long fiber

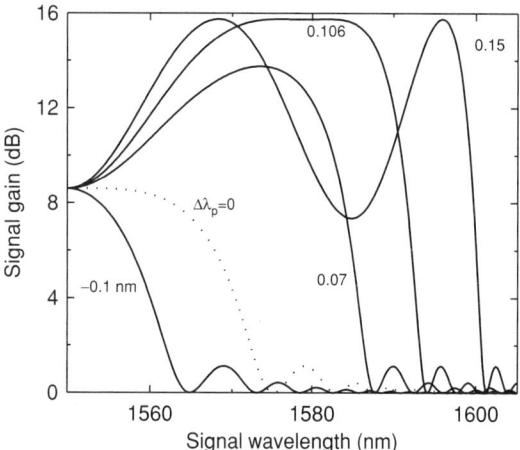

FIGURE 7.2 Gain spectra for a single-pump FOPA for several values of pump detuning $\Delta\lambda_p = \lambda_p - \lambda_0$ from the ZDWL λ_0. The parameters used are $\gamma = 2\,\text{W}^{-1}/\text{km}$, $P_1 = 0.5\,\text{W}$, $L = 2.5\,\text{km}$, $\beta_3 = 0.1\,\text{ps}^3/\text{km}$, and $\beta_4 = 10^{-4}\,\text{ps}^4/\text{km}$.

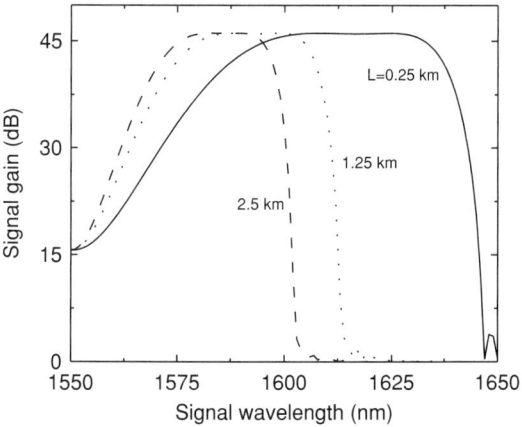

FIGURE 7.3 Gain spectra for single-pump FOPAs of three different lengths. The product $\gamma P_1 L = 6$ is kept constant for all curves. Other parameters are the same as those used for Fig. 7.2.

exhibits a 50-nm region on each side of the ZDWL over which the gain is nearly flat. Therefore, a simple rule of thumb for single-pump FOPAs is to use as high a pump power as possible together with a fiber with as large a nonlinearity as possible. Even though n_2 is fixed for silica fibers, the nonlinear parameter γ can be increased by reducing the effective core area. Such fibers have become available in recent years and are called highly nonlinear fibers (HNLFs) even though it is not the material nonlinearity n_2 that is enhanced in such fibers. Values of $\gamma > 10\,\mathrm{W^{-1}/km}$ can be realized in such fibers.

HNLFs have been used to make FOPAs with a large bandwidth. In a 2001 experiment, a 200-nm gain bandwidth was realized by using Raman-assisted parametric amplification inside a 20-m-long HNLF with $\gamma = 20\,\mathrm{W^{-1}/km}$ [17]. The required pump power ($\sim$10 W) was large enough that the signal was also amplified by stimulated Raman scattering when its wavelength exceeded the pump wavelength. Recent advances in designing microstructure fibers also make it possible to use short fiber lengths. A net peak gain of 22.5 dB over a bandwidth of 20 nm has been realized inside a 12.5-m-long microstructure fiber pumped by high-energy pulses [18]. In another 2003 experiment, a peak gain of 43 dB with an 85-nm bandwidth was obtained by pumping the FOPA with pulses at a repetition rate of 20 Gb/s [19]. However, a pulse-pumped FOPA requires either synchronization between the pump and signal pulses or pumping at a repetition rate much higher than that of the signal.

Another scheme for mitigating the phase-matching problem manages fiber dispersion along the fiber length, resulting in the so-called quasi-phase matching. This can be realized either through periodic dispersion compensation [20, 21] or by carefully arranging different sections of fiber with different dispersion properties [22]. Because quasi-phase matching can be maintained along a fairly long length, continuous-wave (CW) pumps can be used and still realize a considerable amount of gain. Figure 7.4 shows the experimental results for such a single-pump FOPA [23] where both the net signal gain and the net conversion efficiency at the idler wavelength are shown at several pump-power levels. At a pump power of 31.8 dBm (about 1.5 W) at 1563 nm, the FOPA provided a 49-dB peak gain. It was designed using a 500-m-long HNLF ($\gamma = 11\,\mathrm{W^{-1}/km}$) with low dispersion (dispersion slope $S = 0.03\,\mathrm{ps \cdot nm^2 \cdot km}$). The fiber was composed of three sections with ZDWLs 1556.8, 1560.3, and 1561.2 nm.

With its high gain over a wide spectrum, FOPAs have many practical applications. Simultaneous amplification of seven channels has been realized using a CW-pumped FOPA made of HNLF [24]. The experiment showed that dominant degradation stems from gain saturation and FWM-induced crosstalk among channels. FOPAs have also been used as a stable source of pulses at a high repetition rate (40 Gb/s) in long-haul transmission [25]. In another experiment, a transform-limited Gaussian-shape pulse train could be generated at a 40-Gb/s repetition when a weak CW signal wasz amplified using a FOPA whose pump power was sinusoidally modulated at 40 GHz [26].

As discussed earlier, all FOPAs generate the idler wave during signal amplification. Because the idler is a phase-conjugated version of the signal, it carries all the information associated with a signal and thus can be used for wavelength conversion. Indeed, FOPAs can act as highly efficient wavelength converters with a wide bandwidth [3]. As early as 1998, peak conversion efficiency of 28 dB was realized

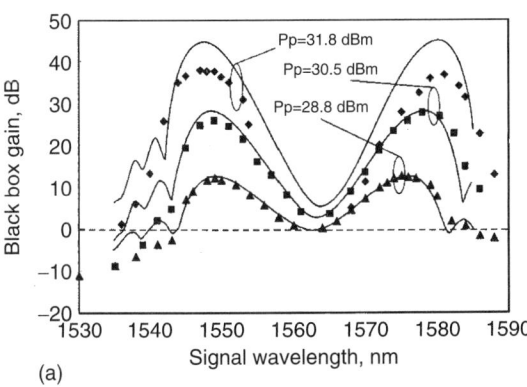
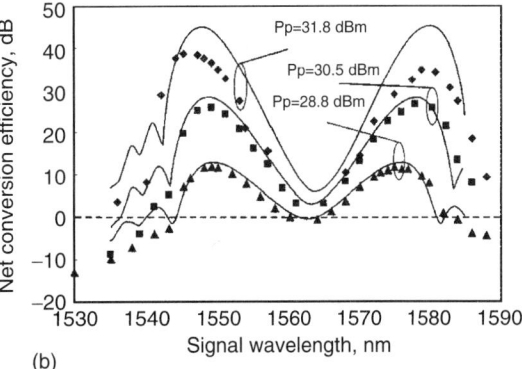

FIGURE 7.4 (a) Measured signal gain and (b) idler conversion efficiency for a single-pump FOPA at several pump powers. Solid curves show the theoretically expected results. (Adapted from [23].)

over a 40-nm bandwidth (full width of the gain spectrum) using a pulsed pump [27]. More recently, transparent wavelength conversion (conversion efficiency > 0 dB) over a 24-nm bandwidth (entire pump tuning range) was realized using a single-pump FOPA made with just 115 m of HNLF [28]. The ultrafast nature of nonlinear response of FOPAs is also useful for many other applications such as optical time division demultiplexing [29] and optical sampling [30]. FOPAs can be used to mitigate noise associated with an input signal when operating in the saturation regime [31]. A similar scheme can be used for all-optical signal regeneration using a higher order idler [32, 33]. FOPAs can also work in the pump-depleted region and can transfer as much as 92% of the pump power to the signal and idler fields [34]. Such FOPAs can be used to realize CW-pumped optical parametric oscillators with 30% internal conversion efficiency and a tuning range of 80 nm [35].

The performance of single-pump FOPAs is affected by several factors that must be considered during the design of such devices. For example, although FOPAs benefit from an ultrafast nonlinear response of silica, they also suffer from it because any fluctuations in the pump power are transferred to the signal and idler fields. As a result, noise in FOPAs is dominated by transfer of relative intensity noise associated with the pump laser [36, 37]. Moreover, because the pump beam is typically amplified using one or two erbium-doped fiber amplifiers (EDFAs) to achieve high powers necessary for pumping an FOPA, amplified spontaneous emission (ASE) from EDFAs can degrade the FOPA considerably. In fact, it is essential to block such ASE noise using optical filters. A noise figure of 4.2 dB with a maximum gain of 27.2 dB has been realized for parametric amplifiers [38]. Similarly, a noise figure of 3.8 dB with 40-dB conversion efficiency has been reported for FOPA-based wavelength converters [39] by blocking the ASE noise through narrowband fiber Bragg gratings. Such values of noise figure are close to the fundamental quantum limit of 3 dB [40].

The second factor that affects FOPAs is the nonlinear phenomenon of stimulated Brillouin scattering (SBS). The SBS threshold is around 10 mW for long fibers (>10 km) and increases to ~0.1 W for fiber lengths of 1 km or so [1]. Because FOPAs require pump-power levels approaching 1 W, a suitable technique is needed that raises the threshold of SBS and suppresses it over the FOPA length. The techniques used in practice include control of temperature distribution along the fiber length [28] and phase modulation of the pump either at several fixed frequencies [23] or over a broad frequency range [39]. The latter technique suppresses SBS by broadening the pump spectrum. Although the amplified signal is affected little by spectral broadening of the pump, the idler suffers from it considerably. In fact, the spectrum of the idler can become twice as broad as the pump if the signal spectrum is narrow. This follows directly from the energy conservation of Eq. (1). The broadening of the idler is not of concern when FOPAs are used for signal amplification but becomes a serious issue when they are used as wavelength converters. This problem cannot be solved for single-pump FOPAs. As discussed in the next section, dual-pump FOPAs can be designed to provide an idler with the same bandwidth as that of the signal.

The third issue associated with the single-pump FOPA is that its gain spectrum is not as uniform as one would like. In practice, only the pump wavelength can be adjusted to optimize the gain spectrum. As discussed before, the phase mismatch κ should be zero at the gain peak. However, Eq. (9) shows that it is hard to maintain this phase-matching condition over a wide bandwidth in a single-pump FOPA. Because $\Delta\beta \to 0$ when the signal wavelength approaches the pump wavelength, $\kappa \to 2\gamma P_1$. This value of κ is quite large and results in only a linear growth of signal ($G = 1 + \gamma P_1 L$). The net result is that the signal gain is considerably reduced in the vicinity of the pump wavelength, and the gain spectrum exhibits a dip. Figure 7.5 shows how variations in the phase mismatch κ affect the FOPA gain. Although κ can be close to zero in the spectral region where the FOPA gain peaks, it changes

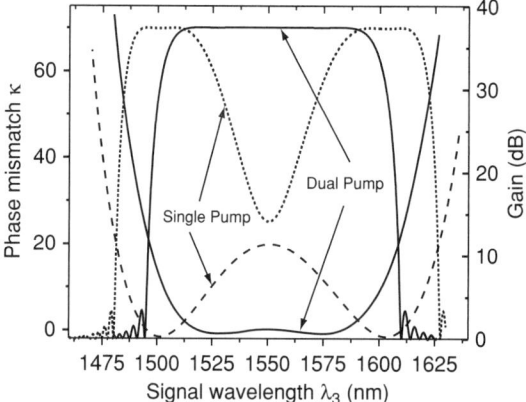

FIGURE 7.5 Optimized gain spectra for single-pump and dual-pump FOPAs and corresponding phase mismatch κ. The same amount of total pump power was used in both cases.

over a large range within the whole gain spectrum. The nonuniform gain of single-pump FOPAs is a consequence of such κ variations. Although amplification over a range as wide as 200 nm is possible, the gain spectrum remains highly nonuniform [17]. In practice, the usable bandwidth is limited to a much smaller region of the whole gain bandwidth. This problem can be solved to some extent by manipulating fiber dispersion [20–22]. It is predicted theoretically that a fairly flat gain spectrum is possible by using several fiber sections of suitable lengths and properly selecting dispersive properties of these fiber sections [41]. However, such a scheme is difficult to implement in practice because dispersive properties of HNLFs are rarely known with sufficient precision. The current practical solution is to make use of dual-pumped FOPAs, discussed in the next section.

The fourth serious issue is the polarization dependence of the FOPA gain. The theoretical analysis in this section is based on the assumption that all optical fields are linearly polarized initially and maintain their state of polarization (SOP) during propagation inside the fiber. In practice, the SOP of the input signal can be arbitrary. The FWM process is highly polarization dependent because it requires angular momentum conservation among the four interacting photons [14]. Polarization-independent operation of single-pump FOPAs can be realized by using a polarization diversity loop [42, 43]. In this approach, the pump beam is split into its orthogonally polarized components with equal amounts of power, which counterpropagate inside a Sagnac loop. When the signal enter the loop, it is also split into its orthogonally polarized components, each of which copropagates with the identically polarized pump. The two polarization components of the signal are then recombined after the polarization diversity loop. Such polarization diversity loops have been used for optical sampling at 80 Gb/s with a residual polarization dependence of only 0.7 dB [44]. By using a polarization maintaining HNLF inside such a loop, the wavelengths of 32 channels, each operating at 10 Gb/s, were converted simultaneously with a polarization dependence of only 0.2 dB [45].

4. DUAL PUMP PARAMETRIC AMPLIFIERS

Dual-pump FOPAs use the nondegenerate FWM process using two pumps with dissimilar frequencies [46]. The properties of such FOPAs have been analyzed in recent years [4–6]. The most interesting aspect is that they can provide relatively flat gain over a much wider bandwidth than that possible for single-pump FOPAs. In the case of nondegenerate FWM, two distinct photons, one from each pump, are used to create the signal and idler photons as shown in Eq. (1). The parametric gain coefficient in this case can be obtained from the simple theory of Section 2 and is found to be [1]

$$g = \sqrt{(2\gamma)^2 P_1 P_2 - (\kappa/2)^2} \quad (11)$$

where the phase mismatch $\kappa = \Delta\beta + \gamma(P_1 + P_2)$ and P_1 and P_2 are the input pump powers, assumed to remain undepleted.

Similar to the single-pump case, one can expand the linear phase mismatch $\Delta\beta = \beta_3 + \beta_4 - \beta_1 - \beta_2$ in a Taylor series. A simple solution is to introduce $\omega_c = (\omega_1 + \omega_2)/2$ as the mean frequency of the two pumps and $\omega_d = (\omega_1 - \omega_2)/2$ as their frequency difference and expand $\Delta\beta$ around ω_c [47]:

$$\Delta\beta = 2\sum_{m=1}^{\infty} \frac{1}{(2m)!}\left(\frac{d^{2m}\beta}{d\omega^{2m}}\right)_{\omega=\omega_c}[(\omega_3 - \omega_c)^{2m} - \omega_d^{2m}] \quad (12)$$

This equation differs from Eq. (9) by the last term. This ω_d term contributes only when two pumps are used and is independent of the signal and idler frequencies. This difference provides the main advantage of dual-pump FOPAs over single-pump FOPAs as the ω_d term can be used to control the phase mismatch. By properly choosing the pump wavelengths, it is possible to use this term for compensating the nonlinear phase mismatch $\gamma(P_1 + P_2)$ stemming from SPM and XPM. As a result, the total phase mismatch κ can be maintained close to zero over quite a wide spectral range after the first term is made small by balancing carefully different orders of fiber dispersion.

The most commonly used configuration of dual-pump FOPAs uses a relatively large wavelength difference between the two pumps for realizing flat gain over a wide spectral range. This increases the magnitude of the ω_d term. At the same time, the mean frequency of the two pumps ω_c is set close to the ZDWL of the fiber so that the linear phase mismatch in Eq. (12) can be maintained at an amount just enough to compensate for nonlinear phase mismatch. Therefore, to achieve a fairly wide phase-matching range, the two pump wavelengths should be located on the opposite sides of the ZDWL in a symmetric fashion, but should be reasonably far from it [4]. Figure 7.5 shows how κ can be reduced to zero over a wide wavelength range using such a scheme, resulting in a flat broadband gain spectrum. Comparing the single-pump and dual-pump cases, it can be seen that although single-pump FOPAs may provide nonuniform gain over a wider bandwidth under certain conditions, the dual-pump FOPAs provide much more uniform gain in general.

The preceding discussion is based on the assumption that only the nondegenerate FWM process contributes to FOPA gain. However, the situation is much more complicated for dual-pump FOPAs because the degenerate FWM process associated with each intense pump always occurs simultaneously with the nondegenerate one. In fact, it turns out that the combination of degenerate and nondegenerate FWM processes can create eight other idler fields besides the one at the frequency ω_4 [2]. Only four among these idlers, say at frequencies ω_5, ω_6, ω_7, and ω_8, are significantly relevant for describing the gain spectrum of FOPA because they are related to the signal frequency through the relations:

$$2\omega_1 \rightarrow \omega_3 + \omega_5, \quad 2\omega_2 \rightarrow \omega_3 + \omega_6 \quad (13)$$

$$\omega_1 + \omega_3 \rightarrow \omega_2 + \omega_7, \quad \omega_2 + \omega_3 \rightarrow \omega_1 + \omega_8 \quad (14)$$

Although these degenerate and nondegenerate FWM processes look as simple as Eq. (1) at the first glance, they do not occur independently because energy conversion is also maintained among the following processes: (15)

$$2\omega_1 \rightarrow \omega_4 + \omega_7, \quad 2\omega_2 \rightarrow \omega_4 + \omega_8 \quad (15)$$

$$\omega_1 + \omega_2 \rightarrow \omega_5 + \omega_8, \quad \omega_1 + \omega_2 \rightarrow \omega_6 + \omega_7 \quad (16)$$

$$\omega_1 + \omega_4 \rightarrow \omega_2 + \omega_5, \quad \omega_1 + \omega_6 \rightarrow \omega_2 + \omega_4 \quad (17)$$

All these processes involve at least two photons from one or both intense pumps and occur in the same order as the process in Eq. (1) as long as their phase-matching conditions are satisfied. As a result, a complete description of the FWM processes inside dual-pump FOPA becomes quite complicated [4]. Fortunately, a detailed analysis shows that the phase-matching conditions associated with these processes are quite different. When the two pumps are located symmetrically far from the ZDWL of the fiber, the 10 FWM processes shown in Eqs. (13)–(17) can only occur when the signal is in the vicinity of the two pumps. Thus, they leave unaffected the central flat part of the parametric gain spectrum resulting from the process shown in Eq. (1). Figure 7.6 compares the FOPA gain spectrum obtained numerically using a complete analysis that includes all five idlers model (solid curve) with that obtained using the sole nondegenerate FWM process of Eq. (1).

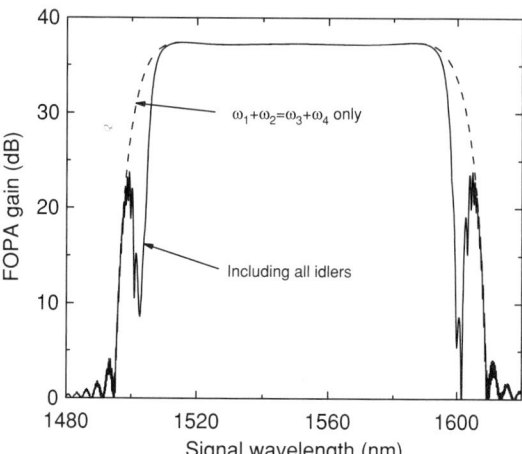

FIGURE 7.6 Gain spectra for a dual-pump FOPA including the contribution of all idlers (solid curve). The dotted curve shows gain spectrum when only a single idler corresponding to the dominant nondegenerate FWM process is included. The parameters used are $L = 0.5$ km, $\gamma = 10$ W^{-1}/km, $P_1 = 0.5$ W, $P_2 = 0.5$ W, $\beta_3 = 0.1$ ps^3/km, $\beta_4 = 10^{-4}$ ps^4/km, $\lambda_1 = 1502.6$ nm, $\lambda_2 = 1600.6$ nm, and $\lambda_0 = 1550$ nm.

It can be seen clearly that the flat portion of the gain spectrum has its origin in the single FWM process of Eq. (1). Other 10 processes [Eqs. (13)–(17)] only affect the edges of gain spectrum and reduce the gain bandwidth by 10–20%. We thus conclude that a model based on Eq. (1) is sufficient to describe the performance of dual-pump FOPAs as long as the central flat gain region is used experimentally.

Dual-pump FOPAs provide several degrees of freedom to realize a flat gain spectrum using just a single piece of fiber. By symmetrically assigning the two pumps on the opposite side of the ZDWL, a flat spectrum with high gain value has been obtained [48]. Because the two pumps are at the edge of the gain spectrum, pump blocking is no longer necessary, unlike the case of single-pump FOPAs. Moreover, because the pump power is distributed over two lasers in a dual-pump FOPA, the required launch power for each pump laser is only half that of the single-pump case. Figure 7.7 shows the data obtained in a recent experiment [49]. By using two pumps with powers of 600 mW at 1559 nm and 200 mW at 1610 nm, a gain of more than 40 dB over a 33.8-nm bandwidth was obtained inside a 1-km-long HNLF for which $\gamma = 17$ W^{-1}/km, ZDWL $= 1583.5$ nm, $\beta_3 = 0.055$ ps^3/km, and $\beta_4 = 2.35 \times 10^{-4}$ ps^4/km. The solid curve shows the theoretical prediction. It was necessary to include multiple idlers as well as the Raman gain to fit the experimental data well.

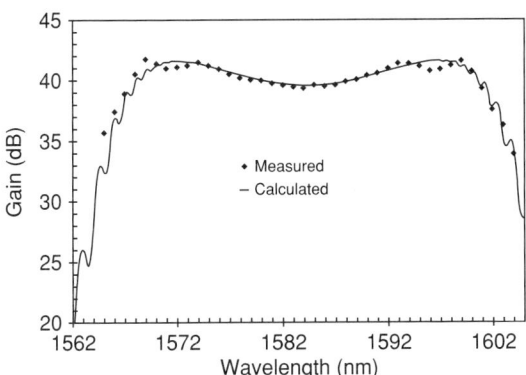

FIGURE 7.7 Measured (diamonds) and calculated (solid) gain spectrum as a function of signal wavelength for a dual-pump FOPA.

Although phase modulation of the pumps is still necessary to suppress SBS, spectral broadening of the idler is no longer a problem in dual-pump FOPAs because the phases of the two pumps can be manipulated such that a specific idler is not broadened, depending on which idler is used for wavelength conversion. If ω_4 in Eq. (1) is used for wavelength conversion, the two pumps should be modulated out of phase [50]. However, if ω_7 or ω_8 in Eq. (14) is used, the two pumps should be modulated in phase [51]. Idler spectrum broadening can also be eliminated by modulating the signal phase at a rate twice that used for modulating the pump phase [52, 53]. In the case of counterphase modulation, higher order idler generation in a dual-pump FOPA is shown to provide optical regeneration with a high extinction ratio and without spectral broadening [54].

Similar to the single-pump case, the gain in dual-pump FOPAs is also strongly polarization dependent if no precaution is taken to mitigate the polarization effects [6, 55]. Apart from the polarization diversity loop used for single-pump FOPAs, polarization independent operation of a dual-pump FOPA can also be realized by using orthogonally polarized pumps [56–59]. When the two pumps are linearly but orthogonally polarized, the nondegenerate FWM process becomes completely polarization independent. In one experiment, a small polarization-dependent gain (PDG) of only 1 dB was observed when the signal was amplified by 15 dB over a bandwidth of 20 nm [59].

A practical issue associated with dual-pump FOPA is the Raman-induced power transfer between the two pumps. As shown in Eq. (5), the FWM efficiency ξ is proportional to $\sqrt{P_1 P_2}$ for a nondegenerate process and is maximized when the two pump powers are the same ($P_1 = P_2$). However, as the two pumps are far from each other but still within the bandwidth of the Raman-gain spectrum, stimulated Raman scattering can transfer energy from the pump of high frequency to that of low frequency. Because the two pumps cannot maintain equality in their powers along the fiber, a significant reduction occurs in the FWM efficiency even though the total power of the two pumps remains constant. To reduce this effect, the power of the high-frequency pump is chosen to be higher than that of the low-frequency pump at the input end of the fiber. With this scheme, the two pumps can maintain their powers close to each other over most of the fiber. Although Raman-induced pump power transfer reduces the FOPA gain by a considerable amount, it does not affect the shape of the gain spectrum because phase matching depends on the total power of the two pumps that is conserved inside FOPA as long as the two pumps are not depleted too much.

5. FLUCTUATIONS OF ZDWL

In the preceding sections, the fiber used to make an FOPA was assumed to be free from any fluctuations in its material properties. However, it is difficult to realize such ideal conditions. In practice, optical waves in a realistic fiber undergo random perturbations originating from imperfections in the fiber. Two such imperfections are related to random variations along the fiber length in the ZDWL and residual birefringence, both of which originate partly from random changes in the core shape and size. In this section, we focus on ZDWL variations and consider the effects of residual birefringence in the next section. As dual-pump FOPAs provide much flatter gain spectra and are more likely to be used for telecommunication applications, we consider such FOPAs but limit our attention to the sole nondegenerate FWM process given in Eq. (1). As pointed out in the last section, this process is sufficient to describe the main flat portion of FOPA gain as long as the two pumps are located far from each other.

As seen clearly in Fig. 7.2, FOPA gain spectrum is extremely sensitive to dispersion parameters of the fiber. Changes in the ZDWL by as small as 0.05 nm change the gain spectrum considerably. Broad and flat gain spectra for dual-pump FOPAs were obtained in Section 4 by assuming that the dispersion characteristics of the fiber do not change along the fiber. However, this is not the case in reality. Fluctuations in the core shape and size along the fiber length make the ZDWL of the fiber change randomly. As such perturbations typically occur during the drawing process, they are expected to have a small correlation length (∼1 m). Long-term variations may also cause the ZDWL to vary over length scales comparable with fiber lengths used for FOPAs [60]. In general, ZDWL fluctuates only by a few nanometers, and the standard deviation of such fluctuations is a small fraction (<0.1%) of the mean ZDWL of the fiber.

In the presence of random ZDWL variations along the fiber, the growth of signal and idler waves is still governed by Eqs. (3) and (4), but the linear phase mismatch $\Delta\beta$ becomes a random function of z. As discussed before, a considerable amount of linear phase mismatch should be maintained over the main portion of the bandwidth to optimize and flatten the FOPA gain spectrum. Because the contribution of ZDWL fluctuations $\delta\lambda_0$ is much smaller than the average value of the linear phase mismatch, we can expand $\Delta\beta$ in a Taylor series to the first order as

$$\Delta\beta \approx K_1 + K_2 \delta\lambda_0 \qquad (18)$$

where $K_1 = \langle \Delta\beta \rangle$ is the average value and K_2 can be obtained from Eq. (9) or (12). The random variable $\delta\lambda_0$ can be modeled as a Gaussian stochastic process. If the correlation length l_c of ZDWL fluctuations is much smaller than the fiber length used for FOPA, the first-order and second-order moments of $\delta\lambda_0$ are given by

$$\langle \delta\lambda_0 \rangle = 0, \quad \langle \delta\lambda_0(z)\delta\lambda_0(z') \rangle = D_\lambda^2 \delta(z-z') \qquad (19)$$

where the diffusion coefficient D_λ is related to the standard deviation σ_λ of ZDWL fluctuations and their correlation length l_c as $D_\lambda^2 = \sigma_\lambda^2 l_c$.

The main question is how the signal/idler power is affected by ZDWL fluctuations. It turns out that the average value of signal or idler power can be found analytically [60]. After averaging Eqs. (3) and (4) over random ZDWL fluctuations [61], the evolution of the average signal/idler power is governed by the following three equations:

$$\frac{d\langle S_0 \rangle}{dz} = -2\xi \langle S_3 \rangle \qquad (20)$$

$$\frac{d\langle S_2 \rangle}{dz} = -\frac{1}{2} D_\lambda^2 K_2^2 \langle S_2 \rangle + K_1 \langle S_3 \rangle \qquad (21)$$

$$\frac{d\langle S_3 \rangle}{dz} = -2\xi \langle S_0 \rangle - \frac{1}{2} D_\lambda^2 K_2^2 \langle S_3 \rangle - K_1 \langle S_2 \rangle \qquad (22)$$

where $S_0(z) = P_3(z) + P_4(z) = 2P_3(z) - P_3(0)$ represents the sum of signal and idler powers and the auxiliary variables S_2 and S_3 are introduced using $S_2 - iS_3 = 2B_3 B_4$. By solving Eqs. (20)–(22) numerically, we obtain the average gain spectrum under the impact of ZDWL fluctuations.

The dashed curve in Fig. 7.8 shows the average gain spectrum obtained by solving the averaged Eqs. (20)–(22) for a FOPA operating under the conditions of Fig. 7.6 using $\sigma_\lambda = 1$ nm and $l_c = 5$ m. The solid curves show for comparison the results obtained numerically by solving the stochastic Eqs. (3) and (4) and averaging over 100 random realizations. Clearly, the agreement is quite good. However, the analytical theory becomes less accurate for larger correlation lengths. As an example, the dot-dashed and dotted curves show how the results change for a correlation length of 50 m. All these curves should be compared with the thin solid curve obtained in the case of a constant value of ZDWL. Clearly, ZDWL fluctuations are detrimental for FOPAs as they affect mainly the flat portion of the gain spectrum. As discussed in the preceding section, the flat gain region is obtained by carefully optimizing the linear phase mismatch $\Delta\beta$ so that it compensates the nonlinear part and results in the total phase mismatch $\kappa \approx 0$ over a fairly broad spectral range. Thus, it is not surprising that ZDWL fluctuations deteriorate mainly the flat portion of FOPA gain spectrum. Figure 7.8 shows that ZDWL variations of even ±1 nm eliminate the flat portion of the gain spectrum and produce two narrow peaks in the vicinity of each pump wavelength because ZDWL fluctuations do not affect the gain around the two

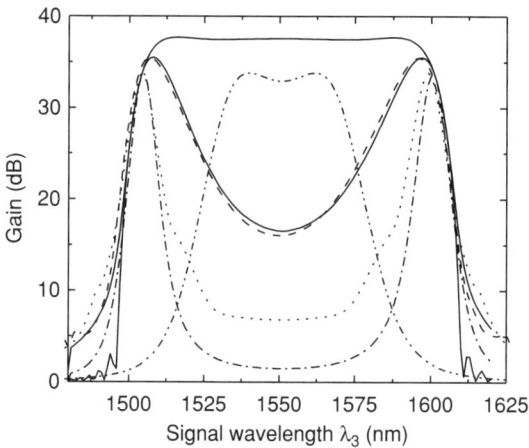

FIGURE 7.8 Average gain spectra for a dual-pump FOPA in which ZDWL varies randomly along the fiber with a standard deviation of 1 nm. Analytical and numerical results are shown by dashed and solid curves for $l_c = 5$ m and by dot-dashed and dashed curves for $l_c = 50$ m. The thin solid curve shows the gain in the absence of fluctuations. The innermost curve shows the results for $l_c = 5$ m when pumps are spaced apart by only 30 nm. All other parameters are identical to those used for Fig. 7.6.

pumps. The analytical theory based on Eqs. (20)–(22) agrees well with numerical simulation when the ZDWL correlation length is much shorter than the total fiber length but begins to deviate when the two are comparable. In the latter case, analytical theory overestimates the degradation caused by random ZDWL variations.

The gain spectra shown in Fig. 7.8 do not show the spectrum expected for a specific FOPA but rather represent an ensemble average. In practice, the gain spectrum varies over a wide range for an ensemble of FOPAs that are otherwise identical. Figure 7.9a shows the individual gain spectra for 100 realizations obtained numerically by solving the stochastic Eqs. (3) and (4). The parameters are identical to those used for the solid curve in Fig. 7.8 that represents the average of these 100 spectra. It is evident that amplified signals can fluctuate over a wide range for different members of the ensemble even when $\sigma_\lambda = 1$ nm. The important question is how one can design FOPAs that can tolerate ZDWL variations of ~ 1 nm? The answer turns out to be that the wavelength separation between the two pumps should be reduced significantly so that the second term in Eq. (12) does not play a major role. Of course, the whole gain spectrum is then much narrower, and the FOPA bandwidth is significantly reduced. However, this reduced-bandwidth gain spectrum is much more tolerant of ZDWL fluctuations. This is evident in Fig. 7.9b obtained under conditions identical to those of Fig. 7.9a except that the two pump wavelengths are separated by 30 nm rather than by 98 nm. The innermost curve in Fig. 7.8 shows the average spectrum. Flatness of the average gain spectrum is nearly maintained under such conditions, but the spectrum is much narrower. We conclude this section by emphasizing that ZDWL fluctuations limit in practice the usable bandwidth of an FOPA.

6. EFFECT OF RESIDUAL FIBER BIREFRINGENCE

Most fibers exhibit residual birefringence that fluctuates randomly along the fiber length. Such birefringence fluctuations induce PMD and randomize the SOP of any optical wave propagating through the fiber [62–65]. They change the relative orientation of the four waves, affect the angular momentum conservation among the four photons during the FWM process, and thus seriously degrade the performance of FOPAs [6]. Because the nonlinear processes contributing to parametric amplification depend on the SOPs of the fields, a vector theory of FWM is needed. In this section we present such a theory and discuss its impact on FOPA performance.

Adopting the Jones vector notation for vector fields used in [64], the four equations governing the propagation of two pumps, signal, and idler in the presence of randomly fluctuating birefringence can be written as follows

$$\frac{\partial |A_1\rangle}{\partial z} = (b_0 \sigma_1 + \beta_1)|A_1\rangle + \frac{i\gamma}{3}\left(|A_1^*\rangle\langle A_1^*| + 2|A_2\rangle\langle A_2| + 2|A_2^*\rangle\langle A_2^*|\right)|A_1\rangle \quad (23)$$

$$\frac{\partial |A_2\rangle}{\partial z} = [b_0 \sigma_1 + b_1(\omega_2 - \omega_1)\sigma_1 + \beta_2]|A_2\rangle + \frac{i\gamma}{3}\left(|A_2^*\rangle\langle A_2^*| + 2|A_1\rangle\langle A_1| + 2|A_1^*\rangle\langle A_1^*|\right)|A_2\rangle \quad (24)$$

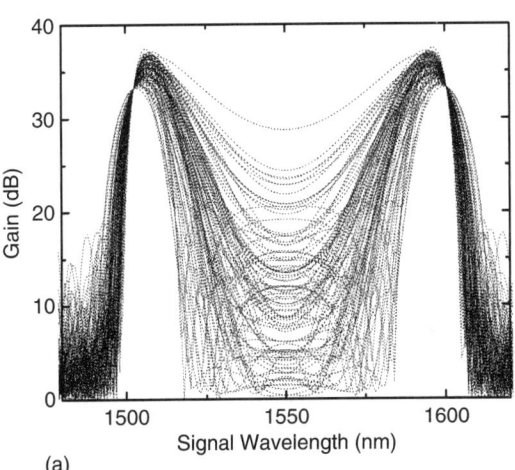 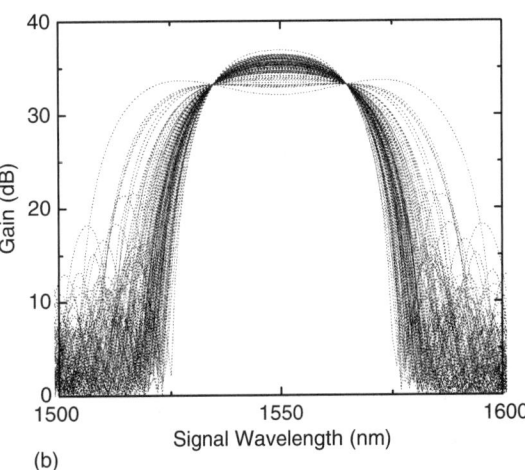

FIGURE 7.9 Changes expected in the gain spectra from fiber to fiber because of ZDWL fluctuations ($\sigma_\lambda = 1$ nm and $l_c = 10$ m) when pump spacing is 98 nm (a) or 30 nm (b). All other parameters are identical to those used for Fig. 7.6.

$$\frac{\partial |A_3\rangle}{\partial z} = [b_0 \sigma_1 + b_1(\omega_3 - \omega_1)\sigma_1 + \beta_3]|A_3\rangle$$
$$+ \frac{2i\gamma}{3} \left[\sum_{j=1}^{2} (|A_j\rangle\langle A_j| + |A_j^*\rangle\langle A_j^*|) \right] |A_3\rangle$$
$$+ \frac{2i\gamma}{3} (|A_1\rangle\langle A_2^*| + |A_2\rangle\langle A_1^*| + \langle A_1^*|A_2\rangle)|A_4^*\rangle \quad (25)$$

$$\frac{\partial |A_4\rangle}{\partial z} = [b_0 \sigma_1 + b_1(\omega_4 - \omega_1)\sigma_1 + \beta_4]|A_4\rangle$$
$$+ \frac{2i\gamma}{3} \left[\sum_{j=1}^{2} (|A_j\rangle\langle A_j| + |A_j^*\rangle\langle A_j^*|) \right] |A_4\rangle$$
$$+ \frac{2i\gamma}{3} (|A_1\rangle\langle A_2^*| + |A_2\rangle\langle A_1^*| + \langle A_1^*|A_2\rangle)|A_3^*\rangle \quad (26)$$

where $|A_j\rangle$ ($j = 1-4$) represents the Jones vector apart from a constant phase factor that depends on the total pump power and σ_1 is a 2×2 diagonal Pauli matrix with elements 1 and -1. The effects of residual birefringence are included through two random variables b_0 and b_1. They are defined using the Taylor expansion $b(\omega) \approx b_0 + b_1(\omega - \omega_1)$, where $b(\omega) = \omega \delta n(\omega)/c$ and $\delta n(\omega)$ denotes refractive index fluctuations. The vector equations for the case of a single pump can be deduced from these equations after minor modifications.

It is clear from Eqs. (23)–(26) that in the absence of b_1, all the fields change their SOP in the same way. These equations are written using the first pump as a reference. As a result, the other three fields change their SOPs around the first pump at a rate given by the frequency difference $\omega_j - \omega_1$, where $j = 2, 3$, or 4. The random variable b_1 is also responsible for the difference in the group velocity between pulses of different polarization and leads to PMD. In fact, the variance of b_1 is related to the PMD parameter D_p through the expression $\langle b_1^2 \rangle = D_p^2/l_c$ [66], where l_c is the correlation length over which b_0 and b_1 change. Both b_0 and b_1 obey Gaussian statistics with zero mean.

Similar to the case of ZDWL fluctuations, one can use the stochastic Eqs. (23)–(26) to calculate the average gain spectrum in the presence of PMD. In the case of single-pump FOPA, the average can be performed analytically. Figure 7.10 shows the analytic results (solid curves) and compares them with the direct numerical simulations (dashed curves) for $D_p = 0.05$ and $0.15\,\mathrm{ps}/\sqrt{\mathrm{km}}$. The dotted curve represents the expected gain in the absence of birefringence fluctuations. It is evident that random fluctuations of birefringence severely impact the FOPA gain spectrum. Random variables b_0 and b_1

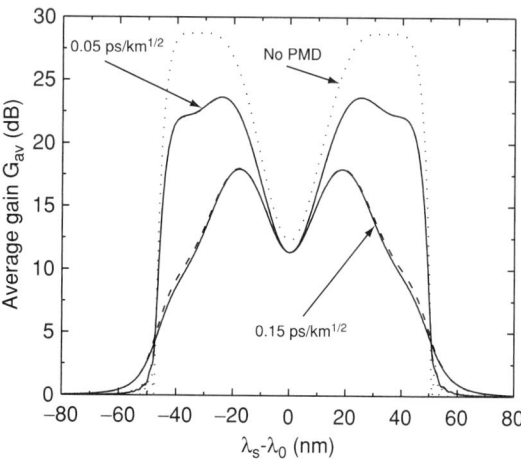

FIGURE 7.10 Average gain spectrum for a single-pump FOPA for two values of the PMD parameter D_p. Solid and dashed curves compare the analytical and numerical results. Dotted curve shows the no-PMD case. Parameters values used are $\gamma = 2\,\mathrm{W}^{-1}/\mathrm{km}$, $L = 2\,\mathrm{km}$, $\lambda_0 = 1550\,\mathrm{nm}$, $\beta_3 = 0.1\,\mathrm{ps}^3/\mathrm{km}$, $\beta_4 = 10^{-4}\,\mathrm{ps}^4/\mathrm{km}$, and $P_1 = 1\,\mathrm{W}$.

affect the FOPA through two different mechanisms. Whereas b_0 rotates the SOP of all four fields in the same manner and thus reduces the available average gain by roughly the same amount at all frequencies, b_1 causes pump and signal SOPs to drift from each other at a rate that depends on their frequency difference.

It is useful to define a diffusion length L_d as $L_d = 3/(D_p \Delta \omega)^2$; it quantifies the length scale over which two fields that have the same SOPs initially develop random SOPs with respect to each other [67]. When L_d is much larger than the total fiber length, the effects of b_1 are negligible. In a typical fiber, b_0 changes the SOP of fields over a length scale of 10 m. Because such changes take place over a much shorter scale than the nonlinear length and are same for all the fields, Eqs. (23)–(26) can be averaged over b_0. The main effect of b_0 is that the nonlinear parameter γ is reduced by a factor of 8/9. Because the efficiency of both FWM and XPM is reduced, the gain is lowered at the peaks from 28 to 24.5 dB but by only 1 dB in the central region. In Fig. 7.10, the gain curve for $D_p = 0.05\,\mathrm{ps}/\sqrt{\mathrm{km}}$ very closely follows this prediction. This is not surprising because for this low value of D_p $L_d = 2.1\,\mathrm{km}$ is comparable with the total fiber length of 2 km even for as large as 30 nm separation between pump and signal. However, L_d is reduced to 0.24 km when $D_p = 0.15\,\mathrm{ps}/\sqrt{\mathrm{km}}$, and the effects of b_1 takes over. As seen in Fig. 7.10, the gain is reduced by more than 10 dB and the spectrum is distorted considerably.

In the case of dual-pump FOPAs, the complexity of Eqs. (23)–(26) hinders an analytic treatment. For this reason, we solve these equations numerically for three different values of D_p. Figure 7.11 shows fluctuations in gain for $D_p = 0.1\,\text{ps}/\sqrt{\text{km}}$ for 50 different realizations of random birefringence. Note that the birefringence parameters can change with time for a given fiber on a time scale ranging from seconds to hours. For this reason, gain fluctuations seen in Fig. 7.11 can also be viewed as fluctuations with time for a given FOPA. The average gain obtained from 50 realizations is shown in Figure 7.12 for three different values of D_p. The ideal case of isotropic fiber is also shown for comparison as a solid curve. Similar to the case of single-pump FOPA, the effect of b_0 is just to reduce the nonlinear coefficient γ by a factor 8/9 [66]. For the parameters used in Fig. 7.12, the lower value of γ reduces the peak gain from 37 to 33 dB but keeps the spectrum flat. The central dip seen in Fig. 7.12 results from b_1. The reason behind this dip can be understood as follows. When the signal frequency is close to one of the pumps, that pump provides the dominant contribution. However, as signal frequency moves toward the center of the spectrum, neither of the two pumps remains oriented parallel to the signal, and the gain is reduced.

As discussed earlier, it is important that FOPAs provide gain that does not depend on the SOP of the input signal, and two methods are commonly used for this purpose. However, the random residual birefringence prevents these methods from working perfectly. The polarization diversity method relies critically on the assumption that the signal and pump maintain their identical SOPs throughout the

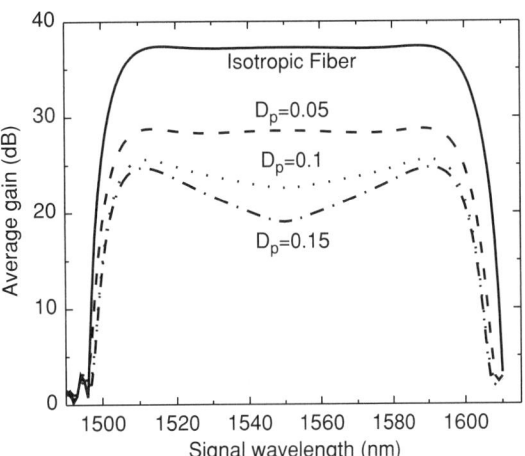

FIGURE 7.12 Average gain spectra for a dual-pump FOPA for three values of the PMD parameter with the same parameter values used in Fig. 7.6. Solid curve shows for comparison the no-PMD case.

fiber. The second method utilizes orthogonal pumps and works only under the assumption that the pumps maintain their orthogonal SOPs throughout the fiber. It was already noted that fields with different frequencies change their SOP at different rates in the presence of PMD. It is thus evident that both of these schemes introduce PDG. This was also observed experimentally when the second method was implemented [63].

To illustrate the performance of the second method in the presence of PMD, we initially performed numerical simulations using two pumps with linear but orthogonal SOPs. For each realization, the signal is launched at an angle of $\theta = 0$, 45 and $90°$ from the pump at the shorter wavelength. Figure 7.13 shows the results for a PMD parameter $D_p = 0.1\,\text{ps}/\sqrt{\text{km}}$. The expected gain curve in the absence of PMD is also shown as a dotted curve. For certain signal wavelengths, PDG can be as large as 12 dB, where PDG is defined as the difference between the maximum and minimum gain as the signal SOP is changed. PDG increases as the signal wavelength gets closer to either of the pumps. The same behavior was also observed in a 2003 experiment [6]. In physical terms, the reason why the largest PDG occurs for a signal close to pump wavelength can be understood as follows. If the signal has a wavelength close to one pump, its relative orientation does not change along the fiber but it decorrelates with the other pump rapidly because they are at the different edges of the spectrum. As a result, whatever the initial polarization of the second pump is, the signal can only sense its average effect. However, because

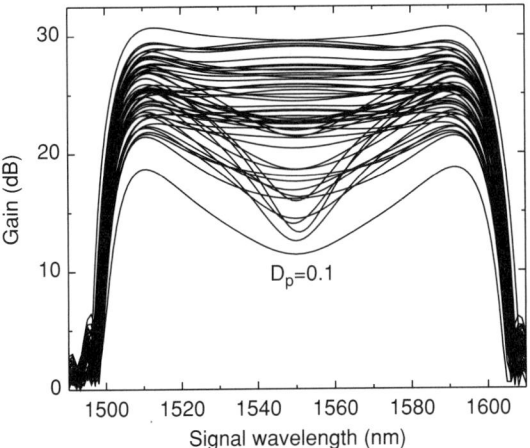

FIGURE 7.11 Changes in gain spectra with birefringence fluctuations for a dual-pump FOPA for $D_p = 0.1\,\text{ps}/\sqrt{\text{km}}$. Other parameters are same as used in Fig. 7.6. Both pumps and the signal have the same SOP initially.

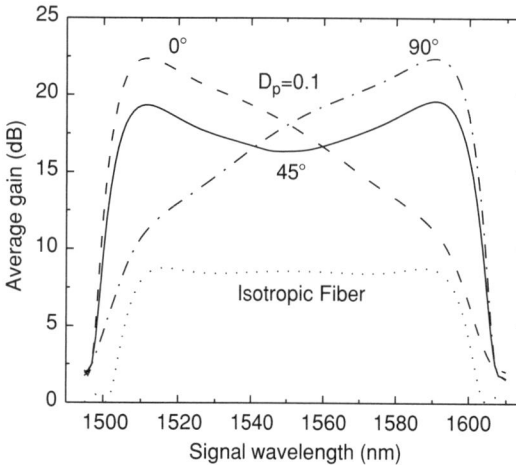

FIGURE 7.13 Average gain vs. signal wavelength for three different initial linear SOP of the signal when $D_p = 0.1\,\text{ps}/\sqrt{\text{km}}$; θ represents the angle between the linear SOPs of signal and the shorter wavelength pump. The other pump is orthogonally polarized. Dotted curve shows for comparison the no-PMD case.

it keeps its relative orientation with the first pump, the signal experiences the highest or smallest gain depending on if it is initially parallel or orthogonal to the first pump. The reason why the overall gain is smallest in the case of isotropic fiber is that the FWM efficiency is the smallest when the pumps are orthogonal. PMD can make the pumps nonorthogonal (and even parallel occasionally) and hence increases the gain.

7. SUMMARY

In this review on FOPAs we focused on some of the recent advances that have considerably improved the state-of-the-art for parametric amplifiers. The well-known simple theory behind the nonlinear phenomenon of FWM was discussed first to provide the background material. It was then used to discuss the performance of single-pump FOPAs and reveal the important role played by the nonlinear contribution to the phase-matching condition. The same scalar theory was used to discuss the more general case of dual-pump FOPAs and show that the gain spectrum in such amplifiers results from several degenerate and nondegenerate FWM processes. However, it turns out that only a single nondegenerate process dominates in the case of FOPAs pumped at two wavelengths relatively far apart from each other and located on opposite sides of the ZDWL. We discussed the design of such FOPAs and showed how they can be designed to provide a gain spectrum that is relatively uniform over a bandwidth larger than 100 nm.

The experiments on dual-pump FOPAs have shown that flat gain over a bandwidth of 40 nm or so can be realized in practice. To resolve this discrepancy between the theoretical and experimental results, we studied the impact of two unavoidable phenomena that affect the performance of all FOPA-based devices. First, the ZDWL of a fiber can vary along its length in a random fashion due to core-diameter variations that invariably occur during fiber manufacturing. We showed that ZDWL fluctuations as small as ±1 nm severely degrade the gain spectrum of FOPAs because they mainly affect its central flat part. This degradation can be avoided to a large extent by moving pump wavelengths closer so that they are separated by about 30–40 nm instead of 100 nm or more. This appears to be the main reason why experiments have realized flat gain only over 40 nm or so. Second, birefringence fluctuations that occur in all practical fibers and lead to PMD also affect the gain spectrum of FOPAs. Their inclusion requires the development of a vector theory based on the Jones matrix formalism that can also be related to the rotation of the SOP of each optical field on the Poincaré sphere. We discussed how PMD affects the gain spectrum and make the FOPA gain polarization dependent even when orthogonally polarized pumps are used.

8. ACKNOWLEDGMENTS

This work is supported by the U.S. National Science Foundation under grants ECS-0320816 and ECS-0334982.

9. REFERENCES

1. G. P. Agrawal, *Nonlinear Fiber Optics*, 3rd ed. (Academic Press, San Diego, 2001).
2. J. Hansryd, P. A. Andrekson, M. Westlund, J. Li, and P. O. Hedekvist, "Fiber-based optical parametric amplifiers and their applications," IEEE J. Sel. Top. Quantum Electron., **8**, pp. 506–520 (2002).
3. M. N. Islam and Ö. Boyraz, "Fiber Parametric Amplifiers for wavelength band conversion," IEEE. J. Sel. Top. Quantum Electron. **8**, 527–537 (2002).
4. C. J. McKinstrie, S. Radic, and A. R. Chraplyvy, "Parametric amplifiers driven by two pump waves," IEEE J. Sel. Top. Quantum Electron. **8**, 538–547 (2002).
5. S. Radic, C. J. McKinstrie, A. R. Chraplyvy, G. Raybon, J. C. Centanni, C. G. Jorgensen, K. Brar, and C. Headley, "Continuous-wave parametric gain synthesis using nondegenerate pump four-wave mixing," IEEE Photon. Technol. Lett. **14**, 1406–1408 (2002).

6. S. Radic and C. J. McKinstrie, "Two-pump fiber parametric amplifiers," Opt. Fiber. Technol. **9**, 7–23 (2003).
7. A. Yariv, D. Fekete, and D. M. Pepper, "Compensation for channel dispersion by nonlinear optical phase conjugation," Opt. Lett. **4**, 52–54 (1979).
8. S. Watanabe and M. Shirasaki, "Exact compensation for both chromatic dispersion and Kerr effect in a transmission fiber using optical phase conjugation," J. Lightwave Technol. **14**, 243–248 (1996).
9. J. Santhanam and G. P. Agrawal, "Reduced timing jitter in dispersion-managed lightwave systems through parametric amplification," J. Opt. Soc. Am. B **20**, 284–291 (2003).
10. C. J. McKinstrie, S. Radic, and C. Xie, "Reduction of soliton phase jitter by in-line phase conjugation," Opt. Lett. **28**, 1519–1561 (2003).
11. J. A. Levenson, I. Abram, T. Rivera, and P. Grangier, "Reduction of quantum noise in optical parametric amplification," J. Opt. Soc. Am. B **10**, 2233–2238 (1993).
12. I. H. Deutsch and I. Abram, "Reduction of quantum noise in soliton propagation by phase-sensitive amplification," J. Opt. Soc. Am. B **11**, 2303–2313 (1994).
13. W. Imajuku, A. Takada, and Y. Yamabayashi, "Inline coherent optical amplifier with noise figure lower than 3 dB quantum limit," Electron. Lett. **36**, 63–64 (2000).
14. R. W. Boyd, *Nonlinear Optics*, 2nd ed. (Academic Press, San Diego, 2003).
15. K. Inoue and T. Mukai, "Signal wavelength dependence of gain saturation in a fiber optical parametric amplifier," Opt. Lett. **26**, 10–12 (2001).
16. M. E. Marhic, N. Kagi, T. K. Chiang, and L. G. Kazovsky, "Broadband fiber optical parametric amplifiers," Opt. Lett. **21**, 573–575 (1996).
17. M. Ho, K. Uesaka, M. Marhic, Y. Akasaka, L. G. Kazovsky, "200-nm-Bandwidth fiber optical amplifier combining parametric and Raman gain," J. Lightwave Technol. **19**, 977–981 (2001).
18. R. Tang, P. Devgan, J. Sharping, P. Voss, J. Lasri, and P. Kumar, "Microstructure fiber based optical parametric amplifier in the 1550-nm Telecom band," Proc. Opt. Fiber Commun. (Optical Society of America, Washington, DC, 2003), pp. 562–563.
19. K. Shimizu, K. Wong, G. Kalogerakis, K. Uesaka, M. Marhic, and L. Kazovsky, "High repetition-rate pulsed-pump fiber OPA for 10Gb/s NPZ modulated signals," Proc. Opt. Fiber Commun. (Optical Society of America, Washington, DC, 2003), pp. 566–567.
20. M. E. Marhic, F. S. Yang, M. Ho, and L. G. Kazovsky, "High-nonlinearity fiber optical parametric amplifier with periodic dispersion compensation," J. Lightwave Technol. **17**, 210–215 (1999).
21. J. Kim, Özdal Boyraz, J. H. Lim, and M. N. Islam, "Gain enhancement in cascaded fiber parametric amplifier with quasi-phase matching: theory and experiment," J. Lightwave Technol. **19**, 247–251 (2001).
22. K. Inoue, "Arrangement of fiber pieces for a wide wavelength conversion range by fiber four-wave mixing," Opt. Lett. **19**, 1189–1191 (1994).
23. J. Hansryd and P. A. Andrekson, "Broad-band continuous-wave-pumped fiber optical parametric amplifier with 49-dB gain and wavelength-conversion efficiency," IEEE Photon. Technol. Lett. **13**, 194–196 (2001).
24. T. Torounidis, H. Sunnerud, P. O. Hedekvist, and P. A. Andrekson, "Amplification of WDM signals in Fiber-Based optical parametric amplifiers," IEEE Photon. Technol. Lett. **15**, 1061–1063 (2003).
25. T. Torounidis, H. Sunnerud, P. O. Hedekvist, and P. A. Andrekson, "40Gb/s transmission using RZ-pulse source based on fiber optical parametric amplification," IEEE Photon. Technol. Lett. **15**, 1159–1161 (2003).
26. J. Hansryd and P. A. Andrekson, "Wavelength tunable 40-GHz pulse source based on fiber optical parametric amplifier," Electron. Lett. **37**, 584–585 (2001).
27. G. A. Nowak, Y. Hao, T. J. Xia, and M. N. Islam, and D. Nolan, "Low-power high-efficiency wavelength conversion based on modulational instability in high-nonlinearity fiber," Opt. Lett. **23**, 936–938 (1998).
28. M. Westlund, J. Hansryd, P. A. Andrekson, and S. N. Knudsen, "Transparent wavelength conversion in fiber with 24 nm pump tuning range," Electon. Lett. **38**, 85–86 (2002).
29. J. Hansryd and P. A. Andrekson, "O-TDM demultiplexer with 40-dB gain based on a fiber optical parametric amplifier," IEEE Photon. Technol. Lett. **13**, 732–734 (2001).
30. J. Li, J. Hansryd, P. O. Hedekvist, P. A. Andrekson, and S. N. Knudsen, "300-Gb/s eye-diagram measurement by optical sampling using fiber-based parametric amplification," IEEE Photon. Technol. Lett. **13**, 987–989 (2001).
31. Y. Su, L. Wang, A. Agarwal, and P. Kumar, "All-optical limiter using gain flattened fiber parametric amplifier," Electron. Lett. **36**, 1103–1104 (2000).
32. E. Ciaramella and S. Trillo, "All-optical signal reshaping via four-wave mixing in optical fibers," IEEE Photon. Technol. Lett. **12**, 849–851 (2000).
33. K. Inoue, "Suppression of level fluctuation without extinction ratio degradation based on output saturation in higher order optical parametric interaction in fiber," IEEE Photon. Technol. Lett. **13**, 338–340 (2001).
34. M. E. Marhic, K. K. Y. Wong, M. C. Ho, and L. G. Kazovsky, "92% pump depletion in a continuous-wave

35. M. E. Marhic, K. K. Y. Wong, L. G. Kazovsky, and T. E. Tsai, "Continuous-wave fiber optical parametric oscillator," Opt. Lett. **27**, 1439–1441 (2002).
36. X. Zhang and B. F. Jørgensen, "Noise characteristics and optimization fiber length of spectral inversion using four-wave mixing in a dispersion-shifted fiber," Opt. Fiber Technol. **3**, 28–43 (1997).
37. P. O. Hedekvist and P. A. Andrekson, "Noise characteristics of fiber-based optical phase conjugators," J. Lightwave Technol. **17**, 74–79 (1999).
38. J. L. Blows and S. E. French, "Low-noise-figure optical parametric amplifier with a continuous-wave frequency-modulated pump," Opt. Lett. **27**, 491–493 (2002).
39. K. K. Y. Wong, K. Shimizu, M. E. Marhic, K. Uesaka, G. Kalogerakis, and L. G. Kazovsky, "Continuous-wave fiber optical parametric wavelength converter with 40 dB conversion efficiency and a 3.8 dB noise figure," Opt. Lett. **28**, 692–694 (2003).
40. G. P. Agrawal, *Fiber-Optic Communication Systems*, 3rd ed. (Wiley, New York, 2002).
41. L. Provino, A. Mussot, E. Lantz, T. Sylvestre, and H. Maillotte, "Broadband and flat parametric amplifiers with a multisection dispersion-tailored nonlinear fiber arrangement," J. Opt. Soc. Am. B **20**, 1532 (2003).
42. T. Hasegawa, K. Inoue, and K. Oda, "Polarization independent frequency conversion by fiber four-wave mixing with a polarization diversity technique," IEEE Photon. Technol. Lett. **5**, 947–949 (1993).
43. K. K. Y. Wong, M. E. Marhic, K. Uesaka, and L. G. Kazovsky, "Polarization-independent one-pump fiber-optical parametric amplifier," IEEE Photon. Technol. Lett. **14**, 1506–1508 (2002).
44. A. Tersigni, V. Calle, A. T. Clausen, L. K. Oxenløwe, J. Mørk, and P. Jeppesen, "Polarization independent optical sampling using four-wave mixing," Conf. Lasers & Electro-Optics (Optical Society of America, Washington, DC, 2003), paper CMR2.
45. S. Watanabe, S. Takeda, and T. Chikama, "Interband wavelength conversion of 320 Gb/s (32 × 10Gb/s) WDM signal using a polarization-insensitive four-wave mixer," ECOC98 (1998) pp. 85–86.
46. C. J. McKinstrie and S. Radic, "Parametric amplifiers driven by two pump waves with dissimilar frequencies," Opt. Lett. **27**, 1138–1140 (2002).
47. M. E. Marhic, Y. Park, F. S. Yang, and L. G. Kazovsky, "Broadband fiber-optical parametric amplifiers and wavelength converters with low-ripple Chebyshev gain spectra," Opt. Lett. **21**, 1354–1356 (1996).
48. S. Radic, C. J. McKinstrie, A. R. Chraplyvy, G. Raybon, J. C. Centanni, C. G. Jorgensen, K. Brar, and C. Headley, "Continuous-wave parametric gain synthesis using nondegenerate pump four-wave mixing," IEEE Photon. Technol. Lett. **14**, 1406–1408 (2002).
49. S. Radic, C. J. McKinstrie, R. M. Jopson, J. C. Centanni, Q. Lin, and G. P. Agrawal, "Record performance of parametric amplifier constructed with highly nonlinear fiber," Electron. Lett. **39**, 838–839. (2003)
50. M. Ho, M. E. Marhic, K. Y. K. Wong, and L. G. Kazovsky, "Narrow-linewidth idler generation in fiber four-wave mixing and parametric amplification by dithering two pumps in opposition of phase," J. Lightwave Technol. **20**, 469–476 (2002).
51. S. Radic, C. J. McKinstrie, R. M. Jopson, J. C. Centanni, A. R. Chraplyvy, C. G. Jorgensen, K. Brar, and C. Headley, "Selective suppression of idler spectral broadening in two-pump parametric architectures," IEEE Photon. Technol. Lett. **15**, 673–675 (2003).
52. K. Torii and S. Yamashita, "Cancellation of spectral spread by pump frequency modulation in optical fiber wavelength converter," OFC01, WW5-1 (2001).
53. K. Torii and S. Yamashita, "Efficiency improvement of optical fiber wavelength converter without spectral spread using synchronous phase/frequency modulations," J. Lightwave Technol. **21**, 1039–1045 (2003).
54. S. Radic, C. J. McKinstrie, R.M. Jopson, J. C. Centanni, and A. R. Chrapyvy, "All-optical regeneration in one and two-pump parametric amplifier using highly nonlinear optical fiber," IEEE Photon. Technol. Lett. **15**, 957–959 (2003).
55. A. Guimaraes, W. A. Arellano, M. O. Berendt, and H. L. Fragnito, "Measurement of polarization dependent gain in a dual pump fiber optical parametric amplifier," Conf. Lasers & Electro-Optics (Optical Society of America, Washington, DC, 2001), pp. 447–448.
56. R. M. Jopson and R. E. Tench, "Polarization-independent phase conjugation of lightwave signals," Electron. Lett. **29**, 2216–2217 (1993).
57. K. Inoue, "Polarization independent wavelength conversion using fiber four-wave mixing with two orthogonal pump lights of different frequencies," J. Lightwave Technol. **12**, 1916–1920 (1994).
58. S. Yamashita and K. Torii, "Polarization-independent highly-efficient optical fiber wavelength converter," Conf. Lasers & Electro-Optics (Optical Society of America, Washington, DC, 2001), pp. 384–385.
59. K. K. Y. Wong, M. E. Marhic, K. Uesaka, and L. G. Kazovsky, "Polarization-independent two-pump fiber optical parametric amplifier," IEEE Photon. Technol. Lett. **14**, 911 (2002).
60. M. Karlsson, "Four-wave mixing in fibers with randomly varying zero-dispersion wavelength," J. Opt. Soc. Am. B **15**, 2269–2275 (1998).
61. C. W. Gardiner, *Handbook of Stochastic Methods*, 2nd ed. (Springer, New York, 1985).

62. P. O. Hedekvist, M. Karlsson, and P. A. Andrekson, "Polarization dependence and efficiency in a fiber four-wave mixing phase conjugator with orthogonal pump waves," IEEE Photon Technol. Lett. **8**, 776–778 (1996).
63. S. Radic, C. McKinstrie, R. Jopson, C. Jorgensen, K. Brar, and C. Headley, "Polarization dependent parametric gain in amplifiers with orthogonally multiplexed optical pumps," Optical Fiber Commun. Conf. (2003) pp. 508–509.
64. J. P. Gordon and H. Kogelnik, "PMD fundamentals: polarization mode dispersion in optical fibers," Proc. Natl. Acad. Sci. USA **97**, 4541–4550 (2000).
65. H. Kogelnik, R. M. Jopson, and L. E. Nelson, "Polarization-mode dispersion," in *Optical Fiber Telecommunications IV B*, I. P. Kaminow and T. Li, Eds. (Academic Press, San Diego, 2002) pp. 725–861.
66. P. K. A. Wai and C. R. Menyuk, "Polarization mode dispersion, decorrelation, and diffusion in optical fibers with randomly varying birefringence," J. Lightwave Technol. **14**, 148–157 (1996).
67. M. Karlsson and J. Brentel, "Autocorrelation function of the polarization-mode dispersion vector," Opt. Lett., **24**, 939–941 (1999).

Erbium-Doped Fiber Amplifiers

K. Thyagarajan*
Department of Physics
Indian Institute of Technology Delhi
New Delhi, India

1. INTRODUCTION

In traditional long-distance optical fiber communication systems, compensation of loss and dispersion is usually accomplished by using electronic regenerators in which the optical signals are first converted into electrical signals, then processed in the electrical domain, and then reconverted into optical signals. Whenever the system limitation is due to insufficient optical power rather than dispersion, amplification of the signal and optical amplifiers can indeed perform this job. Optical amplifiers are devices that amplify the incoming optical signals in the optical domain itself without any conversion to the electrical domain; these devices have truly revolutionized long-distance fiber optic communications.

Compared with electronic regenerators, optical amplifiers do not need any high-speed electronic circuitry, are transparent to bit rate and format, and most importantly can amplify multiple optical signals at different wavelengths simultaneously. Thus, their development has ushered in the tremendous growth of communication capacity using wavelength division multiplexing (WDM) in which multiple wavelengths carrying independent signals are propagated through the same single-mode fiber, thus multiplying the capacity of the link. Of course, compared with electronic regenerators, they have some drawbacks too: They do not compensate for dispersion accumulated in the link and they also add noise to the optical signal. As we see later, this noise leads to a maximum number of amplifiers that can be cascaded so that the received signal-to-noise ratio (SNR) is within the limits.

Optical amplifiers can be used at many points in a communication link. Figure 8.1 shows some typical examples. A booster amplifier is used to boost the power of the transmitter before launching into the fiber link. The increased transmitter power can be used to attain longer reach in a link. The preamplifier placed just before the receiver is used to increase the receiver sensitivity. In-line amplifiers are used at intermediate points in the link to overcome fiber transmission and other losses. Optical amplifiers can also be used for overcoming splitter losses, for example, for distribution of cable televisions (see Chapter 11).

The three main types of optical amplifiers are the erbium-doped fiber amplifier (EDFA), the Raman fiber amplifier (RFA) (see Chapter 9), and the semiconductor optical amplifier (SOA). Today, most optical fiber communication systems use EDFAs because of their advantages in terms of bandwidth, high power output, and noise characteristics. RFAs and SOAs are also becoming important in many applications. In the following we attempt a basic introduction to the characteristics of EDFAs; detailed discussions on EDFAs can be found in many texts [1–3].

2. EDFA

Optical amplification by EDFA is based on the process of stimulated emission, which is the basic principle behind laser operation. In fact, a laser without any optical feedback is just an optical amplifier.

Figure 8.2 shows two levels of an atomic system: the ground level with energy E_1 and an excited level with energy E_2. Under thermal equilibrium, most of the atoms are in the ground level. Thus, if light corresponding to an appropriate frequency ($\nu = (E_2 - E_1)/h$, h being Planck's constant) falls on this collection of atoms, then it will result in a greater number of absorptions than stimulated emissions and the light beam suffers attenuation. On the other hand, if the number of atoms in the upper level could be made more than those in the lower level, then an incident light beam at the appropriate frequency could induce more stimulated emissions than absorptions, thus leading to optical amplification.

---

*E-mail: ktrajan@physics.iitd.ernet.in

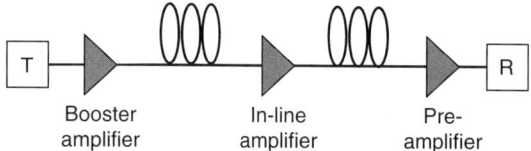

FIGURE 8.1 Optical amplifiers in different forms for boosting the transmitter power (booster amplifier), for use along the link (in-line amplifiers), and for use before the receiver (preamplifier). T, transmitter; R, Receiver.

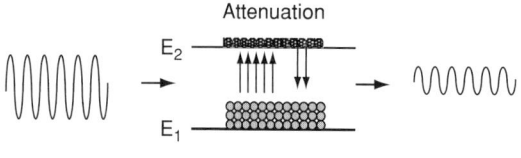

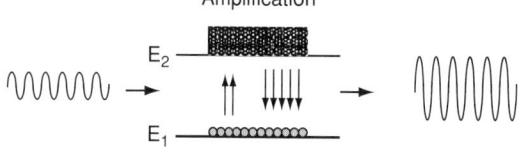

FIGURE 8.2 Schematic representations of signal attenuation and amplification. Under thermal equilibrium, the lower levels are more populated than the upper levels, which leads to absorption of an incoming beam. On the other hand, if the upper level has greater population than the lower level, then optical amplification can take place.

This is the basic principle behind optical amplification by an EDFA.

Figure 8.3 shows the three lowest lying energy levels of erbium ion in silica matrix. A pump laser at 980 nm excites the erbium ions from the ground state to the level marked E_3. The E_3 level is a short-lived level, and the ions fall back to the level marked E_2 after less than a microsecond. The life-time of level E_2 is much larger and is about 12 ms. Hence, ions brought to level E_2 stay there for a long time.

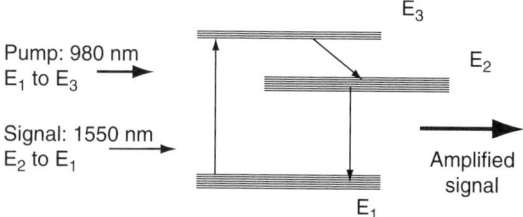

FIGURE 8.3 Energy levels corresponding to the three lowest lying levels of erbium ion. The external pump laser at 980 nm induces excitation from level E_1 to E_3 and ions drop down quickly from E_3 to E_2. Level E_2 is a long lived level and with enough pump rate, one can achieve population inversion between E_2 and E_1, leading to optical amplification.

Thus, by pumping hard enough, the population of ions in the E_2 level can be made larger than the population of E_1 level and thus achieve population inversion between the E_1 and E_2 levels. In such a situation, a light beam at a frequency $\nu_0 = (E_2 - E_1)/h$ incident on the collection of erbium ions gets amplified. For erbium ions, the frequency ν_0 falls in the 1550-nm band and thus is an ideal amplifier for signals in the 1550-nm window, which is the lowest-loss window of silica-based optical fibers. In the case of erbium ions in a silica matrix, the energy levels are not sharp but are broadened due to interaction with other ions in the silica matrix. Hence, the system is capable of amplifying optical signals over a band of wavelengths.

3. POPULATION INVERSION AND OPTICAL AMPLIFICATION

If N_1 is the number of erbium ions per unit volume in the ground level and if light of intensity I_ν at a frequency ν interacts with the ions, then the number of absorptions per unit time per unit volume from level E_1 to E_2 can be written as

$$\frac{dN_1}{dt} = -\sigma_a(\nu)\frac{I_\nu}{h\nu}N_1 \quad (1)$$

where σ_a is called the absorption cross-section (has dimensions of area) and is a function of the frequency. Similarly, if N_2 is the number of ions per unit volume in the level E_2, then the number of stimulated emissions per unit volume per unit time from level E_2 to E_1 can be written as

$$\frac{dN_2}{dt} = -\sigma_e(\nu)\frac{I_\nu}{h\nu}N_2 \quad (2)$$

where σ_e is called the emission cross section. The absorption and emission cross sections depend on the specific ion as well as on the pair of levels for a given ion.

Similarly, an atom in the excited state E_2 can emit a photon spontaneously and get de-excited to the level E_1. The number of such emissions per unit volume per unit time is given by

$$\frac{dN_2}{dt} = -\frac{N_2}{t_{sp}} \quad (3)$$

where t_{sp} is called the spontaneous lifetime of the level E_2.

Consider the propagation of a light beam at frequency ν through a collection of these ions. Let $I_\nu(z)$ be the intensity of the light beam at the plane P_1 (of cross-sectional area S) and let $I_\nu(z + \Delta z)$ be the intensity of the light beam on the plane P_2 (Fig. 8.4).

Erbium-Doped Fiber Amplifiers

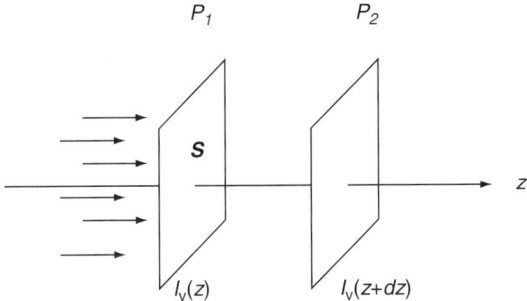

FIGURE 8.4 Light propagating through a collection of ions induces absorption and emission in the ions, leading to a change of intensity of the beam.

The intensity of the propagating light beam changes due to absorption and stimulated emission. The difference in energy entering and leaving the volume lying between planes P_1 and P_2 must be due to the above processes. Thus, we can write

$$I_\nu(z)S - I_\nu(z + \Delta z)S = -\frac{dI_\nu}{dz} = (\sigma_a N_1 - \sigma_e N_2) I_\nu S dz$$

which gives us the following equation for the evolution of light intensity with distance:

$$\frac{dI_\nu}{dz} = -(\sigma_a N_1 - \sigma_e N_2) I_\nu \qquad (4)$$

Hence, for optical amplification we must have $\sigma_e N_2 > \sigma_a N_1$. If $\sigma_a = \sigma_e$ (which is usually the case), then for optical amplification we need to have $N_2 > N_1$, that is, have population inversion.

Figure 8.5 shows the spectral variation of absorption and emission cross-sections in the spectral region of 1550 nm of erbium ions in silica matrix. We notice that the absorption and emission spectra are broad, and this is what leads to the broad gain spectrum of EDFAs.

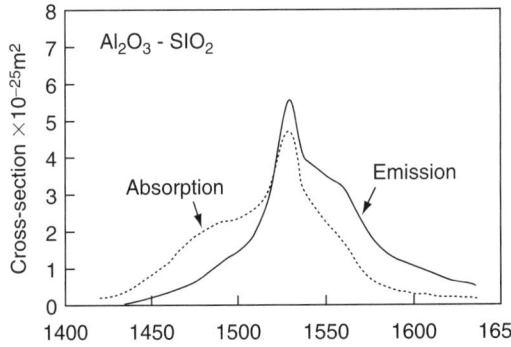

FIGURE 8.5 Absorption and emission cross-section spectra of erbium ion in silica matrix. (After [4] © IEEE 2005)

4. OPTICAL AMPLIFICATION IN EDFAs

We now consider the evolution of pump and signal in an erbium-doped fiber. We follow the analysis of [5]. Let us consider an erbium-doped fiber and let $I_p(z)$ and $I_s(z)$ represent the variation of intensity of the pump at frequency ν_p (assumed to be at 980 nm) and the signal at frequency ν_s assumed to be in the region of 1550 nm. As the beams propagate through the fiber, the pump would induce absorption from E_1 to E_3, whereas the signal would induce absorption and stimulated emissions between levels E_2 and E_1. The population of the various levels would depend on the fiber length, that is, z, because the intensities of the beams would be z dependent. We assume that the lifetime of level E_3 is very small so that we can neglect the population density N_3 of level E_3 and put $N_3 = 0$. Thus, the ions are either in level E_1 or in level E_2. We can write the rate of change of population of level E_2 as

$$\frac{dN_2}{dt} = -\frac{N_2}{t_{sp}} + \frac{\sigma_{pa} I_p}{h\nu_p} N_1 - (\sigma_{se} N_2 - \sigma_{sa} N_1) \frac{I_s}{h\nu_s} \quad (5)$$

The various terms on the right-hand side of the above equation correspond to the following: term 1, spontaneous emission; term 2, pump absorption; and term 3, absorption and stimulated emission due to signal.

Subscripts p and s correspond to pump and signal, respectively. Because we are neglecting the population of level E_3, we have

$$\frac{dN_1}{dt} = -\frac{dN_2}{dt} \qquad (6)$$

The pump and signal intensity variations with z are caused due to absorption and stimulated emission and can be described by the following equations [see Eq. (4)]:

$$\frac{dI_p}{dz} = -\sigma_{pa} N_1 I_p \qquad (7)$$

$$\frac{dI_s}{dz} = -(\sigma_{sa} N_1 - \sigma_{se} N_2) I_s \qquad (8)$$

In the case of optical fibers, because the pump and signal beams propagate in the form of modes, we should describe amplification in terms of powers rather than in terms of intensities. The optical powers at the pump and signal are given by

$$P_{p,s}(z, t) = \int I_{p,s}(r, z, t) 2\pi r\, dr \qquad (9)$$

We separate the r dependence of pump and signal intensities through the following equations (see, e.g., [3]):

$$I_{p,s}(r,z,t) = P_{p,s}(z,t) f_{p,s}(r) \qquad (10)$$

where the function $f_{p,s}(r)$ is normalized through the following equation:

$$2\pi \int_0^\infty f_{p,s}(r) r\, dr = 1 \qquad (11)$$

The functions $f_{p,s}(r)$ describe the transverse intensity distributions at the pump and signal wavelengths. This can be approximated as a Gaussian function (referred to as the Gaussian envelope approximation) defined by

$$f_{p,s}(r) = \frac{1}{\pi \Omega_{p,s}^2} e^{-r^2/\Omega_{p,s}^2} \qquad (12)$$

where $\Omega_{p,s}$ give the widths at the pump and signal wavelengths and depend on the fiber parameters as well as the wavelength. For step-index fibers they can be approximated by

$$\Omega_i = a J_0(U_i) \frac{V_i}{U_i} \frac{K_1(W_i)}{K_0(W_i)}; \quad i = p,s \qquad (13)$$

where U_i, W_i, and V_i are defined through the following equations:

$$\begin{aligned} U_i &= a(k_i^2 n_1^2 - \beta_i^2)^{1/2} \\ W_i &= a(\beta_i^2 - k_i^2 n_2^2)^{1/2} \\ V_i &= a k_i (n_1^2 - n_2^2)^{1/2} \end{aligned} \qquad (14)$$

where a is the core radius, n_1 and n_2 are the core and cladding refractive indices of the fiber, and $k_i = 2\pi/\lambda_i$.

Under steady state, Eqs. (5)–(8) can be solved to get the following equation:

$$\bar{N}_2 = -\frac{1}{L\varsigma} \sum Q_j^{in} \left[\exp\left\{ (\gamma_j + \alpha_j) \bar{N}_2 - \alpha_j \right\} L - 1 \right] \qquad (15)$$

In the above equation the sum over j on the right-hand side is over different pump and signal wavelengths (assuming propagation of multiple pump and multiple signal wavelengths through the amplifier). The various quantities appearing in the above equation are defined below:

Average fractional upper level population:

$$\bar{N}_2 = \frac{1}{LN_t} \int_0^L N_2(z) dz$$

Number of photons incident per unit time at the jth pump or signal:

$$Q_j^{in} = \frac{P_j^{in}}{h\nu_j}$$

Absorption (α) and emission (γ) constants at pump and signal wavelengths:

$$\alpha_p = \sigma_{pa} \Gamma_p N_t; \quad \alpha_s = \sigma_{sa} \Gamma_s N_t; \quad \gamma_s = \sigma_{se} \Gamma_s N_t$$

Pump and signal confinement factors:

$$\Gamma_{p,s} = 2\pi \int f_{Er}(r) f_{p,s}(r) r\, dr$$

Saturation parameter:

$$\varsigma = \frac{SN_t}{t_{sp}}$$

Effective area of erbium doped region:

$$S = \int f_{Er}(r) 2\pi r\, dr$$

where $f_{Er}(r)$ is the transverse distribution of erbium doping, L is the length of the erbium doped fiber, and N_t is the total erbium ion density in the fiber core.

For a given doped fiber and input pump and signal powers, Eq. (15) can be solved for $\bar{N}_2$ and from which we get the gain coefficient $g(\lambda)$ as

$$g(\lambda) = \{\gamma(\lambda) + \alpha(\lambda)\} \bar{N}_2 - \alpha(\lambda) \qquad (16)$$

The amplifier gain in dB is thus given as

$$G(\lambda) = 10 g(\lambda) L \log e \qquad (17)$$

A typical erbium-doped fiber would have a numerical aperture of 0.2 and a cutoff wavelength around 900 nm so that at the pump wavelength of 980 nm, the fiber would be single moded. This corresponds to a core radius of about 1.7 µm. The erbium concentration is usually about 100 to 500 parts per million (concentration of $5.7 \times 10^{24}\,\mathrm{m}^{-3}$ to $2.9 \times 10^{25}\,\mathrm{m}^{-3}$).

We now present results of simulations of an EDFA with the following parameters:

$a = 1.64\,\mu\mathrm{m}$, NA = 0.21
λ_p, Pump wavelength = 980 nm
λ_s, Signal wavelength = 1550 nm
$t_{sp} = 12 \times 10^{-3}\,\mathrm{s}$
$N_t = 0.68 \times 10^{25}\,\mathrm{m}^{-3}$
$L = 7\,\mathrm{m}$

Figure 8.6 shows the variation of gain versus length of EDFA for different pump powers. For any given pump power there is an optimum length for achieving maximum gain. This is because as the pump propagates through the fiber it is absorbed, and beyond a certain length the pump power is unable to produce population inversion. This leads to attenuation of the signal rather than amplification beyond the optimum length.

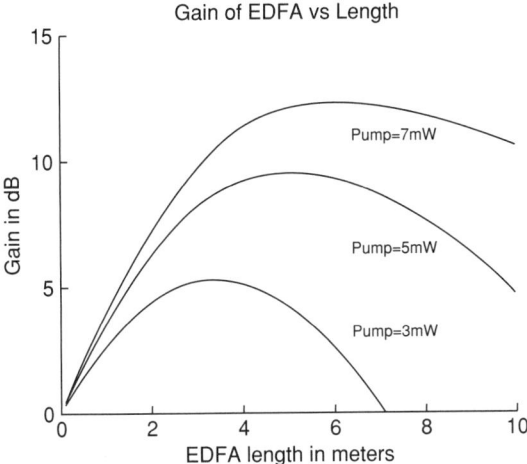

FIGURE 8.6 A typical simulated signal gain vs. the length of an erbium-doped fiber. For any pump power there is an optimum length for which the gain is maximum.

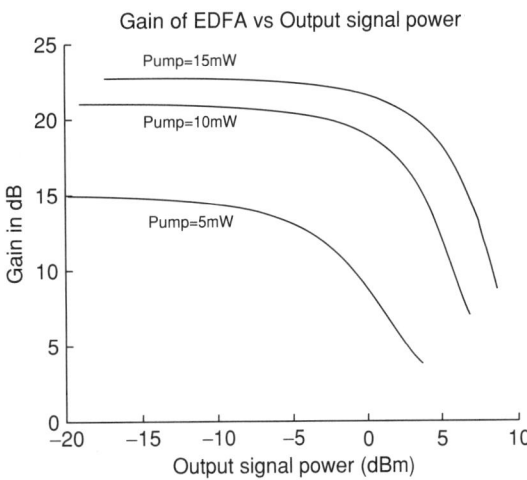

FIGURE 8.8 Variation of gain with output signal power showing gain saturation.

Figure 8.7 shows the variation of signal gain with pump power for different lengths of the doped fiber. For low pump powers the gain is negative, that is, the fiber is absorbing. At a certain pump power the fiber has zero loss or gain and becomes transparent for the signal. As the pump power increases further, the gain saturates because that increase in the pump power leads to inversion of even greater numbers of erbium ions in the fiber, and ultimately after inverting all the erbium ions no more increase in inversion is possible.

Figure 8.8 shows the variation of signal gain with output signal power. As the input signal power increases from low values, the output also increases with almost constant gain. But as the input power increases further, the inversion starts to reduce due to increased stimulated emission by the signal; the increase in output power is then not that high,

resulting in a drop of the gain. This is referred to as gain saturation and is a characteristic of all amplifiers. The output power for which the gain is 3 dB lower than the small signal gain is referred to as the saturation output power and is a very important parameter of EDFAs.

Figure 8.9 shows the spectral variation of gain per unit length with the level of pumping. In the absence of the pump, the fiber is attenuating at all signal wavelengths. As the pump power increases, the fiber first starts to have gain at higher wavelengths, and at significantly large pump powers, the gain spectrum peaks at around the wavelength of 1530 nm.

Figure 8.10 shows the gain spectra at various input signal power levels. It can be seen that the gain spectrum is in general not flat, and as the input signal power increases, the gain decreases (gain saturation) and the gain spectrum becomes more and more flat. The gain spectrum also becomes modified as the input signal power changes. This feature has

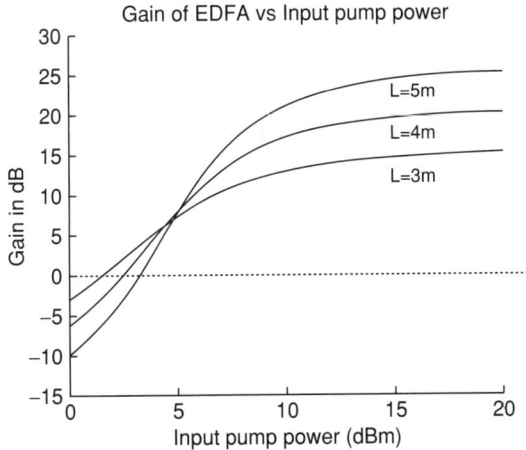

FIGURE 8.7 Variation of gain with input pump power.

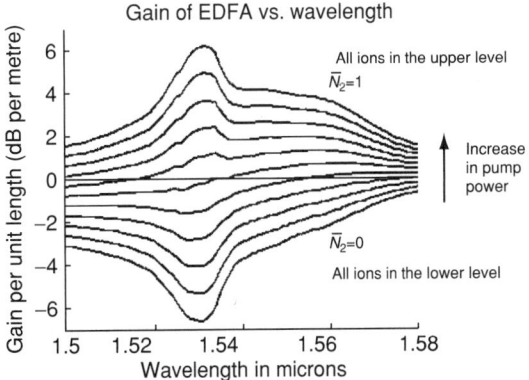

FIGURE 8.9 Spectral variation of gain per unit length for different pump power levels.

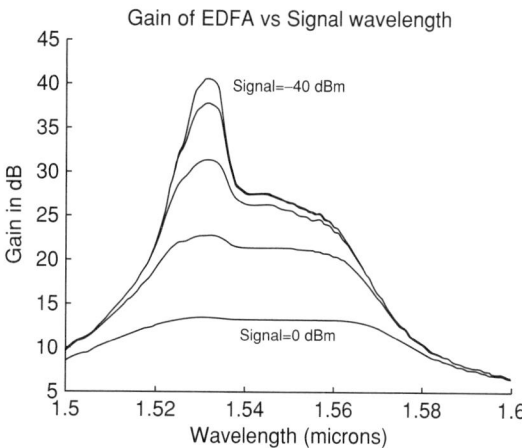

FIGURE 8.10 Variation of gain spectrum with input signal power.

important consequences in the design of gain flattening filters, discussed in Section 5.

Figure 8.11 is a schematic of an EDFA that consists of a short piece (~20 m in length) of erbium doped fiber and is pumped by a 980-nm pump laser through a WDM coupler. The WDM coupler multiplexes light of wavelength 980 and 1550 nm from two different input arms to a single output arm. The 980-nm pump light is absorbed by the erbium ions to create population inversion between levels E_2 and E_1. Thus, incoming signals in the 1550-nm wavelength region get amplified as they propagate through the population inverted doped fiber.

Figure 8.12 shows typical measured gain spectra of an EDFA for various input signal powers. As can be seen in the figure, EDFA can provide amplifications of greater than 20 dB over the entire band of 40 nm from 1525 nm to about 1565 nm. This wavelength band is referred to as the C-band (conventional band) and is the most common wavelength band of operation. We also note that the gain decreases and

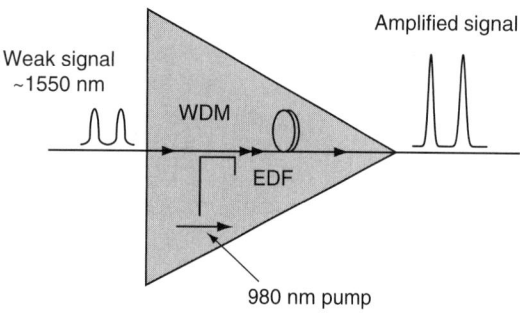

FIGURE 8.11 An EDFA consists of a short piece of erbium-doped fiber that is pumped by a 980-nm laser diode through a WDM coupler. Signals around the 1550-nm wavelength get amplified as they propagate through the pumped doped fiber.

the spectrum flattens as the input signal power increases showing signal saturation.

With proper amplifier optimization, EDFAs can also amplify signals in the wavelength range of 1570–1610 nm; this band of wavelengths is referred to as the L-band (long wavelength band). From Fig. 8.9 it can be noticed that the gain profile of the EDFA depends on the average degree of inversion. For the average degree of inversion of around 0.4, the fiber is absorbing in the C-band, whereas in the L-band the fiber is amplifying. Thus, by controlling the average inversion in the fiber, it is possible to achieve amplification in the L-band.

The C-band and L-band amplifiers together can be used to simultaneously amplify 160 wavelength channels. Such systems are indeed now commercially available.

It can be seen from Fig. 8.12 that although EDFAs can provide gains over an entire band of 40 nm, for low input signal powers the gain is not flat, that is, the gain depends on the signal wavelength. Thus, if multiple wavelength signals with the same power are input into the amplifier, then their output powers will be different. In a communication system using a chain of amplifiers, a differential signal gain among the various signal wavelengths (channels) from each amplifier results in a significant difference in signal power levels and hence in the SNR among the various channels. In fact, signals for which the gain in the amplifier is greater than the loss suffered in the link will keep on increasing in power level, whereas those channels for which the amplifier gain is less than the loss suffered will keep on reducing in power. The former channels will finally saturate the amplifiers and will also lead to increased nonlinear effects in the link, whereas the latter will have reduced SNR leading to increased errors in detection. Thus, such a differential amplifier gain is not desirable in a dense WDM communication system, and it is very important to have gain flattened amplifiers.

5. GAIN FLATTENING OF EDFAs

There are basically two main techniques for gain flattening: One uses external wavelength filters to flatten the gain, whereas the other relies on modifying the amplifying fiber properties to flatten the gain.

5.1. Gain Flattening Using External Filters

Figure 8.13 shows the principle behind gain flattening using external filters. In this, the output of the amplifier is passed through a special wavelength filter whose transmission spectrum is exactly the inverse of the gain spectrum of the amplifier. Thus,

Erbium-Doped Fiber Amplifiers

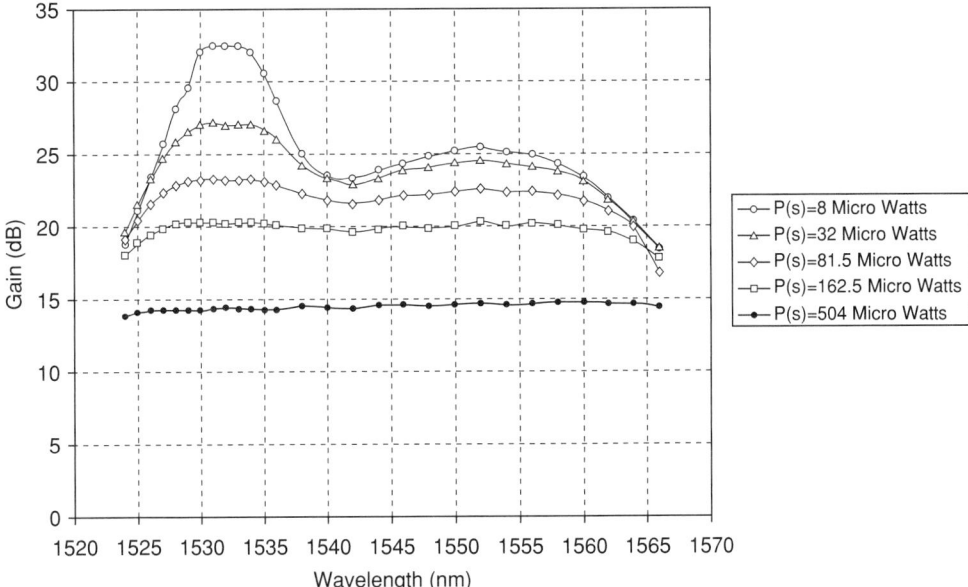

FIGURE 8.12 Typical measured gain spectra for different input signal power levels of an EDFA designed using an erbium-doped fiber fabricated at CGCRI, Kolkata.

channels that have experienced greater gain in the amplifier will suffer greater transmission loss, whereas channels that experience smaller gain will suffer smaller loss. By appropriately tailoring the filter transmission spectrum profile, it is possible to flatten the gain spectrum of the amplifier. Although Fig. 8.13 shows the gain flattening filter to be placed after the amplifier, in actual practice the filter is located within the amplifier between two sections of erbium-doped fiber so as to optimize both the gain and noise characteristics of the EDFA.

Transmission filters with specific transmission profiles can be designed and fabricated using various techniques. These include thin film interference filters and filters based on long period fiber gratings (LPG). Figure 8.14 shows a typical gain flattening achieved by using an LPG (LPG is discussed in detail in Chapter 15). The LPG is a transmission filter with adjustable spectral variation, and being all-fiber is attractive for use as a gain-flattening filter. Typical gain flatness of better than 0.5 dB can be achieved, and commercially available EDFAs are gain flattened.

5.2. Intrinsically Flat Gain Spectrum

We note that the gain of the EDFA is not flat due to the spectral dependence of the absorption and emission cross-sections and also due to the variation of the modal overlap between the pump, signal, and the erbium-doped region of the fiber. Thus, it is, in principle, possible to flatten the gain of the amplifier by appropriately choosing the transverse refractive index profile of the fiber and the doping profile of

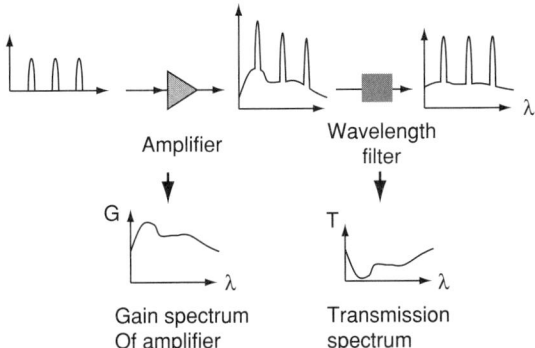

FIGURE 8.13 The principle of gain flattening of EDFA using an external wavelength filter.

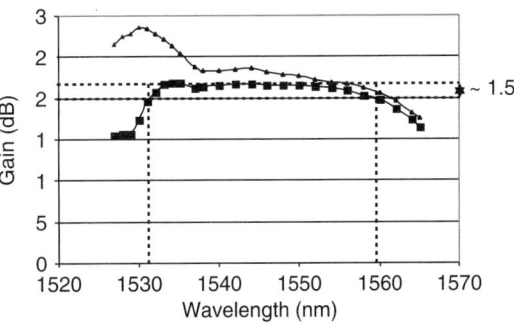

FIGURE 8.14 Gain flattening of an EDFA pumped by a 980-nm laser, using an LPG gain-flattening filter.

the fiber to achieve flatter gain. Figure 8.15 shows a schematic of a refractive index profile distribution and the corresponding erbium-doped region that can provide gain flattening by appropriately optimizing the various parameters. Figure 8.16 shows the comparison of the gain profile of an EDFA with a conventional fiber and the gain profile of an optimized EDFA with the proposed designs based on coaxial and staircase profiles [6, 7]. As is evident, much flatter gain profiles can be achieved using proper optimization of the refractive index profile and the doping profile of an erbium-doped fiber.

6. NOISE IN AMPLIFIERS

In this section we briefly discuss noise characteristics of EDFAs. As discussed earlier, in an EDFA population inversion between two energy levels of erbium ion leads to optical amplification by the process of stimulated emission. Erbium ions occupying the upper energy level can also make spontaneous transitions to the ground level and emit radiation. This radiation appears over the entire fluorescent band of emission of erbium ions and travels both in the forward and in the backward directions along the fiber. Part of the spontaneous emission generated at any point along the fiber gets coupled into the propagating mode of the fiber and can also get amplified, just like the signal as it propagates through the population inverted fiber. The resulting radiation is called amplified spontaneous emission (ASE). This ASE is the basic mechanism leading to noise in the optical amplifier (see, e.g., [1]).

ASE appearing in the wavelength region not coincident with the signal can be filtered using an optical filter. On the other hand, the ASE that appears in the signal wavelength region cannot be separated and constitutes the minimum added noise from the amplifier. Figure 8.17 shows a typical output from an EDFA. The uppermost curve corresponds to the output ASE spectrum in the absence of any input signal, and the two lower curves correspond to the output spectra with signal input at two different wavelengths. Notice that the amplified signal rides on a background ASE and also note the reduction in the ASE power due to the presence of signal that consumes a part of the upper level population.

If P_{in} represents the signal input power (at frequency ν) into the amplifier and G represents the gain of the amplifier, then the output signal power is given by GP_{in}. Along with this amplified signal, there is also ASE power that can be shown to be given by (see, e.g., [1])

$$P_{ASE} = 2n_{sp}(G-1)h\nu B_o \quad (18)$$

where B_o is the optical bandwidth over which the ASE power is being measured (which must be at least equal to the optical bandwidth of the signal),

$$n_{sp} = \frac{N_2}{(N_2 - N_1)} \quad (19)$$

Here N_2 and N_1 represent the population densities in the upper and lower amplifier energy levels of

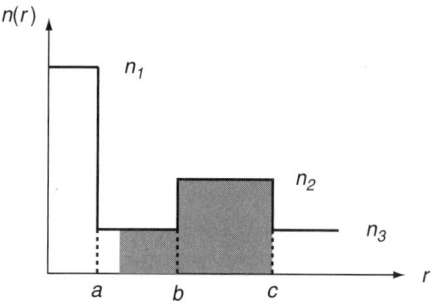

FIGURE 8.15 A schematic of a refractive index profile of an erbium-doped fiber that exhibits intrinsic gain flattening.

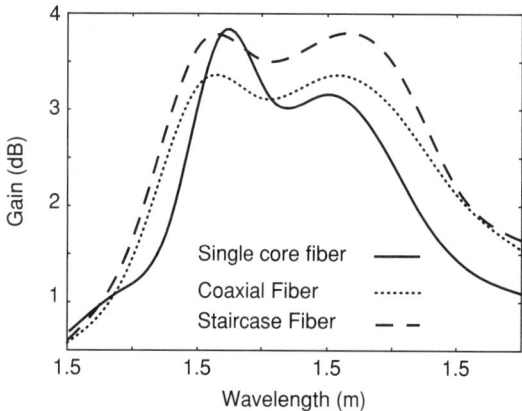

FIGURE 8.16 Comparison of the gain spectrum of an EDFA with a conventional erbium-doped fiber and the proposed erbium-doped fiber. (Adapted from [6, 7].)

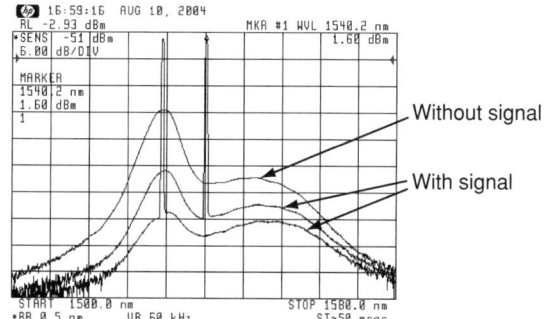

FIGURE 8.17 ASE spectrum of an EDFA with and without input signals. In the presence of a signal, the ASE power reduces.

erbium in the fiber. The minimum value for n_{sp} corresponds to a completely inverted amplifier for which $N_1 = 0$ and thus $n_{sp} = 1$.

As a typical example, we have

$$n_{sp} = 2$$
$$G = 100 \ (20 \, \text{dB})$$
$$\lambda = 1550 \, \text{nm}$$
$$B_o = 12.5 \, \text{GHz} \ (= 0.1 \, \text{nm at } 1550 \, \text{nm})$$

which gives

$$P_{ASE} = 0.6 \, \mu\text{W} (= -32 \, \text{dBm})$$

which corresponds to an ASE noise spectral density of $-22 \, \text{dBm/nm}$. We can define the optical SNR (OSNR) as the ratio of the output optical signal power to the ASE power:

$$\text{OSNR} = \frac{P_{out}}{P_{ASE}} = \frac{GP_{in}}{2n_{sp}(G-1)h\nu B_o} \quad (20)$$

where P_{in} is the average power input into the amplifier (which is about half of the peak power in the bit stream, assuming equal probability of ones and zeroes). For large gains $G \gg 1$, and assuming $B_o = 12.5 \, \text{GHz}$, for a wavelength of 1550 nm, we obtain

$$\text{OSNR(dB)} \approx P_{in}(\text{dBm}) + 58 - F \quad (21)$$

where

$$F(\text{dB}) = 10 \log(2n_{sp}) \quad (22)$$

is the noise figure of the amplifier (for large gains). For $n_{sp} = 2$ and $P_{in} = -30 \, \text{dBm}$, we obtain an OSNR of 22 dB. In system designs, typically one looks for an OSNR of greater than 20 dB for the detection to have low bit error rates.

Each amplifier in a chain adds noise, and thus in a fiber optic communication system consisting of multiple spans of optical fiber links with amplifiers, OSNR will keep falling (Fig. 8.18). At some point in the link when the OSNR falls below a certain value, the signal will need to be regenerated. If we assume a noise of $0.6 \, \mu\text{W}$ added by each amplifier then after, say, 10 amplifiers, the signal power would still be the same as at the beginning (assuming the amplifier gain exactly compensates for the attenuation in the span) but the noise power would be $6 \, \mu\text{W}$. Thus, as the signal passes through multiple spans and amplifiers, there is a reduction in the OSNR. Hence, there is a maximum number of amplifiers that can be placed in a link beyond which the signal needs to be regenerated.

For a link consisting of multiple spans of transmission fiber and EDFAs compensating the loss of each span, the OSNR is given by

$$\text{OSNR(dB)} \approx P_{out}(\text{dBm}) - 10 \log(n) + 58 - F(\text{dB})$$
$$- 10 \log(N+1) - L_{sp}(\text{dB}) \quad (23)$$

where P_{out} is the total output power from the amplifier in dBm, n represents the number of wavelength channels in the link, F represents the noise figure of each EDFA (assumed to be the same), N represents the number of amplifiers, and L_{sp} the loss of each span. As a typical application of the above equation, let us consider a link consisting of EDFAs with the following specifications:

$$P_{out} = 17 \, \text{dBm}$$
$$n = 32$$
$$F = 5 \, \text{dB}$$
$$L_{sp} = 20 \, \text{dB}$$

If we require an OSNR of 22 dB at the end of the link, then using Eq. (23) the maximum number of amplifiers that can be used in the link comes out to be about 18. If more than this number of amplifiers are used, then the OSNR will fall below the required value of 22 dB. Thus, for proceeding further along the length, the signal needs to be regenerated.

It is also interesting to note from Eq. (23) that for achieving the same OSNR at the output of the link, the number of amplifiers in the chain can be increased by reducing the noise figure of each amplifier, increasing the output power of the amplifiers, or decreasing the span loss. Indeed, by choosing smaller span loss, the number of amplifiers can be increased

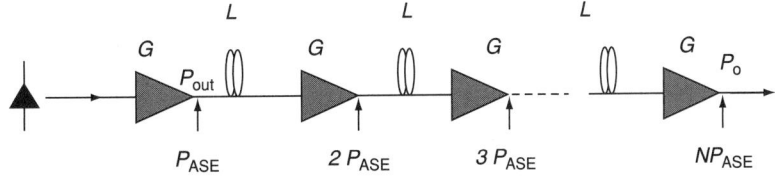

FIGURE 8.18 A long fiber optic link with EDFAs for compensation of loss of each span. The signal power returns to the original value after each span, whereas the ASE power keeps on building up, leading to reduction in OSNR as the number of spans increases.

significantly so that the distance for regeneration can be made very large. Thus, reducing each span loss by 3 dB would result in a doubling of the maximum allowed number of amplifiers (all other parameters being the same). Of course, in this case we would have to use a larger number of amplifiers.

The above discussion was based on the OSNR of the amplifier. When the amplified output is eventually received by a detector, the detector converts the optical signal into an electric current and the noise characteristics of the generated electrical signal are of importance. Apart from the optical signal, the amplified spontaneous emission within the bandwidth of the signal is also received by the photodetector. However, the ASE noise is completely random and contains no information. The photodetector would convert the total optical power received into electrical current; if the electric fields of the signal and noise are E_s and E_n, respectively, then the generated electric current would be proportional to $(E_s + E_n)^2$. The expansion of this would consist of terms proportional to $(E_s)^2$, $(E_n)^2$, and $E_s E_n$. The first term leads to the signal current, whereas the other two terms correspond to noise. The second term leads to a beating between noise components at various frequencies lying within the signal bandwidth and is referred to as spontaneous-spontaneous beat noise. The last term leads to beating between the signal and spontaneous emission and is referred to as signal-spontaneous beat noise. Under normal circumstances, the signal-spontaneous noise and the signal-shot noise are the important noise terms, and assuming the input to the amplifier to be shot noise limited we can calculate the output SNR from the noise terms. We define the noise figure of the amplifier by the following relation:

$$F = \frac{(SNR)_{in}}{(SNR)_{out}} \quad (24)$$

By calculating the input and output SNR, we can obtain an expression for the noise figure of the amplifier which is given by

$$F = \frac{1 + 2n_{sp}(G-1)}{G} \quad (25)$$

Thus, the noise figure depends on the inversion through n_{sp} and on the amplifier gain through G. For large gains $G \gg 1$, the noise figure is approximately given by $2 n_{sp}$. Because the smallest value of n_{sp} is unity, the smallest noise figure is given by 2 or in decibel units as 3 dB.

The noise figure is a very important characteristic of an amplifier and determines the overall performance of any amplified link. Noise figures of typical commercially available EDFAs are about 5 dB.

7. EDFAs FOR THE S-BAND

Opening up of newer bands for transmission, namely the L-band (1565–1625 nm) and the S-band (1460–1530 nm), is a potential solution for increasing the capacity of current dense WDM systems. It has been shown that efficient S-band EDFAs require high inversion levels along the fiber and C-band ASE suppression, which otherwise depletes the population inversion. S-band amplifiers based on silica-based erbium-doped fibers have also been recently reported based on a W-index fiber, with fundamental mode cutoff at 1525 nm [8]. An efficient design for single-stage S-band EDFA based on a coaxial core fiber, wherein distributed ASE filtering is achieved by winding the fiber with an optimally chosen bend radius, has recently been reported [9]. Bend loss usually has a strong spectral variation because of the variation of mode field diameter with wavelength. Coaxial fiber design provides extra degrees of freedom in terms of tuning the bend loss variation. Hence, by optimizing the fiber parameters and the bend radius, high bend loss is ensured at wavelengths above 1525 nm (>6 dB/m), whereas wavelengths below 1525 nm suffer minimal loss. In the new fiber design, the bend loss at 1530 nm is 100 times greater than that at 1490 nm, which leads to a high net gain in the S-band.

Figure 8.19 shows the refractive index profile of the coaxial core fiber, and Fig. 8.20 shows the gain spectrum of the S-band amplifier (designed with a bend radius of 3 cm) under unsaturated and saturated signal input conditions and also the noise figure of the amplifier. As can be seen, an average gain of 26 dB with a gain variation of ±2.9 dB over the wavelength region 1495–1525 nm is achievable.

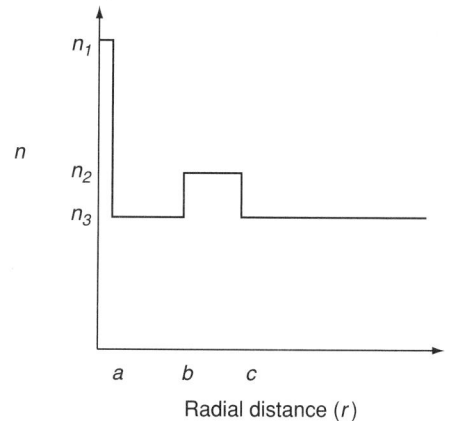

FIGURE 8.19 Refractive index profile of a coaxial fiber for an S-band EDFA. (After [9] © IEEE 2005)

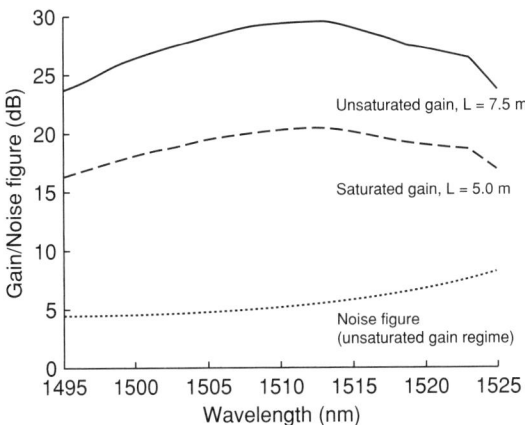

FIGURE 8.20 Gain and noise figure spectrum of an S-band EDFA (After [9] © IEEE 2005)

The pump is at 980 nm, and the pump power is 500 mW.

8. CONCLUSIONS

Optical fiber amplifiers have truly revolutionized optical fiber communications and have made possible practical implementation of dense WDM for increased bandwidth. EDFAs can be made to operate in S-, C-, and L-bands, and although EDFAs are the most popular amplifiers, RFAs and SOAs are also becoming very important. With the availability of optical amplifiers capable of spanning the entire wavelength region of 1250–1650 nm, the entire low-loss window (53 THz) of optical fibers will become available for exploitation and use by the telecommunications engineers.

9. ACKNOWLEDGMENTS

I thank Dr. M. N. Satyanarayan and Mr. Mandip Singh for their help in experiments on gain flattening of EDFAs using LPGs and Mr. Deepak Gupta for help in simulations of many of the results presented here. Thanks are also due to Dr. S. K. Bhadra of CGCRI, Kolkata for providing samples of EDF manufactured by them.

10. REFERENCES

1. Becker, P.C., Olsson, N.A. and Simpson, J.R. (1999) *Erbium Doped Fiber Amplifiers*, Academic Press, San Diego.
2. Desurvire, E. (1994) *Erbium Doped Fiber Amplifiers*, Academic Press, New York.
3. Ghatak, A.K. and Thyagarajan, K. (1998) *Introduction to Fiber Optics*, Cambridge University Press, UK.
4. Barnes, W.L., Laming, R.I., Tarbox, E.J., and Morkel, P. (1991) Absorption and emission cross section of Er^{3+} doped silica fibers, *IEEE J Quant. Electron.* **27**, pp. 1004–1010.
5. Sun, Y., Zyskind, J.L. and Srivastava, A. (1997) Average inversion level, modeling, and physics of erbium doped fiber amplifiers, *IEEE J. Sel. Topics Quant. Electron.* **3**, pp. 991–1007.
6. Thyagarajan, K. and Kaur, J. (2000) A novel design of an intrinsically gain flattened erbium doped fiber, *Opt. Commun.* **183**, pp. 407–413.
7. Thyagarajan, K. and Kaur, J. (2003) Intrinsically gain-flattened staircase profile erbium doped fiber amplifier, *Opt. Commun.* **222**, pp. 227–233.
8. Arbore, M.A., Zhou, Y., Thiele, H., Bromage, J. and Nelson, L. (2003) 'S-band erbium doped fiber amplifiers for WDM transmission between 1488 and 1508 nm, in *Proceedings of Optical Fiber Communications*, Georgia, USA, paper WK 2.
9. Thyagarajan, K. and Kakkar, C. (2004), S-band single stage EDFA with 25 dB gain using distributed ASE suppression, *IEEE Photonics Tech. Letts.* **16** pp. 2448–2450.

Fiber Optic Raman Amplifiers

Govind P. Agrawal*
Institute of Optics
University of Rochester
Rochester, New York, USA

1. INTRODUCTION

Stimulated Raman scattering (SRS) is the fundamental nonlinear process that turns optical fibers into broadband Raman amplifiers. Although Raman amplification in optical fibers was observed as early as 1972, until recently SRS was mainly viewed as a harmful nonlinear effect because it could also severely limit the performance of multichannel lightwave systems by transferring energy from one channel to its neighboring channels (see Chapter 6).

This chapter focuses on the SRS phenomenon from the standpoint of Raman amplifiers. Section 2 discusses the basic physical mechanism behind SRS, with emphasis on the Raman gain spectrum. It also presents the coupled pump and Stokes equations that one needs to solve for predicting the performance of Raman amplifiers. The simplest case in which both the pump and Stokes are in the form of continuous-wave (CW) beams is considered in Section 2. This section also introduces forward, backward, and bidirectional pumping configurations.

Section 3 focuses on modern broadband Raman amplifiers pumped by multiple lasers, whose powers and wavelengths are optimized to produce uniform gain over a wide spectral range in the 1.55-μm region. Section 4 deals with several practical issues that are relevant for Raman amplifiers. The topics covered include spontaneous Raman scattering, double Rayleigh backscattering, pump-noise transfer, and polarization mode dispersion (PMD). A vector theory of the SRS process shows that PMD can induce large fluctuations in the amplified signal whose magnitude depends not only on the PMD parameter but also on the pumping configuration. Fluctuations are reduced significantly when a backward-pumping configuration is used. The amplification of short pulses is discussed in Section 5 with emphasis on the role of group-velocity mismatch and dispersion.

*E-mail: gpa@optics.rochester.edu

2. FUNDAMENTAL CONCEPTS

In any molecular medium, spontaneous Raman scattering can transfer a small fraction of laser power operating in one frequency to another optical field whose frequency is downshifted by an amount determined by the vibrational modes of the molecules. This phenomenon was discovered by C. V. Raman in 1928 and is known as the *Raman effect* [1]. Classically, it is described as an inelastic scattering process in which light frequency shifts toward the red side.

As shown schematically in Fig. 9.1, the Raman scattering process can be viewed quantum mechanically as absorption of a photon of energy $\hbar\omega_p$ by a molecule that ends up in a virtual state, shown by the dashed line. Virtual states do not correspond to a real electronic state but are allowed in quantum mechanics as long as they exist for a time interval shorter than that dictated by Heisenberg's uncertainty relation. For this reason, a molecule in the virtual state decays immediately to a real vibrational state by emitting a lower frequency photon of energy $\hbar\omega_s$. The energy difference $\hbar(\omega_p - \omega_s)$ is used by an optical phonon generated during this process. From a practical standpoint, the incident light acts as a pump for generating the red-shifted radiation called the *Stokes line*. A blue-shifted component, known as the *anti-Stokes line*, is also generated but its intensity is much weaker than that of the Stokes line because the anti-Stokes process requires the vibrational state to be initially populated with a phonon of right energy and momentum. In what follows, we ignore the anti-Stokes process because it plays virtually no role in fiber amplifiers.

Although spontaneous Raman scattering takes place in any molecular medium, it is weak enough that it can be ignored when an optical beam propagates through an optical fiber. In 1962, it was observed that for intense optical fields, the nonlinear phenomenon of SRS can occur in which the Stokes wave grows rapidly inside the medium such that

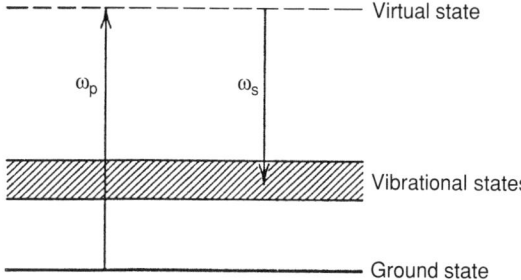

FIGURE 9.1 Schematic illustration of the Raman-scattering process from a quantum-mechanical viewpoint. A Stokes photon of reduced energy $\hbar\omega_s$ is created spontaneously when a pump photon of energy $\hbar\omega_p$ is lifted to a virtual level shown as a dashed line.

2.1. Raman Gain Spectrum

The most important parameter characterizing Raman amplifiers is the Raman gain coefficient γ_R, which is related to the cross-section of spontaneous Raman scattering [7]. It describes how the Stokes power grows as pump power is transferred to it through SRS. On a more fundamental level, γ_R is related to the imaginary part of the third-order nonlinear susceptibility [8]. In a simple approach, valid under the CW or quasi-CW conditions, the initial growth of a weak optical signal is governed by

$$\frac{dI_s}{dz} = \gamma_R(\Omega) I_p I_s \qquad (1)$$

where $\Omega \equiv \omega_p - \omega_s$ represents the Raman shift, and ω_p and ω_s are the optical frequencies associated with the pump and signal fields having intensities I_p and I_s, respectively.

The Raman gain spectrum has been measured for silica glasses as well as silica-based fibers [10–18]. Figure 9.3 shows the Raman gain coefficient for bulk silica as a function of the frequency shift Ω when the pump and signal are copolarized (solid curve) or orthogonally polarized (dotted curve). The peak gain is normalized to 1 in the copolarized case so that the same curves can be used for any pump wavelength λ_p. The peak value scales inversely with λ_p and is about 6×10^{-14} m/W for a pump near 1.5 μm.

The most significant feature of the Raman gain spectrum for silica fibers is that the gain exists over a large frequency range (up to 40 THz) with a broad peak located near 13.2 THz. This behavior is attributable to the noncrystalline nature of silica glasses. In amorphous materials such as fused silica, molecular vibrational frequencies spread out into bands that overlap and create a continuum. As a result, in contrast to most molecular media, for which the Raman gain occurs at specific well-defined frequencies, it extends continuously over a broad range in silica fibers. Optical fibers can act as broadband Raman amplifiers because of this feature. Another

most of the power of the pump beam is transferred to it [2]. Since 1962, SRS has been studied extensively in a variety of molecular media and has found a number of applications [3–9]. SRS was observed in silica fibers in 1972 after losses of such fibers had been reduced to acceptable levels [10]. Since then, the properties of the Raman scattering process have been quantified for many optical glasses, in both the bulk and the fiber form [11–18].

From a practical standpoint, SRS is not easy to observe in optical fibers using CW pump beams because of its relatively high threshold ($\sim$1 W). However, if a Stokes beam of the right frequency is launched together with the pump beam, as shown in Fig. 9.2, it can be amplified significantly using a CW pump beam with power levels of $\sim$100 mW. The pump and signal can even be launched in the opposite directions because of the nearly isotropic nature of SRS. In fact, as discussed later, the backward-pumping configuration is often preferred in practice because it leads to better performance of Raman amplifiers. Although fiber-based Raman amplifiers attracted considerable attention during the 1980s [19–37], it was only with the availability of appropriate pump lasers in the late 1990s that their development matured for telecommunications applications [38–62].

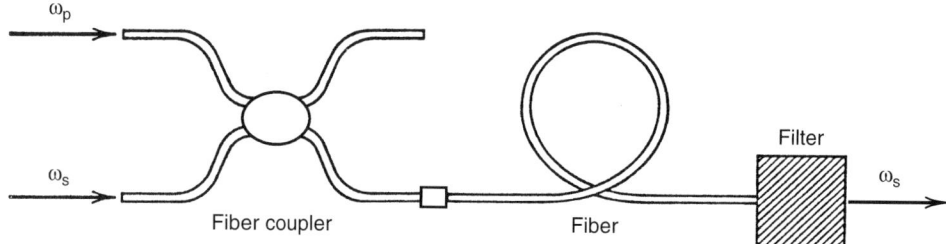

FIGURE 9.2 Schematic of a fiber-based Raman amplifier in the forward-pumping configuration. The optical filter passes the signal beam but blocks the residual pump.

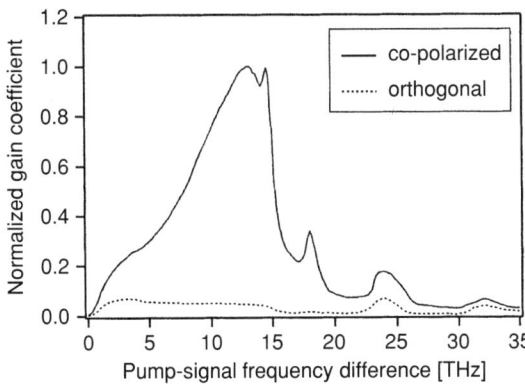

FIGURE 9.3 Raman gain spectrum for bulk silica measured when the pump and signal are copolarized (solid curve) or orthogonally polarized (dotted curve). The peak gain is normalized to 1 in the copolarized case. (After [62]; ©2004 IEEE.)

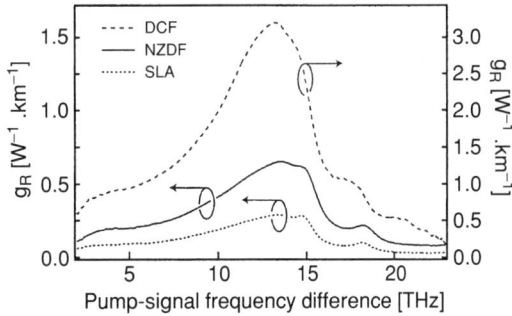

FIGURE 9.4 Measured Raman gain spectra for three kinds of fibers pumped at 1.45 μm. Both the effective area and GeO_2 doping levels are different for three fibers. SLA, super-large area. (After [62]; ©2004 IEEE.)

important feature of Fig. 9.3 is the polarization dependence of the Raman gain; the gain nearly vanishes when pump and signal are orthogonally polarized. The polarization dependence of the Raman gain affects the performance of Raman amplifiers in several different ways.

In single-mode fibers, the spatial profile of both the pump and signal beams is dictated by the fiber design and does not change along the entire fiber length. For this reason, one deals with the total optical power defined as $P(z) = \iint_{-\infty}^{\infty} I(x, y, z)dxdy$. Equation (1) can be written in terms of optical powers as

$$\frac{dP_s}{dz} = (\gamma_R/A_{\text{eff}})P_pP_s \equiv g_R P_p P_s \quad (2)$$

where the effective core area is related to the fiber-mode profile $F(x,y)$ as

$$A_{\text{eff}} = \frac{(\iint_{-\infty}^{\infty} |F(x, y)|^2 dxdy)^2}{\iint_{-\infty}^{\infty} |F(x, y)|^4 dxdy} \quad (3)$$

if we assume that the mode profile is nearly the same for both the pump and the Stokes. If we further approximate the mode profile by a Gaussian function of the form $F(x, y) = \exp[-(x^2 + y^2)/w^2]$, where w is the mode-field radius (MFR) [9], and perform the integrations in Eq. (3), we obtain the simple well-known result $A_{\text{eff}} \approx \pi w^2$. Because the MFR w is specified for any fiber, A_{eff} is a known parameter whose values can range from 10 to 100 μm² depending on fiber design; low values of A_{eff} occur for dispersion compensating fibers (DCFs) for which the core diameter is relatively small.

Figure 9.4 shows $g_R \equiv \gamma_R/A_{\text{eff}}$ (sometimes called the Raman gain efficiency) for a DCF, a nonzero dispersion fiber (NZDF), and a super-large area fiber with $A_{\text{eff}} = 15$, 55, and 105 μm², respectively. In all cases, the fiber was pumped at 1.45 μm and provided gain near 1.55 μm. The main point to note is that a DCF is nearly 10 times more efficient for Raman amplification. An increase by a factor of 7 is expected from its reduced effective core area. The remaining increase is due to a higher doping level of germania in DCFs (GeO_2 molecules exhibit a larger Raman gain peaking near 13.1 THz). Spectral changes seen in Fig. 9.4 for three fibers can be attributed to GeO_2 doping levels.

It is evident from Fig. 9.4 that when a pump beam is launched into the fiber together with a weak signal beam, it is amplified because of the Raman gain as long as the frequency difference $\Omega = \omega_p - \omega_s$ lies within the bandwidth of the Raman gain spectrum. The signal gain depends considerably on the frequency difference Ω and is maximum when the signal beam is downshifted from the pump frequency by 13.2 THz (about 100 nm in the 1.5-μm region). The Raman gain exists in all spectral regions, that is, optical fibers can be used to amplify any signal provided an appropriate pump source is used. This remarkable feature of Raman amplifiers is quite different from erbium-doped fiber amplifiers, which can amplify only signals whose wavelength is close to the atomic transition wavelength occurring near 1.53 μm.

The nonuniform nature of the Raman gain spectrum in Fig. 9.4 is of concern for wavelength division multiplexed (WDM) lightwave systems because different channels are amplified by different amounts. This problem is solved in practice by using multiple pumps at slightly different wavelengths. Each pump provides nonuniform gain but the gain spectra associated with different pumps overlap partially. With a suitable choice of wavelengths and powers for each pump laser, it is possible to realize in nearly flat gain profile over a considerably wide wavelength range. We discuss the single-pump scheme first, because it allows us to introduce the basic concepts in a simple manner, and then focus on the multiple-pump configuration of Raman amplifiers.

2.2. Simple Theory

Consider the simplest situation in which a single CW pump beam is launched into an optical fiber used to amplify a CW signal. Even in this case, Eq. (2) should be modified to include fiber losses before it can be used. Moreover, the pump power does not remain constant along the fiber. When these effects are included, the Raman-amplification process is governed by the following set of two coupled equations:

$$\frac{dP_s}{dz} = g_R P_p P_s - \alpha_s P_s \quad (4)$$

$$\xi \frac{dP_p}{dz} = -\frac{\omega_p}{\omega_s} g_R P_p P_s - \alpha_p P_p \quad (5)$$

where α_s and α_p account for fiber losses at the Stokes and pump wavelengths, respectively. The parameter ξ takes values ± 1 depending on the pumping configuration; the minus sign should be used in the backward-pumping case.

Equations (4) and (5) can be derived rigorously from Maxwell's equations. They can also be written phenomenologically by considering the processes through which photons appear in or disappear from each beam. The frequency ratio ω_p/ω_s appears in Eq. (5) because the pump and signal photons have different energies. One can readily verify that in the absence of losses,

$$\frac{d}{dz}\left(\frac{P_s}{\omega_s} + \xi \frac{P_p}{\omega_p}\right) = 0 \quad (6)$$

Noting that P_j/ω_j is related to photon flux at the frequency ω_j, this equation merely represents the conservation of total number of photons during the SRS process.

Equations (4) and (5) are not easy to solve analytically because of their nonlinear nature. In many practical situations, pump power is so large compared with the signal power that pump depletion can be neglected by setting $g_R = 0$ in Eq. (5), which is then easily solved. As an example, $P_p(z) = P_0 \exp(-\alpha_p z)$ in the forward-pumping case ($\xi = 1$), where P_0 is the input pump power at $z = 0$. If we substitute this solution in Eq. (4), we obtain

$$\frac{dP_s}{dz} = g_R P_0 \exp(-\alpha_p z) P_s - \alpha_s P_s \quad (7)$$

This equation can be easily integrated to obtain

$$P_s(L) = P_s(0) \exp(g_R P_0 L_{\text{eff}} - \alpha_s L) \equiv G(L) P_s(0) \quad (8)$$

where $G(L)$ is the net signal gain, L is the amplifier length, and L_{eff} is an effective length defined as

$$L_{\text{eff}} = [1 - \exp(-\alpha_p L)]/\alpha_p \quad (9)$$

The solution (8) shows that because of pump absorption, the effective amplification length is reduced from L to L_{eff}.

The backward-pumping case can be considered in a similar fashion. In this case, Eq. (5) should be solved with $g_R = 0$ and $\xi = -1$ using the boundary condition $P_p(L) = P_0$; the result is $P_p(z) = P_0 \exp[-\alpha_p(L-z)]$. The integration of Eq. (4) yields the same solution given in Eq. (8), indicating that the amplified signal power at a given pumping level is the same in both the forward and the backward pumping configurations.

The case of bidirectional pumping is slightly more complicated because two pump lasers are located at the opposite fiber ends. The pump power in Eq. (4) now represents the sum $P_p = P_f + P_b$, where P_f and P_b are obtained by solving (still ignoring pump depletion)

$$dP_f/dz = -\alpha_p P_f, \qquad dP_b/dz = \alpha_p P_b \quad (10)$$

Solving these equations, we obtain total pump power $P_p(z)$ at a distance z in the form

$$P_p(z) = P_0\{r_f \exp(-\alpha_p z) + (1 - r_f) \exp[-\alpha_p(L-z)]\} \quad (11)$$

where P_0 is the total pump power and $r_f = P_L/P_R$ is the fraction of pump power launched in the forward direction. The integration of Eq. (4) yields the signal gain

$$G(z) = \frac{P_s(z)}{P_s(0)} = \exp\left(g_R \int_0^z P_p(z) dz - \alpha_s z\right) \quad (12)$$

Figure 9.5 shows how the signal power changes along a 100-km-long distributed Raman amplifier as r_f is varied from 0 to 1. In all cases, the total pump power is chosen such that the Raman gain is just sufficient to compensate for fiber losses, that is, $G(L) = 1$.

One may ask which pumping configuration is the best from the system standpoint. The answer is not so simple because it depends on many factors. As discussed in Section 3.2, forward pumping is superior from the noise viewpoint. However, for a long-haul system limited by fiber nonlinearities, backward pumping may offer better performance because the

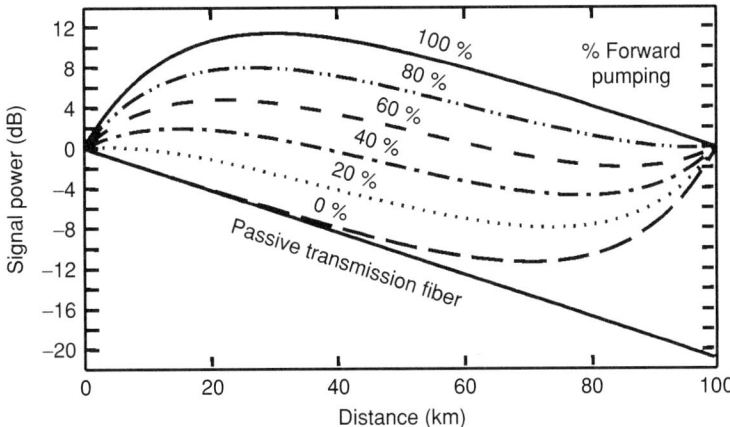

FIGURE 9.5 Evolution of signal power in a bidirectionally pumped 100-km-long Raman amplifier as the contribution of forward pumping is varied from 0 to 100%. The straight line shows for comparison the case of a passive fiber with no Raman gain. (After [57]; ©2003 Springer.)

signal power is the smallest throughout the link length in this case. The total accumulated nonlinear phase shift induced by self-phase modulation (SPM) can be obtained from [9]

$$\Phi_{NL} = \gamma \int_0^L P_s(z)dz = \gamma P_s(0) \int_0^L G(z)dz \quad (13)$$

where $\gamma = 2\pi n_2/(\lambda_s A_{eff})$ is the nonlinear parameter responsible for SPM. Using $G(z)$ from Eq. (12), it is easy to show that Φ_{NL} is smaller in the case of backward pumping.

The quantity $G(L)$ represents the net signal gain and can be even <1(net loss) if the Raman gain is not sufficient to overcome fiber losses. It is useful to the introduce the concept of the on–off Raman gain using the definition

$$G_A = \frac{P_s(L) \text{ with pump on}}{P_s(L) \text{ with pump off}} = \exp(g_R P_0 L_{eff}) \quad (14)$$

Clearly, G_A represents the total amplifier gain distributed over a length L_{eff}. If we use a typical value of $g_R = 3$ W^{-1}/km for a DCF from Fig. 9.4 together with $L_{eff} = 1$ km, the signal can be amplified by 20 dB for $P_0 \approx 1.5$ W. The exponential dependence of G_A on pump power has been observed in many experiments. In a 1981 experiment, a 1.3-km-long fiber was used to amplify the 1.064-μm signal by using a 1.017-μm pump [19]. The output power increased exponentially at low signal powers but began to saturate for $P_0 > 1$ W when input signal power was relatively large.

2.3. Gain Saturation

An approximate expression for the saturated gain G_s in Raman amplifiers can be obtained by solving Eqs. (4) and (5) analytically [9] with the assumption $\alpha_s = \alpha_p \equiv \alpha$. This approximation is not always valid but can be justified for optical fibers in the 1.55-μm region. Assuming forward pumping ($\xi = 1$) and making the transformation $P_j = \omega_j F_j \exp(-\alpha z)$ with $j = $ s or p, we obtain two simple equations:

$$\frac{dF_s}{dz} = \omega_p g_R F_p F_s, \quad \frac{dF_p}{dz} = -\omega_p g_R F_p F_s \quad (15)$$

Noting that $F_p(z) + F_s(z) = C$, where C is a constant, the differential equation for F_s can be integrated over the amplifier length to obtain the following result:

$$G_s = \frac{F_s(L)}{F_s(0)} = \left(\frac{C - F_s(L)}{C - F_s(0)}\right) \exp(\omega_p g_R C L_{eff}) \quad (16)$$

Using $C = F_p(0) + F_s(0)$ in the preceding equation, the saturated gain of the amplifier is given by

$$G_s = \frac{(1+r_0)G_A^{1+r_0}}{1 + r_0 G_A^{1+r_0}} \quad (17)$$

where r_0 is related to the signal-to-pump power ratio at the fiber input as

$$r_0 = \frac{F_s(0)}{F_p(0)} = \frac{\omega_p}{\omega_s} \frac{P_s(0)}{P_p(0)} \quad (18)$$

and $G_A = \exp(g_R P_0 L_{eff})$ is the unsaturated gain introduced in Eq. (14). Typically, $P_s(0) \ll P_p(0)$. For example, $r_0 < 10^{-3}$ when $P_s(0) < 1$ mW, whereas $P_p(0) \approx 1$ W. Under such conditions, the saturated gain of the amplifier can be approximated as

$$G_s = \frac{G_A}{1 + r_0 G_A} \quad (19)$$

The gain is reduced by a factor of 2 or 3 dB when the Raman amplifier is pumped hard enough that

$r_0 G_A = 1$. This can happen for $r_0 = 10^{-3}$ when the on–off Raman gain approaches 30 dB.

Figure 9.6 shows the saturation characteristics by plotting G_s/G_A as a function of $G_A r_0$ for several values of G_A. The saturated gain is reduced by a factor of 2 when $G_A r_0 \approx 1$. This condition is satisfied when the power in the amplified signal starts to approach the input pump power P_0. In fact, P_0 is a good measure of the saturation power of Raman amplifiers. As typically $P_0 > 1$ W, the saturation power of Raman amplifiers is much larger compared with that of erbium-doped fiber amplifiers.

3. MODERN RAMAN AMPLIFIERS

As early as 1981, it was found that Raman amplifiers could amplify an input signal by a factor of 1000 when the pump power exceeded 1 W [19]. Most of the early experiments used a Nd:YAG laser operating at 1.06-μm for pumping because it provided such high CW power levels. This laser can also operate at 1.32 μm. In a 1983 experiment [21], a 1.4-μm signal was amplified using such a laser, and gain levels of up to 21 dB were obtained at a pump power of 1 W. The amplifier gain was nearly the same in both the forward and backward pumping configurations. Signal wavelengths of most interest from the standpoint of optical fiber communications are near 1.5 μm. A Nd:YAG laser can still be used if a higher order Stokes line is used as a pump. For example, the first-order Stokes line at 1.4 μm from a 1.32-μm laser can act as a pump to amplify an optical signal near 1.5 μm. As early as 1984, amplification factors of more than 20 dB were realized by using such schemes [24–26]. These experiments also indicated the importance of matching the polarization directions of the pump and signal waves because SRS nearly ceases to occur in the case of orthogonal polarization. The use of a polarization-preserving fiber with a high-germania core resulted in a 20-dB gain at 1.52 μm when such a fiber was pumped with 3.7 W of pump power.

The main drawback of Raman amplifiers from the standpoint of lightwave system applications is that they require a high-power CW laser for pumping. Most experiments performed in the 1980s in the 1.55-μm spectral region used tunable color-center lasers as a pump; such lasers are too bulky for telecommunications applications. For this reason, with the advent of erbium-doped fiber amplifiers around 1989, Raman amplifiers were rarely used in the 1.55-μm wavelength region.

The situation changed with the availability of compact high-power semiconductor and fiber lasers. Indeed, the field of fiber optic Raman amplifiers experienced a virtual renaissance during the 1990s [38–56]. In a 1995 experiment, three pairs of fiber gratings were inserted at two ends of the fiber used for Raman amplification [39]. The Bragg wavelengths of these gratings were chosen such that they formed three cavities for three Raman lasers operating at wavelengths of 1.117, 1.175, and 1.24 μm. These wavelengths correspond to first-, second-, and third-order Stokes generated in succession when a 1.06-μm laser is used. All three lasers were pumped through cascaded SRS using a single, diode-pumped, Nd-fiber laser. The 1.24-μm laser then pumped the Raman amplifier to provide signal amplification in the 1.3-μm region. The same idea of cascaded SRS was used to obtain 39-dB gain at 1.3 μm by using WDM couplers in place of fiber gratings [38]. In a different approach, the core of a silica fiber was doped heavily with germanium. Such a fiber provided 30-dB gain at a pump power of only 350 mW [40], power levels that can be realized using one or more semiconductor lasers. A dual-stage configuration was also used, in which a 2-km-long germanium-doped fiber was placed in series with a 6-km-long dispersion shifted fiber in a ring geometry [46]. Such a Raman amplifier, when pumped with a 1.24-μm Raman laser, provided 22-dB gain in the 1.3-μm wavelength region with a noise figure of about 4 dB.

3.1. Broadband Raman Amplifiers

Starting in 1998, the use of multiple pumps for Raman amplification was pursued for developing broadband optical amplifiers required for WDM lightwave systems operating in the 1.55-μm region [47–56]. Massive WDM systems (80 or more channels) typically require optical amplifiers capable of providing

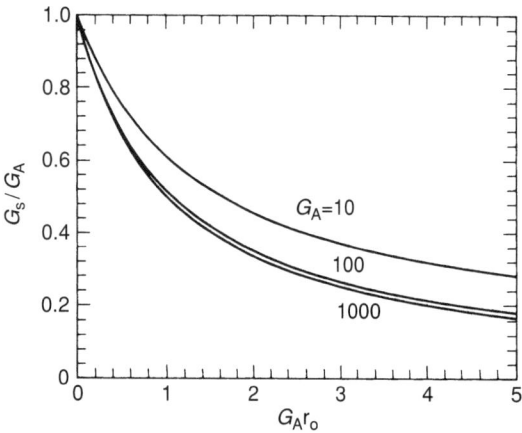

FIGURE 9.6 Gain-saturation characteristics of Raman amplifiers for several values of the unsaturated amplifier gain G_A.

uniform gain over a 70- to 80-nm wavelength range. In a simple approach, hybrid amplifiers made by combining erbium doping with Raman gain were used. In one implementation of this idea [49], a nearly 80-nm bandwidth was realized by combining an erbium-doped fiber amplifier with two Raman amplifiers pumped simultaneously at three different wavelengths (1471, 1495, and 1503 nm) using four pump modules, each module launching more than 150 mW of power into the fiber. The combined gain of 30 dB was nearly uniform over the wavelength range 1.53–1.61 µm.

Broadband amplification over 80 nm or more can also be realized by using a pure Raman-amplification scheme. In this case, a relatively long span (typically > 5 km) of a fiber with a relatively narrow core (such as a DCF) is pumped using multiple pump lasers. Alternatively, one can use the transmission fiber itself as the Raman gain medium. In the latter scheme, the entire long-haul fiber link is divided into multiple segments (60–100 km long), each one pumped backward using a pump module consisting of multiple pump lasers. The Raman gain accumulated over the entire segment length compensates for fiber losses of that segment in a distributed manner.

Multiple-pump Raman amplifiers make use of the fact that the Raman gain exists at any wavelength as long as the pump wavelength is suitably chosen. Thus, even though the gain spectrum of a single pump is not very wide and is flat only over a few nanometers (Fig. 9.4), it can be broadened and flattened considerably by using several pumps of different wavelengths. Each pump creates a gain profile that mimics the spectrum shown in Fig. 9.4. Superposition of several such spectra can produce a relatively constant gain over a wide spectral region when pump wavelengths and power levels are chosen judiciously. Figure 9.7 shows a numerical example when six pump lasers operating at wavelengths in the range of 1420–1500 nm are used [62]. The individual pump powers (vertical bars) are chosen to provide individual gain spectra (dashed curves) such that the total Raman gain of 18 dB is nearly flat over an 80-nm bandwidth (solid trace). Pump powers range from 40 to 200 mW and are larger for shorter wavelength pumps because all pumps interact through SRS, and some power is transferred to longer wavelength pumps within the amplifier. This technique can provide a gain bandwidth of more than 100 nm with a suitable design [50–54]. In a 2000 demonstration, 100 WDM channels with 25-GHz channel spacing, each operating at a bit rate of 10 Gb/s, were transmitted over 320 km [53]. All channels were amplified simultaneously by pumping each 80-km fiber span in the backward direction using four semiconductor lasers. Such a distributed Raman amplifier provided 15-dB gain at a total pump power of 450 mW.

3.2. Design of Raman Amplifiers

Design of broadband Raman amplifiers requires attention to many details. As seen in Fig. 9.3, Raman gain is polarization sensitive in the sense that a signal copolarized with the pump experiences much

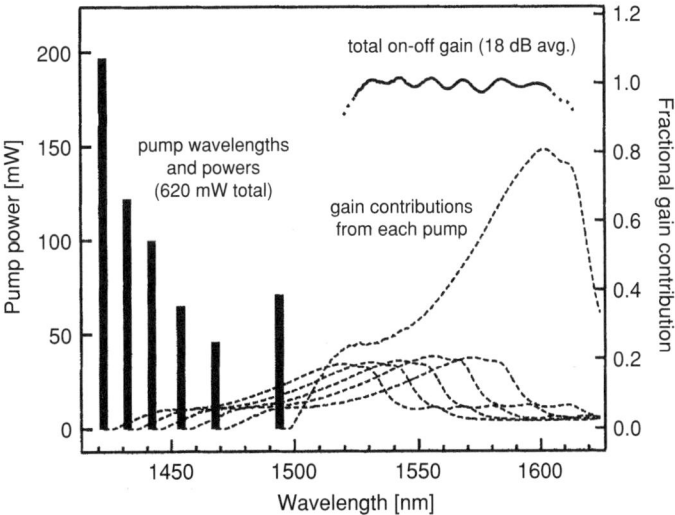

FIGURE 9.7 Numerically simulated composite Raman gain (solid trace) of a Raman amplifier pumped with six lasers with different wavelengths and input powers (vertical bars). Dashed curves show the Raman gain provided by individual pumps. (After [62]; ©2004 IEEE.)

more gain than the one polarized orthogonal to the pump. This creates a problem in practice because signal polarization is unpredictable in most lightwave systems. The polarization problem is solved in practice by pumping a Raman amplifier such that two orthogonally polarized lasers are used at each pump wavelength. Alternatively, one can depolarize the output of each pump laser using a suitable technique that reduces the degree of polarization close to zero. It should be stressed that the state of polarization of the pump and signal fields changes randomly in any realistic fiber because of birefringence variations along the fiber length.

Several other issues must be considered while designing a Raman amplifier. For example, when multiple pump lasers are used, each shorter wavelength laser can amplify the remaining pump beams through Raman gain because, as seen in Fig. 9.4, some gain exists even when pump wavelengths differ by only a few nanometers. Another issue that must be addressed is related to *double Rayleigh backscattering*. Rayleigh scattering occurs in all fibers and is the fundamental source of loss in them. A part of signal is backscattered by it and is amplified by the Raman gain. If this field is backscattered a second time by Rayleigh scattering, it ends up propagating with the signal and acts as noise. Without Raman gain, the contribution of double Rayleigh backscattering is negligible. However, it is amplified by the Raman process in both directions and can build up to sufficiently high levels to be of concern, especially when the amplifier gain is large.

Modern Raman amplifiers are designed using a numerical model that includes pump–pump interactions, Rayleigh backscattering, and spontaneous Raman scattering. Such a model considers each frequency component separately and requires the solution of a large set of coupled equations of the form [52]

$$\frac{dP_f(\nu)}{dz} = \int_{\mu>\nu} g_R(\mu,\nu)[P_f(\mu)+P_b(\mu)]$$
$$\times [P_f(\nu)+2h\nu n_{sp}(\mu-\nu)]d\mu$$
$$- \int_{\mu<\nu} g_R(\nu,\mu)[P_f(\mu)$$
$$+ P_b(\mu)] \times [P_f(\nu)+4h\nu n_{sp}(\nu-\mu)]d\mu$$
$$- \alpha(\nu)P_f(\nu)+f_r\alpha_r P_b(\nu) \quad (20)$$

where μ and ν denote optical frequencies and the subscripts f and b denote forward- and backward-propagating waves, respectively. The parameter n_{sp} is defined as

$$n_{sp}(\Omega) = [1-\exp(-\hbar h\Omega/k_B T)]^{-1} \quad (21)$$

where $\Omega = |\mu-\nu|$ is the Raman shift and T denotes absolute temperature of the amplifier. In Eq. (20), the first and second terms account for the Raman-induced power transfer into and out of each frequency band. The factor of 2 in the first term accounts for the two polarization modes of the fiber. An additional factor of 2 in the second term includes spontaneous emission in both the forward and the backward directions [54]. Fiber losses and Rayleigh backscattering are included through the last two terms and are governed by the parameters α and α_r, respectively; f_r represents the fraction of backscattered power that is recaptured by the fiber mode. A similar equation holds for the backward-propagating waves.

To design broadband Raman amplifiers, the entire set of such equations is solved numerically to find the channel gains, and input pump powers are adjusted until the gain is nearly the same for all channels (Fig. 9.7). Figure 9.8 shows an experimentally measured gain spectrum for a Raman amplifier made by pumping a 25-km-long dispersion shifted fiber with 12 diode lasers. The frequencies and powers of pump lasers are also indicated in a tabular form. Notice that all powers are under 100 mW. The amplifier provided about 10.5-dB gain over an 80-nm bandwidth with a ripple of less than 0.1 dB. Such amplifiers are suitable for dense WDM systems covering both the C- and L-bands. Several experiments have used broadband Raman amplifiers to demonstrate transmission over long distances at high bit rates. In a 2001 experiment, 77 channels, each operating at 42.7 Gb/s, were transmitted over 1200 km by using the C- and L-bands simultaneously [55]. Since then, many demonstrations have used Raman amplification for a wide variety of WDM systems [58–61].

4. PERFORMANCE LIMITING FACTORS

The performance of modern Raman amplifiers is affected by several factors that need to be controlled. In this section we focus on spontaneous Raman scattering, double Rayleigh backscattering, and pump-noise transfer. The impact of PMD on the performance of Raman amplifier is also considered in this section.

4.1. Spontaneous Raman Scattering

Spontaneous Raman scattering adds to the amplified signal and appears as noise because of random phases associated with all spontaneously generated photons. This noise mechanism is similar to the

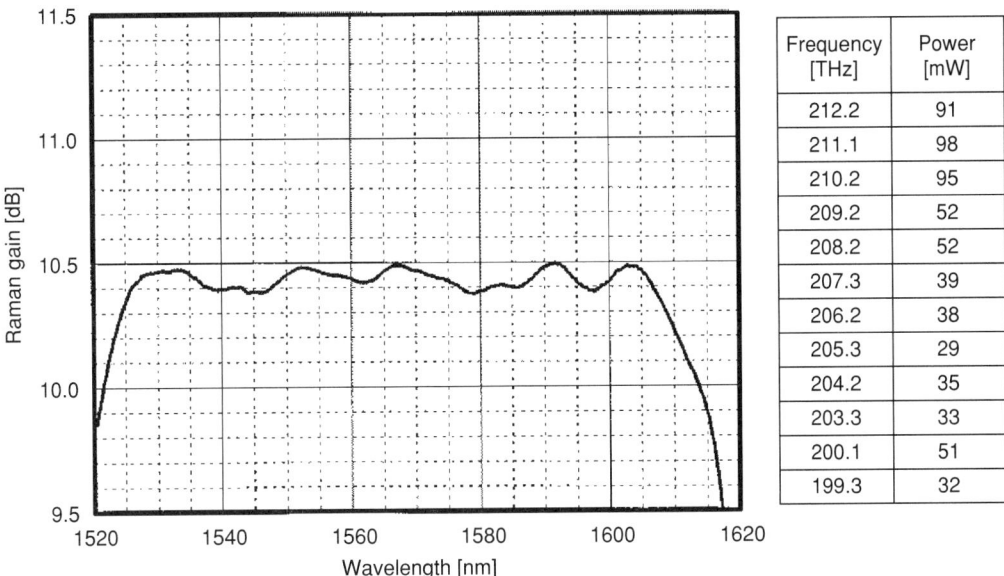

FIGURE 9.8 Measured Raman gain as a function of signal wavelength for a 25-km-long amplifier pumped with 12 lasers. Pump frequencies and power levels used are indicated on the right. (After [54]; ©2001 IEEE.)

spontaneous emission that affects the performance of erbium-doped fiber amplifiers except that, in the Raman case, it depends on the phonon population in the vibrational state, which in turn depends on the temperature of the Raman amplifier. On a more fundamental level, one should consider the evolution of signal with the noise added by spontaneous Raman scattering. However, because noise photons are not in phase with the signal photons, we cannot use the rate equations firm Eq. (4) and (5) satisfied by the signal and pump powers. Rather, we should write equations for the two optical fields. If we neglect pump depletion, it is sufficient to replace Eq. (4) with

$$\frac{dA_s}{dz} = \frac{g_R}{2} P_p(z) A_s - \frac{\alpha_s}{2} A_s + f_n(z, t) \quad (22)$$

where A_s is the signal field defined such that $P_s = |A_s|^2$, P_p is the pump power, and the Langevin noise source $f_n(z, t)$ takes into account the noise added through spontaneous Raman scattering. Because each scattering event is independent of others, this noise can be modeled as a Markovian stochastic process with Gaussian statistics such that $\langle f_n(z, t) \rangle = 0$ and its second moment is given by [63]

$$\langle f_n(z, t) f_n(z', t') \rangle = n_{sp} h\nu_0 g_R P_p(z) \delta(z - z') \delta(t - t') \quad (23)$$

where n_{sp} is the spontaneous-scattering factor introduced earlier and $h\nu_0$ is the average photon energy. The two delta functions ensure that all spontaneous events are independent of each other.

Equation (22) can be easily integrated to obtain $A_s(L) = \sqrt{G(L)} A_s(0) + a_{ASE}(t)$, where $G(L)$ is the amplification factor defined earlier in Eq. (8), ASE is the amplified spontaneous emission, and the total accumulated noise from spontaneous Raman scattering is given by

$$a_{ASE}(t) = \sqrt{G(L)} \int_0^L \frac{f_n(z, t)}{\sqrt{G(z)}} dz,$$
$$G(z) = \exp\left(\int_0^z [g_R P_p(z') - \alpha_s] dz'\right) \quad (24)$$

This noise is often referred to as ASE because of its amplification by the distributed Raman gain. It is easy to show that it vanishes on an average ($\langle a_{ASE}(t) \rangle = 0$) and its second moment is given by

$$\langle a_{ASE}(t) a_{ASE}(t') \rangle = G_L \int_0^L dz \int_0^L dz' \frac{\langle f_n(z, t) f_n(z', t') \rangle}{\sqrt{G(z) G(z')}}$$
$$= S_{ASE} \delta(t - t') \quad (25)$$

where $G_L \equiv G(L)$ and the ASE spectral density is defined as

$$S_{ASE} = n_{sp} h\nu_0 g_R G_L \int_0^L \frac{P_p(z)}{G(z)} dz \quad (26)$$

The presence of the delta function in Eq. (25) is due to the Markovian assumption implying that S_{ASE} is constant and exists at all frequencies (white noise). In practice, the noise exists only over the amplifier bandwidth and can be further reduced by placing an optical filter at the amplifier output. Assuming this

Frequency [THz]	Power [mW]
212.2	91
211.1	98
210.2	95
209.2	52
208.2	52
207.3	39
206.2	38
205.3	29
204.2	35
203.3	33
200.1	51
199.3	32

to be the case, we can calculate the total ASE power after the amplifier using

$$P_{\text{ASE}} = 2 \int_{-\infty}^{\infty} S_{\text{ASE}} H_f(\nu) d\nu = 2 S_{\text{ASE}} B_{\text{opt}} \quad (27)$$

where B_{opt} is the bandwidth of the optical filter. The factor of 2 in this equation accounts for the two polarization modes of the fiber. Indeed, the ASE can be reduced by 50% if a polarizer is placed after the amplifier. Assuming that a polarizer is not used, the optical signal-to-noise ratio (SNR) of the amplified signal is given by

$$\text{SNR}_o = \frac{P_s(L)}{P_{\text{ASE}}} = \frac{G_L P_{\text{in}}}{P_{\text{ASE}}} \quad (28)$$

It is evident from Eq. (26) that both P_{ASE} and SNR_o depend on the pumping scheme through pump-power variations $P_p(z)$ occurring inside the Raman amplifier. As an example, Fig. 9.9 shows how the spontaneous power per unit bandwidth $P_{\text{ASE}}/B_{\text{opt}}$, and the optical SNR vary with the net gain $G(L)$ for several different pumping schemes assuming that a 1-mW input signal is amplified by a 100-km-long, bidirectionally pumped, distributed Raman amplifier. The fraction of forward pumping varies from 0 to 100%. The other parameters were chosen to be $\alpha_s = 0.21$ dB/km, $\alpha_p = 0.26$ dB/km, $n_{\text{sp}} = 1.13$, $h\nu_0 = 0.8$ eV, and $g_R = 0.68$ W^{-1}/km. The optical SNR is highest in the case of purely forward pumping (about 54 dB or so) but degrades by as much as 15 dB as the fraction of backward pumping is increased from 0 to 100%. This can be understood by noting that the spontaneous noise generated near the input end experiences losses over the full length of the fiber in the case of forward pumping, whereas it experiences only a fraction of such losses in the case of backward pumping. Mathematically, $G(z)$ in the denominator in Eq. (26) is larger in the forward-pumping case, resulting in reduced S_{ASE}.

4.2. Effective Noise Figure

The preceding discussion shows that spontaneous Raman scattering degrades the SNR of the signal amplified by a Raman amplifier. The extent of SNR degradation is generally quantified through a parameter F_r, called the *amplifier noise figure*, and defined as [64–66]

$$F_n = \frac{(\text{SNR})_{\text{in}}}{(\text{SNR})_{\text{out}}} \quad (29)$$

In this equation, SNR is not the optical SNR but refers to the *electric power* generated when the optical signal is converted into an electric current. In general, F_n depends on several detector parameters that govern thermal noise associated with the detector. A simple expression for F_n can be obtained by considering an ideal detector whose performance is limited by shot noise only. The electrical SNR of the input signal is then given by [67]

$$(\text{SNR})_{\text{in}} = \frac{\langle I_d \rangle^2}{\sigma_s^2} = \frac{(R_d P_{\text{in}})^2}{\sigma_s^2} \quad (30)$$

where I_d is the current and $R_d = q/h\nu_0$ is the responsivity of a detector with 100% quantum efficiency.

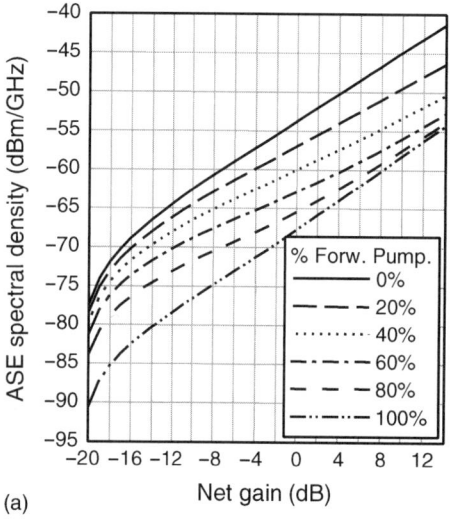
(a)

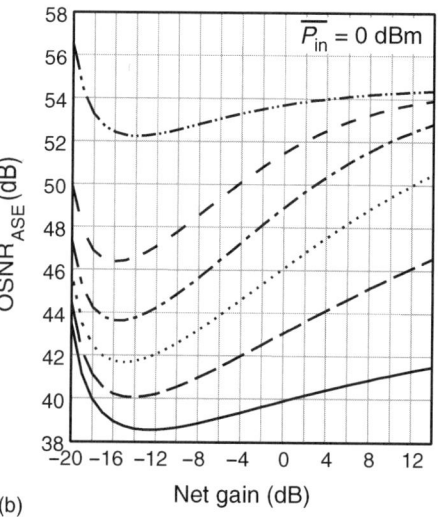
(b)

FIGURE 9.9 (a) Spontaneous spectral density and (b) optical SNR as a function of net gain $G(L)$ at the output of a 100-km-long, bidirectionally pumped, distributed Raman amplifier assuming $P_{\text{in}} = 1$ mW. (After [57]; ©2003 Springer.)

The variance of shot noise over the receiver bandwidth Δf can be written as $\sigma_s^2 = 2q(R_d P_{in})\Delta f$, resulting in an input SNR of $P_{in}/(2h\nu_0\Delta f)$.

To calculate the electrical SNR of the amplified signal, we should add the contribution of ASE to the receiver noise. When all noise sources are included, the detector current takes the form

$$I_d = R_d[|\sqrt{G_L}E_s + E_{cp}|^2 + |E_{op}|^2] + i_s + i_T \quad (31)$$

where i_s and i_T are current fluctuations induced by shot and thermal noises, respectively, E_s is the signal field, E_{cp} is the part of ASE copolarized with the signal, and E_{op} is its orthogonally polarized part. It is necessary to separate the ASE into two parts because only its copolarized part can beat with the signal. Because ASE occurs over a broader bandwidth than the signal bandwidth $\Delta\nu_s$, it is common to divide the ASE bandwidth B_{opt} into M bins, each of bandwidth $\Delta\nu_s$, and write E_{cp} in the form [64–66]

$$E_{cp} = \sum_{m=1}^{M} \sqrt{S_{ASE}\Delta\nu_s}\exp(i\Phi_m - i\omega_m t) \quad (32)$$

where Φ_m is the phase of the noise component at the frequency $\omega_m = \omega_f + m(2\pi\Delta\nu_s)$. An identical form should be used for E_{op}.

Using $E_s = \sqrt{P_{in}}\exp(i\Phi_s - i\omega_s t)$ in Eq. (31) and including all the beating terms, the current I_d can be written in the form

$$I_d = R_d G_L P_{in} + i_s + i_{ASE} + i_s + i_T \quad (33)$$

where i_b and i_{ASE} represent current fluctuations resulting from signal–ASE and ASE–ASE beating, respectively, and are given by

$$i_b = 2R_d(G_L P_{in} S_{ASE}\Delta\nu_s)^{1/2}$$
$$\sum_{m=1}^{M}\cos[(\omega_s - \omega_m)t + \Phi_m - \Phi_s] \quad (34)$$

$$i_{ASE} = 2R_d S_{ASE}\Delta\nu_s \sum_{m=1}^{M}\sum_{n=1}^{M}\cos[(\omega_n - \omega_m)t + \Phi_m - \Phi_n] \quad (35)$$

Because these two noise terms fluctuate with time, we need to find their variances. Because details are available in several texts, we write the final result directly [64–66]:

$$\sigma_b^2 = 4R_d^2 G_L P_{in} S_{ASE}\Delta f \quad (36)$$

$$\sigma_{ASE}^2 = 4R_d^2 S_{ASE}^2 \Delta f(B_{opt} - \Delta f/2) \quad (37)$$

The total variance of current fluctuations can be written from Eq. (33) as

$$\sigma^2 = \sigma_b^2 + \sigma_{ASE}^2 + G_L\sigma_s^2 + \sigma_T^2 \quad (38)$$

We can neglect the thermal-noise contribution σ_T^2 because as it is relatively small. The σ_{ASE}^2 term is also small in comparison with σ_b^2. For this reason, the electrical SNR of the amplified signal is approximately given by

$$(\text{SNR})_{out} = \frac{(R_d G_L P_{in})^2}{G_L\sigma_s^2 + \sigma_b^2} \quad (39)$$

The noise figure can now be obtained by substituting Eqs. (30) and (39) in Eq. (29). The result is found to be

$$F_n = \frac{1}{G_L}\left(1 + \frac{\sigma_b^2}{G_L\sigma_s^2}\right) \quad (40)$$

Using σ_b^2 from Eq. (36) and S_{ASE} from Eq. (26), the noise figure becomes

$$F_n = 2n_{sp}g_R\int_0^L \frac{P_p(z)}{G(z)}dz + \frac{1}{G_L} \quad (41)$$

This equation shows that the noise figure of a Raman amplifier depends on the pumping scheme. It provides reasonably small noise figures for "lumped" Raman amplifiers for which the fiber length is $\sim$1 km and the net signal gain exceeds 10 dB.

When the fiber within the transmission link itself is used for distributed amplification, the length of fiber section typically exceeds 50 km, and pumping is such that net gain $G(z) < 1$ throughout the fiber length. In this case, F_n predicted by Eq. (41) can be very large and can exceed 15 dB depending on the span length. This does not mean distributed amplifiers are more noisy than lumped amplifiers. To understand this apparent contradiction, consider a 100-km-long fiber span with a loss of 0.2 dB/km. The 20-dB span loss is compensated for by using a hybrid scheme in which a lumped amplifier with a 5-dB noise figure is combined with the Raman amplification through backward pumping. The on–off gain G_A of the Raman amplifier can be varied in the range of 0–20 dB by adjusting the pump power. Clearly, $G_A = 0$ and 20 dB correspond to the cases of pure lumped and distributed amplifications, respectively.

The solid line in Fig. 9.10 shows how the noise figure of such a hybrid amplifier changes as G_A is varied from 0 to 20 dB [62]. When $G_A = 0$, Eq. (41) shows that the passive fiber has a noise figure of 20 dB. This is not surprising because any fiber loss reduces signal power and thus degrades the SNR [56]. When the signal is amplified by the lumped amplifier, an additional 5-dB degradation occurs,

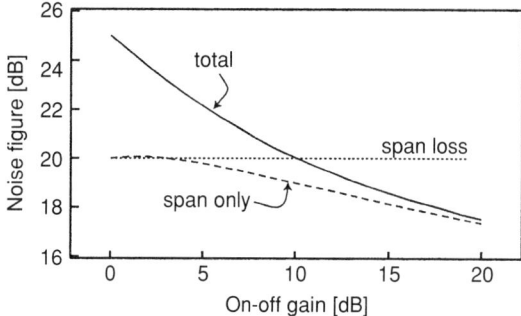

FIGURE 9.10 Total noise figure as a function of the on–off gain of a Raman amplifier when 20-dB loss of a 100-km-long fiber span is compensated for by using a hybrid amplification scheme. Dashed curve shows the noise figure of the Raman-pumped fiber span alone. Dotted line shows 20-dB span loss. (After [62]; ©2004 IEEE.)

resulting in a total noise figure of 25 dB in Fig. 9.10. This value decreases as G_A increases, reaching a level of about 17.5 dB for $G_A = 20$ dB (no lumped amplification). The dashed line shows the noise figure of the Raman-pumped fiber span alone as predicted by Eq. (41). The total noise figure is higher because the lumped amplifier adds some noise if span losses are only partially compensated for by the Raman amplifier. The important point is that total noise figure drops below the 20-dB level (dotted line) when the Raman gain exceeds a certain value.

To emphasize the noise advantage of distributed amplifiers, it is common to introduce the concept of an *effective* noise figure using the definition $F_{\text{eff}} = F_n \exp(-\alpha L)$, where α is the fiber-loss parameter at the signal wavelength. In decibel units, $F_{\text{eff}} = F_n - \mathcal{L}$, where $\mathcal{L}$ is the span loss in dB. As seen from Fig. 9.10, $F_{\text{eff}} < 1$ (or negative on the decibel scale) by definition. It is this feature of distributed amplification that makes it so attractive for long-haul WDM lightwave systems. For the example shown in Fig. 9.10, $F_{\text{eff}} \approx -2.5$ dB when pure distributed amplification is used. Note, however, the noise advantage is almost 7.5 dB when the noise figures are compared for lumped and distributed amplification schemes.

As seen from Eq. (41), the noise figure of a Raman amplifier depends on the pumping scheme used because, as discussed in Section 2.2 $P_p(z)$ can be quite different for forward, backward, and bidirectional pumping. In general, forward pumping provides the highest SNR and the smallest noise figure, because most of the Raman gain is then concentrated toward the input end of the fiber where power levels are high. However, backward pumping is often used in practice because of other considerations such as the transfer of pump noise to signal.

4.3. Rayleigh Backscattering

The phenomenon that most limits the performance of distributed Raman amplifiers turned out to be Rayleigh backscattering [68–73]. Rayleigh backscattering affects the performance of Raman amplifiers in two ways. First, a part of backward propagating ASE can appear in the forward direction, enhancing the overall noise. This noise is relatively small and is not of much concern for Raman amplifiers. Second, *double Rayleigh backscattering* of the signal creates a cross-talk component in the forward direction that has nearly the same spectral range as the signal (in-band cross-talk). It is this Rayleigh-induced noise, amplified by the distributed Raman gain, that becomes the major source of power penalty in Raman-amplified lightwave systems.

Because Rayleigh backscattering is a statistical process governed by random density fluctuations within the fiber core, in general one must follow a complicated approach outlined, for example, in [57]. However, it is relatively easy to calculate the fraction of *average* signal power that ends up propagating in the forward direction after double Rayleigh scattering (DRS). In this simple approach, one assumes that depletion of the signal through Rayleigh scattering is negligible and supplements the signal Eq. (7) with two more equations of the form

$$-\frac{dP_1}{dz} = [g_R P_p(z) - \alpha_s]P_1 + f_r \alpha_r P_s \quad (42)$$

$$\frac{dP_2}{dz} = [g_R P_p(z) - \alpha_s]P_2 + f_r \alpha_r P_1 \quad (43)$$

where P_1 and P_2 represent the average power levels of the noise components created through single and double Rayleigh backscattering, respectively.

Equations (7), (42), and (43) need to be integrated over the fiber length to find the fraction f_{DRS} that ends up coming out with the signal at the output end at $z = L$. The signal equation has already been integrated and has the solution $P_s(z) = G(z)P_s(0)$, where $G(z)$ is given in Eq. (24). If we introduce the new variables using $P_1 = Q_1/G(z)$ and $P_2 = G(z)Q_2$, Eqs. (42) and (43) reduce to

$$-\frac{dQ_1}{dz} = f_r \alpha_r G^2(z) P_s(0), \quad \frac{dQ_2}{dz} = f_r \alpha_r G^{-2}(z) Q_1(z) \quad (44)$$

These equations can be readily integrated to obtain

$$Q_2(L) = f_r \alpha_r \int_0^L G^{-2}(z) Q_1(z) dz \quad (45)$$

$$Q_1(z) = f_r \alpha_r \int_z^L G^2(z) P_s(0) dz \quad (46)$$

The fraction of input power that ends up coming out with the signal at the output end is thus given by [70]

$$f_{\text{DRS}} = \frac{P_2(L)}{P_s(L)} = (f_r \alpha_r)^2 \int_0^L dz \, G^{-2}(z) \int_z^L G^2(z') dz \quad (47)$$

Figure 9.11 shows how f_{DRS} increases with the on–off Raman gain G_A using $f_r \alpha_r = 10^{-4}$, $g_R = 0.7 \, \text{W}^{-1}/\text{km}$, $\alpha_s = 0.2 \, \text{dB/km}$, $\alpha_p = 0.25 \, \text{dB/km}$, and backward pumping for a 100-km-long Raman amplifier. The cross-talk begins to exceed the -35-dB level for the 20-dB Raman gain needed for compensating fiber losses. Because this cross-talk accumulates over multiple amplifiers, it can lead to large power penalties in long-haul lightwave systems.

A simple way to estimate the degradation induced by double Rayleigh backscattering is to consider how the noise figure of a Raman amplifier increases because of it. The noise field of power $P_2(L)$ is incident on the photodetector together with the signal field with power $P_s(L)$. Similar to the case of spontaneous Raman scattering, the two fields beat and produce a noise component in the receiver current of the form

$$\Delta I_r = 2R_d(G_L P_{\text{in}}) f_{\text{DRS}}^{1/2} \cos\theta \quad (48)$$

and θ is a rapidly varying random phase. If σ_r^2 represents the variance of this current noise, we should add it to the other noise sources and replace Eq. (39) with

$$(\text{SNR})_{\text{out}} = \frac{(R_d G_L P_{\text{in}})^2}{(G_L \sigma_s^2 + \sigma_b^2 + \sigma_r^2)} \quad (49)$$

where $\sigma_r^2 = 2 f_{\text{DRS}} (R_d G_L P_{\text{in}})^2$ after using $\langle \cos^2 \theta \rangle = \frac{1}{2}$.

The noise figure of a Raman amplifier can now be calculated following the procedure discussed earlier. In place of Eq. (40), we obtain

$$F_n = \frac{1}{G_L} \left(1 + \frac{\sigma_b^2}{G_L \sigma_s^2} + \frac{\sigma_r^2}{G_L \sigma_s^2} \right) \quad (50)$$

where the last term represents the contribution of double Rayleigh backscattering. This contribution has been calculated including the depolarization effects as well as the statistical nature of Rayleigh backscattering [57]. In the case of a Gaussian pulse, Gaussian-shaped filters, and a relatively large optical bandwidth, it is found to be

$$\frac{\sigma_r^2}{G_L \sigma_s^2} = \frac{5 P_{\text{in}} f_{\text{DRS}}}{9 h \nu_0 G_L \Delta f} \quad (51)$$

where P_{in} is the input channel power, f_{DRS} is the fraction given in Eq. (47), and we assumed that signal bandwidth $\Delta \nu_s \approx \Delta f$. As an example, for $P_{\text{in}} = 1 \, \text{mW}$, $f_{\text{DRS}} = 10^{-4}$, and $\Delta f = 40 \, \text{GHz}$, the factor of 8 is large enough to increase the noise figure.

4.4. Pump-Noise Transfer

All lasers exhibit some intensity fluctuations. The situation is worse for semiconductor lasers used for pumping a Raman amplifier because the level of power fluctuations in such lasers can be relatively high because of their relatively small size and a large rate of spontaneous emission. As seen from Eq. (12), the gain of a Raman amplifier depends on the pump power exponentially. It is intuitively expected from

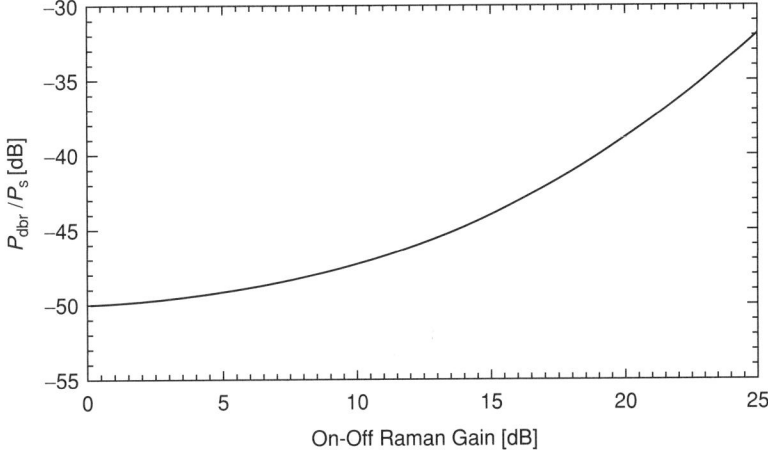

FIGURE 9.11 Fraction of signal power that is converted to noise by double Rayleigh backscattering (DRB) plotted as a function of Raman gain for a 100-km-long backward-pumped distributed Raman amplifier. (After [56]; ©2002 Elsevier.)

this equation that any fluctuation in the pump power would be magnified and result in even larger fluctuations in the amplified signal power. This is indeed the case, and this source of noise is known as pump-noise transfer. Details of the noise transfer depend on many factors, including amplifier length, pumping scheme, and the dispersion characteristics of the fiber used for making the Raman amplifier.

Power fluctuations of a semiconductor laser are quantified through a frequency-dependent quantity called the *relative intensity noise* (RIN). It represents the spectrum of intensity of power fluctuations and is defined as

$$\frac{\sigma_p^2}{\langle P_0 \rangle^2} = \int_0^\infty \text{RIN}_p(f)df \qquad (52)$$

where σ_p^2 is the variance of pump-power fluctuations and $\langle P_0 \rangle$ is the average pump power. When pump noise is included, the amplified signal also exhibits fluctuations, and one can introduce the RIN of the amplified signal as

$$\frac{\sigma_s^2}{\langle P_s(L) \rangle^2} = \int_0^\infty \text{RIN}_s(f)df \qquad (53)$$

The pump-noise transfer function represents the enhancement in the signal noise at a specific frequency f and is defined as

$$H(f) = \text{RIN}_s(f)/\text{RIN}_p(f) \qquad (54)$$

To calculate the noise transfer function $H(f)$, one must include the fact that the pump and signal do not travel along the fiber at the same speed because of their different wavelengths. The speed difference depends on the dispersion of the fiber. For this reason, the pump-noise transfer process depends on the pumping scheme, the amplifier length, and the dispersion parameter D of the fiber used to make the amplifier. Mathematically, Eqs. (4) and (5) are modified to include an additional term and take the form

$$\frac{\partial P_s}{\partial z} + \frac{1}{v_{gs}} \frac{\partial P_s}{\partial t} = g_R P_p P_s - \alpha_s P_s \qquad (55)$$

$$\xi \frac{\partial P_p}{\partial z} + \frac{1}{v_{gp}} \frac{\partial P_p}{\partial t} = -\frac{\omega_p}{\omega_s} g_R P_p P_s - \alpha_p P_p \qquad (56)$$

where v_{gs} and v_{gp} are the group velocities for the signal and pump, respectively. These equations can be solved with some reasonable approximations [74] to calculate the noise transfer function $H(f)$.

Figure 9.12 compares the calculated frequency dependence of $H(f)$ in the forward and backward pumping configurations for two types of Raman amplifiers pumped at 1450 nm to provide gain near 1550 nm [62]. Solid curves are for a discrete Raman amplifier made with a 5-km-long DCF with $\alpha_p = 0.5$ dB/km and a dispersion of $D = -100$ ps/km·nm at 1.55 μm. Dashed curves are for a distributed Raman amplifier made with a 100-km-long NZDF with $\alpha_p = 0.25$ dB/km and a dispersion of 4.5 ps/km·nm at 1.55 μm. Dashed curves are for a distributed Raman amplifier made with a 100-km-long NZDF with $\alpha_p = 0.25$ dB/km and a dispersion of 4.5 ps/km·nm at 1.55 μm. The on–off Raman gain is 10 dB in all cases. In the case of forward pumping, RIN is enhanced by a factor of nearly 6 in the frequency range of 0–5 MHz for the DCF and the frequency range increases to 50 MHz for the NZDF because of its relatively low value of D. However,

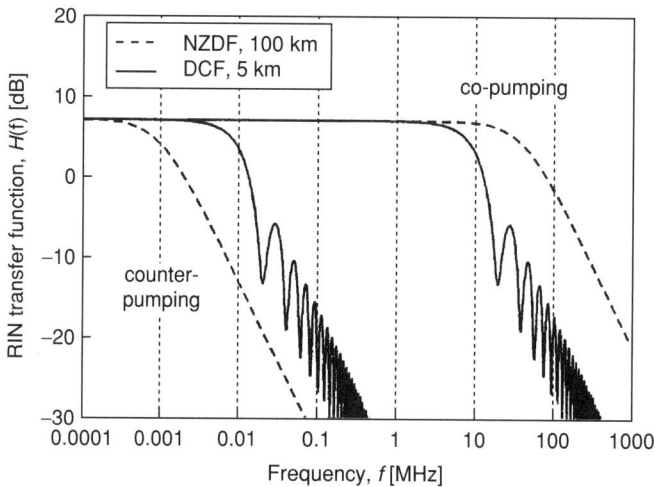

FIGURE 9.12 Calculated RIN transfer function for two types of Raman amplifiers when pump and signal copropagate or counterpropagate. (After [62]; ©2004 IEEE.)

when backward pumping is used, RIN is actually reduced at all frequencies except for frequencies below a few kilohertz. Notice that $H(f)$ exhibits an oscillatory structure in the case of a DCF. This feature is related to a relatively short length of discrete amplifiers.

In general, the noise enhancement is relatively small in the backward pumping configuration and for large values of D [74]. This behavior can be understood physically as follows. The distributed Raman gain builds up as the signal propagates inside the fiber. In the case of forward pumping and low dispersion, pump and signal travel at nearly the same speed. As a result, any fluctuation in the pump power stays in the same temporal window of the signal. In contrast, when dispersion is large, the signal moves out of the temporal window associated with the fluctuation and sees a somewhat averaged gain. The averaging is much stronger in the case of backward pumping because the relative speed is extremely large (twice that of the signal group velocity). In this configuration, the effects of pump-power fluctuations are smoothed out so much that almost no RIN enhancement occurs. For this reason, backward pumping is often used in practice even though the noise figure is larger for this configuration. Forward pumping can only be used if fiber dispersion is relatively large and pump lasers with low RIN are used.

4.5. Effects of PMD

In the scalar approach used so far, it has been implicitly assumed that both pump and signal are copolarized and maintain their state of polarization (SOP) inside the fiber. However, unless a special kind of polarization-maintaining fiber is used for making Raman amplifiers, residual fluctuating birefringence of most fibers changes the SOP of any optical field in a random fashion and also leads to PMD, a phenomenon that has been studied extensively in recent years [75]. Indeed, the effects of PMD on Raman amplification have been observed in several experiments [76–78]. A vector theory of Raman amplification has also been developed, and it can explain the experimental features quite well [79–81]. The theory shows that the amplified signal fluctuates over a wide range if PMD changes with time, and the average gain is significantly lower than that expected in the absence of PMD.

Figure 9.13 shows how the average gain and σ_s change with the PMD parameter D_p when the input signal is copolarized (solid curves) or orthogonally polarized (dashed curves) with respect to the pump. The curves are shown for both the forward and backward pumping schemes for a 10-km-long Raman amplifier pumped with 1 W of power using a single 1.45-μm laser. Fiber losses are taken to be 0.273 and 0.2 dB/km at the pump and signal wavelengths, respectively. When D_p is zero, the two beams maintain their SOP, and the copolarized signal experiences a maximum gain of 17.6 dB while the orthogonally polarized signal has a 1.7-dB loss, irrespective of the pumping configuration. The loss is not exactly 2 dB because a small gain exists for the orthogonally polarized input signal. As the PMD parameter increases, the gain difference between the copolarized and orthogonally polarized cases decreases and disappears eventually.

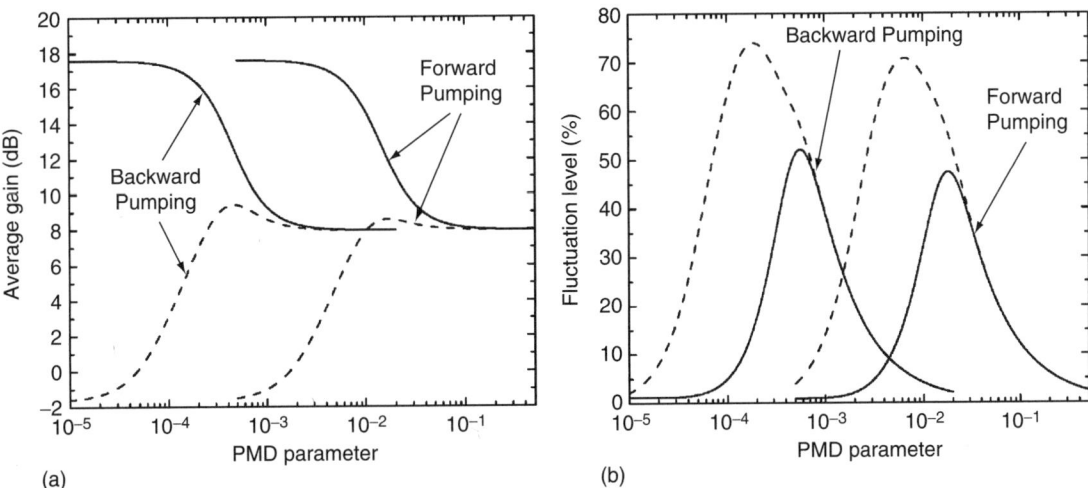

FIGURE 9.13 (a) Average gain and (b) standard deviation of signal fluctuations at the output of a Raman amplifier as a function of PMD parameter in the cases of forward and backward pumping. The solid and dashed curves correspond to the cases of copolarized and orthogonally polarized signals, respectively.

The level of signal fluctuations in Fig. 9.13 increases quickly with the PMD parameter, reaches a peak, and then decreases slowly to zero with further increase in D_p. The location of the peak depends on the pumping scheme as well as on the initial polarization of pump. The noise level can exceed 20% for $D_p = 0.05\,\text{ps}/\sqrt{\text{km}}$ in the case of forward pumping. If a fiber with low PMD is used, the noise level can exceed 70% under some conditions. These results suggest that forward-pumped Raman amplifiers perform better if a fiber with $D_p > 0.1\,\text{ps}/\sqrt{\text{km}}$ is used. The curves for backward pumping are similar to those for forward pumping but shift to smaller D_p values and have a higher peak. In spite of an enhanced peak, the backward pumping produces the smallest amount of fluctuations for all fibers for which $D_p > 0.01\,\text{ps}/\sqrt{\text{km}}$.

In the case of backward pumping, curves in Fig. 9.13 are nearly identical to those obtained for the case of forward pumping except that they are shifted to the left. As a result, the solid and dashed curves merge at a value of D_p that is smaller by about a factor of 30. This difference is related to the relative speed with which pump and signal propagate through the fiber. In the case of backward pumping, the quantity $|\Omega| = \omega_p + \omega_s$ is about 30 times larger than $\Omega = \omega_p - \omega_s$ in the forward-pumping case [81]. For this reason, the level of signal fluctuations depends strongly on the relative directions of pump and signal propagation. In the case of forward pumping, σ_s grows monotonically with distance, reaching 24% at the end of the 10-km fiber. In contrast, σ_s is only 0.8% even for a 10-km-long amplifier in the case of backward pumping, a value 30 times smaller than that occurring in the forward-pumping case.

The PMD effects in Raman amplifiers are often quantified using the concept of polarization-dependent gain (PDG), a quantity defined as the difference between the maximum and the minimum values of G realized while varying the SOP of the input signal. The gain difference $\Delta = G_{\max} - G_{\min}$ is itself random because both $G_{\max}$ and $G_{\min}$ are random. It is useful to know the statistics of Δ and its relationship to the operating parameters of a Raman amplifier because they can identify the conditions under which PDG can be reduced to acceptable levels. The statistical properties of PDG have been found using a vector theory of Raman amplification [81]. For $D_p \leq 0.05\,\text{ps}/\sqrt{\text{km}}$, $p(\Delta)$ becomes Maxwellian and its peaks shifts to smaller values as D_p increases. This behavior has been observed experimentally [78].

Figure 9.14 shows how the mean and the variance of PDG fluctuations vary with the PMD parameter for the same Raman amplifier used for Fig. 9.13. As expected, the mean PDG decreases monotonically as D_p increases. The mean PDG, $\langle\Delta\rangle$, is not exactly $2G_{av}$ when $D_p = 0$ because the gain is not zero when pump and signal are orthogonally polarized. Note however that $\langle\Delta\rangle$ can be as large as 30% of the average gain for $D_p = 0.05\,\text{ps}/\sqrt{\text{km}}$ in the case of forward pumping, and it decreases slowly with D_p after that, reaching a value of 8% for $D_p = 0.2\,\text{ps}/\sqrt{\text{km}}$. In the case of backward pumping, the behavior is nearly identical to that in the case of forward pumping except that the curve shifts to a value of D_p smaller by about a factor of 30. These conclusions are in agreement with recent experiments [78].

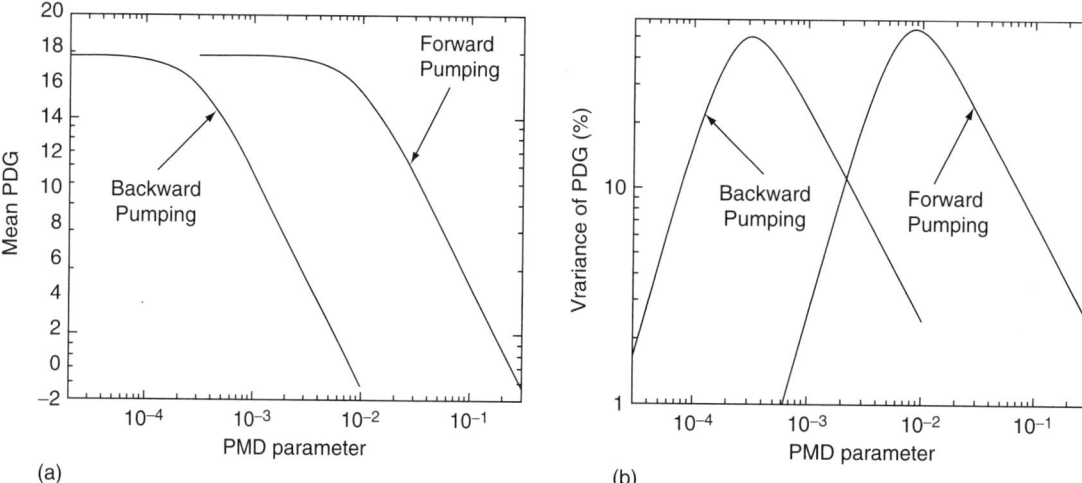

FIGURE 9.14 (a) Mean PDG and (b) variance σ_Δ (both normalized to the average gain G_{av}) as a function of PMD parameter under forward and backward pumping conditions.

5. AMPLIFICATION OF OPTICAL PULSES

The CW regime considered so far assumes that both the pump and signal fields are in the form of CW beams. This is rarely the case in practice. Although pumping is often continuous in lightwave systems, the signal is invariably in the form of a pulse train consisting of a random sequence of 0 and 1 bits. Fortunately, the CW theory can be applied to such systems, if the signal power P_s is interpreted as the average channel power, because the Raman response is fast enough that the entire pulse train can be amplified without any distortion. However, for pulses shorter than 10 ps or so, one must include the dispersive and nonlinear effects that are likely to affect the amplification of such short pulses in a Raman amplifier.

The situation is quite different when a Raman amplifier is pumped with optical pulses and is used to amplify a pulsed signal. Because of the dispersive nature of silica fibers, the pump and signal pulses travel with different group velocities, v_{gp} and v_{gs}, respectively, because of their different wavelengths. Thus, even if the two pulses were overlapping initially, they separate after a distance known as the *walk-off length*. The CW theory can still be applied for relatively wide pump pulses (width $T_0 > 1$ ns) if the walk-off length L_W, defined as

$$L_W = T_0 / \left| v_{gp}^{-1} - v_{gs}^{-1} \right| \quad (57)$$

exceeds the fiber length L. However, for shorter pump pulses for which $L_W < L$, Raman amplification is limited by the group-velocity mismatch and occurs only over distances $z \sim L_W$ even if the actual fiber length L is considerably larger than L_W. At the same time, the nonlinear effects such as SPM and cross-phase modulation (XPM) become important and affect considerably the evolution of the pump and signal pulses [9]. This section discusses the pulsed case assuming that pulse widths remain larger than the Raman response time ($\sim$50 fs) so that transient effects are negligible. The PMD effects are ignored to simplify the following discussion.

5.1. Pulse-Propagation Equations

To study the propagation of a short pulse in a dispersive nonlinear medium, one must consider both the electronic and nuclear motion while calculating the material polarization induced by the pump field. If we consider the third-order susceptibility [8], neglect fiber birefringence, but account for the frequency dependence of the dielectric constant, we arrive at the following set of two coupled propagation equations [82]:

$$\frac{\partial A_s}{\partial z} + \frac{\alpha_s}{2} A_s + \beta_{1s} \frac{\partial A_s}{\partial z} + \frac{i\beta_{2s}}{2} \frac{\partial^2 A_s}{\partial t^2} - \frac{\beta_{3s}}{6} \frac{\partial^3 A_s}{\partial t^3}$$
$$= i\gamma_s \left(1 + \frac{i}{\omega_s} \frac{\partial}{\partial t}\right) \left[(1 - f_R)(|A_s(t)|^2 + 2|A_p(t)|^2) A_s(t) \right.$$
$$+ f_R A_s(t) \times \int_{-\infty}^{t} h_R(t - t')(|A_s(t')|^2 + |A_p(t')|^2$$
$$\left. + A_s(t') A_p^*(t') e^{-i\Omega(t-t')}) dt' \right] \quad (58)$$

$$\frac{\partial A_p}{\partial z} + \frac{\alpha_p}{2} A_p + \beta_{1p} \frac{\partial A_p}{\partial z} + \frac{i\beta_{2p}}{2} \frac{\partial^2 A_p}{\partial t^2} - \frac{\beta_{3p}}{6} \frac{\partial^3 A_p}{\partial t^3}$$
$$= i\gamma_p \left(1 + \frac{i}{\omega_p} \frac{\partial}{\partial t}\right) \left[(1 - f_R)(|A_p(t)|^2 + 2|A_s(t)|^2) A_p(t) \right.$$
$$+ f_R A_p(t) \times \int_{-\infty}^{t} h_R(t - t')(|A_p(t')|^2 + |A_s(t')|^2$$
$$\left. + A_p(t') A_s^*(t') e^{i\Omega(t-t')}) dt' \right] \quad (59)$$

where $h_R(t)$ is the Raman response function and f_R represents the fractional through which nuclei contribute to the third-order susceptibility ($f_R \approx 0.18$); $h_R(t)$ has been normalized such that $\int_0^\infty h_R(t) dt = 1$. The nonlinear parameter $\gamma_j = n_2 \omega_j/(cA_{\text{eff}})$, where n_2 is the Kerr coefficient and ω_j is the optical frequency for the field with amplitude A_j ($j = $ s and p for signal and pump, respectively).

The three dispersion parameters, β_1, β_2, and β_3, include the effects of fiber dispersion to the first three orders. They become increasingly important for shorter pulses with a wider spectrum. The first-order dispersion parameter is related to the group velocity v_{gj} of a pulse as $\beta_{1j} = 1/v_{gj}$, where $j = $ p or s. The second-order parameter β_{2j} represents the dispersion of group velocity and is known as the group-velocity dispersion (GVD) parameter [9]. For a similar reason, β_{3j} is called the third-order dispersion parameter.

The Raman response function $h_R(t)$ plays an important role for short optical pulses [83–87]. The Raman-gain spectrum is related to the imaginary part of its Fourier transform $\tilde{h}_R(\omega)$. The real part of $\tilde{h}_R(\omega)$ can be obtained from the imaginary part by using the Kramers–Kronig relations. Figure 9.15 shows the real and imaginary parts of $\tilde{h}_R(\omega)$ obtained by using the experimentally measured Raman gain spectrum [16]. Physically, the real part $\tilde{h}_R(\omega)$ leads to Raman-induced index changes.

Although Eqs. (58) and (59) should be solved for femtosecond pulses, they can be simplified considerably for picosecond pulses [82]. For relatively broad pulse widths (>1 ps), A_s and A_p can be treated as constants compared with the time scale over which $h_R(t)$ changes (<0.1 ps). The integrals in Eqs. (58) and (59) can then be performed analytically.

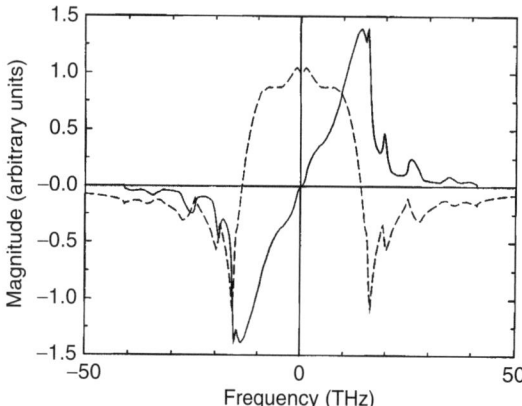

FIGURE 9.15 Imaginary (solid curve) and real (dashed curve) parts of the Raman response as a function of frequency shift. (After [82]; ©1996 OSA.)

Neglecting the index changes associated with the real part of $\tilde{h}_R(\omega)$ and introducing a reduced time $T = t - z/v_{gp}$ in a frame of reference moving with the pump pulse, these equations reduce to [9]

$$\frac{\partial A_s}{\partial z} - d\frac{\partial A_s}{\partial T} + \frac{i\beta_{2s}}{2}\frac{\partial^2 A_s}{\partial T^2} = i\gamma_s[|A_s|^2 + 2|A_p|^2]A_p \\ + \frac{g_s}{2}|A_p|^2 A_s - \frac{\alpha_s}{2}A_s \quad (60)$$

$$\frac{\partial A_p}{\partial z} + \frac{i\beta_{2p}}{2}\frac{\partial^2 A_p}{\partial T^2} = i\gamma_p[|A_p|^2 + 2|A_s|^2]A_p \\ - \frac{g_p}{2}|A_s|^2 A_p - \frac{\alpha_p}{2}A_p \quad (61)$$

where the Raman gain coefficients are defined as

$$g_j = 2f_R\gamma_j|\text{Im}[\tilde{h}_R(\Omega)]| \quad (j = p, s) \quad (62)$$

and the walk-off parameter d accounts for the group-velocity mismatch:

$$d = \beta_{1p} - \beta_{1s} = v_{gp}^{-1} - v_{gs}^{-1} \quad (63)$$

Typical values of d are in the range of 2–6 ps/m. The GVD parameter β_{2j}, the nonlinearity parameter γ_j, and the Raman gain coefficient g_j are slightly different for the pump and signal pulses because of their different wavelengths. In terms of the wavelength ratio λ_p/λ_s, these parameters are related as

$$\beta_{2s} = \frac{\lambda_p}{\lambda_s}\beta_{2p}, \quad \gamma_s = \frac{\lambda_p}{\lambda_s}\gamma_p, \quad g_s = \frac{\lambda_p}{\lambda_s}g_p \quad (64)$$

Equations (60) and (61) reduce to Eqs. (4) and (5) when the dispersive and nonlinear terms are neglected and pump and signal powers are introduced as $P_j = |A_j|^2 (j = p, s)$.

Four length scales can be introduced to determine the relative importance of various terms in Eqs. (60) and (61). For pump pulses of duration T_0 and peak power P_0, these are defined as

$$L_D = \frac{T_0^2}{|\beta_{2p}|}, \quad L_W = \frac{T_0}{|d|}, \quad L_{NL} = \frac{1}{\gamma_p P_0}, \quad L_G = \frac{1}{g_p P_0} \quad (65)$$

The dispersion length L_D, the walk-off length L_W, the nonlinear length L_{NL}, and the Raman gain length L_G provide, respectively, the length scales over which the effects of GVD, walk-off, nonlinearity (both SPM and XPM), and Raman gain become important. The shortest length among them plays the dominant role. Typically, $L_W \approx 1$ m for $T_0 < 10$ ps, whereas L_{NL} and L_G become smaller or comparable with walk-off length L_W for $P_0 > 100$ W. In contrast, $L_D \approx 1$ km for $T_0 = 10$ ps. Thus, the GVD effects are generally negligible for pulses as short as 10 ps. The situation changes for pulse widths ~1 ps or less because L_D decreases faster than L_W with a decrease in the pulse width. The GVD effects can then affect the amplification process significantly, especially in the anomalous-dispersion regime.

5.2. Effects of Group-Velocity Mismatch

When the second-derivative terms in Eqs. (60) and (61) are neglected by setting $\beta_{2p} = \beta_{2s} = 0$, these equations can be solved analytically [88–92]. The analytic solution takes a simple form if the signal pulse is assumed to be relatively weak and pump depletion is neglected by setting $g_p = 0$. We also neglect fiber losses for simplicity. Equation (60) for the pump pulse then yields the solution

$$A_p(z, T) = A_p(0, T)\exp[i\gamma_p|A_p(0, T)|^2 z] \quad (66)$$

where the XPM term has been neglected assuming $|A_s|^2 \ll |A_p|^2$. For the same reason, the SPM term in Eq. (61) can be neglected. The solution of Eq. (61) is then given by [88]

$$A_s(z, T) = A_s(0, T + zd)\exp[(g_s/2 + 2i\gamma_s)\psi(z, T)] \quad (67)$$

where

$$\psi(z, T) = \int_0^z |A_p(0, T + zd - z'd)|^2 dz' \quad (68)$$

Equation (66) shows that the pump pulse of initial amplitude $A_p(0, T)$ propagates without change in its shape. However, the SPM-induced phase shift imposes a frequency chirp on the pump pulse that broadens its spectrum. The signal pulse, in contrast, changes both its shape and spectrum as it propagates through the fiber; temporal changes occur due to

Raman gain, whereas spectral changes have their origin in XPM. Because of pulse walk-off, both kinds of changes are governed by an overlap factor $\psi(z, T)$ that takes into account the relative separation between the two pulses along the fiber. This factor depends on the pulse shape. For a Gaussian pump pulse with the input amplitude

$$A_p(0, T) = \sqrt{P_0} \exp(-T^2/2T_0^2) \quad (69)$$

the integral in Eq. (68) can be performed in terms of error functions with the result

$$\psi(z, \tau) = [\text{erf}(\tau + \delta) - \text{erf}(\tau)](\sqrt{\pi}P_0 z/\delta) \quad (70)$$

where $\tau = T/T_0$ and δ is the propagation distance in units of the walk-off length, that is, $\delta = zd/T_0 = z/L_W$. An analytic expression for $\psi(z, \tau)$ can also be obtained for pump pulses having "sech" shape [89]. In both cases, the signal pulse compresses initially, reaches a minimum width, and then begins to rebroaden as it is amplified through SRS. It also acquires a frequency chirp through XPM. This qualitative behavior persists even when pump depletion is included [89–91].

The analytic solution (67) can be used to obtain both the shape and the spectrum of the signal pulse during the initial stages of Raman amplification. The spectral evolution is governed by the XPM-induced frequency chirp. Note, however, that the signal pulse travels faster than the pump pulse in the normal GVD regime. As a result, chirp is induced mainly near the trailing edge in the case of normal dispersion and near the leading edge in the case of anomalous dispersion. It should be stressed that both pulse shapes and spectra are considerably modified when pump depletion is included because the growing signal pulse affects itself through SPM and the pump pulse through XPM.

When fiber length is comparable with the dispersion length L_D, it is important to include the GVD effects. Such effects cannot be described analytically, and a numerical solution of Eqs. (60) and (61) is necessary to understand the Raman amplification process. A generalization of the split-step Fourier method can be used for this purpose [9]. The results depend on whether the signal being amplified experiences anomalous or normal dispersion. We consider the two cases separately.

5.3. Anomalous Dispersion Regime

When the wavelengths of the pump and signal pulses lie inside the anomalous dispersion regime of the optical fiber used for Raman amplification, both of them experience the soliton-related effects [93–103]. Under suitable conditions, almost all the pump-pulse energy can be transferred to the signal pulse, which propagates undistorted as a fundamental soliton. Numerical results show that this is possible if most of the energy transfer occurs at a distance at which the pump pulse, propagating as a higher order soliton, achieves its minimum width [93]. In contrast, if energy transfer is delayed and occurs at a distance where the pump pulse has split into its components, the signal pulse does not form a fundamental soliton, and its energy rapidly disperses.

Equations (60) and (61) can be used in the anomalous GVD regime by simply choosing the negative sign for the second-derivative terms. As expected, energy transfer to the signal pulse occurs near $z \approx L_W$. This condition implies that L_W should not be too small in comparison with the dispersion length L_D. Typically, L_W and L_D become comparable for femtosecond pulses of widths $T_0 \approx 100$ fs. For such ultrashort pump pulses, the distinction between pump and signal pulses gets blurred as their spectra begin to overlap considerably. This can be seen by noting that the Raman gain peak in Fig. 9.3 corresponds to a spectral separation of about 13 THz, whereas the spectral width of a 100-fs pulse is ~10 THz. Equations (60) and (61) do not provide a realistic description for femtosecond pump pulses, and one should solve Eqs. (58) and (59) in their place.

An interesting situation occurs when a single ultrashort pulse propagates inside the fiber without a pump pulse. Figure 9.16 shows the evolution of an ultrashort pulse in the anomalous GVD regime of the fiber by solving Eq. (58) numerically with an input of the form $A(0,t) = \sqrt{P_0} \text{ sech}(t/T_0)$ using $T_0 = 50$ fs. Input peak power P_0 of the pulse was chosen such that the pulse forms a second-order soliton in the absence of higher order effects. As seen in Fig. 9.16, pulse is initially distorted but most of its energy is carried by a pulse whose center shifts to the right. This temporal shift indicates a reduction in the group velocity of the pulse.

To understand why group velocity of the pulse changes, we need to consider modifications in the pulse spectrum. As seen in Fig. 9.16, the pulse spectrum shifts toward the red side, and this red shift continues to increase as the pulse propagates further down the fiber. This shift has its origin in the integral appearing in Eq. (58) and containing the Raman response function. Physically speaking, high-frequency components of an optical pulse pump the low-frequency components of the same pulse through

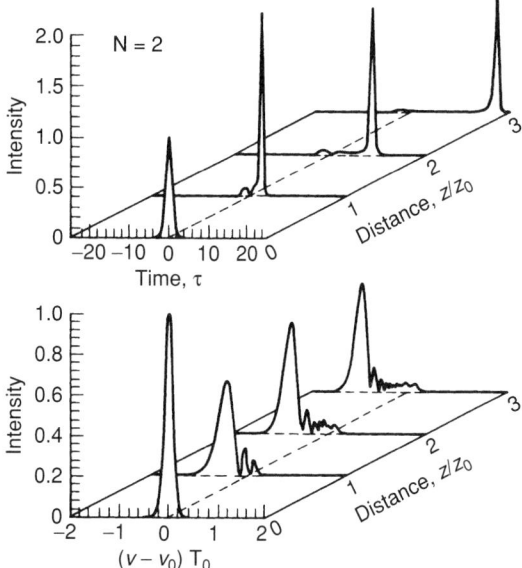

FIGURE 9.16 Pulse shapes and spectra of a 50-fs pulse in the anomalous dispersion regime showing Raman-induced spectral shift and temporal delay. The solition period z_0 is related to the dispersion length as $z_0 = (\pi/2)L_D$.

SRS, a phenomenon referred to as *intrapulse* Raman scattering. Because energy in the blue part of the original pulse spectrum is used to amplify the red-shifted components, the pulse spectrum shifts toward lower frequencies, as the pulse propagates down the fiber. This spectral shift is known as the soliton self-frequency shift [95] and was first observed in a 1986 experiment [96]. As the pulse spectrum shifts toward the red side, the speed of pulse changes because group velocity in a dispersive medium depends on the central frequency at which pulse spectrum is located. More specifically, the pulse is delayed because the red-shifted pulse travels slower in the anomalous GVD regime of an optical fiber. However, it should be stressed that anomalous dispersion is not a prerequisite for the spectrum to shift. In fact, the Raman-induced spectral shift can occur even in the normal dispersion regime, although its magnitude is relatively small because of a rapid broadening of the pulse in the case of normal dispersion [104]. We turn to this case next.

5.4. Normal Dispersion Regime

In this section we discuss what happens to a short pulse amplified inside a Raman amplifier while experiencing normal dispersion so that it cannot form a soliton. We expect the pulse to undergo some reshaping through the amplification process. It turns out that the pulse acquires a parabolic shape and its width is always the same irrespective of the shape of input pulses. Parabolic pulses have attracted considerable attention in recent years because it was observed experimentally and discovered theoretically that such a shape exists for pulses being amplified inside an optical fiber while experiencing normal dispersion [105–112]. Parabolic pulses were first observed in a 1996 experiment using a fiber amplifier in which gain was provided by erbium ions [105]. They were later observed in ytterbium-doped fiber amplifiers [106–108].

There is no fundamental reason why parabolic pulses should not form in a distributed Raman amplifier provided the dispersion is normal for the signal pulse. Indeed, such pulses were observed in a Raman amplifier in a 2003 experiment [111]. The amplifier was pumped using a CW laser operating at 1455 nm and capable of delivering up to 2 W of power. Signal pulses of 10-ps width were obtained from a mode-locked fiber laser. The pump and signal were copropagated inside a 5.3-km-long fiber whose GVD parameter was about 5 ps^2/km (normal GVD) at the signal wavelength. The amplified pulses were characterized using the frequency-resolved optical gating technique. Figure 9.17 shows the frequency-resolved optical gating trace and the pulse shape and frequency chirp deduced from it when a pulse with 0.75 pJ of energy was amplified by 17 dB. The chirp is nearly linear and the pulse shape is approximately parabolic.

One can predict the pulse shape and chirp profiles by solving Eqs. (58) and (59) numerically. For a CW pump, the dispersive effects can be neglected but the nonlinear effects should be included. Figure 9.18 shows the evolution toward a parabolic shape when a "sech" input pulse is amplified over the 5.3-km length of the Raman amplifier. The shape of the output pulse can be fitted quite well with a parabola, but the fit is relatively poor when the shape is assumed to be either Gaussian (dashed curve) or hyperbolic secant (dotted curve). The chirp is approximately linear. Both the predicted pulse shape and the chirp profile agree well with the experimental data, as shown in Figure 9.17. It was also confirmed experimentally that the final pulse characteristics were independent of input pulse parameters and were determined solely by the Raman amplifier. Both the width and the chirp changed when the gain of the amplifier was changed, but the pulse shape remained parabolic and the frequency chirped across the pulse remained nearly linear. These properties are in agreement with the theoretical solution of the underlying equation. Such solutions are referred to as

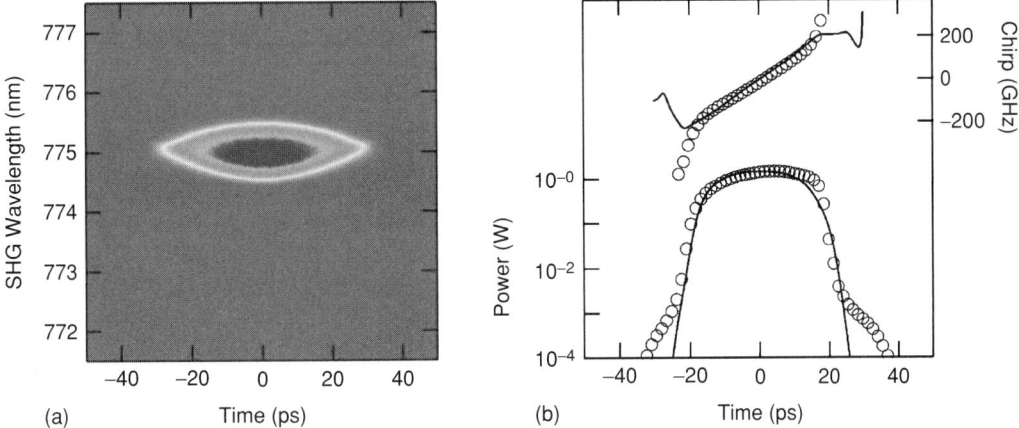

FIGURE 9.17 (a) Frequency-resolved optical gating (FROG) trace of a 0.75-pJ signal pulse amplified by 17 dB using Raman gain in a 5.3-km-long fiber. SHG, second harmonic (b) Intensity and chirp profiles (circles) retrieved from the FROG trace. Solid lines show the results of numerical simulations. (After [111]; ©2003 OSA.)

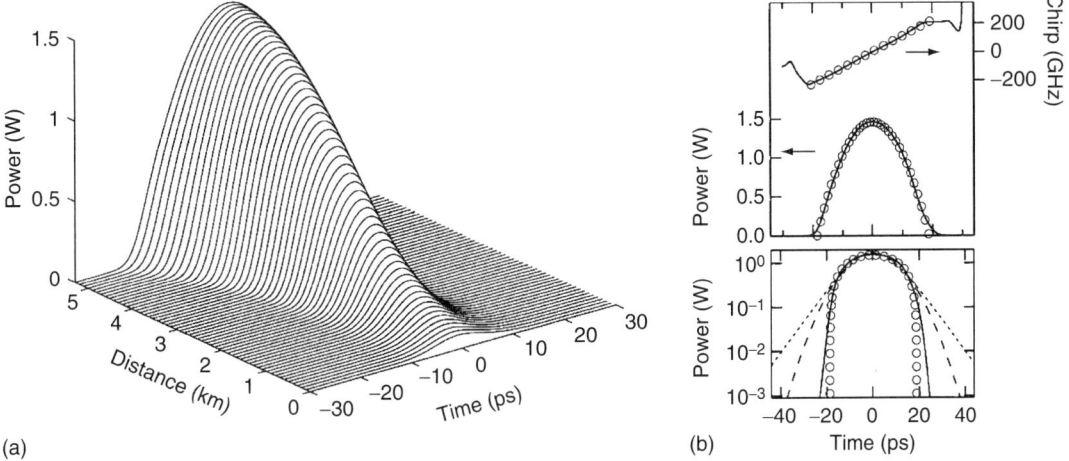

FIGURE 9.18 (a) Evolution toward a parabolic shape of a "sech" input pulse over the 5.3-km length of the Raman amplifier. (b) Intensity and chirp (solid lines) profiles of the output pulse together with parabolic and linear fits, respectively (circles). The bottom part shows the same results on a logarithmic scale together with the Gaussian (dashed curve) and "sech" (dotted curve) fits. (After [111]; ©2003 OSA.)

being self-similar and are of considerable interest both from the fundamental and the applied points of view [106–110].

6. REFERENCES

1. C. V. Raman, *Indian J. Phys.* **2**, 387 (1928).
2. E. J. Woodbury and W. K. Ng, *Proc. IRE* **50**, 2347 (1962).
3. R. W. Hellwarth, *Phys. Rev.* **130**, 1850 (1963); *Appl. Opt.* **2**, 847 (1963).
4. E. Garmire, E. Pandarese, and C. H. Townes, *Phys. Rev. Lett.* **11**, 160 (1963).
5. Y. R. Shen and N. Bloembergen, *Phys. Rev.* **137**, A1786 (1965).
6. W. Kaiser and M. Maier, in *Laser Handbook*, Vol. 2, F. T. Arecchi and E. O. Schulz-Dubois, Eds. (North-Holland, Amsterdam, 1972). Chap. E2.
7. Y. R. Shen, *The Principles of Nonlinear Optics* (Wiley, New York, 1984), Chap. 10.
8. R. W. Boyd, *Nonlinear Optics*, 3rd ed. (Academic Press, San Diego, 2003), Chap. 9.
9. G. P. Agrawal, *Nonlinear Fiber Optics*, 3rd ed. (Academic Press, San Diego, 2001), Chap. 8.
10. R. H. Stolen, E. P. Ippen, and A. R. Tynes, *Appl. Phys. Lett.* **20**, 62 (1972).
11. R. H. Stolen and E. P. Ippen, *Appl. Phys. Lett.* **22**, 276 (1973).
12. R. H. Stolen, *IEEE J. Quantum Electron.* **15**, 1157 (1979).

13. F. L. Galeener, *Phys. Rev. B* **19**, 4292 (1979).
14. R. H. Stolen, *Proc. IEEE* **68**, 1232 (1980).
15. N. Shibata, M. Horigudhi, and T. Edahiro, *J. Noncrys. Solids* **45**, 115 (1981).
16. R. H. Stolen, J. P. Gordon, W. J. Tomlinson, and H. A. Haus, *J. Opt. Soc. Am. B* **6**, 1159 (1989).
17. D. J. Dougherty, F. X. Kartner, H. A. Haus, and E. P. Ippen, *Opt. Lett.* **20**, 31 (1995).
18. D. Mahgerefteh, D. L. Butler, J. Goldhar, B. Rosenberg, and G. L. Burdge, *Opt. Lett.* **21**, 2026 (1996).
19. M. Ikeda, *Opt. Commun.* **39**, 148 (1981).
20. G. A. Koepf, D. M. Kalen, and K. H. Greene, *Electron. Lett.* **18**, 942 (1982).
21. Y. Aoki, S. Kishida, H. Honmou, K. Washio, and M. Sugimoto, *Electron. Lett.* **19**, 620 (1983).
22. E. Desurvire, M. Papuchon, J. P. Pocholle, J. Raffy, and D. B. Ostrowsky, *Electron. Lett.* **19**, 751 (1983).
23. A. R. Chraplyvy, J. Stone, and C. A. Burrus, *Opt. Lett.* **8**, 415 (1983).
24. M. Nakazawa, M. Tokuda, Y. Negishi, and N. Uchida, *J. Opt. Soc. Am. B* **1**, 80 (1984).
25. K. Nakamura, M. Kimura, S. Yoshida, T. Hikada, and Y. Mitsuhashi, *J. Lightwave Technol.* **2**, 379 (1984).
26. M. Nakazawa, T. Nakashima, and S. Seikai, *J. Opt. Soc. Am. B* **2**, 215 (1985).
27. L. F. Mollenauer, J. P. Gordon, and M. N. Islam, *IEEE J. Quantum Electron.* **22**, 157 (1986).
28. N. A. Olsson and J. Hegarty, *J. Lightwave Technol.* **4**, 391 (1986).
29. K. Vilhelmsson, *J. Lightwave Technol.* **4**, 400 (1986).
30. E. Desurvire, M. J. F. Digonnet, and H. J. Shaw, *J. Lightwave Technol.* **4**, 426 (1986).
31. Y. Aoki, S. Kishida, and K. Washio, *Appl. Opt.* **25**, 1056 (1986).
32. S. Seikai, T. Nakashima, and N. Shibata, *J. Lightwave Technol.* **4**, 583 (1986).
33. K. Mochizuki, N. Edagawa, and Y. Iwamoto, *J. Lightwave Technol.* **4**, 1328 (1986).
34. N. Edagawa, K. Mochizuki, and Y. Iwamoto, *Electron. Lett.* **23**, 196 (1987).
35. R. W. Davies, P. Melman, W. H. Nelson, M. L. Dakss, and B. M. Foley, *J. Lightwave Technol.* **5**, 1068 (1987).
36. L. F. Mollenauer and K. Smith, *Opt. Lett.* **13**, 675 (1988).
37. M. J. O'Mahony, *J. Lightwave Technol.* **6**, 531 (1988).
38. S. V. Chernikov, Y. Zhu, R. Kashyap, and J. R. Taylor, *Electron. Lett.* **31**, 472 (1995).
39. S. G. Grubb, *Proc. Conf. on Optical Amplifiers and Applications* (Optical Society of America, Washington, DC, 1995).
40. E. M. Dianov, *Laser Phys.* **6**, 579 (1996).
41. D. I. Chang, S. V. Chernikov, M. J. Guy, J. R. Taylor, and H. J. Kong, *Opt. Commun.* **142**, 289 (1997).
42. P. B. Hansen, L. Eskilden, S. G. Grubb, A. J. Stentz, T. A. Strasser, J. Judkins, J. J. DeMarco, J. R. Pedrazzani, and D. J. DiGiovanni, *IEEE Photon. Technol. Lett.* **9**, 262 (1997).
43. A. Bertoni, *Opt. Quantum Electron.* **29**, 1047 (1997).
44. A. Bertoni and G. C. Reali, *Appl. Phys. B* **67**, 5 (1998).
45. E. M. Dianov, M. V. Grekov, I. A. Bufetov, V. M. Mashinsky, O. D. Sazhin, A. M. Prokhorov, G. G. Devyatykh, A. N. Guryanov, and V. F. Khopin, *Electron. Lett.* **34**, 669 (1998).
46. D. V. Gapontsev, S. V. Chernikov, and J. R. Taylor, *Opt. Commun.* **166**, 85 (1999).
47. H. Masuda, S. Kawai, K. Suzuki, and K. Aida, *IEEE Photon. Technol. Lett.* **10**, 516 (1998).
48. J. Kani and M. Jinno, *Electron. Lett.* **35**, 1004 (1999).
49. H. Masuda and S. Kawai, *IEEE Photon. Technol. Lett.* **11**, 647 (1999).
50. Y. Emori, K. Tanaka, and S. Namiki, *Electron. Lett.* **35**, 1355 (1999).
51. S. A. E. Lewis, S. V. Chernikov, and J. R. Taylor, *Electron. Lett.* **35**, 1761 (1999).
52. H. D. Kidorf, K. Rottwitt, M. Nissov, M. X. Ma, and E. Rabarijaona, *IEEE Photon. Technol. Lett.* **12**, 530 (1999).
53. H. Suzuki, J. Kani, H. Masuda, N. Takachio, K. Iwatsuki, Y. Tada, and M. Sumida, *IEEE Photon. Technol. Lett.* **12**, 903 (2000).
54. S. Namiki and Y. Emori, *IEEE J. Sel. Topics Quantum Electron.* **7**, 3 (2001).
55. B. Zhu, L. Leng, L. E. Nelson, Y. Qian, L. Cowsar, S. Stulz, C. Doerr, L. Stulz, S. Chandrasekhar, S. Radic, D. Vengsarkar, Z. Chen, J. Park, K. S. Feder, H. Thiele, J. Bromage, L. Gruner-Nielsen, and S. Knudsen, *Electron. Lett.* **37**, 844 (2001).
56. K. Rottwitt and A. J. Stentz, in *Optical Fiber Telecommunications*, Vol. 4A, I. Kaminow and T. Li, Eds. (Academic Press, San Diego, CA, 2002), Chap. 5.
57. J. Bromage, P. J. Winzer, and R. J. Essiambre, in *Raman Amplifiers for Telecommunications*, M. N. Islam, Ed. (Springer, New York, 2003), Chap. 15.
58. H. Suzuki, N. Takachio, H. Masuda, and K. Iwatsuki, *J. Lightwave Technol.* **21**, 973 (2003).
59. M. Morisaki, H. Sugahara, T. Ito, and T. Ono, *IEEE Photon. Technol. Lett.* **15**, 1615 (2003).
60. B. Zhu, L. E. Nelson, S. Stulz, A. H. Gnauck, C. Doerr, J. Leuthold, L. Gruner-Nielsen, M. O. Pedersen, J. Kim, and R. L. Lingle, Jr., *J. Lightwave Technol.* **22**, 208 (2004).
61. D. F. Grosz, A. Agarwal, S. Banerjee, D. N. Maywar, and A. P. Küng, *J. Lightwave Technol.* **22**, 423 (2004).
62. J. Bromage, *J. Lightwave Technol.* **22**, 79 (2004).
63. G. P. Agrawal, *Applications of Nonlinear Fiber Optics* (Academic Press, San Diego, 2001), Chap. 7.

64. E. Desuvire, *Erbium-Doped Fiber Amplifiers: Principles and Applications* (Wiley, New York, 1994).
65. P. C. Becker, N. A. Olsson, and J. R. Simpson, *Erbium-Doped Fiber Amplifiers: Fundamentals and Technology*, Academic Press, San Diego, CA, 1999.
66. E. Desuvire, D. Bayart, B. Desthieux, and S. Bigo, *Erbium-Doped Fiber Amplifiers: Device and System Development* (Wiley, New York, 2002).
67. G. P. Agrawal, *Fiber-Optic Communication Systems*, 3rd ed. (Wiley, New York, 2002).
68. P. Wan and J. Conradi, *J. Lightwave Technol.* **14**, 288 (1996).
69. P. B. Hansen, L. Eskilden, A. J. Stentz, T. A. Strasser, J. Judkins, J. J. DeMarco, R. Pedrazzani, and D. J. DiGiovanni, *IEEE Photon. Technol. Lett.* **10**, 159 (1998).
70. M. Nissov, K. Rottwitt, H. D. Kidorf, and M. X. Ma, *Electron. Lett.* **35**, 997 (1999).
71. S. R. Chinn, *IEEE Photon. Technol. Lett.* **11**, 1632 (1999).
72. S. A. E. Lewis, S. V. Chernikov, and J. R. Taylor, *IEEE Photon. Technol. Lett.* **12**, 528 (2000).
73. C. H. Kim, J. Bromage, and R. M. Jopson, *IEEE Photon. Technol. Lett.* **14**, 573 (2002).
74. C. R. S. Fludger, V. Handerek, and R. J. Mears, *J. Lightwave Technol.* **19**, 1140 (2001).
75. H. Kogelnik, R. M. Jopson, and L. E. Nelson, in *Optical Fiber Telecommunications*, Val. 4B, I. P. Kaminow and T. Li, Eds. (Academic Press, San Diego, CA, 2002), Chap. 15.
76. P. Ebrahimi, M. C. Hauer, Q. Yu, R. Khosravani, D. Gurkan, D. W. Kim, D. W. Lee, and A. E. Willner, *Proc. Conf. on Lasers and Electro-Optics* (Optical Society of America, Washington, DC, 2001), p. 143.
77. S. Popov, E. Vanin, and G. Jacobsen, *Opt. Lett.* **27**, 848 (2002).
78. A. B. dos Santos and J. P. von der Weid, *Opt. Lett.* **29**, 1324 (2004).
79. Q. Lin and G. P. Agrawal, *Opt. Lett.* **27**, 2194 (2002).
80. Q. Lin and G. P. Agrawal, *Opt. Lett.* **28**, 227 (2003).
81. Q. Lin and G. P. Agrawal, *J. Opt. Soc. Am. B* **20**, 1616 (2003).
82. C. Headley, III and G. P. Agrawal, *J. Opt. Soc. Am. B* **13**, 2170 (1996).
83. K. J. Blow and D. Wood, *IEEE J. Quantum Electron.* **25**, 2665 (1989).
84. P. V. Mamyshev and S. V. Chernikov, *Opt. Lett.* **15**, 1076 (1990).
85. S. V. Chernikov and P. V. Mamyshev, *J. Opt. Soc. Am. B* **8**, 1633 (1991).
86. P. V. Mamyshev and S. V. Chernikov, *Sov. Lightwave Commun.* **2**, 97 (1992).
87. R. H. Stolen and W. J. Tomlinson, *J. Opt. Soc. Am. B* **9**, 565 (1992).
88. J. T. Manassah, *Appl. Opt.* **26**, 3747 (1987); J. T. Manassah and O. Cockings, *Appl. Opt.* **26**, 3749 (1987).
89. J. Hermann and J. Mondry, *J. Mod. Opt.* **35**, 1919 (1988).
90. R. Osborne, *J. Opt. Soc. Am. B* **6**, 1726 (1989).
91. D. N. Cristodoulides and R. I. Joseph, *IEEE J. Quantum Electron.* **25**, 273 (1989).
92. Y. B. Band, J. R. Ackerhalt, and D. F. Heller, *IEEE J. Quantum Electron.* **26**, 1259 (1990).
93. V. A. Vysloukh and V. N. Serkin, *JETP Lett.* **38**, 199 (1983).
94. E. M. Dianov, A. M. Prokhorov, and V. N. Serkin, *Opt. Lett.* **11**, 168 (1986).
95. J. P. Gordon, *Opt. Lett.* **11**, 662 (1986).
96. F. M. Mitschke and L. F. Mollenauer, *Opt. Lett.* **11**, 659 (1986).
97. P. Beaud, W. Hodel, B. Zysset, and H. P. Weber, *IEEE J. Quantum Electron.* **23**, 1938 (1987).
98. A. S. Gouveia-Neto, A. S. L. Gomes, J. R. Taylor, and K. J. Blow, *J. Opt. Soc. Am. B* **5**, 799 (1988).
99. A. Höök, D. Anderson, and M. Lisak, *J. Opt. Soc. Am. B* **6**, 1851 (1989).
100. E. A. Golovchenko, E. M. Dianov, P. V. Mamyshev, A. M. Prokhorov, and D. G. Fursa, *J. Opt. Soc. Am. B* **7**, 172 (1990).
101. C. Headley, III and G. P. Agrawal, *J. Opt. Soc. Am. B* **10**, 2383 (1993).
102. R. F. de Souza, E. J. S. Fonseca, M. J. Hickmann, and A. S. Gouveia-Neto, *Opt. Commun.* **124**, 79 (1996).
103. K. Chan and W. Cao, *Opt. Commun.* **158**, 159 (1998).
104. J. Santhanama and G. P. Agrawal, *Opt. Commun.* **222**, 413 (2003).
105. K. Tamura and M. Nakazawa, *Opt. Lett.* **21**, 68 (1996).
106. M. E. Fermann, V. I. Kruglov, B. C. Thomsen, J. M. Dudley, and J. D. Harvey, *Phys. Rev. Lett.* **26**, 6010 (2000).
107. V. I. Kruglov, A. C. Peacock, J. D. Harvey, and J. M. Dudley, *J. Opt. Soc. Am. B* **19**, 461 (2002).
108. J. Limpert, T. Schreiber, T. Clausnitzer, K. Zollner, H. J. Fuchs, E. B. Kley, H. Zellmer, and A. Tunnermann, *Opt. Exp.* **10**, 628 (2002).
109. A. C. Peacock, N. G. R. Broderick, and T. M. Monro, *Opt. Commun.* **218**, 167 (2003).
110. V. I. Kruglov, A. C. Peacock, and J. D. Harvey, *Phys. Rev. Lett.* **90**, 113902 (2003).
111. C. Finot, G. Millot, C. Billet, and J. M. Dudley, *Opt. Exp.* **11**, 1547 (2003).
112. T. Hirooka and M. Nakazawa, *Opt. Lett.* **29**, 498 (2004).

10
CHAPTER

Application of Numerical Analysis Techniques for the Optimization of Wideband Amplifier Performances

Namkyoo Park*, Pilhan Kim, Hansuek Lee, and Jaehyoung Park

Optical Communication Systems Laboratory
School of Electrical Engineering, Seoul National University
Seoul, South Korea

1. FOREWORD

Increasing demands for the high-capacity wavelength division multiplexed (WDM) transmission system now require the development of new transmission windows exceeding the gain bandwidth formerly supported by Erbium-doped fiber amplifiers (EDFA). With intensive development efforts on new rare-earth dopants, high power laser diodes, and highly-nonlinear fiber in the past years, today's wideband optical amplifiers now can easily support gain bandwidths that are over 4-5 times wider than those formerly possible with the conventional EDFAs. With various breeds of optical amplifier, there exist three distinct approaches for the gain bandwidth expansion, available in the commercial market. These includes: Thulium-doped fiber amplifiers (TDFA) for S+ band (1450-1480 nm) and S band (1480-1530 nm), EDFAs for C band (1530-1560nm) and L band (1570-1610nm), and Raman amplifiers with 100nm's of gain bandwidth (with flexible location from S+ to L band). Even though there have been much increased experimental reports for all of these amplifiers, the complexity of the amplification dynamics due to the number of involved energy levels, and difficulties in the measurement of experimental parameters have made it difficult to predict the performance of wideband amplifiers in general. In this respect, a reliable and efficient numerical analysis techniques that can predict and estimate the optical amplifier performances for different applications would be useful, minimizing the possibility of future system impairment and design cost. In this chapter, we present some of the successful examples for the application of the numerical analysis in the amplifier design, concentrating on three different main aspects of amplifier performances—power efficiency, gain engineering, and transient—for three different types of optical amplifiers (EDFA, Raman, and TDFA). For this purpose, we concentrate in this chapter for the specific issues on the table leaving out the fundamentals, because many excellent works on the fundamentals already exist, including previous chapters in this book detailing the physics of the optical amplifiers for each breed.

2. POWER EFFICIENCY: L-BAND EDFA

2.1. Introduction

With their potential for doubling the current transmission gain bandwidth, the long-wavelength band erbium-doped fiber amplifiers (L-band EDFAs) have become a tempting option for aggressive system integrators. Nevertheless, as reported in previous studies, the L-band EDFA suffers from a well-recognized drawback in terms of its power conversion efficiency (PCE) when compared with its hybrid, the C-band EDFA. Briefly stated, this problem stems from the somewhat faster absorption rate of the pump wave (especially for the 980 nm pump) in the L-band EDFA, and the resulting development of a strong C-band amplified spontaneous emission (ASE), which wastes pump photons that could otherwise be used to amplify the L-band signal. Among the various attempts to circumvent this problem [1-4], EDFAs with wavelength-detuning of the pump [3, 4], or structural detuning of EDF absorption [5] distinguish themselves from other methods because they do not require any modifications for the amplifier structure (which would otherwise be necessary, e.g., For an additional source, for extra wavelength division multiplexing [WDM] components, in prior articles [1, 2]). In this section, we apply a numerical analysis technique to explain the internal

*E-mail: nkpark@plaza.snu.ac.kr

dynamics of PCE for the pump wavelength detuned L-band EDFAs and provide a clue for a generalized explanation on the PCE of an amplifier.

2.2. Pump Wavelength Detuning

As formerly addressed in [6], by selectively doping those to-be-excited erbium ions only at those regions where the pump power is sufficiently intense (a subset of the core), the higher inversion level (>60%) that is required for efficient C-band amplification can be achieved. Although this approach seems natural for the C-band, it needs to be modified for the L-band EDFA, which requires a much lower inversion level of 30–40%. Considering the current resolution required for achieving this lower average inversion for the L-band amplifier, which is using longer lengths of doped fiber, the amplifier develops fiber regions of low-inversion states, where detrimental ASE starts to develop as well as regions of extremely high-inversion (>90%) states, where C-band ASE amplification dominates. To avoid this problem, the formerly proposed pump-detuning method is used, in which the wavelength-dependent lower pump absorption coefficient delivers pump power deeper into the fibers, and eliminates these regions of extreme low/high-inversion levels and expands the useful L-band amplification regions [3, 4].

To understand and confirm the internal dynamics of PCE improvement, we show the results of numerical simulations for pump wavelength detuned L-band EDFAs. The amplifier was modeled as a homogeneously broadened three level system including both spatial and spectral variations, ASE, and the Rayleigh scattering effect [6], with a spectral grid of 1 nm within the spectral range of interest (1500–1620 nm). Absorption and emission cross-sections in this range were obtained from the fiber manufacturer's data sheets, including other additional parameters shown in Table 10.1. The pump absorption cross-section between 950 and 1010 nm was also obtained from a prior report [7]. The accuracy of the simulator was also tested intensively over most of the operating conditions before the following analysis, against other commercial, manufacturer-supplied programs or against the experimental results. For the sample analysis, we assumed a forward-pumped single stage EDFA with an input signal of 0 dBm (1585 nm), and a pump power of 100 mW (950–1010 nm).

Figure 10.1 shows the variation of the output signal power and noise figure (NF) as a function of pump

TABLE 10.1 EDF parameters used in the study.

Core radius	1.16 μm
Cutoff wavelength (λ_c)	876 nm
Doping concentration	6.1×10^{-24} m^{-3}
Scattering capture fraction	0.0025
Background loss at 980 nm	10 dB/km
Background loss at 1550 nm	3.86 dB/km
Pairing percentage	1%
Peak absorption cross-section	8.15×10^{-25} m^2
980 nm absorption cross-section	3.17×10^{-25} m^2
1480 nm absorption cross-section	3.42×10^{-25} m^2

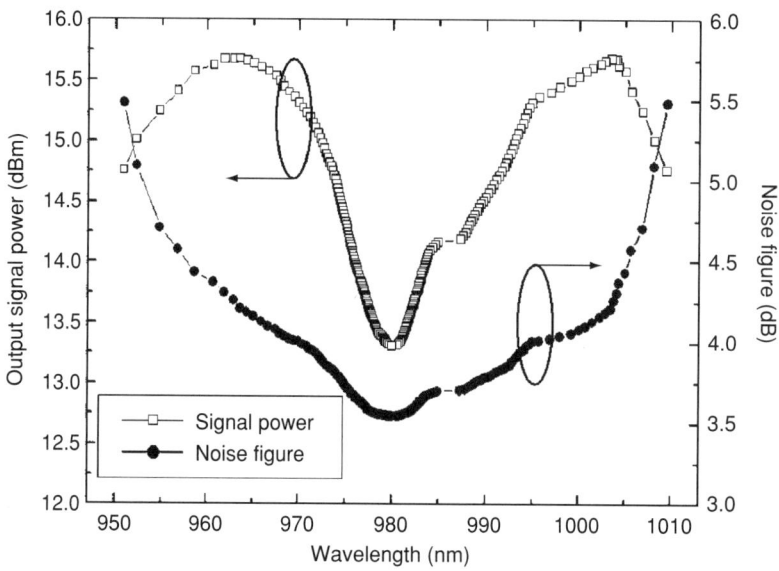

FIGURE 10.1 Output power and NF of L-band EDFA at different pump wavelengths (950–1010 nm).

wavelength (950–1010 nm). The output signal power and NF were obtained at the *optimal length* of EDF, which achieved the maximum PCE (at a given pump wavelength). As can be seen from Fig. 10.1, it was possible to observe the enhancement of the PCE as we increase or decrease the pump wavelength from 980 to 963 nm or 1003 nm. At the PCE-optimized pump wavelengths (963 or 1003 nm), the observed PCE enhancement was 65% larger than that of the 980-nm pumped EDFA. Figure 10.1 also shows the monotonic increase of the NF from the pump wavelength detuning effect. The NF degradation at the PCE optimum pump wavelength (963 and 1003 nm) was about 0.67 dB.

To better explain these behaviors at the fundamental level, Fig. 10.2a illustrates the pump power evolution and inversion distribution for 963/1003 nm (optimally pump wavelength detuned design) and 980-nm (conventional design) pumped EDFAs. As explained earlier, the strong absorption of the pump power and higher inversion level for the 980 nm pumped EDFA could be observed, especially at the pumping end, where significant development of backward ASE occurs. In contrast, an overall reduction in the pump power absorption rate/inversion levels for 963/1003-nm pumped EDFAs was evident, enabling the higher power efficiency in terms of the L-band amplification. Still, it is worth noting that this overall reduction of the inversion level (especially at the input end) from the wavelength detuning effect results in an increased NF (Fig. 10.1). To quantitatively show the effect of the absorption cross-section change by pump detuning on the PCE enhancement, Fig. 10.2b plots the amount of output signal, backward ASE, and wasted pump (unabsorbed output pump) power at various pump absorption cross-sections normalized by a peak pump absorption cross-section at 980 nm (3.17×10^{-25} m^{-3}). As we decrease the pump cross-section value (or physically as we change the pump wavelength away from 980 nm), the suppression of the backward ASE power and the increase in the unabsorbed pump power was evident. Even though the smaller pump cross-section enables a reduced pump depletion rate, thus suppressing the backward ASE development at the input end, this result also implies that it can result in the increase in unabsorbed pump output. This observation explains the existence of the optimum pump wavelength for L-band signal amplification shown in Fig. 10.1, which is determined by the trade-off between the suppression of ASE and the wasting of the pump power.

2.3. Fiber Structural Detuning

Based on the observation from the previous section, we now transfer the effect of pump absorption control into the fiber structure. Figure 10.3a shows the simplest fiber structure (type L) that can provide an equivalent effect on the inversion distribution inside the EDF without the need for a pump wavelength change [5]. With the annular doping structure, the pump wave would experience relatively higher reduction in its absorption rate than would signal waves, as a result of the wavelength-dependent field distribution inside the fiber (Fig. 10.3b). To prove the efficiency of the fiber structural detuning method, we show the result of the numerical simulation below. Fiber parameters from conventional C-band EDFs (Table 10.1) were used to calculate the solutions of the Bessel functions and to obtain precise radial distributions of the pump/signal waves so that the solutions could be applied to the full numerical analysis of the EDFA. Because erbium is also doped in the

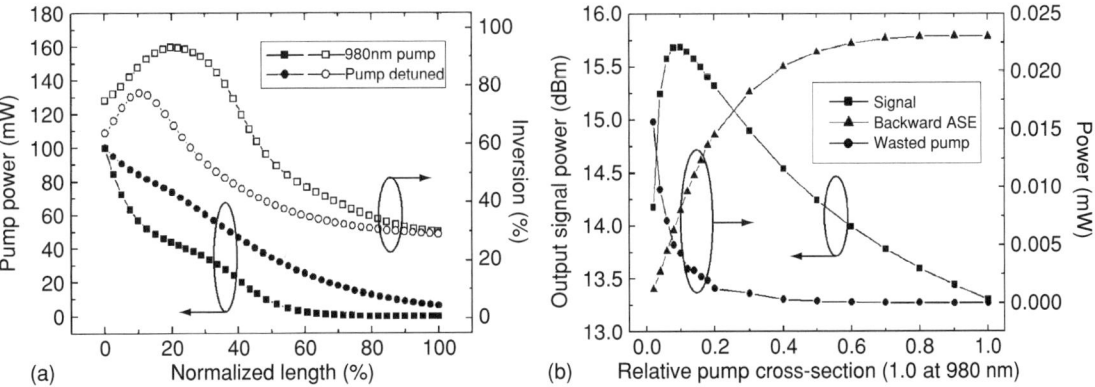

FIGURE 10.2 (a) Pump power/inversion evolution as a function of normalized fiber length (with 980/963 nm pump). (b) Output signal/ASE/wasted pump powers plot as a function of pump absorption cross-sections (1.0 at 980 nm peak).

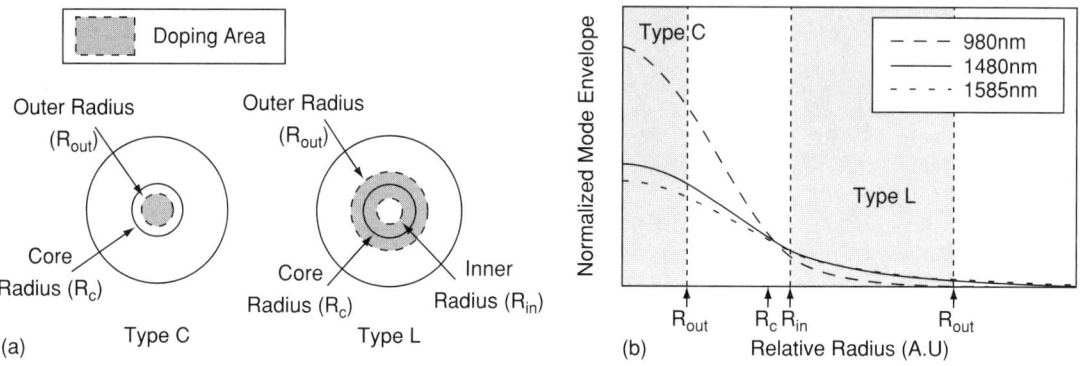

FIGURE 10.3 (a) Doping profile of type C/L EDF; (b) Normalized mode envelope/doping regions of type C/L EDF.

cladding area, the spatial models of EDFAs have to correctly include the effect of inversion distribution along the radial direction [8].

Figure 10.4 shows the distribution of inversion as a function of the radial axis (r) as well as for the longitudinal (normalized) direction of the EDF (type C/L). Most of the simulation conditions were identically set to those of the pump wavelength detuning analysis for fair comparison of the results. For the sample analysis, we also assumed a forward-pumped single-stage EDFA with an input signal of 0 dBm (1585 nm) and a pump power of 100 mW (both for 980 and 1480 nm, respectively). The output power and NF for various EDF structures were then obtained at the optimal length of EDF, for which we achieved the maximum PCE.

Figure 10.5a shows the variation of the output signal power with various values of the outer/inner normalized doping radius ($R_{out,in} = r_{out,in}/r_{core}$). In this figure, the $R_{in} = 0$ line corresponds to a type C EDF, and the other lines represent various EDFs of type L (with different core-filling factors). From this analysis, it was possible to observe the enhancement of the PCE from the structural detuning effect. The optimum hole radius (R_{in}) under the given fiber parameters was found to be about the same as the size of the core. The other observation is the proportional increase of PCE with the size of the doping area, at a fixed hole radius. At the optimum design point ($R_{out} = 2.2$, $R_{in} = 1.0$), the observed PCE enhancement was 58.5% greater (with 980-nm pumping; and 9.1% with 1480 nm) than that of the commercial C-band EDF (Type C, $R_{out} = 0.5$, $R_{in} = 0$). The amount of NF degradation for the EDF at this optimum PCE design stayed well below 0.32 dB and −0.07 dB, depending on the pump wavelength (980 and 1480 nm). It is worth noting that the suggested type-L doping structure results in a slightly lower (6%) PCE than the type C EDF with the pump detuning method ($R_{out} = 0.5$, and $R_{in} = 0$, with the 963-nm pump) but results in a higher (12.6%) PCE when compared with a type C EDF optimized specifically for the L band ($R_{out} = 3$, $R_{in} = 0$). Figure 10.5b also illustrates the pump evolution and inversion distribution for the type L (optimal design) and type C EDFs. Overall reduction in the pump power

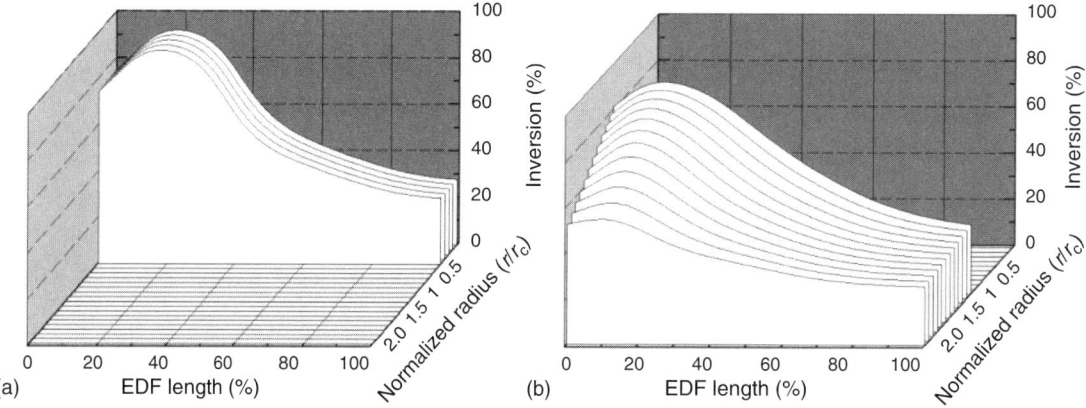

FIGURE 10.4 Inversion distribution along the radial/propagation axis: (a) type C EDF; (b) type L EDF.

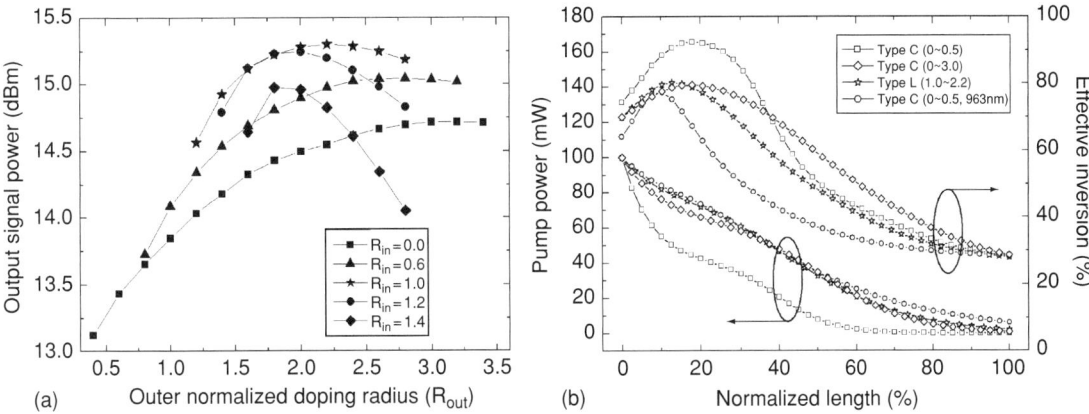

FIGURE 10.5 (a) EDFA output power as a function of normalized outer/inner doping radius. (b) Pump propagation/inversion map as a function of normalized fiber length for different doping profiles.

absorption rate/inversion levels for the 980-nm pumped optimum designed type-L EDF and pump-detuned (963 nm) type-C EDF was evident.

To quantitatively assess the effect of the annular-doping position on the absorption rate changes, we also show the overlap factor (between the doped region and the mode field, for the signal and pump waves) as a function of the outer normalized doping radius, under a delta doping condition ($R_{out} - R_{in} = 0.1 \ll R_c = 1$, background loss neglected to observe the net effect). After the expected increase in the overlap factor between the annular-doped region and the propagating waves (contribution from the radial factor: $2\pi r$), it was possible to observe a wavelength-dependent decrease (contribution from the mode-envelope shape, Fig. 10.3b) in the overlap factor both for the signal and pump waves, especially at much higher rates for the short-wavelength waves.

As a result of this change in the mode overlap factor, a greater reduction in the absorption rate of the pump waves than that of the signal waves can be obtained, especially for the 980 nm pumps, thus enabling deeper penetration of the pump waves inside the EDF. The relatively smaller changes in the overlap factor ratio for 1585/1480 nm waves to the delta-doping radius (Fig. 10.6a) also explains the somewhat smaller values of PCE enhancement observed for the 1480 nm pumps with this structural detuning approach. In contrast to the 980-nm pumping case, the 1480-nm pump and the 1585-nm signal experience almost the same amount of changes in the absorption coefficient, virtually equivalent to the effect from the PCE-independent doping concentration adjustment.

As an additional measure of the power efficiency, Fig. 10.6b, we plotted the amount of output signal, total backward ASE (1500–1610 nm), and wasted pump (unabsorbed 980 nm) powers at various delta-doping radii. As the radius is increased, the suppression of the backward ASE power and the

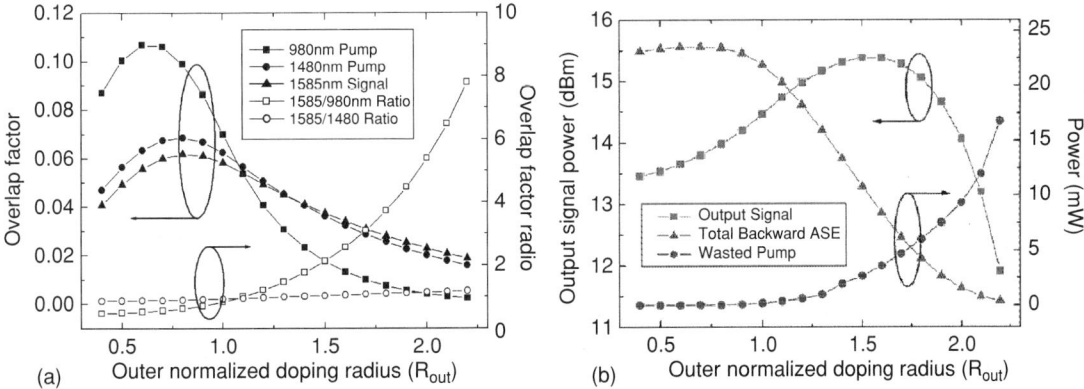

FIGURE 10.6 (a) Doped area - mode field overlap factor, and (b) output signal/ASE/pump powers obtained from the EDFs at different radius of delta doping ($R_{out} - R_{in} = 0.1 \ll R_c = 1$) conditions.

increase in wasted pump power were evident. This pattern is very similar to Fig. 10.2b because the pump absorption control mechanism is common in both cases. Even though the smaller pump overlap factor enables a reduced pump depletion rate, which suppresses the backward ASE development, this result also implies that it can result in the increase of the unabsorbed pump output. This phenomenon explains the existence of the optimum doping profile, which is determined by the trade-off between the suppression of ASE and the wasting of pump power.

2.4. Conclusion

In this section we explained the underlying physics and internal dynamics of PCE enhancement for L-band EDFA in terms of the inversion distribution and pump propagation map, which is a quantity hard to access with the experiment, for both the pump wavelength detuning and the fiber structure detuning approaches. With the help of a powerful numerical tool, it was possible to design a highly efficient EDF structure by applying the concept learned from the numerical analysis of the pump detuning method. Results showed that a properly designed, structurally detuned EDF could increase the PCE of L-band EDFAs by more than 58.5% when compared with the PCE of a conventional one using a C-band EDF, with a negligible increase in the NF.

3. GAIN ENGINEERING: RAMAN AMPLIFIER

3.1. Introduction

The fiber Raman amplifier (FRA) has become an indispensable technology with its distinctive advantages, such as flexible gain bandwidth and intrinsically lower noise characteristics. To enable the optimal design of Raman amplifier/amplified transmission systems, various approaches have been proposed for the efficient modeling of the Raman amplifier [9–11]. Although different in their detailed algorithms, levels of convergence speed and accuracies, all these approaches share the common platform of coupled ordinary differential equations (ODE) for the Raman equation set; they are solved along the long length of fiber propagation axis.

This section introduces an alternative highly efficient modeling method for FRA analysis. Under this platform, a Raman amplifier is solved along the iteration axis rather than the fiber axis, enabling orders of faster convergence speed at the equivalent accuracy achievable with the methods based on coupled ODEs. Furthermore, we show that the traditionally hard-to-access problem of multi-pumped FRA gain design can be treated in a much simpler and faster way with this method.

3.2. Implementation of the Closed Form Raman Equation

To construct such an algorithm or platform stated in the Introduction, we treat the Raman gain coefficient as the perturbation factor in the adiabatic process and (1) derive a recursion relation of Raman integral equation, (2) construct a matrix formalism for the solution of the given FRA problem, and (3) find the output power of FRA (the final target solution) while in the iteration process, considering the power integral as the interim target solution to achieve faster convergence. The following shows the step-by-step implementation method of the above idea.

3.2.1. Construction of the Formalism

First, ignoring the negligible effect of ASE and Rayleigh scattering (see the end of Section 2.3 for the latter calculation/justification), the coupled nonlinear Raman process in the fiber can be expressed as [12, 13]:

$$\pm \frac{dP_i}{dz} = -\alpha_i P_i + \sum_{j=1}^{M+N} g_{ji} P_j P_i \quad (1)$$

where P_i is the power at ith wavelength, α_i is the attenuation coefficient, g_{ji} is the scaled Raman gain coefficient [14], M is the pump number, and N is the number of signal waves. After dividing Eq. (1) by P_i and integrating over z, we get:

$$P_i(z) = P_i(0) \exp\left[\mp \alpha_i z \pm \left(\sum_{j=1}^{M+N} g_{ji} \int_0^z P_j(\zeta) d\zeta\right)\right] \quad (2)$$

With an additional integration progress, we get the following integral form of Raman wave equations:

$$\frac{\int_0^z P_i(\zeta) d\zeta}{P_i(0)} = \int_0^z \exp\left[\mp \alpha_i \zeta \pm \left(\sum_{j=1}^{M+N} g_{ji} \int_0^\zeta P_j(\xi) d\xi\right)\right] d\zeta \quad (3)$$

Now, using the definition of $L_{\text{eff}\_i}(z) = \int_0^z P_i(\zeta)/P_i(0) dz'$, we can then rewrite Eq. (3) into a closed integral form for the effective length.

$$L_{\text{eff}\_i}(z) = \int_0^z \exp\left[\mp \alpha_i \zeta \pm \left(\sum_{j=1}^{M+N} g_{ji} P_j(0) L_{\text{eff}\_j}(\zeta)\right)\right] d\zeta \quad (4)$$

To solve this equation, we apply Picard's iteration method to Eq. (4) taking L_{eff_i} s as the interim target solution. At the nth iteration step, Eq. (4) becomes

$$L^{n^{th}}_{\text{eff}\_i}(z) = \int_0^z \exp\left[\mp \alpha_i \zeta \pm \left(\sum_{j=1}^{M+N} g_{ji} P^{(n-1)^{th}}_j(0) L^{(n-1)^{th}}_{\text{eff}\_j}(\zeta)\right)\right] d\zeta \quad (5)$$

For the implementation of the above equations into the numerical domain, we now construct a vector $\tilde{L}^{n^{th}}_{\text{eff}\_i}(\vec{z}_k)$ to assign the value of the effective length at the ith wavelength at the position z_k (the discrete position element covering the whole fiber link with step size of Δz, as shown in Fig. 10.7) with the vector elements

$$L^{n^{th}}_{\text{eff}\_i}(\vec{z}_k) = \exp\left[\mp \alpha_i \vec{z}_k * \right.$$
$$\left. \pm \vec{g}_{ji} \vec{P}^{(n-1)^{th}}_j(0) \cdot \tilde{L}^{(n-1)^{th}}_{\text{eff}\_j}(\vec{z}_k)\right] \cdot \tilde{T}_{\text{trig}} \cdot \Delta z \quad (6)$$

where

$\vec{z}_k = [0 \ \Delta z \ 2 \cdot \Delta z \ 3 \cdot \Delta z \ \cdots \ L - \Delta z \ L]$,
$L^{n^{th}}_{\text{eff}\_i}(\vec{z}_k) = \lfloor L^{n^{th}}_{\text{eff}\_i}(0) \ L^{n^{th}}_{\text{eff}\_i}(\Delta z) \ L^{n^{th}}_{\text{eff}\_i}(2\Delta z) \ \cdots \ L^{n^{th}}_{\text{eff}\_i}(L)\rfloor$
$\vec{g}_{ji} \vec{P}^{(n-1)^{th}}_j(0) = \left[g_{1i} P^{(n-1)^{th}}_1(0) \ g_{2i} P^{(n-1)^{th}}_2(0) \right.$
$\left. g_{3i} P^{(n-1)^{th}}_3(0) \ \cdots \ g_{M+N i} P^{(n-1)^{th}}_{M+N}(0)\right]$, and

$$\tilde{L}^{(n-1)^{th}}_{\text{eff}\_j}(\vec{z}_k) = \begin{bmatrix} L^{(n-1)^{th}}_{\text{eff}\_1}(0) & L^{(n-1)^{th}}_{\text{eff}\_1}(\Delta z) & L^{(n-1)^{th}}_{\text{eff}\_1}(2\Delta z) & \cdots & L^{(n-1)^{th}}_{\text{eff}\_1}(L) \\ L^{(n-1)^{th}}_{\text{eff}\_2}(0) & L^{(n-1)^{th}}_{\text{eff}\_2}(\Delta z) & L^{(n-1)^{th}}_{\text{eff}\_2}(2\Delta z) & \cdots & L^{(n-1)^{th}}_{\text{eff}\_2}(L) \\ L^{(n-1)^{th}}_{\text{eff}\_3}(0) & L^{(n-1)^{th}}_{\text{eff}\_3}(\Delta z) & L^{(n-1)^{th}}_{\text{eff}\_3}(2\Delta z) & \cdots & L^{(n-1)^{th}}_{\text{eff}\_3}(L) \\ \vdots & \vdots & \vdots & \vdots & \vdots \\ L^{(n-1)^{th}}_{\text{eff}\_M+N}(0) & L^{(n-1)^{th}}_{\text{eff}\_M+N}(\Delta z) & L^{(n-1)^{th}}_{\text{eff}\_M+N}(2\Delta z) & \cdots & L^{(n-1)^{th}}_{\text{eff}\_M+N}(L) \end{bmatrix},$$

$$\tilde{T}_{\text{trig}} = \begin{bmatrix} 1 & 1/2 & 1/2 & 1/2 & \cdots & 1/2 \\ 0 & 1/2 & 1 & 1 & \cdots & 1 \\ 0 & 0 & 1/2 & 1 & \cdots & 1 \\ 0 & 0 & 0 & 1/2 & \cdots & 1 \\ \vdots & \vdots & \vdots & \vdots & & \vdots \\ 0 & 0 & 0 & 0 & \cdots & 1 \\ 0 & 0 & 0 & 0 & \cdots & 0 \end{bmatrix}$$

Meanwhile, utilizing the rule of superposition with fractional Raman gains [12], the Raman gain at a certain wavelength can be expressed as:

$$G_{dB\_t} = \sum_{j=1}^{M+N} G_{dB\_j}$$
$$= \sum_{j=1}^{M+N} \left(10 \log(e) \frac{g_R}{2 A_{\text{eff}}} \left(\frac{\lambda_R}{\lambda_j}\right)^k\right) \cdot L_{\text{eff}\_j} \cdot P_j(0) \quad (7)$$

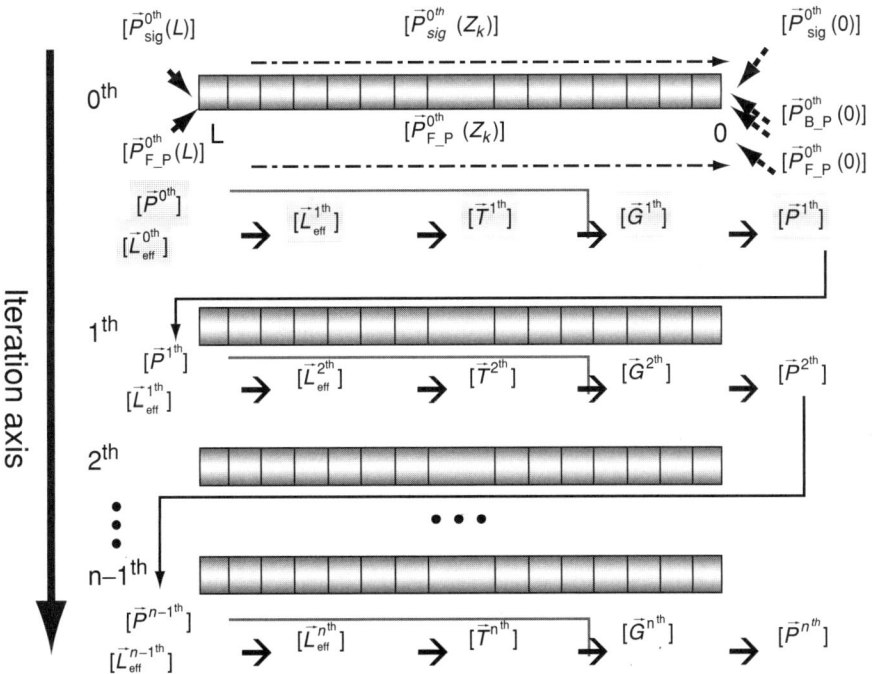

FIGURE 10.7 Flow diagram of the suggested adiabatic iteration algorithm along the iteration axis. ($\vec{P}^{0^{th}}_{\text{sig}}(0)$, $\vec{P}^{0^{th}}_{F_P}(0)$, $\vec{P}^{0^{th}}_{B_P}(0)$;0th order signal, forward-pump, and backward-pump power at $z_k = 0$)

where, k is the frequency scale factor and g_R is the gain at the reference wavelength [15]. Extending Eq. (7) for every pump/signal wave and treating the power $P_j(0)$ of signal/pump waves as the individual vector element, we now transform the gain equation into following matrix form

$$\vec{G}_{dB\_t} = \tilde{T} \times \vec{P} \qquad (8)$$

where

$$\tilde{T} = \begin{bmatrix} G_R^{(1,1)} L_{\text{eff}\_1} & G_R^{(1,2)} L_{\text{eff}\_2} & \cdots & G_R^{(1,M+N)} L_{\text{eff}\_M+N} \\ G_R^{(2,1)} L_{\text{eff}\_1} & G_R^{(2,2)} L_{\text{eff}\_2} & \cdots & G_R^{(2,M+N)} L_{\text{eff}\_M+N} \\ \vdots & \vdots & \vdots & \vdots \\ G_R^{(M+N,1)} L_{\text{eff}\_1} & G_R^{(M+N,2)} L_{\text{eff}\_2} & \cdots & G_R^{(M+N,M+N)} L_{\text{eff}\_M+N} \end{bmatrix},$$

$$\vec{P} = \begin{bmatrix} P_1(0) \\ P_2(0) \\ \vdots \\ P_{N+N}(0) \end{bmatrix}, \text{ and } G_R^{(i,j)} = 10 \log(e) \frac{g_R(\lambda_i,\lambda_j)}{2A_{\text{eff}}} \left(\frac{\lambda_R}{\lambda_j}\right)^k$$

Utilizing these formulations, the Raman gain values for every pump/signal wave can be obtained with a matrix multiplication process without relying on a complex solution search procedure of coupled ODEs.

Following the steps described, the actual procedures to practically implement the above-derived analytic formulations.[Eqs. (6) and (8)] into the numerical domain are as described in the following steps:

3.2.2. Step 1: Initial Condition Setup

Treating the $g_{ji} \times P_j(0) \times L_{\text{eff}\_j}$ as a perturbation parameter in the adiabatic process, we first assume it to be zero for the 0th iteration step to get the 0th order initial effective length, $\tilde{L}_{\text{eff}\_j}^{0^{\text{th}}} = (1 - \exp(-\alpha_j L))/\alpha$; At the same time, to ease the application of transfer matrix for all forward/backward-propagating waves, we treat the forward-propagating waves injected at the input end ($z = L$) as the backward-propagating waves assigned to the output end [12] ($z = 0$, as illustrated in Fig. 10.7). For this, we set $\vec{P}_{\text{sig}}^{0^{\text{th}}}(0)$ and $\vec{P}_{F\_P}^{0^{\text{th}}}(0)$ to be equal to $(\vec{P}_{\text{sig}}^{0^{\text{th}}}(L) - |\text{fiber loss}|)$ and $(\vec{P}_{F\_P}^{0^{\text{th}}}(L) - |\text{fiber loss}|)$, respectively, for the forward-propagating waves in the counter-/bidirectional pumping configuration and use the predetermined initial input signal power and forward pump power $\vec{P}_{\text{sig}}^{0^{\text{th}}}(0)$ and $\vec{P}_{F\_P}^{0^{\text{th}}}(0)$, respectively, for codirectional waves. It is worth mentioning here that the effect of all this approximation is reasonable in the initial iteration process, because the perturbation term $g_{ji} \times P_j(0) \times L_{\text{eff}\_j}$ of the forward propagating signal and pump is much smaller than that of backward-propagating pump ($\vec{P}_{\text{sig}}^{0^{\text{th}}}(0), \vec{P}_{F\_P}^{0^{\text{th}}}(0) \ll \vec{P}_{B\_P}^{0^{\text{th}}}(0)$).

3.2.3. Step 2: Calculation of Higher Order Terms

Substituting $\vec{P}^{0^{\text{th}}}(0)$ and $\tilde{L}_{\text{eff}\_j}^{0^{\text{th}}}$ into Eq. (6), first-order approximation for the effective length $\tilde{L}_{\text{eff}\_j}^{1^{\text{th}}}$ can be obtained. The value of $\vec{G}^{1^{\text{th}}}$ from $\vec{P}^{0^{\text{th}}}(0)$ and $\tilde{T}^{1^{\text{th}}}$ (calculated from $\tilde{L}_{\text{eff}\_j}^{1^{\text{th}}}$) can also be obtained by using Eq. (8). Finally, to get the first-order iteration value of the target solution $\vec{P}^{1^{\text{th}}}(0)$, we set $\vec{P}_{\text{sig}}^{1^{\text{th}}}(0)$ and $\vec{P}_{F\_P}^{1^{\text{th}}}(0)$ equal to $(\vec{P}_{\text{sig}}^{0^{\text{th}}}(L) - |\text{fiber loss}| + \vec{G}^{1^{\text{th}}})$ and $(\vec{P}_{F\_P}^{0^{\text{th}}}(L) - |\text{fiber loss}| + \vec{G}^{1^{\text{th}}})$, respectively, for the forward-propagating waves, while using the predetermined initial input signal power and forward pump power $\vec{P}_{\text{sig}}^{0^{\text{th}}}(0)$ and $\vec{P}_{F\_P}^{0^{\text{th}}}(0)$ for the codirectional waves.

3.2.4. Step 3: Reiteration

Repeating the iteration procedure described in step 2 at the higher order values of $\tilde{L}_{\text{eff}\_j}$, $\tilde{T}$, $\vec{G}$, $\vec{P}(0)$, the final solution set can be obtained.

3.3. Application Example 1: Gain Prediction

Figures 10.8 and 10.9 show the comparison result between the described algorithm and the conventional average power method [11]. Seventy-one signals were used at wavelengths of 1530–1600 nm with 1 nm spacing at signal power of −13 dBm per channel. The co-/counterdirectional pumping scheme, we used 14 pumps (1420–1480 nm with 5-nm spacing and an additional pump at 1495 nm), whereas for the bidirectional scheme we used 4 pumps in the forward direction (1420–1435 nm with 5-nm spacing) and 10 pumps in the backward direction (1440–1480 nm with 5-nm spacing and 1495 nm). Thirty-seven kilometers of dispersion shifted fiber which has a larger Raman gain coefficient than that of a single-mode fiber was used as the test gain medium.

As can be seen from Fig. 10.8, the suggested method shows an excellent convergence accuracy; after six iteration steps, the accuracy improvement from the next order of iteration becomes less than 0.002 dB. The relative gain difference between the proposed method and the average power method remained less than 0.03 dB over the whole 71-nm gain spectral range. Meanwhile, the required computation time to get the exact Raman gain and signal/pump power distribution was less than a few tenths of a second for all the co-, counter- and bidirectional pumping configurations, with a conventional PC (2.0-GHz CPU clock). To compare, the measured convergence speed with the average power analysis method for the same system configuration was

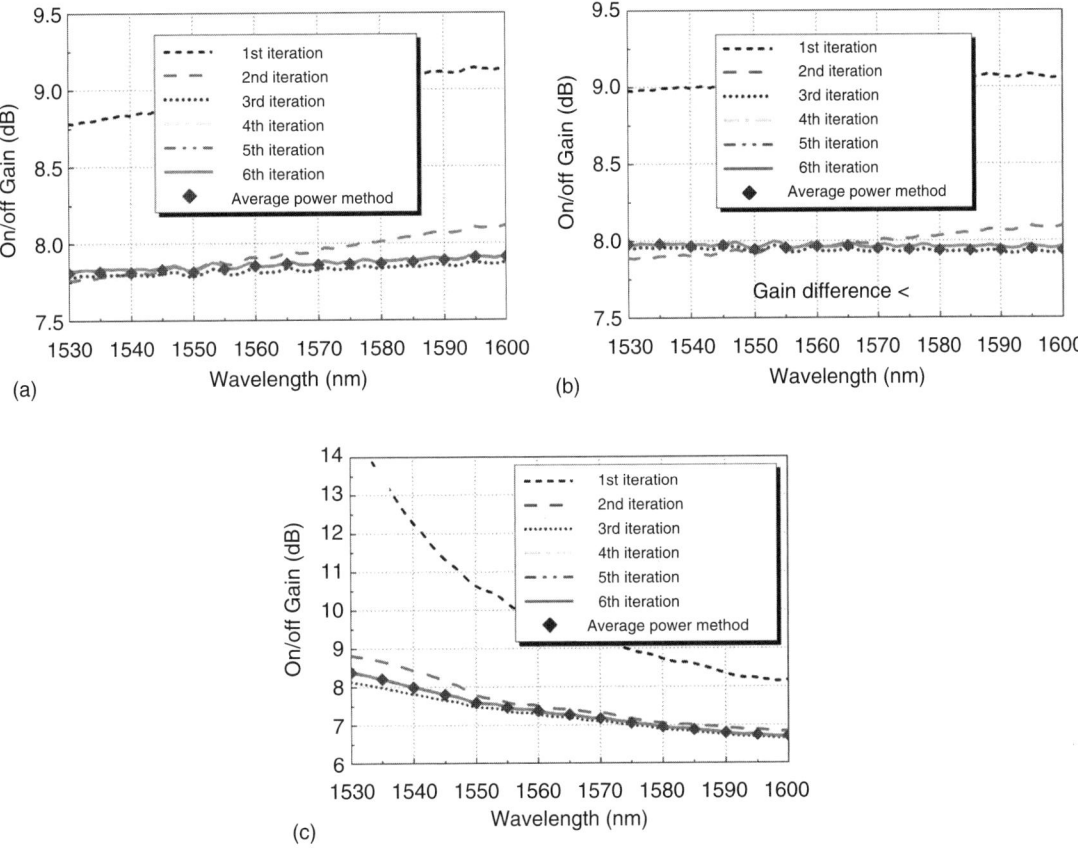

FIGURE 10.8 Raman on–off gain for different pumping configurations: (a) co-, (b) counter-, and (c) bidirectional pumping.

much longer (an order of 10–100). Figure 10.9 also shows the signals evolution along the fiber. The excellent agreement between the two simulation results shows the stability of the proposed algorithm. Using the acquired signal distribution, various FRA noise also can be obtained, if necessary (Fig. 10.10), from the known formula described in [16].

3.4. Application Example 2: Raman Gain Engineering—The Inverse Scattering Problem

Of the many distinctive advantages that the Raman fiber amplifier provides, the flexibility in gain engineering can be considered one of the key advantages that other types of amplifiers cannot provide. Still, there is an additional difficulty in the gain profile design of the FRA, when compared with other types of optical amplifiers: The strong interactions of pump waves inside the FRA make it difficult to predict the required pump power sets for the target gain profile [17,18]. To back trace the pump powers required for the target design, one has to iteratively adjust the pump input powers by using either the amount of gain error between the signal gain and target spectrum (equivalent to the shooting method in one form or another) with different tracking algorithms [12] or rescaled pump power integrals obtained from the numerical integration of coupled differential equations [15]. Even if it were possible to obtain the required set of pump powers/wavelengths after either of these different optimization procedures, all these aforementioned approaches require time-consuming process of repetitive numerical integrations for coupled ODE, including all the involved signal/pump waves.

With the constructed efficient formulation of the Raman equation described in the previous section, we show here that the highly accurate estimation of pump powers and fast construction of gain profiles can be achieved with much less effort while still keeping all the Raman interaction effects in the calculation. For this, we first rewrite the expression for Raman on–off gain in Eq. (8), keeping the signal gain term only:

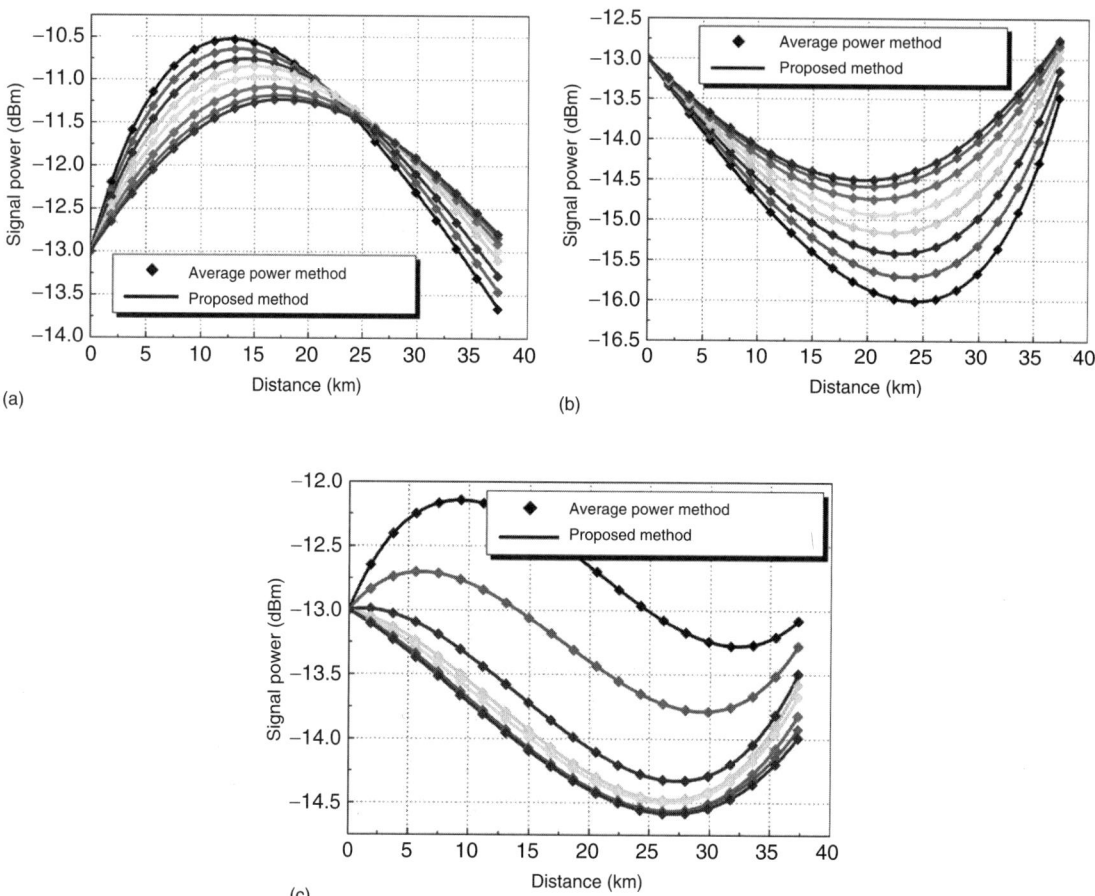

FIGURE 10.9 Signal evolution (at 1530–1600 nm, 10-nm spacing) for (a) co-, (b) counter-, and (c) bidirectional pumping.

$$\begin{aligned}
G_{dB\_t} &= \sum_{j=1}^{M+N} G_{dB\_j} \\
&= \sum_{j=1}^{M+N} \left(10 \ln \frac{g_R}{2A_{\text{eff}}} \left(\frac{\lambda_R}{\lambda_j}\right)^k\right) \cdot P_j(0) \cdot L_{\text{eff}\_j} \\
&= \tilde{T}_P \cdot \vec{P}_P + \tilde{T}_S \cdot \vec{P}_S,
\end{aligned} \quad (9)$$

where

$$\tilde{T}_{I=P,S} = \begin{bmatrix} G_{RI}^{(1,1)} L_{\text{eff}\_r1} & G_{RI}^{(1,2)} L_{\text{eff}\_r2} & \cdots & G_{RI}^{(1,K)} L_{\text{eff}\_lK} \\ G_{RI}^{(2,1)} L_{\text{eff}\_r1} & G_{RI}^{(2,2)} L_{\text{eff}\_r2} & \cdots & G_{RI}^{(2,K)} L_{\text{eff}\_lK} \\ \vdots & \vdots & \vdots & \vdots \\ G_{RI}^{(N,1)} L_{\text{eff}\_r1} & G_{RI}^{(N,2)} L_{\text{eff}\_r2} & \cdots & G_{RI}^{(N,K)} L_{\text{eff}\_lK} \end{bmatrix}_{K=M,N},$$

$$\vec{P}_{I=P,S} = \begin{bmatrix} P_{r1}(0) \\ P_{r2}(0) \\ \vdots \\ P_{lK}(0) \end{bmatrix}_{K=M,N}, \quad G_{RI}^{(i,j)} = 10 \ln \frac{g_R(\lambda_i, \lambda_j)}{2A_{\text{eff}}} \left(\frac{\lambda_R}{\lambda_j}\right)^k$$

with index $I = P, S$ for pump and signal waves, respectively.

From this, we start the gain design procedure for multi pumped Raman fiber amplifiers by defining the target gain G_{dB_t} given set of constraints (bandwidth, flatness, available pump numbers, etc.) and thus determine the strength of the gain contribution components G_{dB_j} from each pump/signal wavelength. Note that G_{dB_j} is related to the product of injected power and effective length $(P_j(0) \cdot L_{\text{eff}\_j})$, and from this relation the required injection pump power $P_{Pj}(0)$ can be obtained by using the information on effective length L_{eff_j} (or equivalently, pump integrals in [12]) at each wavelength.

The following steps are used to practically acquire the solution set for a given problem stated above:

Step 1. First, considering the nature of adiabatic expansion where we treat $g_{ji} \cdot P_j(0) \cdot L_{\text{eff}\_j}$ as a perturbation parameter, $L_{\text{eff}\_Ij}^{(0)}$ should be equal to $(1 - \exp(-\alpha_j L))/\alpha_j$. With $L_{\text{eff}\_Ij}^{(0)}$, it is possible to get initial estimations for transfer matrices $T_P^{(0)}$ and $T_S^{(0)}$. In good approximation, $T_S^{(0)}$ also can be ignored

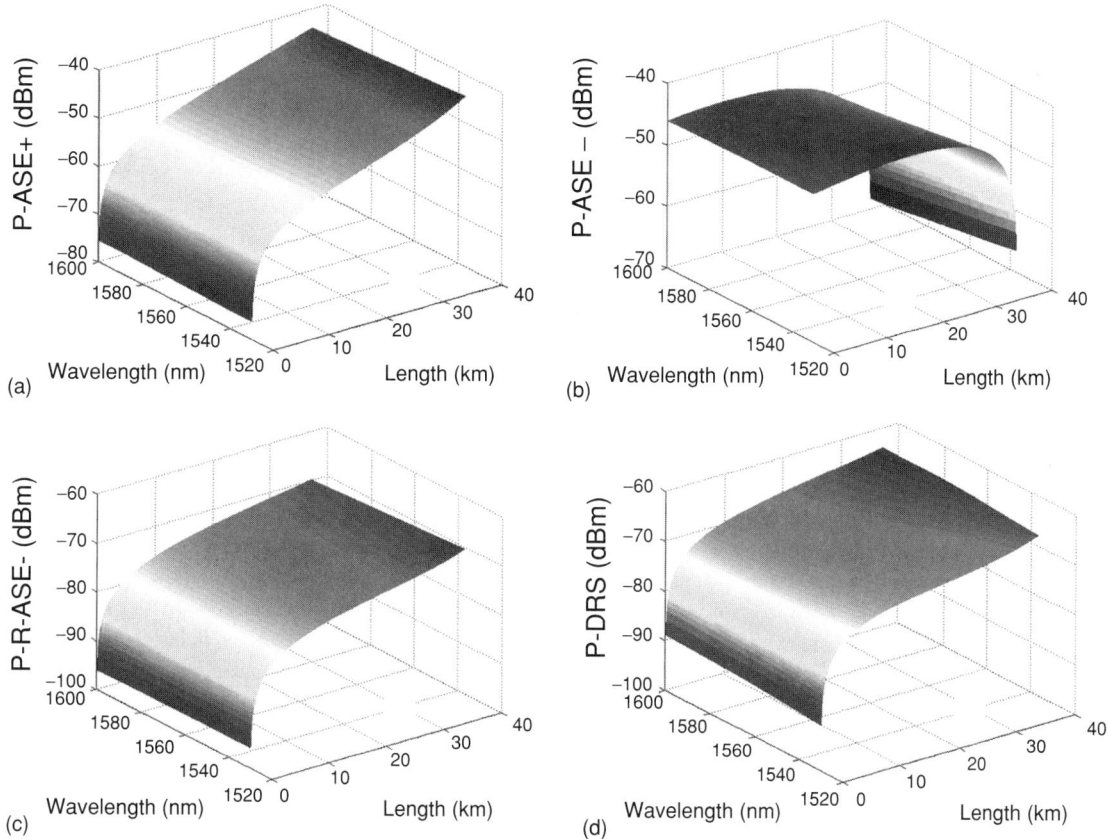

FIGURE 10.10 FRA noise (at $T = 300\,\text{K}$) in the counterpumping configuration: (a) forward ASE power, (b) backward ASE power, (c) Rayleigh scattered backward ASE power, and (d) double Rayleigh backscattered signal power.

in this first step of the iteration procedure, because $g_R(\nu_{si} - \nu_{sj}) \ll g_R(\nu_{pi} - \nu_{sj})$, and $P_S \ll P_P$. This last step enables us to pseudo-invert the transfer matrix in Eq. (9), to give $P_{Pj}^{(0)}(0) = (T_P^{(0)})^{-1} \cdot G_{dB\_t}$. It is worth noting that the effective length evolution (over z) data $L_{\text{eff}\_lj}^{(0)}(z) = (1 - \exp(-\alpha_j z))/\alpha_j$ needs to be generated in this step, so that they can be used in the next stage of iteration process.

Step 2. Substituting the $P_{Pj}^{(0)}(0)$ and $L_{\text{eff}\_lj}^{(0)}(z)$ into Eq.(5), effective lengths are updated to new values $L_{\text{eff}\_lj}^{(1)}(z)$ and $L_{\text{eff}\_lj}^{(1)}(L) = L_{\text{eff}\_lj}^{(1)}$, which are then used to get updated matrices $T_P^{(1)}$ and $T_S^{(1)}$. From this, $P_{Pj}^{(1)}(0)$ can be obtained using $(T_P^{(1)})^{-1} \cdot (G_{dB\_t} - T_S^{(1)} P_S)$, where $P_S = P_{Sj}^{(n)}(0)$ is the fixed boundary value inversely determined from the ideal Raman gain, G_{dB_t} and predetermined input signal power.

Step 3. Reiterating step 2 for higher orders, the final solution set can be obtained.

Several design examples are shown in Figs. 10.11 and 10.12. For the example in Fig. 10.11, we set G_{dB_t} (ideal Raman on/off gain profile) to $8 \pm 0.04\,\text{dB}$ over the 70 nm (1530–1600 nm) spectral range. Consider-

ing practical restrictions in the pump WDM wavelength allocation, 16 equally spaced pump waves were used between 1420 and 1495 nm at 5-nm spacings. Seventy signal waves at $-3\,\text{dBm}$ channel input power (1-nm spacing, 1530–1600 nm) were also assumed. Thirty-seven kilometers of dispersion shift fiber with large Raman gain coefficients was used to test the robustness of the algorithm.

Figure 10.11 shows the ideal target gain (horizontal solid line, with open symbol) G_{dB_t}, and gain profiles generated from the full Raman differential equations using the pump powers obtained at each stage of the iteration process (steps 1–3). The right side of the figure also illustrates the typical trace of the pump evolution at each iteration step. As can be seen from the figure, after 4th iteration steps, the pump evolution converges to that of a real value, and the gain curve exactly overlaps that of an ideal target gain. The convergence error from the iteration procedure becomes almost negligible after only four iteration steps. The observed absolute error between ideal target gain and the iteratively obtained gain profile after the four iteration steps was smaller than 0.08 dB

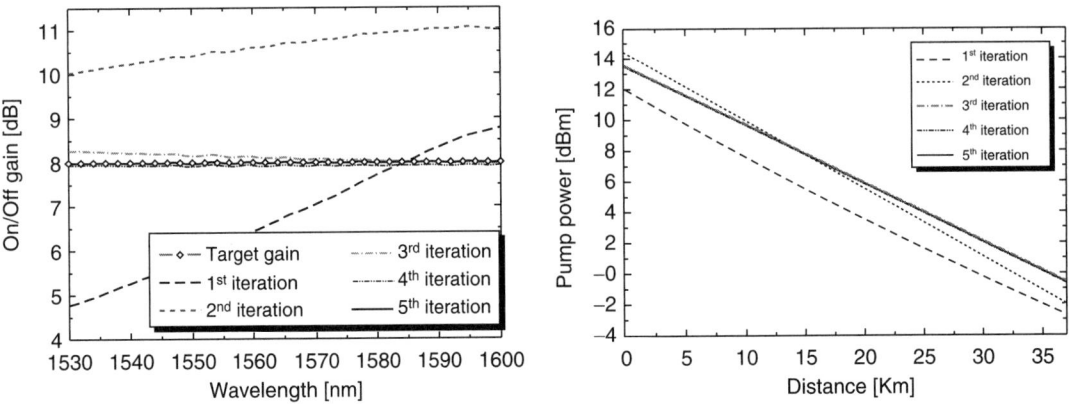

FIGURE 10.11 Gain profiles and pump evolutions of FRA obtained from the results at each iteration process.

over the whole 70-nm gain spectral range. The level of gain accuracy could be further improved with additional iteration steps by monitoring the convergence of $L_{\text{eff}\_ij}^{(1)}(z)$ or $<\Delta G>_{AVG}$, but considering most of the practical applications under experimental error, nine iteration steps were sufficient (for $<\Delta G>_{AVG}$ improvement < 0.002 dB). The required computation time to obtain this result after 10 steps of iteration was less than a few tenths of a second with a conventional PC (2.0-GHz CPU clock). For comparison, the common approach using direct integration of the Raman amplifier differential equation usually takes several minutes [11,12], with insufficient compensation of the signal stimulated Raman scattering (SRS) effects.

This unprecedented level of convergence speed is attributed to the following: (1) removing time-consuming coupled ODE equations in the optimization process; (2) transforming coupled Raman equations into closed a integral equation form, which enables the solution to be found along the iteration axis, not the fiber propagation axis; (3) using a matrix formulation for both the signal SRS effect and using an effective length iteration equation, enabling fast numerical process in the software; and (4) simultaneously applying the signal SRS effect and a pump integral update in every iteration step. Figure 10.12 illustrates design examples applied with this algorithm for various shapes of ideal target gain profiles (under an identical setting as that of Fig. 10.11).

3.5. Application Example 3: Channel Reconfiguration

Along with technological advancements in the optical link equipped with a dynamic lambda channel routing function, there have been many efforts to develop the control method for an optical amplifier

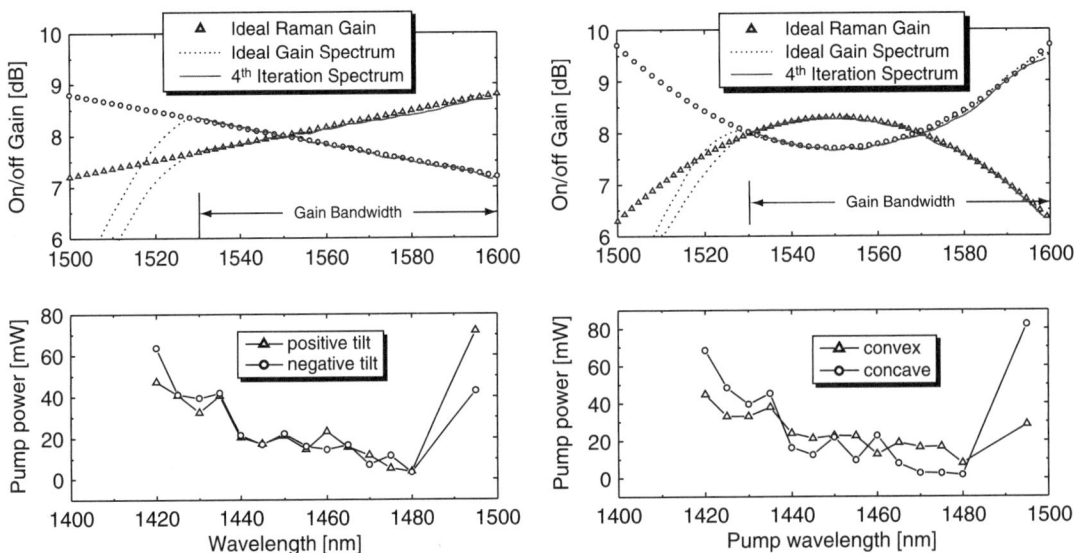

FIGURE 10.12 Examples of gain spectrum design (with corresponding pump powers).

to cope with the input power variation induced by WDM channel add/drop. Here, we apply the gain design algorithm described in the previous section to the gain clamping problem as another application example to find out the required pump power set under the dynamic network reconfiguration environment. To apply the technique described in the previous section and to obtain a stable and reasonable solution, the algorithm should be processed with surviving signal channels and with dropped signal channels, while naturally treating the power of the dropped signal channel as zero.

As an example, a wide-band FRA providing 11.8 dB average on–off gain to the 80-channel C/L-band WDM signal has been constructed using the algorithm presented before (C-band, 40 channels, 1529.2–1560.4 nm; L-band, 40 channels, 1570.0–1601.2 nm with 100 GHz spacing −10 dBm per channel). The required pump powers for the seven pumps (1420–1470 nm with 10-nm spacing and one additional pump at 1490 nm) were 135, 110, 55, 48, 51, 9, and 84 mW, respectively.

Figures 10.13 and 10.14 show the output spectrum of FRA with/without the pump adjustment. As illustrated, successful gain clamping has been achieved with the proper adjustment of pump powers, which were found by applying the algorithms described in the previous section. To compare, 0.5 dB of excursion in the gain was observed for surviving channels (40 channel drops-out of 80 channels), when no pump control was applied.

3.6. Application Example 4: Analytic Solution for the Gain Clamping Problem

Even if the gain clamping problem could be easily addressed with the methods described in the previous section, the dynamic control of FRA requires an algorithm that is much faster than the order of

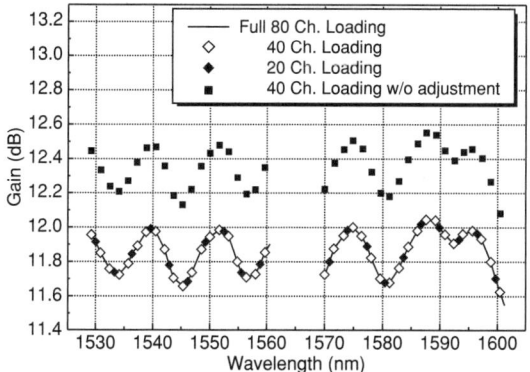

FIGURE 10.13 Gain spectrum with/without the adjustment of pump power (channels dropped interleaving).

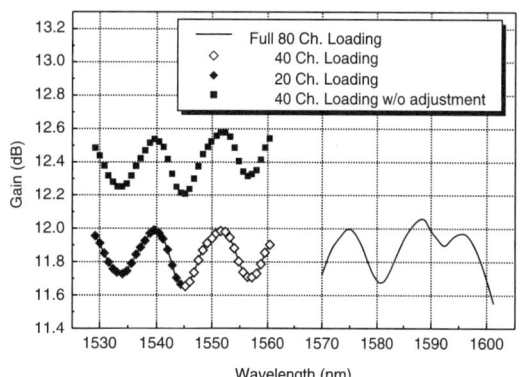

FIGURE 10.14 Gain spectrum with/without the adjustment of pump power (L-band channels all dropped out).

seconds. Restricting the given problem to practical applications where we use a small signal for the input of the Raman fiber amplifier, we now show that the simplified analytic solution enabling a faster calculation of required pump power set could be obtained. To begin, we rewrite the FRA gain in Eq. (8) to the following matrix form by splitting the pump/signal part:

$$\begin{bmatrix} G_p \\ G_s \end{bmatrix} = \begin{bmatrix} C_{pp} & C_{sp} \\ C_{ps} & C_{ss} \end{bmatrix} \begin{bmatrix} I_p \\ I_s \end{bmatrix} \quad (10)$$

where $G_p(G_s)$ is an $M \times 1$ ($N \times 1$) vector representing the Raman gain for pumps (signals), and I_p (I_s) is the vector constructed from the power integral of pumps (signals). Comparing with Eq. (8), it is easy to see that C is a constant matrix composed of elements $G_R^{(i,j)}$ (or, equivalently, the same as matrix T without L_{eff}). Under this formulation, the physical meanings of the submatrix C_{ij} become clear: C_{pp} is the pump-to-pump interaction, C_{ss} is the signal-to-signal interaction (signal SRS), C_{sp} is the signal-to-pump interaction (pump depletion), and C_{ps} is the pump-to-signal interaction (signal amplification).

On the other hand, we note that the following equation should be satisfied if one wants to keep the gain of the surviving channel unchanged under the channel reconfiguration environment:

$$\Delta G_s = C_{ps}\Delta I_p + C_{ss}\Delta I_s = 0 \quad (11)$$

where ΔG_s, ΔI_s, and ΔI_p are the change in the signal gain and change in the integral of the signal and pump power, respectively. If we require that ΔG_s be equal to zero under channel reconfiguration, we then obtain,

$$\Delta I_p = -(C_{ps})^{-1} C_{ss}\Delta I_s \quad (12)$$

Applying Eq. (12) to the gain equation of the pump, the relation between ΔG_p and ΔI_s becomes

$$\Delta G_{\mathrm{p}} = C_{\mathrm{pp}}\Delta I_{\mathrm{p}} + C_{\mathrm{sp}}\Delta I_{\mathrm{s}} = \left\{-C_{\mathrm{pp}}(C_{\mathrm{ps}})^{-1}C_{\mathrm{ss}} + C_{sp}\right\}\Delta I_{\mathrm{s}} \quad (13)$$

Meanwhile, using the definition of effective length, the variation of pump power integral and pump gain can be related to the variation of the pump power $\Delta P_p(0)$ (note that is this the final target solution) as follows:

$$L_{\mathrm{eff,\,p}} + \Delta L_{\mathrm{eff,\,p}} = \frac{\int_0^L P_{\mathrm{p}}(z) + \Delta P_{\mathrm{p}}(z)dz}{P_{\mathrm{p}}(0) + \Delta P_{\mathrm{p}}(0)} \quad (14)$$
$$= \frac{I_{\mathrm{p}} + \Delta I_{\mathrm{p}}}{P_{\mathrm{p}}(0) + \Delta P_{\mathrm{p}}(0)}$$

$$\therefore \Delta P_{\mathrm{p}}(0) = \frac{\Delta I_{\mathrm{p}} - \Delta L_{\mathrm{eff,\,p}} P_{\mathrm{p}}(0)}{L_{\mathrm{eff,\,p}} + \Delta L_{\mathrm{eff,\,p}}} \quad (\because, I_{\mathrm{p}} = L_{\mathrm{eff,\,p}} P_{\mathrm{p}}(0)) \quad (15)$$

where ΔI_p is a known quantity from Eq. (12), $P_p(0)$ is the input pump power before the reconfiguration, and $L_{\mathrm{eff,\,p}}$ is the effective length (vector) from the initial setup (which we know).

Now, to get the (undetermined) value for the variation in the effective length $\Delta L_{\mathrm{eff,\,p}}$, we express the effective length formula in terms of the pump gain and its differential value which we know from Eq. (13), and finally get

$$L_{\mathrm{eff,\,p}} + \Delta L_{\mathrm{eff,\,p}}$$
$$= \int_0^L \left[\frac{(P_{\mathrm{p}}(0) + \Delta P_{\mathrm{p}}(0))\exp\{-\alpha_{\mathrm{p}}z + (G_p(z) + \Delta G_p(z))\}}{P_{\mathrm{p}}(0) + \Delta P_{\mathrm{p}}(0)}\right] dz \quad (16)$$

and

$$\Delta L_{\mathrm{eff,\,p}} = -L_{\mathrm{eff,\,p}} + \int_0^L \exp\{-\alpha_{\mathrm{p}}z + (G_p(z) + \Delta G_p(z))\} dz \quad (17)$$

Now, restricting the given problem to the undepleted pump regime, the signal power evolutions along the fiber should not be different from that of original operating condition even after the channel reconfiguration. For this case, the given problem can be easily solved, as we can determine the vector ΔI_s which can be used in Eq. (13) along with Eq. (17) to calculate $\Delta L_{\mathrm{eff,\,p}}$, and in Eq. (12) to get ΔI_p, and finally to solve for the required adjustment in the pump power $\Delta P_p(0)$ using Eq. (15). (Note that ΔI_s of dropped channels are naturally $-I_s$, and ΔI_s of surviving channels are zero in this approximation. The tested error $\Delta I_s/I_s$ was less than 0.2 dB for 78 channels dropping out of 80 in all examples.

Because this analytic approach does not require any iteration process, in contrast to the examples of the previous section, but only simple calculation from the known parameters of FRA (before the reconfiguration) and changes in the signal power, the whole procedure for the search of the proper pump power sets required for the gain control can be achieved with orders of higher speed (approximately a few tenths of a microsecond).

As an example, for the identical reconfiguration scenario used in the previous section, we calculated the pump power sets from the above equations. The gain spectrums from this adjusted pump power set are illustrated in Figs. 10.15 and 10.16. Up to the input power change of 6 dB (60 from 80 channel drop-out), the residual gain excursion was suppressed within 0.2 dB with this method. It is worth noting that even if the small signal approximation were applied to get the above analytic result, the observed gain error was still smaller than 0.5 dB when the same test was carried out with an increased signal power of −5 dBm per channel.

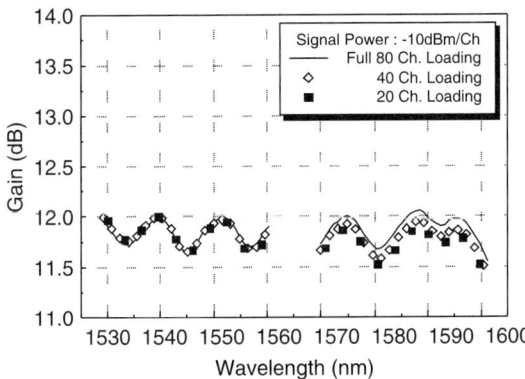

FIGURE 10.15 Gain spectrum of surviving channels, with the adjustment of pump power (analytic algorithm).

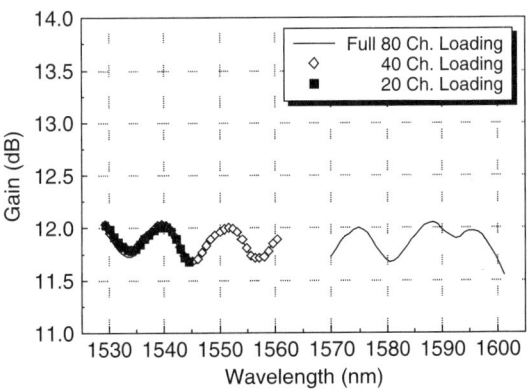

FIGURE 10.16 Gain spectrum with the adjustment of pump (analytic algorithm, L-band channel all dropped).

3.7. Conclusion

In this section we introduced a novel FRA modeling framework by transforming the complicated differential Raman equation into a closed integral/matrix form. By taking the effective length as the interim target solution and updating its value along the iteration axis, Raman gain and pump/signal power evolution can be obtained with orders of magnitude increases in convergence speed and excellent accuracy when compared with the previous coupled ODE-based approaches. Application of this formalism to the problems of (1) gain spectrum engineering for the search of optimum pump power set under the given constraints, (2) gain clamping for the channel reconfiguration, and (3) derivation of analytic formula for the faster FRA dynamic control have been addressed, with excellent accuracy and conversion speed to assist the high-level practical design of the FRA.

4. TRANSIENT THULIUM-DOPED FIBER AMPLIFIER

4.1. Introduction

To prepare for the increased data traffic in the future, there have been serious efforts to expand the gain bandwidth of the amplifier, extending to the S/S+-band (1530–1430 nm). Among various approaches to achieve this mission (including FRA and EDFA), thulium-doped fiber amplifiers (TDFAs) are distinguished by the excellent positioning of the Tm luminescence band and a much higher power efficiency. Even with the increased reports for S- and S+-band TDFAs [19–23], complexities in the dynamics and difficulties in the characterization of experimental parameters—from the lack of materials and characterization tools—has restricted the prediction of TDFA performances to a static domain that still has limited accuracy.

In this section, we use a reduced form of differential equations for TDFA to analyze its transient responses—a parameter of serious importance in an optically switched network. Through a numerical analysis, we show that the transient response of the TDFA should differ from that of an EDFA in terms of the numbers and magnitudes of time constants involved with the transition. Experimental verification with a 1.4/1.5 μm dual wavelength pumped TDFA confirms the existence of two different time constants (∼60 μs, vs 8 ms) involved with the transient, with a close relationship to the characteristics lifetimes ($^3F_4$ and $^3H_4$) of the thulium ion.

4.2. Average Inversion Analysis of TDFA Transient: Comparison with Experiment

The complete set of TDFA rate equations [19] involves too many transition processes over extremely wide wavelength windows to be applied to transient analysis. To use established time domain approaches such as average inversion analysis [24] to TDFA with reasonable accuracy and computation speed, we first need to reduce the full set of equations into a simpler form. At first, Fig. 10.17 shows the energy levels of the Tm ion, the transition of 1.4/1.55 μm dual-wavelength pumping, stimulated emission for S-Band amplification, and cross-relaxation (CR) [25]. Writing N_i as the population density of energy level i, A_{ij} as the spontaneous emission probability from energy levels i to j given in [19, 23], and W_{ij} as the absorption/emission transition rate between levels i and j, we get the following set of equations after ignoring relatively weaker effects, such as the up-conversion and multiphoton decay related with 2,4,5 levels.

$$\frac{dN_0}{dt} = -W_{01} \cdot N_0 + A_{10} \cdot N_1 + A_{30} \cdot N_3$$

$$\frac{dN_1}{dt} = W_{01} \cdot N_0 - (A_{10} + W_{13}) \cdot N_1 + (A_{31} + W_{31}) \cdot N_3 + 2CR$$

$$\frac{dN_3}{dt} = W_{13} \cdot N_1 - (A_{30} + A_{31} + W_{31}) \cdot N_3 - CR$$

$$CR = k_{3101} N_3 N_0 - k_{1310}(N_1)^2$$

Here, N is the total population density ($N = N_0 + N_1 + N_3$) and W_{ij} is given by

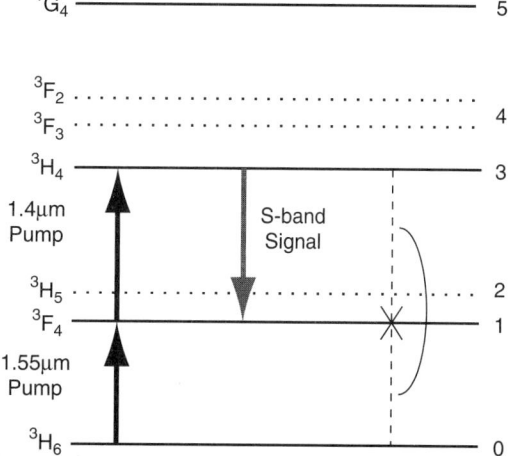

FIGURE 10.17 Diagram of energy level of Tm ion and transition processes for 1.4/1.55 μm dual-wavelength pumped TDFA. CR: cross-relaxation

$$W_{ij} = \int_\lambda \frac{\Gamma(\lambda)\sigma_{ij}(\lambda)P(\lambda)}{h\upsilon\, A_{\text{eff}}(\lambda)} d\lambda$$

where $\Gamma(\lambda)$ is the overlap factor, $\sigma_{ij}(\lambda)$ is the cross-section, $P(\lambda)$ is the optical intensity, and $A_{\text{eff}}(\lambda)$ is the effective interaction area at the wavelength of λ.

For the analysis, we used four signals channels (1460, 1470, 1480, and 1490 nm) and 20m of Tm-doped Zr-based fluoride fiber, with a doping concentration of 2000 ppm and n value of 2.5%, respectively. Figure 10.18 shows the transient response curves for the surviving channel at 1490 nm, under the 1480 nm channel amplitude on–off modulation. Numerical analysis then was carried out, assuming fixed main pump power (210 mW) and different levels of subsidiary pump power. For simulation parameters, we used A, S, and from [20] and emission/absorption across-section from [26]. The k_{3101} and $k_{3101}\, k_{1310}$ are set as 3.6×10^{-23} and 3×10^{-25} m^3/s, respectively normalized from [27]. As can be seen from Fig. 10.18, a relatively slow (4–10 ms) transient time constant related with the $^3F_4$ level was observed, following the fast overshoot from the $^3H_4$ level.

Also observed is the variation of excursion levels (at steady state), with strong dependence on the amount of subsidiary pump power. When the subsidiary pump power is (relatively) smaller than the main pump power, 1.5-μm pump photons excite just enough ions to reach the first excited state (or signal ground state, $^3F_4$), where the relatively larger number of main pump photons generate participating ions ($^3F_4 \to {}^3H_4$) for the amplification process. In this regime [e.g., state (v) (iv) → (iii) → (ii) in Fig. 10.18], with the larger number of excited ions in the amplifier, the surviving channel begins to experience a higher level of excursions as the subsidiary pump power increases. Increasing further the number of subsidiary pump photons (beyond what can be consumed with main pump photons, via $^3F_4$ level ions), the signal inversion level begins to decrease, resulting in the reduction of the final gain value after the transient [state (ii) → (i) in Fig. 10.17] [21]. In this regime, the TDFA response curve becomes similar to that of the EDFA.

Simulation results have been compared with the experiment using the set up in Fig. 10.19. Four laser diodes (1460, 1470, 1480, and 1490 nm; −6 dBm each) were used as TDFA saturating signal sources, and three high power laser diodes (stabilized with fiber Bragg grating, 1420, 1427, 1435 nm:total power of 210 mW) were used as main pumping sources at the 1.4 μm band [20, 22, 23]. As a subsidiary pump, a 1.5-μm tunable laser with a high power Er/Yb codoped fiber amplifier (pumped with a high power fiber laser at 1060 nm) was used.

While applying an on–off modulation to the 1480-nm signal laser diode with a function generator, a tunable band-pass filter was also tuned to the

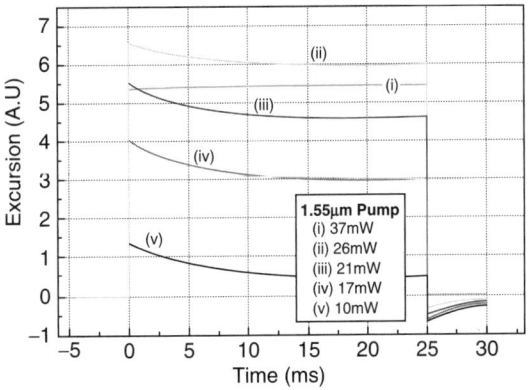

FIGURE 10.18 Surviving channel (1490 nm) response under on–off modulation of 1480-nm signal (simulation).

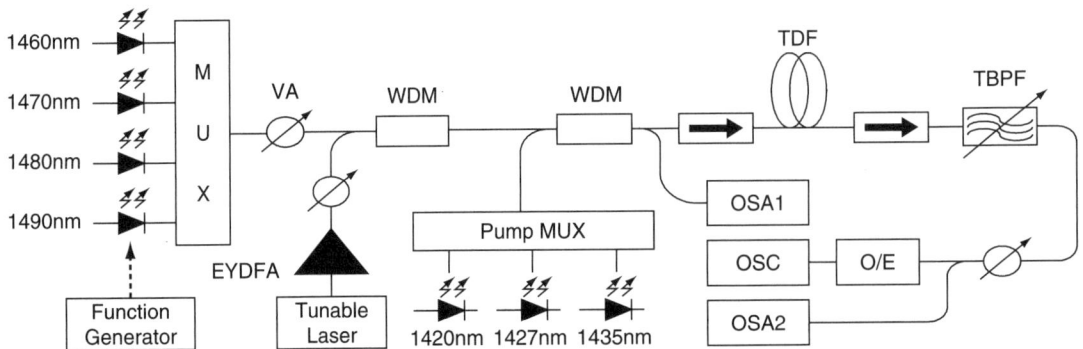

FIGURE 10.19 Setup (VA, variable attenuator; EYDFA, Er/Yb codoped fiber amplifier; TDF, thulium-doped fiber; TBPF, tunable bandpass filter; O/E, O-E converter; OSC, oscilloscope, OSA, optical spectrum analyzer).

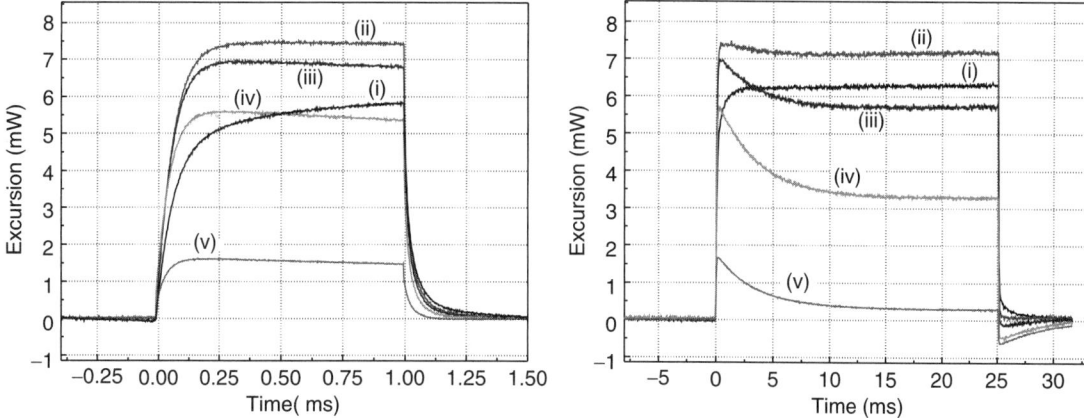

FIGURE 10.20 Surviving channel (1490 nm) transient response under on–off modulation (a) at 500 Hz and (b) at 20 Hz, with 1550 nm subsidiary pump/signal power (mW): (i) 37/14.9, (ii) 26/15.2, (iii) 21/14.1, (iv) 17/13.4, (v) 10.5/8.9

surviving channel wavelength (1490 nm) to monitor the transient responses. Figure 10.20a shows the measured transient response of the 1490-nm signal channel under 500 Hz modulation, with different levels of 1550-nm subsidiary pump power. Looking similar to the transient behavior from a conventional EDFA, the measured TDFA response curve showed a characteristic time of about 100 µs, for both rising and falling edges. We attribute this time constant to a spontaneous lifetime for the second excited state ($^3H_4$) of TDFA (Fig. 10.17. $\tau_{31} \approx$ 1 ms; to compare, EDFA $\approx$ 10 ms). However, when viewed, in detail with a slower (20-Hz) modulation frequency to the 1480nm signal LD (Fig. 10.20b), it was possible to observe another time constant (~8 ms) after the large overshoot, as predicted from the numerical analysis (Fig. 10.18).

4.3. Conclusion

In this section, we characterized the transient responses of TDFA as a function of a 1.55-µm subsidiary pump power by using a reduced form of the rate equation. Results show the signature of the first excited state ($^3F_4$) lifetime, making transient response characteristics differ from that of an EDFA. We also showed that the transient response of the TDFA depends on the relative strength between main/subsidiary pump power and signal power. Different contributions for the transition response curve from $^3H_4$ and $^3F_4$ level of the TDFA with ~60 µs and ~8 ms characteristic times have been found to explain these phenomena. These results imply the need of a different algorithm for the transient control of TDFA when compared with that of the EDFA, which involves just one time constant and a fixed number of ions for the inversion process.

5. CONCLUSION

In this chapter, we showed application examples of numerical analysis techniques for optical amplifiers, at different levels of investigation and for different types of amplifiers (EDFA, FRA, and TDFA), which all have different amplification dynamics and gain bands. Different numerical analysis techniques, suitable for each subject of key interest have been used either to gain useful insights, or to address important practical applications, which otherwise are difficult to approach with experimental techniques. Even if the demonstration was provided for individual types of amplifiers, its application to hybrid-type amplifiers should be straightforward. These wavelength domain data, obtained from the above described techniques, also can be further applied to the time domain transmission simulator to find correct Q factors and to estimate the power margin/bit error rate of transmission signal under the WDM environment.

6. REFERENCES

1. B. Min, et al., "Coupled structure for wideband EDFA with gain and noise figure improvements from C to L-band ASE injection," *IEEE Photon. Technol. Lett.*, **12**, No. 5, January 1999.
2. Y. Zhang, et al., "Wavelength and power dependence of injected C-band laser on pump conversion efficiency of L-band EDFA," *IEEE Photon. Technol. Lett.*, **14**, No. 3, March 2002.
3. F. A. Flood, et al., "980nm Pump-band wavelengths for long-wavelength-band erbium-doped fiber

amplifiers," *IEEE Photon. Technol. Lett.*, **11**, No. 10, October 1999.

4. R. Di Muro, et al., "Dependence of L-band amplifier efficiency on pump wavelength and amplifier design," *Optical Fiber Communication Conference*, **2**, 2000.

5. H. Lee, et al., "Structural Detuning of Absorption Rate in Doped Fiber for the Enhancement of Power Efficiency," *IEEE Photon. Technol. Lett.*, **16**, No. 6, pp. 1468–1470, June 2004

6. E. Desurvire, *Erbium Doped Fiber Amplifier: Principles and Applications*, New York: Wiley, 1994.

7. R. M. Percival, et al., "Erbium-doped fiber amplifier with constant gain for pump wavelengths between 966 and 1004 nm," *Electron. Lett.*, **27**, No. 14, July 1991.

8. C. Giles, et al., "Modeling erbium-doped fiber amplifiers," *IEEE J. Lightwave Technol.*, **9**, No. 2, February 1991.

9. M. N. Islam, "Raman amplifier for telecommunications," *IEEE J. Lightwave Technol.*, **8**, pp. 548–559, May–June 2002.

10. H. Kidorf, K. Rottwitt, M. Nissov, M. Ma, and E. Rabarijaona, "Pump interactions in a 100-nm bandwidth Raman amplifier," *IEEE Photon. Technol. Lett.*, **11**, pp. 530–532, May, 1999.

11. B. Min, W. J. Lee, and N. Park, "Efficient formulation of Raman amplifier propagation equations with average power analysis," *IEEE Photon. Technol. Lett.*, **12**, pp. 1486–1488, November, 2000.

12. V. E. Perlin and H. G. Winful, "Optimal design of flat-gain wide-band fiber Raman amplifiers," *IEEE J. Lightwave Technol.*, **20**, pp. 250–252, February, 2002.

13. L. K. Choi, P. Kim, J. Park, J. Park, and N. Park "Adiabatic, closed form approach for the highly efficient analysis of fiber Raman amplifier problem," *Optics Lett.*, **30**, pp. 126–128, January, 2005.

14. J. Park, P. Kim, J. Park, H. Lee, and N. Park, "Closed integral form expansion of Raman equation for efficient gain optimization process," *IEEE Photon. Technol. Lett.*, **16**, pp. 1649–1651, July 2004.

15. N. Newbury, "Full wavelength dependence of Raman gain in optical fibers: measurements using a single pump laser," *Opt. Fiber Commun. Tech. Dig.*, **1**, 309, (2003).

16. K. Rottwitt and A. J. Stentz, "Raman amplification in lightwave communication systems," in *Opt. Fiber Telecommun. IVA*, I. P. Kaminow and L. Tingye, Eds. 217, Academic, San Diego, 2002.

17. Y. Emori, S. Kado, and S. Namiki, "Simple gain control method for broadband Raman amplifiers gain-flattened by multi-wavelength pumping," *European Conference on Optical Communication 2001*, **2**, pp. 158–159, 2001.

18. P. Kim, J. Park, H. Yoon, J. Park, and N. Park, "In-situ Design Method for Multi-channel Gain of Distributed Raman Amplifier with the multi-wave OTDR," *IEEE Photon. Technol. Lett.*, **14**, pp. 1683–1685, December 2002.

19. T. Komukai, T. Yamamoto, T. Sugawa, and Y. Miyajima, "Upconversion pumped thulium-doped fluoride fiber amplifier and laser operating at 1.47 μm," *IEEE J. Quantum Electron.*, **31**, No. 11, pp. 1880–1889, 1995.

20. T. Kasamatsu, Y. Yano, and T. Ono, et al., "Laser-diode-pumped highly-efficient gain-shifted thulium-doped fiber amplifier operating in the 1480–1510-nm band," *Optical Fiber Communication Conference*, TuQ4, 2001.

21. W. J. Lee, et al., "Study on the pumping wavelength dependency of s+ band fluoride based thulium doped fiber amplifiers," *IEEE Optical Fiber Communication Conference*, TuQ5, OFC 2001.

22. J. Byun, et al., "Analysis on the Transient Response of 1.4um/1.5um dual wavelength pumped thulium doped fiber amplifiers," *Photon. Tech. Lett.*, **14**, No.11, pp. 1503–1505, 2002.

23. W. J. Lee, et al., "Study on the gain excursion & tilt compensation for 1.4/1.5 um dual wavelength pumped TDFA," *IEEE Photon. Tech. Lett.*, **14**, No. 6, pp. 786–788, 2002.

24. T. G. Hodgkinson, "Improved average power analysis technique for erbium-doped fiber amplifiers," *IEEE Photon. Technol. Lett.*, **4**, pp. 1273–1275, November 1992.

25. P. Kim, Y. J. Jung, H. S. Suh, and N. Park, "Transient analysis of 1.4/1.55 μm dual-wavelength pumped TDFA with average inversion method," paper 7P-044, *OptoElectronics and Communications Conference*, Seoul, South Korea, July, 2005.

26. H. Inoue, K. Soga, and A. Makishima, "Simulation of the optical properties of Tm:ZBLAN glass," *J. Non-Crystalline Sol.*, **306**, No. 1, pp. 17–29, 2002.

27. S. D. Jackson and T. A. King, "Theoretical modeling of Tm-doped silica fiber lasers," *J. Lightwave Techonol.*, **17**, pp. 948–956, May, 1999.

11
CHAPTER

Analog/Digital Transmission with High-Power Fiber Amplifiers

Puneit Dua, Kunzhong Lu, Niloy K. Dutta,* and James Jacques
Department of Physics
University of Connecticut
Storrs, Connecticut, USA

James Jaques
Lucent Technologies
Government Communications Lab
Murray Hill, New Jersey, USA

1. INTRODUCTION

The optically transmitted multichannel microwave signals, like those in the community antenna television (CATV) systems, use the subcarrier multiplexing (SCM) technique. The fiber optic analog SCM systems are amplitude modulated with vestigial sideband (AM-VSB), and the VSB modulation technique helps eliminate fiber dispersion effect [1] and provides dispersion compensation [2]. Also, these systems provide flexibility and the upgradability required in the design of broadband networks [3]. Such CATV systems can be analog and/or digitally modulated to transmit multiple voice, data, and video signals to users; they are evolving from analog broadcast systems to communications networks supporting analog/digital video, two-way, fully interactive multimedia services platforms and a variety of other targeted services, such as video-on-demand to homes and business in a much larger area [4, 5] than before.

Because the waveform of an analog signal must be preserved during transmission, analog SCM systems require a high carrier-to-noise ratio (CNR) at the receiver and impose strict linearity requirements on the optical source and the communication channel. Thus, the optical CATV applications at 1.5 μm require very high powers for overcoming large splitting losses and the relatively large input power necessary at the receiver. High-power fiber optic amplifiers, which are compatible with fiber optical communication system and also offer other advantages such as high efficiency, compactness, and reliability, are used to compensate for the increases in power budget. The cladding pumped double-clad fiber amplifier (DCFA) provides such a unique solution. The performance of an analog and a hybrid analog/digital transmission system using an Er/Yb co-doped high-power cladding pumped fiber amplifier was investigated. The system performance parameters, namely bit error rate (BER), CNR, composite second order (CSO) and composite triple beat (CTB) distortion without and with the DCFA, were measured and compared. In such AM-SCM cable TV systems with the optical fiber amplifiers, the CSO degradation arises largely due to the AC (modulated signal) gain tilt in the fiber amplifier. We also experimentally measured the gain tilt of the high-power erbium/ytterbium co-doped DCFA, based on the method proposed by K. Kikushima [6] for the erbium-doped fiber amplifier (EDFA). The effect of the gain tilt on the CSO of the AM-SCM transmission system using the DCFA is calculated and compared with the direct measurement of the second-order distortion in the system.

2. EXPERIMENT

2.1. Analog Transmission

The setup of a subcarrier multiplexed CATV distribution experiment is shown in Fig. 11.1. The 60-channel NTSC (National Telecommunication Standards Committee), AM-VSB (amplitude modulated vestigal side band) video signals were generated

*E-mail:nkd@phys.uconn.edu

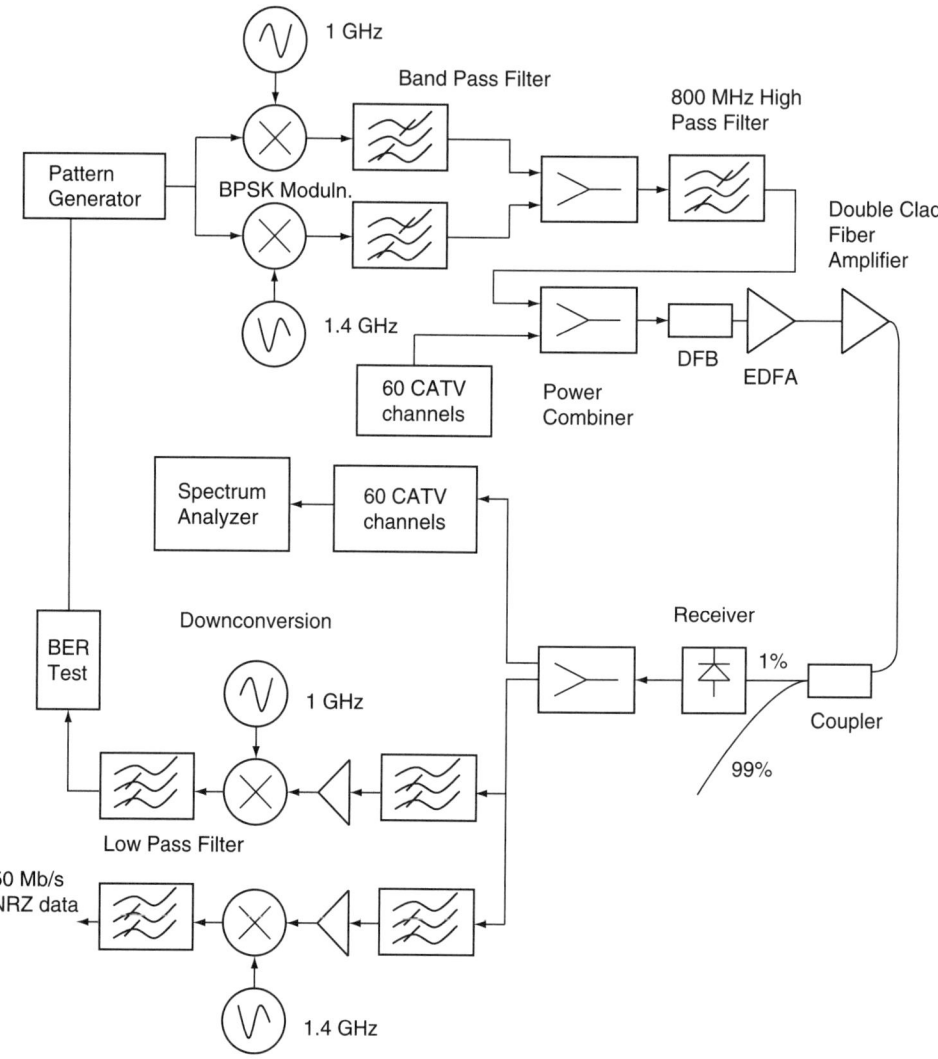

FIGURE 11.1 Experimental setup of subcarrier multiplexed CATV distribution experiment. NRZ, nonreturn to zero.

using Matrix Test Equipment Inc. ASX-16B and had a frequency range of 55.25–445.25 MHz with a 6-MHz separation. Two nonreturn to zero data stream channels at 50 Mb/s were mixed using binary phase shift keying (BPSK) modulation techniques with two carriers at 1.0 and 1.4 GHz, respectively [4]. In the BPSK modulation scheme, the digital data and an RF analog signal generated by a local oscillator were mixed in a double-balance mixer. After the mixing, the phase is shifted 180 degrees between transmission of 0 and 1, and the optical intensity remains constant during all bits and the signal appears to have a continuous waves format. The up-conversion expands the bandwidth of the data. Two bandpass filters with 300 MHz bandwidth and 1 and 1.4 GHz center frequencies, respectively, were placed after each mixer to remove the higher harmonics. The two up-con-

verted signals were then combined and passed through another high-pass filter. The transmitted signal was obtained by combining the two digital subscriber signals and the 60-channel AM-VSB analog signals and was used to directly modulate a distributed feedback laser. RF amplifiers and attenuators (not shown) were added to enhance the mixing and to obtain the required RF spectrum for transmission.

The output from the distributed feedback (DFB) laser was preamplified by an EDFA and then boosted by the high-power erbium/ytterbium co-doped DCFA; ~1 W of output power was obtained from the DCFA. To simulate the distribution system, a 1% coupler was used to simulate the distribution to 100 users. A linear p-i-n InGaAs photodiode followed by a transimpedance amplifier was used to convert the optical signal back to electrical form. The recovered electrical signal was split into two branches, one for digital channels and the other for analog channels. Down-conversion of the digital subcarrier channels was performed to retrieve the data from the subcarrier channel. The local oscillators for the down-conversion (1.0 and 1.4 GHz) must be phase locked to the incoming RF carrier; therefore, we used the same local oscillators as upconversion. The BER test as a function of the modulation depth was obtained by inserting a tunable attenuator between the mixer (modulator) of the digital channel under test and the combiner for combining the two digital channels. The recovered analog channels were fed to a standard CATV test system consisting of a tunable filter and a spectrum analyzer. The measurement of CNR, CSO, and CTB of selected channels was carried out. The experiment was repeated without the DCFA and the 1% coupler as a comparison. Both tests were performed with the same received power.

The BER test results of the digital channels, at 1.0 and 1.4 GHz, are shown in Fig. 11.2. The nonlinearity due to the BPSK signal can be calculated using the Gaussian approximation of Saleh [7]. The change in carrier-to-nonlinear distortion ratio (C/NLD) is dependent on the number of channels N and the optical modulation depth m and is given by

$$\Delta\left(\frac{C}{NLD}\right) = -20\log\left(1 + \frac{m_{BPSK}^2}{4Nm^2}\right) \quad (1)$$

For $m_{BPSK} = 40\%$, $N = 60$, and $m = 3.5\%$, the change in distortion is approximately 3 dB, which is in qualitative agreement with the data measured.

The performance of a multichannel transmission system is evaluated on the basis of the measured values of CNR, CSO, and CTB. The system CNR after the in-line amplifier can be written as [8]

$$CNR = \frac{m^2}{RIN + \frac{2h\nu F_n}{P_i} + \left(\frac{h\nu F_n}{P_i}\right)^2 B_0 + \frac{2e}{RP_r} + \frac{ith^2}{(RP_r)^2}} \frac{1}{2B_e} \quad (2)$$

where m is the optical modulation depth (OMD) per channel, RIN is the relative intensity noise of the laser transmitter, F_n is the noise figure of the in-line amplifier, R is the receiver responsivity, B_0 is the optical bandwidth of the received signal, B_e is the electrical bandwidth of each video channel which is 4 MHz in an AM-VSB system, ith^2 is the receiver spectral noise power density, P_i is the amplifier input power, and P_r is the optical power at the receiver.

The second and third term of the denominator represent signal–spontaneous and spontaneous–spontaneous beat noise in the amplifier. The system CNR after the in-line amplifier is governed mainly by the amplifier signal–spontaneous beat noise at lower input levels. When the input level is increased until reaching the amplifier saturation region, the CNR is limited by the noise of the receiver. Therefore, it is desirable to have the DCFA operate in the saturation region to minimize the CNR penalty due to the spontaneous noise. The optical power should be as large as possible [9]. The optical power at the receiver was kept at 3 dBm in our experiment to ensure the receiver was shot noise limited. The upper limit of the system CNR is set by transmitter RIN and RF loading (i.e., OMD). A large OMD is desired to obtain large CNR and reduce the laser RIN, but the expression for CNR, Eq. (2), does not take into account the distortion noise. A large OMD produces nonlinear distortions from resonance distortion, clipping effects, and spatial hole burning [10]. In our experiment the laser bias was 35 mA, the point of minimum second-order distortion, and the OMD

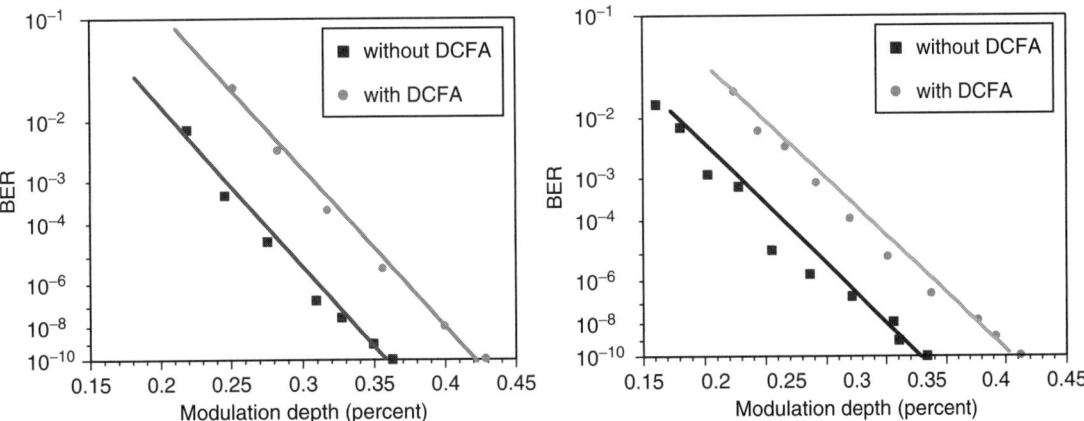

FIGURE 11.2 BER for the 50-Mb/s data channel with subcarrier at 1.0 GHz (left) and 1.4 GHz (right).

was 3.5%. The total RMS modulation index was $\mu = \sqrt{(Nm^2/2)} = 19.2\%$, where N is the total number of channels loaded, which was 60 in our experiment. Figure 11.3 shows the CNR test results of 60 CATV channels with and without the DCFA.

The composite signal obtained by summing all the modulated subcarriers is used to modulate the intensity of the DFB laser. The transmitted power can be written as [3]

$$P(t) = P_b \left[1 + \sum_{j=1}^{N} m_j a_j \cos(2\pi f_j t + \phi_j) \right] \quad (3)$$

where P_b is the output power at the bias level and $m_j, a_j, f_j,$ and ϕ_j are, respectively, the modulation index, amplitude, frequency, and phase associated with the jth microwave subcarrier. The power received is also given by Eq. (3), except for an overall power reduction occurring because of transmission losses if the communication channel is perfectly linear. In practice, the signal is distorted because of a deviation from linearity. Such a distortion is referred to as intermodulation distortion. Any nonlinearity in the response of the laser or in the propagation characteristics of fibers generates new frequencies of the form $f_i + f_j$ and $f_i + f_j \pm f_k$, some of which lie within the transmission bandwidth and distort the analog signal.

With up to 60 channels, tens of second order and over 1000 third-order distortion products are generated within each channel [9]. The optical CATV systems with fiber optic amplifiers experience second order distortion due to the interaction between the gain tilt of the amplifier and the laser chirp. The third order generation in a directly modulated DFB is due to the dynamic characteristic curve of the laser and the CATV hybrid used to drive it [11]. The usual measurement technique when quantifying this type of distortion is to use a filter bandwidth of 30 KHz, which is wide enough to contain the entire range of significant products. The power level of the second- and third-order distortion products, known as CSO and CTB, for a specific channel are normalized to the carrier power of that channel and expressed in dBc units (i.e., decibel relative to the carrier power level in dB, which is always a negative number). The experimental results obtained are shown in Fig. 11.4.

2.2. Hybrid Digital/Analog Transmission

A hybrid transmission system allows for increased communication services over the existing network, and digital signals are transmitted along with the AM video channels. This is achieved by substituting a digital channel in the place of one or more of the existing AM channels [4]. The experimental setup for the hybrid AM-VSB/digital fiber optic system is shown in Fig. 11.5. The digital channel at 2 Mb/s that was BPSK modulated is substituted in place of channel no. 23 at 223.25 MHz by using a double-balanced mixer. A tunable 4-MHz filter was used to isolate the BPSK data channel from the adjacent CATV channels. The BPSK data stream was then combined with the other 59 standard CATV channels and used to directly modulate a 1.5-μm DFB laser. As before, the output from the DFB laser was preamplified by an EDFA and then boosted to about 1 W by the high-power DCFA and a 1% coupler was used to simulate the distribution to 100 users. The receiver and demodulation network was used to convert the optical signal back to the elec-

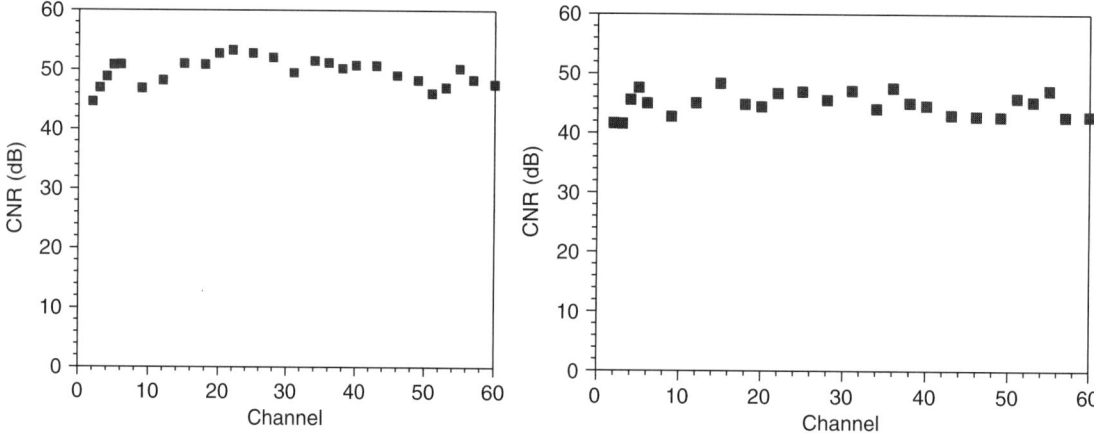

FIGURE 11.3 CNR for the various CATV channels without and with the DCFA.

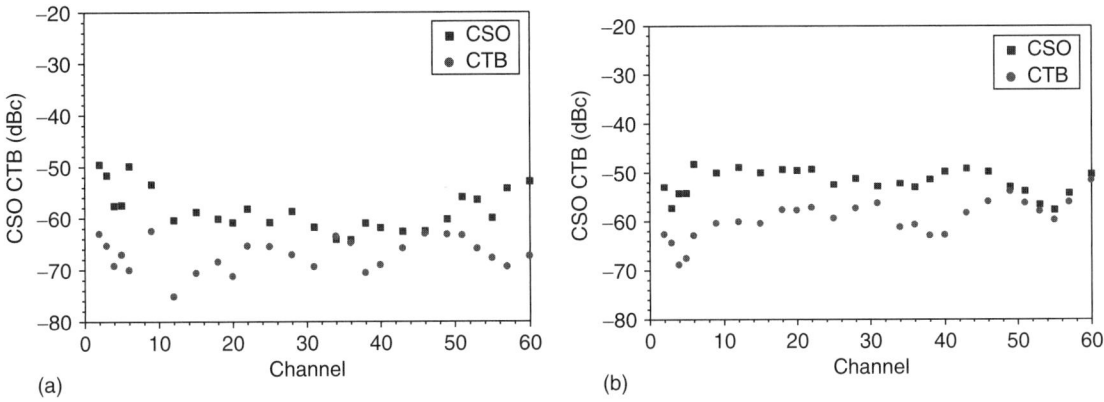

FIGURE 11.4 (a) CSO and CTB of various CATV channel without the DCFA. (b) CSO and CTB of various CATV channel with the DCFA.

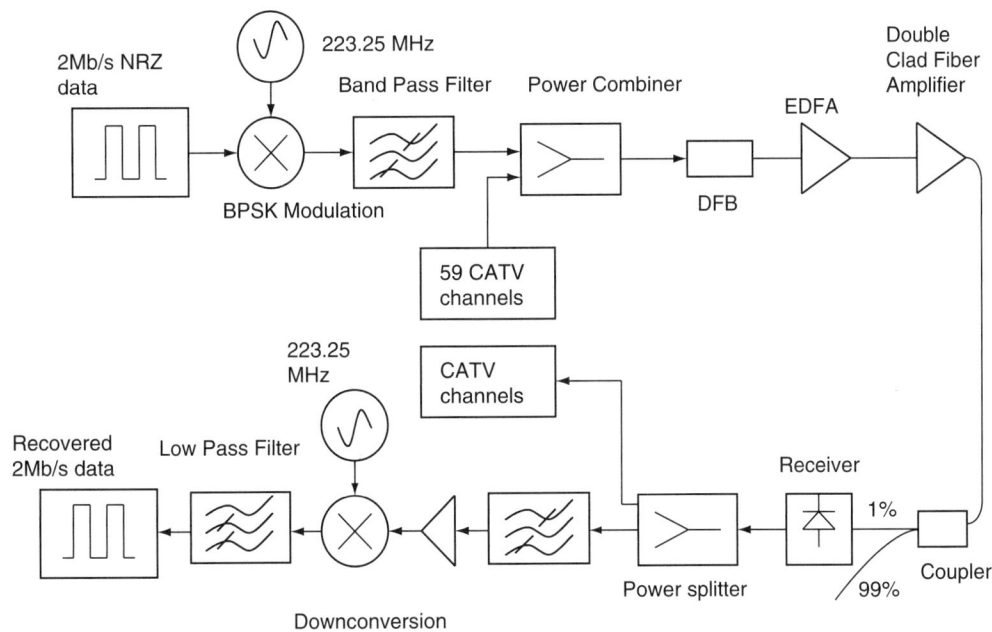

FIGURE 11.5 Experimental setup for hybrid AM-VSB/digital fiber optic transmission system. NRZ, nonreturn to zero.

trical form. The digital data was reconstructed by isolating the BPSK carrier using a bandpass filter, amplifying the signal, and then down-converting the data by mixing with the signal identical to the original carrier frequency.

The measured BER as a function of the modulation depth of the BPSK in presence of the AM carriers with modulation depth of 2.8% is shown in Fig. 11.6. Figure 11.7 shows the measurement of CNR, and Fig. 11.8 shows the measurements of intermodulation distortions CSO and CTB as a function of the various CATV channels in the presence of the 2 Mb/s digital carrier and the DCFA.

2.3. Gain Tilt Measurement of the Er/Yb Co-Doped DCFA

The frequency chirp inherent to a directly modulated transmitter can result in distortion due to interaction with the system components that may have properties dependent on the variation of optical frequency [12]. The dynamic gain tilt of a doped fiber amplifier is one such parameter. The doped fiber amplifiers have nonuniform gain that varies with wavelength of the input signal. The distortion introduced by a doped fiber amplifier for a transmitter with chirp arises due to the departure from flatness of the gain spectrum, and different signal levels

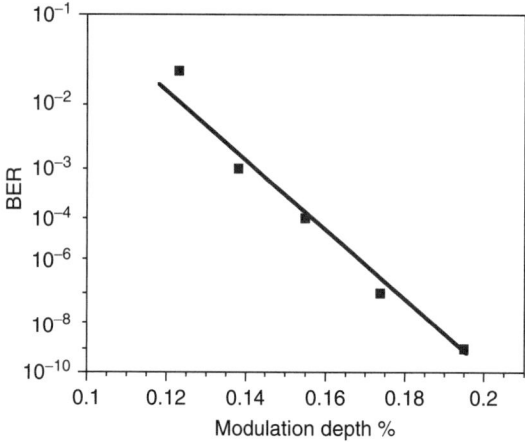

FIGURE 11.6 BER of digital channel at 2 Mb/s as a function of the modulation depth (%).

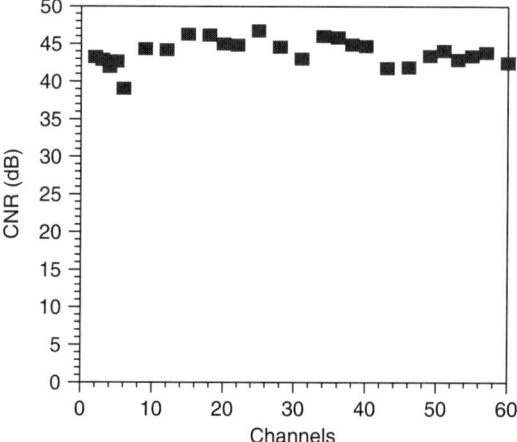

FIGURE 11.7 CNR as a function of CATV channels.

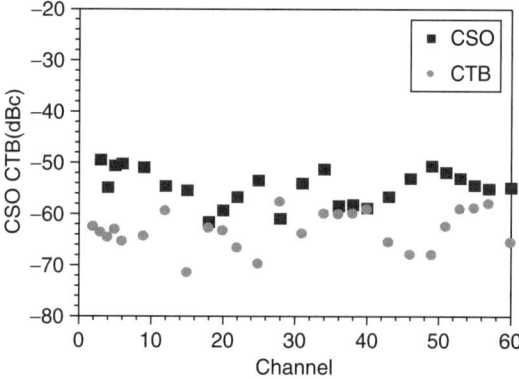

FIGURE 11.8 CSO and CTB as a function of CATV channels.

then experience different gains and the original waveform is distorted [13]. The CSO distortion is such a nonlinear effect, and new frequencies of the form $f_i \pm f_j$ are generated due to the presence of the channels of the system, some of which lie within the transmission bandwidth and distort the signal.

The experimental setup used for gain tilt measurement of the high-power Er/Yb co-doped DCFA is shown in Fig. 11.9; the theoretical and experimental details are described elsewhere [14]. The gain tilt G_t can be calculated by [6]

$$G_t = \frac{G(\lambda_2) - G(\lambda_1)}{\Delta \lambda} \qquad (4)$$

Here, $G(\lambda_1)$ and $G(\lambda_2)$ are the Er-Yb doped fiber amplifier gain values at wavelengths of λ_1 and λ_2, respectively. The experimentally obtained gain difference under AC conditions, where the modulation frequency is 10 MHz, is almost proportional to the wavelength difference $\Delta\lambda$ as shown in Fig. 11.10. The slope of the fitted linear curve is $\sim$0.42 dB/nm. Figure 11.11 shows the measured dependence of gain tilt on the wavelength of input light, and the horizontal axis shows the center wavelength of λ_1 and λ_2. The gain tilt values range from -0.47 to 0.5 dB/nm in the measured wavelength range of 1528–1564 nm. This large AC gain tilt causes CSO distortion. The CSO for an AM-SCM 60-channel transmission system is calculated from the gain tilt and laser chirp as follows [9]:

$$\text{CSO} = 10\text{Log}\left\{\sqrt{n}\frac{G_t(\lambda)}{G(\lambda)}\frac{\lambda_{\text{chirp}}}{\sqrt{N}}\right\} \qquad (5)$$

where n is the number of components, 40 in our experiment, and N is the total number of channels, 60 in our setup, and λ_{chirp} is the laser light chirp due to modulation. Figure 11.12 shows calculated values of CSO in a 60-channel AM-SCM system with an Er/Yb co-doped fiber amplifier and their variation with the input wavelength. Here we used $\lambda_{\text{chirp}} = 0.0012$ nm (wavelength chirp per channel), $n = 40$ and $N = 60$. The gain tilt values are obtained from Fig. 11.11. The calculated values for the second-order distortion are in the range of -45.5 to -53 dBc.

3. RESULTS

The performance of an analog and a hybrid digital/analog transmission system using a high-power cladding pumped fiber amplifier was investigated. A BER $< 10^{-9}$ was obtained. The BPSK modulated digital channels require a lower signal-to-noise ratio than the AM-VSB analog channels; a smaller modulation depth is required to obtain an error-free transmission. A modulation depth of 0.4% and 0.44% for the 1.0 and 1.4-GHz channel, respectively, and 0.2% for the 2 Mb/s channel was enough to

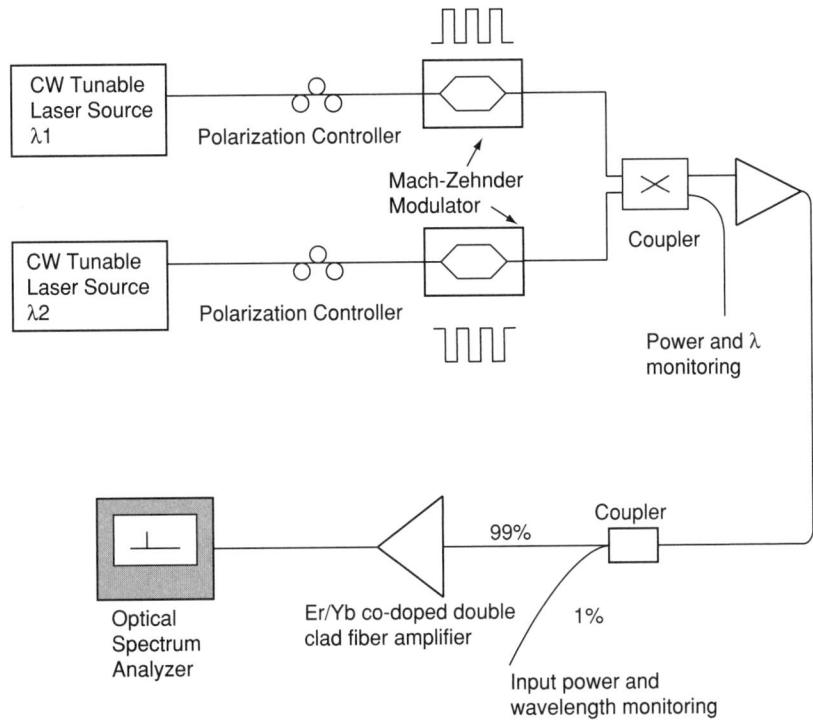

FIGURE 11.9 Experimental setup for the gain tilt measurement of the Er/Yb co-doped fiber amplifier. CW, continuous wave.

obtain a BER $< 10^{-9}$. Without the DCFA, slightly smaller modulation depths were required. A slight increase in the modulation depth with the DCFA should not be a problem because other experimental results [4] showed that only a much higher modulation depth (~10%) induces significant noise and distortion to affect the system performance.

The AM-VSB video signals require a high CNR. With 3.5% of the optical modulation depth, the CNR ranges from 45 to 52 dB without the DCFA. By introducing the DCFA and the distribution coupler, the CNR ranges from 42 to 48 dB for the 60 AM-VSB channels. Increasing the modulation depth increases the CNR, but the noise caused by distortion increases too. Although the CNR of most channels after the DCFA was still in the acceptable range, the CNR results can be improved by using a better DFB laser. The DFB laser used in this experiment has a high spontaneous level, resulting in a high RIN (relative intensity noise) of the laser, which set the upper limit of the CNR. There is actually only about a 4-dB penalty added to the CNR of the system because of the DCFA. The CSO and CTB measurements show that the DCFA

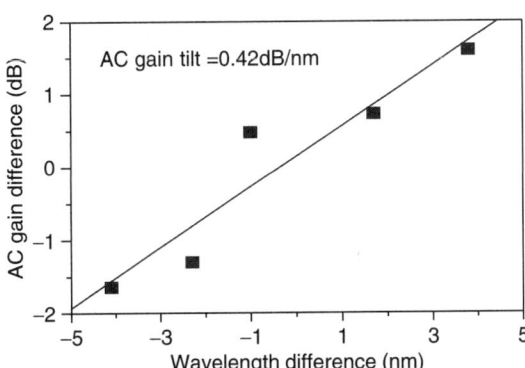

FIGURE 11.10 Dependence of AC gain difference on the wavelength variation $\Delta\lambda$. The center wavelength is 1545 nm.

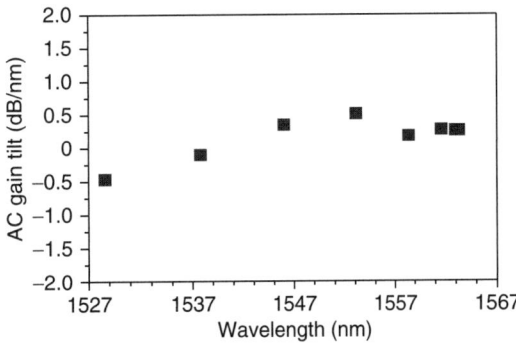

FIGURE 11.11 Measured wavelength dependence of AC gain tilt on the input wavelength under 10 MHz modulation.

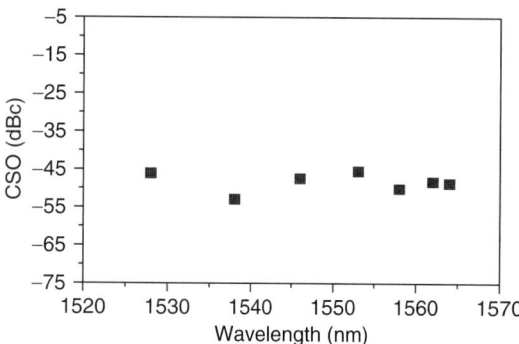

FIGURE 11.12 Calculated values of CSO for a 60-channel system with the DCFA and their variation with the input wavelength.

caused a penalty of only 4–7 dB, and most channels can satisfy the CSO and CTB requirements for communications systems.

We also experimentally measured the AC gain tilt of a high-power Er/Yb DCFA. The gain tilt value is ~0.42 dB/nm in the measured wavelength range of 1528–1564 nm. The calculated values of CSO degradation (Fig. 11.12) due to the presence of the DCFA in the AM-SCM transmission system with 60 channels are in good agreement to the experimentally measured values as shown in Fig. 11.4b. The gain tilt of the DCFA couples with the transmitter chirp and causes composite second-order distortion degradation that is also experimentally observed for the AM-SCM analog/digital transmission system with the amplifier.

4. REFERENCES

1. H. Kim and A. H. Gnuack, "10 Gb/s 177 km transmission over conventional singlemode fiber using a vestigial side-band modulation format," Electron. Lett., **37**, 1533–1534, 2001.
2. H. Lu, "CSO/CTB performance improvement by using optical VSB modulation technique," IEEE Photon. Technol. Lett., **14**, no. 10, 1478–1480, 2002.
3. Govind P. Agrawal, *Fiber optic communication systems*, 2nd ed. Wiley-Interscience, New York, 1997.
4. P. N. Freeman and N. K. Dutta, "Intermodulation distortion for a hybrid AM-VSB/digital system using a 1.55 μm laser and an optical amplifier," IEEE Photon. Technol. Lett., **8**, no. 11, 1558–1560, 1996.
5. E. Schweitzer, "WDM provides targeted services in CATV networks," in *WDM solutions*, Metro/Access Networks, PennWell publications, January 2003.
6. K. Kikushima, "AC and DC gain tilt of Erbium doped fiber amplifiers," J. Lightwave Technol., **12**, no. 3, 463–470, 1994.
7. A. Saleh, "Fundamental limit on number of channels in subcarrier multiplexed lightwave CATV system," Electron. Lett., **25**, no. 12, 770–771, 1989.
8. E. Desurvire, *Erbium doped fiber amplifiers*, Wiley Publications, New York, 1994.
9. T. E. Darcie and G. E. Bodeep, "Lightwave subcarrier CATV transmission systems," IEEE Trans. Microwave Theory Tech., **38**, 524–533, 1990.
10. H. Lin and Y. Kao, "Nonlinear distortions and compensations of DFB laser diode in AM-VSB lightwave CATV applications," J. Lightwave Technol., **14**, no. 11, 2567–2574, 1996.
11. K. D. LaViolette, "CTB performance of cascaded externally modulated and directly modulated CATV transmitters," IEEE Photon. Technol. Lett., **8**, 281–283, 1996.
12. C. Y. Kuo and E.E. Bergmann, "Erbium-doped fiber amplifier second order distortion in analog links and electronic compensation," IEEE Photon. Technol. Lett., **3**, 829–831, 1991.
13. K. Kikushima and H. Yoshinaga, "Distortion due to gain tilt of erbium-doped fiber amplifiers," IEEE Photon. Technol. Lett., **3**, no. 10, 945–947, 1991.
14. P. Dua, N. K. Dutta, and J. Jacques, "Measurement of gain tilt and its contribution to CSO in the analog/digital transmission systems with Er/Yb fiber amplifiers," Appl. Opt., **43**, no. 8, 1747–1751, 2004.

12
CHAPTER

Erbium-Doped Fiber Amplifiers for Dynamic Optical Networks

Atul Srivastava[*] and Yan Sun
Onetta/Bookham Technology
Morganville, New Jersey, USA

1. INTRODUCTION

Lightwave networks are evolving from centrally planned provisioned circuits to intelligent dynamic network elements supporting point and click provisioning and simplified operations with reduced capital and operational costs. In recent years, the emphasis of lightwave networks has shifted from point-to-point line system to a more two-dimensional architecture involving multiwavelength networking capability [1] via optical add/drop multiplexers and optical cross-connects, as illustrated in Fig. 12.1. Erbium-doped fiber amplifiers (EDFAs) are a key element of these wavelength division multiplexing (WDM) systems. WDM systems require high output power, low noise EDFAs with a high level of gain flatness over a wide and well-managed bandwidth.

The capacity of lightwave communication systems has undergone enormous growth during the last few years due to huge capacity demand for data transmission. Several laboratory demonstrations of high capacity transmission exceeding 10 Tb/s capacity [2] have been reported, and terabit capacity commercial systems are now available from system vendors. These impressive strides in the capacity of lightwave systems have been made possible due to several key technological advances in optical amplifiers, such as wide bandwidth EDFAs [3] having both low noise and high output power characteristics. In addition to wide bandwidth, the EDFAs have evolved to become a subsystem that incorporates significant amounts of loss in the midstage to accommodate dispersion compensation or optical add/drop modules needed in high capacity systems. There has also been considerable progress in the understanding of the time-dependent behavior of the EDFA, which is key in the design of suitable amplifiers needed for dynamic networks.

In this chapter, following the Introduction, we include a section on the EDFAs that covers the fundamentals of EDFAs including design considerations, followed by system and network issues related to EDFAs. In the next section we describe the issues related to EDFAs in dynamic WDM networks. After an introduction to the gain dynamics of single EDFA, we describe the phenomena of fast power transients in EDFA chains constituting optical networks. This section is followed by a discussion on the system impairments caused by the channel power transients in a WDM system. Finally, we describe several control schemes to limit the impairments caused by fast power transients. These include fast pump control, link control via insertion of compensating signals, gain clamping by lasing in an all-optical feedback loop, and three schemes, pump control, link control, and laser control, suitable for controlling the signal channel power transients.

2. EDFAs FOR HIGH CAPACITY NETWORKS

In traditional optical communication systems, optoelectronic regenerators are used between terminals to convert signals from optical to electrical and then back to the optical domain. Since the first reports in 1987 [4, 5], the EDFA has revolutionized optical communications. Unlike optoelectronic regenerators, the EDFAs do not need high-speed electronic circuitry and are transparent to data rate and format, which dramatically reduces cost. EDFAs also provide high gain, high power, and low noise. More importantly, all the optical signal channels can be amplified simultaneously within the EDFA in one fiber, thus enabling WDM technology.

In recent years, tremendous progress has been made in the development of EDFAs, including the erbium-doped fiber, semiconductor pump lasers, passive components, and splicing and assembly technology. Recent research in EDFAs has led to the development of an EDFA with a bandwidth of 84 nm. With these amplifiers, long distance transmission at 1 Tb/s was

---

[*]E-mail: Atul.Srivastava@bookham.com

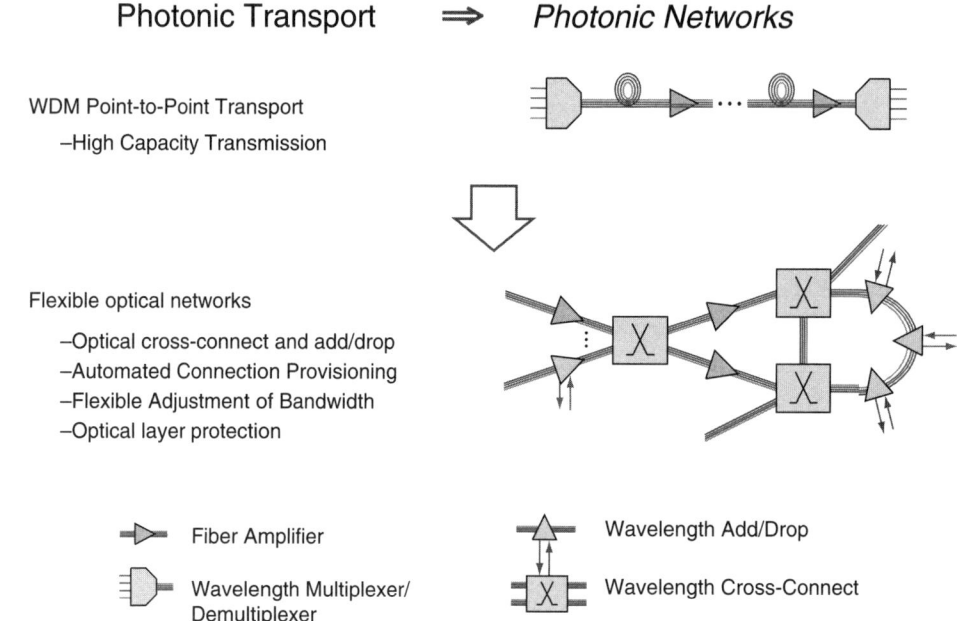

FIGURE 12.1 The evolution of optical networks.

achieved for the first time [6]. These developments have led to the application of EDFAs to commercial optical communication systems. The first field trial of WDM optical communication system was carried out by AT&T in 1989 [7], and subsequently the first commercial WDM system was deployed in 1995. Since the first deployment, the capacity of WDM systems has been increasing at a very fast pace. Today, optical amplifiers and WDM technology are offering an unprecedented cost-effective means to meet the ever-increasing demand for transport capacity, networking functionality, and operational flexibility.

2.1. Basic Characteristics of EDFAs

EDFAs consist of erbium-doped fiber having silica glass host core doped with active Er ions as the gain medium. Erbium-doped fiber is usually pumped by semiconductor lasers operating at the 980-nm or 1480-nm wavelength. Basic elements of an EDFA are shown schematically in Fig. 12.2. The gain medium in the amplifier is a specially fabricated optical fiber having its core doped with erbium (Er). The erbium-doped fiber is pumped by a semiconductor laser, which is coupled by using a wavelength selective coupler, also known as the WDM coupler, that combines the pump laser light with the signal light. The pump light propagates either in the same direction as the signal (copropagation) or in the opposite direction (counterpropagation). Optical isolators are used to prevent oscillations and excess noise due to unwanted reflection in the assembly. More advanced amplifier architecture consists of multiple stages

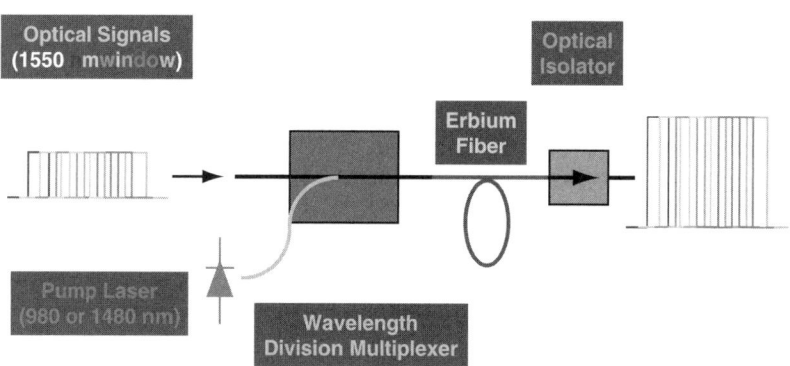

FIGURE 12.2 Schematic diagram of an erbium-doped fiber amplifier.

designed to optimize the output power and noise characteristics while incorporating additional loss elements in the midstage.

The energy level scheme of the erbium ion and the associated spontaneous lifetime in the glass host are shown in Fig. 12.3. The atomic levels of erbium ions are broadened by local field variations at the microscopic level in the glass host. The light emission due to optical transitions from the first excited state ($^4I_{13/2}$) to the ground level are perfectly matched with the transmission window of silica transmission fiber (1525–1610 nm). The erbium ions can be excited to the upper energy levels by 980- or 1480-nm pumps. In both cases, it is the first excited state, $^4I_{13/2}$, that is responsible for the amplification of optical signals. The amplification is achieved by the signal photons causing stimulated emission from the first excited state.

A three-level model can be used to describe the population of energy levels in the case of 980-nm pumping; whereas a two-level model usually suffices for the 1480-nm pumping case [8, 9]. Nearly complete inversion of erbium ions can be achieved with 980-nm pumping, whereas due to the stimulated emission at the pump wavelength the inversion level is usually lower in the case of 1480-nm pumping [9]. A higher degree of inversion leads to a lower noise level generated from the spontaneous emission process and is therefore highly desirable for the preamplifier stage. The quantum efficiency of the amplifier is higher for 1480 pumping due to the closer match between the signal and pump energies. The spontaneous lifetime of the metastable energy level ($^4I_{13/2}$) is about 10 ms, which is much slower than the signal bit rates of practical interest. The slow dynamics is responsible for the key advantage of EDFAs because of their negligible intersymbol distortion and interchannel cross-talk.

2.1.1. Amplifier Gain

The gain of the amplifier is defined as the ratio of the signal output and input powers. It is related to the gain factor $g(z)$, which changes along the length z of the erbium-doped fiber. The gain factor dependence on z arises due to pump depletion and gain saturation.

$$G \equiv P_{\text{out}}/P_{\text{in}} = \exp\left(\int_0^L g(z)dz\right) \quad (1)$$

The gain factor $g(z)$ is a measure of the local growth of optical power $P(z)$.

$$g(z) = \frac{1}{p(z)}\frac{dP(z)}{dz} = \rho\Gamma(\sigma_e N_2 - \sigma_a N_1) = \frac{g_0}{1+P/P_{sat}} \quad (2)$$

where N_2 and N_1 are the populations of the upper and lower energy levels of the active erbium ions, ρ is the active ion density, and Γ is a confinement factor that is a measure of the overlap of the signal field with the doped core. σ_e and σ_a are the emission and absorption cross-sections of the two-level system at frequency ν.

The rightmost term in equation (2) expresses the homogeneous saturation of the gain. When the signal power P reaches the saturation power P_{sat}, the gain reduces to half of the small signal gain g_0. The small signal gain factor g_0 is related to the pump power P_p as described by the following equation:

$$g_0 = \frac{\rho\Gamma\sigma_e(P_p - P_{\text{TH}})}{P_p + P_{\text{TH}}\sigma_e/\sigma_a} \quad (3)$$

where the threshold pump power P_{TH} is given by

$$P_{\text{TH}} = (\sigma_a/\sigma_e)(h\nu_p A/\Gamma_p \tau \sigma_p) \quad (4)$$

where τ is the spontaneous lifetime of the upper energy level, A is the core area, and h is Planck's constant. For pump powers lower than the threshold value, the gain is negative and it reaches a maximum value $\rho\Gamma\sigma_e$ for very high pump powers. Γ_p is the pump confinement factor that is usually designed in the Er-doped fiber to maximize the overlap.

The amplifier saturation behavior is depicted in Fig. 12.4. For small input signal power the gain is constant and starts to decrease as the input power is increased further. In the saturation region, the output power is approximately proportional to the pump power. In this condition, the pump absorption from the ground state is balanced by stimulated emission from the first excited state induced by the signal. The balance shifts to the higher signal power level as the

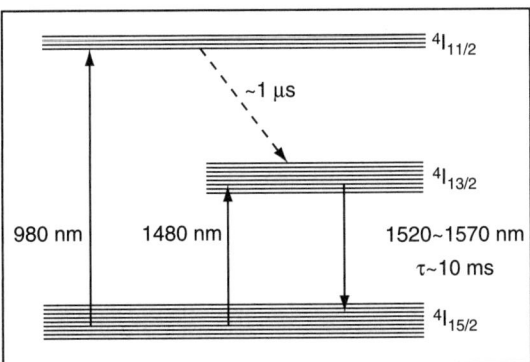

FIGURE 12.3 Erbium ion energy-level scheme.

- Usually operated in deep saturation regime
- Gain remains constant at high data rate
  - OA is slow, it does not respond to the modulation (> kHz), it only "sees" the average power, and sets its gain at G_{ave}

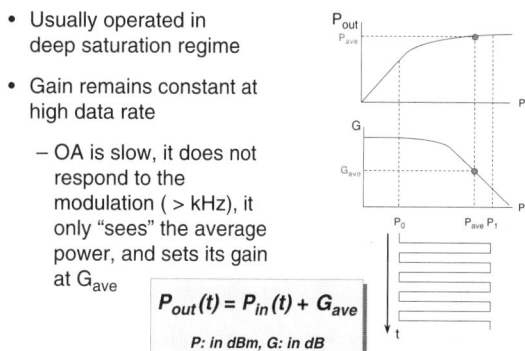

$$P_{out}(t) = P_{in}(t) + G_{ave}$$
P: in dBm, G: in dB

FIGURE 12.4 Saturation characteristics of EDFAs.

pump power is increased. The output signal power at which the gain reduces to half of the small signal gain value is the saturation power P_{sat} and is given by

$$P_{sat} = \frac{h\nu A}{(\sigma_e + \sigma_a)\Gamma\tau}\left[1 + \frac{\sigma_a}{\sigma_e}\frac{P_p}{P_{TH}}\right] \quad (5)$$

For $P_p \gg P_{TH}$, P_{sat} is proportional to P_p/P_{TH}. The above equations describe the gain saturation at a particular value of z. This local description is usually quite adequate in determining the amplifier saturation behavior because the saturation occurs primarily near the output end of the amplifier where the signal power is at its highest value.

The gain and loss coefficient spectra are shown in Fig. 12.5 at different inversion levels for erbium-doped fiber with Al and Ge co-doping [10, 11]. Under a homogeneous broadening approximation,

the overall gain spectrum of any piece of Er-doped fiber always matches one of the curves after scaling and does not depend on the details of pump power, signal power, and saturation level along the fiber [9]. The gain spectrum is very important for amplifier design.

2.1.2. Amplifier Noise

Amplification generates optical noise through the spontaneous emission process. The spontaneous emission arises when light emitted by spontaneous decay of excited erbium ions is coupled to the optical fiber waveguide. The spontaneous emission is amplified by the gain in the medium and therefore increases with the gain G. The total amplified spontaneous emission noise power N_{ASE} at the amplifier output is given by

$$N_{ASE} = 2n_{sp}(G-1)h\nu B \quad (6)$$

where the factor of 2 takes into account the two modes of polarization supported by the fiber waveguide, B is the optical noise bandwidth, and n_{sp} is the spontaneous emission factor given by

$$n_{sp} = \sigma_e N_2/(\sigma_e N_2 - \sigma_a N_1) \quad (7)$$

The spontaneous emission factor indicates the relative strengths of the spontaneous and stimulated emission process. It reaches its minimum value of 1 for complete inversion ($N_1 = 0$) when all the erbium ions are in the excited state. The ASE power can be related to the noise figure (NF) of the amplifier,

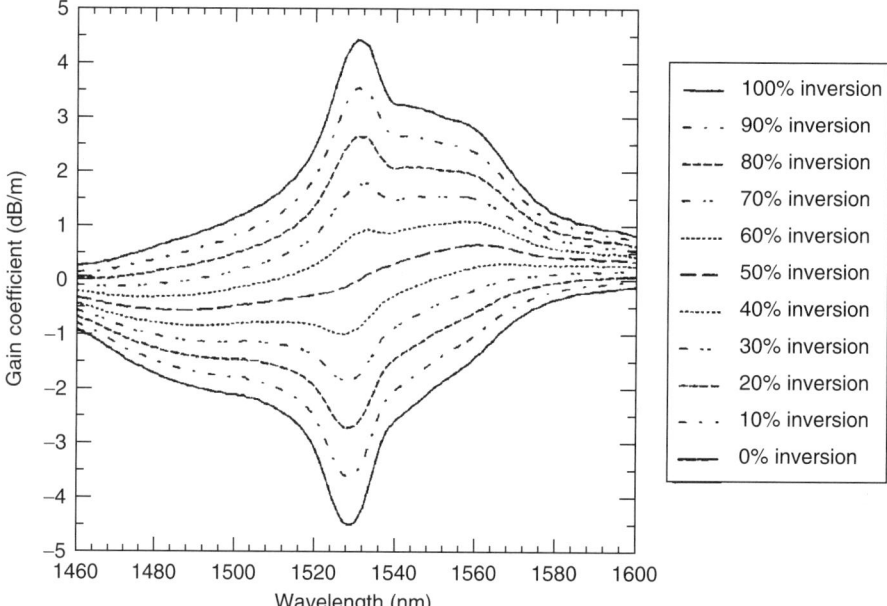

FIGURE 12.5 The gain and loss coefficient spectra at different inversion levels for erbium-doped fiber with Al and Ge co-doping.

which is defined as the signal-to-noise ratio (SNR) corresponding to the shot noise of the signal at the input divided by SNR at the output. The NF is a measure of the degradation of the optical signal by the ASE noise added by the amplifier.

$$\text{NF} \equiv \text{SNR}_{\text{in}}/\text{SNR}_{\text{out}} \tag{8}$$

which is given by

$$\text{NF} = N_{\text{ASE}}/h\nu BG = 2n_{\text{sp}}(G-1)/G \tag{9}$$

As discussed earlier, for an EDFA with full inversion, the spontaneous emission factor reaches a value of unity at high gain.

A high inversion level provides low NF, whereas a low inversion level yields high efficiency in the conversion of photons from pump to signal. To achieve both low NF and high efficiency, two or more gain stages are typically used where the input stage is kept at a high inversion level and the output stage is kept at a low inversion level [12, 13]. An ASE filter is usually inserted in the middle stage to prevent gain saturation caused by the ASE peak around 1530 nm. For optical amplifiers with two or more gain stages, the overall NF is mainly decided by the high gain input stage and the output power is basically determined by the strongly saturated output stage. The passive components have minimal impact on the NF and the output power when they are in the midstage.

2.2. System Issues

The EDFA is an optical amplifier that amplifies optical signals in the optical domain. Unlike the optoelectronic regenerators used in earlier systems, the optical amplifier does not need high-speed electronic circuitry and is transparent to data rate and format, which dramatically reduces cost. All the optical signal channels can be amplified simultaneously within the EDFA, which provides high gain, high power, and low NF. In lightwave communication systems it can be used as a booster to increase the power of the transmitter, an in-line amplifier, or a repeater to enhance the overall reach of the system, or as a preamplifier to increase the receiver sensitivity. Because of slow gain dynamics in EDFAs, the high-speed data modulation characteristics of optical signals are preserved during the amplification process. Application of EDFAs as repeaters in WDM systems is particularly important because they offer a cost-effective means of faithfully amplifying all the signal wavelengths within the amplifier band simultaneously, thereby eliminating the need of costly optoelectronic regenerators. WDM-based lightwave systems were deployed by AT&T in 1996 for the first time. In present-day lightwave systems, the EDFA has replaced the optoelectronic regenerator as the repeater of choice in both submarine and terrestrial systems.

Currently, optical communications technology is moving from point-to-point systems to optical networking, where EDFAs can play a role at many places in WDM optical networks. The EDFAs are used mostly in optical transport line systems, which consists of a transmitter, fiber spans with in-line amplifiers, and a receiver. On the transmitter end, multiple optical channels are combined in an optical multiplexer, and the combined signal is amplified by a power amplifier before being launched into the first span of the transmission fiber. At the receiving end, the incoming WDM signals are amplified by a preamplifier before being demultiplexed into individual channels that are fed into the respective receivers. Long-haul applications also require in-line repeater amplifiers to extend the total system reach [14].

2.2.1. Optical SNR

In an optically amplified system, channel power reaching the receiver at the end of the link is optically degraded by the accumulated ASE noise from the optical amplifiers in the chain. At the front end of receiver, ASE noise is converted to electrical noise, primarily through signal–ASE beating, adding to the bit error rate (BER) floor. System performance therefore places an important requirement on optical signal-to-noise ratio (OSNR) of each of the optical channels. OSNR therefore becomes the most important design parameter for an optically amplified system. Other optical parameters in system design consideration are channel power divergence, which is generated primarily due to the spectral gain nonuniformity in EDFAs (described in the next section), and maximum channel power relative to the threshold levels of optical nonlinearities, such as self-phase modulation, cross-phase modulation, and four-photon mixing [15].

Although optical amplifiers are conventionally classified into power, in-line, and preamplifiers, state-of-the-art WDM systems require all three types of amplifiers to have low NF, high output power, and uniform gain spectrum. We do not distinguish these three types of amplifiers in the discussion presented in this section. The nominal OSNR for a 1.55-μm WDM system with N optical transmission spans can be given by the following formula [11]:

$$OSNR_{nom} = 58 + P_{out} - 10\log_{10}(N_{ch}) - L_{sp} - NF$$
$$- 10\log_{10}(N) \quad (10)$$

where OSNR is normalized to the 0.1-nm bandwidth, P_{out} is the optical amplifier output power in dBm, N_{ch} is the number of WDM channels, L_{sp} is the fiber span loss in dB, and NF is the amplifier noise figure in dB. For simplicity, it has been assumed here that both optical gain and NF are uniform for all channels.

The above equation shows how various system parameters contribute to OSNR; for example, the OSNR can be increased by 1 dB, by increasing the amplifier output power by 1 dB or decreasing NF by 1 dB, or reducing the span loss by 1 dB. This equation indicates that we can make trade-offs between number of channels and number of spans in designing a system. However, the trade-off may not be straightforward in a practical system because of the mutual dependence of some of the parameters. Other system requirements also impose additional constraints; for example, optical nonlinearities place an upper limit on channel power and this limit depends on number of spans, fiber type, and data rate.

The simple formula in Eq. (10) highlights the importance of two key amplifier parameters: NF and output power. Although it provides valuable guidelines for amplifier and system design, it is always necessary to simulate the OSNR evolution in a chain of amplifiers when designing a practical WDM system. The amplifier simulation is usually based on an accurate mathematical model of amplifier performance. Amplifier modeling is also a critical part of the end-to-end system transmission performance simulation that incorporates various linear and nonlinear transmission penalties.

2.3. Dynamic Network Related Issues

Another important control function is amplifier power adjustment. In a WDM system, there is a need to adjust the total amplifier output as a function of number of equipped channels. The total output power must be adjusted such that while the per-channel power is high enough to ensure sufficient OSNR at the end of the chain, it is low enough not to exceed the nonlinear threshold. In addition, per-channel power must be maintained within the receiver dynamic range as the system channel loading is changed. Such power adjustment has traditionally been achieved through a combination of channel monitoring and software-based pump power adjustment. Fast transient control is becoming increasingly important, as the EDFAs need to maintain a constant channel gain during channel load changes. Each component in the optical transmission layer is subject to failure and degradation. The long-term degradation of transmission components results in an increase of insertion loss, lower isolation, higher back-reflection, or lower output power. Active monitoring and slowly adjusting power levels and optical amplifier gain can compensate for these effects. Proper systems are designed to accommodate those changes. On the other hand, catastrophic component failure can be very fast by nature. If optical power transients are not handled properly, these failures can considerably affect the quality of service of the optical channels in the network. The transmitter module can fail due to electronic or optical problems. Fortunately, a catastrophic transmitter failure, which could have a sub-millisecond time frame, does not affect the other transmitters. Catastrophic failure of any receiver module should be limited to that module if properly designed.

Other network elements and transmission fiber catastrophic failures have the potential of severely affecting a large number of channels. Fiber handling at the central office can result in up to sub-microsecond optical power changes. Examples of fiber mishandling are the removal of wrong connectors, stepping on the fiber, and heavy objects sectioning the fiber. The other network elements that can be placed between the transmitter and the receiver are optical amplifiers, reconfigurable or fixed optical add-drop, optical cross-connects, and dynamic gain equalization filters. The power changes depend on the technology used and the failure mode. MEMS-based devices have the potential for insertion loss changes on the order of a microsecond. Mechanical, magnetooptic, and thermal-based switches and variable optical attenuators have transition speeds of hundreds of microseconds to tens of milliseconds.

The network provisioning can be designed such that connections are added and removed slowly (several milliseconds), one at the time, and would only marginally disturb the rest of the channels in the network. When a failure is detected, the service is rerouted. If optical shared protection is used, power transients are observed and depend on the switching technology used. A large number of channels can be switched at the same time, imposing large power changes. The amplifier is affected by the magnitude and the speed of the changes. Therefore, the amplifier should be designed to handle fast (sub-microsecond) and large power changes.

The recent advances in WDM optical networking has called for fast gain control to minimize the channel power excursion when a large number of channels are changed due to, for example, catastrophic partial system failure. Various techniques that will be detailed in Section 3 have been demonstrated to stabilize amplifier gain, thereby achieving the goal of maintaining per channel power. In addition to amplifier dynamics control, practical implementation in a system also requires a receiver design that can accommodate power change on a very short time scale.

3. EDFAs FOR DYNAMIC NETWORKS

Dynamic reconfigurable WDM networks consist of intelligent and dynamic network elements supporting point and click provisioning and simplified operations with reduced capital and operational costs. Optical amplifiers are a key element of the present-day WDM systems. The WDM systems require high output power, low NF EDFAs with high level of gain flatness over a wide and well-managed bandwidth. In addition to high performance, the implementation of the new features such as dynamic wavelength provisioning and automated reconfiguration can lead to optical power transients in a conventional gain-controlled EDFA, which is usually operated under saturation. These signal transients can compromise the flexibility of network architecture and impact overall system performance. Should a network be reconfigured or a component fail, the number of WDM signals passing along a chain of amplifiers changes, thus causing the power of surviving channels to either increase or decrease. The service quality of the surviving channels can be impaired through four mechanisms when channel loading changes. First, optical nonlinear effects in transmission fibers occur or increase if the power excursions are large enough when signal channels are lost. For example, self-phase modulation (SPM) has been observed to affect the performance of the surviving channels [16]. Second, when channels are added, the optical power at the receiver can be reduced during the transient period, which would cause eye closure and severe degradation of BER. Third, OSNR may be degraded due to the change of inversion level and therefore the change in gain spectrum during the transient period. Fourth, the received power at the receiver is varying, which requires that the threshold of the receiver be optimized at high speed, which can be a problem for certain receivers.

To effectively control the power transients in DWDM systems, it is important to understand the factors that influence the speed or the rate the surviving channels' power change during transients. The speed of transients depends on both the EDFA characteristics and the number of EDFAs constituting the system. In a single EDFA, the time constant of the power transients decreases with increases in saturated power output. In a system with long chains of EDFAs the amplifiers strive to maintain the saturated output power levels. The time varying output of the first EDFA (after a fiber span loss) appears as an input to the second EDFA, which has time-dependent gain, and therefore the output power of the second EDFA changes at a faster rate. Consequently, with increasing numbers of amplifiers in the chain, the speed of transients becomes faster and faster, thereby requiring control on shorter and shorter time scales to be able to limit the power excursion of surviving channels [17]. Proper transient control design needs to take into account the fastest single event in the network as well as the acceleration of this event due to the cascading of amplifiers.

In this section we provide an overview of the physics of gain dynamics of EDFAs and the impact of EDFA gain dynamic on the WDM transmission system performance. Recent advances in techniques for mitigating system impairments caused due to channel power excursions in a dynamic network are also reviewed.

3.1. Gain Dynamics of Single EDFA

After years of extensive study on steady state problems, recently much attention has been focused on dynamics. The dynamics of EDFAs are generally considered to be slow as a result of the long spontaneous lifetime of around 10 ms. For transmission of high-speed data, the gain of EDFAs is undisturbed by signal modulation. Furthermore, EDFAs, in contrast to semiconductor laser amplifiers, do not introduce intersymbol interference in single channel systems or cross-talk among channels in WDM systems. This is one of the chief advantages of EDFAs and the reason why steady-state models have been used to model transmission systems. The results of measurement of cross-talk between two wavelength signals traversing an unsaturated single-stage EDFA is shown in Fig. 12.6. The frequency response of the cross-talk shows a corner frequency to be in the few kHz range with corresponding gain recovery time constant between 110 and 340 μs [18].

The speed of gain dynamics in a single EDFA is in general much faster than the spontaneous lifetime because of the gain saturation effect. EDFAs with high output powers for multichannel optical networks are strongly saturated, and the resulting effective time constants are reduced to the order of tens of microseconds. In a recent report, the characteristic transient times were reported to be tens of microseconds in a two-stage EDFA [19]. The measurement

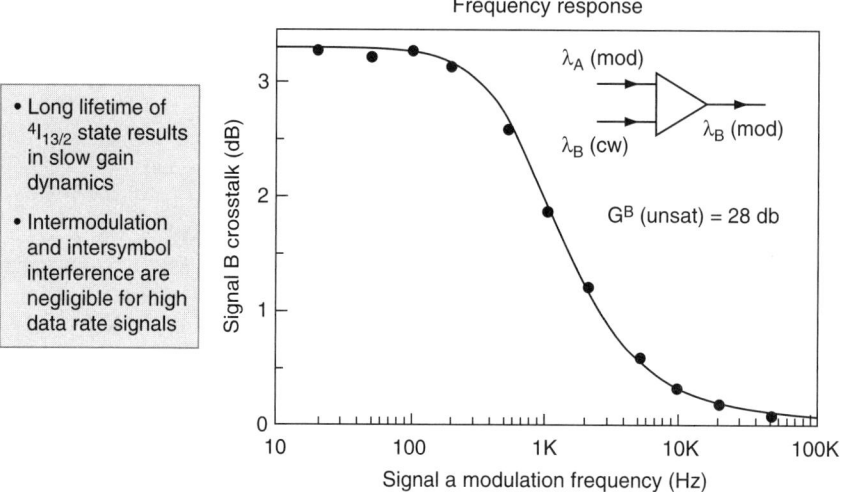

FIGURE 12.6 Frequency response of cross-talk between two channels in an EDFA [18].

of transient behavior of surviving channel power in an EDFA operating under saturation, suitable for an eight-channel multiwavelength network, was presented. In the experiment, the total input power to the EDFA corresponded to 7-dBm input power, as would result from eight WDM channels with an input power of −2 dBm each. The amplifier gain was 9 dB with a corresponding total output power for all channels of 16 dBm. The transient behavior of surviving channels when seven channels are dropped is shown in Fig. 12.7. The upper trace in the figure shows the modulation signal applied to the DFB laser power supply. The transition time for the laser light output to reach 90% of the steady-state value is less than 3 μs. The current supply to the laser diode has a few-microsecond electronic delay in the response of the surviving channels after the channels are dropped. After this short delay, a transient in the power of the surviving channels is observed. During the transient the power rises to 90% of the steady-state value in nearly 80 μs. The transient becomes slower as the number of dropped channels increases. The transient behavior of surviving channel power for the cases of one, four, and seven dropped channels is shown in Fig. 12.8. In the case of seven dropped channels, the transient time constant is nearly 52 μs. As can be seen, the transient becomes faster as the number of dropped channels decreases. The time constant decreases to 29 μs when only one out of eight channels is dropped. The rate equations [20] for the photons and the populations of the upper ($^4I_{13/2}$) and lower ($^4I_{15/2}$) states can be used to derive the following approximate formula for the power transient behavior [21]:

$$P(t) = P(\infty)[P(0)/P(\infty)]^{(-t/\tau_e)} \quad (11)$$

where $P(0)$ and $P(\infty)$ are the optical powers at time $t = 0$ and $t = \infty$, respectively. The characteristic time τ_e is the effective decay time of the upper level averaged over the fiber length. It is used as a fitting parameter to obtain best fit with the experimental data. The experimental data are in good agreement (cf. Fig. 12.8) with the model for the transient response. The model has been used to calculate the fractional power excursions in decibels of the surviving channels for the cases of one, four, and seven dropped channels. The times required to limit the power excursion to 1 dB are 18 and 8 μs when four or seven channels are dropped, respectively. The time constant of gain dynamics is a function of the saturation caused by the pump power and

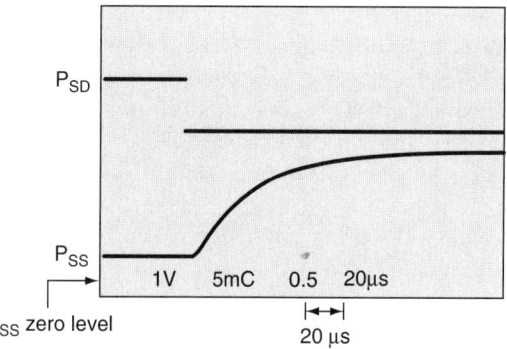

FIGURE 12.7 Oscilloscope trace showing the time dependence of the surviving chanel power (P_{SS}) along with the dropped channel power (P_{SD}) [19].

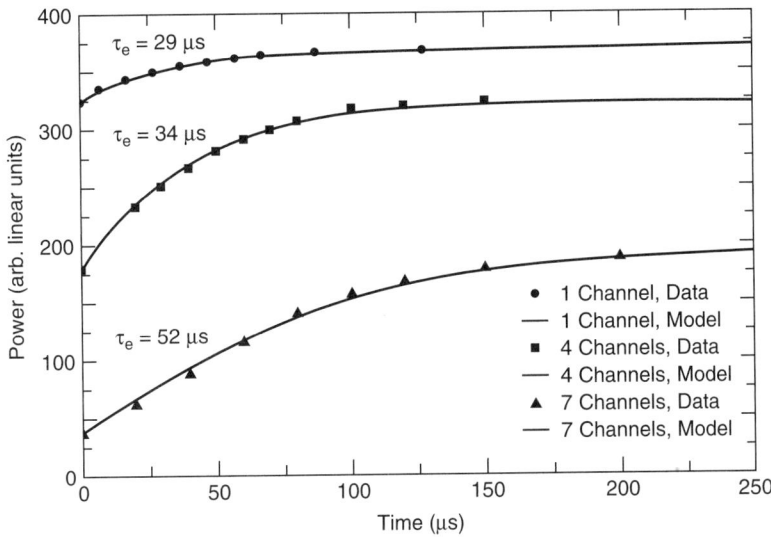

FIGURE 12.8 Measured and calculated surviving power transients for the cases of one-, four-, and seven-channel loss out of eight WDM channels.

the signal power. Present-day WDM systems with 40–100 channels require high-power EDFAs in which the saturation factor becomes higher, leading to a shorter transient time constant. Dynamic gain control of the EDFAs with faster response times is necessary to control the signal power transients.

A model of EDFA dynamics is needed to understand the transient behavior in large system or networks. A simple model has been developed for characterizing the dynamic gain of an EDFA. The time dependent gain is described by a single ordinary differential equation for an EDFA with an arbitrary number of signal channels having arbitrary power levels and propagation directions. Most previous EDFA models are represented by sets of coupled partial differential equations [22, 23], which can be solved only through iterative computationally intensive numerical calculations, especially for multichannel WDM systems with counterpropagating pump or signals. The time-dependent partial differential equations can, however, be dramatically reduced to a single ordinary differential equation. The mathematical details of the model are provided in [24]. Here the simulation results from the model are compared with the measured time-dependent power excursions of surviving channels when one or more input channels to an EDFA are dropped. The structure of the two-stage EDFA used in the experiment [19] and simulation is shown in the inset of Fig. 12.9. The experimentally measured power of the surviving channel when one, four, or seven out of out of eight WDM channels are dropped is plotted in Fig. 12.9. It is seen from the figure that the simulation results agree reasonably well with the experimental data without any fitting parameters. The exception is the 0.9-dB difference at the most for the seven-channel drop case. This discrepancy is believed to arise from pump excited state absorption at high pump intensity. The model can be very useful in the study of power transients in amplified optical networks. A summary of the time constants for both the unsaturated and saturated EDFAs is given in Fig. 12.10. The saturation factor increases with both the signal input

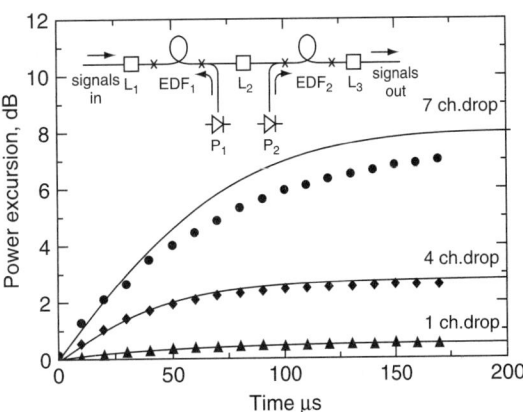

FIGURE 12.9 Comparison between theory and experiment for output power excursions of surviving channels from a two-stage amplifier when one, four, and seven input channels are dropped.

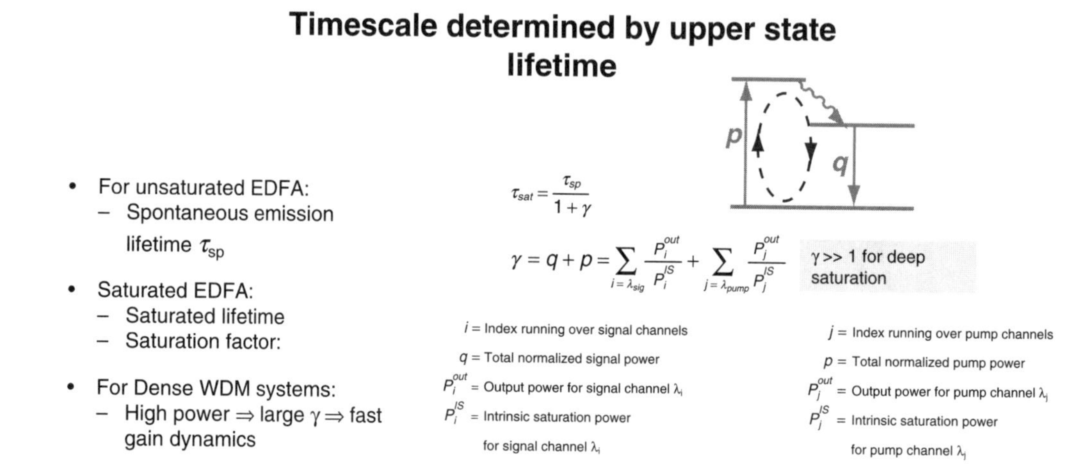

Timescale determined by upper state lifetime

- For unsaturated EDFA:
  - Spontaneous emission lifetime τ_{sp}
- Saturated EDFA:
  - Saturated lifetime
  - Saturation factor:
- For Dense WDM systems:
  - High power $\Rightarrow$ large γ $\Rightarrow$ fast gain dynamics

$$\tau_{sat} = \frac{\tau_{sp}}{1+\gamma}$$

$$\gamma = q + p = \sum_{i=\lambda_{sig}} \frac{P_i^{out}}{P_i^{IS}} + \sum_{j=\lambda_{pump}} \frac{P_j^{out}}{P_j^{IS}}$$

$\gamma \gg 1$ for deep saturation

i = Index running over signal channels
q = Total normalized signal power
P_i^{out} = Output power for signal channel λ_i
P_i^{IS} = Intrinsic saturation power for signal channel λ_i

j = Index running over pump channels
p = Total normalized pump power
P_j^{out} = Output power for pump channel λ_j
P_j^{IS} = Intrinsic saturation power for pump channel λ_j

FIGURE 12.10 Lifetimes of saturated and unsaturated EDFAs.

power and pump power and gives rise to a shorter gain dynamics time constant as governed by the lifetime of the upper state.

A dynamic behavior study of L-band EDFA has been carried out [25]. In this work, the transient response of the surviving channels in a two-stage L-band EDFA under different channel loading conditions was reported. The observed dynamic behavior of an L-band EDFA shows similar dependence on the output power and the number of dropped channels to that in a C-band EDFA. However, the magnitude of the response time is very different. The response time constants as a function of the number of dropped channels under different saturation conditions is shown in Fig. 12.11. The time constants are about 105 and 260 μs when one and seven channels, respectively, out of eight channels are dropped and the amplifier is well saturated. These values are about four to five times larger than that observed in a C-band EDFA. The difference can be explained by the different intrinsic saturation power in these two bands.

3.2. Fast Power Transients in EDFA Chains

The phenomenon of fast power transients in EDFA chains has been reported [17, 26]. The effect of dropped channels on surviving powers in an amplifier chain is illustrated in Fig. 12.12. When four of eight WDM channels are suddenly lost, the output power of each EDFA in the chain drops by 3 dB and the power in each surviving channel then increases toward double the original channel power to conserve the saturated amplifier output power. Even though the gain dynamics of an individual EDFA is unchanged, the increase in channel power at the end of the system becomes faster for longer amplifier chains. Fast power transients result from the effects of the collective behavior in chains of amplifiers. The output of the first EDFA attenuated by the fiber span loss acts as the input to the second EDFA. Because both the output of the first EDFA and the gain of the second EDFA increase with time, the output power of the second amplifier increases at a faster rate. This cascading effect results in faster and faster transients as the number of amplifiers increase in the chain. To prevent performance penalties in a large-scale WDM optical network, surviving channel power excursions must be limited to certain values depending on the system margin. Take the MONET network as an example: The power swing should be within 0.5 dB when channels are added and 2 dB when channels are dropped [27].

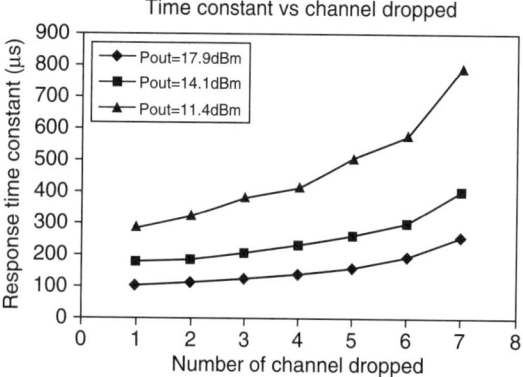

FIGURE 12.11 Response time constant vs. the number of dropped channels under different saturation conditions in a two-stage L-band EDFA [25].

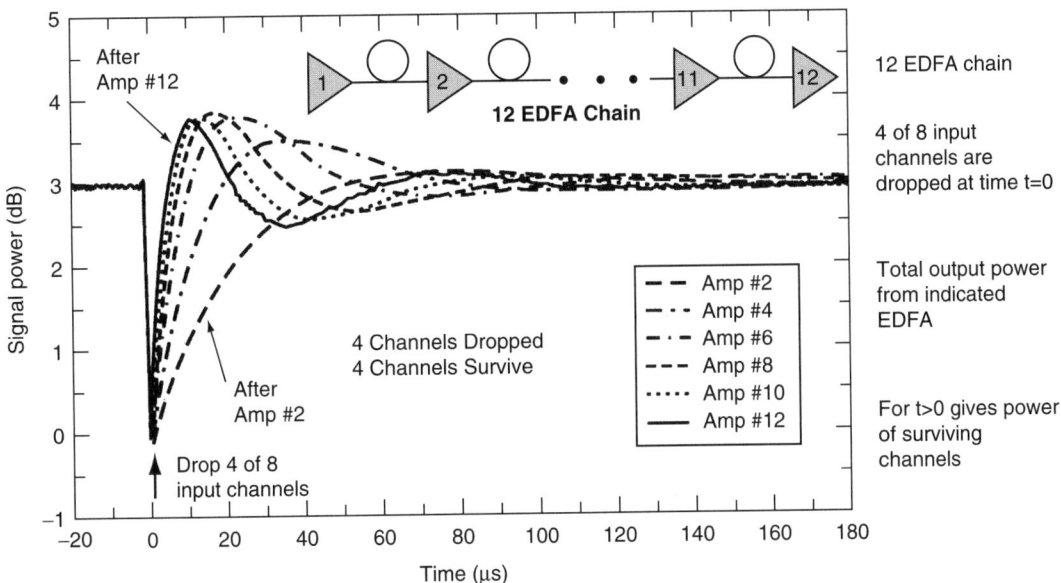

FIGURE 12.12 Measured output power as a function of time after 0, 2, 4, 6, 8, 10, and 12 EDFAs (at time $t = 0$, four of eight WDM channels are dropped) [17].

In a chain consisting of 10 amplifiers, the response times required to limit the power excursions to 0.5 and 2 dB would be 0.85 and 3.75 μs, respectively. The response times are inversely proportional to the number of EDFAs in the transmission system.

The response time of EDFAs can be divided into three regions: the initial perturbation region, the intermediate oscillation region, and the final steady-state region. In the initial perturbation region, the gain of the EDFA increases linearly with time and the system gain and output power increases at a rate proportional to the number of EDFAs. The time delays for a channel power excursion of 2 dB are measured from Fig. 12.12, and the inverse of time delays, which are the power transient slope in the perturbation region, are plotted in Fig. 12.13. Assuming that the amplifiers operate under identical conditions, the rate of change of gain at each EDFA is the same and is proportional to the total lost signal power. The slope plotted in Fig. 12.13 therefore increases linearly with the number of EDFAs in the chain. These experimental results have been confirmed by modeling and numerical simulation from a dynamic model [9, 24].

In the intermediate region an overshoot spike can be observed after two EDFAs shown in Fig. 12.12. The first overshoot peak is the maximum power excursion because the oscillation peaks that follow are smaller than the first one. From the results of both experimental measurements and numerical simulation on a system with N EDFAs, the time to reach the peak is found to be inversely proportional to N and the slope to the peak is proportional to $N - 1$ [28]. This indicates that the overshoot peaks are bound by a value determined by the dropped signal power and the operating condition of the EDFAs. These properties in the perturbation and oscillation regions can be used to predict the power excursions in large optical networks.

Better understanding of the physical phenomena occurring in EDFA chains during transients was achieved by numerical simulation. Calculated optical channel power excursion when channels are added and lost under very similar conditions as the above

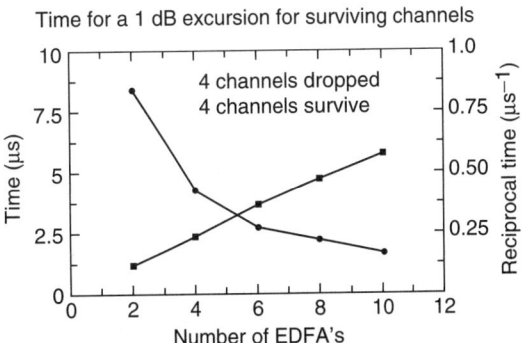

FIGURE 12.13 Delay and reciprocal of delay for surviving channel power excursion to reach 2 dB after the loss of four of eight WDM channels.

experiment is shown in Fig. 12.14. The numerical simulation results are qualitatively in agreement with the experimental data described earlier. From both the experimental and simulation results, the time response of the surviving channels in general can be divided into four sequential regions and the final steady-state region. In the linear region the output power increases linearly with time. The slope increases linearly with the number of EDFAs in the system as witnessed in the experimental results (cf. Fig. 12.12). After the initial linear period, overshoot spikes are observed in all the EDFAs except the first one. The delay before the overshoot peak is inversely proportional to the number of EDFAs, and the slope to the peak, that is, the slope of the line connecting 0 dB at $t = 0$ to the overshoot peak, is proportional to the number of EDFAs, as shown in Fig. 12.15. These two properties indicate that the overshoot peak asymptotically approaches but never exceeds a certain value. The overshoot peak is followed by damped oscillations. Finally, a new steady state is reached, which, for EDFAs after the first one, show a 3-dB increase in channel output power. This is because the EDFAs are deeply saturated and therefore the total output power is essentially the same before and long after the channel drop.

The impact of transients in a WDM system is to cause error bursts in surviving channels. The error bursts in channels in service are unacceptable to service providers; it is therefore essential that fast transients are controlled. Although typical time scales for gain change in a single amplifier are tens of microseconds, the time constants for a chain of N amplifiers is $1/N$ times that of a single amplifier. Thus, systems with chains of amplifiers require faster control to limit the undesirable power excursions, so that long chains of amplifiers present a greater gain stabilization challenge.

3.3. System Impairments due to Transients

The quality of service of the channels in the network can be severely affected through four mechanisms when channel loading changes. First, optical nonlinear effects in transmission fibers occur if the power excursions are large enough when signal channels are lost. Nonlinear effects, such as Brillouin scattering, SPM, cross-phase modulation, and the Raman effect, could degrade the signal quality depending on the number of WDM channels and their spacing and type of transmission fiber. SPM has been demonstrated to affect the performance of the surviving channels in the link control experiment described in a later section [16]. Second, when channels are added, the optical power at the receiver can be reduced during the transient period, which would cause eye closure and possibly inadvertent protection switching. If the optical power at the receiver is lowered by more than the system margin and crosses the receiver sensitivity, the BER could be severely degraded. Third, OSNR may be lowered

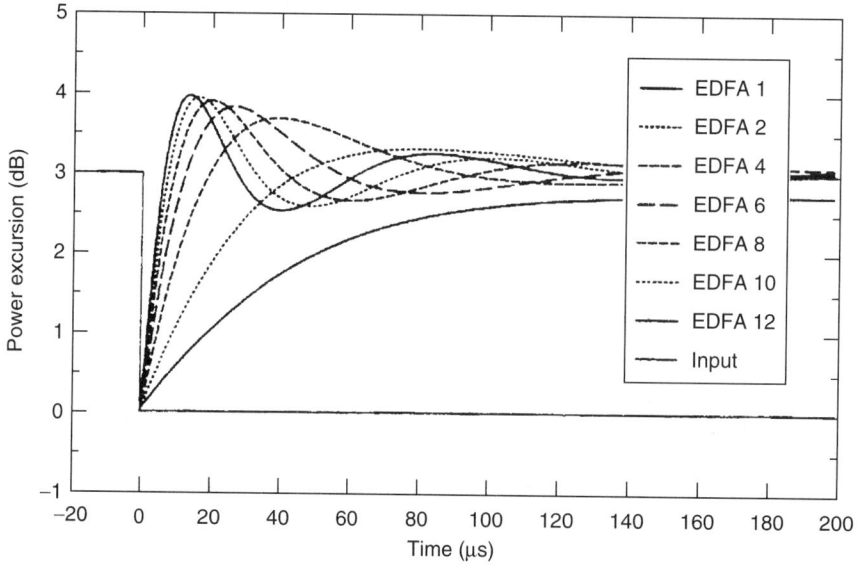

FIGURE 12.14 Numerical simulation of total signal power excursion as a function of time. There are four regions: initial linear region, power overshoot peak, decayed oscillation region, and final steady-state region [24].

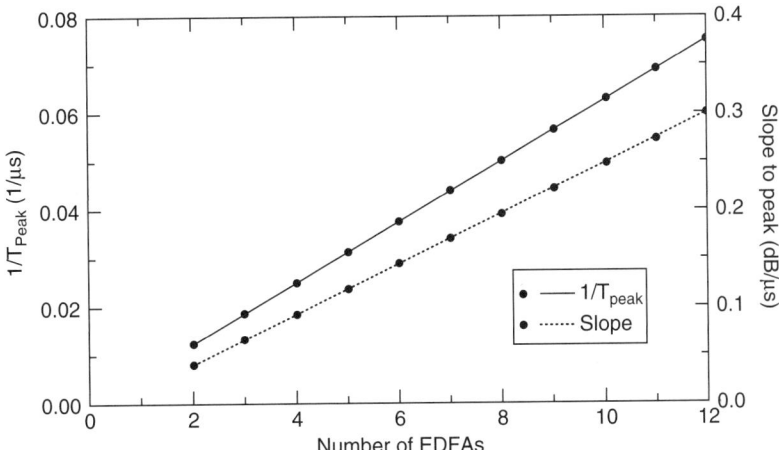

FIGURE 12.15 Time to reach overshoot peak and slopes to the peaks (i.e., the slopes of the lines connecting 0 dBm at $t = 0$ to the overshoot peaks) measured from Fig. 12.14. Both reciprocal of time and slope increase linearly as a function of the number of EDFAs [24].

due to the change of inversion level and therefore the change in gain spectrum during the transient period. The gain spectrum change is a combination of both the linear tilt due to inversion level change and other variations due to spectral hole burning, which could be very strong, near 1532 nm. Fourth, because the received power at the receiver is varying with time during transients, the receiver needs to have a high-speed threshold adjustment capability, which can be a problem for some receivers.

In a recent study the impact of transients on a WDM transmission system with channels at 2.5 Gb/s was reported [29]. The effects of both the power excursions and the transient pulse duration were investigated in the transmission system. Optical power levels in transmission systems are typically adjusted to steady-state levels that reflect optimum network performance. It is therefore convenient to define transient magnitudes in terms of the power excursion from the planned steady-state levels. The definition of transient parameters, including the maximum power excursions above and below the initial steady-state value for a channel drop event, is shown in Fig. 12.16. One possible impact of positive optical power spikes is the potential for damage to optical receivers and other components. The damage threshold of most optical components is duration dependant and is typically higher for shorter excursions. The susceptibility to damage is therefore a function of both the amplitude and speed of positive power transients. Another consideration is data integrity, which is limited by the sensitivity of the detection circuit to changes in average optical power. The maximum magnitude of optical transients that can accumulate in a transmission system must be considered when designing the network to ensure that the optical power at the receiver remains well within the dynamic range of the receiver to ensure error-free operation. Optical power excursions that are significantly slower than the data transmission rate can be prepared for by adjusting the detection threshold according to changes in the average power level. However, the accommodation of large optical transients can be difficult. In the experiment, performance of a typical of 2.5 Gb/s WDM transmission system with 100-µs duration transient pulses showed that the received optical power remains well within the regime of error-free steady-state operation for small excursions (<4 dB). However, larger power variations of 100-µs duration prevent successful data transmission.

In addition to the amplitude, the speed of the transient pulse can be an important factor. The detection circuitry design is usually limited by the slowest data rate that may be carried by the system, to avoid distortion of the data pattern and a degraded eye diagram. This limit restricts the speed of average power changes that can be compensated for by the detection circuit. The measurement of performance degradation for a 2.5 Gb/s receiver circuit as a function of transient speed showed significant degradation for transient rise times exceeding 200 µs for 4-dB transient amplitude. The study concluded that for a given detection circuit and receiver, the system performance is always determined both by the amplitude and speed of incoming power transients.

In another report [30], the maximum power excursion of one of the surviving channels was

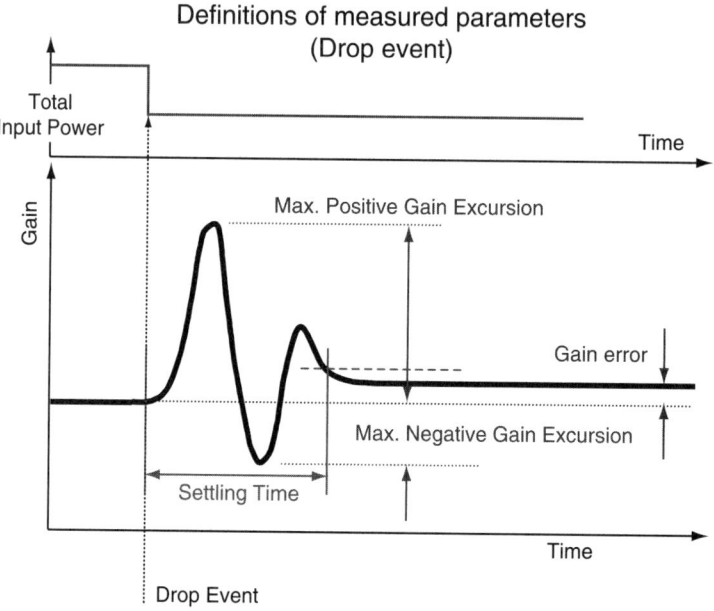

FIGURE 12.16 Definitions of different parameters used to characterize signal power transients for a channel drop event.

measured to study the effect of transient propagation along a chain of EDFAs. The power of one of the surviving channels at the output of each span for a 3-dB add/drop (20 of 40 channels) is plotted in Fig. 12.17. Without transient control, a large excursion was measured after the first EDFA. On the other hand, with pump control, a maximum excursion of 0.1 dB was measured per EDFA. However, it was observed that the excursion increases after every span. After 12 EDFAs the measured excursion is just under 1 dB. The reason for the transient increase along the EDFA chain is as follows. After the first EDFA, the surviving channel waveform acquires a momentary power deviation after an add/drop event. This deviation is now an inherent component of the signal power and cannot be corrected by EDFAs farther down the link. Thus, the transient at the output of the second EDFA is the combination of the transient at the output of the first EDFA and any transient acquired at the second EDFA due to the add/drop. Clearly, even with transient-control EDFAs, the transient increases in speed and magnitude after each EDFA in the link. These results show that fast transient control in multispan links is essential for next generation networks.

As discussed earlier, the EDFA transients degrade BER performance and quality of service of surviving channels. During an *add* event, the OSNR decreases, and the signal power can fall below the threshold level of the receiver (i.e., the BER tester, in this case). During an overshoot caused by a drop event, BER performance may be degraded due to nonlinearities caused by the high channel power or by a suboptimal receiver threshold. Naturally, the severity increases for longer chains of EDFAs. Figure 12.18 shows the time-dependent BER performance at the end of the link for add and drop events. During the *add* event, the BER increases to a level higher than 10–4, though the chain of transient-control EDFAs brings the system performance back to the original level within 30 μs. The BER increase during the *drop* event is much less significant, as shown in Fig. 12.18. These data show how quickly BER can degrade during a transient event as well as the large impact on performance that even a controlled transient can have. However, the data also suggest that for a 3-dB add/drop in a long-haul link, fast transient control

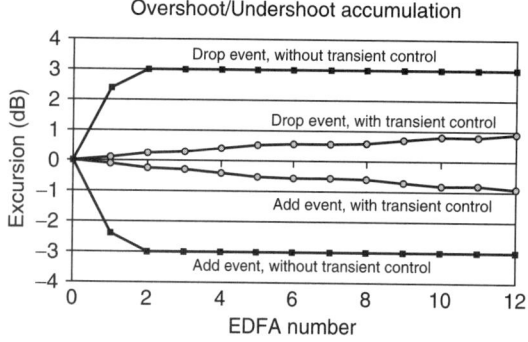

FIGURE 12.17 Accumulation of maximum channel power excursion along the link after a 3-dB add/drop.

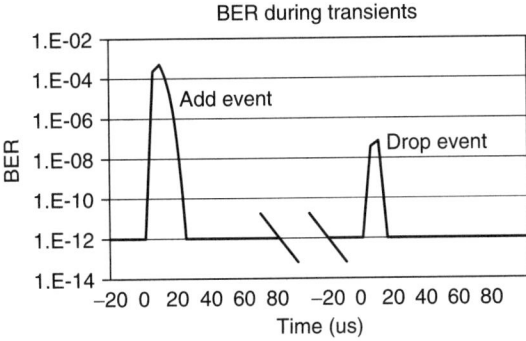

FIGURE 12.18 BER measurement for 3-dB add/drop at the end of the 800-km link.

EDFAs can quickly bring the link performance back to the proper level. To get a sense of the improvement over the case with no transient control, it is much larger than the 0.1-dB excursion after one amplifier in the transient-controlled case. This strongly suggests that transient-control EDFAs must be evaluated in a field-deployed long-haul link to properly assess performance. The BER degradation measured in the experiment demonstrates the need for fast transient control. After propagation through 12 EDFAs, the signal transient was dramatic enough to cause severe degradation to the BER. However, because the commercial EDFAs used in this experiment control the transient on a fast time scale, the BER degradation lasted less than 30 μs. Altogether, the experimental results verify that high-speed transient control is necessary for next generation networks, and the results prove that system makers *must* evaluate the performance of transient-control EDFAs across the *entire link* to correctly assess the performance of a system.

A new technique for making time-resolved measurements of BER and OSNR was reported to assess the BER of surviving channels during transients [31]. In the experiment, the OSNR and the BER of a surviving channel were measured for different channel loading. A severe drop of 5 dB in OSNR was observed despite the fact that the channel power decreased by only 3 dB. After measuring the OSNR as a function of time during an add/drop, its degradation was quantified in the electrical domain using BER measurements. The measurement resolution in time is limited by (1) the duration of the gating interval and by (2) the duty cycle where data is collected. For example, in the case of a sampling pulse of 20 ms with a repetition rate of 500 Hz, the number of errors in a measurement interval of 10 seconds at 10 Gb/s at a BER of 1×10^{-8} is 10, which is barely enough statistically. After 340 km of transmission, the BER of the surviving channel is shown in Fig. 12.19 when transient control of the EDFAs is disabled. BER of 10^{-12} or lower is inferred from Q factor measurements. The worst BER occurs at the channel power undershoot that follows immediately after a channel-add event. A corresponding dip in the OSNR can also be found at the identical time delay. When transient control is enabled, the BER and the Q factor are found to remain unchanged during the add event.

3.4. Channel Protection Schemes

Several schemes have been demonstrated to control the unwanted power excursions of surviving channels in EDFAs. These include fast pump control, link control via insertion of a compensating signals, and gain clamping by an all-optical feedback loop.

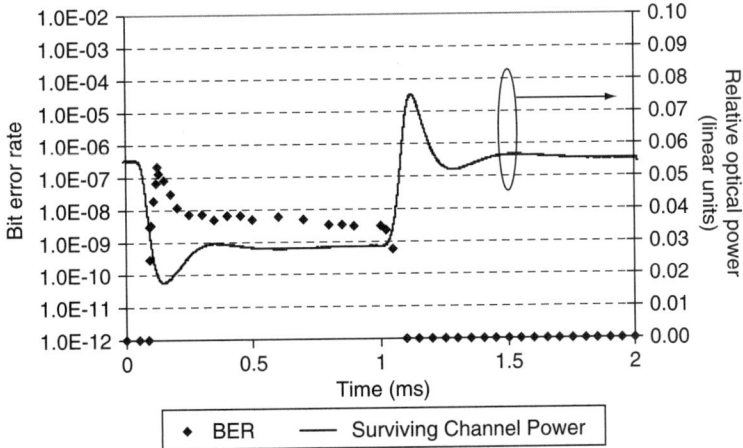

FIGURE 12.19 Plot of the time-resolved BER during a channel-add event when transient control is off. When transient control is enabled, the BER and the Q factor are found to remain unchanged during the add event [31].

Most of these schemes have focused on achieving a common goal: The maximum value of the power excursions of the surviving channel should be less than a fraction of a decibel for any possible change in channel loading to minimize the error bursts caused by signal power transients resulting from a line failure or a network reconfiguration. Such error bursts in surviving channels represent a service impairment that is absent in electronically switched networks and is unacceptable to service providers. In this section we describe the previously-mentioned three schemes to control the gain of amplifier.

3.4.1. Pump Control

The gain of an EDFA can be controlled by adjusting its pump current. The basic scheme for the pump control is shown in Fig. 12.20 and involves making measurements of input and output power of the EDFA through signal taps and monitor photodiodes. Early reported work addressed pump control on time scales of the spontaneous lifetime in EDFAs [22]. One study demonstrated low frequency feed-forward compensation with a low frequency control loop [18]. Results of pump power control on time scales much shorter than the erbium spontaneous lifetime demonstrated to arrest the power excursion in the surviving channels are shown in Fig. 12.21. The necessary response time was characterized by monitoring the power of the surviving channel as a function of the delay after the cutoff of the dropped channels. The second-stage pump power of the amplifier is then decreased by an amount suitable to restore the gain of the surviving channels. The experiment demonstrates that the dynamic time scales for changes in signal power and pump power are comparable and the power excursion of the surviving channels can be arbitrarily limited if the pump power is decreased with sufficiently short delay. For example, in the last trace, negligible power excursion occurs when a correction is applied after a delay of 7 μs. This shows that with standard pumps if the decision to take the corrective action can be reached in time, the pump power can be turned down quickly enough to control the excursions of surviving channels. These measurements demonstrate that for the pump control to minimize the variations in the power of the surviving channels in case of channel loss, the response of the control scheme must be at the most a few tens of microseconds.

After the discovery of fast power transients, pump control on short time scales [32] was demonstrated

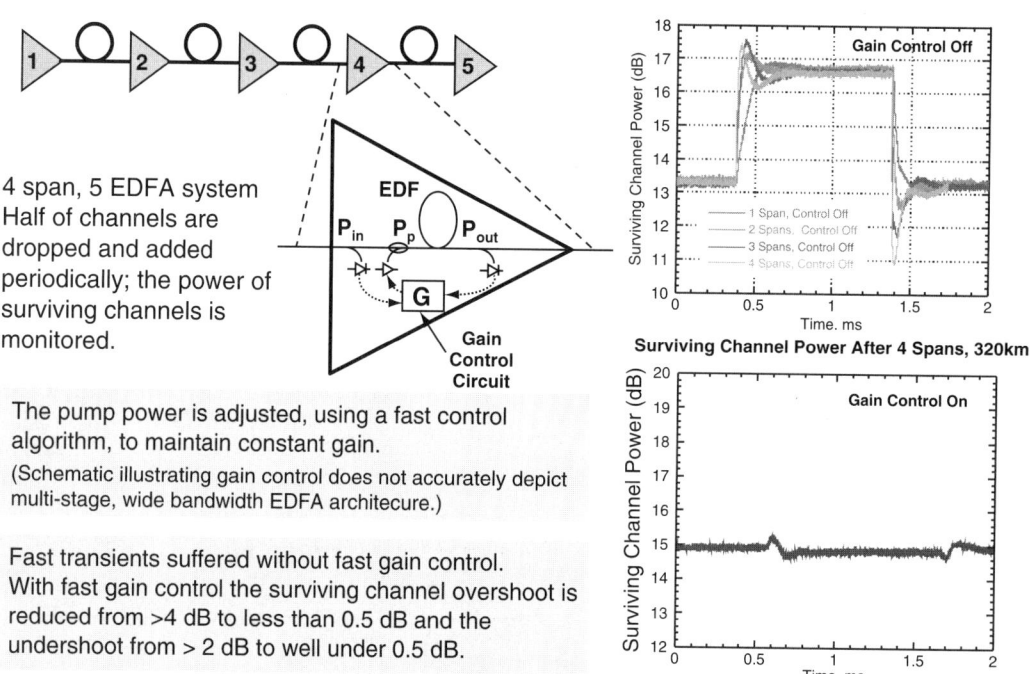

FIGURE 12.20 A schematic of the pump control scheme for EDFAs.

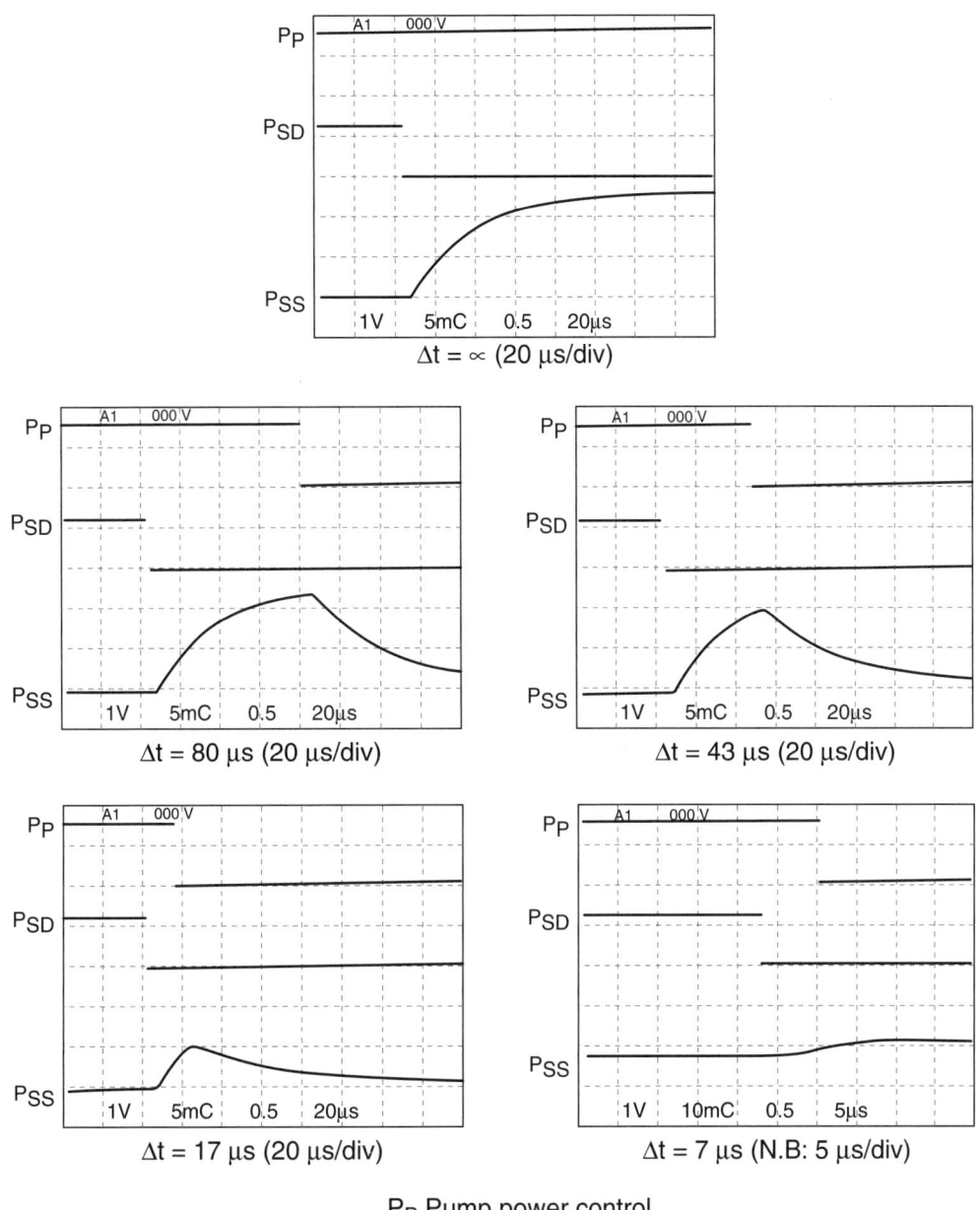

FIGURE 12.21 Oscilloscope traces showing the behavior of the surviving channel power (P_{SS}) transient when the pump power (P_P) is reduced after indicated delay time Δt following the channel loss [18].

to limit the power excursion of surviving channels. In the experiment, automatic pump control in a two-stage EDFA operating on a time scale of microseconds was demonstrated. The changes in the surviving channel power in the worst case of seven-channel drop/add in an eight-channel WDM system are shown in Fig. 12.22. In the absence of gain control, the change in surviving channel signal power exceeds 6 dB. When the pump control on both stages is active, the power excursion is less than 0.5 dB both for drop and add conditions. The control circuit acts to correct the pump power within 7–8 μs, and this effectively limits the surviving channel power excursion. In practical EDFAs, a suitable combination of the control of the two stages is necessary to optimize the NF, mainly determined by the pumping of the first stage. A careful calibration of the ASE in the amplifier under different operating

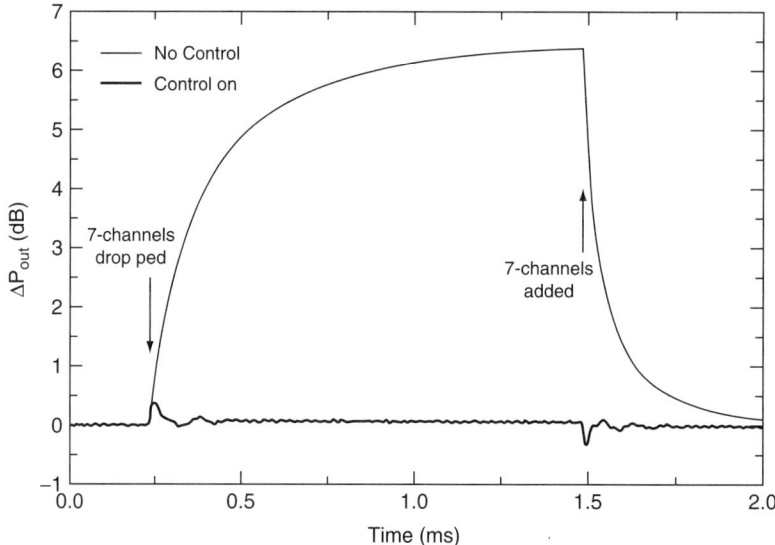

FIGURE 12.22 Impact of pump control on surviving channel power transient in a two-stage EDFA when seven of eight channels are dropped and added.

conditions is necessary in order to subtract it to obtain the actual signal strength and accurate gain measurement.

Conventional synchronous optical networks and synchronous digital hierarchy ring networks have built-in protection switching and a service restoration mechanism that has a 50-ms response time. Alternatively, physical layer protection switching can be achieved in nearly 15 ms in an advanced WDM system. In both cases, transients are not detected quickly enough to prevent system performance degradation, as illustrated by the example in Fig. 12.23. The signal power variation of a surviving channel in a four fiber span system is shown when half of the channels are dropped and added (Fig. 12.23), and the corresponding eye diagram (Fig. 12.23) shows a BER degradation to 1×10^{-4}. When the fast transient control is switched, the surviving channel power variation is restricted to within 0.5 dB. Under these conditions, the eye diagram shows significant improvement with BER $< 1 \times 10^{-12}$. This example shows the necessity of fast transient control in a WDM network that guarantees quality of service.

3.4.2. Link Control

The pump control scheme described above would require protection at every amplifier in the network. Another technique makes use of a control channel in the transmission band to control the gain of amplifiers. Earlier work demonstrated gain compensation in an EDFA at low frequencies (<1 kHz) using an idle compensation signal [33]. Link control, which provides surviving channel protection against fast transients, has been demonstrated [16]. The scheme, as illustrated in Fig. 12.24, protects surviving channels on a link-by-link basis. A control channel is added before the first optical amplifier in a link (commonly the output amplifier of a network element). The control channel is stripped off at the next network element (commonly after its input amplifier) to prevent improper loading of downstream links. The power of the control channel is adjusted to hold constant the total power of the signal channels and the control channel at the input of the first amplifier. This maintains constant loading of all EDFAs in the link.

In the experimental demonstration of link control surviving channel protection, the setup consists of seven signal channels and one control channel. A fast feedback circuit with a response time of 4 μs is used to adjust the line control channel's power to maintain total power at a constant level. The signal channels and control channels are transmitted through seven amplified spans of fiber, and BER performance of one of the signal channels is monitored. The results of system measurements are summarized in Fig. 12.25. When five of seven signal channels are added/dropped at a rate of 1 kHz, the surviving channel suffers a power penalty exceeding 2 dB and exhibits a severe BER floor. An even worse BER floor is observed when five of six channels are added/dropped, resulting from cross-saturation induced by change in channel loading. With the fast link control in operation, the power excursions are

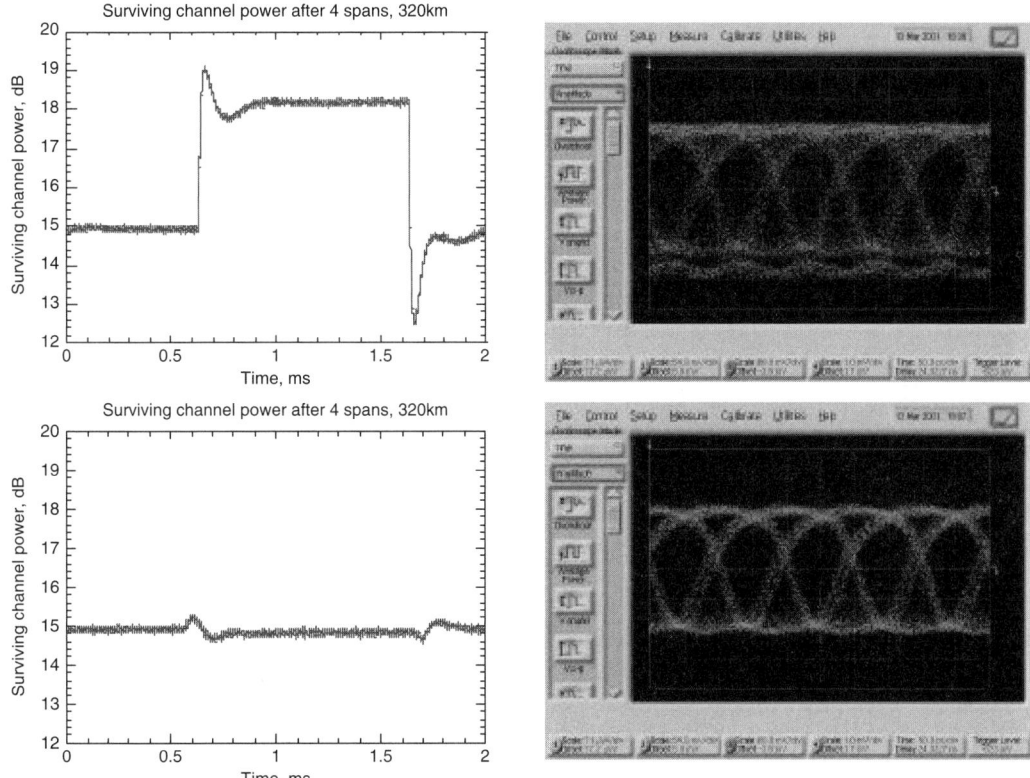

FIGURE 12.23 Impact of transient control on system performance. The top two figures show surviving channel power excursion (left) and eye diagram (right) in absence of axis transient control. The bottom two figures show the corresponding characteristics with transient control.

mitigated and BER penalties are reduced to a few tenths of a decibel and error floors disappear.

The link control technique is fast; changes in channel loading result in prompt changes in a link's total power permitting much faster detection and response than schemes that rely on detecting the much slower changes in channel output power, gain, or ASE in individual EDFAs (which are much slower than the fast transients resulting from collective effects in EDFA chains and networks). Protection on a per link basis rather than a per amplifier basis does not increase the complexity of the network's EDFAs and is simple, less expensive, and well suited to the architecture of wavelength routed networks.

3.4.3. Laser Control

In this section we present an all-optical gain stabilization technique for multichannel EDFAs and the impact on the WDM network performance requirement. The laser automatic gain control (AGC) is achieved by placing the amplifier in a laser cavity

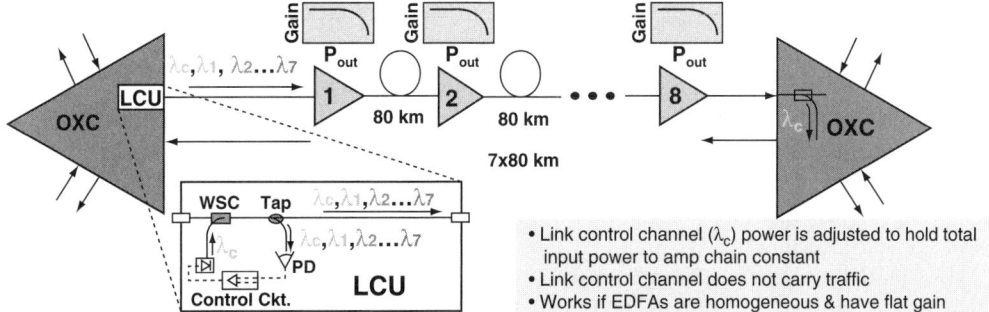

FIGURE 12.24 Schematic representation of link control for surviving channel protection in optical networks: Link Control Unit (LCU), Optical Cross Connect (OXC).

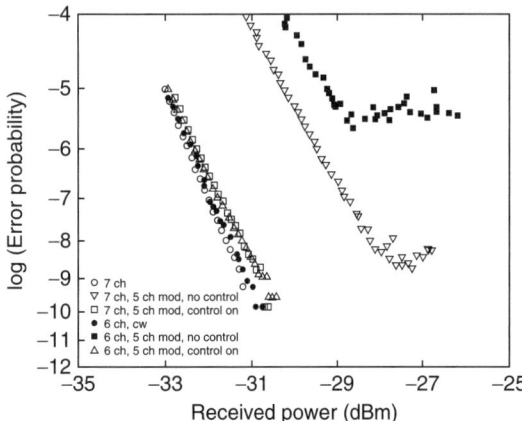

FIGURE 12.25 BERs measured for transmission of six and seven channels without modulation and with modulation of channels 1, 2, 3, 5, and 8 on/off at 1 kHz with and without link control [16].

to clamp its gain [34–36]. The gain of the amplifier is clamped by the laser, and its power changes in response to channel loading in the amplifier. It is possible to achieve the same steady-state signal gain in an all-optical gain controlled EDFA configuration using different lasing wavelengths. The steady-state behavior in terms of output powers may be the same, but the transients are wavelength dependent. Figure 12.26 shows transient response of the surviving signal output power for three different lasing wavelengths for the following scenario: addition and loss of seven of the eight WDM channels in a two-stage EDFA. At the 1532-nm lasing wavelength shown in Fig. 12.26a, the laser AGC suppresses transients in the surviving channel, but not completely. The signal experiences a power offset of 0.62 dB compared with the steady-state value. This offset arises from spectral hole burning (SHB), that is, inhomogeneity of the erbium gain medium. The transition between the two gain levels is from 100 to 200 μs, reflecting the slow gain dynamics of the erbium gain medium. The spikes arising due to relaxation oscillation in the laser do not undershoot and/or overshoot the steady-state values for lasing wavelengths near 1530 nm. This indicates that when the lasing wavelength is close to the spectral band occupied by signal wavelengths, steady-state power excursions arising from SHB are dominant. At 1540 nm, as can be seen from Fig. 12.26b, relaxation oscillations observed in the laser give rise to fast oscillations in the surviving channel power, resulting in transients that undershoot the lower gain level and overshoot the higher gain level. On the other hand, the residual power excursions due to SHB become smaller. At the 1555-nm wavelength (Fig. 12.26c), relaxation oscillations imposed by the laser give rise to the dominant residual power excursions in the surviving signal, whereas the power excursions due to SHB are completely absent. As can be seen from the figure, both the frequency and amplitude of the transient power excursions of the surviving channel are lower in the case of adding channels versus that of losing channels. Dropping channels results in a much more severe effect on the surviving channels.

From this report, it is clear that there is a trade-off in selecting the lasing control wavelength to minimize impairments from SHB and a dynamic contribution due to the relaxation oscillations in the laser, and a lasing wavelength cannot be chosen that avoids both impairments. These effects are shown to be small but compound in large networks of concatenated EDFAs in which laser AGC is used in each EDFA and the performance of surviving channels maybe be impaired. Residual power excursions related to SHB resulting from inhomogeneity are less significant as the lasing wavelength approaches the spectral band occupied by the signal wavelengths centered around 1550 nm. However, in this case, it is the dynamic power excursion imposed on the amplified signals due to relaxation oscillations in the laser that becomes the dominant effect.

Laser automatic gain control has been extensively studied since it was experimentally demonstrated [34]. A new scheme for link control based on laser gain control has been proposed [37]. In this work, a compensating signal in the first amplifier is generated using an optical feedback laser loop and then propagates down the link. Stabilization is reached within a few tens of microseconds, and output power excursion after six EDFAs is reduced by more than an order of magnitude to a few tenths of a decibel. For laser gain control, the speed is limited by laser relaxation oscillations [38,39], which are generally on the order of tens of microseconds or slower. Inhomogeneous broadening of EDFAs and the resulting SHB can cause gain variations at the signal wavelength, which limit the extent of control from this technique. The same is true for the link control scheme.

4. ACKNOWLEDGMENTS

Much of the work discussed in this chapter was done at the Crawford Hill Laboratory of Lucent Technologies in Holmdel, New Jersey. We are grateful to all our colleagues for very fruitful collaborations. Special thanks go to J. L. Zyskind who helped

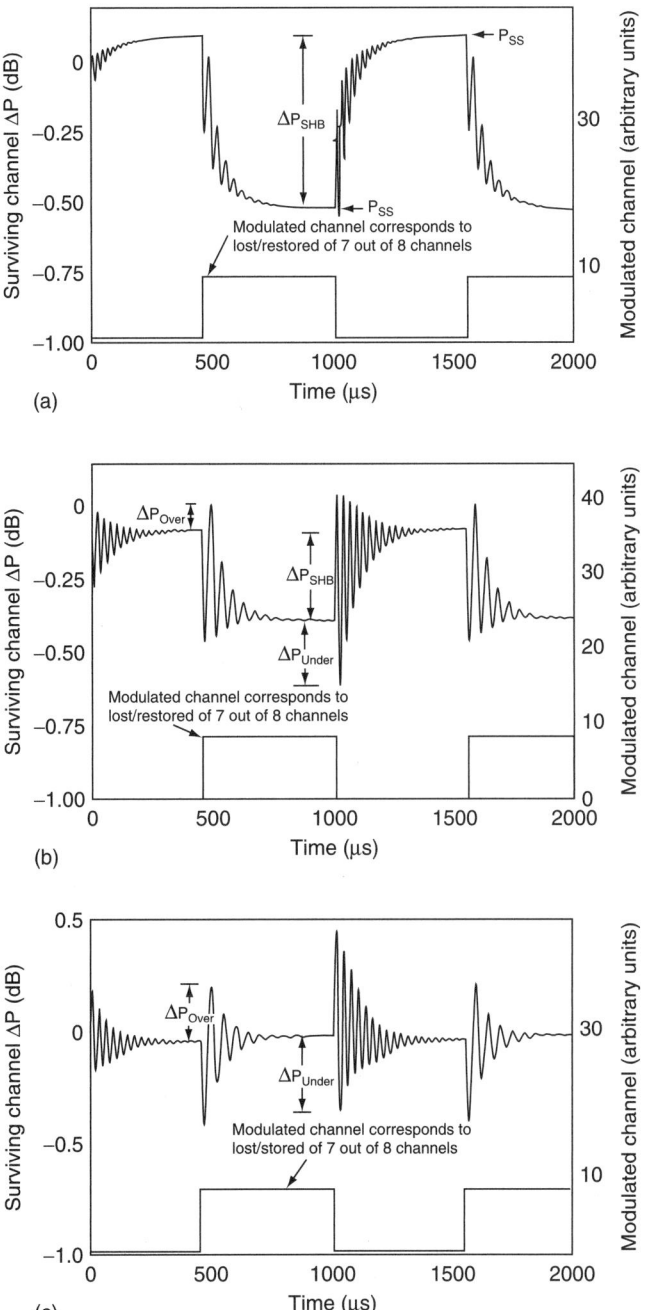

FIGURE 12.26 Transient response of surviving channel output power to dropping/adding seven of eight channels. (a) The amplifier operated at 15 dB gain compression. The lasing wavelength is 1532 nm. (b) The same as (a) except that the lasing wavelength is 1540 nm (c) The same as (a) except the lasing wavelength is 1555 nm [38].

the authors and made significant contribution to erbium-doped fiber amplifier research at Crawford Hill Laboratory. We are also thankful to several colleagues at Onetta for sharing their results and many exciting discussions. Collaboration with the Optical Group at Sprint Labs is also gratefully acknowledged.

5. REFERENCES

1. R. Ramaswami and K. N. Sivarajan, *Optical Networks*, Morgan Kaufman, San Francisco (1998).
2. K. Fukuchi, "Wideband and ultra-dense WDM transmission technologies toward over 10 Tb/s capacity," Paper THX5, Proc. Optical Fiber Communications Conf., Anaheim, CA (March 2002).

3. A. K. Srivastava and Y. Sun, "Advances in erbium-doped fiber amplifiers," *Optical Fiber Telecommunications IVA*, Ed. I. Kaminow and T. Li, pp. 174–212. Academic Press, London (2002).

4. R. J. Mears, Y. L. Reekie, J. M. Jauncey and D. N. Payne. "Low noise erbium-doped fiber amplifier operating at 1.54 μm," *Electron. Lett.*, **23**, September 10, pp. 1026–1028 (1987).

5. E. Desurvire, J. R. Simpson, and P. C. Becker, "High gain erbium-doped traveling wave fiber amplifier," *Opt. Lett.*, **12**, pp. 888–890 (1987).

6. A. K. Srivastava, Y. Sun, J. W. Sulhoff, C. Wolf, M. Zirngibl, R. Monnard, A. R. Chraplyvy, A. A. Abramov, R. P. Espindola, T. A. Strasser, J. R. Pedrazzani, A. M. Vengsarkar, J. L. Zyskind, J. Zhou, D. A. Ferrand, P. F. Wysocki, J. B. Judkins, and Y. P. Li, "1 Tb/s transmission of 100 WDM 10 Gb/s channels over 400 km of TrueWave™ fiber," OFC '98 Technical Digest, Postdeadline Paper, PD 10, San Jose, CA, February 22–27 (1998).

7. D. Fishman, J. A. Nagel, T. W. Cline, R. E. Tench, T. C. Pleiss, T. Miller, D. G. Coult, M. A. Milbrodt, P. D. Yeates, A. Chraplyvy, R. Tkach, A. B. Piccirilli, J. R. Simpson, and C. M. Miller, "A high capacity non-coherent FSK lightwave field experiment using Er-doped fiber optical amplifiers," *Photonic Technol. Lett.*, **2**, No. 9, p. 662 (1990).

8. C. R. Giles and E. Desurvire, "Modeling erbium-doped fiber amplifiers," *J. Lightwave Technol.*, **9**, Feb., pp. 271–283 (1991).

9. Y. Sun, J. L. Zyskind, and A. K. Srivastava, "Average saturation level, modeling, and physics of erbium-doped fiber amplifiers," *IEEE J. Selected Areas in Quantum Elect.*, **3**, No. 4, pp. 991–1007 (August 1997).

10. P. F. Wysocki, J. R. Simpson, and D. Lee, "Prediction of gain peak wavelength for Er-doped fiber amplifiers and amplifier chains," *IEEE J. Photon. Technol. Lett.*, **6**, No. 9 pp. 1098–1100 (1994).

11. J. L. Zyskind, J. A. Nagel, and H. D. Kidorf, "Erbium-doped fiber amplifiers," *Optical Fiber Telecommunications*, Vol. IIIB, Ed. I. P. Kaminow and T. L. Koch, Academic Press, pp. 13–68 (1997).

12. R. G. Smart, J. L. Zyskind, and D. J. DiGiovanni, "Two-stage erbium-doped fiber amplifiers suitable for use in long-haul soliton systems," *Electron. Lett.*, **30**, No. 1, pp. 50–52 (1994).

13. J.-M. P. Delavaux and J. A. Nagel, "Multi-stage erbium-doped fiber amplifier designs," *J. Lightwave Technol.*, **13**, No. 5, pp. 703–720 (1995).

14. A. R. Chraplyvy, J.-M. Delavaux, R. M. Derosier, G. A. Ferguson, D. A. Fishman, C. R. Giles, J. A. Nagel, B. M. Nyman, J. W. Sulhoff, R. E. Tench, R. W. Tkach, and J. L. Zyskind, "1420-km Transmission of sixteen 2.5-Gb/s channels using silica-fiber-based EDFA repeaters," *Photonic Technol. Lett.*, **6**, No. 11, pp. 1371–1373 (1994).

15. F. Forgieri, R. W. Tkach, and A. R. Chraplyvy, "Fiber nonlinearities and their impact on transmission systems," *Optical Fiber Communications*, Vol. IIIA, Ed. I. P. Kaminow and T. L. Koch, Academic Press, pp. 196–264, (1997).

16. A. K. Srivastava, J. L. Zyskind, Y. Sun, J. Ellson, G. Newsome, R. W. Tkach, A. R. Chraplyvy, J. W. Sulhoff, T. A. Strasser, C. Wolf, and J. R. Pedrazzani, "Fast-link control protection of surviving channels in multiwavelength optical networks," *IEEE Photon. Technol. Lett.*, **9**, No. 11, pp. 1667–1669 (1997).

17. J. L. Zyskind, Y. Sun, A. K. Srivastava, J. W. Sulhoff, A. L. Lucero, C. Wolf, and R. W. Tkach, "Fast power transients in optically amplified multiwavelength optical networks," Proc. Optical Fiber Communications Conference, Postdeadline Paper PD 31 (1996).

18. C. R. Giles, E. Desurvire, and J. R. Simpson, "Transient gain and cross-talk in erbium-doped fiber amplifier," *Opt. Lett.*, **14**, No. 16, pp. 880–882 (1989).

19. A. K. Srivastava, Y. Sun, J. L. Zyskind, and J. W. Sulhoff, "EDFA transient response to channel loss in WDM transmission system," *IEEE Photonics Technol. Lett.*, **9**, No. 3, pp. 386–388 (1997).

20. E. Desurvire, "Analysis of transient gain saturation and recovery in erbium-doped fiber amplifiers," *IEEE Photonics Technol. Lett.*, **1**, pp. 196–199 (1989).

21. Y. Sun, J. L. Zyskind, and A. K. Srivastava, "An analytical formula on the transient response of an erbium-doped fiber amplifier." (unpublished work).

22. E. Desurvire, *Erbium-doped fiber amplifiers—principles and applications*, John Wiley, NY, pp. 469–480 (1994).

23. P. R. Morkel and R. I. Laming, "Theoretical modeling of erbium-doped fiber amplifiers with excited state absrorption," *Opt. Lett.*, **14**, pp. 1062–1064, (1989).

24. Y. Sun, G. Luo, J. L. Zyskind, A. A. M. Saleh, A. K. Srivastava, and J. W. Sulhoff, "Model for gain dynamics in erbium-doped fiber amplifiers," *Electron. Lett.*, **32**, No. 16, pp. 1490–1491 (1996).

25. S. J. Shieh, J. W. Sulhoff, K. Kantor, Y. Sun, and A. K. Srivastava, "Dynamic behavior in L-band EDFA," OFC'00 Paper PD-27, Baltimore, MD, March 7–10 (2000).

26. Y. Sun, J. L. Zyskind, A. K. Srivastava, J. W. Sulhoff, C. Wolf, and R. W. Tkach, "Fast Power Transients in WDM Optical Networks with Cascaded EDFA's" *Electron. Lett*, **33**, No. 4, February 13, 1997, pp. 313–314.

27. MONET, 2nd Quarterly Report, Payable Milestone #2 (1995).

28. Y. Sun and A. K. Srivastava, "Dynamic effects in optically amplifed networks," Proc. Optical Amplifiers and their Applications, Victoria, Canada, July (1997).
29. P. Lundquist, M. Levesque, J. Morrier, and D. Zaccarin, "Optical transients in cascaded EDFA's: effects on transmission system performance," Proc. Optical Amplifiers and their Applications, Otaru, Japan, Paper MD03, pp. 70–72 (2003).
30. I. M. White, S. Chen, S. Yam, D. Harris, G. Kalogerakis, J. Pan, A. K. Srivastava, Y. Sun, Yuh-Jen Cheng, and R. Monnard, "Optical transient control in an 800 km field deployed link" (unpublished).
31. W. S. Wong, H.-S. Tsai, C.-J. Chen, H. K. Lee, and M.-C. Ho, "Novel time-resolved measurements of bit-error-rate and optical-signal-to-noise-ratio degradations due to EDFA gain dynamics in a WDM network," Proc. Optical Fiber Communications Conference, Anaheim, CA, Paper THR4, pp. 515–516 (2002).
32. A. K. Srivastava, Y. Sun, J. L. Zyskind, J. W. Sulhoff, C. Wolf, and R. W. Tkach, "Fast gain control in erbium-doped fiber amplifiers," Proc. Optical Amplifiers and their Applications, Postdeadline Paper PDP4 (1996).
33. E. Desurvire, M. Zirngibl, H. M. Presby, and D. DiGiovanni, "Dynamic gain compensation in saturated erbium-doped fiber amplifiers," *IEEE Photonic Technol. Lett.*, **3**, pp. 453–455 (1991).
34. M. Zirngibl, "Gain control in erbium-doped fiber amplifiers by an all-optical feedback loop," *Electron. Lett.*, **27**, pp. 560–561 (1991).
35. E. Delavaque, T. Georges, J. F. Bayon, M. Monerie, P. Niay, and P. Bernage, "Gain control in erbium-doped fiber amplifiers by lasing at 1480 nm with grating written on fiber ends," *Electron. Lett.*, **29**, pp. 1112–1113 (1993).
36. B. Landousies, T. Georges, E. Delavaque, R. Lebref, and M. Monerie, "Low power transient in multi-channel equalized and stabilized gain amplifier using passive gain control," *Electron. Lett.*, **32**, pp. 1912–1913 (1996).
37. J. L. Jackel and D. Richards, "All-optical stabilization of cascaded multi-channel erbium-doped fiber amplifiers with changing numbers of channels," OFC '97 Technical Digest, pp. 84–85, Dallas, TX, February 22–27 (1997).
38. Luo, J. L. Zyskind, Y. Sun, A. K. Srivastava, J. W. Sulhoff, and M. A. Ali, "Relaxation oscillations and spectral hole burning in laser automatic gain control of EDFA's," OFC '97 Technical Digest, pp. 130–131, Dallas, TX, February 22–27 (1997).

13
CHAPTER

Fused Fiber Couplers: Fabrication, Modeling, and Applications

Bishnu P. Pal,* P. Roy Chaudhuri, M. R. Shenoy, and Naveen Kumar

Physics Department
Indian Institute of Technology Delhi
New Delhi, India

1. INTRODUCTION

Development of broadband optical fiber amplifiers in the form of erbium-doped fiber amplifiers (EDFAs), which led to the birth of the era of wavelength division multiplexing (WDM) technology in the mid-1990s, ushered in a revolution in optical fiber communication. WDM technology, more precisely known as dense WDM (DWDM) technology, involving simultaneous transmission of at least four wavelengths in the 1550-nm low-loss wavelength window through one single-mode fiber has indeed resulted in an enormous increase in available bandwidth for high-speed telecommunications and data transfer. Terabit transmission of signals has already been demonstrated at least on a laboratory scale through DWDM [1]. The advent of DWDM technology has created a demand for a host of devices and components. Some of these are directional couplers, fiber Bragg gratings, silicon optical bench/planar lightwave circuits, micro-optoelectromechanical systems for switching, and other optical signal processing functions.

The optical networks require a host of branching components that distribute optical signals among the fibers in a predetermined fashion. Several of these components are passive and data-format transparent and are able to combine/split optical power or multiplex wavelengths, regardless of the information content of the signal. If configured in an all-fiber form, such devices could be introduced in a communication link through splicing to the transmission fiber with relatively low insertion loss. Fused biconically tapered fiber coupler technology is a versatile platform to realize a variety of such components for an optical network. Often referred to as a fused fiber coupler (FFC), it forms the building block for a variety of branching components used in various applications; these range from passive telephony, fiber amplifiers, and cable television networks, to distributed sensing [2, 3]. The basic device in FFC technology is a 2 × 2 fused fiber coupler, which can be configured to realize power splitter/combiners, wavelength division multiplexers and demultiplexers, bandpass/bandstop filters, broadband and wavelength selective couplers, polarization splitters, signal taps, and so on. In this chapter we present an overview of the fabrication process, developed in-house, and a supermode analysis to model an FFC. A summary description follows of a variety of all-fiber components realized by us that are relevant to optical fiber communication.

2. FABRICATION

As is well known, the basic technique of fabrication of an FFC involves lateral flame fusion and simultaneous stretching of a pair of unjacketed single-mode fibers over a short length in a high temperature microflame (or electric furnace), as shown in Fig. 13.1a. The fused region transforms into a biconical-tapered junction (Fig. 13.1b), and hence it is also called "fused biconical tapered" coupler. Because the input and output ends are identical, either side may be used as an "input" end, and hence these components are also referred to as bidirectional couplers. During the process of "fusion and elongation," light is launched into one of the input ports and output light from the two exit fiber ports are constantly monitored. This enables one to appropriately control the precise time for withdrawal of the flame and stoppage of fiber pull to realize a targeted distribution of light between the two output ports. For a given pair of fibers that are optimized for a given wavelength of operation, the conditions of fusion and pulling (flame condition and pulling speed) dictate the performance of the fabricated coupler in terms of excess loss, splitting ratio, spectral response, and back reflection. Because of the smallness of the fiber dimensions, the

---

*E-mail: bppal@physics.iitd.ernet.in

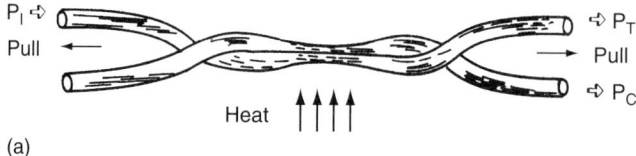

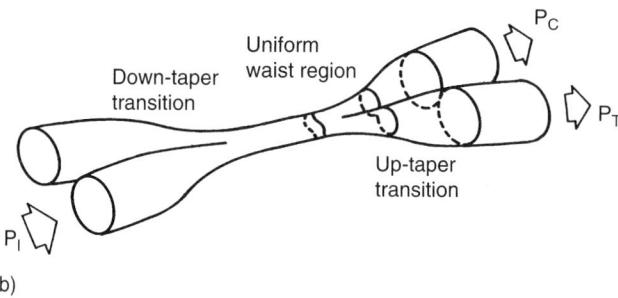

FIGURE 13.1 (a) Fabrication of fused coupler: the basic "fuse-pull-taper" method. (b) The basic structure of a fused biconical tapered (FBT) fiber coupler.

process of fabrication requires a high degree of precision and control on the process parameters, and even a minute perturbation in any of the process variables from the designed ones may significantly affect the performance characteristics of the device [4].

Figure 13.2 shows a schematic of the integrated fabrication assembly developed in-house, which comprise the mechanical hardware and the associated electronic control system. The mechanical system consists of a pair of translational pulling stages and a flame-holder assembly. These are designed around a set of motorized precision movements.

The fiber pulling stages are identically spring loaded to impart a uniform tension to the to-be-fused pair of fibers. The initial set positions of these translation stages are ensured by the use of limiter switches. The flame holder unit is a motorized three-dimensional translation stage, interfaced with microswitches that enables insertion and withdrawal of the micro-flame burner, adjustment of flame height, and realization of a "flame-brush" to impart a uniform distribution of temperature in the fiber fusion zone during fusion. The entire setup is mounted on a vibration-free platform. All the electrical terminals are

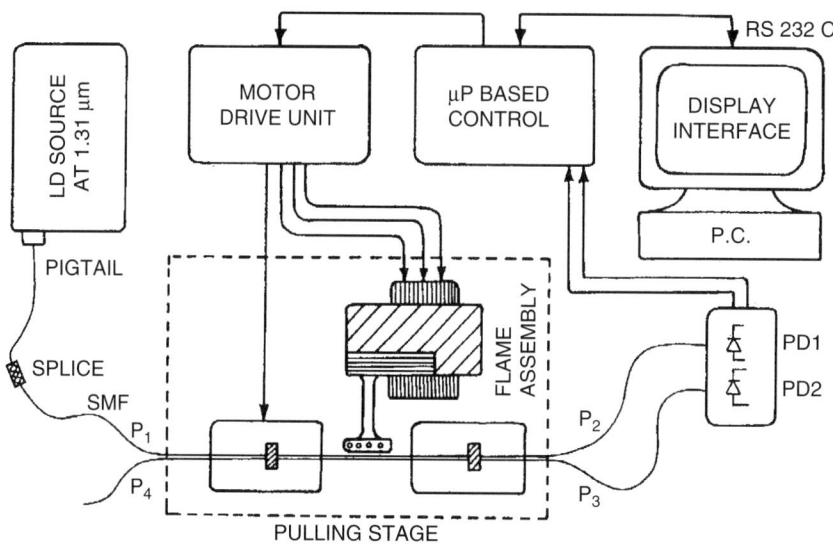

FIGURE 13.2 Schematic of the integrated fabrication with motorized mechanical stages, microprocessor interface, and a PC-based control system. SMF, single mode fiber.

connected to the drive circuits of a microprocessor (μP) interface, which controls these translational movements at different stages of the fabrication process [5]. A useful attribute of the software-driven fabrication is that it records all instantaneous status information in a backup file location. The data stored for the entire process is extremely relevant to analyze and investigate the coupling phenomenon, in particular, as a function of process parameters. The real-time generated pulling signature[†] at the monitoring wavelength is perhaps the most important signature of the target device (Fig. 13.3). This signature enables subsequent analysis of the coupling behavior for a given set of process parameters that could be fine tuned for precise realization of a targeted component.

Excess loss is one of the most important figures of merit for a fused fiber coupler. It is defined as (Fig. 13.1)

$$\text{Excess loss} = -10\, \log_{10}\left(\frac{P_T + P_C}{P_I}\right)$$

where P_T and P_C represent the output powers at the throughput and coupled ports, respectively, and P_I is the power coupled into the fiber at the input port. Figure 13.4 depicts the histogram for the measured excess loss of a large number of FFCs fabricated in-house over various periods and optimized for 1310 nm. It can be seen that the median of excess loss is 0.30 dB, with a standard deviation of 0.16 dB. Typical optical characteristics of various power splitters purported for operation at 1310, 1550, and 632.8 nm that were realized by using our machine are tabulated in Table 13.1. The fiber used in the fabrication of these couplers for operation at 1310- and 1550-nm wavelengths was an SMF-28™ (Corning, NY, USA) (optimized for single-mode operation at the 1310-nm wavelength and having a cutoff wavelength of 1270 nm). Couplers for single-mode operation at 632.8 nm were fabricated using an SM-600™ single-mode fiber from Fibrecore (Southampton, UK), which has a cutoff wavelength of 511 nm.

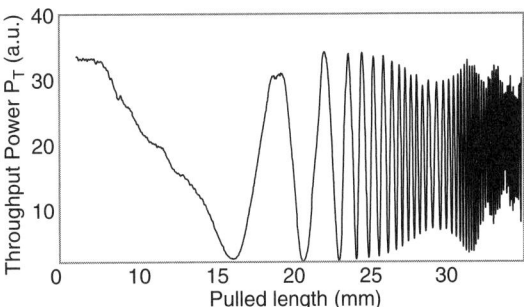

FIGURE 13.3 Recorded pulling signature during the fabrication of a strongly fused long over-coupled coupler; the monitoring wavelength was 1550 nm.

3. MODELING

Fused fiber couplers involve an axially-varying complex waveguide geometry [2, 3], and a rigorous mathematical analysis of the fused fiber coupler has been a great challenge to theorists for years. The most common route of analysis has been through coupled mode theory (CMT) [6, 7], which fails for modeling strongly interacting cladding-mode coupling. Supermode analysis adequately describes the propagation characteristics of a fused fiber coupler, even though it owes its origin to the coupled mode equations of the CMT. The beating between the supermodes along the fused region depends on the transverse geometry and the longitudinal taper profile of the coupler structure. Several authors have treated the cross-section of the fused region by approximating the structure as being rectangular [8–11], dumbbell

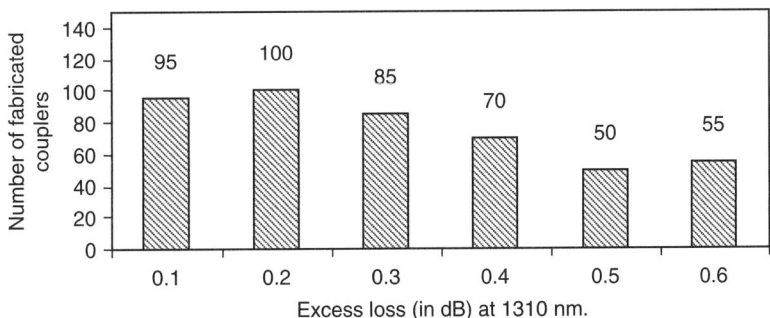

FIGURE 13.4 Histogram of excess loss for the fabricated couplers, measured at 1310 nm.

[†]A software entitled "Pullsig"© has been copyrighted by Foundation for Innovation and Technology Transfer (FITT), IIT Delhi.

TABLE 13.1 Typical optical characteristics of power splitters optimized for operation at three different wavelengths.

Characteristics	SMF at 1310 nm	SMF at 1550 nm	SMF at 632.8 nm
Splitting ratio	10–90%	10–90%	20–80%
Excess loss	0.1–0.3 dB	0.1–0.5 dB	0.6–1.1 dB
Directivity	40–60 dB	40–60 dB	45–55 dB

SMF, single-mode fiber.

shaped, or as two touching/intersecting [12–15] tapered fibers. The longitudinal taper profile, in most of the reported works, has been approximated by exponential [11, 13, 16], parabolic [17, 18], or sinusoidal/polynomial [9, 19] functions. In all cases, determination of the propagation characteristics of the (two lowest order) supermodes formed a key step in the modeling.

Methods well known in the literature dealing with this problem are effective index [20], finite element [14, 21–23], finite difference (FD) [24], beam propagation [25], point matching [26], boundary element [27], the method of Eyges et al. [28], circular harmonics expansion [29], and variational [30]. These methods, in some cases, yield reasonably accurate results but require a large computation time for calculation of each propagating mode. Some of the mode-solving routes, on the other hand, are efficient but do not yield the accuracy required for design and analysis. In most cases, the algorithms rely on involved algebra or a complex eigen-matrix formulation and thus need considerable programming effort. In addition, these approaches were unable to fully explain all the characteristics of fused couplers.

A very successful model was presented [31] using the scalar field correction method (FCM) [32], which could interpret most of the experimental characteristics of fused couplers. Though it is effectively equipped with exponentially decaying boundary fields, simple field optimization logic, and progressive grid multiplication scheme, the technique still requires an extensive computation time and memory, because the field distribution initiates from a uniform or a Gaussian profile. Furthermore, FCM uses an approximate polarization correction to the propagation constants of scalar modes for calculating the polarization dependence of mode coupling. Thus, there is room for improvement of the basic FCM to make it fast, precise, and computationally user friendly for design and analysis of fused fiber coupler–based branching components.

With this goal, we developed a method that is relatively simple, semianalytical, and fast enough to precisely yield the coupling characteristics of the propagating local supermodes. Also, the algorithm could be embedded with input process variables in a computer-controlled fabrication environment. The basic philosophy of this approach is to first calculate an approximate mode by applying the perturbation method (or Kumar method [16]) to an appropriately chosen pseudo-waveguide structure that replaces the true waveguide [33]. The effective index and field distribution so obtained are then modified through a discretized wave equation until they converge to a true mode. For birefringence analysis, the perturbation method being amenable to transverse electric (TE) and transverse magnetic (TM) polarizations, the formalism is adapted through semivectorial FD equations. In this analysis, we incorporated the shape of realistic taper profile and the resulting cross-sections for a given "degree of fusion" to model the characteristics exhibited by fabricated coupler components, namely 3-dB, tree and tap couplers, and classical WDM couplers for 980/1550- and 1310/1550-nm wavelengths.

3.1. Mode Analysis Algorithm

To determine the approximate modes of the waveguide using the perturbation method [33], we chose a rectangular core pseudo-waveguide, which has axial dimensions the same as those of the actual waveguide (or which closely resembles the latter in index profile) [34]. Thus, the fused coupler geometry simplifies to one with two square cores embedded in a hypothetical rectangular-cladding structure. The first-order correction of the perturbation recipe applied to this structure yields a modal field distribution and a corrected mode index as [33]

$$e(x, y) = e(x) \cdot e(y) \quad (1)$$

$$n_{effo}^2 = n_{effx}^2 + n_{effy}^2 - n_{core}^2$$
$$n_{effp}^2 = n_{effo}^2 + \Delta n_{eff}^2 \quad (2)$$

Where $e(x)$ and $e(y)$ are the field distributions of the x- and y-slab guides constituting the pseudo waveguide structure, while n_{effo} and n_{effp} represent mode effective indices of the pseudo waveguide and that of the perturbed waveguide after incorporation of correction to it (calculated through the perturbation method described below). The resulting modal field of the perturbed waveguide $e(x, y)$ and the corrected effective index $n_{\text{eff}p}$ of the mode, given by Eqs. (1) and (2), are then treated as an approximate solution for the mode of the true coupler structure.

For calculation of modes of different orders (namely, the fundamental and first order), the appropriate orders of the constituent slab modes are evaluated. For example, to calculate the even supermode of the perturbed waveguide, we calculate the fundamental modes of both x- and y-slabs, respectively, whereas the odd supermode is generated through the fundamental and first higher order modes of the y- and x-slabs, respectively. Thus, the field distribution and effective indices for the even and odd modes are given by

$$e_e(x, y) = e_0(x) \cdot e_0(y)$$
$$e_o(x, y) = e_1(x) \cdot e_0(y)$$

and

$$\beta_e^2 = \beta_{x0}^2 + \beta_{y0}^2 - k_0^2 n_{\text{core}}^2 + \Delta\beta_{00}^2$$
$$\beta_o^2 = \beta_{x1}^2 + \beta_{y0}^2 - k_0^2 n_{\text{core}}^2 + \Delta\beta_{01}^2 \quad (3)$$

where the subscripts 0 and 1 denote the fundamental and the first higher order modes.

By following the usual discretization scheme, namely $x = i \cdot \Delta x$, $y = j \cdot \Delta y$, the index profile $n^2(x, y) = n_{ij}^2$ and the field $e(x, y) = e_{ij}$ are then placed in the array of computational domain. Using the above approximate field distribution and mode index as the initial conditions, we then verify discretized Helmholtz equation at all grid points by the basic mode convergence equation of the FCM [32] through

$$e_{i,j} = \frac{e_{i,j+1} + e_{i,j-1} + e_{i+1,j} + e_{i-1,j}}{4 - (\Delta x)^2 k_0^2 (n_{i,j}^2 - n_{\text{eff}}^2)} \quad (4)$$

Here, we assumed $\Delta x = \Delta y$ for simplicity. After a few convergence scans, the resulting field distribution yields a more accurate value for the mode index $n_{\text{eff}m}$ through the equation [32, 35]

$$n_{\text{eff}m}^2 = \frac{\int_{-\infty}^{+\infty}\int_{-\infty}^{+\infty} \left[\begin{array}{c}(e_{i,j+1}+e_{i,j-1}+e_{i+1,j}+e_{i-1,j}) \\ +\{(\Delta x)^2 k_0^2 n_{i,j}^2 - 4\} e_{i,j}\end{array}\right] dxdy}{(\Delta x)^2 k_0^2 \int_{-\infty}^{+\infty}\int_{-\infty}^{+\infty} e_{i,j}^2 dxdy} \quad (5)$$

The above process of repetitive convergence scan and subsequent evaluation of the mode effective index is continued until a desired precision in the value of $n_{\text{eff}m}$ is achieved. For even faster convergence, the algorithm could be modified with a relaxation factor by which the change in the field before and after convergence scans is added in the next step of convergence.

The above described method yields the scalar fields of the waveguide modes. In the analysis of polarized modes, the approximate modes can be calculated by considering the appropriate TE and TM modes of the constituent slab guides. For example, the TE mode of x-slab and the TM mode of y-slab combined together through Eqs. (1) and (2) correspond to the x-polarized mode and vice versa. For modal correction, the FD discretization can be simply adapted through the semivectorial Helmholtz equation. For TE modes, the continuity condition leaves the scalar Helmholtz equation unchanged to represent the semivectorial form. For TM modes, the corresponding equation is [36]

$$\left(\frac{\partial^2}{\partial y^2} + k_0^2 n^2 - \beta^2\right) E_y + \frac{\partial}{\partial y}\left(E_y \frac{\partial}{\partial y} \log n^2\right) = 0 \quad (6)$$

Built around a three-point centered difference approximation for the operator $\partial^2/\partial y^2$ and using a uniform sampling grid, the field in this case is converged to an x-polarized mode through the scan equation [9,16,19]

$$e_{i,j} = \frac{e_{i+1,j} + e_{i-1,j} + \frac{2\varepsilon_{i,j-1}}{\varepsilon_{i,j-1}+\varepsilon_{i,j}}e_{i,j-1} + \frac{2\varepsilon_{i,j+1}}{\varepsilon_{i,j+1}+\varepsilon_{i,j}}e_{i,j+1}}{4 - k_0^2(\Delta y)^2(\varepsilon_{i,j} - n_{\text{eff}}^2) + \frac{\varepsilon_{i,j}-\varepsilon_{i,j-1}}{\varepsilon_{i,j-1}+\varepsilon_{i,j}} + \frac{\varepsilon_{i,j}-\varepsilon_{i,j+1}}{\varepsilon_{i,j+1}+\varepsilon_{i,j}}} \quad (7)$$

In Eq. (7) $\varepsilon_{i,j} = (n_{i,j})2$ where n is the local refractive index. From the field distribution, the n_{eff} is then calculated using the integral

$$n_{\text{eff}}^2 = \frac{\int_{-\infty}^{+\infty}\int_{-\infty}^{+\infty}\left[\begin{array}{c}e_{i+1,j}+e_{i-1,j}+\frac{2\varepsilon_{i,j-1}}{\varepsilon_{i,j-1}+\varepsilon_{i,j}}e_{i,j-1}+\frac{2\varepsilon_{i,j+1}}{\varepsilon_{i,j+1}+\varepsilon_{i,j}}e_{i,j+1} \\ -\left\{4+\frac{\varepsilon_{i,j}-\varepsilon_{i,j-1}}{\varepsilon_{i,j-1}+\varepsilon_{i,j}}+\frac{\varepsilon_{i,j}-\varepsilon_{i,j+1}}{\varepsilon_{i,j+1}+\varepsilon_{i,j}}-k_0^2(\Delta y)^2\varepsilon_{i,j}\right\}e_{i,j}\end{array}\right]e_{i,j}dx.dy}{k_0^2(\Delta y)^2 \int_{-\infty}^{+\infty}\int_{-\infty}^{+\infty} e_{i,j}^2 dx.dy} \quad (8)$$

For the y-polarized mode, the above two equations are once again implemented simply by interchanging indices i, j.

In the semivectorial formulation [36], the terms in the wave equation corresponding to the interaction between x-directed electric field components E_x and the y-directed ones,

$$\frac{\partial}{\partial x}\left(\frac{1}{n^2}\frac{\partial n_{ij}^2}{\partial y}E_y\right) \text{ and } \frac{\partial}{\partial y}\left(\frac{1}{n^2}\frac{\partial n^2}{\partial x}E_x\right)$$

are usually small, and therefore these terms are neglected to decouple the vectorial wave equations.

Thus, the wave equation is reduced to semivectorial form. Solving this equation is numerically efficient and could be widely used when designing optical waveguide devices for which the coupling between the x- and y-polarizations is negligible. Therefore, for birefringent waveguide structures, for example, the fused coupler and the elliptic core fiber/high-birefringence photonic crystal fiber, semivectorial analysis has been shown to be highly efficient and accurate [37]. Above all, the formulation is relatively easy to implement.

Importantly, the perturbation correction method itself yields reasonably good results and has been applied successfully to practical structures, namely rectangular [33, 38] and elliptical core [39, 40] waveguides, anisotropic channel waveguides [41], strip and rib waveguides [42], and rectangular core couplers [43]. However, the field correction scheme works as a fine-tuning step in the modal convergence. Thus, the two-step algorithm not only yields accuracy but also substantially reduces the computation time for convergence of each of the local modes. The perturbation method of mode calculation described above applies to a step-index profile waveguide; for analysis of couplers made from graded profile fibers, the approximate mode of the hypothetical waveguide is generated by evaluating the corresponding axial slab modes, that is, $e(x)$ and $e(y)$, and multiplying them using Eq. (1). For this purpose, any numerical technique for solving graded-index planar waveguides can be used [44].

In any finite difference algorithm, the differential operators and continuous derivatives are approximated by a difference between neighboring points on a numerical grid. Because the index distribution $n^2(x, y)$ is only known at the mesh points, there is an uncertainty in the effective index n_{eff} of the structure due to the assumptions that the algorithm takes care of $n^2(x, y)$ between the grid nodes. Therefore, it is only natural that the smaller the grid size, the more the finite difference method approaches the differential formulation of the rigorous Helmholtz equation. However, due to the finite precision that is inherent in a computer's handling of real numbers, continued smallness of the grid size does not significantly improve the order of accuracy, even though the computation time could become unusually large.

3.2. Modeling the Propagation in the Coupling Region

The structure of the fused region of a coupler is represented by the "model of fusion" and "model of elongation" to take account of the local transverse geometry and the longitudinal taper profile. In practice, the longitudinal taper profile and the nature of fusion together determine the local cross-section of the coupling region, which dictate the characteristics of the local supermodes at a given point along the coupler length. The taper profile of the fused coupler depends on the specific technique used to form the taper [45]. For a flame burner of extent Δz, the exponentially decaying taper profile results in the local radius to evolve with z as (Fig. 13.5) [29, 45]

$$b(z) = b \cdot e^{-\frac{z}{\Delta z}} \text{ and } b_0 = b \cdot e^{-\frac{L}{2\Delta z}} \quad (9)$$

where b_0 is the radius at the coupler's waist and b is that of the untapered fiber. Most exponential profiles are seen to match well with the following form [13,16]:

$$b(z) = b - (b - b_0) \times \exp\left[-\left(\frac{z}{z_0}\right)^2\right] \quad (10)$$

Here z_0 is the scalar taper length and is approximately equal to one-third of the total elongation.

Figure 13.6 shows the typical cross-section of a 2×2 fused coupler. In practice, a coupler may assume any intermediate shape of cross-section between the two extreme cases of weakly fused (resembling that of two touching cylinders) and strongly fused (near-elliptical) couplers depending on the speed of taper elongation and the "degree of fusion" f defined by [32]

$$f = \frac{d_{\text{weakest}} - d_{\text{arbitrary}}}{d_{\text{weakest}} - d_{\text{strongest}}} \quad d_{\text{strongest}} < d_{\text{arbitrary}} < d_{\text{weakest}}$$

$$= \frac{2b - d}{2b \cdot (2 - \sqrt{2})} \quad (11)$$

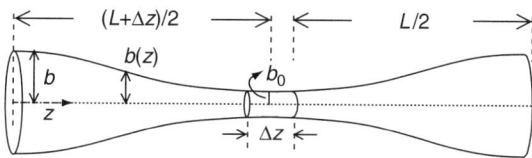

FIGURE 13.5 Schematic of the longitudinal taper profile formed by the fuse-pull-taper method used in coupler fabrication.

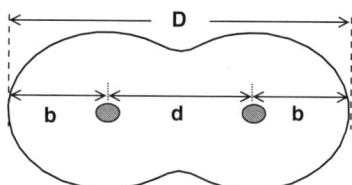

FIGURE 13.6 Modeling the cross-section of a coupler.

where $2b$ is the diameter of the original untapered fiber and d is the center-to-center difference between the two cores. Therefore, the weakest fused coupler has $d = 2b$. Accordingly, $0 \leq f \leq 1$; $f = 0$ and $f = 1$ would correspond to weakest and strongest fusion, respectively. To estimate the degree of fusion of a fabricated coupler, the width $D_0 (= d_0 + 2b_0)$ of the cross-section at the waist is measured under a traveling microscope, obtaining an estimate of d_0. The interaction length of the coupler is divided into a large number of elemental slices each of thickness dz with a staircase-like concatenation of varying transverse dimensions to simulate the longitudinal taper profile [16]. In each of these elemental waveguides, we obtain the local even and odd modes of this waveguide following the procedure described in Section 3.1. Thus, by specifying the initial condition for the electric field of x-polarization or y-polarization at a location well before the coupling region, we calculate the local supermodes of each section along the entire length of interaction.

3.3. Supermodes and Beating

Figure 13.7 depicts the dimensions and index distribution of the pseudo-waveguide in the backdrop of the coupler cross-section. Here $\delta n_1^2 = n_1^2 - n_2^2$ and $\delta n_2^2 = n_1^2 - n_a^2$ represent small index differences between the rectangular core and the hypothetical waveguide used in the calculation of the corrected mode index [33]. Though the structure appears complex in geometry, its symmetry about x- and y-axes makes the algebra quite simple. Examples of field distributions of the fundamental and first-order modes (local modes) of this fused coupler generated by the perturbation method, using Eq. (1), are shown in Fig. 13.8. In Fig. 13.9, the true modal fields after execution of the convergence process are shown. These modes correspond to the cross-sectional point of the coupler waist at which the inverse taper ratio (= radius of tapered fiber/radius

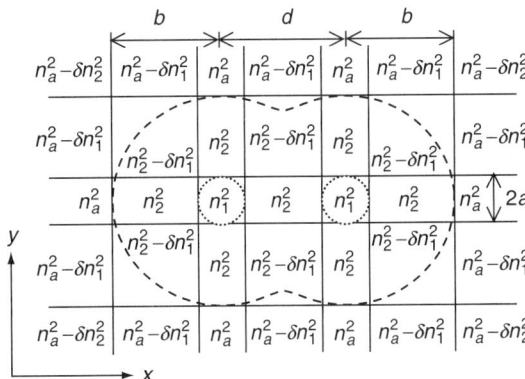

FIGURE 13.7 The index profile and dimensions of the pseudo-waveguide in the backdrop of the coupler cross-section for perturbation analysis.

of untapered fiber), $\rho = 0.35$. Evidently, the approximate fields differ more at the corner regions (sharply falling fields) where the index difference between the coupler and the pseudo-waveguide is high. However, the shape of the approximate field distribution as evaluated by Eq. (1) is close to that of the corresponding true mode of the real waveguide and has the correct signs for the field across the waveguide cross-section (Fig. 13.9).

The propagation constants of the even and odd modes of the structure so determined are then used to obtain the cumulative phase difference between them after traveling through the entire length of the coupling region through

$$\Delta\phi = \int_{-w}^{w} (\beta_e - \beta_0) dz \qquad (12)$$

Here w represents the extent of taper on either side of the center of taper waist up to a point beyond, and the power coupling between the two fibers is insignificant as seen during simulation. The powers at the throughput and coupled ports are then evaluated using the well-known expressions

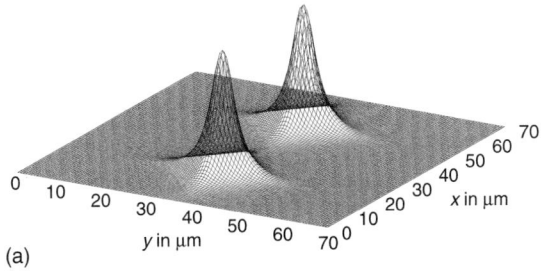
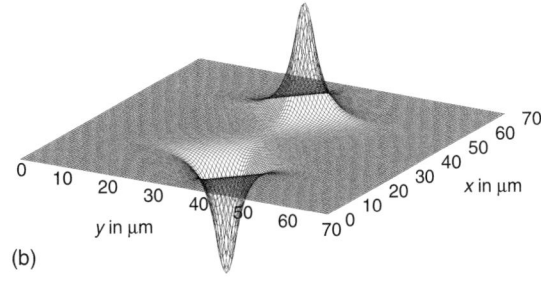

FIGURE 13.8 Approximate field distributions of the (a) fundamental and (b) first-order local modes of the fused coupler generated by the perturbation method, that is, the modes of the waveguide under the perturbation approximation.

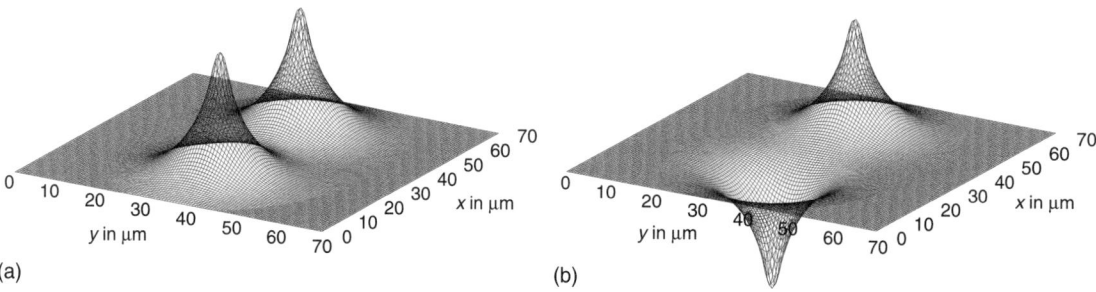

FIGURE 13.9 The true modal fields of the fused coupler as obtained after the convergence of the approximate fields (shown in Fig. 13.4) corresponding to a $\rho = 0.35$ and $f = 0.5$: the local modes (a) fundamental and (b) first order.

$$P_T = P_0 \cos^2\left(\frac{\Delta\phi}{2}\right) \quad \text{and} \quad P_C = P_0 \sin^2\left(\frac{\Delta\phi}{2}\right) \quad (13)$$

3.4. Polarization Characteristics

Because of the relatively large Δn encountered at the cladding–air interface and physical asymmetry in x- and y-dimensions, sensitivity to the states of polarization of the propagating fields in long couplers is quite prominent. The intensity of the output power periodically reaches its maximum for an unpolarized input light as the fused pair of fibers is pulled. The cumulative phase difference for the x-polarized mode $\Delta\phi^X$ is determined through

$$\Delta\phi^X = \int_{-w}^{+w} \left(\beta_e^X - \beta_0^X\right) dz \quad (14)$$

$\Delta\phi^Y$ for the y-polarized mode is calculated in a similar manner. Assuming that the power injected into the input port is unity, the power in the transmitted port evolves as

$$P_T = \eta \cos^2\left(\frac{\Delta\phi^X}{2}\right) + (1-\eta) \cos^2\left(\frac{\Delta\phi^Y}{2}\right) \quad (15)$$

where η is the fraction of injected power into the x-polarization mode (i.e., $\eta = 0.5$, for equal excitation of both the polarizations).

3.5. Results and Discussions

To check the effectiveness of our analysis and the corresponding algorithm, as an initial step we consider the coupling behavior in many coupling-cycle long couplers. To define the geometry of the coupler structure for simulation, we estimate b and d from measured data, and for the degree of fusion, we approximate f as 0.3, 0.5, and 0.8, respectively, mimicking sample examples for weakly, moderately, and strongly fused couplers fabricated under a given fusion condition. The power variation in the output port as a function of elongation at a given monitoring wavelength, that is, the so-called pulling signature, represents the real-time coupling characteristics. Figure 13.10a shows the experimentally recorded real-time power variation in the transmitted port at the monitoring wavelength of 1310 nm when the coupler was drawn under a weakly fused condition. Because of the relatively high form birefringence and large index contrast, the weakly fused couplers exhibit a strong polarization modulation over the drawn length of the coupler. To make a quantitative comparison of this, the theoretically estimated signature using the fiber parameters and fabrication condition as used in the experiment is shown in Fig. 13.10b. The measured

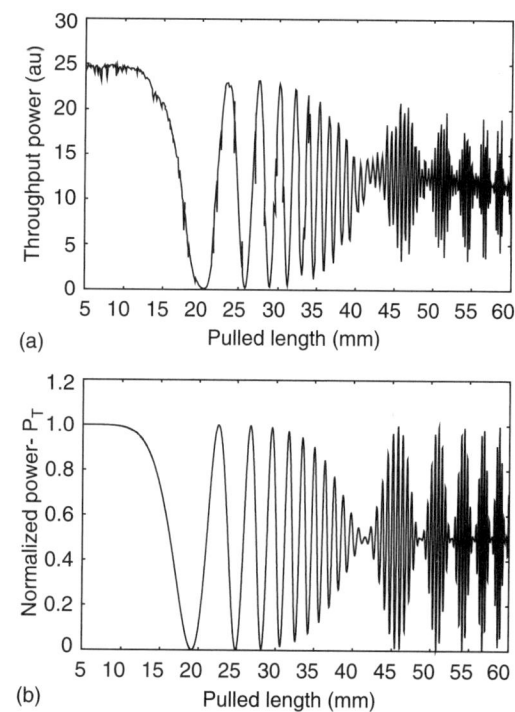

FIGURE 13.10 (a) Experimental pulling signature at the monitoring wavelength of 1310 nm recorded during fabrication of a 1310/1550-nm WDM (moderately fused condition). (b) Theoretical pulling signature at 1310-nm wavelength for $f = 0.5$.

TABLE 13.2 Fiber (SMF 28®, Corning Inc.) parameters as measured by the far-field technique [46] and used in the calculation.

Core radius, a	4.1 μm
Overall fiber diameter, $2b$	125 μm
Δn	0.0036
Mode field diameter (MFD)	9.2 μm @ 1310 nm
Cladding refractive index, n_2	1.447 @ 1310 nm

parameters of the fiber used in the fabrication of couplers are given in Table 13.2. The sinusoidal power variation, modulation in the power amplitude, and the "bunch-up" in periodicity as the characteristics of polarization dependent coupling nature of fused coupler could be easily explained from these figures.

In many coupling-cycle long couplers where the taper becomes as thin as ~ 10–$20\,\mu m$, the guiding at the tapered waist is mainly due to the air–cladding interface and the core does not play any significant role in modal guidance. The analysis, in that case, can be simplified even more (Fig. 13.7), and the results of this "no-core model" can explain the trend of coupling characteristics quite well. However, during the fabrication process, at the beginning coupling initiates between the core modes in the taper transition region. As the pulling continues, which results in progressively reduced fiber cross-section, the coupling due to core modes is gradually taken over by that due to cladding modes. In practical coupler devices, for example, for WDM mapping, only a few coupling cycles (e.g., 1.0, 1.5, and 2.0) are required; for 3-dB couplers or signal splitters/tap couplers, the required number of coupling cycles is only 0.5 or less. Thus, intuitively, it is conceivable that interacting core modes indeed contribute to coupling [20, 29, 47].

Figures 13.11 and 13.12 show a comparison of the theory and experiment pertaining to typical WDM couplers intended for operation at 1310- and 1550-nm wavelengths. The dashed-dotted curves labeled as (a) in these figures represent the estimated pulling signatures at 1310-nm for WDMs corresponding to two different degrees of fusion, $f = 0.5$ and $f = 0.8$, respectively. The corresponding pulling signatures recorded during fabrication are shown as curves (b). Evidently, the agreement between experimental results and the theoretical prediction is seen to be very good. For the case of $f = 0.5$, the measured (using a monochromator and lock-in amplifier arrangement) and estimated spectral response of the WDM are shown in Fig. 13.13 as curves (a) and (b), respectively. Clearly, under the given fabrication condition with standard fibers, the WDM character-

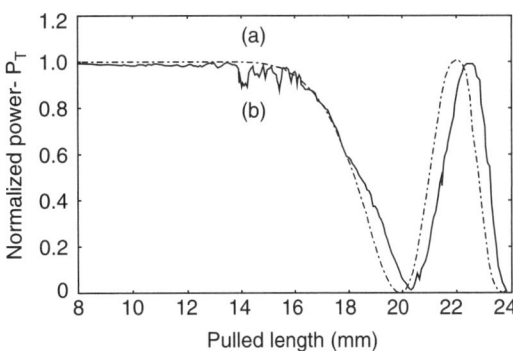

FIGURE 13.11 (a) Theoretically estimated elongation signature at 1310 nm wavelength with $f = 0.5$. (b) The same experimentally recorded during fabrication of a moderately fused 1310/1550-nm WDM with 1.5 coupling cycle at 1310 nm.

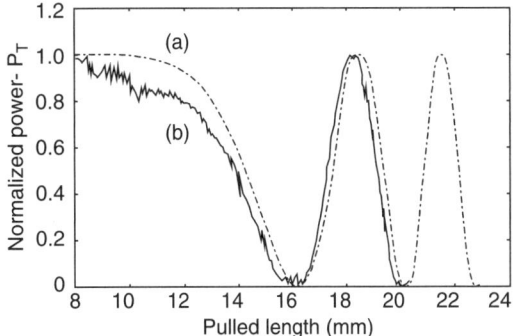

FIGURE 13.12 (a) Computed elongation response of the coupler at the 1310-nm wavelength for $f = 0.8$ compared with (b) the recorded pulling signature of a fabricated 1310/1550-nm WDM coupler under the strongly fused condition.

istics are mapped with 1.5 coupling cycles at the monitoring wavelength of 1310 nm (exiting through the coupled port). For 1550 nm as the monitoring wavelength, the same mapping requires 2.0 coupling cycles (exiting through the throughput port). However, because the rate of power splitting at the

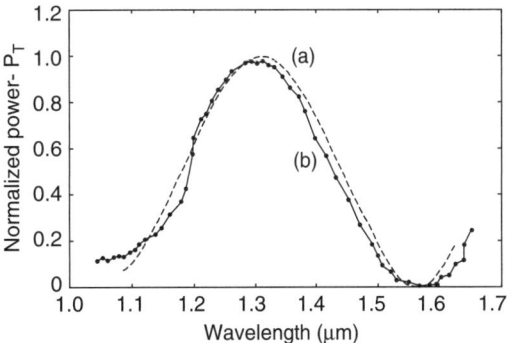

FIGURE 13.13 (a) Computed wavelength response compared with (b) measured response of a typical fabricated 1310/1550-nm WDM.

longer wavelength is faster, we find that monitoring the fabrication at lower wavelength ensures a better control to stop the process of elongation at a desired power-splitting ratio.

To estimate the elongation required for fabrication of a 980/1550 nm WDM, we followed the preceding analysis. Figure 13.14 shows the elongation response of the coupler at the monitoring wavelengths of (a) 1550 nm and (b) 980 nm; these results were computed for a Fibercore SM-980™ fiber. Mapping the process parameters with this estimate, we could fabricate 980/1550-nm WDMs within a few trials, after fine tuning the fabrication condition in line with the above analysis. The measured typical wavelength response of a fabricated WDM as obtained through an optical spectrum analyzer is depicted in Fig. 13.15. The excess loss, channel isolation, and isolation bandwidth of these WDMs were comparable with state-of-the-art specifications.

4. APPLICATIONS: FFC-BASED ALL-FIBER COMPONENTS

4.1. Beam Splitter/Combiner

The beam splitter/combiner forms the most common class of fused biconical taper (FBT) coupler-based branching components, which are widely used in optical networks. A splitter/combiner is a symmetric coupler in which the constituent fibers are identical. The commonly used components of this class are 3-dB couplers, tap/access couplers, and tree couplers.

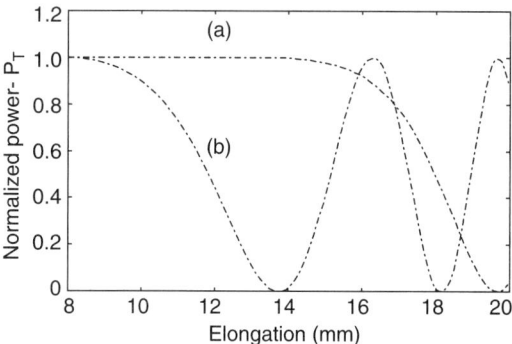

FIGURE 13.14 Theoretical elongation response for mapping a 980/1550-nm WDM coupler, which corresponds to (a) 0.5 coupling cycle when monitoring at 980 nm, whereas (b) 2.0 coupling cycles for 1550 nm, indicating that under the given fabrication condition the coupler has to be pulled through ~20 mm.

4.1.1. 3-dB Coupler

A 3-dB coupler is a 2×2 symmetric fiber coupler, which equally splits input optical signal at the desired wavelength between the throughput (T) and coupled (C) ports; the effective interaction length of the device is $L_C/2$, where L_C is the coupling length [48]. The 50:50 splitting of the input power at a given wavelength (say, 1310 or 1550 nm) is ensured by real-time monitoring of powers from the T- and C-ports during the fabrication. Because the spectral splitting ratio of these components is approximately sinusoidal, the splitting ratio varies most rapidly at the 50:50 point of the curve. These couplers find

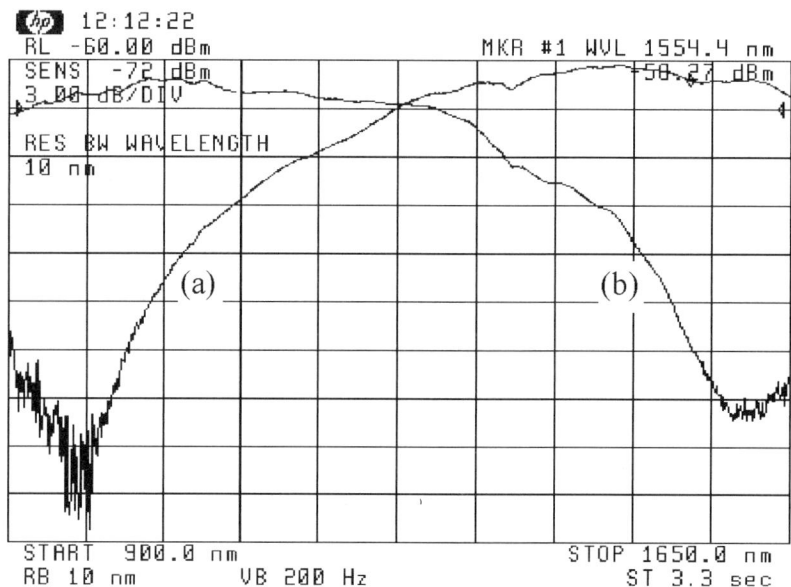

FIGURE 13.15 Wavelength response of a typical 0.5 cycle WDM coupler, fabricated for operation at 980/1550 nm, as measured by optical spectrum analyser (OSA). The curves show the power variation (in dBm) with wavelength in the (a) throughput ports and (b) coupled port.

extensive use in optical telecommunication networks, test equipments, optical fiber sensors (e.g., interferometric fiber sensors [49, 50]), and cable television networks. These are bi-directional devices, and could be used to either split or combine signals.

4.1.2. Tap/Access Couplers

A tap coupler is essentially a beam splitter with a desired splitting ratio, usually in the range of 5–50%. Such a branching component permits passive add/drop of a signal at an optical network node. These are used as power taps for monitoring the stability of an optical source, optimizing power budgets in optical networks, accessing Rayleigh backscattered signal from the test fiber in an optical time domain reflectometer (OTDR), monitoring signal level status in a fiber transmission link, and coupling at nodes in bus networks.

4.1.3. Tree Coupler

The tree coupler is a bidirectional branching component, configurable through integration of a number of FFCs, which distributes the optical signal from one fiber to several others ($1 \times n$ ports) or combines signals from several fibers to one (Fig. 13.16). Typically, it splits input power equally to all output ports. The number of output ports could range from 2 to 16. However, in an all-fiber form this route is more practical for lower order arrays (for $n \leq 8$). A tree coupler is also realized by fabricating 3-dB couplers on each of the T- and C-arms of the previously fabricated 3-dB couplers or through splicing of 3-dB couplers in tandem (Fig. 13.17).

4.2. WDM Coupler

A WDM coupler is also a symmetric 2×2 FFC in which two inputs at two different wavelengths, say λ_2 and λ_2, from the two input ports are combined at one output port. On the other hand, if these two signals are injected into the same input port, they would separate out at the two output ports (Fig. 13.18). Such a design owes its origin to the fact that for a given coupler, the coupling coefficients and the effective lengths of interaction at two different wavelengths, say 1310 and 1550 nm, are different. Therefore, the splitting ratios at these wavelengths are usually different. For a coupler to function as a WDM at these two operating wavelengths, the fabrication process has to be tailored such that the splitting ratio of the device is a maximum at one wavelength and a minimum at the other wavelength [51]. This implies that all the input power at λ_1 emerges at one output port and all the input power at λ_2 emerges at the second output port.

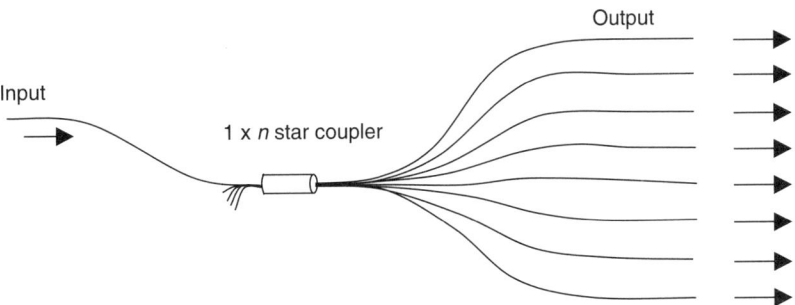

FIGURE 13.16 Schematic of fused fiber coupler-based $1 \times n$ star coupler having one input and several outputs.

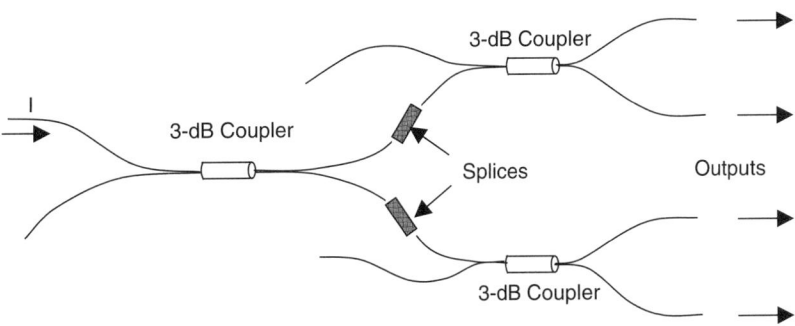

FIGURE 13.17 Schematic of a $1 \times$ four-power splitter configured through three 3-dB couplers.

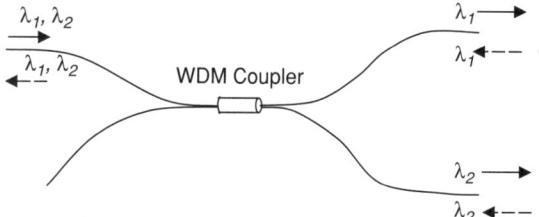

FIGURE 13.18 Schematic of a WDM at wavelengths λ_1 and λ_2, designed around a 2 × 2 fused coupler.

A WDM that is extensively used for telecommunications applications is the 1310/1550-nm WDM and is referred to as "classical WDM" to distinguish it from DWDMs [52]. The two specific wavelengths at which a WDM is required to operate are often referred to as the channels. When configured around a 2 × 2 FFC coupler, the separation between the two wavelengths, known as channel spacing, requires the coupler to exhibit a specific wavelength response. Based on the channel spacing, classical WDMs are classified as either narrow channel spacing (<100 nm) or wide channel spacing devices.

4.3. Principle of Operation of Classical WDM

It is well known that the transmission characteristics of an FFC are wavelength dependent, because the coupling coefficient κ depends on the operating wavelength and the V number [3, 48]. Thus, if the length of interaction L (which is fixed for a given coupler) is such that for a given wavelength λ_1, $L = pL_c(\lambda_1)$, and for another wavelength λ_2, $L = (p+1)L_c(\lambda_2)$, where p is an integer, the entire power at λ_1 escapes from one of the output ports, whereas the entire power at λ_2 is available from the complementary port (Fig. 13.19). Physically, p corresponds to the number of coupling cycles. Such a coupler would act as a WDM for these two wavelengths, λ_1 and λ_2.

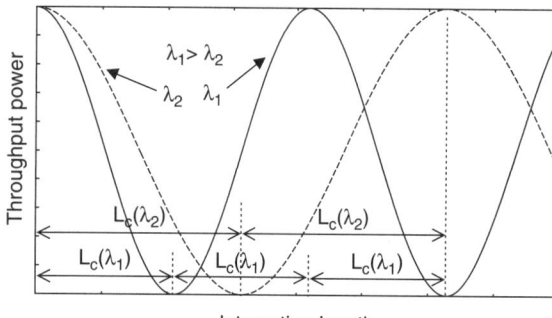

FIGURE 13.19 Qualitative variation of power in the throughput port with interaction length, illustrating the requirement of a WDM for two wavelengths λ_1 and λ_2 ($\lambda_1 > \lambda_2$).

In general, for the coupler to function as a narrow channel-spacing WDM, one is required to go through a relatively larger number of power-coupling cycles. These couplers may exhibit a relatively high excess loss and stronger polarization dependence [53, 54]. Wide channel-spacing WDMs are usually characterized by a larger bandwidth due to a relatively large wavelength period in their spectral response. A very important wide channel-spacing WDM is the 980/1550-nm WDM, which forms an integral part of an EDFA, where the pump and signal at these two wavelengths are multiplexed within the amplifying (erbium-doped) fiber [55]. An alternative 1480/1550-nm WDM, which is a narrow channel-separation WDM, could also be used to combine pump and signal in an EDFA [56].

4.4. Wavelength Interleaver

A key figure of merit for a DWDM optical transmission system is known as spectral efficiency, which is defined as the ratio of the bit rate per channel to the channel spacing, and it is expressed in bits/s/Hz. Due to the well-known difficulties encountered in attaining high-speed electronic components at speeds beyond 40 Gb/s, a decrease in the channel spacing has evolved as the best near-term option to achieve high spectral efficiency in a DWDM link. In contrast to classical WDM, DWDM involves simultaneous transmission of at least four wavelengths in the EDFA bands (e.g., C-band, 1530–1565 nm; L-band, 1570–1610 nm); typically, the number of wavelengths could be 8, 16, 32, 64, and so on. To avoid haphazard growth of DWDM fiber links, the International Telecommunication Union (ITU) has fixed certain wavelength standards, referred to as ITU wavelength grids, for DWDM optical transmission systems (see Chapter 1). The recommended standard channel spacings are 200 GHz ($\equiv$1.6 nm), 100 GHz ($\equiv$0.8 nm), and 50 GHz ($\equiv$0.4 nm) with 1550 nm as the reference wavelength. As the demand for more and more bandwidth grows, the systems designers are expected to pack in more wavelength channels within an amplifier band. This would necessarily lead to smaller channel spacings, which would entail tighter specifications for components such as optical source (laser diode), multiplexers/demultiplexers (mux/demux), and wavelength filters.

A wavelength interleaver is a device that combines two input streams of wavelength channels having a constant spacing in the frequency domain, $\Delta\nu$ (e.g., as per, the ITU grid), into a single dense stream of channels with separation $\Delta\nu/2$ at the output (Fig. 13.20). The device being reciprocal [57], it

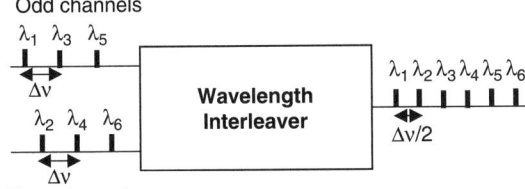

FIGURE 13.20 Schematic representation of a wavelength interleaver that interleaves the input channels with spacing in frequency domain $\Delta\nu$ into a dense packed output having inter-channel spacing $\Delta\nu/2$; here $\lambda_1, \lambda_2, \lambda_3, \lambda_4\ldots$ represent central wavelengths corresponding to channels 1, 2, 3, 4, and so on.

functions as a wavelength slicer when operated in the reverse direction, thereby greatly relaxing the tight tolerance otherwise required on components required in detection of associated DWDM signals, because the adjacent channels would then be spaced apart by double the original frequency spacing. Of the few alternative routes followed to realize a wavelength interleaver, the one based on unbalanced Mach-Zehnder Interferometer (MZI) is perhaps the most popular [58–60], which can be realized either in an integrated optic form [61] or in an all-fiber form [60].

4.4.1. MZI-Based All-Fiber Wavelength Interleaver

In an all-fiber form, a single-stage unbalanced MZI can be formed by joining two 2 × 2 fiber couplers in such a way that the lengths of the two arms are slightly unequal [60, 62] by an amount ΔL. A schematic of an all-fiber MZI is shown in Fig. 13.21. The differential path length (ΔL) is equivalent to a delay line. The corresponding differential phase delay $\Delta\varphi$ suffered by any wavelength λ while traversing the extra path length ΔL is given by

$$\Delta\varphi = 2\pi n_{\text{eff}} \Delta L/\lambda \qquad (16)$$

The transmission characteristics of an interleaver based on MZIs can easily be described in terms of transfer matrices of individual couplers and delay lines. From the coupled mode equations for the electric field amplitudes, the transfer matrix [60,62] for the coupler of splitting ratio $\sin^2(\kappa z)$ can be written in the form

$$M_{\text{coupler}} = \begin{pmatrix} c & -js \\ -js & c \end{pmatrix} \qquad (17)$$

where $c = \cos(\kappa z)$ and $s = \sin(\kappa z)$. κ is the coupling coefficient and z is the interaction length. The transfer matrix corresponding to the differential delay between the two arms is given by

$$M_{\Delta L} = \begin{pmatrix} e^{j\Delta\varphi} & 0 \\ 0 & 1 \end{pmatrix} \qquad (18)$$

Thus, the transfer function of an unbalanced MZI is given by

$$M_{\text{MZI}} = M_{\text{coupler2}} M_{\Delta L} M_{\text{coupler1}} \qquad (19)$$

If E_1 and E_2 are input fields to the MZI at its port 1 and port 2, respectively, then the output fields E_T and E_C at the ports 3 and 4 can be expressed using Eq. (19) as

$$\begin{pmatrix} E_T \\ E_C \end{pmatrix} = M_{\text{MZI}} \begin{pmatrix} E_1 \\ E_2 \end{pmatrix} \qquad (20)$$

Thus, if $P_1(\lambda_1)$ and $P_2(\lambda_2)$ represent powers at two input channels corresponding to wavelengths λ_1 and λ_2, respectively, then the outputs are given by

$$P_T = P_1(\lambda_1)\sin^2\frac{\Delta\varphi(\lambda_1)}{2} + P_2(\lambda_2)\cos^2\frac{\Delta\varphi(\lambda_2)}{2} \qquad (21)$$

$$P_C = P_1(\lambda_1)\cos^2\frac{\Delta\varphi(\lambda_1)}{2} + P_2(\lambda_2)\sin^2\frac{\Delta\varphi(\lambda_2)}{2} \qquad (22)$$

Here, we have assumed that the splitting ratios of the coupler 1 and coupler 2 in the MZI configuration are 50:50. To achieve wavelength interleaving using an unbalanced MZI configuration, Eqs. (21) and (22) have to satisfy the following condition:

$$\Delta\varphi(\lambda_1) = (2n+1)\pi \quad \text{and} \quad \Delta\varphi(\lambda_2) = 2n\pi \qquad (23)$$

or

$$\Delta\varphi(\lambda_1) = 2n\pi \quad \text{and} \quad \Delta\varphi(\lambda_2) = (2n+1)\pi \qquad (24)$$

where n is an integer. From Eqs. (23) and (24) we can observe that $|\Delta\varphi(\lambda_1) - \Delta\varphi(\lambda_2)| = \pi$. Thus, substituting the values of $\Delta\varphi(\lambda_1)$ and $\Delta\varphi(\lambda_2)$ from Eq. (16), we get

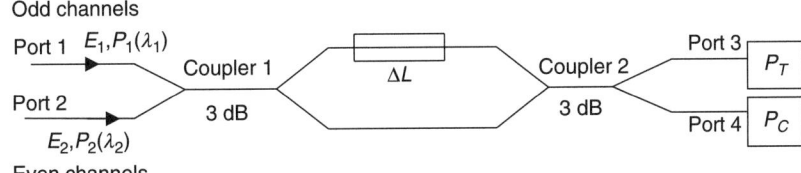

FIGURE 13.21 Schematic of an all-fiber Mach-Zehnder interferometer based interleaver realized using two 3-dB couplers.

$$\Delta L = \frac{\lambda_1 \lambda_2}{2 n_{\text{eff}} \Delta \lambda}. \qquad (25)$$

If the DWDM signal channels are input at port 1, the signal wavelengths that suffer a differential phase delay of $(2n+1)\pi$ exit through port 3, and the wavelengths that suffer a differential phase delay of $2n\pi$ exit from port 4. These wavelengths correspond to the peaks in the spectral response of the MZI. Such a configuration could separate/interleave the wavelength channels with a sinusoidal response function. The wavelength separation between the consecutive peaks at any output port is known as free spectral range (FSR) of the configuration. The spectral response of such a configuration with an FSR of 25 GHz is shown in Fig. 13.22. If we change ΔL by an amount $\lambda/2 n_{\text{eff}}$ (which introduces an additional phase change of π), the spectral response at the output ports gets reversed as can be observed from Fig. 13.23.

When evaluating interleavers for application in optical fiber communication, the most important parameters of concern are insertion loss, shape of the pass-band, and cross-talk performance for the desired channel spacing. A uniform flattop pass-band, besides lowering insertion losses, is also very desirable to minimize channel power variations as the signal wavelength drifts. Any initial imbalance in channel power cascades as the signal passes through multiple amplifiers and affects the signal-to-noise performance of the system. Often a compromise is made between achieving a flattop pass-band and low insertion loss, because additional filtering elements are necessary to achieve a uniform flattop response from a pass-band characterized by sinusoidal response. To meet these requirements, a two-stage MZI configuration has been proposed [58, 59, 62]. Two-stage MZI configuration consists of cascading three couplers in tandem with two delay lines in their arms; the schematic of such a configuration is shown in Fig. 13.24. A typical simulated spectral response of this configuration is shown in Fig. 13.25 [62].

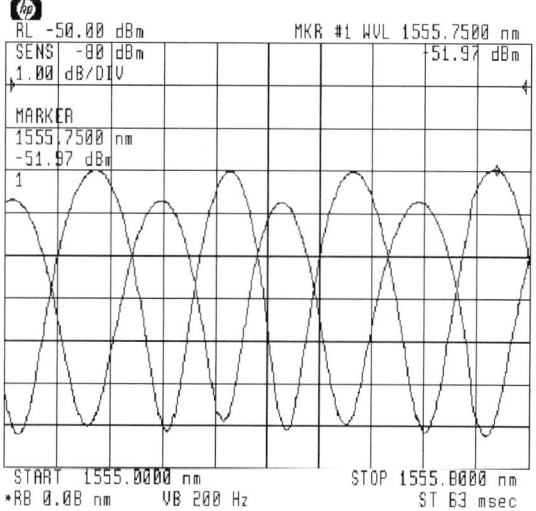

FIGURE 13.22 Measured spectral response of the all-fiber wavelength interleaver, tuned for an FSR of 25 GHz, based on single-stage MZI; the marker at one of the peaks corresponds to $\lambda = 1555.75$ nm.

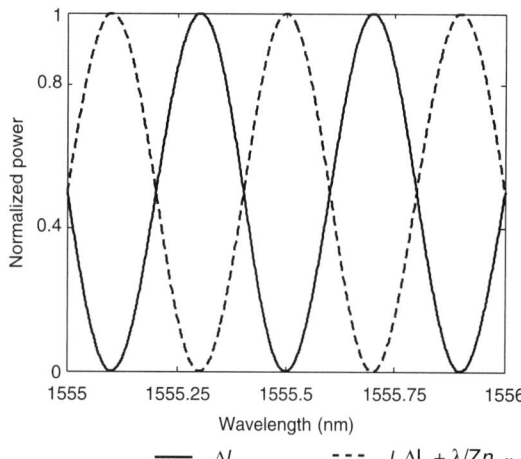

FIGURE 13.23 Typical simulated spectral response at port 3 of the MZI-based wavelength interleaver when differential delay is ΔL and $\Delta L \pm \lambda/2 n_{\text{eff}}$.

4.5. Fiber Loop Reflector

An all-fiber loop reflector, also called a fiber loop mirror (FLM), is generally constructed by joining/splicing the two output ports of a 3-dB fiber coupler; the schematic is shown in Fig. 13.26. FLMs find applications as a resonant cavity in a fiber laser, passive fiber Fabry-Perot devices, and also in duplex transmission along a single fiber with a light source

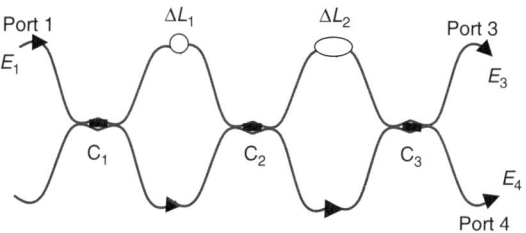

FIGURE 13.24 Schematic of all-fiber two-stage MZI configuration used as wavelength interleaver with flattop response; ΔL_1 and ΔL_2 denote the two delay lines, and C_1, C_2, and C_3 represent three couplers.

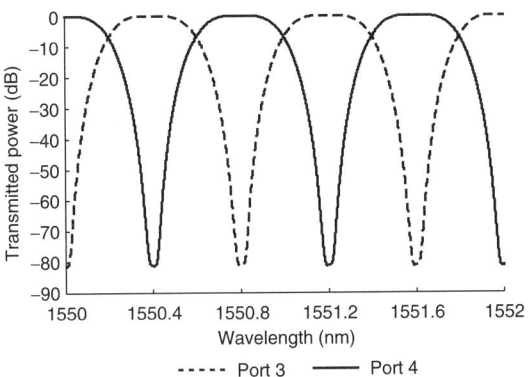

FIGURE 13.25 Typical flattop spectral response of a two-stage MZI configuration; corresponding optimized splitting ratios [64] of couplers C_1, C_2, and C_3 are 50:50, 68:32, and 4:96, respectively.

at only one end of the link [63, 64]. More specifically, it has been exploited in soliton switching with a fiber nonlinear mirror [65], in ultrafast all-optical demultiplexing [66], in all-optical demultiplexing of TDM data at 250 Gb/s [67], as a nonlinear amplifying loop mirror [68], as an all-optical loop mirror switch using an asymmetric amplifier/attenuator configuration [69], and so on.

In an FLM configuration, polarization plays an important role because some amount of inherent birefringence is induced in the fiber loop due to the presence of bends and twist. To take into account the effect of this birefringence, we assume the presence of a hypothetical waveplate [64,70] at some arbitrary location within the loop. The amount of birefringence introduced by the fiber loop is equivalent to that of this waveplate [64,70], with retardation ϕ and orientation angle θ, and depends on the orientation of the plane of the fiber loop. This birefringence modulates the phases of the fields traveling around the loop in opposite directions.

The input light is totally or partially reflected and transmitted, depending on the nature of interference between the clockwise (cw) and counterclockwise (ccw) propagating fields through the loop. Thus, rotating the waveplate can alter the reflectance and the transmittance of the FLM. The Jones matrix elements for an equivalent wave plate with retardation ϕ and oriented at an angle θ can be written as [70]

$$J_{xx} = e^{j\phi}\sin^2\theta + \cos^2\theta, \quad J_{xy} = J_{yx}$$
$$= (e^{j\phi} - 1)\sin\theta\cos\theta, \quad \text{and} \quad J_{yy} = e^{j\phi}\cos^2\theta + \sin^2\theta \quad (26)$$

The reflectance and the transmittance at ports 1 and 2, respectively, of the FLM (Fig. 13.26) are given by

$$R = E'_{1x}E'^*_{1x} + E'_{1y}E'^*_{1y} \quad (27)$$

$$T = E'_{2x}E'^*_{2x} + E'_{2y}E'^*_{2y} \quad (28)$$

where * denotes the complex conjugate. Please note that $E_{1x}E^*_{1y} = E'_{1y}E'^*_{1x} = 0$ because there is no correlation between x- and y-polarized input lights, E_{1x} and E_{1y}. Thus, by use of the formulation given by Morishita and shimamoto [70] and Mortimore [64], the expressions for T and R are obtained as

$$T = |J_{xx}|^2 \cos^2(2\kappa_x z)P_{xi} + |J_{yy}|^2 \cos^2(2\kappa_y z)P_{yi}$$
$$+ |J_{xy}|^2 \cos^2(\kappa_x z - \kappa_y z)(P_{xi} + P_{yi}), \quad \text{and}$$
$$R = 1 - T \quad (29)$$

where

$$|J_{xx}|^2 = |J_{yy}|^2 = 1 - \sin^2(2\theta)\sin^2\left(\frac{\phi}{2}\right), \quad |J_{xy}|^2$$
$$= 1 - |J_{xx}|^2, \quad P_{xi} = E_{1x}E^*_{1x}, \quad \text{and} \quad P_{yi} = E_{1y}E^*_{1y}$$

In the case of $\theta = 0$, the birefringence caused by fiber bends, i.e. ϕ, has no effect on the performance of the FLM. Thus, $T = \cos^2(2\kappa_{x,y}z)$ and $R = \sin^2(2\kappa_{x,y}z)$. In such cases, if the coupler forming the FLM is a

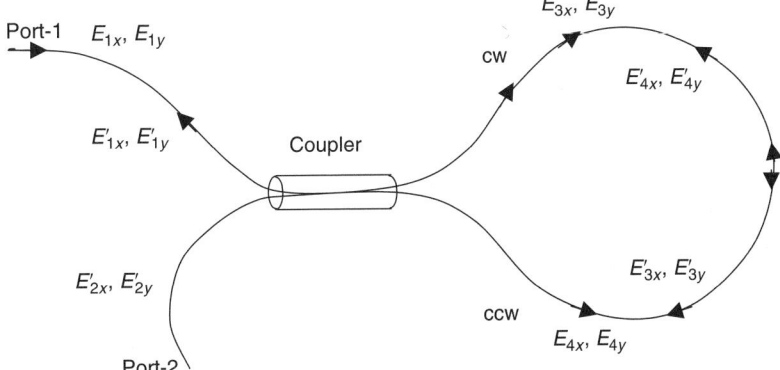

FIGURE 13.26 Schematic of a fiber loop mirror reflector. E_{1x}, E_{1y} are the output fields at parts 1 and 2, respectively; here cw and ccw represent clockwise and counterclockwise propagating beams.

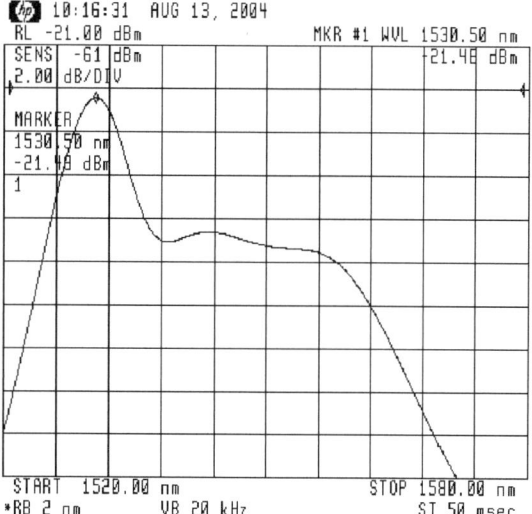

FIGURE 13.27 Typical amplified spontaneous emission spectral response of an erbium-doped fiber.

3-dB one, then all the light is reflected back to the input port, making it a perfect mirror.

4.5.1. Gain Flattening of EDFA

A modern DWDM system operates in the low-loss window around 1550 nm with EDFA forming an integral part of it (see Chapter 8). The gain spectrum of a typical EDFA [48] (Fig. 13.27) shows a peak around 1530 nm, which reduces the available flat gain band. This imposes a serious limitation on the number of channels that can be incorporated in an efficient DWDM system. Use of a high birefringence (HiBi) FLM has been proposed [71] for gain flattening of EDFA. A HiBi-FLM can be realized by incorporating a number of sections of HiBi fiber in the loop mirror configuration. By appropriately adjusting the orientations of the polarization controllers placed in the fiber loop of the HiBi-FLM, fan-Solc filters [72] and Lyot-type filters [73] have also been reported. But the involvement of a HiBi fiber leads to higher losses and makes the technique less attractive.

If the coupler forming the FLM is fabricated with communications grade fiber (such as SMF-28), the inherent birefringence present in the loop is not sufficient to produce wavelength-filtering action as in the case of HiBi-FLM. However, if over-coupled couplers are fabricated, with wavelength filtering properties and low losses, it makes them suitable candidates for achieving gain flattening of EDFA. The FSR of the transmittance/reflectance of the loop reflector formed with an over-coupled coupler is half the FSR of the coupler. Thus, the optimized FSR for the coupler to flatten the amplified spontaneous emission spectrum of an EDFA over the range of 35 nm (from 1525 to 1560 nm) is equal to 70 nm. Typical simulation of the wavelength response of an FLM, assumed to be formed with an over-coupled coupler of FSR 70 nm and a dip at wavelength 1530 nm, for different retardation and orientation ($\theta = \pi/4$), is shown in Fig. 13.28. Based on this study, we realized an FLM formed with over-coupled coupler for experimentation. The schematic of the experimental setup is shown in Fig. 13.29. We introduced an appropriate bend-induced birefringence inside the loop layout, and we could flatten [74] the ASE spectrum of an EDFA (Fig. 13.30). By adjusting the orientation of the two polarization controllers (PC_1 and PC_2), flatness within ± 0.5 dB has been achieved over a range of 30 nm in the C-band.

4.5.2. Nonlinear Optical Loop Mirror

By incorporating a nonlinear optical element, located asymmetrically within the loop, one can configure a switch through such a loop mirror, which is referred to as a nonlinear optical loop mirror (NOLM) in the literature [75]. An NOLM can function as a switch for optical time domain multiplexed signals if the symmetry between the two signals is broken through introduction of a relative phase shift of π between them, so that entire input power at port 1 is transmitted through port 4. By inserting an EDFA within the loop at a location on the path of the continuous wave beam shortly after it leaves the coupler, one may induce a nonlinear phase shift to the beam as it would encounter the rest of the loop as a high power beam, and hence the CW beam would undergo a larger phase shift as

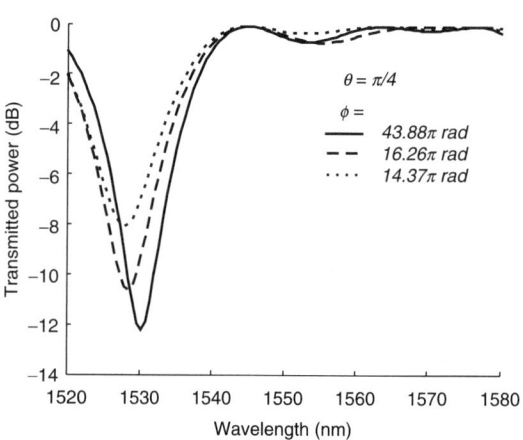

FIGURE 13.28 Wavelength response of a fiber loop mirror (FLM) at a transmitted port with $\theta = \pi/4$.

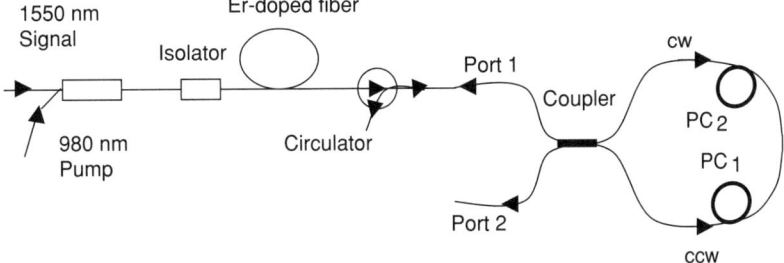

FIGURE 13.29 Schematic of the experimental setup used in gain flattening of the amplified spontaneous emission spectrum of EDFA. cw, clockwise; ccw, counter-clockwise.

compared with the unamplified ccw beam. Because nonlinearity in a silica fiber is rather weak, one requires a relatively long (~ 5 km) loop. By this we means if a relative phase difference of π is introduced between the two counterpropagating beams, all the input power would appear in the transmitted port (port 4). NOLM is the time division multiplexer (TDM) counterpart of WDM add drop multiplexer. This basic idea has been used in [76], in which the EDFA has been replaced by a semiconductor optical amplifier (SOA) to construct a terahertz optical asymmetric demultiplexer (TOAD). In a TOAD, a clock signal in the form of a train of optical pulses of sufficiently high intensity at the signal channel bit rate is injected by means of an additional tap coupler at an appropriate location within the loop (Fig. 13.31) so that it encounters the SOA by propagating along ccw path after the continuous wave signal pulse but before the ccw signal pulse reaches the SOA. Thus, ccw propagating signal pulse experiences a saturated gain and hence it accumulates an extra phase in comparison with the continuous wave pulse, which experiences an unsaturated gain in the SOA. The net effect would lead to bar- and cross-states of the port 4, depending on the timing sequence of the clock pulse. The clock pulse is normally chosen to be centered at a wavelength, that is different from the signal pulse wavelength. Thus, by means of a filter it can be separated from the signal pulse.

5. SUMMARY

We described the fabrication, modeling, and some applications of fused fiber couplers. The basic fuse-pull-taper method has been described in detail. A simple and accurate modeling algorithm has been presented to analyze 2×2 fused fiber coupler–based branching components. The analysis is based on the perturbation correction technique of mode calculation, devised with an FD field convergence approach. We used realistic models for "fusion and elongation" to take into account the axially varying coupler cross-section and evaluated the mode-coupling properties in a semivectorial form of supermodes between. Elongation response of many coupling-cycle long couplers exhibiting polarization modulation and the "bunch-up" effect, as recorded during real-time monitoring, is accurately predicted through this model. For practical components with a few coupling cycles, our analysis, through a mapping of coupling cycles with

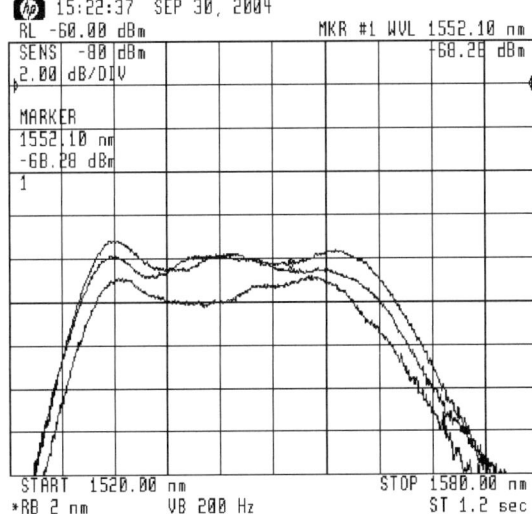

FIGURE 13.30 Typical gain flattened spectra of EDFA for different orientations of two polarization controllers in the loop mirror configuration.

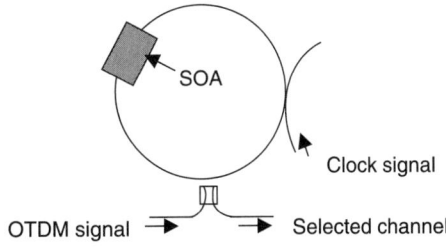

FIGURE 13.31 Schematic configuration of a loop mirror-based TOAD (after [75]). OTDM, optical time division multiplexing.

the spectral-splitting ratio, yields an important design parameter, namely pulled length that the coupler has to be pulled through to realize a component with desired characteristics at the given monitoring wavelength. Thus, our analysis successfully explains all the experimentally observed characteristics of fabricated couplers and should be very useful as a design guideline to realize components having desired specifications. We also presented several applications of fused fiber couplers relevant to optical fiber communications. In particular, use of all-fiber MZI as a wavelength interleaver and an FLM for gain flattening of EDFAs has been detailed. For the sake of completeness, the essential theory behind the working of these devices has also been presented.

6. ACKNOWLEDGMENTS

This work was partially supported by OPTEL Telecommunications Ltd., Bhopal India. We gratefully acknowledge the constant encouragement and constructive criticism received from our senior colleague Prof. A. K. Ghatak. We also thank a number of colleagues from the Fiber Optics Group, IIT Delhi for very fruitful academic interactions, and also several past students who have worked on the FFC fabrication rig from time to time as part of their master's dissertations. Part of the modeling work was carried out in the Institute of Infocomm Research (I^2R), Singapore. One of the authors (P. R. C.) thanks Prof. Cheng Tee Hiang and Prof. Lu Chao of the Lightwave Department, I^2R, Singapore, for their encouragement in this work.

7. REFERENCES

1. S. Bigo, Y. Frignac, G. Charlet, S. Borne, P. Tran, C. Simonneau, D. Bayert, A. Jourdan, J. P. Hamaide, W. Idler, R. Dischler, G. Veith, H. Gross, and W. Poehlmann, Technical Digest Optical Fiber Communication Conference (2001), Opt. Soc. Am., Washington, Post-deadline paper PD25.
2. B. S. Kawasaki, K. O. Hill, and R. G. Lamont, Opt. Lett. **6** (1981) 327.
3. B. P. Pal, "Electromagnetics of all-fiber components," in *Electromagnetic Fields in Unconventional Structures and Materials*, Eds. O. N. Singh and A. Lakhtakia, pp. 359–432, John Wiley, New York, 2000.
4. I. Yokohama, J. Node, and K. Okamoto, IEEE J. Lightwave Technol. LT-5 (1987) 910.
5. P. Roy Chaudhuri, M. R. Shenoy, B. P. Pal, Proc. SPIE (USA), **67** (1999) 3666.
6. D. Marcuse, Bell Syst. Tech. J., 50 (1971) 1791.
7. A. Yariv, IEEE J. Quantum Electron. QE-9 (1973) 919.
8. F. P. Payne, C. D. Hussey, and M. S. Yataki, Electron. Lett. 21(1985) 461.
9. J. D. Love and M. Hall, Electron. Lett. **21** (1985) 519.
10. J. M. P. Rodriguez, T. S. M. Maclean, B. K. Gazey, and J. F. Miller, Electron. Lett. **22** (1986) 402.
11. M. Eisenmann and E. Weidel, J. Lightwave Technol. **8** (1988) 113.
12. F. P. Payne, J. Instit. Electron. Telecom. Engineers (India), **32** (1986) 319.
13. J. V. Wright, Electron. Lett. **21** (1985) 1064.
14. X. H. Zheng, Electron. Lett. **22** (1986) 804.
15. J. L. Zhang, Z. M. Mao, and Z. Q. Lin, Appl. Opt. **28** (1989) 2026.
16. K. Okamoto, *Fundamentals of Optical Waveguides*, Academic Press, San Diego, 2000.
17. J. Bures, S. Lacroix, and J. Lapierre, Appl. Opt. **22** (1983) 1918.
18. W. K. Burns and M. Abebe, Appl. Opt. **26** (1987) 4190.
19. T. A. Birks and Y. W. Li, J. Lightwave Technol. **10** (1992) 432.
20. K. S. Chiang, App. Opt. 25 (1986) 348; Electron Lett. **22** (1986) 1221; Opt. Lett. **12** (1987) 431.
21. C. Yeh, K. Ha, S. B. Dong, and W. P. Brown, Appl. Opt. **18** (1979) 1490.
22. B. M. A. Rahman and J. B. Davies, IEEE Trans. Microwave Theory and Tech. **32** (1984) 20.
23. K. S. Chiang, Opt. Quantum Elect. **17** (1985) 381.
24. E. Schweig, and W. B. Bridges, IEEE Trans. Microwave Theory Tech. **32** (1984) 531.
25. P. Kaczmarski, P. Lagasse, and J. Vandewge, IEE Proc. Pt. J. **143** (1987).
26. J. R. James and I. N. L. Gallett, Radio Elect. Engineering **42** (1972) 103.
27. T.-L. Wu, J. Lightwave Technol. **18** (2000) 1024.
28. L. P. Eyges, P. Wintersteiner, and P. D. Gianino, J. Opt. Soc. Am. **69** (1976) 1226.
29. H. S. Huang and H. C. Chang, J. Lightwave Technol. **8** (1990) 823.
30. S. She, L. Qiqo, and J. Wang, Opt. Comm. **65** (1988) 415.
31. S. Lacroix, F., Gonthier, and J. Bures, App. Opt. **33** (1994) 8361.
32. F. Gonthier, S. Lacroix, and J. Bures, Opt. Quantum Electron. **26** (1994) S135.
33. A. Kumar, K. Thyagarajan, and A. K. Ghatak, Opt. Lett. **8** (1983) 63.
34. P. Roy Chaudhuri, Proc. Optoelectronics and Communication Conference, Yokohama, Japan, 8–12 July 2002, 354.
35. A. W. Snyder and J. D. Love, *Optical Waveguide Theory*, Chapman and Hall, London, (1983).

36. K. Kawano and T. Kitoh, *Introduction to Optical Waveguide Analysis*, John Wiley & Sons, Inc., New York (2001).
37. P. Roy Chaudhuri, C. Zhao, V. Paulose, J. Hao, and C. Lu, Asia Pacific Optoelectronics Conference SPIE, Wuhan, China, 2–6 November 2003, paper 5279–02.
38. A. K. Ghatak and K. Thyagarajan, "Electromagnetics of integrated optical waveguides," in *Fundamentals of Fiber Optics in Telecommunication and Sensor Systems*, Ed. B. P. Pal, John Wiley, New York and Wiley Eastern, New Delhi, 1992.
39. A. Kumar and R. K. Varshney, Opt. Quantum Electron. **16** (1984) 349.
40. A. Kumar and R. K. Varshney, Opt. Lett. **14** (1989) 817.
41. A. Kumar, M. R. Shenoy, and K. Thyagarajan, IEEE Trans. Microwave Theory & Tech. **32** (1984) 1415.
42. R. K. Varshney and A. Kumar, J. Lightwave Technol. **6** (1988) 601.
43. A. Kumar, A. N. Kaul, and A. K. Ghatak. Opt. Lett. **10** (1985) 86.
44. V. Ramaswamy and R. K. Lagu, J. Lightwave Technol. **1** (1983) 408.
45. R. P. Kenny, T.A. Birks, and K. P. Oakley, Elect. Lett. **27** (1991) 1654.
46. S. Roy, R. Tewari, and K. Thyagarajan, J. Opt. Commun. **12** (1991) 26.
47. K. S. Chiang, Opt. Lett. **12** (1987) 431.
48. See, for example A. Ghatak and K. Thyagarajan, *Introduction to Fiber Optics*, f. 362, Cambridge University Press, Cambridge, 1999.
49. B. Culshaw, Chapter in *Fundamentals of Fiber Optics in Telecommunication and Sensor Systems*, Ed. B. P. Pal, John Wiley, New York and Wiley Eastern, New Delhi, p. 584, 1992.
50. J. D. C. Jones, Chapter in *Fundamentals of Fiber Optics in Telecommunication and Sensor Systems*, Ed. B. P. Pal, John Wiley, New York and Wiley Eastern, New Delhi, p. 657, 1992.
51. C. M. Lawson, P. M. Kopera, T. Y. Hsu, and V. J. Tekippe, Electron. Lett. **20** (1984) 963.
52. S. S. Orlov, A. Yariv, and S. V. Essen, Opt. Lett. **22** (1997) 688.
53. I. J. Wilkinson and C. J. Rowe, Electron. Lett. **26** (1990) 382.
54. M. N. McLandrich, R. J. Orazi, and H. R. Marlin, J. Lightwave Technol. **9** (1991) 442.
55. F. Gonthier, D. Ricard, S. Lacroix, and J. Bures, Electron. Lett. **27** (1991) 42.
56. J. D. Minelly and M. Suyama, Electron. Lett. **26** (1990) 523.
57. S. Bourgeois, "Fused-fiber developments offer passive foundation for optical slicing," Special Report in Lightwave (PennWell Corporation) 2000, **17**, No. 3.
58. H. Arai, T. Chiba, H. Uetsuka, H. Okano, and L. Mara, Tech. Digest NFOEC- 2000, 444.
59. J. Chon, A. Zeng, P. Peters, B. Jian, A. Luo, and K. Sullivan, Tech. Digest NFOEC-2001, 1410.
60. H. Yonglin, L. Jie, M. Xiurong, K. Guiyun, Y. Shuzhong, and D. Xiaoyi, Opt. Commun. **222**, (2003) 191.
61. B. H. Verbeek, C. H. Henry, N. A. Olosson, K. J. Orlowski, R. F. Kazarinov, and B. H. Johnson, J. Lightwave Technol. **6** (1988) 1011.
62. N. Kumar, M. R. Shenoy, and B. P. Pal, Int. Conf. On Fiber Optics and Photonics—PHOTONICS-2002, Mumbai, Dec. 14–18, 2002, Paper FBRP-15.
63. I. D. Miller, D. B. Mortimore, W. P. Urquhart, B. J. Ainslie, S. P. Craig, C. A. Millar, and D. B. Payne, Appl. Opt. **26** (1987) 2197.
64. D. B. Mortimore, J. Lightwave Technol. **6** (1988) 1217.
65. M. N. Islam, E. R. Sunderman, R. H. Stolen, W. Pleibal, and J. R. Simpson, Opt. Lett. **15** (1989) 811.
66. K. J. Blow, N. J. Doran, and B. P. Nelson, Electron. Lett. **26**, (1990) 262.
67. I. Glask, J. P. Sokoloff, and P. R. Prucnal, "Demonstration of all-optical of TDM data at 250 Gb/s," Electron. Lett. **30** (1990) 339.
68. M. E. Fermann, F. Haberl, M. Hofer, and H. Hochreiter, Opt. Lett. **15** (1990) 752.
69. A. W. Oneill and R. P. Woff, Electron. Lett. **26** (1990) 2008.
70. K. Morishita and K. Shimamoto, J. Lightwave Technol. **13** (1995) 2276.
71. S. Li., K. S. Chiang, and W. A. Gambling, IEEE Photon. Technol. Lett. **13** (2001) 942.
72. X. Fang and R. O. Claus, Opt. Lett. **20** (1995) 2146.
73. X. Fang, H. Ji, C. T. Allen, K. Demarest, and L. Pelz, IEEE Photon. Technol. Lett. **9** (1997) 458.
74. B. P. Pal, G. Thursby, N. Kumar, and M. R. Shenoy, in Proc. Photonics-2004, Dec. 8–11, Cochin, India, 2004.
75. R. Ramaswamy and K. N. Sivarajan, *Optical Networks: A Practical Perspective*, Morgan Kaufmann, San Francisco, 1998.
76. T. Morioka, H. Takara, S. Kawonishi, T. Kitoh, and M. Saruwatori, Electron. Lett. **32** (1996) 833.

14
CHAPTER

Side-Polished Evanescently Coupled Optical Fiber Overlay Devices: A Review

Walter Johnstone*
Department of Electronic and Electrical Engineering
University of Strathclyde
Glasgow, Scotland

1. INTRODUCTION

Access to the evanescent field of an optical fiber can yield a number of useful optical functions involving coupling of the field to structures formed directly on the fiber. In such devices the fiber remains continuous, obviating the need for expensive fiber-to-device interfacing with its requirements for high precision alignment. Continuous fiber devices offer advantages of intrinsically low insertion loss, low back reflections, and excellent mechanical stability relative to their micro-optic or integrated optic counterparts where the fiber needs to be broken for device insertion.

The most enduring technique for accessing the evanescent field of single-mode optical fiber is that of side polishing to partially remove the cladding on one side of the fiber, thus forming what has become known as the side-polished fiber block coupler or the fiber half coupler. This was first demonstrated by Bergh et al. [1], who also demonstrated the polished block tunable coupler [1] in which two polished fiber blocks were assembled to give evanescent field coupling from one to the other with the coupled power adjusted by variation of the intercore spacing. After the demonstration of evanescent field access by Bergh, a number of researchers reported side-polished fiber devices involving the interaction of the field with various bulk and thin film materials to form variable attenuators/modulators [2, 3] and polarizers [4, 5]. If the refractive index of any material in direct contact with the evanescent field is greater than the fiber effective index, then the guided fiber mode becomes a leaky mode and the attenuation of the structure is high. If it is less, the mode remains strongly guided with low loss.

Some materials have a dispersion characteristic that crosses that of the fiber mode at a particular wavelength, providing an edge filter characteristic with low loss below that wavelength and high loss beyond it [2]. In liquid crystals the molecular alignment and the refractive index may be changed by applying an electric field. Using an appropriate liquid crystal, the refractive index of an overlay may be varied above and below the fiber effective index by controlling the applied field. This technique has been used to demonstrate a variable attenuator/modulator [3].

Birefringent material with ordinary and extraordinary indices above and below the fiber effective index has also been used to extract one of the linear polarization states in the fiber while leaving the orthogonal state unaffected. Extinction ratios in excess of 30 dB were achieved with this approach [4]. Differential absorption between the transverse magnetic (TM) and transverse electric (TE) modes into a metal surface in contact with the exposed evanescent field region has also been used to yield extinction ratios of about 15 dB. However, the best polarizers were realized by making use of the resonant excitation of TM-only surface plasmon waves at a single metal dielectric interface [6, 7] and in thin metal films [8, 9] in evanescent contact with the fiber mode. The TE mode cannot excite the surface plasmon waves and passes through with low loss. For the thin metal film device, the design enabled precise phase matching between the TM mode in the fiber and the symmetric leaky surface plasmon mode. The result was TE/TM extinction ratios in excess of 50 dB with losses of less than 0.5 dB for the TE mode.

Perhaps the most interesting devices based on the side-polished fiber block are those involving evanescent field coupling to a planar dielectric overlay

*E-mail: w.johnstone@eee.strath.ac.uk

waveguide [10]. These are variously referred to as fiber to planar waveguide couplers or side-polished fiber overlay devices. Their intrinsic response is that of a bandstop filter [10, 11]. However, the basic structure has been modified and developed to yield tunable filtering functions in both bandstop and bandpass formats. They have also led to the development of electro-optic routing switches, modulators, fiber laser mode lockers, and spectral flattening filters. In the following sections, the technology of fiber to planar waveguide couplers and their applications is reviewed and some new results on filters for flattening the wavelength response of erbium-doped fiber (EDF) devices are presented.

2. PRINCIPLES OF OPERATION

Figure 14.1 shows the structure of a fiber to planar waveguide coupler comprising a side-polished fiber block with the single-mode fiber evanescently coupled to a multimode planar waveguide overlay with a refractive index greater than the fiber core. The device is fundamentally a directional coupler, and light propagating in the fiber only couples efficiently to the overlay when the effective index of one of the modes of the overlay, n_{eo}, is matched to that of the fiber, n_{ef}. The effective indices of the overlay modes are strongly wavelength dependent, and hence the wavelength response of the device is that of a quasi-periodic bandstop filter (Fig. 14.2) as the individual modes of the overlay are tuned in and out of resonance with the fiber mode with variation of the input wavelength, λ_o. The mode effective indices n_{eo} of an overlay of thickness d, and index n_o are given by the standard planar waveguide equation as follows:

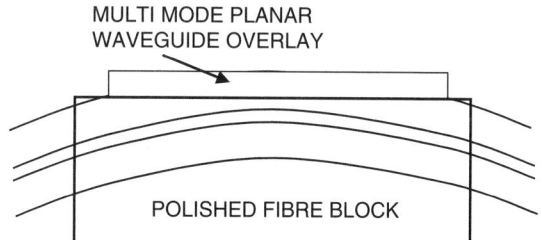

FIGURE 14.1 Schematic diagram of a fiber to planar waveguide bandstop coupler.

$$\frac{2\pi d(n_o^2 - n_{eo}^2)^{1/2}}{\lambda_o} = m\pi + \phi_1 + \phi_2 \quad (1)$$

where m is an integer and ϕ_1 and ϕ_2 are the evanescent field phase shifts at the boundaries with the fiber cladding and the superstrate, respectively. The phase shifts, ϕ_1 and ϕ_2, are given by

$$\phi_i = \tan^{-1} \xi \left(\frac{n_{eo}^2 - n_i^2}{n_0^2 - n_{eo}^2} \right)^{1/2} \quad \text{for } i = 1, 2 \quad (2)$$

where n_1 is the fiber cladding index, n_2 is the superstrate index, and ξ is a polarization-dependent term equal to 1 for TE modes and n_o^2/n_i^2 for TM modes.

Substituting the condition for coupling, $n_{ef} = n_{eo}$, into Eq. (1) and reorganizing for wavelength, we get an expression for the bandstop resonant wavelengths, λ_m, associated with modes m for which the transmission is minimum

$$\lambda_m = \frac{2\pi d(n_o^2 - n_{ef}^2)^{1/2}}{m\pi + \phi_1 + \phi_2} \quad (3)$$

The key design parameters of the device are the refractive index and the thickness of the overlay

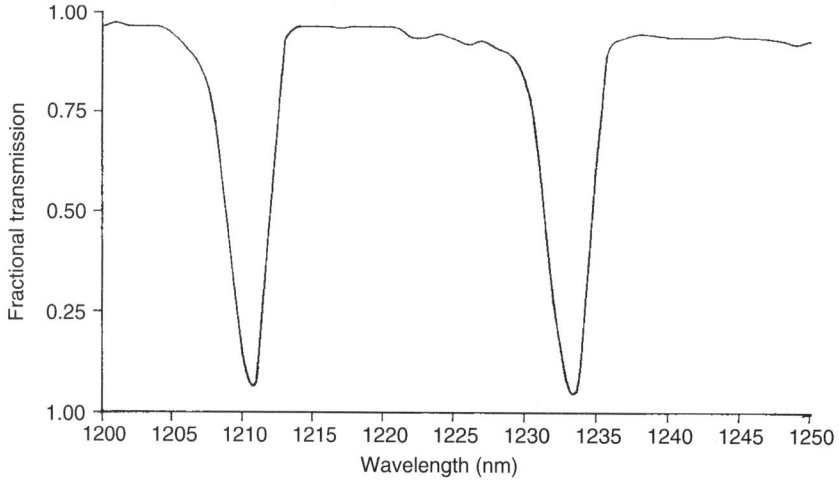

FIGURE 14.2 Wavelength response of a bandstop overlay filter with a 20-μm-thick polished lithium niobate planar waveguide overlay.

and the remaining cladding thickness as set by the polishing process. The overlay thickness and index control the resonance positions and spacing. The remaining cladding thickness controls the coupling strength between the two waveguides and hence fundamentally determines the resonance width in terms of n_{eo}. The linewidth of the bandstop resonances in the wavelength response are thus determined by the remaining cladding thickness and the mode dispersion, $dn_{eo}/d\lambda$, which should be large for narrow linewidth. Simple numerical analysis of Eq. (3) reveals that the spacing between resonances decreases and the mode dispersion $dn_{eo}/d\lambda$ increases with increasing overlay index and thickness [12]. For a given overlay index, we can thus adjust the free spectral range of the filter by varying the thickness, d. For a given index and thickness, the mode dispersion is fixed and the linewidth is then determined by the remaining cladding thickness/coupling strength.

It is here that we must consider the differences between the overlay devices and conventional directional couplers comprising two-channel waveguides. In the conventional directional couplers, power oscillates between the waveguides with a period determined by the coupling length. In the overlay devices, power coupled into the planar waveguide diverges rapidly by diffraction and is lost as regards the coupling process. Hence, there is little or no oscillation of power with length and, provided the coupling length is shorter than the interaction length, the device exhibits a high extinction ratio. However, we can control the remaining cladding thickness, thereby tuning the coupling strength and the bandstop linewidth. The coupling strength may be decreased to the limit where the coupling length increases to equal the interaction length. This produces the narrowest possible linewidth.

With a simple modification to the above device, it is possible to create a bandpass filter response [13]. This device (Fig. 14.3) is based on a side-polished fiber block for which the polishing process has been continued to partially remove the core to such an extent that the fiber mode is cut off and transmission is negligible throughout the intended operating range of wavelengths. Application of the overlay then restores transmission at wavelengths close to the resonant wavelengths of a bandstop device with the same overlay index and thickness. At these wavelengths, given approximately by Eq. (3), power couples resonantly to a composite mode of the overlay/fiber core structure at the input to the "cutoff" region. That mode traverses the cutoff region with low loss and then reexcites the fiber mode at the output. Power at off-resonance wavelengths propagates into the cutoff region and radiates into the cladding, providing extinction between the bandpass resonant peaks. Hence, the wavelength response is that of a quasi-periodic bandpass filter (Fig. 14.4). The overlay index and thickness influence the spacing and linewidth of the bandpass peaks in the same way as for the bandstop filter. Again the polishing depth, this time as regards the extent of core removal, is crucial in determining the linewidth of the resonances, the on-resonance loss and the off-resonance extinction ratio. Sufficient removal of the core is required to achieve narrow linewidths and good off-resonance extinction, but if too much is removed the on-resonance loss becomes very high.

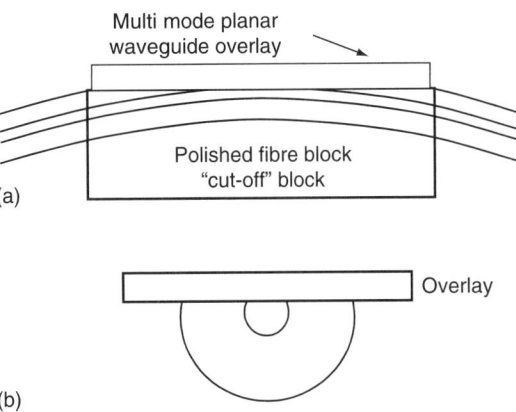

FIGURE 14.3 Schematic diagrams showing the axial (a) and radial (b) cross-sections of a fiber to planar waveguide bandpass coupler.

Equation (3) indicates that the resonance wavelengths for the bandstop device (similarly for the bandpass filter) may be tuned by variation of the overlay index. Hence, simply by using an electro-optic material as the overlay and applying a suitable electrode structure (Fig. 14.5), we can realize electrically tunable bandstop and bandpass filters or modulators.

3. DEVICES

The polished fiber blocks are fabricated according to the method described by Bergh et al. [1]. However, as discussed above, the polishing depth of the fiber half coupler is crucial to achieving the required filter linewidth, loss, and extinction ratio. Precise and repeatable control of the polishing depth is achieved by periodic measurement of the change in loss of the device when an oil of a particular refractive index is applied to the exposed core surface [12]. Removal of

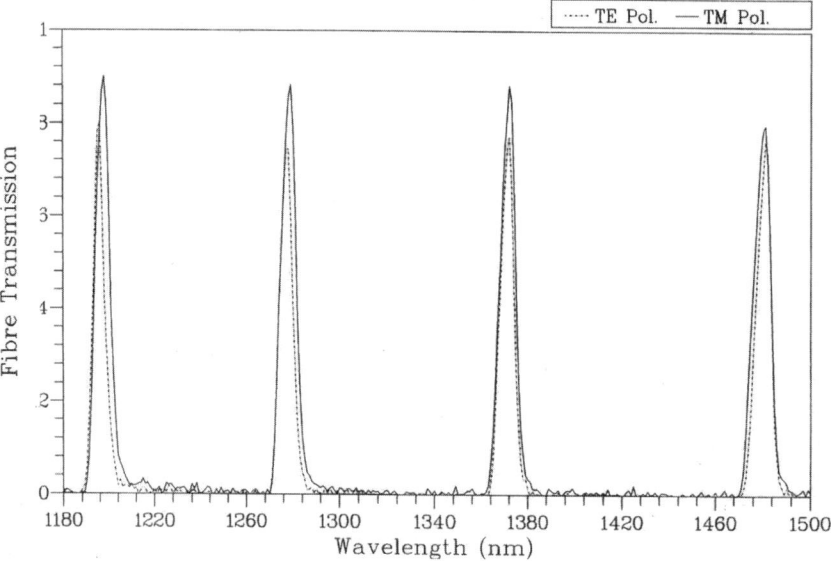

FIGURE 14.4 Wavelength response of a bandpass overlay filter with a 6-μm-thick sputtered zinc oxide overlay for which stress birefringence has been engineered to offset waveguide birefringence, resulting in coincident TE and TM resonances.

the cladding is characterized by measurement of the transmission before and after application of an oil of refractive index 1.60. The extinction ratio (relative to no oil) for the leaky waveguide formed by applying the oil is a repeatable and reliable parameter to characterize the polishing depth and the coupling strength of the device up to the point at which core removal becomes dominant. An oil of refractive index equal to 1.60 is chosen because it provides the best compromise between its sensitivity to polishing depth and the range it needs to cover. Core removal is best characterized by measurement of the transmission before and after application of an oil of refractive index equal to the fiber cladding index (1.447 for quartz fiber claddings). As the core is removed, the stand-alone fiber half coupler exhibits increasing loss. Application of an oil of refractive index equal to the cladding restores strong waveguiding and low-loss transmission. Again, the transmission ratio is a repeatable and reliable parameter to characterize the extent of core removal depth for the purposes of fabricating devices to a given specification.

The first bandstop filters were fabricated using liquid overlays introduced by capillary action into the cavity formed by a quartz glass superstrate separated from the polished fiber surface by precision mica spacers [10]. Later, more practical solid-state devices were formed by bonding high refractive index glasses and other materials such as lithium niobate [14] to the fiber block surface and then polishing them to the desired thickness to achieve the required wavelength response (Fig. 14.3). Devices formed by vacuum evaporation or sputter deposition of the overlay were also demonstrated [11, 12]. In this work, extinction ratios of >20 dB with off-resonance losses of <0.5 dB were readily achieved and linewidths of less than 2 nm were realized using polished high index lithium niobate [14]. In later work, higher index semiconductors such as silicon and longer interaction lengths were used to yield linewidths of considerably less than 1 nm [15]. Resonance spacings were readily variable from a few nanometers up to >200 nm as controlled by the overlay thickness.

Bandpass filters have been demonstrated using oils, polished overlays (as above), and vacuum-deposited overlays [12, 13, 16]. In the bandpass configuration, linewidths down to 3 nm have been achieved with on-resonance losses of less than 0.5 dB and with off-resonance extinction ratios of greater than 20 dB. Figure 14.4 shows the wavelength response, for both the TE and TM polarization states, of a bandpass device [17] formed by sputtering of zinc oxide directly onto a polished fiber block. Stress birefringence was engineered to offset the waveguide birefringence to yield coincident TE and TM resonances.

Wavelength tunable devices and modulators have been realized in both device types by using electro-optic materials in the role of the overlay. Figure 14.5 shows two variants of electrode structure used to

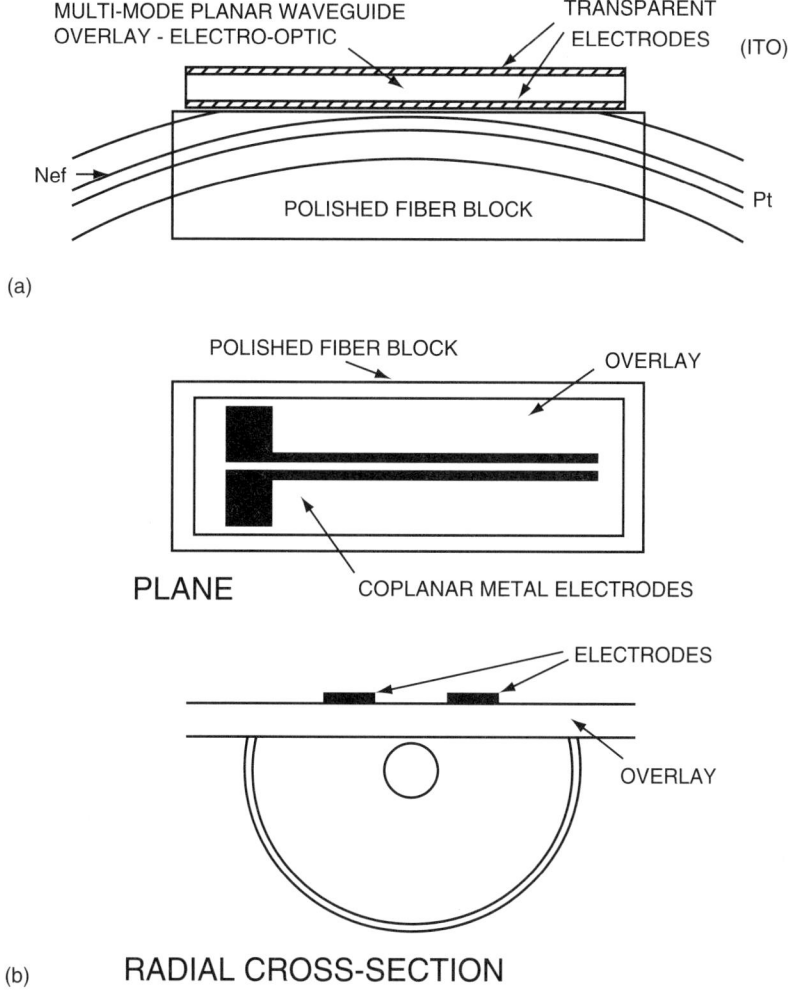

FIGURE 14.5 Electrode structures for fiber to planar waveguide couplers showing (a) transparent plane parallel electrodes and (b) coplanar microstrip metal electrodes (in plane view and radial cross-section).

address the active material. In one option, transparent indium tin oxide electrodes are applied by sputtering below and then above the overlay to realize a field vertically through the plane of the overlay on application of a voltage. Such a configuration exhibits high impedance, and the electrical bandwidth is limited. In an alternative approach, coplanar electrodes, defined photolithographically, are applied onto the top surface of the overlay on either side of the fiber core. Here an applied voltage leads to a field in the plane of the overlay and the device impedance can be made very low, implying high electrical bandwidth.

Tunable bandstop devices have been fabricated with polished lithium niobate [14], electro-optic polymers [18], and liquid crystals [19], all using the indium tin oxide plane parallel electrodes of Fig. 14.5a.

Tuning voltages of 100, 34, and 0.2 V/nm, respectively, were demonstrated. One of the tunable lithium niobate bandstop devices was later converted into a routing switch by aligning a second polished fiber block onto the top surface to capture the resonantly coupled field [20]. Lithium niobate was also used to form a tunable bandpass device with plane parallel indium tin oxide electrodes [16]. Finally, a high-speed bandpass modulator [21] was realized using polished lithium niobate and coplanar microstrip electrodes (Fig. 14.5b). It demonstrated a 3-dB bandwidth of 4 GHz.

4. APPLICATIONS

The original intended application of the overlay devices was filtering in dense wavelength division multiplexing (DWDM). However, with DWDM channel spacing of 1.6 or 0.8 nm, the filter linewidths required proved to be at or less than the absolute lower limit of the single layer devices that were demonstrated. Other applications soon emerged though. Tunable lithium niobate bandpass overlay filters were used to define and tune the wavelength of EDF ring lasers configured as shown in Fig. 14.6 [22, 23]. Single-mode operation of the laser was demonstrated with electro-optic tuning ranges of a few nanometers. In another experiment using the same laser structure as that in Fig. 14.6, a high-speed lithium niobate modulator with coplanar electrodes was used to provide mode locking at a pulse rate up to 4 GHz [21].

Further applications were found in the technology of EDF amplifiers and super fluorescent sources. Here, bandstop overlay filters were used to flatten the wavelength spectra of both EDF amplifiers and super fluorescent sources [24]. Figure 14.7 shows the configuration of a double-pass forward EDF super fluorescent source with an overlay bandpass filter inserted between the EDF and the fiber loop mirror. Figure 14.8 shows the output spectrum with and without the filter. The filter was designed to have the necessary polishing depth (core removal) to provide a response that approximated the inverse of the EDF spectral peak at 1530 nm. It had an on-resonance extinction ratio of 7 dB and a 3-dB linewidth of 5 nm. The overlay was formed of a thermoplastic material, and the resonant wavelength was tuned by Peltier temperature control to exactly coincide with the 1530-nm peak of the EDF source. Insertion into the super fluorescent sources, as shown in Fig. 14.8, removed the 1530-nm peak, flattened the output to a ripple of 1 dB, and resulted in a 30% increase in the spectral power density at 1550 nm.

5. CONCLUSIONS

Fiber to planar waveguide coupling has been used to realize useful solid-state bandstop and bandpass filters, offering all the advantages of continuous fiber devices. Linewidths down to a few nanometers have been demonstrated. Active devices using electro-optic overlays and appropriate electrode configurations have been realized to provide wavelength tuning in both filter configurations and high-speed modulation in a bandpass arrangement. Active bandpass filters have found useful applications in the tuning and high-speed mode locking of EDF ring lasers, and a bandstop filter has been used to flatten the output spectrum of an EDF super fluorescent sources. It is proposed that in the future, multilayer overlays may be used to sharpen and

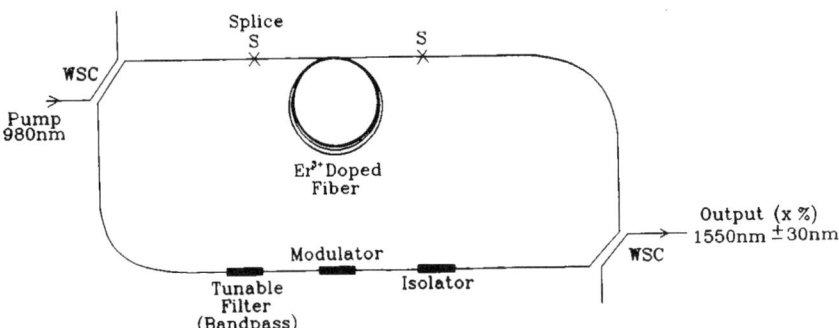

FIGURE 14.6 Schematic diagram of the EDF ring laser containing the tunable filter and the high-speed modulator for mode locking. WSC, wavelength selective coupler.

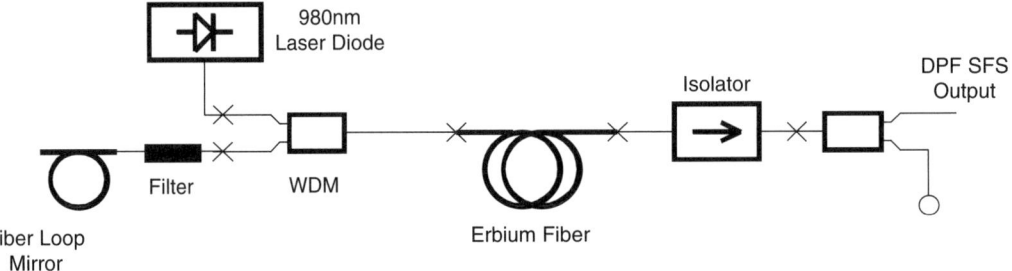

FIGURE 14.7 Schematic diagram of a double-pass forward (DPF) EDF super fluorescent source (SFS) with an overlay filter providing spectral flattening.

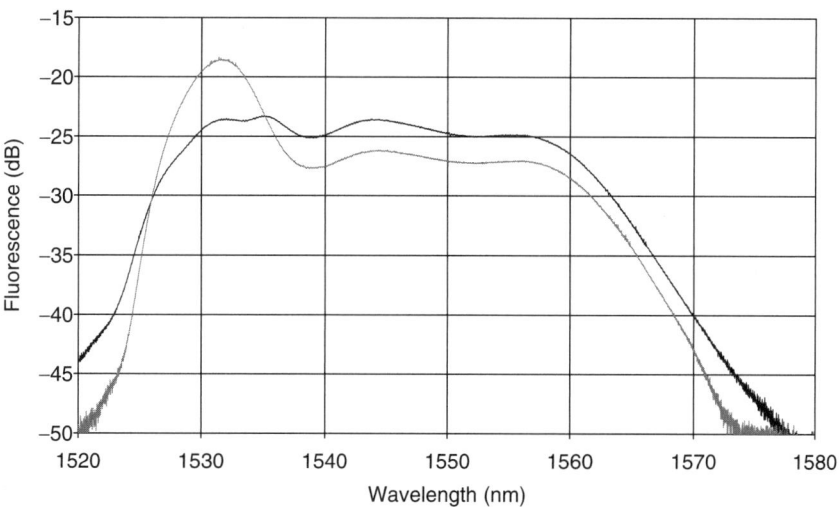

FIGURE 14.8 Output spectra of the super fluorescent source before and after insertion of the spectral flattening overlay filter.

shape the linewidth to realize filters suitable for DWDM applications as originally intended. However, coarse WDM systems, with 20-nm channel spacing, may provide a much more readily accessible application with demands for filter linewidths in the region of 10 nm.

6. REFERENCES

1. R. A. Bergh, G. Kotler and H. J. Shaw, "Single mode fibre optic directional couplers," Electron. Lett., **16**, pp. 260–261 (1980).
2. R. Zergele and O. Leminger, "Optical edge filters made of single mode fibre," J. Opt. Comms., **6**, pp. 150–152 (1985).
3. B. K. Nayar, R. Kashyap and K. R. White, "Electro-optic monomode fibre devices with liquid crystal overlays," Proc. Euro. Conf. Opt. Comms., Barcelona, Spain, pp. 177–180 (1986).
4. R. A. Bergh, H. C. Lefevre and H. J. Shaw, "Single mode fibre optic polarizer," Opt. Lett., **5**, pp. 479–481 (1980).
5. W. Eichkoff, "In line fibre optic polariser," Electron. Lett., **16**, pp. 762–763 (1980).
6. O. Parriaux, S. Gidon and S. Cochet, "Fibre optic polariser using plasmon guided wave resonance," Conf. Proc. 7th ECOC, Copenhagen, Denmark, paper 6–1 (1981).
7. D. Grunchmann, K. Petermann, L. Staudigel and E. Weidel, "Fibre optic polarisers with high extinction ratio," Conf. Proc. 9th ECOC, Amsterdam, The Netherlands, pp. 305–307 (1983).
8. W. Johnstone, G. Stewart, T. Hart and B. Culshaw, "Fibre optic polarisers and polarising couplers," Electron. Lett., **24**, 14, pp. 866–868 (1988).
9. W. Johnstone, G. Stewart, T. Hart and B. Culshaw, "Surface plasmon polaritons in thin metal films and their role in fibre optic polarising devices," J. Lightwave Technology, **8**, 4, pp. 538–544 (1990).
10. C. A. Millar, M. C. Brierley and S. R. Mallinson, "Exposed core single mode fibre channel dropping filter using a high index overlay waveguide," Opt. Lett., **12**, pp. 284–286 (1987).
11. W. Johnstone, G. Thursby, D. G. Moodie, R. Varshney and B. Culshaw, "Fibre optic wavelength channel selector with high resolution," Electron. Lett., **28**, 14, pp. 1364–1365 (1992).
12. K. McCallion, "Investigation of continuous fibre devices based on coupling to high index planar overlays," Ph.D. Thesis, EEE Dept., University of Strathclyde, Glasgow, Scotland, UK (1993).
13. K. MacCallion, W. Johnstone and G. Fawcett, "A tunable in-line fibre optic bandpass filter," Opt. Lett., **19**, 8, pp. 542–544 (1994).
14. W. Johnstone, S. Murray, M. Gill, A. McDonach, G. Thursby, D. Moodie and B. Culshaw, "Fibre optic modulators using a multi-mode lithium niobate waveguide overlay," Electron. Lett., **27**, 11, pp. 894–896 (1991).
15. K. McCallion, Private communication (1997).
16. K. McCallion, S. Creaney, I. Madden and W. Johnstone, "A tunable fibre optic bandpass filter based on polished fibre to planar waveguide coupling techniques," Opt. Fibre Technol., **1**, 3, pp. 271–277 (1995).
17. G. Fawcett, "Fibre optic components using passive and active polymer overlays," Ph.D. Thesis, EEE Dept., University of Strathclyde, Glasgow, Scotland, UK (1993).

18. G. Fawcett, W. Johnstone and I. Andonovic, "In-line fibre optic intensity modulator using electro-optic polymer," Electron. Lett., **28**, 11, pp. 985–986 (1992).
19. I. S. Mauchline, W. I. Madden and W. Johnstone, "Low voltage tunable in-line channel dropping filter using liquid crystal waveguide overlays," Electron. Lett., **33**, 11, pp. 985–986 (1997).
20. K. McCallion, W. Johnstone and G. Thursby, "Investigation of an optical fibre switch using electro-optic interlays," Electron. Lett., **28**, 4, pp. 410–411 (1992).
21. S. C. Creaney and W. Johnstone, "Continuous-fibre modulator with high bandwidth co-planar microstrip electrodes," Photon. Technol. Lett., **8**, 3, pp. 355–357 (1996).
22. A. Gloag, K. McCallion, W. Johnstone and N. Landford, "A tunable erbium doped fibre laser using a novel overlay bandpass filter," Opt. Lett., **19**, 11, pp. 801–803 (1994).
23. A. Gloag, K. McCallion, W. Johnstone and N. Langford, "Tuneable single frequency erbium fibre laser using an overlay bandpass filter," Appl. Phys. Lett., **66**, 24, pp. 3263–3265 (1995).
24. D. MacMillan, "Spectrum engineering in erbium doped fibre devices," Ph.D. Thesis, EEE Dept., University of Strathclyde, Glasgow, Scotland, UK (2003).

15
CHAPTER

Optical Fiber Gratings

K. Thyagarajan*
Physics Department
Indian Institute of Technology Delhi
New Delhi, India

1. INTRODUCTION

Optical fiber gratings consist of a periodic modulation of refractive index along the core of an optical fiber. There are two main types of fiber gratings, namely short period gratings and long period gratings. Short period gratings, also referred to as fiber Bragg gratings (FBG), have periods on the order of half a micrometer for operation around the 1550-nm window, whereas long period gratings (LPG) have periods of a few hundred micrometers. Both types of gratings exhibit strong wavelength dependent characteristics and hence have important applications in wavelength division multiplexed (WDM) optical fiber communication systems and networks.

Short period gratings couple light from a guided core mode propagating in the forward direction to another guided core mode propagating in the backward direction. If Λ represents the period of the short period grating, then an optical wavelength satisfying the relation $\lambda = 2n_{\text{eff}}\Lambda$ (referred to as the Bragg condition), where n_{eff} is the effective index of the guided core mode, gets strongly reflected while other wavelengths get transmitted with almost negligible loss (Fig. 15.1). The peak reflectivity, wavelength of peak reflectivity, and the bandwidth depend on the period, length of the grating, and on the magnitude of the index modulation. Thus, by an appropriate choice of the grating parameters, one can realize gratings with required characteristics. Short period gratings find applications in add/drop multiplexers, wavelength lockers, wavelength filters, dispersion compensators, and sensors.

LPGs are gratings that couple light from a guided core mode to a cladding mode (modes guided by the cladding–air boundary) propagating in the same direction as the core mode or to another guided core mode. The specific cladding mode to which a core mode gets coupled depends on the fiber parameters as well as on the period of the grating. Because cladding modes are lossy, this results in a loss peak in the transmission spectrum of the grating (Fig. 15.2). The magnitude of the loss peak, the spectral width, and position of the loss peak can all be controlled by an appropriate choice of the grating period, the length, and the index modulation. Unlike FBGs, there is no reflected wave in the case of LPGs. Thus, LPGs can be used to induce specific loss at a specific wavelength and mainly find applications as gain flattening filters in erbium-doped fiber amplifiers (EDFAs), in sensors, and so on.

In this chapter we briefly describe the coupled-mode theory used for the analysis of both FBG and LPG and then discuss the properties of gratings and some of their applications. For more details, readers are referred to [1, 2].

2. FIBER BRAGG GRATINGS

Consider a periodic dielectric medium consisting of a number of alternate layers of higher $(n_0 + \Delta n)$ and lower $(n_0 - \Delta n)$ refractive indices with $\Delta n \ll n_0$ (Fig. 15.3). When a lightwave enters this medium, it undergoes minute reflections from every interface. If all the individual reflections are in phase, then the medium strongly reflects the incident wave. If the reflected waves are not in phase, then the net reflection is weak. Because the phase difference between adjacent reflections is dependent on the wavelength, this implies that the overall reflection from such a medium is very strongly wavelength dependent.

Assuming the thickness of each medium to be $\Lambda/2$, the phase difference between reflections 1 and 2 (Fig. 15.3) is (assuming $\Delta n \ll n_0$)

$$\Delta\phi = \pi - \frac{2\pi}{\lambda_0}n_0\Lambda \quad (1)$$

The extra phase difference of π comes about due to the phase change on reflection suffered during

---
*E-mail: ktrajan@physics.iitd.ernet.in

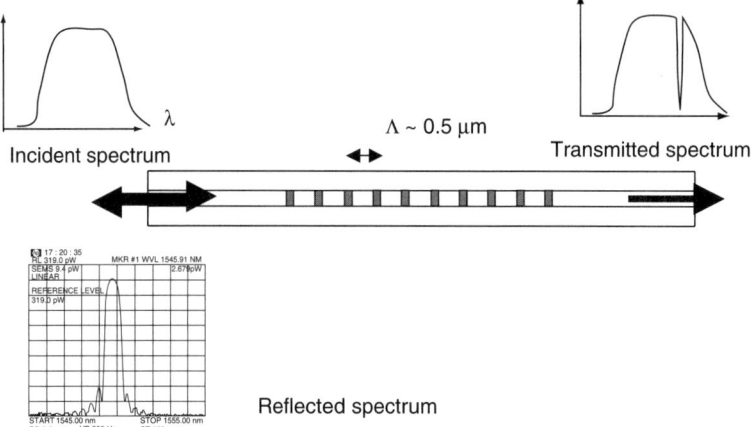

FIGURE 15.1 When light from a broadband source is incident on an FBG, peak reflection occurs at the wavelength λ_B, satisfying the Bragg condition.

reflection from a denser medium. Similarly, the phase difference between waves 2 and 3 is given by

$$\Delta\phi = \pi + \frac{2\pi}{\lambda_0 n_0} \qquad (2)$$

For wavelengths satisfying the following condition, constructive interference between waves 1, 2, 3, etc. takes place:

$$\Delta\phi = 2\pi$$

that is, whenever the wavelength satisfies the following equation:

$$\lambda_B = 2n_0\Lambda \qquad (3)$$

The above equation is referred to as the Bragg condition (reminiscent of x-ray diffraction from atomic planes in crystals) and the specific wavelength λ_B satisfying Eq. (3) is referred to as the Bragg wavelength. As the wavelength deviates from the wavelength specified by Eq. (3), the waves reflected from the layers are not in phase and thus the reflection would drop down. For the condition $\Delta\phi = 2\pi$ we could have chosen any integral multiple of 2π on the right-hand side. The wavelength given by Eq. (3) is the longest wavelength for constructive interference and other wavelengths satisfying the Bragg condition would be smaller.

If we now consider a sinusoidal refractive index modulation with a period Λ within the core of a single-mode optical fiber, when a guided mode propagates through such a grating, the forward propagating mode can get strongly coupled to a backward propagating mode when the following Bragg condition is satisfied:

$$\lambda_B = 2\Lambda n_{\text{eff}} \qquad (4)$$

where n_{eff} is the effective index of the propagating mode and λ_B is the Bragg wavelength. Such a periodic index modulation within the fiber is referred to as an FBG. Thus, when broadband light or a set of wavelengths are incident on an FBG, then only the wavelength corresponding to λ_B gets reflected; the

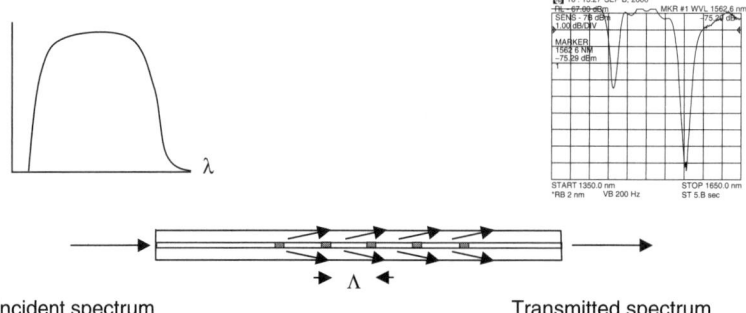

FIGURE 15.2 In the case of an LPG, the grating couples light from a forward propagating core mode to a forward propagating cladding mode, and this results in dips in the transmitted spectrum.

Optical Fiber Gratings

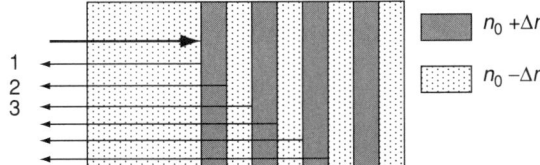

FIGURE 15.3 A periodic structure consisting of alternate layers of high and low refractive indices. When reflections from every interface (1, 2, 3..., etc.) add constructively, then it leads to a strong reflection from the periodic structure.

other wavelengths just get transmitted to the output (Fig. 15.1).

To understand in a more rigorous way the properties of FBG, we now discuss the coupled mode theory.

2.1. Coupled-Mode Theory for FBG

One of the standard methods of analysis of FBG is using the coupled-mode theory [3]. In this, the total field at any value of z is written as a superposition of the two interacting modes and the coupling process results in a z-dependent amplitude of the two coupled modes. We assume that at any point along the grating within the single-mode fiber, we have a forward propagating mode and a backward propagating mode. Thus, the total field within the core of the fiber is given by

$$\Psi(x, y, z, t) = A(z)\psi(x, y)e^{i(\omega t - \beta z)} + B(z)\psi(x, y)e^{i(\omega t + \beta z)} \quad (5)$$

where $A(z)$ and $B(z)$ represent the amplitudes of the forward and backward propagating modes (assumed to be the same order mode), $\psi(x, y)$ represents the transverse modal field distribution, and β represents the propagation constant of the mode. The total field given by Eq. (5) has to satisfy the wave equation given by

$$\nabla^2 \Psi + k_0^2 n_g^2(x, y, z)\Psi = 0 \quad (6)$$

where $n_g^2(x, y, z)$ represents the refractive index variation along the fiber. For an FBG it is given by

$$n_g^2(x, y, z) = n^2(x, y) + \Delta n^2(x, y)\sin(Kz) \quad (7)$$

where $K = 2\pi/\Lambda$ represents the spatial frequency of the grating and Δn^2 represents the index modulation of the grating. For uniform gratings K is independent of z; when K depends on z, such gratings are referred to as chirped gratings. We consider uniform gratings.

Substituting Eq. (5) and Eq. (7) in Eq. (6) and making some simplifying approximations, we obtain the following coupled-mode equations (see, e.g., [3]):

$$\frac{dA}{dz} = \kappa B e^{i\Gamma z}$$
$$\frac{dB}{dz} = \kappa A e^{-i\Gamma z} \quad (8)$$

where $\Gamma = 2\beta - K$ and κ represents the coupling coefficient given by

$$\kappa = \frac{\omega \varepsilon_0}{8} \iint \psi^* \Delta n^2(x, y) \psi \, dx \, dy \quad (9)$$

If the perturbation in the refractive index is constant and finite only within the core of the fiber, then

$$\Delta n^2(x, y) = \Delta n^2, \quad r < a$$
$$= 0, \quad r > a \quad (10)$$

and we obtain

$$\kappa \approx \frac{\pi \Delta n}{\lambda_B} I \quad (11)$$

where

$$I = \frac{\int_0^a \psi^2 r \, dr}{\int_0^\infty \psi^2 r \, dr} \quad (12)$$

Under the Gaussian approximation for the fundamental fiber mode, we can assume ψ to be given by a Gaussian function of width w_0 and we obtain the following approximate expression for the coupling coefficient:

$$\kappa \approx \frac{\pi \Delta n}{\lambda_B}\left(1 - \exp\{-2a^2/w_0^2\}\right) \quad (13)$$

The coupled-mode Eqs. (8) can be solved easily with the boundary conditions

$$A(z = 0) = 1 \quad \text{and} \quad B(z = L) = 0 \quad (14)$$

where L is the length of the grating. Equations (14) imply that the incident wave has unit amplitude at $z = 0$ and the amplitude of the reflected wave at $z = L$ is zero because there is no reflected wave beyond $z = L$. We define the reflectivity of the FBG by the ratio of the reflected power at $z = 0$ to the incident power at $z = 0$. Solving the coupled-mode equations and using the boundary conditions [Eq. (14)] we obtain for the reflectivity of the grating

$$R = \frac{\kappa^2 \sinh^2(\Omega L)}{\Omega^2 \cosh^2(\Omega L) + \frac{\Gamma^2}{4}\sinh^2(\Omega L)} \quad (15)$$

where

$$\Omega^2 = \kappa^2 - \frac{\Gamma^2}{4} \quad (16)$$

2.2. Phase Matched Interaction

The phase matched case corresponds to $\Gamma = 0$, that is, $2\beta = K$, or in terms of effective index n_{eff} and the grating period Λ,

$$2\frac{2\pi}{\lambda_B} n_{\text{eff}} = \frac{2\pi}{\Lambda}$$

or

$$\lambda_B = 2\, n_{\text{eff}}\Lambda \tag{17}$$

which is nothing but the Bragg condition. In such a case the reflection coefficient given by Eq. (15) becomes

$$R = \tanh^2 \kappa L \tag{18}$$

Hence, as κL increases, the grating reflectivity increases and in the limit of κL approaches unity (Fig. 15.4).

Figure 15.5 shows the vector diagram for the case of reflection from FBG when the Bragg condition is satisfied. The wave vectors of the incident wave and the reflected waves are denoted by $+\beta$ and $-\beta$, respectively, and that of the grating by the spatial frequency vector K. As seen from the figure, for phase matched interaction, the three vectors add to give zero.

2.3. Nonphase Matched Interaction

If $\Gamma \neq 0$ then we have the nonphase matched case, and the reflectivity given by Eq. (15) can be shown to be smaller than that given by Eq. (18). Thus, peak reflectivity appears when the Bragg condition is satisfied.

Figure 15.6 shows the simulated variation of reflection coefficient of a typical FBG with a length of 5 mm, a grating period of 0.545 μm, and an index modulation $\Delta n = 4.93 \times 10^{-4}$. The peak of the re-

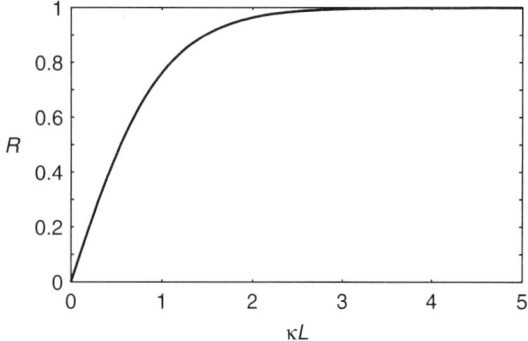

FIGURE 15.4 Length variation of the reflectivity at the Bragg wavelength of an FBG. For large values of κL, the reflectivity approaches unity.

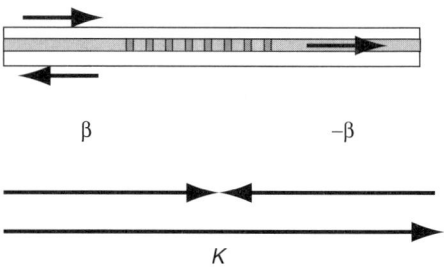

FIGURE 15.5 Vector diagram showing coupling of forward propagating mode to a backward propagating mode by an FBG when the Bragg condition is satisfied.

flection corresponds to $\Gamma = 0$, and as Γ deviates from 0, the reflectivity drops from its peak value. One can define the spectral bandwidth of the grating in terms of the grating parameters as (see, e.g., [3])

$$\Delta\lambda \approx \frac{\lambda_B^2}{n_{\text{eff}}L}\left(1 + \frac{\kappa^2 L^2}{\pi^2}\right)^{1/2} = \frac{\lambda_B^2}{n_{\text{eff}}L}\left(1 + \frac{(\Delta n L I)^2}{\lambda_B^2}\right)^{1/2} \tag{19}$$

If we assume the propagating mode to be well confined within the core, then for small reflectivity, the approximate reflection coefficient at λ_B is given by

$$R(\lambda_B) \approx \kappa^2 L^2 \approx \frac{\pi^2 \Delta n^2 L^2}{\lambda_B^2} \tag{20}$$

where Δn is the peak refractive index change in the grating and L is the length. As we move away from λ_B, the reflection coefficient drops on either side.

We now consider two limiting cases. For small $\Delta n (\Delta n \ll \lambda_B/L)$, the bandwidth defined as the wavelength spacing between the two zeroes in reflection on either side of λ_B is given approximately by [see Eq. (19)]

$$\Delta\lambda \approx \frac{\lambda_B^2}{n_{\text{eff}}L} \tag{21}$$

In this case, the bandwidth of the grating is inversely proportional to the length of the grating.

On the other hand, for $\Delta n \gg \lambda_B/L$, we have [see Eq. (19)]

$$\Delta\lambda \approx \frac{\Delta n \lambda_B I}{n_{\text{eff}}} \tag{22}$$

independent of the grating length.

As an example, let us consider a grating made on a fiber with $a = 3$ μm, numerical aperture (NA) = 0.1, and $n_2 = 1.45$. The fiber is single moded at 850 nm and the fundamental mode has an effective index of 1.4517. If we wish to strongly reflect a wavelength of 850 nm, the corresponding spatial period required is

Optical Fiber Gratings

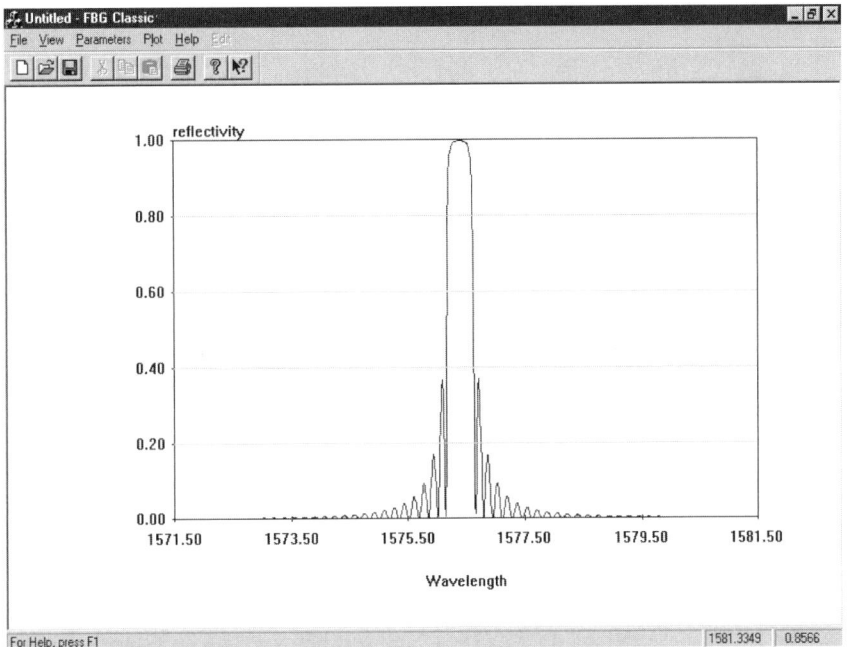

FIGURE 15.6 Simulated transmission spectrum of an FBG.

$$\Lambda = \frac{\lambda_B}{2n_{\text{eff}}} \approx 0.293 \ \mu m$$

Note that the required spatial period is less than the optical wavelength. If the grating has a length of 1 cm, then the required value of Δn for 20% peak reflectivity is

$$\Delta n = \frac{\lambda_B}{\pi L}\sqrt{R} \approx 1.2 \times 10^{-5}$$

The corresponding bandwidth is $\Delta\lambda \approx 0.05$ nm. Note also the extremely small bandwidth of the grating.

From Eqs. (18) and (19) it also follows that one can design gratings having the same peak reflectivity but different bandwidths by appropriate choice of the peak index modulation and grating length. A given peak reflectivity implies a given value of κL (i.e., ΔnL). The bandwidth can be increased or reduced by decreasing or increasing the grating length while keeping the product ΔnL constant.

3. SOME APPLICATIONS OF FBGS

FBGs find many applications. These include applications in add/drop multiplexers, fiber grating sensors, to provide external feedback for laser diode wavelength locking, dispersion compensation, and so on.

3.1. Add/Drop Multiplexers

Figure 15.7 shows a typical configuration of an add/drop multiplexer based on an FBG. An FBG that reflects light at the desired wavelength that needs

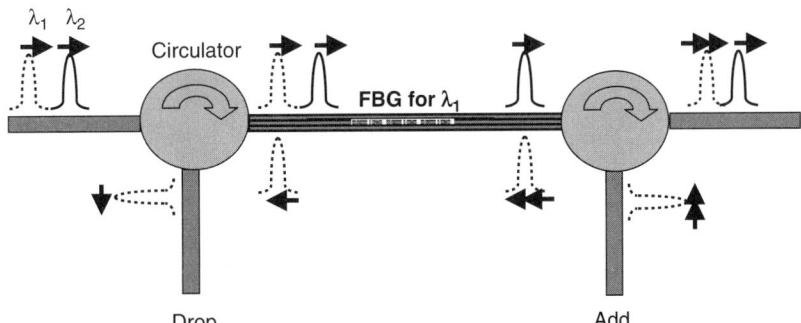

FIGURE 15.7 Application of an FBG in an add/drop multiplexer. The FBG reflects the desired wavelength that is then dropped by the circulator. Information at the same wavelength can be added using the same FBG.

to be dropped is placed between two circulators. From among the incoming wavelength channels, the wavelength matching the FBG wavelength gets reflected and routed to the dropped port. The same time signal at the same wavelength can be added using the second circulator with the same grating reflecting the added channel into the link. Because an FBG reflects only the chosen wavelength and transmits all other channels, it is possible to add and drop multiple channels by using multiple FBGs at the desired wavelengths instead of a single FBG. The gratings used here need to have high reflectivity so that there is no residual signal leading to crosstalk among the dropped and added channels. Also, the gratings need to be designed so that there is no coupling to the backward propagating cladding modes at wavelengths other than the chosen wavelength.

3.2. Dispersion Compensation

Another important application of FBGs is in dispersion compensation in fiber optic communication systems. When pulses of light propagate through an optical fiber link, they suffer from dispersion due to the dependence of group velocity of the mode on the wavelength. This dispersion needs to be compensated in the link, and this can be achieved either through the use of dispersion compensating fibers (see, e.g., [3]) or with the help of chirped FBGs. In chirped FBGs the period of the grating varies with the position along the fiber length, and this leads to a variation of Bragg wavelength along the grating (Fig. 15.8). Thus, different wavelengths get reflected at different positions along the grating, leading to different time delays for different wavelengths. By using an appropriate chirped FBG, one can compensate for the differential delay of different wavelengths accumulated while propagating through the fiber. As shown in Fig. 15.8, wavelength λ_1 suffers a larger delay than wavelength λ_2 while propagating through the fiber. In the anamolous dispersion region of a single-mode fiber (wavelength greater than the zero dispersion wavelength), $\lambda_1 > \lambda_2$. The chirped grating is designed so that λ_1 reflects in the near end of the grating whereas λ_2 reflects from the far end so as to compensate for the differential delay between λ_1 and λ_2, leading to dispersion compensation.

Let us consider a linearly chirped FBG with the following spatial frequency variation:

$$K(z) = K_0 + \frac{Fz}{L_g^2} \qquad (23)$$

Here K_0 is the spatial frequency at $z = 0$ (the input position of the grating), L_g is the length of the grating, and F is the chirp parameter. If Λ_f and Λ_b represent the spatial periods of the FBG at $z = 0$ (front) and $z = L_g$ (back), then the wavelengths that undergo reflections at $z = 0$ and $z = L_g$ are approximately given by

$$\lambda_f = 2n_{\text{eff}}\Lambda_f, \quad \lambda_b = 2n_{\text{eff}}\Lambda_b$$

where we have neglected the wavelength dependence of n_{eff}. Thus, from Eq. (23), we can write

$$\Delta\left(\frac{1}{\Lambda}\right) = \frac{1}{\Lambda_b} - \frac{1}{\Lambda_f} = \frac{F}{2\pi L_g} \approx \frac{2n_{\text{eff}}\Delta\lambda}{\lambda_a^2}$$

where $\Delta\lambda = \lambda_b - \lambda_f$ represents the bandwidth of the chirped grating, and $\lambda_f = (\lambda_b + \lambda_b)/2$ represents the average wavelength. Thus, the bandwidth of the FBG is given by

$$\Delta\lambda = \frac{F\lambda_a^2}{4\pi n_{\text{eff}} L_g} \qquad (24)$$

Now, the time delay between the wavelengths λ_f and λ_b is

$$\Delta\tau = \frac{2L_g}{c/n_{\text{eff}}} = \frac{2n_{\text{eff}} L_g}{c}$$

Hence, the dispersion introduced by the grating is given by

$$\frac{d\tau}{d\lambda} = \frac{8\pi n_{\text{eff}}^2}{c\lambda_a^2} \frac{L_g^2}{F} \qquad (25)$$

Thus, to compensate for accumulated dispersion of $D \cdot L_f$ of the link fiber with dispersion coefficient D and length L_f, we require

$$D \cdot L_f = -\frac{8\pi n_{\text{eff}}^2}{c\lambda_a^2} \frac{L_g^2}{F} \qquad (26)$$

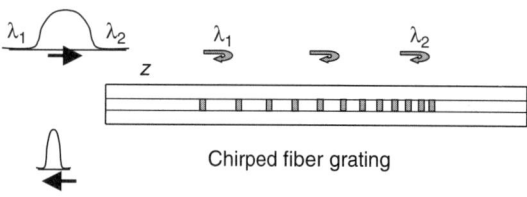

FIGURE 15.8 In a chirped FBG, the grating period varies with position. Thus, different wavelengths are reflected at different positions of the FBG, leading to differential delay between wavelengths. Such chirped FBGs find applications in dispersion compensation.

The bandwidth over which compensation takes place is given by Eq. (24), which in terms of frequency is

$$\Delta \nu = \frac{c}{4\pi n_{\text{eff}}} \frac{F}{L_g} \quad (27)$$

As an example, let us consider a chirped grating of length 11 cm with the chirp parameter $F = 640$, operating at an average wavelength of 1550 nm with the fiber effective index of 1.45. Using Eqs. (24) and (25), we obtain for the dispersion of the grating as 1380 ps/nm operating over a bandwidth of 0.77 nm. This grating can compensate for dispersion accumulated over a fiber with a dispersion coefficient of 17 ps/km·nm and of length 81 km over a bandwidth of 0.77 nm, which is approximately equal to 96 GHz.

Chirped dispersion compensating gratings are commercially available for compensation of accumulated dispersion of up to 80 km of G.652 fiber for up to 32 wavelength channels. Unlike dispersion compensating fibers, chirped FBGs provide the possibility of tweaking the required dispersion compensation, especially for 40-Gbps systems where the margin of dispersion available is rather small. Also, by using nonlinearly chirped FBGs it has been shown that delay variation from -200 ps to -1200 ps is possible [4]. Polarization mode dispersion is another important effect that degrades pulse propagation and becomes important, especially at high bit rates. Compensation of polarization mode dispersion can be achieved using chirped FBGs fabricated in high birefringence fibers [4].

Apart from the applications detailed above, FBGs have a great potential for applications as sensors for sensing mechanical strain, temperature, acceleration, and so forth [5]. They are especially attractive for quasi-distributed sensing applications wherein localized FBGs with different center wavelengths can be written along a fiber's length to sense signals at different points along the same fiber. Because different FBGs reflect light at different wavelengths, the signals coming from different points along the fiber can easily be differentiated.

4. LONG PERIOD GRATINGS

LPGs are periodic perturbations along the length of the fiber with periods of greater than 100 μm that induce coupling between two co-propagating modes. This coupling could be between two guided core modes or between a guided core mode and a cladding mode. Coupling due to periodic perturbation being wavelength selective, these gratings act as wavelength-dependent loss components. This makes them attractive candidates for applications in wavelength filters with specific application in gain flattening of EDFA, band rejection filters, WDM isolation filters, or as polarization filtering components, sensors, and so on [6–8].

4.1. Coupled-Mode Theory for LPG

An LPG corresponds to a periodic perturbation of the core refractive index as given by Eq (7). In the coupled-mode approximation, the total field at any value of z can be written as

$$\Psi(x, y, z, t) = A(z)\psi_1(x, y)e^{i(\omega t - \beta_1 z)} + B(z)\psi_2(x, y)e^{i(\omega t - \beta_2 z)} \quad (28)$$

where $A(z)$ and $B(z)$ represent the amplitudes of the two interacting modes propagating along the same direction, $\psi_1(x, y)$ and $\psi_2(x, y)$ represent the normalized field distributions of the interacting modes, and β_1 and β_2 represent their propagation constants. Substituting the expressions for Ψ and n^2 in the wave equation and after simplification, we obtain the following pair of coupled-mode equations describing the variation of $A(z)$ and $B(z)$:

$$\frac{dA}{dz} = \kappa B e^{i\Gamma z}$$
$$\frac{dB}{dz} = -\kappa A e^{-i\Gamma z} \quad (29)$$

where the detuning parameter is given by

$$\Gamma = \beta_1 - \beta_2 - K \quad (30)$$

and the coupling coefficient κ is given by

$$\kappa = \frac{\omega \varepsilon_0}{8} \iint \psi_1 \Delta n^2 \psi_2 \, dx \, dy \quad (31)$$

The coupled-mode equations can be solved and assuming the following boundary conditions

$$A(z = 0) = 1, \quad B(z = 0) = 0 \quad (32)$$

that is, unit power is incident only in mode ψ_1 at $z = 0$, we obtain for the fractional power remaining in the core and the cladding after a length of interaction z (see, e.g., [3])

$$P_{\text{co}} = 1 - \frac{\kappa^2}{\gamma^2} \sin^2 \gamma z$$
$$P_{\text{cl}} = 1 - P_{\text{co}} \quad (33)$$

where

$$\gamma = \sqrt{\kappa^2 + \frac{\Gamma^2}{4}} \quad (34)$$

and determines the strength of coupling between the modes. Knowing the field distributions of the interacting modes, one can evaluate the value of κ.

It can be seen from Eq. (33) that unlike the case of FBG, the exchange of energy between the core mode and the cladding mode is periodic with distance and complete transfer of energy from the guided mode to the cladding mode is possible if $\Gamma = 0$, that is, if the following condition is satisfied:

$$\beta_{co} = \beta_{cl} + K \quad (35)$$

If n_{eff}^{co} and n_{eff}^{cl} represent the effective indices of the interacting core and cladding modes, respectively, then from Eq. (35) we obtain for the required period of the LPG

$$\Lambda = \frac{\lambda_0}{n_{eff}^{co} - n_{eff}^{cl}} \quad (36)$$

If we consider coupling from a well-guided LP_{01} mode of a single-mode fiber into a low order cladding mode, then the difference between the effective indices of the core and the cladding mode would be approximately equal to the index difference between the core and the cladding, and for typical telecommunications fibers, this is approximately 0.0035. Thus, for an LPG to operate at a wavelength of 1550 nm, the required grating period is approximately 440 μm, which is much larger than that of an FBG. Thus, LPGs require much longer grating periods than FBGs.

The two important parameters of the grating are its period (Λ) and the peak index modulation. If β_{co} is the propagation constant of the core mode of a fiber and $\beta_{cl}^{(n)}$ that of the cladding mode of nth order, then the two modes are required to satisfy the phase matching condition given by $\beta_{co} - \beta_{cl}^{(n)} = 2\pi/\Lambda$ to exchange energy efficiently. Figure 15.9 shows the vector diagram corresponding to phase matched interaction between a core mode with propagation constant β_{co} and a cladding mode with propagation constant β_{cl} propagating along the same direction.

Figure 15.10 shows a comparison of the spectral dependence of the transmitted power before and

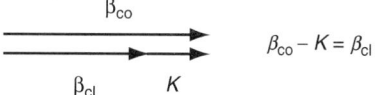

FIGURE 15.9 Vector diagram showing coupling from the core mode to a cladding mode propagating in the same direction. Notice that the grating vector in this case is much smaller than the case of FBG (see Figure 15.5), thus implying much larger periods.

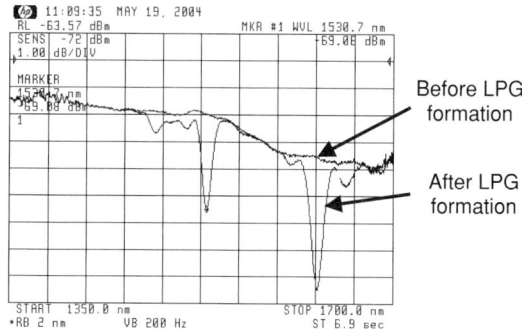

FIGURE 15.10 Spectral dependence of transmitted power before and after formation of an LPG using a fiber fusion splice machine in SMF-28. The LPG induces transmission loss at specific wavelengths corresponding to coupling to different cladding modes.

after formation of an LPG in SMF-28 fiber using a fiber fusion splice machine [9], and Fig. 15.11 shows the comparison of the measured and simulated transmission spectra. One can see multiple dips in the transmission spectrum corresponding to the coupling of the LP_{01} guided mode to individual cladding modes. Note that if the perturbation is azimuthally symmetric, then coupling from the LP_{01} core mode takes place only to LP_{0m} cladding modes.

5. SOME APPLICATIONS OF LPGs

5.1. WDM Filter

Two concatenated chirped LPGs act as a Mach-Zehnder interferometer (MZI) for the range of wavelengths for which coupling is enabled by the gratings (Fig. 15.12). The first grating couples part of the core mode into the cladding, whereas the second grating recombines them. Depending on the phase difference accumulated between the core and the cladding mode, the output power is either in the core or in the cladding. Hence, a periodic

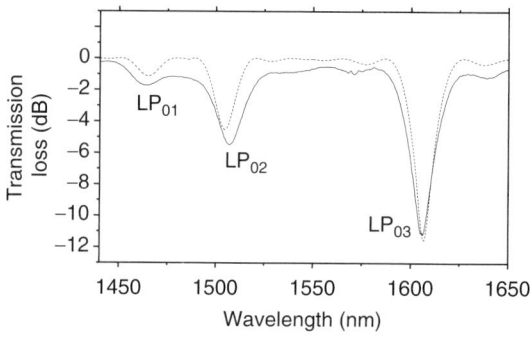

FIGURE 15.11 Simulated transmission spectrum (dashed curve) of an LPG in SMF-28 showing multiple dips due to coupling from the core LP_{01} mode to various cladding modes. The solid curve is the corresponding experimental plot.

Optical Fiber Gratings

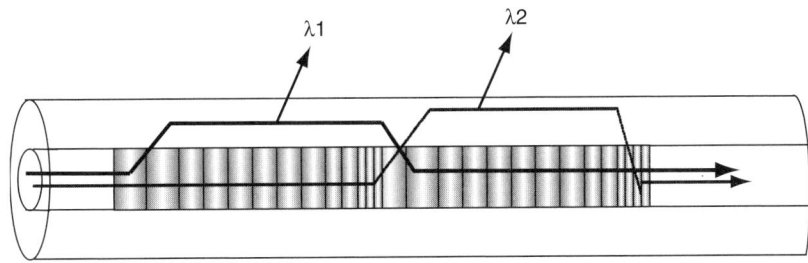

FIGURE 15.12 Two concatenated chirped LPGs act as a Mach-Zehnder interferometer (MZI).

transmission spectrum is expected from the output of the device [10]. Figure 15.13 shows a typical transmission spectrum of two concatenated chirped LPGs. There are about 32 pass bands with an average wavelength spacing of 1.175 nm and an isolation at the stop bands of greater than 15 dB. Such filters can find applications in various DWDM-based devices [11].

5.2. Broadband LPGs

Uniform LPGs usually have a spectral bandwidth of only a few nanometers because as the wavelength changes, the modes get out of phase matching and the power exchange reduces. However, many applications such as polarization dependent loss compensators and WDM filters require broad spectral bandwidth of the transmission dips [12]. It is indeed possible to tailor the refractive index profile of the fiber in which the LPG is fabricated to achieve large spectral bandwidths. This is achieved by having the phase matching condition satisfied over a large wavelength range and at the same time having a coupling coefficient that does not vary significantly over the wavelength range. Figure 15.14 shows the transmission spectrum of an LPG on a fiber having a novel refractive index profile [13]. As is evident, for this design the 20-dB spectral bandwidth is very large and is about 100 nm.

5.3. Gain Flattening of EDFAs

The spectral variation of gain of an EDFA is not flat, and this causes problems in their applications in DWDM optical fiber communication systems (see Chapter 10). Because LPGs can induce spectrally dependent loss, they find applications as gain flattening filters for EDFAs. By designing LPGs with the desired loss at the desired wavelength (~ 1532 nm for EDFA), they can flatten the gain of an EDFA. Figure 8.14 of Chapter 8 shows the gain spectrum of an EDFA without a filter and with an LPG filter (fabricated using a fiber fusion splice machine) showing gain flattening. Such gain flattening LPG filters are also available commercially.

6. CONCLUSIONS

FBGs and LPGs have strong spectral characteristics and are finding application in many devices such as add/drop multiplexers, wavelength lockers, dispersion compensators, sensors, and optical amplifier gain flatteners. By tailoring the grating properties,

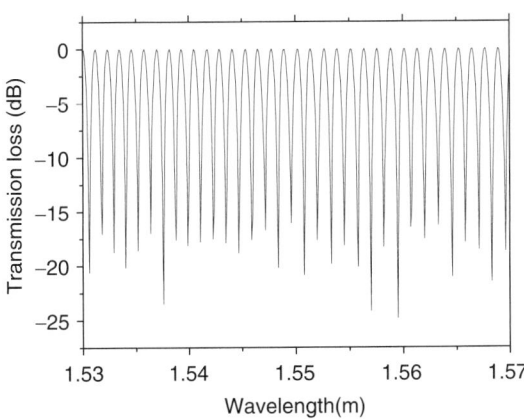

FIGURE 15.13 Transmission spectrum of two concatenated chirped LPGs (After [10]).

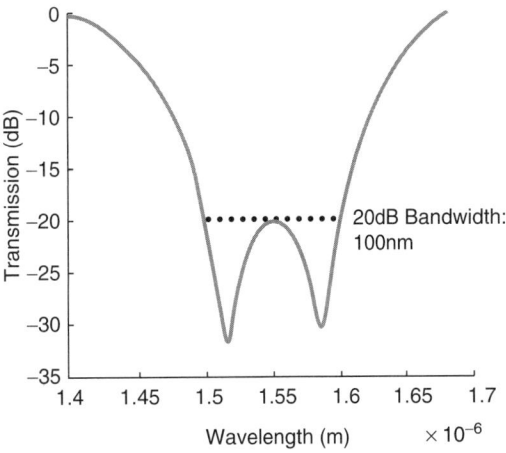

FIGURE 15.14 Simulated transmission spectrum of an LPG in a novel fiber design showing a 20-dB spectral bandwidth of about 100 nm (After [13]).

different spectral characteristics can be achieved. In this chapter we discussed the basic analysis of fiber gratings and some applications.

7. REFERENCES

1. R. Kashyap, *Fiber Bragg Gratings*, Academic Press, San Diego, 1999.
2. A. Othonos and K. Kalli, *Fiber Bragg Gratings*, Artech House, Boston, 1999.
3. A. Ghatak and K. Thyagarajan, *Introduction To Fiber Optics*, Cambridge University Press, UK (1998).
4. A. E. Willner, K. M. Feng, J. Cai, S. Lee, J. Peng and H. Sun, "Tunable compensation of channel degrading effects using nonlinearly chirped passive fiber Bragg gratings," IEEE J. Sel. Topics in Quant. Electron., **5** pp. 1298–1311 (1999).
5. A. D. Kersey, M. A. Davis, H. J. Patrick, M. LeBlac, K. P. Koo, C. G. Askins, M. A. Puman and E. J. Friebele, "Fiber grating sensors," J. Lightwave Technol. **15** pp. 1442–1463 (1997).
6. A. M. Vengsarkar, P. J. Lemaire, J. B. Judkins, V. Bhatia, T. Erdogan and J. E. Sipe, "Long period fiber gratings as band rejection filters," J. Lightwave Technol., **14** pp. 58–65 (1996).
7. S. W. James and R. P. Tatam, "Optical fiber long period grating sensors: characteristics and applications," Meas. Sci. Tech., **14** pp. R49–R61 (2003).
8. A. M. Vengsarkar, J. Renee Pedrazzani, J. B. Judkins, P. J. Lemaire, N. S. Bergano, and C. R. Davidson, "Long period fiber grating based gain equalizers," Opt. Letts., **21** pp. 336–338 (1996).
9. P. Palai, M. N. Satyanarayan, M. Das, K. Thyagarajan, and B. P. Pal, "Characterization and simulation of long period gratings fabricated using electric discharge," Opt. Comm., **193** p. 181 (2001).
10. M. Das and K. Thyagarajan, "Wavelength-division multiplexing isolation filter using concatenated chirped long period gratings," Opt. Comm., **197** pp. 67–71 (2001).
11. K. Thyagarajan and J. K. Anand, "Gain enahanced EDFA for WDM applications," Proc. International Conference On Optical Communication and Networks (ICOCN) 2002, Singapore, Paper 12D2, 73 (2002).
12. S. Ramachandran, S. Ghalmi, Z. Wang, and M. Yan, "Band selection filters with concatenated long period gratings in few mode fibers," Opt. Lett., **27** pp. 1678–1680 (2002).
13. C. Kakkar and K. Thyagarajan, "Novel fiber design for broadband long period gratings," Opt. Comm., **220** pp. 309–314 (2003).

16
CHAPTER

Enhancing Photosensitivity in Optical Fibers

Nirmal K. Viswanathan*
Photonics Division
IRDE
Dehradun, India

1. INTRODUCTION

Photosensitivity in optical fiber enables fabrication of gratings and grating-based devices that are critical components in optical communications networks. The intrinsic photosensitivity in standard single-mode (SSM) telecommunication fiber is typically low ($\Delta n \approx 3\text{E}^{-5}$). As a result, several methods of enhancing the fiber photosensitivity have been actively researched over the past decade [1, 2]. One of the major areas of research has been to develop intrinsically photosensitive fiber by increasing the germanium (Ge) content in the fiber core and also by co-doping with materials such as boron, phosphorous, erbium, and tin [1]. Using these specialty fibers results in large (≥ 1 dB) splice losses with the SSM fiber due to mode-field diameter mismatch, an increase in ultraviolet (UV)-induced impairments such as absorption loss and birefringence due to grating fabrication, and a reduction in thermal stability of the gratings due to dopant migration [1, 2]. Developing new methods to increase the intrinsic photosensitivity in the SSM fiber is still attractive because of their wide use, excellent optical characteristics, low polarization mode dispersion, low cost, and minimal insertion loss. Here I discuss results using two methods—UV sensitization and dilute hydrogen sensitization—of enhancing photosensitivity in SSM optical fibers for manufacturing fiber Bragg grating (FBG)–based devices.

UV-sensitizing fibers to write Bragg gratings [3–5] is an attractive alternative to the standard hydrogen (H_2)/deuterium (D_2) loading methods because this technique offers several advantages, including better thermal stability [6] and low UV-induced absorption losses [7], and offers the flexibility to use different laser wavelengths for sensitization and grating writing processes [8]. Since first demonstrated in silicon oxynitride (SiON) waveguide [3], researchers have used this technique to increase the photosensitivity and write strong and stable gratings in different specialty fibers with excess germanium, boron, and phosphorus doping [5, 8, 9]. Our demonstration of larger than expected refractive index changes ($\Delta n > 5 \times 10^{-4}$) obtained in UV-sensitized low-Ge content SSM fibers was explained by extending the two-step process to include recirculation of catalyst during the grating writing process [10].

Also, for fibers with low intrinsic photosensitivity, saturating the fibers with molecular hydrogen (H_2) before grating inscription is commonly practiced [11]. In the standard method, fibers are exposed to pure H_2 at high pressure (>100 atmosphere [atm]) and above room temperature for several days or even weeks to increase their photosensitivity to fabricate gratings with useful refractive index changes ($\Delta n > 1 \times 10^{-4}$). Alternately, we demonstrated the use of dilute H_2 (1 part per thousand) in very high-pressure gas mixtures to enhance photosensitivity in SSM optical fibers by more than an order of magnitude compared with the standard H_2-loading method [12] and explained using the excluded volume effect and the fugacity of gases.

2. UV SENSITIZATION OF FIBERS

One of the promising methods to sensitize standard optical fiber uses UV sensitization [3–5], where the fiber is saturated with H_2, exposed to UV light, the H_2 is removed, and then the standard FBG writing technique is followed. This method permanently photosensitizes the fiber, even after hydrogen desorption. It produces FBGs that do not require annealing for most applications [6] and has reduced optical loss [7] and possibly produced low birefringence as well. We used the UV sensitization method to study (1) the effect of core Ge concentration, (2) the stability of FBGs using isothermal and continuous isochronal annealing techniques, and (3) the

*E-mail: Nirmal@irde.res.in

mechanism to explain the increase in fiber photosensitivity involving recirculation of H_2 during the grating writing process [10, 13].

2.1. Effect of Ge Concentration

Standard telecommunication fibers with low Ge content in the range of about 3–5 mol% are typically less photosensitive ($\Delta n_{max} \approx 3 \times 10^{-5}$) than fibers with high Ge content and specialty fibers [1]. This limits the use of SSM fiber, because the desirable refractive index changes for manufacturing FBG-based devices are typically $\geq 1 \times 10^{-4}$. Here we present results on strong ($\Delta n \geq 5 \times 10^{-4}$) Bragg gratings written in UV-sensitized SSM fibers with low Ge content and photosensitive fibers with high Ge content. The maximum strength of Bragg gratings that can be written in these fibers appears to depend strongly on the fiber type (perhaps Ge content), the UV-sensitization fluence, and the grating writing time. With UV sensitization, strong gratings can be written and index change can be tailored along the grating length in fibers with low Ge in the core without any co-dopants, opening potential applications including wavelength division multiplexing components and dispersion compensating gratings (DCGs).

Three different types of fibers were used in this study with an average core Ge concentration of 3.5, 5, and 10 mol%, referred to as fiber A, B, and C, respectively. A frequency doubled continuous wave Ar^+ laser operating at 244 nm was the UV source used for both the sensitization exposure and grating writing. The fibers were hydrogen loaded at 60°C and at 2000 psi for 3 days. The concentration of H_2 diffused into the fiber can be calculated by measuring the absorbance at 1.24 μm. The measurement known as the cutback method involves coupling the sample fiber to the source and measuring the power out of the far end. The fiber is then cut near the source, reconnected, and the power measured again. By knowing the power at the source (P_s), at the end of the fiber (P_e), and the length of the fiber (L), the absorbance (in dB/m) can be determined by calculating $[(P_e - P_s)/L]$.

Figure 16.1 is a plot of the wavelength dependence of absorbance (α) for fiber B treated under different conditions. A motorized linear stage was used to scan the UV beam across the buffer-stripped fiber to increase its photosensitivity. Adjusting the scan rate and the laser intensity the fibers were exposed to varied fluence levels. The UV-sensitized fibers were put into then kept inside an oven at 70°C for 3 days to remove the excess hydrogen from the fiber, which was verified by the absence of the 1.24-μm hydrogen

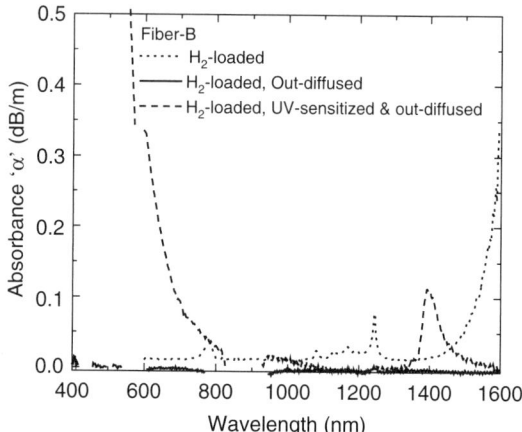

FIGURE 16.1 Absorption spectra of H_2-loaded, out-diffused, and UV-sensitized fiber B measured using the cutback method.

absorption peak (dashed line in Fig. 16.1). An FBG fabrication system based on Talbot interferometer was used to write gratings in the test fibers at a fixed UV power of 180 mW with a spot size of 3×0.1 mm. The grating growth was monitored in transmission as a function of time during the grating inscription.

The refractive index modulation Δn induced in fibers during grating inscription was calculated as a function of time t from the Bragg wavelength λ_B, the grating length L, and the transmission minimum of the grating T_{min} as

$$\Delta n(t) = (\lambda_B/\pi L)\tanh^{-1}\left(\sqrt{1 - T_{min}(t)}\right) \quad (1)$$

Figure 16.2 shows index growth in fiber B exposed to increasing UV sensitization fluence (0.56–5.6 kJ/cm^2) as a function of FBG fabrication time. For a sensitization fluence of ~5.6 kJ/cm^2, we wrote gratings with >25 dB in transmission in ~30 minutes,

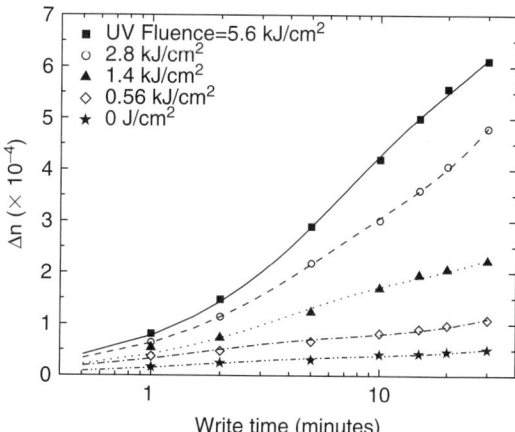

FIGURE 16.2 Refractive index modulation (Δn) growth curves during FBG fabrication in fiber B, UV sensitized to different fluence levels.

corresponding to a maximum refractive index modulation of 7.34×10^{-4}.

The maximum refractive index modulation reached for a 30-minute grating writing as a function of UV-sensitization fluence is shown in Figure 16.3 for all three test fibers under identical exposure conditions. The Δn in the low Ge content fibers, fibers A and B, increases with sensitization fluence until saturation at a relatively high fluence of $>5.6 \, \text{kJ/cm}^2$, whereas the Δn saturates at a relatively low sensitization fluence of $\sim 1.9 \, \text{kJ/cm}^2$ in the high Ge content fiber, fiber C, and decreases thereafter. Other researchers reported the saturation of photosensitivity at a lower sensitization fluence of $\sim 50 \, \text{J/cm}^2$ for $\sim 3 \, \text{mol}\%$ boron co-doped high Ge ($\sim 12 \, \text{mol}\%$) specialty photosensitive fiber [5].

We also calculated the average refractive index change (Δn_{avg}) induced in the fibers during the grating writing process using

$$\Delta n_{\text{avg}} = (n_{\text{eff}} \Delta \lambda / \lambda_{\text{B}}) \quad (2)$$

where n_{eff} is the effective refractive index of the fiber core and $\Delta \lambda$ is the Bragg wavelength shift. Bragg gratings written in UV-sensitized fibers were compared with H_2-loaded and pristine fibers. All the gratings written in the UV-sensitized fibers have a refractive index modulation contrast of $\Delta n / \Delta n_{\text{avg}} \approx 1.13$.

2.2. Thermal Stability of Bragg Gratings

For an FBG to be useful in most telecommunications applications, the relative strength of the FBG is expected to decay less than 2% when operated in the temperature range of -20 to $+80°C$ for 25 years [14]. The strength of gratings written in hydrogen-loaded fibers decays rapidly, which affects the long-term stability of the grating. In comparison, although gratings written in photosensitive fibers have higher refractive index changes, they may suffer from large-splice loss (>1 dB) with SSM and thermal stability issues due to higher germanium concentration and co-dopants.

Experimental results are presented here to establish the long-term and high-temperature stability of gratings written in UV-sensitized fibers A, B, and C using isothermal and continuous isochronal annealing studies. The results are compared with gratings written in H_2-loaded fibers. Analysis based on the stretched exponential model estimates a decay of $\sim 0.2\%$ for gratings in UV-sensitized fibers kept at $80°C$ for 25 years. Further, the gratings written in UV-sensitized fibers show better stability at even higher temperatures [15].

The thermal decay property of the UV-induced index change (Δn) and the long-term stability of the Bragg gratings written in UV-sensitized and H_2-loaded fibers were first measured using isothermal anneal experiments [14]. Here the FBGs are annealed at different temperatures (above room temperature) for a long period of time, while monitoring the changes in the grating strength. To characterize the FBGs for high-temperature stability, we carried out continuous isochronal annealing experiments, where the gratings are heated to high temperatures ($\sim 600°C$) at a fixed ramp rate and the corresponding decrease in the grating strength is measured. The gratings written in UV-sensitized and H_2-loaded fibers were kept at room temperature for 7 days before the annealing measurements were carried out. The gratings were isothermally annealed at 80 and $160°C$ to study their long-term stability and were subsequently heated to $580°C$ from room temperature ($25°C$) to study their high-temperature behavior.

The integrated coupling constant (ICC) of the gratings can be calculated from the grating's strength (T_{min}) using

$$\text{ICC} = \tanh^{-1}\left(\sqrt{1 - T_{\text{min}}}\right) \quad (3)$$

and normalized to its initial value at time $t = 0$. The annealing behavior of the gratings written in H_2-loaded and UV-sensitized fibers was modeled using the stretched exponential function [16]

$$\text{ICC}(t) = \text{ICC}(0) \exp\left[-(t/\tau)^\beta\right] \quad (4)$$

where t is the time, $1/\tau$ is the rate of relaxation, and β is the degree of nonexponentiality characteristic of the relaxation time distribution.

Gratings written to similar strengths in H_2-loaded and UV-sensitized fibers A–C were used in our

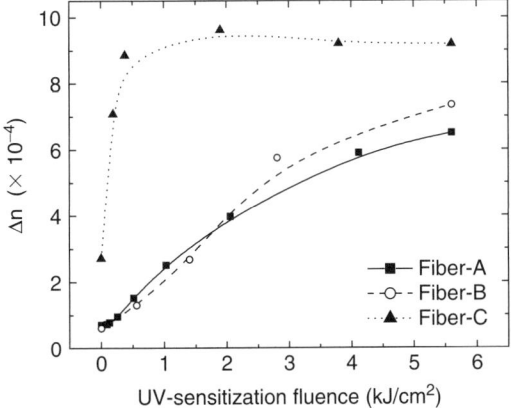

FIGURE 16.3 Maximum refractive index modulation plotted as a function of UV-sensitization fluence for fibers A, B, and C.

analysis. Figure 16.4 shows the behavior of the normalized ICC (NICC) as a function of time for gratings written in H_2-loaded and UV-sensitized fiber B annealed at 80 and 160°C. Figure 16.4 also shows the corresponding theoretical predictions based on the stretched exponential model [Eq. 4] to fit the isothermal annealing behavior of the gratings. For fiber A, the stretch factor β is estimated to be 0.48 and 0.28 for the H_2-loaded and UV-sensitized fibers, respectively, independent of their annealing temperatures. Similar measurements were done in fibers with higher Ge concentrations in the core and resulted in β values of 0.43 and 0.27 for fiber B and 0.40 and 0.15 for photosensitive fiber C for H_2-loaded and UV-sensitized fibers, respectively.

Subsequently, continuous isochronal annealing experiments were carried out on the gratings written in H_2-loaded and UV-sensitized fibers. The temperature of gratings fixed in a clamshell annealer was increased linearly at a fixed ramp rate (480°C/hr) from room temperature (25°C) to ~560°C. Figure 16.5 shows a plot of the continuous isochronal annealing data and the corresponding theoretical fits for gratings written to similar strengths in different fibers. The gratings written in UV-sensitized fibers lose <25% of their initial grating strength, independent of the sensitization fluence, as compared with >50% strength loss in H_2-loaded fiber at these high temperatures. When the experimental data were fitted with the stretched exponential model, the onset of different decay mechanisms was predicted for gratings subjected to high temperatures [13]. The gratings written in UV-sensitized fiber C were more stable at temperatures < 100°C but lost their strength rapidly at higher temperatures, possibly due to diffusion of

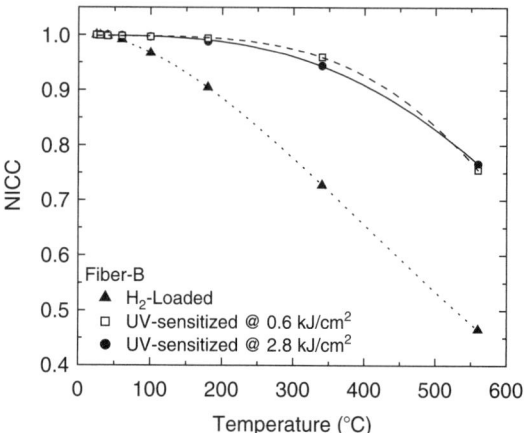

FIGURE 16.5 Continuous isochronal annealing data for gratings written in H_2-loaded and UV-sensitized fiber B. NICC normalized integrated coupling constant.

higher Ge content in the fiber core [13]. Interestingly, no significant differences were noticed in the behavior of the gratings written to different strengths in UV-sensitized fibers.

Thermal stability of the fibers subjected to UV-sensitization exposure and H_2-loading were then compared for write gratings suitable for high-temperature applications. Here the fibers were annealed at a higher temperature for different lengths of time before the grating fabrication. As expected, photosensitivity in the UV-sensitized fiber remained stable even after annealing at 300°C for 6 minutes, enabling the fabrication of ~10-dB strong gratings (Fig. 16.6). In comparison, only very weak (~0.1 dB) gratings could be fabricated in H_2-loaded fibers using the same annealing and writing processes. Such high thermal stability of UV-sensitized fibers opens up the

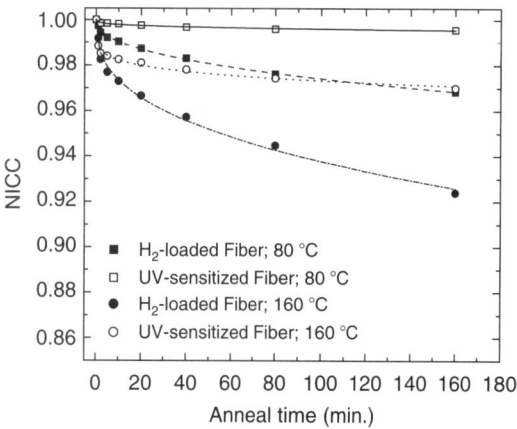

FIGURE 16.4 Isothermal annealing data and the corresponding stretched exponential fit for H-loaded and UV-sensitized fiber B. NICC, normalized integrated coupling constant.

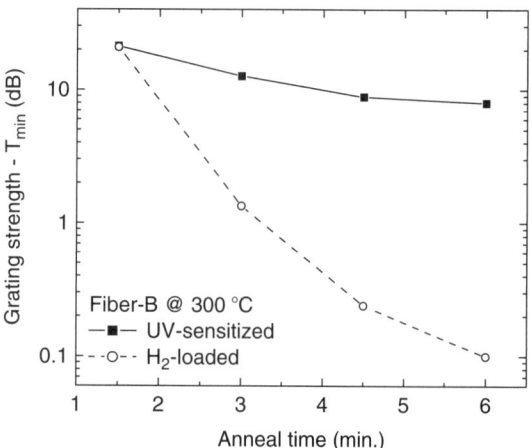

FIGURE 16.6 Strength of gratings written in H_2-loaded and UV-sensitized fiber B after annealing at 300°C for various times.

potential to fabricate gratings suitable for high-temperature applications using standard optical fibers.

2.3. Indication of Recirculating Catalyst

A two-step process described by the reaction equation $A \xrightarrow{k_1} B \xrightarrow{k_2} C$ has been proposed to explain the UV-sensitization process, where k_1 and k_2 are the rate coefficients describing the transformation of species A to species B and subsequently of species B to species C [5]. The nature of these species is uncertain, but species C, which is related to the final index change, may be related to GeE centers, and its rate of formation depends on the consumption of species B. It is assumed that hydrogen plays a catalytic role in this process by transforming species A to species B, which may be Ge-H radicals. The two-step model has qualitatively explained most phenomena encountered with the UV-sensitization processes.

In our investigation of the UV-sensitization process, we completely removed the hydrogen from the UV-exposed hydrogen-loaded fibers and then wrote FBGs into them. Assuming that the pathway from species A to species B is eliminated, or reduced significantly, without hydrogen and that the index change in the fiber during grating writing is due to species B transforming into species C, the growth of the index change should follow a simple saturating exponential function [17], but this was not the case. The index change growth continues past the point where the original concentration of species B should be exhausted, and the additional growth is much greater than that which can be explained by intrinsic fiber photosensitivity. Several reaction kinetic models, based primarily on either cooperative binding or freed catalysts, were explored to explain these findings. One of the more appealing models assumes that a catalyst, possibly hydrogen, is being freed as species B is transformed into species C and that more species B is made from species A by the freed catalyst.

The change of index data in Fig. 16.2 does not fit the saturated exponential function well [17]. So alternate reaction models were investigated, including those for catalyzed reactions. Assuming the two-step process is valid and that species B is a hydrogen radical formation, hydrogen may be freed during the transformation of species B to species C and enable the generation of more species B from species A. Such a reaction can be described by the Michaelis-Menten equation [18], and a time-integrated solution of this equation may be used to explain the growth dynamics of the index change in experimental data and may be expressed as

$$\Delta n \propto n_1 t + n_2 \left(1 - e^{-k_2 t}\right) \quad (5)$$

where Δn is directly proportional to the amount of species C created during the grating writing process. In the absence of a recirculating catalyst, Eq. (5) reduces to a saturated exponential function, where the parameter n_2 is proportional to the initial concentration of species B after sensitization. The speed of the reaction k_2 is proportional to the photon flux and the photoreactive cross-section of species B. In the steady state at long exposure times, Eq. (5) may be approximated by a linear function of the form $(n_1 t + n_2)$, where the slope of the line n_1 is proportional to the available hydrogen, which in turn is proportional to the original concentration of species B. The initial growth rate at the short exposure time can be approximated as $(n_1 + n_2 k_2)t$, where the product of k_2 and n_2 increase the growth rate. Indeed, the initial growth rates are much greater than the steady-state rates [10].

We fit Eq. (5) to data similar to that in Fig. 16.2 for the three test fibers to determine the parameters n_1, n_2, and k_2. The fits to the data for fiber B are shown by the curves in Fig. 16.2. By plotting these parameters as a function of sensitization fluence for the three test fibers, n_1 and n_2 increase with sensitization fluence until the maximum amount of species B formation is reached, as with fiber C [10]. The rate coefficient k_2 remains constant within the fit error, independent of the fiber type, for a sensitization fluence $\geq 1\,\text{kJ/cm}^2$ but deviates for lower fluence.

3. DILUTE H_2 SENSITIZATION OF FIBERS

In contrast to the standard H_2-loading method—where the optical fibers are exposed to pure H_2 at high pressures (<100 atm) and above room temperatures for several days or even weeks to increase their photosensitivity [11]—we used dilute hydrogen (0.1%) in very high-pressure gas mixtures with different diluent gases [12]. Ultra-low concentrations of H_2 mixed with a gas such as argon, carbon dioxide, methane, or ethane and pressurized to very high pressures of ~1000 atm were found to significantly enhance the diffusion of H_2 into the fiber core and hence its photosensitivity beyond that achievable by the standard loading methods at equivalent partial pressures [12]. The diluent gas dependence of the resulting enhancement in fiber photosensitivity explained using the excluded volume effect and the fugacity of gases also allows for useful sensitization while keeping partial pressure of H_2 near or below 1 atm.

3.1. Comparison with Standard H$_2$-Loaded Fibers

Two different types of SSM fibers with average core Ge concentrations of 3.5 and 5 mol%, referred to as fiber A and B, respectively, were used in this study. Spools of these fibers were placed in a very high-pressure vessel capable of withstanding gas pressures as large as 2041 atm [19]. A schematic of the loading apparatus used is shown in Fig. 16.7. The gas mixtures were formed either by pressurizing the chamber with premixed gas or by forming the mixture from the components in situ. The loading vessel was connected to a booster pump and can be heated to the required temperatures using heater coils. The fibers were typically loaded at 80°C for 1 day with different total pressures of the gas mixtures.

The concentration of molecular H$_2$ diffused into the fibers at different loading conditions were quantified by measuring (1) the peak absorbance (α) at 1.24 µm, due to molecular H$_2$ in the fiber core, using the cutback method and (2) the refractive index modulation (Δn) due to UV-written FBGs. The first overtone absorption peak of molecular H$_2$ at 1.24 µm, measured using the cutback method (see Section 2.1 for details), provides an accurate measure of its concentration in the fiber core [20, 21]. An increase in the H$_2$ concentration in the fiber core, related to an increase in photosensitivity in the fiber to UV radiation, was further verified by fabricating Bragg gratings using the method discussed in Section 2.1. The grating growth was monitored in transmission (T_{min}) as a function of time during the grating inscription from which the refractive index modulation is calculated using Eq. (1).

Samples for the control experiment were prepared with fibers A and B using the standard H$_2$-loading process [11]. The fibers were placed inside the loading vessel and heated up to 80°C. The heated vessel was then filled with pure H$_2$ up to the desired pressure, and the fiber was exposed to the hydrogen-containing atmosphere for 24 hr. After 24 hr, the vessel was vented and the fibers were removed quickly from the vessel and cooled rapidly by placing in a freezer at −40°C where the fiber was stored until further experimental measurements were performed. The linear dependence of both the peak absorbance (α) at 1.24 µm and the maximum index change (Δn) achieved in 10 minutes of grating writing as a function of total H$_2$ pressure for fiber A [12] confirms the diffusion process [21, 22]. Similar results were obtained for fiber B.

To demonstrate the ability of dilute H$_2$ to enhance the sensitivity of fibers, spools of fibers A and B were loaded with 1 ppt concentration of compressed hydrogen premixed with argon (Ar). The premixed gas was pressurized to various levels at 80°C for 24 hr. The concentration of H$_2$ that diffused into the fiber was measured from the peak absorbance (α) at 1.24 µm using the cutback method (see Section 2.1) and compared with fibers prepared in the control experiment [12, 23]. The partial pressure of H$_2$ was calculated as the molar fraction of H$_2$ in the mixture multiplied by the total pressure of the mixture. The maximum Δn values measured by writing FBGs in fiber A loaded with 1 ppt of H$_2$ premixed in Ar are compared with fibers loaded with pure H$_2$ using the standard method (Fig. 16.8). Enhancement in the H$_2$ concentration, measured by absorbance and photosensitivity, is calculated as the ratio of the pressure of pure H$_2$ that would result in the same level of H$_2$ content in the fiber divided by the initial partial pressure of H$_2$ in the mixed gas sample.

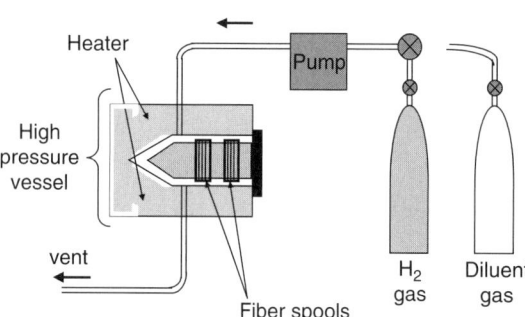

FIGURE 16.7 Schematic of the experimental setup used to hydrogen load fibers using the method described in this chapter.

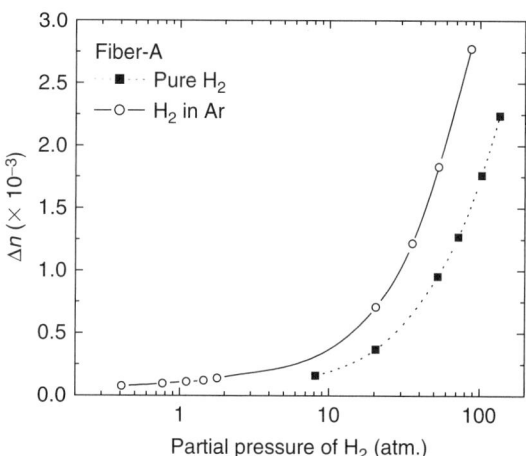

FIGURE 16.8 Partial pressure dependence of refractive index changes (Δn) in fiber B loaded with pure H$_2$ and with Ar as diluent gas.

An average enhancement in excess of two times in both the absorbance (α) and the resulting photosensitivity (Δn) were measured for fiber A, loaded with a premixed gas of 1 ppt H_2 in Ar as compared with the pure H_2 samples [12, 13].

3.2. Effect of Diluent Gas

We investigated the enhancement further by using different diluent gases either premixed or mixed in situ with low-concentration H_2. Diluent gases enhance the sensitivity of fibers through the increase in the fugacity of H_2 in mixtures relative to pure H_2 at an equivalent partial pressure. The fugacity depends on the temperature, pressure, and selection of diluent gas [23]. Loading fibers with a premixed gas of 1 ppt H_2 in carbon dioxide (CO_2) results in an enhancement of ~30 in absorbance and ~41 in the maximum refractive index change (Fig. 16.9). In other experiments, H_2 was mixed in situ with diluent gases such as nitrous oxide (N_2O), methane (CH_4), or ethane (C_2H_6) using the apparatus shown in Fig. 16.7. Here, the vessel was filled with H_2 to about 10 atm and vented to atmospheric pressure. The vessel was then pressurized to about 1000 atm total pressure with the diluent gas. The Peng-Robinson equations of state [24] and Aspen plus software (Aspen Technology, Cambridge, MA, USA) were used to calculate the partial pressure of the hydrogen in the vessel. An enhancement in absorbance (α) for fibers A and B in excess of 50 times was measured with ethane as the diluent gas in comparison with the pure H_2 at an equivalent partial pressure (Fig. 16.9). The fiber photosensitivity, measured using the grating writing process, also shows similar enhancements (Fig. 16.9).

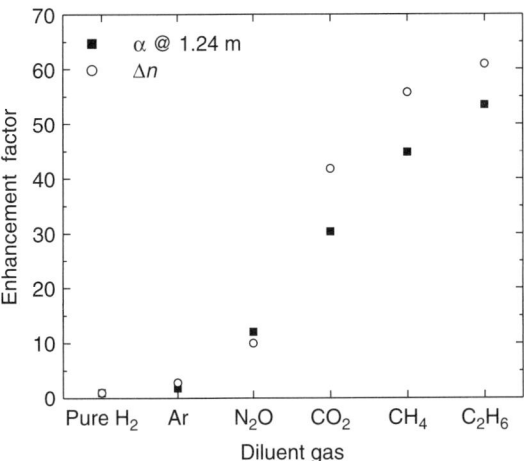

FIGURE 16.9 Enhancements in the absorbance (α) and the photosensitivity (Δn) achieved for different diluent gases.

In the context of this work, fugacity can be understood as the tendency for the H_2 to escape the supercritical fluid mixture of hydrogen and diluent gas by dissolving in the fiber, which is to be sensitized. Two effects dominate the increase, which we observed. The first effect is simply the failure of the ideal gas law due to excluded volume. The ideal gas law states that pressure times volume is equal to a constant times the number of molecules divided by absolute temperature. This is strictly true only under very low pressures, because it ignores the decrease in available volume due to the displacement of the gas molecules themselves. Under conditions of high pressure, where the density of the supercritical fluid approaches that of a conventional liquid, the volume occupied by molecules is a substantial fraction of the total volume of the chamber. This excluded volume effect is taken into account by the partial pressure calculations and is not the source of the enhancement discussed in this chapter.

The second effect is due to an interaction among the molecules in the supercritical fluid. As the density increases, these interactions become more important. At densities over 0.2 g/cc, these interactions can dominate the fugacity. At the temperatures and pressures used for this work, the interaction of hydrogen with argon allows for a more stable solution (and thus lower fugacity hydrogen) than that of hydrogen with methane or ethane. The interaction of hydrogen with carbon dioxide is intermediate. These trends are consistent with simulations based on thermodynamic properties.

At conditions of very low hydrogen concentration and high fugacity, the loss of hydrogen to the metal in the vessel becomes an appreciable fraction of the total, and the experiment must be designed to take these losses into account. Preloading the chamber walls with hydrogen was typically done, and hydrogen concentration was measured in the exhaust gas, which was vented at the end of the experiments. The hydrogen measured in the exhaust gas was consistently less than that placed in the chamber, even with the chamber pretreatment. The values reported in this chapter, which were used to calculate the enhancements, were those of the initial hydrogen loading and ignore this decrease.

4. CONCLUSION

The UV-sensitization technique allows the inscription of ≥ 25 dB ($\Delta n \geq 1 \times 10^{-3}$) strong Bragg gratings in SSM fibers with low Ge content with no hydrogen present during grating writing. The fiber type and the germanium content in the fiber core have an impact on the characteristics of gratings written in UV-sensitized fibers.

FBGs written in UV-sensitized SSM fibers are found to be more stable (expected to decay only ~0.2% when kept at 80°C for 25 years) and withstand higher temperatures (~600°C) as compared with the gratings written in H_2-loaded fibers. Thermal stability analyses of these gratings suggest the possible elimination of the postfabrication annealing step.

The dynamics of Bragg grating growth in UV-sensitized, hydrogen saturated, germanosilicate fibers are explained using a transient kinetic model. Recirculation of hydrogen, functioning as a catalyst in the photosensitive reactions, is believed to be responsible for an increase in the steady-state value of the refractive index changes.

The choice of the diluent gas has a significant influence in increasing the concentration of hydrogen diffused into the fibers and hence the photosensitivity of the fibers used to fabricate FBG-based devices. Such amplification can be explained using the excluded volume effect and fugacity of gases in a mixture. This method allows us to use either hydrogen or deuterium in low concentrations and hence low partial pressures, near or below 1 atm, where the sensitization with pure H_2 is negligible.

5. ACKNOWLEDGMENTS

I thank James F. Brennan, III, William V. Dower, and Dora M. Paolucci for their contributions, Janet A. Kling for critical reading of the manuscript, and the management of the 3M Company for permission to publish this work.

6. REFERENCES

1. A. Othonos, *Fiber Bragg Gratings*, Artech House, Inc., MA (1999).
2. K.O. Hill, IEEE J. Selected Topics in Quant. Electron., 6, 1186 (2000); Technical Digest of *Bragg Gratings Photosensitivity and Poling in Glass Waveguides* (2003).
3. J. Hubner, D. Wiesmann, R. Germann, B. J. Offrein and M. Kristensen, in *Conference on Bragg Gratings, Photosensitivity, and Poling in Glass Waveguides*, Vol. 33 of OSA Trends in Optics and Photonics Series, paper PD 6-1 (1997).
4. G. E. Kohnke, D. W. Nightingale and P. G. Wigley, in *Optical Fiber Communication Conference* (OFC '99), paper PD20.
5. J. Canning, Opt. Fiber Technol. **6**, 275 (2000).
6. M. Aslund and J. Canning, Opt. Lett. **25**, 692 (2000).
7. J. Canning, M. Aslund and P.-F. Hu, Opt. Lett. **25**, 1621 (2000).
8. K. P. Chen and P. R. Herman, IEEE J. Lightwave Technol. **21**, 1958 (2003).
9. J. Canning and P.-F. Hu, Opt. Exp. **9**, 476 (2001).
10. N. K. Viswanathan and J. F. Brennan, in *Optical Fiber Communication Conference Technical Digest* (OFC '02), paper TuQ1, p. 107.
11. P. J. Lemaire, R. M. Atkins, V. Mizrahi and W. A. Reed, Electron. Lett. **29**, 1191 (1993).
12. N. K. Viswanathan, W. V. Dower, D. M. Paolucci and M. D. Barrera, in *Technical Digest Bragg Gratings, Photosensitivity, and Poling in Glass Waveguides* (BGPP 2003) paper MC5, pp. 49–52.
13. N. K. Viswanathan and J. F. Brennan, 3M Technical Reports (2001).
14. S. R. Baker, H. N. Rourke, V. Baker and D. Goodchild, IEEE J. Lightwave Technol. **15**, 1470 (1997).
15. B. O. Guan, H. Y. Tam, X. M. Tao and X. Y. Dong, IEEE Photon. Technol. Lett., **12**, 1349 (2000).
16. R. J. Egan, H. G. Inglis, P. Hill, P. A. Krug and F. Ouellette, OFC, Paper TuO3F (1996).
17. T. A. Strasser, T. Erdogan, A. E. White, V. Mizrahi and P. J. Lemaire, Appl. Phys. Lett. **26**, 3308 (1994).
18. J. H. Espenson, *Chemical Kinetics and Reaction Mechanisms*, McGraw-Hill, New York (1981).
19. J. F. Brennan, D. Sloan, J. Dent and D. L. LaBrake, in *Optical Fiber Communication Conference* (OFC '99), ThD4-1/59.
20. J. Stone, J. Lightwave Technol. **5**, 712 (1987).
21. J. F. Shackleford, P. L. Studt, and R. M. Fulrath, J. Appl. Phys. **43**, 1619 (1972).
22. P. L. Swart, A. A. Chtcherbakov, W. L. Joubert and M. G. Shlyagin, Opt. Commun. **217**, 189 (2003).
23. W. V. Dower and N. K. Viswanathan, 3M Technical Report (2002).
24. D. Y. Peng, and D. B. Robinson, Ind. Eng. Chem. Fundam. **15**, 59 (1976).

17
CHAPTER

Solitons in a Fiber Bragg Grating

K. Porsezian*
Department of Physics
Pondicherry University
Pondicherry, India

Krishnan Senthilnathan
Department of Physics
Anna University
Chennai, India

1. INTRODUCTION

The search for a new communication technology is a quest in human society for the betterment of our living conditions. Communication technology is also at the forefront of all other sciences and technologies. Ever since the invention of telephone at the beginning of the 18th century, the dissemination of information has become faster and the volume of information being transferred has grown exponentially [1–7]. Hence, communication network services ultimately demand high bandwidth information transmission networks. A better way to cope with these demands is by means of optical fiber communication (OFC) systems. A worldwide communication network is now in place but is continuously extended, and the OFC systems act as "information highways."

1.1. Optical Fiber Communications

Optical fibers have distinct advantages over their metallic counterparts, copper cables (both twisted pair and coaxial). Optical fibers are compact and lightweight and have the ability to transfer large amounts of information. Apart from the economic advantages, optical fibers also exhibit technological superiority. Optical fibers are made up of silica-based glass or plastic that are insulators and therefore have no currents flowing in them. As a result, fibers are immune to electromagnetic interference. In addition, no tapping of information is possible with fiber communications. Hence, optical fibers are extremely suitable for deployment in defense services where secrecy is given the highest priority.

Another advantage of optical fibers is that they do not corrode and can be placed near a chemical environment. Apart from all these merits, bandwidth and speed of this communication system are large, compared with any other communication systems that exist today. Consequently, there is an increased interest in fiber-based communication systems. Despite the fact that fiber communication is a wired communication, it offers several other advantages compared with other communications, as discussed.

1.2. Soliton-Based OFC

During the last quarter of the previous century, there have been numerous dramatic and exciting discoveries in many areas of pure and applied physics, chemistry, medical sciences, and astronomy [1–7]. Remarkably, at the same time there have been considerable developments in mathematics, particularly in nonlinear mathematics. In addition, the advent of sophisticated computer technology has made many previously impossible tasks a reality. These developments are responsible for the present status of nonlinear science, which includes both soliton and chaotic phenomena. Our intention here is to concentrate on the soliton theory, mainly *optical solitons*. In 1834, John Scott-Russell, a British naval architect, introduced the concept of a solitary wave after observing a single isolated wave on the surface of water in a canal by following it while riding on horseback. He described this remarkable phenomenon in "Report on Waves" as a large solitary elevation, a rounded heap of water, that traversed along the canal without any change in its shape or speed (8 or 9 miles per hour) and gradually diminished until it got lost in the windings of the canal [8, 9].

---

*E-mail: ponzsol@yahoo.com

A theoretical description for this type of solitary wave propagation in shallow water was provided by Korteveg and de Vries, in terms of a nonlinear partial differential equation, now popularly called the K-dV equation. One solution to this equation describes Russell's solitary water wave. In 1965, Zabusky and Kruskal [9] performed a computer simulation of the collision of two solitary waves described by the K-dV equation. The surprising result was that these waves reemerged after collision with unvaried shape and speed, as though they were particles. To emphasize the fact that they do behave like particles, they coined the word "soliton," which sounds like particles such as proton, electron, and photon. To be precise we can say that a solitary wave is a wave that travels without change in shape after an interaction with another wave and a soliton is a solitary wave whose shape and speed are not altered by the collision. An immense amount of research soon followed, and solitons were speculated to have vast applications in many branches of science, an intuition that was subsequently confirmed.

Solitons can be propagated without any distortion, if the nonlinear characterstics (like amplitude and hence intensity, depending on velocity) and dispersion characterstics (frequency depending on velocity) of the media are balanced. In the absence of nonlinearity, dispersion can destroy a solitary wave as various spectral components of the wave would propagate with different velocities, called group-velocity dispersion (GVD). In the absence of dispersion, the nonlinearity simply sharpens the solitary wave, finally leading to the breaking of pulses.

The silica (SiO_2) glass with which the optical fibers are made meets the above two requirements. For very intense light, the refractive index of such a medium increases with intensity of light. Keiser [7] explained the development of OFC. In glass fibers the wave guiding is based on the principle of total internal reflection. Depending on the magnitude of intensity of the pulse that is propagating through the fiber, one could anticipate nonlinear effects or only linear effects. The propagating pulse width and the amplitude are affected by the linear or nonlinear behavior of the fiber media. If the intensity of the pulse is very low, the linear effects such as dispersion and optical losses come into play.

Though the OFC systems have several advantages, to exploit them undesirable effects such as optical loss, dispersion, amplifier-induced noise, and nonlinearity need to be resolved (see Chapters 1, 6, 8, and 9). In short distance communication, the impact of these effects is hardly worrisome. In the following, we briefly discuss the various effects that influence the soliton and its propagation in the fibers.

2. LINEAR EFFECTS

2.1. Optical Loss

Optical loss, during transmission of optical signals inside a fiber, is an important parameter to reckon with. There is an exponential loss of launched power in the fiber. If P_o is the power launched at the input point of the fiber of length L, the transmitted power P_T is given by [1, 2, 5, 7],

$$P_T = P_o \exp(-\alpha L)$$

where α is the attenuation constant, commonly referred to as fiber loss.

Among the several factors that contribute to the losses, material absorption and Rayleigh scattering are the major ones. Material absorption arises due to the presence of several impurities inside the core medium. The major impurity constituent is the presence of hydroxyl ions. These hydroxyl ions are absorbed by the silica material during the manufacturing process. Depending on the vibrational absorption characteristics of hydroxyl ion impurities, the loss profile due to material dispersion is formed. Care is to be taken to minimize the density of hydroxyl ions during manufacture [1, 2, 5, 7].

Rayleigh scattering arises due to random density fluctuations of frozen silica materials formed at the time of manufacturing. It is a well-known fact that loss due to Rayleigh scattering varies inversely as the fourth power of the propagating optical wavelength. Other losses are those that occur due to bending and coupling that, in turn, can be minimized by taking suitable care during installation [1, 2, 5, 7]. For SiO_2 glass fiber, the minimum loss is found to be around 1.55 μm.

2.2. Chromatic Dispersion

Depending on the type of fiber used, dispersion can be broadly classified into two types: intramodal dispersion and intermodal dispersion. In the case of a single mode fiber (SMF), which can handle only one mode, there is only intramodal dispersion. SMFs are mainly used only for long distance communications [1, 2, 5, 7]. Multimode fibers carry many modes of different phases and group velocities. The differences in group velocities of different modes cause spreading of the temporal envelopes of the individual pulse-excited modes and thus cannot maintain high bit-rate pulse streams over long distances.

Soliton type propagation is possible only with SMFs. Here, intramodal dispersion comprises material dispersion and waveguide dispersions. Material dispersion is due to the frequency dependence of the index of refraction and waveguide dispersion is due to the frequency dependence of the propagation constant of a mode. Material dispersion can produce devastating effects in a SMF.

When an optical pulse is transmitted in a waveguide, it suffers from broadening due to dispersion. The optical pulse has a spectrum of Fourier frequency components. Because the index of refraction of any optical system is a function of frequency, various Fourier components of the pulse experience different indices of refraction in a dielectric medium. Because refractive index is a measure of the velocity of the wave propagating in the dielectric, different Fourier components travel with different velocities, called the group velocity. Because of this, the optical pulse spreads in the time domain during the course of propagation. This is called GVD or chromatic dispersion [1, 2, 5, 7].

If dispersion exists in a fiber, the index of refraction n becomes a function of the wave frequency of the lightwave. As a result, derivatives of the propagation constant k (also called wave number, $k = n\omega/c$) with respect to the frequency ω; $\partial k/\partial \omega$, $\partial^2 k/\partial \omega^2 \ldots$, also have finite values. By making use of this fact, one can find the propagation property of the side band (which is the Fourier component around the carrier frequency ω_o produced by the modulation).

The wave number k around the carrier frequency ω_o can be written using the index of refraction n as a function of ω,

$$k - k_0 = \left.\frac{\partial k}{\partial \omega}\right|_{\omega_0}(\omega - \omega_0) + \frac{1}{2}\left.\frac{\partial^2 k}{\partial \omega^2}\right|_{\omega_0}(\omega - \omega_0)^2 + \frac{1}{6}\left.\frac{\partial^3 k}{\partial \omega^3}\right|_{\omega_0}(\omega - \omega_0)^3 + \cdots$$

This equation gives the wave number of the frequency component of the modulated wave whose frequency deviates slightly from the carrier frequency ω_o. The group velocity v_g of the modulated wave is given by

$$v_g = \frac{\partial \omega}{\partial k} = \frac{1}{k'}$$

and the group velocity parameter k'' is given by

$$k'' = \frac{\partial}{\partial \omega}\left(\frac{1}{v_g}\right) = -\frac{1}{v_g^2}\frac{\partial v_g}{\partial \omega}$$

The third derivative $k''' \equiv (\partial^3 k/\partial \omega^3)$ is called higher order dispersion (HOD) or third-order dispersion parameter. Here, pulse broadening is mainly due to the GVD parameter (k''). When compared with k'', the contribution of k''' is negligible. But for the case of ultrashort pulses, the effect of k''' cannot be ignored. GVD broadens the pulse symmetrically in the cases of Gaussian and hyperbolic secant pulses, whereas HOD disperses asymmetrically.

It can be seen that at one point k'' vanishes ($\lambda = 1.33\,\mu$m) for SiO$_2$. This particular wavelength is called the zero dispersion wavelength (λ_D) that can be varied by doping with dopants like GeO$_2$ or P$_2$O$_5$. Today, optical communication is usually based on this zero dispersion wavelength because the pulse broadening is very low, called nonreturn to zero communication. Near λ_D, the dispersion is due to the HOD parameter (k'''). From the soliton point of view, λ_D plays a very important role.

For wavelengths below λ_D, that is, $\lambda < \lambda_D$, redshifted components of a pulse travel faster than blue-shifted components. This is called the normal or the positive dispersion regime. In contrast, if it is such that the wavelengths are greater than λ_D, that is, $\lambda > \lambda_D$, then it is called anomalous or negative dispersion regime. Soliton type pulse propagation in the anomalous and the normal dispersion regimes are called bright and dark solitons, respectively.

3. NONLINEAR EFFECTS

It is clear from the discussion of linear effects that all dielectric properties depend only on the frequency of the optical pulse. But for very intense electromagnetic fields, all dielectrics behave nonlinearly, that is, the properties also depend on the pulse intensity. This can be very easily seen from the relation connecting the induced polarization P and the electric field E as

$$P = \varepsilon_0[\chi^{(1)}E + \chi^{(2)}E^2 + \chi^{(3)}E^3 + \cdots]$$

where ε_o is the vacuum permittivity and $\chi^{(j)}$ ($j = 1, 2, \ldots$) is the jth order susceptibility. As discussed in the previous section, the linear effects are due to $\chi^{(1)}$. In the above equation $\chi^{(2)}$ represents the second-order susceptibility. In centrosymmetric molecules, all the even order susceptibilities vanish. In silica fibers, the lowest dominant nonlinearity is due to $\chi^{(3)}$, because they are made up of SiO$_2$ molecules. For further details about nonlinear effects in fiber, see Chapter 6.

3.1 Kerr Nonlinearity

An intense pulse propagating through an optical fiber gives rise to many nonlinear effects. The intensity dependent refractive index can be represented as

$$n(\omega; |E|^2) = n_o(\omega) + n_2|E|^2$$

where $n_o(\omega)$ is the linear index of refraction and n_2 is the Kerr nonlinearity parameter. Usually, the index of refraction gives the pulse, a phase delay given by

$$\phi = nk_0L = (n_0 + n_2|E|^2)k_oL$$

where $k_0 = 2\pi/\lambda$ and L is the length of the fiber. The intensity dependent nonlinear phase shift, $\phi_{NL}(= n_2k_oL|E|^2)$, is due to the Kerr effect. This nonlinear effect makes the index of refraction a function of intensity. This is called Kerr nonlinearity, which induces self-phase modulation (SPM). That is, in any signal propagation there is a generation of phase shift between different frequency components. This phase shift depends on the refractive index of the medium. As the refractive index depends on the intensity of the pulse, which is a time varying quantity, the induced phase shift also varies with time. This can be considered as a generation of new frequency components on both sides of the bandwidth. As this phase modulation of the pulse is due to its own intensity, it is called SPM. This can be considered as a spread in the frequency domain. This is an important nonlinear effect as it balances the effect of GVD in the anomalous dispersion regime and produces bright solitons. In this case the pulse dynamics is governed by the nonlinear Schrodinger [NLS] equation [1–5]. The nonlinear phase shift is a function of time as it depends on the time varying intensity of the pulse.

The situation is different if one pairs the negative GVD with Kerr nonlinearity. Here the solitary waves exist but do not change their shape and spectrum as they propagate along the fiber. The chirp produced by the SPM at high frequencies in back and at low frequencies in front is combated by the dispersion that retards the low frequencies in front of the pulse and advances the high frequencies in the back of the pulse. The result is a pulse that does not change its shape as it travels along the fiber. The dispersion is kept in balance by the nonlinearity and vice versa. Another way to describe the counteraction of these two effects is to note that the nonlinearity increases the index locally (where the pulse intensity is high). This increase in index counteracts the dispersion (in time) of the pulse of finite duration. When a continuous wave propagates through a silica fiber, with intensity capable of inducing Kerr effect, the optical field experiences an amplitude modulation called modulation instability [1–4].

3.2. Self-Steepening

The channel handling capacity of a fiber system can be increased by propagating ultrashort pulses. When the width of the pulse is too narrow, the higher order nonlinear effects are to be considered. Two different phenomena come under these higher order nonlinear effects, namely Kerr dispersion, or self-steepening, and stimulated inelastic scattering. They are all considered as dominant perturbations in soliton propagation.

Self-steepening is due to the intensity dependent group velocity. This causes the peak of the pulse to travel slower than the wings. So during propagation, the pulse becomes steeper and steeper because of its intensity, hence the name self-steepening. There is also a possibility of breaking of the pulse due to instability. Because this phenomenon is basically due to the variation in the group velocity (with respect to the intensity), this is also called the Kerr dispersion. In the frequency domain, self-steepering gives asymmetrical broadening to the pulse [1–4].

3.3 Stimulated Inelastic Scattering

Stimulated inelastic scattering is inelastic in the sense that there is an energy transfer between the interacting field and the dielectric medium. There are two types of inelastic scattering observed: stimulated Raman scattering (SRS) and stimulated Brillouin scattering (SBS). The main difference between the two is that optical phonons participate in SRS, whereas acoustic phonons participate in SBS.

In a simple quantum-mechanical picture, SRS can be described as follows: A photon of the incident field (often called the pump) is annihilated to create a photon at the down shifted Stokes frequency. In the case of SBS, the interaction brings out a phonon with appropriate energy and momentum to conserve the energy and the momentum of the system.

Of course, a higher energy photon at the so called anti-Stokes frequency can also be created, provided a phonon of appropriate energy and momentum is available for absorption. Even though SRS and SBS are very similar in their origin, different dispersion relations for acoustic and optical phonons lead to some basic differences between the two. A fundamental difference is that SBS in optical fibers can occur only in the backward direction, whereas SRS dominates in the forward direction.

The effect of SRS produces a self-frequency shift, which can cause decay in the soliton pulse. This causes splitting of the pulse. The phenomenon of SRS is used for the amplification of soliton pulses.

Light pulses of higher frequency than that of soliton pulses are periodically pumped into the fiber at constant intervals. The SRS of the pump wave gives the necessary amplification for the propagating pulses [1, 2, 5].

SRS also gives asymmetrical broadening to the pulse in the frequency domain. Self-frequency shift is a self-induced red shift in the pulse spectrum arising from SRS, that is, the long wavelength components of the pulse experience Raman gain at the expense of the short wavelength components, resulting in an increasing red shift as the pulse progresses. It has been recognized that self-frequency shift is potentially a detrimental effect in soliton communication systems because the power fluctuations at the source translate into frequency fluctuations in the fiber through the power dependence of the soliton self-frequency shift and hence into timing jitter at the receiver [1, 2].

3.4. Effect of Birefringence

As already discussed in Section 2.2, though a SMF can support only one mode of propagation, it can support two degenerate modes that are dominantly polarized in two mutually orthogonal directions. In the case of an ideal fiber made of isotropic material and of perfect geometry, a mode excited with its polarization along the x-direction would not couple to the mode with orthogonal y-polarization state. But in practice, any deviation in the fiber geometry or even small fluctuations in material anisotropy may result in mixing of the two polarization axes, which would lead to breaking of degeneracy mode. The mode propagation constants for the modes polarized along x- and y-directions are slightly different. This property is called modal birefringence measured as the degree of modal birefringence B. It is defined as

$$B = |(k_x - k_y)|/k_0 = |n_x - n_y|$$

Here, n_x and n_y are the effective mode indices in the two orthogonal polarization states. It can be shown that for a given value of B, the power between the two modes is exchanged periodically as they propagate inside the fiber with the period L_B defined by

$$L_B = 2\pi/|(k_x - k_y)| = \lambda/B$$

Here, L_B is called the beat length. The axis along which the effective mode index is smaller is called the fast axis because the group velocity is higher for light propagating through this axis. Similarly, for the same reason the axis with the larger mode index is called the slow axis. In conventional SMFs, B changes randomly because of fluctuations in the core shape and stress-induced anisotropy. Because of this, when light is launched into the fiber with linear polarization, it quickly reaches a state of arbitrary polarization. But for some applications, it is desirable to have fibers transmitting light without changing its state of polarization. Such polarization is called polarization-maintaining or polarization-preserving fibers. With the inclusion of this effect, for high intensity pulses, the dynamics is governed by the coupled nonlinear Schrödinger (NLS) equation [1, 2].

4. OPTICAL SOLITONS IN PURE SILICA FIBER

The electromagnetic field vectors that characterize the lightwave are related by Maxwell's equations as [1, 2]

$$\nabla \cdot \vec{D} = \rho, \quad \nabla \cdot \vec{B} = 0, \quad \nabla \times \vec{E} = -\frac{\partial \vec{B}}{\partial t} \quad (1)$$

$$\text{and} \quad \nabla \times \vec{H} = \vec{J} + \frac{\partial \vec{D}}{\partial t}$$

where $\vec{E}$ and $\vec{H}$ are electric and magnetic field vectors, $\vec{D}\,(=\varepsilon_0\vec{E}+\vec{P})$ is the displacement vector, and $\vec{B}\,(=\mu_0\vec{H})$ is the magnetic flux density. As we consider the medium of propagation as homogeneous, which is charge neutral, we set $\vec{J}=0$, $\rho=0$. The only source for the lightwave in the medium is then the polarization term $\vec{P}$. The lightwave that satisfies a wave equation may be obtained through the usual mathematical methods as

$$\nabla \times \nabla \times \vec{E} = -\frac{1}{c^2}\frac{\partial^2 \vec{E}}{\partial t^2} - \mu_0 \frac{\partial^2 \vec{P}}{\partial t^2}$$

where $\mu_0\varepsilon_0 = 1/c^2$. Picosecond pulse propagation is governed by the NLS equation that would require the following valid physical assumptions:

- Slowly varying envelope approximation

$$\omega^{-2}\frac{\partial^2 q}{\partial t^2} \ll \omega^{-1}\frac{\partial q}{\partial t} \ll q$$

$$k^{-2}\frac{\partial^2 q}{\partial t^2} \ll k^{-1}\frac{\partial q}{\partial t} \ll q$$

- No fiber loss (dielectric constant is real)
- Instantaneous response from the dielectric medium
- Supports only single mode and perfect cylindrical symmetry (polarization preserving)

With the above assumptions in Maxwell's equation, the NLS equation is derived in the form

$$iq_z - \frac{k''}{2}q_{tt} + (n_2\omega_0/cA_{\text{eff}})|q|^2 q + ik'q_t = 0 \quad (2)$$

where q is the slowly varying envelope of the electric field and subscripts z and t denote the spatial and time derivatives. Certain solutions of nonlinear partial differential equations like the NLS equation provide details of the pulse shape, width, and intensity of the pulse to be transmitted for soliton-type pulse propagation. Identification of the linear eigenvalue problem (Lax pair) confirms its complete integrability and soliton-type pulse propagation in the system. The NLS equation has a Lax pair and N-soliton solution. Soliton wave propagation in the NLS equation was first theoretically proposed by Hasegawa and Tappert [10] in 1973 and experimentally verified by Mollenauer et al. in 1980 [11].

Though the NLS equation has been successful in explaining a large number of nonlinear phenomena, it needs to be modified depending on the intensity of the applied pulse. For example, if the peak of the incident pulse is above a threshold level, both SRS and SBS can transfer energy from the pulse to a Stokes pulse. Also, if the spectral width Δw of the incident pulse becomes comparable with the carrier frequency ω_0, then the above-mentioned approximations made while deriving the above NLS equation become questionable. With the inclusion of the above important results, the resulting equation is obtained as [1, 2]

$$q_z = i\alpha_1 q_{tt} + i\alpha_2 |q|^2 q - \alpha_3 q_{ttt} - \alpha_4 \left(|q|^2 q\right)_t \\ - \alpha_5 q \left(|q|^2\right)_t + \Gamma q + \alpha_6 q_t + \cdots \quad (3)$$

where α_1 is related to GVD, α_2 to SPM, α_3 to the HOD (for wide bandwidth even when the wavelength λ is relatively far away from λ_D), α_4 to nonlinear dispersion (self-steepening and shock wave formation at a pulse edge), and α_5 to the stimulated inelastic scattering (self-frequency shift).

As discussed earlier, in a fiber a clever configuration of both linear (dispersion) and the nonlinear (SPM) effect does end up with the generation of a pulse that can keep its shape over a long propagation distance, withstanding minor perturbations. These steady pulses are called optical fiber solitons. The solitons discussed in the anomalous regime are called bright solitons having hyperbolic secant shape for their intensity profile, as shown in Fig. 17.1a. In the normal dispersion regime, the resulting solitons are called dark solitons whose intensity profile contains a dip in its shape with uniform background having hyperbolic tangent shape, as shown in Fig. 17.1b). Because of their short pulse duration and high stability, solitons could form the high-speed communications backbone of tomorrow's information superhighway. From the detailed experimental investigations, it has been shown that soliton-based communication systems have the following remarkable advantages: highly stable and ultrafast, 100 times better than the best-existing communication system, and free from noise and less expensive [1–6].

Having discussed the formation of solitons in fibers, we now turn to discuss the fundamentals of fiber Bragg grating (FBG) and details regarding realization of solitons in FBGs. Thus far, we focused our discussion on the OFC system and then discussed the need for deployment of solitons in OFC systems. Now we present the study of formation of solitons in the FBG structure and their subsequent use in optical communications, which will revolutionize existing communication links. Section 5 deals with the fundamentals of FBG covering both linear and nonlinear effects. Section 6 describes the formation and classification of solitons in FBG. In Section 7, gap solitons, both in quadratic and Kerr media, are discussed using Stokes parameters. In Section 8, using multiple scale analysis, we discuss the formation of Bragg grating solitons at the edges of the photonic bandgap. The solution to the

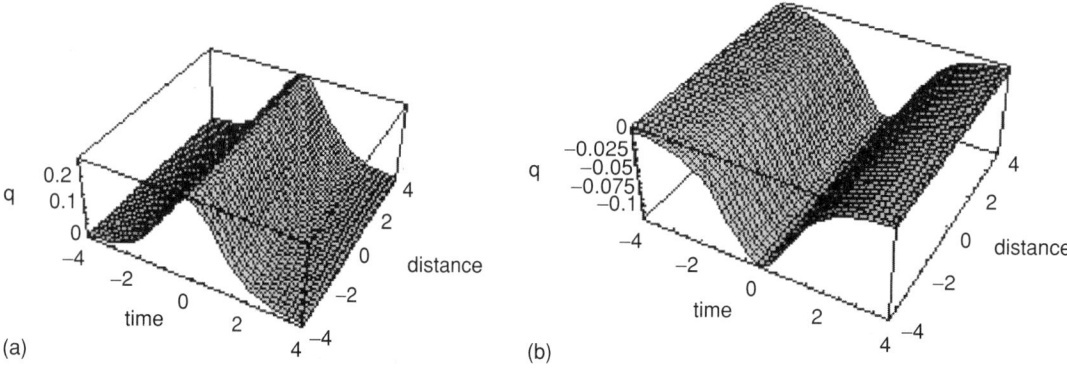

FIGURE 17.1 (a) Bright soliton intensity profile. (b) Dark soliton intensity profile.

nonlinearity of high intensity laser sources is discussed and the results and discussions presented in Section 9.

4.1. Why Solitons in FBG?

Though the use of solitons for communication seems advantageous, they do pose some technical difficulties. The current electronic amplifiers and switches just cannot respond quickly enough for these short soliton pulses of high speed. In addition, at present, there are also some inherent difficulties such as timing jitter that prevent solitons from being used in commercial links. Many experts predict that these difficulties could be overcome in the near future, and hence soliton-based high-speed fiber communication will most probably replace the current major communication links.

To realize solitons in fiber, we need fibers of hundreds of kilometers in length, so that dispersion counteracts with SPM. Thus, the generation of solitons in fibers is economically disadvantageous, and this process ultimately makes the technology more complicated.

In recent times, scientists have been able to generate solitons in so-called FBGs whose length is on the order of a few centimeters (see Chapter 15). This is mainly because the grating-induced dispersion is about six orders of magnitude greater than that due to the conventional standard telecommunication fiber. In FBG, the solitons arise as a result of grating-induced dispersion and Kerr nonlinearity. These grating solitons are more intense than the regular fiber solitons and hence are more practical in realizing a soliton-based optical fiber communication system. In this chapter we discuss the formation of solitons in an FBG.

The *features of FBG* are precision wavelength centering, low adjacent channel cross-talk, low cost and low insertion loss, simplicity and flexibility, high tensile strength, superior spectral characteristics, and compatibility with existing OFC networks. The *applications of FBG* are as follows. FBGs can act as narrowband retro reflectors for providing feedback at a specific wavelength in fiber lasers; filters for multichannel wavelength division multiplying systems; wavelength selective filters and add/drop filters; dispersion compensators in long-distance telecommunication networks; sensing devices, ranging from structural monitoring to chemical sensing; and optical strain gauges in bridges, building structures, elevators, reactors, composites, mines, and smart structures.

These factors ultimately create a great curiosity among researchers to turn to the study of nonlinear laser pulse propagation through the FBG. Though fiber gratings were discussed in Chapter 15, for self-consistency, in the following we discuss the fundamentals of FBG.

5. FUNDAMENTALS OF FBG

5.1. Introduction

An FBG is a periodic variation of the refractive index of the fiber core along the length of the fiber [3, 5, 12–15]. The refractive index variations are formed by exposure of the fiber core to an intense optical interference pattern of ultraviolet (UV) light, as shown in Fig. 17.2. The bright band in the interference pattern corresponds to an increased refractive index region and the dark band corresponds to a constant index region. Thus, when an optical fiber is irradiated by UV light, the refractive index of the fiber is changed permanently. Therefore, the capability of light to induce permanent refractive index changes in the core of an optical fiber due to exposure to a UV light is referred to as *photosensitivity*. The change in refractive index is permanent in the sense that it lasts for several years, that is, a lifetime on the order of 25 years is predicted. Initially, photosensitivity was thought to be a phenomenon that was associated only with germanium doped-core optical fibers. Then, subsequent experiments showed that photosensitivity could be observed in a wide variety of different fibers, many of which do not contain germanium as a dopant. Nevertheless, optical fiber with a germanium-doped core remains the most popular material for the fabrication of Bragg grating-based devices.

The magnitude of the photoinduced refractive index change (Δn) depends on several different factors, namely the irradiation conditions (wavelength, intensity, and total dosage of irradiating light), the composition of glassy material forming the fiber core, and processing of the fiber before and after the irradiation. A wide variety of different continuous

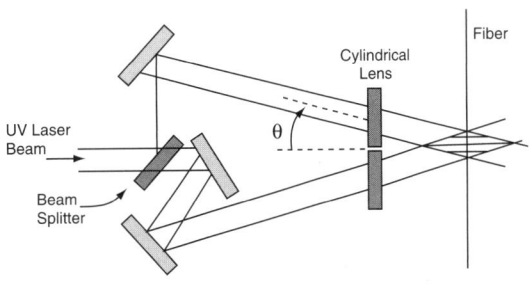

FIGURE 17.2 Fabrication of fiber Bragg grating.

wave and pulsed laser light sources with wavelengths ranging from the visible to the vacuum UV has been used to produce refractive index changes in optical fibers. In practice, the most commonly used light sources are KrF and ArF excimer lasers that generate 248- and 193-nm light pulses, respectively, at pulse repetition rates of 50 to 100 pps. Typically, the fiber core is exposed to laser light for a few minutes at pulse energy levels ranging from 100 to 1000 mJ/cm/pulse. Under these conditions, Δn is positive in germanium-doped SMF with a magnitude ranging between 10^{-3} and 10^{-5}.

The important property of the photoinduced refractive index change is anisotropy. The anisotropy with the photoinduced refractive index change results in fiber becoming birefringent to the light propagating through it. The effect is useful for fabricating polarization mode–converting devices or rocking filters.

This photosensitivity was first discovered by Hill and colleagues [12, 13] in 1978 at the Communications Research Centre in Canada. Thus, the discovery of photosensitivity led to the techniques of fabricating Bragg gratings in the core of an optical fiber and a means for manufacturing a wide range of FBG-based wavelength selective devices that have applications in OFCs and optical sensor systems.

5.2. Types of Grating

Based on the period of the grating, we can classify the FBGs into two types: short period grating (SPG) and long period grating (LPG).

A grating is called an SPG when the period of the grating is less than the optical wavelength, which is on the order of 1.3–1.6 μm. For typical SPG, the period of the grating is $\Lambda = 0.5$ μm. In this grating structure, the interaction of light takes place with a periodic structure, as shown in Fig. 17.3a. As a result, reflection of light occurs due to the Bragg condition, and hence this grating can also be called reflection grating. Because of the occurrence of Bragg reflection, this type of grating can be used as an optical filter. In SPG, coupling occurs between modes traveling in the opposite directions.

When the period of the grating is greater than the optical wavelength, then the grating is referred to as the LPG. In this case, the grating period is around 10–500 μm, as shown in Fig. 17.3b. As a result, the light does not interact with the period of the grating, and hence there will not be any back reflection; instead light travels in the forward direction unlike in an SPG. Thus, the coupling occurs between modes traveling in the same direction in LPG. Thus, such

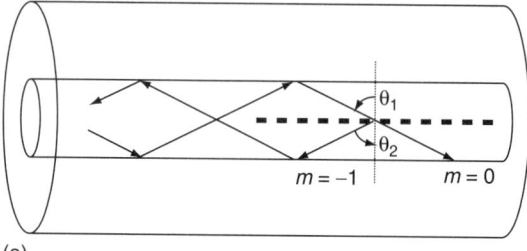

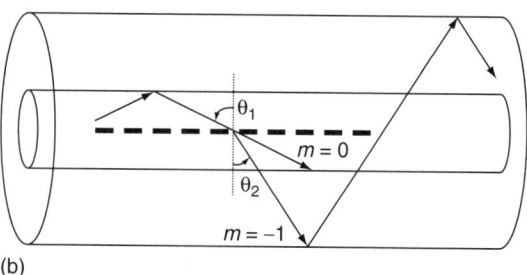

FIGURE 17.3 (a) Pulse propagation in an SPG. (b) Pulse propagation in an LPG.

gratings can be used for mode conversion, and power transfer from one mode to another, polarization conversion, and power transfer between two orthogonally polarization modes. In this chapter, we consider the pulse propagation only through SPG. An extensive basic introduction to FBG can be found in Chapter 15 in this volume.

5.3. Properties of FBG

Before embarking on the investigation on pulse propagation in a fiber Bragg grating structure (FBG), we discuss some of the interesting as well as important linear and nonlinear properties of the FBG structure in the following section. Then in the subsequent section, we focus our attention on pulse propagation in FBG. First, let us discuss the linear properties of FBG and then discuss the nonlinear properties by allowing Kerr-nonlinearity into the system.

5.3.1. Linear Properties

5.3.1.1. Bragg Reflection in FBG. We consider a light signal propagating in a one-dimensional Bragg grating located in an SMF. The linear refractive index n of a uniform grating is given by

$$n = \bar{n} + \Delta n \cos\left(\frac{2\pi}{\Lambda}z\right)$$

where $\bar{n}$ is the average refractive index, Δn is the index modulation depth, and Λ is its period. For typical fiber gratings, $\bar{n} \approx 1.5$ and $\Delta n \approx 0.01$. To simplify the analysis, we usually assume that $\Delta n << \bar{n}$. Such gratings are called shallow gratings.

To answer the question of what happens to a light signal when it enters into an FBG, we consider the following periodic variation of the refractive index structure in an SMF. When a light signal enters into the FBG structure, it gets reflected at each periodic variation of the refractive index, as shown in Fig. 17.4. All the reflected light components combine coherently and form a large reflection at a particular wavelength when the period of the grating is approximately half the wavelength of the input light. This condition is referred to as the Bragg condition, and the wavelength at which this reflection occurs is called the Bragg wavelength, λ_B. Note that the mathematical relation for the Bragg wavelength can be obtained from grating theory. According to the theory of gratings, the relation between the incident angle θ_i and diffracted angle θ_r can be written as

$$\sin\theta_t - \sin\theta_r = m\lambda/(\bar{n}\Lambda) \qquad (4)$$

Here, the light enters perpendicular to the periodic structure and hence the incident and diffracted angles in the above equation take the following values, $\theta_i = \pi/2$ and $\theta_r = -\pi/2$. Further, we assume the order of the grating as $m = 1$. Under these conditions, the Bragg wavelength, λ_B, from the above equation, is found to be

$$\lambda_B = 2\bar{n}\Lambda \qquad (5)$$

Very often, it is convenient to consider the properties of periodic structure in terms of frequency ω, rather than wavelength. In this case, the Bragg condition can be written as

$$\omega_B = \frac{\pi c}{\bar{n}\Lambda}$$

We may emphasize that the grating structure reflects light not only at λ_B, but also in a range of wavelengths around this value. The extent of this range, in general, is tedious to calculate. However, this range of wavelengths depends on the depth of index modulation Δn. The range of wavelengths $\Delta\lambda$ is given by

$$\frac{\Delta\lambda}{\lambda_B} = \frac{\Delta\omega}{\omega_B} \approx \frac{\Delta n}{\bar{n}}$$

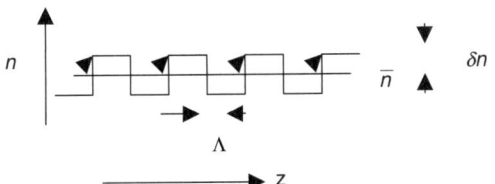

FIGURE 17.4 Fresnel's reflection of light at each interface.

where ω_B is the frequency associated with the Bragg wavelength λ_B and $\Delta\omega$ is the width of the response in frequency. Thus, the depth of index modulation determines the range of reflected wavelengths $\Delta\lambda$. For typical gratings $\frac{\Delta\lambda}{\lambda_B}$ is between 10^{-3} and 10^{-4}. Note that the grating reflects light over a fraction of nanometers because λ_B is typically around $1\,\mu m$. Thus, this property makes the grating very useful in optical fiber telecommunication systems. Such applications of FBG are wide and perhaps beyond the scope of this chapter (see Chapter 15).

Light signals at wavelengths other than the Bragg and neighboring wavelengths are not phase matched. Hence, they are transmitted through the periodic structure. Thus, in an FBG, we have both forward (transmitted) and backward (reflected) propagating modes.

5.3.1.2. Pulse Propagation in FBG. Recently, several methods have been adopted to study the reflection and transmission properties of FBG. To describe pulse propagation in FBG and, more precisely, to analyze how a Bragg grating affects the wave propagation in optical fibers, we use the following two methods: coupled-mode theory (CMT) and Bloch wave analysis. We discuss CMT here and discuss Bloch wave analysis in later sections.

The forward and backward propagating modes in FBG due to Bragg reflection can be described using the CMT as explained by Yariv [16] in the distributed feedback structure.

To describe pulse propagation in FBG, we use CMT. Usually, the governing equation for the pulse propagation in FBG is derived using Maxwell's equations. Here we work in the frequency domain as the nonlinear effects are assumed to be relatively weak. It can easily be shown that Maxwell's equations are reduced to the following wave equation in the form

$$\frac{\partial^2 \vec{E}}{\partial z^2} - \frac{\varepsilon(z)}{c^2}\frac{\partial^2 \vec{E}}{\partial t^2} = 0 \qquad (6)$$

where $\varepsilon(z) = \bar{n}^2 + \tilde{\varepsilon}(z)$, $\bar{n}^2$ is the spatial average of $\tilde{\varepsilon}(z)$, and $\bar{n}$ is the average refractive index of the medium. We consider the term $\tilde{\varepsilon}(z)$ with a period Λ and define $k_0 = \pi/\Lambda$. Using the Fourier series, $\tilde{\varepsilon}(z)$ can be written as

$$\tilde{\varepsilon}(z) = 2\tilde{\varepsilon}\cos(2k_0 z) \qquad (7)$$

The electric field inside the grating can be written as

$$\vec{E}(z,t) = \vec{E}_f(z,t)e^{+i(k_b z-\omega_d t)} + \vec{E}_b(z,t)e^{-i(k_a z-\omega_a t)} + \cdots \qquad (8)$$

where $E_{f,b}(z,t)$ represents the forward and backward propagating waves, respectively, inside the FBG structure. Now inserting Eqs. (7) and (8) into Eq. (6) and considering that the fields $E_{f,b}(z,t)$ are varying slowly with respect to ω_0^{-1} in time and k_0^{-1} in space, the resulting frequency domain coupled mode equations can be written as

$$i\frac{\partial \vec{E}_f}{\partial z} + i\frac{\bar{n}}{c}\frac{\partial \vec{E}_f}{\partial t} + \kappa \vec{E}_b = 0 \\ -i\frac{\partial \vec{E}_b}{\partial z} + i\frac{\bar{n}}{c}\frac{\partial \vec{E}_b}{\partial t} + \kappa \vec{E}_f = 0 \quad (9)$$

In the above equations, κ represents the coupling between the forward and backward propagating waves in the FBG. The set of Eqs. (9) are called linear coupled-mode (LCM) equations in which the non–phase-matched terms have been neglected. In the LCM equations we are dealing with slowly varying amplitudes rather than the electric field itself. Note that CMT is an approximate description that is valid for shallow gratings and for wavelengths close to the Bragg resonance.

5.3.1.3. Dispersion Relation and Photonic Band Gap. To have a general idea about group velocity and GVD in FBG and to understand the concept of photonic bandgap (PBG), we derive the dispersion relation from the LCM equation. To obtain the dispersion relation for an infinite uniform grating structure, we consider the following plane-wave solution of the form

$$E_{f,b} = A_{f,b} e^{i(qz - c\delta/\bar{n})} \quad (10)$$

where $A_\pm$ are constants and the parameter

$$\delta = \frac{\bar{n}}{c}(\omega_0 - \omega_B)$$

represents the detuning from the Bragg frequency, whereas $q = k_0 - k_B$ represents wavenumber. Substituting Eq. (10) into Eq. (9), we get

$$(-q + \delta)A_f + \kappa A_b = 0 \\ (q + \delta)A_b + \kappa A_f = 0$$

The above set of equations can also be written in matrix form as

$$\begin{bmatrix} \delta - q & \kappa \\ \kappa & \delta + q \end{bmatrix} \begin{bmatrix} A_f \\ A_b \end{bmatrix} = 0 \quad (11)$$

and by setting the determinant equal to zero, we obtain

$$\delta = \pm\sqrt{q^2 + \kappa^2} \quad (12)$$

Equation (12) is the dispersion relation for the periodic structure in the SMF. Figure 17.5 represents dispersion curves for both uniform and periodic media. For a uniform medium, the dispersion curve is found to be a straight line that is represented by the dashed lines in Fig. 17.5, and for a periodic medium the dispersion curve is drastically modified.

Note that the parameter q becomes purely imaginary when frequency detuning δ of the incident light signal falls in the range $-\kappa < \delta < \kappa$. That is, the range $|\delta| \leq \kappa$ is referred to as the PBG [17] (in analogy with the electronic energy bands occurring in crystals). For frequencies within the PBG, the grating reflectivity is high, and hence the field envelopes are evanescent. However, for frequencies outside the PBG, the grating reflectivity is lower (smaller). Therefore, for frequencies within the PBG, the grating does not allow the running wave solutions of the form in Eq. (10). It should be noted that the concept of PBG in FBG is analogous to the electronic band gaps in semiconductor crystals. In an electronic crystal, a moving electron experiences a periodic potential produced by the atomic lattice [18]. This produces a gap in the electronic energy band and hence a divide into the lower valence band and upper conduction band. The optical analog is the photon experiencing a periodic potential due to a lattice of different macroscopic dielectric media.

From the above matrix, we calculate the two eigenvectors $v_\pm$ and they are

$$v_\pm = \begin{bmatrix} \kappa \\ q \mp \sqrt{q^2 + \kappa^2} \end{bmatrix} \quad (13)$$

These eigenvectors correspond to the two Bloch functions bordering the PBG under consideration. Note that the Bloch waves are the eigenmodes of the

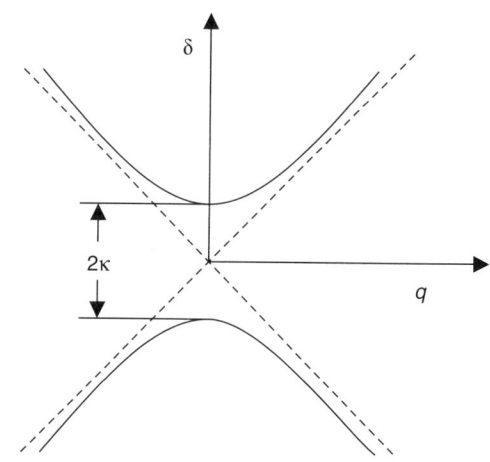

FIGURE 17.5 Dispersion curve.

periodic media in the same way as plane waves are the natural modes of free space propagation. We discuss the Bloch wave analysis in detail in Section 8.

5.3.1.4. Dispersion in FBG.

In the usual manner, we interpret the dispersion relation as follows. The slope of the dispersion curve gives the group velocity, whereas curvature gives the GVD. Note that for a uniform medium, the dispersion relation is a straight line that is represented as dashed lines (cf. Fig. 17.5), with the slope $c/\bar{n}$ (i.e., the slope is constant). Thus, the dispersion of bare fiber is ignored for the range of frequencies considered here. Similarly, now we turn our discussion to the group velocity and GVD in FBG. Using Eq. (12) the group velocity in FBG is calculated as

$$v_g \approx d\omega/dk = (c/\bar{n})d\delta/dq$$
$$v_g = \pm \frac{c}{\bar{n}}\sqrt{1 - \frac{\kappa^2}{\delta^2}} \quad (14)$$

where the positive sign represents the group velocity of the forward propagating wave and the negative sign represents the group velocity of the backward propagating wave. Hence, at the edges of the bandgap, that is, at $\delta = \pm\kappa$, the group velocity $v_g = 0$, and for $|\delta| \to \infty$, the group velocity $v_g = c/\bar{n}$. Thus, the group velocity in FBG varies between 0 and $c/\bar{n}$. This means that light can travel at any velocity from zero to the average speed of light in the medium. Note that on the short-wavelength side of the PBG, the group velocity increases with decreasing wavelength and vice versa.

As discussed earlier, by differentiating Eq. (14), the quadratic dispersion at the edges of the bandgap in FBG is calculated as

$$\omega_2 \equiv \frac{d^2\omega}{dk^2} = \pm \frac{c}{\bar{n}\kappa} \quad (15)$$

In Eq. (15), the positive sign represents the GVD on the upper branch of the dispersion curve where the grating induced dispersion is anomalous, that is, $d^2\omega/dk^2 > 0$, whereas the negative sign represents the GVD on the lower branch of the dispersion curve where the grating induced dispersion is normal, that is, $d^2\omega/dk^2 < 0$. For typical physical parameter values in the above equation, the GVD in FBG is calculated as $\omega_2 = 6 \times 10^5 \, \text{m}^2/\text{s}$, whereas for a typical standard telecommunication fiber the dispersion is given as 20 ps/nm·km, that is, $\omega_2 = 0.22 \, \text{m}^2/\text{s}$. From this fact, it is clear that the grating-induced quadratic dispersion (GVD) in FBG is about six orders of magnitude greater than the dispersion in the conventional standard telecommunication fiber at 1310 nm [19].

As in the case of conventional fiber, cubic dispersion and HOD are important and they play a crucial role in FBG. So, we now turn to calculate the third-order dispersion (TOD) in FBG by differentiating Eq. (15),

$$\omega_3 \equiv \frac{d^3\omega}{dk^3} \approx \frac{c}{\bar{n}\kappa^2} \quad (16)$$

For typical physical parameter values, TOD in FBG is found to be $\omega_3 = 1.9 \times 10^3 \, \text{m}^3/\text{s}$, whereas for conventional bare fiber $\omega_3 = 0.1 \, \text{ps}^3/\text{m} \approx 10^{-7} \, \text{m}^3/\text{s}$. Thus, TOD in FBG is 10 orders of magnitude greater than the TOD in the conventional fiber. Note that the TOD in a bare fiber can be neglected except when close to the zero dispersion wavelength as well as for very short pulses, whereas in the case of FBG, TOD and HODs are much more significant.

From the above analysis, it is clear that the GVD in grating is six orders greater than that due to bare fiber, whereas TOD is 10 times greater. A similar conclusion holds for other HODs in FBG. That is why the magnitude of the grating-induced dispersion is very large when compared with the total chromatic dispersion of SMFs.

5.3.1.5. Reflection and Transmission Coefficients.

The reflectivity (R) of the grating is given by the simple expression $R = \tanh^2(\kappa L)$, where κ is the coupling coefficient at the Bragg wavelength and L is the length of the grating. Figure 17.6 shows the reflected spectrum obtained from the uniform grating. This figure clearly exhibits the range of the PBG.

Note that the product κL measures the grating strength. For $\kappa L = 1$, the grating reflectivity is 58%. Similarly, for $\kappa L = 2, 3$, the grating reflectivity is 93% and 99%, respectively. Therefore, a grating

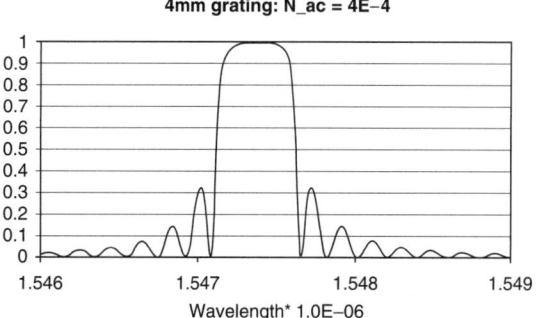

FIGURE 17.6 Reflection spectrum of a uniform FBG.

with κL greater than one ($\kappa L > 1$) is termed a strong grating and a grating with κL less than one ($\kappa L < 1$) is a weak grating.

5.3.1.6. Bandwidth of the Grating. The other important property of the grating is its bandwidth. It is a measure of the wavelength range over which the grating reflects light. The bandwidth of an FBG is easily measured as the full width at half-maximum, $\Delta\lambda_{FWHM}$, of the central reflection peak, which is defined as the wavelength interval between the 3-dB points. That is the separation in the wavelength between the points on either side of the Bragg wavelength where the reflectivity has decreased to 50% of its maximum value. However, a much easier quantity to calculate the bandwidth is $\Delta\lambda_0 = \lambda_0 - \lambda_B$, where λ_0 is the wavelength where the first zero in the reflection spectra occurs. This bandwidth can be found by calculating the difference in the propagation constants, $\Delta\beta_0 = \beta_0 - \beta_B$, where β_0 ($=2\pi N_{eff}/\lambda_0$) is the propagation constant at wavelength λ_0 for which the reflectivity is first zero, and β_B ($=2\pi N_{eff}/\lambda_B$) is the propagation constant at the Bragg wavelength for which the reflectivity is maximum.

In the case of weak gratings ($\kappa L \langle 1$), $\beta_0 - \beta_B = \pi/L$, from which it can be arrived that $\Delta\lambda_{FWHM} \approx \Delta\lambda_0 = \lambda_B^2/N_{eff}L$. From this relation, it is clear that the bandwidth of a weak grating is inversely proportional to the grating length L. Thus, a long weak grating can have very narrow bandwidths. An FBG, written with a fiber length of about 1 m, has a bandwidth less than 100 MHz, which is an astonishingly narrow bandwidth for a reflector of visible light.

On the other hand, in the case of strong gratings ($\kappa L \rangle 1$), $\Delta\beta_0 = \beta_0 - \beta_B = 4\kappa$, and $\Delta\lambda_{FWHM} \approx 2\Lambda\lambda_0 = 4\lambda_B^2\kappa/\bar{n}N_{eff}$. Thus for strong gratings, the bandwidth is directly proportional to the coupling coefficient κ and is independent of the grating length.

5.3.2. Nonlinear Properties

So far we have discussed the different intriguing linear properties of FBG. Now, in this section we discuss the role of nonlinear effects in FBG. We also discuss the derivation of (pulse governing) the nonlinear coupled-mode (NLCM) equations in the presence of Kerr nonlinearity (for the nonlinearity management system).

5.3.2.1. Introduction to Derivation of NLCM Equations. Before the invention of lasers, scientists that all optical media were purely linear. Fortunately, the invention of the laser enabled believed scientists to examine the behavior of light in optical materials at higher intensities, making the study of nonlinear optics possible. Since then, it has been extensively investigated by many researchers, and hence the area of nonlinear optics has progressed steadily.

It is obvious that in a nonlinear medium, the presence of an optical field modifies the properties of the medium, which in turn modifies another optical field or the field itself. In other words, we can say that the properties of the medium are dependent on the intensity of the incident applied optical field. Under this circumstance, the refractive index of the medium can be expressed as

$$n = n_0 + n_2 I \quad (17)$$

We point out that the presence of Kerr nonlinearity may change the width and depth of the PBG described in Eq. (12). Therefore, the transmission of light through a nonlinear periodic structure depends on both wavelength and intensity of the incident light pulse.

In recent times, there has been great interest in considering the nonlinear laser pulse propagation through a medium consisting of alternating oppositely signed (i.e., positive and negative) Kerr coefficients [20, 21]. Let us derive the pulse governing equation in such kind of medium considered above. To do so, it is necessary to have knowledge about the refractive index profile $n(z, |E|^2)$ in this periodic structure. Therefore, as a first step, we calculate the refractive index profile as follows.

5.3.2.2. Calculation of the Refractive Index Profile. If the variations of the refractive index due to nonlinearity and linear gratings are much smaller than the average index, then the refractive index profile $n(z, |E|^2)$ can be approximately viewed as a periodic function along the spatial direction parallel to the layers. That is, the function $n(z, |E|^2)$ may be described analytically over one period as follows:

$$n(z, |E|^2) = \begin{cases} n_{02} + n_{nl2}|E|^2, & \text{if } -n\Lambda/2 < z < -n\Lambda/4 \\ n_{01} + n_{nl1}|E|^2, & \text{if } -n\Lambda/4 < z < n\Lambda/4 \\ n_{02} + n_{nl2}|E|^2, & \text{if } n\Lambda/4 < z < n\Lambda/2, n = 1, \ldots \end{cases} \quad (18)$$

with the periodic medium being an N-layered quarter-waves stack. Here, n_{01} and n_{02} are the linear part of the refractive indices of the (oppositely signed Kerr coefficients) two different media, respectively. Similarly, the coefficients n_{nl1} and n_{nl2} are the oppositely signed Kerr coefficients. The refractive index profile $n(z, |E|^2)$ in the oppositely signed Kerr coefficients is illustrated in Fig. 17.7.

By using the Fourier series, the refractive index profile $n(z, |E|^2)$ can be written as

$$n(z, |E|^2) = a_0 + 2\sum_{n=1}^{\infty} a_n \cos(2\pi n f_0 z) + b_n \sin(2\pi n f_0 z) \quad (19)$$

where $f_0 (=1/\Lambda)$ is the fundamental frequency. The coefficient a_0 is the mean value of the periodic signal over one period and is defined as

$$a_0 = \frac{1}{\Lambda} \int_{-\Lambda/2}^{\Lambda/2} n(z, |E|^2) dz \quad (20)$$

The coefficients of a_n and b_n represent the amplitudes of the cosine and sine functions, and they are

$$a_n = \frac{1}{\Lambda} \int_{-\Lambda/2}^{\Lambda/2} n(z, |E|^2) \cos(2\pi n f_0 z) dz,$$

$$b_n = \frac{1}{\Lambda} \int_{-\Lambda/2}^{\Lambda/2} n(z, |E|^2) \sin(2\pi n f_0 z) dz, \quad n = 1, 2, \ldots \quad (21)$$

For an even function, it is easy to prove that $a_n \neq 0$ and $b_n = 0$. Therefore, it is enough to calculate a_0 and a_n. Substituting Eq. (16) into (18), we obtain

$$a_0 = \frac{1}{\Lambda} \int_{-\Lambda/2}^{\Lambda/2} n(z, |E|^2) dz = \frac{2}{\Lambda} \int_0^{\Lambda/4} (n_{01} + n_{nl1}|E|^2) dz$$

$$+ \int_{\Lambda/4}^{\Lambda/2} (n_{02} + n_{nl2}|E|^2) dz$$

$$= \frac{n_{01} + n_{02}}{2} + \frac{n_{nl1} + n_{nl2}}{2}|E|^2$$

Similarly, the coefficient a_n can be calculated as

$$a_n = \frac{1}{\Lambda} \int_{-\Lambda/2}^{\Lambda/2} n(z, |E|^2) \cos(2\pi n f_0 z) dz$$

$$= \frac{2}{\Lambda} \int_0^{\Lambda/4} (n_{01} + n_{nl1}|E|^2) \cos(2\pi n f_0 z) dz$$

$$+ \frac{2}{\Lambda} \int_{\Lambda/4}^{\Lambda/2} (n_{02} + n_{nl2}|E|^2) \cos(2\pi n f_0 z) dz$$

$$= \frac{1}{\pi n} \left(n_{01} - n_{02} + n_{nl1}|E|^2 - n_{nl2}|E|^2 \right) \sin\left(\frac{\pi n}{2}\right)$$

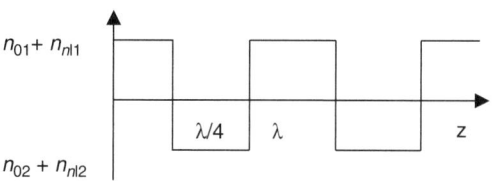

FIGURE 17.7 Refractive index profile in the oppositely signed Kerr coefficients.

Thus, a_n is calculated as follows:

$$a_n = \begin{cases} 0, & \text{if } n \text{ is even} \\ \frac{1}{\pi n}\left(n_{01} - n_{02} + n_{nl1}|E|^2 - n_{nl2}|E|^2\right)\sin\left(\frac{\pi n}{2}\right), & \text{if } n \text{ is odd} \end{cases} \quad (22)$$

Using the values of a_0 and a_n, the refractive index profile $n(z, |E|^2)$ in Eq. (19) can be written as

$$n(z, |E|^2) = \frac{n_{01} + n_{02}}{2} + \frac{n_{nl1} + n_{nl2}}{2}|E|^2$$

$$+ 2 \sum_{n=1, n \text{ odd}}^{\infty} \frac{1}{\pi n} \left(n_{01} - n_{02} + n_{nl1}|E|^2 - n_{nl2}|E|^2 \right) \sin\left(\frac{\pi n}{2}\right) \cos(2\pi n f_0 z)$$

$$n(z, |E|^2) = \frac{n_{01} + n_{02}}{2} + \frac{n_{nl1} + n_{nl2}}{2}|E|^2$$

$$+ 2 \left(\frac{n_{01} - n_{02}}{\pi} + \frac{n_{nl1} - n_{nl2}}{\pi}|E|^2 \right) \cos\left(\frac{2\pi}{\Lambda}z\right) \quad (23)$$

To simplify Eq. (30), we introduce the following four parameters:

$$n_{ln} = \frac{n_{01} + n_{02}}{2}, \quad n_{nl} = \frac{n_{nl1} + n_{nl2}}{2}, \quad n_{0k} = \frac{n_{01} - n_{02}}{\pi},$$

$$\text{and} \quad n_{2k} = \frac{n_{nl1} - n_{nl2}}{\pi} \quad (24)$$

By using these new parameters, the refractive index profile $n(z, |E|^2)$ can now be written in the following simplified form as

$$n(z, |E|^2) = n_{ln} + n_{nl}|E|^2 + 2n_{0k}\cos kz + 2n_{2k}|E|^2 \cos kz \quad (25)$$

where $k(= 2\pi/\Lambda)$ is defined as the wavenumber.

5.3.2.3. Derivation of NLCM Equations. To derive the pulse governing equation oppositely signed Kerr coefficients, we start with Maxwell's equations

$$\nabla \cdot \vec{D} = 0, \quad \nabla \cdot \vec{B} = 0, \quad \nabla \times \vec{E} = -\frac{\partial \vec{B}}{\partial t}$$

$$\text{and} \quad \nabla \times \vec{H} = \vec{J} + \frac{\partial \vec{D}}{\partial t} \quad (26)$$

where $\vec{E}$ and $\vec{H}$ are electric and magnetic field vectors $\vec{D}$ $(=\varepsilon_0\vec{E}+\vec{P})$ is the displacement vector, and $\vec{B}$ $(=\mu_0\vec{H})$ is the magnetic flux density. It can easily be shown that Maxwell's equations are reduced to the following wave equation in the form

$$\frac{\partial^2 \vec{E}}{\partial z^2} - \frac{n(z, |E|^2)}{c^2} \frac{\partial^2 \vec{E}}{\partial t^2} = 0 \qquad (27)$$

where $c = 1/\sqrt{\varepsilon_0\mu_0}$ is the speed of light and $\vec{E}(z,t)$ is the electric field.

The electric field inside the grating can be written as

$$\vec{E}(z,t) = \vec{E}_\text{f}(z,t)e^{+i(k_0 z - \omega_0 t)} + \vec{E}_\text{b}(z,t)e^{-i(k_0 z - \omega_0 t)} + \cdots \qquad (28)$$

where $E_{\text{f,b}}(z,t)$ represents the forward and backward propagating waves, respectively, inside the FBG structure. The central frequency is given by $\omega_0 = ck_0/|n_\text{ln}|$ and the wave number is given by $k_0 = 2\pi|n_\text{ln}|/\lambda$. Note that peak reflectance occurs at the center of a forbidden gap and can be written as $\lambda = 2n_\text{ln}\Lambda$. In other words, resonance in the first bandgap occurs when $k = 2k_0$. Now substituting Eq. (28) into Eq. (27), we obtain the first term as

$$\frac{\partial^2 \vec{E}}{\partial z^2} = \left(\frac{\partial^2 \vec{E}_\text{f}}{\partial z^2} + 2ik_0\frac{\partial \vec{E}_\text{f}}{\partial z} - k_0^2 \vec{E}_\text{f}\right)e^{i(k_0 z - \omega_0 t)}$$
$$+ \left(\frac{\partial^2 \vec{E}_\text{b}}{\partial z^2} - 2ik_0\frac{\partial \vec{E}_\text{b}}{\partial z} - k_0^2 \vec{E}_\text{b}\right)e^{-i(k_0 z + \omega_0 t)} \qquad (29)$$

Here we assume that the forward and backward field components $E_{\text{f,b}}(z,t)$ are varying slowly with respect to ω_0^{-1} in time and k_0^{-1} in space, viz

$$\left|\frac{\partial \vec{E}_\text{f,b}}{\partial t}\right| \ll \omega_0 \left|\vec{E}_\text{f,b}\right|, \quad \left|\frac{\partial \vec{E}_\text{f,b}}{\partial z}\right| \ll k_0 \left|\vec{E}_\text{f,b}\right| \qquad (30)$$

By applying the slowly varying envelope approximation, Eq. (29) transforms to

$$\frac{\partial^2 \vec{E}}{\partial z^2} = \left(2ik_0\frac{\partial \vec{E}_\text{f}}{\partial z} - k_0^2 \vec{E}_\text{f}\right)e^{i(k_0 z - \omega_0 t)}$$
$$+ \left(-2ik_0\frac{\partial \vec{E}_\text{b}}{\partial z} - k_0^2 \vec{E}_\text{b}\right)e^{-i(k_0 z + \omega_0 t)} \qquad (31)$$

Similarly, by applying the slowly varying envelope approximation, the second term in Eq. (27) can be written as

$$\frac{\partial^2 \vec{E}}{\partial t^2} = -\omega_0\left[\left(2i\frac{\partial \vec{E}_\text{f}}{\partial t} + \omega_0 \vec{E}_\text{f}\right)e^{i(k_0 z - \omega_0 t)}\right.$$
$$\left. + \left(2i\frac{\partial \vec{E}_\text{f}}{\partial t} + \omega_0 \vec{E}_\text{f}\right)e^{-i(k_0 z + \omega_0 t)}\right] \qquad (32)$$

Using Eqs. (25) and (32), the second term in Eq. (19) can be written as

$$\frac{n(z, |E|^2)}{c^2}\frac{\partial^2 \vec{E}}{\partial t^2} = \frac{k_0}{\omega_0 n_\text{ln} c}\left[n_\text{ln}^2 + 2n_\text{ln} n_\text{nl}|E|^2\right.$$
$$\left. + (2n_\text{ln} n_{0k} + 2n_\text{ln} n_{2k}|E|^2)(e^{ikz} + e^{-ikz})\right]\frac{\partial^2 \vec{E}}{\partial t^2}$$

Neglecting all the higher order terms in n_{2k}, we have

$$\frac{n(z, |\vec{E}|^2)}{c^2}\frac{\partial^2 \vec{E}}{\partial t^2} =$$
$$\frac{k_0}{\omega_0 c}\left[n_\text{ln} + 2n_\text{nl}|E|^2 + 2(n_{0k} + n_{2k}|E|^2)(e^{i2k_0 z} + e^{-i2k_0 z})\right]$$
$$- \omega_0\left[\left(2i\frac{\partial \vec{E}_\text{f}}{\partial t} + \omega_0 \vec{E}_\text{f}\right)e^{i(k_0 z - \omega_0 t)}\right.$$
$$\left. + \left(2i\frac{\partial \vec{E}_\text{f}}{\partial t} + \omega_0 \vec{E}_\text{f}\right)e^{-i(k_0 z + \omega_0 t)}\right] \qquad (33)$$

The intensity term $|\vec{E}|^2$ in the above equation can be expressed in terms of the field components as

$$|\vec{E}|^2 = \vec{E}\cdot\vec{E}^* = |\vec{E}_\text{f}|^2 + |\vec{E}_\text{b}|^2 + \vec{E}_\text{f}\vec{E}_\text{b}^* e^{i2k_0 z} + \vec{E}_\text{b}\vec{E}_\text{f}^* e^{-i2k_0 z} \qquad (34)$$

Using Eq. (34), Eq. (33) can be simplified to

$$\frac{n(z,|\vec{E}|^2)}{c^2}\frac{\partial^2 \vec{E}}{\partial t^2} = -\frac{k_0}{c}\left[n_\text{ln}\omega_0\vec{E}_\text{f} + 2in_\text{ln}\frac{\partial \vec{E}_\text{f}}{\partial t} + 2n_{0k}\omega_0\vec{E}_\text{b}\right.$$
$$+ 2n_\text{nl}\left(|\vec{E}_\text{f}|^2 + |\vec{E}_\text{b}|^2\right)\omega_0\vec{E}_\text{f} + 2n_\text{nl}\vec{E}_\text{f}\vec{E}_\text{b}^*\omega_0\vec{E}_\text{b}$$
$$+ 2n_{2k}\left(|\vec{E}_\text{f}|^2 + |\vec{E}_\text{b}|^2\right)\omega_0\vec{E}_\text{b} + 2n_{2k}\vec{E}_\text{b}\vec{E}_\text{f}^*\omega_0\vec{E}_\text{f}$$
$$\left. + 2n_{2k}\vec{E}_\text{f}\vec{E}_\text{b}^*\omega_0\vec{E}_\text{f}\right]e^{i(k_0 z - \omega_0 t)} - \frac{k_0}{c}\left[n_\text{ln}\omega_0\vec{E}_\text{b} + 2in_\text{ln}\frac{\partial \vec{E}_\text{b}}{\partial t}\right.$$
$$+ 2n_{0k}\omega_0\vec{E}_\text{f} + 2n_\text{nl}\left(|\vec{E}_\text{f}|^2 + |\vec{E}_\text{b}|^2\right)\omega_0\vec{E}_\text{b} + 2n_\text{nl}\vec{E}_\text{b}\vec{E}_\text{f}^*\omega_0\vec{E}_\text{f}$$
$$+ 2n_{2k}\left(|\vec{E}_\text{f}|^2 + |\vec{E}_\text{b}|^2\right)\omega_0\vec{E}_\text{f} + 2n_{2k}\vec{E}_\text{f}\vec{E}_\text{b}^*\omega_0\vec{E}_\text{b}$$
$$\left. + 2n_{2k}\vec{E}_\text{b}\vec{E}_\text{f}^*\omega_0\vec{E}_\text{b}\right]e^{-i(k_0 z + \omega_0 t)} \qquad (35)$$

Now combining Eqs. (29) and (35) and using them in Eq. (27) and collecting all the terms having the coefficient of $e^{i(k_0 z - \omega_0 t)}$, we obtain

$$2ik_0\frac{\partial \vec{E}_\text{f}}{\partial z} - k_0^2\vec{E}_\text{f} + \frac{k_0}{c}\left[n_\text{ln}\omega_0\vec{E}_\text{f} + 2in_\text{ln}\frac{\partial \vec{E}_\text{f}}{\partial t}\right.$$
$$+ 2n_{0k}\omega_0\vec{E}_\text{b} + 2n_\text{nl}\left(|\vec{E}_\text{f}|^2 + |\vec{E}_\text{b}|^2\right)\omega_0\vec{E}_\text{f}$$
$$+ 2n_\text{nl}\vec{E}_\text{f}\vec{E}_\text{b}^*\omega_0\vec{E}_\text{b} + 2n_{2k}\left(|\vec{E}_\text{f}|^2 + |\vec{E}_\text{b}|^2\right)\omega_0\vec{E}_\text{b}$$
$$\left. + 2n_{2k}\vec{E}_\text{b}\vec{E}_\text{f}^*\omega_0\vec{E}_\text{f} + 2n_{2k}\vec{E}_\text{f}\vec{E}_\text{b}^*\omega_0\vec{E}_\text{f}\right] = 0$$

After simplification, the above equation becomes

$$i\frac{c}{\omega_0}\frac{\partial \vec{E}_f}{\partial z} + i\frac{n_{ln}}{\omega_0}\frac{\partial \vec{E}_f}{\partial t} + n_{0k}\vec{E}_b + n_{nl}\left(\left|\vec{E}_f\right|^2 + \left|\vec{E}_b\right|^2\right)\vec{E}_f$$
$$+ n_{nl}\left|\vec{E}_b\right|^2\vec{E}_f + n_{2k}\left(\left|\vec{E}_f\right|^2 + \left|\vec{E}_b\right|^2\right)\vec{E}_b + n_{2k}\left|\vec{E}_f\right|^2\vec{E}_b$$
$$+ n_{2k}\vec{E}_f^2\vec{E}_b^* = 0 \quad (36)$$

Similarly, by grouping all the terms $e^{-i(k_0 z + \omega_0 t)}$, we obtain

$$-i\frac{c}{\omega_0}\frac{\partial \vec{E}_b}{\partial z} + i\frac{n_{ln}}{\omega_0}\frac{\partial \vec{E}_b}{\partial t} + n_{0k}\vec{E}_f + n_{nl}\left(\left|\vec{E}_f\right|^2 + \left|\vec{E}_b\right|^2\right)\vec{E}_b$$
$$+ n_{nl}\left|\vec{E}_f\right|^2\vec{E}_b + n_{2k}\left(\left|\vec{E}_f\right|^2 + \left|\vec{E}_b\right|^2\right)\vec{E}_f + n_{2k}\left|\vec{E}_b\right|^2\vec{E}_f$$
$$+ n_{2k}\vec{E}_b^2\vec{E}_f^* = 0 \quad (37)$$

To normalize Eqs. (36) and (37), we introduce $\xi = \omega_0 z/c$ and $\tau = \omega_0 t/n_{ln}$. Now the normalized coupled-mode equations can be written as

$$i\frac{\partial \vec{E}_f}{\partial \xi} + i\frac{\partial \vec{E}_f}{\partial \tau} + n_{0k}\vec{E}_b + n_{nl}\left(\left|\vec{E}_f\right|^2 + 2\left|\vec{E}_b\right|^2\right)\vec{E}_f$$
$$+ n_{2k}\left[\left(2\left|\vec{E}_f\right|^2 + \left|\vec{E}_b\right|^2\right)\vec{E}_b + \vec{E}_f^2\vec{E}_b^*\right] = 0$$
$$-i\frac{\partial \vec{E}_b}{\partial \xi} + i\frac{\partial \vec{E}_b}{\partial \tau} + n_{0k}\vec{E}_f + n_{nl}\left(\left|\vec{E}_b\right|^2 + 2\left|\vec{E}_f\right|^2\right)\vec{E}_b$$
$$+ n_{2k}\left[\left(2\left|\vec{E}_b\right|^2 + \left|\vec{E}_f\right|^2\right)\vec{E}_f + \vec{E}_b^2\vec{E}_f^*\right] = 0 \quad (38)$$

Equation (38) represents the nonlinear pulse propagation in a nonlinearity management system. If we consider a medium consisting of periodically varying nonlinear refractive indices in which the Kerr coefficients of two adjacent layers are oppositely signed, that is, $n_{nl1} = -n_{nl2}$, under this condition $n_{nl} = 0$ and $n_{2k} \neq 0$. Under this circumstance, Eq. (38) becomes

$$i\frac{\partial \vec{E}_f}{\partial \xi} + i\frac{\partial \vec{E}_f}{\partial \tau} + n_{0k}\vec{E}_b$$
$$+ n_{2k}\left[\left(2\left|\vec{E}_f\right|^2 + \left|\vec{E}_b\right|^2\right)\vec{E}_b + \vec{E}_f^2\vec{E}_b^*\right] = 0$$
$$-i\frac{\partial \vec{E}_b}{\partial \xi} + i\frac{\partial \vec{E}_b}{\partial \tau} + n_{0k}\vec{E}_f$$
$$+ n_{2k}\left[\left(2\left|\vec{E}_b\right|^2 + \left|\vec{E}_f\right|^2\right)\vec{E}_f + \vec{E}_b^2\vec{E}_f^*\right] = 0 \quad (39)$$

If we consider the medium having the same Kerr coefficients throughout the periodic structure, then we have $n_{nl1} = n_{nl2}$. From Eq. (24), it is clear that $n_{nl} \neq 0$ and $n_{2k} = 0$. Under the above-mentioned condition, Eq. (38) reduces to

$$i\frac{\partial \vec{E}_f}{\partial \xi} + i\frac{\partial \vec{E}_f}{\partial \tau} + n_{0k}\vec{E}_b + n_{nl}\left(\left|\vec{E}_f\right|^2 + 2\left|\vec{E}_b\right|^2\right)\vec{E}_f = 0$$
$$-i\frac{\partial \vec{E}_b}{\partial \xi} + i\frac{\partial \vec{E}_b}{\partial \tau} + n_{0k}\vec{E}_f + n_{nl}\left(\left|\vec{E}_b\right|^2 + 2\left|\vec{E}_f\right|^2\right)\vec{E}_b = 0$$
$$(40)$$

The set of Eqs. (40) are the well-known NLCM equations for the medium having the same positive Kerr coefficients [3].

5.3.2.4. Role of Nonlinearity on the Dispersion Curves.
To investigate the role of nonlinearity on the system under consideration, we derive the nonlinear dispersion relation from the governing equation. That is, by plotting the nonlinear dispersion relation, we can get a rough idea about the impact of nonlinearity on the dispersion curves. From this analysis, one can very easily explain how the positive (negative) nonlinearity affects the upper (lower) branch of the dispersion curves.

To simplify the discussion, here we consider a medium that consists of the same Kerr coefficients. Under this condition we have the system of Eqs. (40). Now, we substitute the stationary solution to the coupled-mode equations by assuming $\vec{E}_{f,b}(\xi,\tau) = A_{f,b}(\xi)e^{-i\hat{\delta}ct/\bar{n}}$, where $\hat{\delta}$ is the detuning parameter that measures detuning from the Bragg condition. The parameter $\omega_B \left(= \frac{\pi c}{n_0 \Lambda} = \frac{2\pi c}{\lambda_B}\right)$ is the Bragg frequency and λ_B is the Bragg free-space wavelength. Using the stationary solution in Eq. (40), we obtain

$$i\frac{d\vec{E}_f}{d\xi} + \hat{\delta}\vec{E}_f + \kappa\vec{E}_b + \Gamma\left(\left|\vec{E}_f\right|^2 + 2\left|\vec{E}_b\right|^2\right)\vec{E}_f = 0$$
$$-i\frac{d\vec{E}_b}{d\xi} + \hat{\delta}\vec{E}_b + \kappa\vec{E}_f + \Gamma\left(\left|\vec{E}_b\right|^2 + 2\left|\vec{E}_f\right|^2\right)\vec{E}_b = 0$$
$$(41)$$

where $\kappa = n_{0k}$ and $\Gamma = n_{nl}$. To solve Eq. (41) in the continuous wave limit, we assume the following form of the solution

$$\vec{E}_f = u_f e^{iqz}, \quad \vec{E}_b = u_b e^{iqz} \quad (42)$$

where u_f and u_b are the constants along the grating length. Now, let us introduce a parameter $f = u_b/u_f$, which describes how the total power $P_0 = u_f^2 + u_b^2$ is divided between the forward and backward propagating waves. The constants u_f and u_b can be written in terms of the total power as follows

$$u_f = \sqrt{\frac{P_0}{1+f^2}}, \quad u_b = \sqrt{\frac{P_0}{1+f^2}}f \quad (43)$$

Now, using Eqs. (42) and (43) in Eq. (41), we get the following simplified equations:

$$-q + \delta + \kappa f + \Gamma P_0 \frac{1+2f^2}{1+f^2} = 0,$$

$$q + \delta + \frac{\kappa}{f} + \Gamma P_0 \frac{2+f^2}{1+f^2} = 0$$

Adding and subtracting the above equations, we obtain the relations for q and δ in terms of the parameters f and P_0, and they are given as

$$q = -\kappa \frac{(1-f^2)}{2f} - \frac{\Gamma P_0}{2} \frac{1-f^2}{1+f^2}, \quad \delta = -\kappa \frac{(1+f^2)}{2f} - \frac{3\Gamma P_0}{2} \tag{44}$$

The set of Eqs. (44) are called the nonlinear dispersion relation for the NLCM equations (40).

Note that for the low-power case, the nonlinear effects are negligible ($\Gamma = 0$) and hence Eq. (44) reduces to $q^2 = \delta^2 - \kappa^2$, which is precisely the dispersion relation [Eq. (12)] obtained in the previous section for the linear case.

To understand the impact of positive (focusing) and negative (defocusing) nonlinearities, we plot the nonlinear dispersion curves using Eq. (44). Figure 17.8, a and b, shows such curves at two power levels.

The positive (focusing) nonlinearity (where $n_2 > 0$) changes the upper branch of the dispersion curve qualitatively and forms a loop beyond a critical power level. This critical value of P_0 can be calculated by looking for the value of f at which q becomes zero while $f \neq 1$. From Eq. (44), we find that this can occur when

$$f \equiv f_c = -(\Gamma P_0/2\kappa) + \sqrt{(\Gamma P_0/2\kappa)^2 - 1} \tag{45}$$

The formation of a loop on the upper branch of the dispersion curve can be explained using the relation $\omega_0 = \pi c/\bar{n}\Lambda$. Here, we consider the positive nonlinearity (focusing) that increases the refractive index. This in turn increases the Bragg wavelength and ultimately shifts the PBG toward the lower frequencies (i.e., the upper branch). Whenever the total power P_0 exceeds the critical power P_c, there is a loop formation on the upper branch of the dispersion curve. Thus, a loop is formed only on the upper branch (where $f < 0$) for a positive nonlinearity. For negative nonlinearity where $n_2 < 0$ (i.e., the self-defocusing effect), loop formation takes place on the lower branch of the dispersion curve. A similar conclusion can be drawn for the general system of NLCM Eqs. (40).

As discussed above, we can also study the impact of nonlinearity for the nonlinearity management system, discussed in Eq. (40). Now, we carry out the same analysis by deriving the dispersion relation for the system of Eqs. (38). To study the role of the nonlinearity management coefficient on the PBG structure, first we intend to study the physics behind the role of nonlinearity on PBG in the absence of the nonlinearity management coefficient. This aspect has been clearly discussed in the previous paragraph. Whenever the applied input power P_0 exceeds the critical power, $P_c (= \frac{2\kappa}{\alpha})$, there is a loop formation on the upper branch of the dispersion curve, as shown in Fig. 17.8, a and b.

An introduction of nonlinearity management into the system ultimately reduces the size of the loop already formed, and we find that the size further decreases as we keep increasing the magnitude of the nonlinearity management coefficient. The reason behind this could be because the more we introduce nonlinearity management, the more the system ceases to hold its nonlinear property. That is why the dispersion curves resemble the case of a PBG of a linear case, as shown in Fig. 17.8, c and d.

So far, we have discussed the linear and nonlinear effects in FBG and also the impact of nonlinearity on the system in terms of dispersion curves. Having discussed the role of nonlinearity on the PBG, in the following section we discuss how the nonlinearity is counterbalanced with the grating-induced dispersion near the PBG structure.

6. SOLITONS IN FBG

It is obvious that the soliton in fibers is formed after the exact balancing of GVD arising as a combination of material and waveguide dispersion with that of the SPM due to Kerr nonlinearity. We can also expect a similar soliton-type pulse formation in FBG where the strong grating-induced dispersion is exactly counterbalanced by the Kerr nonlinearity through the SPM and cross-phase modulation effects. As a result, there is a formation of slowly traveling localized envelope in FBG structures known as, in general, Bragg grating solitions. They are often referred to as gap solitons if their spectra lie well within the frequency (photonic) bandgap. They are referred to in this way because the frequency of the incident pulse exactly matches the Bragg frequency. Thus, based on the pulse frequency spectrum with respect to PBG, solitons in FBG can be classified into two categories, namely Bragg grating solitons and gap solitons. Recent developments in theory as well as experimental investigations of these solitons provide a wide scope for realizing many devices such as optical add/drop multiplexers, optical filters, and sensors that

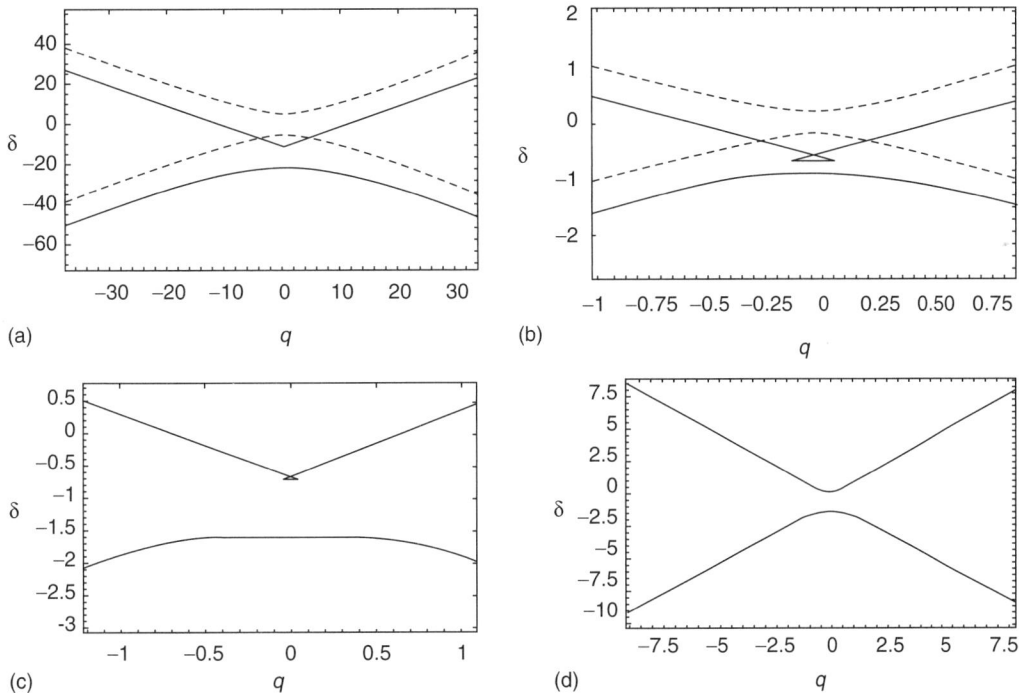

FIGURE 17.8 Nonlinear dispersion curves showing variation of δ and q for $\Gamma P_0/\kappa = 2$. Dashed curves represent the linear case ($\Gamma = 0$). (c) For NM coefficient $= 0.4$. (d) For NM coefficient $= 0.9$.

ultimately could improve the OFC and sensor technology. The observation of solitons in FBG also paves the way for many potential applications, such as all optical-logic gates [22], pulse compression [23, 24], all-optical switching [25], and limiting [26]. Thus, the pulse propagation in a nonlinear periodic structure (FBG) exhibits different kinds of fundamentally unique and technologically interesting regimes. Among these regimes, in this chapter we concentrate only on the intriguing solitary waves existing in FBGs, or gap solitons, and solitons propagating near the bandgap edge, or Bragg grating solitons.

There are basically two conditions under which one can describe the formation of solitons in FBG. The first one is based on the high intensity pulse propagation (refractive index modulation is weak, i.e., $\alpha P_0 \gg \kappa$) in FBG where NLCM equations are used to describe a coupling between forward and backward propagating modes. The second one deals with the low intensity pulse propagation in FBG where we assume the peak intensity of the pulse to be small enough so that the nonlinear index change $n_2 I$ is much smaller than the maximum value of δn, that is, $\alpha P_0 \ll \kappa$. Under the low intensity limit, the NLCM equations can be reduced to the NLS equation by using a technique known as multiple scale analysis.

In the following sections, we investigate in detail these solitons in FBG. We use the Stokes parameter formalism to investigate the gap solitons, and using the Bloch wave (method of multiple scales) analysis, we analyze the Bragg grating solitons propagating near the frequency bandgap. Note that gap solitons are obtained from the NLCM equations and their spectra lie within the PBG structure. As discussed earlier, there is another class of solitons, called Bragg solitons, whose frequencies fall close to but outside the band edge of the PBG. These are obtained from the approximated NLS-type equation that results from reducing the NLCM equations using multiple scale analysis. Generally speaking, the gap solitons are a special class of Bragg grating solitons.

7. MAP SOLITONS

7.1. Gap Solitons in Kerr Media

The stationary properties of one-dimensional Bragg gratings were first analyzed by Winful et al. [27]. Several leading groups investigated and reported the formation of these solitons in FBG [28–40]. Chen and Mills [28] coined the term *gap soliton* in their numerical work covering the nonlinear optical response of super lattices. Mills and Trullinger [29] obtained an analytical solution for stationary gap solitary waves. Sipe and Winful [30], Christodoulides and Joseph [31], Aceves and Wabnitz [32], de Sterke and Sipe [33], and recently our group [34–36] obtained analytical solutions for the Bragg solitons. These solitons in FBG have been extensively reviewed [33, 37–39] and were experimentally verified and demonstrated independently by Eggleton et al. [40, 41] and Taverner et al. [42]. Feng and Kneubuhl [43] found the in-gap and out-gap solitary wave solutions for the NLCM equations and showed the existence of bright and dark solitons in the out-gap solution. Conti and Trillo [44] investigated the bright GS with high amplitude close to the upper edge of the frequency bandgap. Moreover, they showed that this gap soliton acquires a Lorentzian shape and then gets bifurcated into the dark–antidark pairs outside the bandgap. They explained the bifurcation of this gap soliton in terms of catastrophe theory.

Theoretical results have been reported for the self-induced transparency in the extended Bragg reflector (resonant interaction between the counter-propagating waves and two-level atoms). That is, formation of gap solitons have been discovered in a periodic array of thin layers of resonant two-level systems separated by half-wavelength nonabsorbing dielectric layers, that is, a resonantly absorbing Bragg reflector [39]. Opatrny et al. [45] investigated both bright and dark gap solitons depending on the initial conditions in resonantly absorbing gratings.

Here, we discuss the construction of explicit analytical solutions of bright and dark gap solitons under the Bragg resonance condition using Stokes parameters. That is, we show that the exact balance between Kerr nonlinearity and grating-induced dispersion forms the bright and dark localized soliton envelopes whose carrier frequencies lie well within the PBG of the periodic medium. Further, we show that another class of bright soliton exists for arbitrary values of the detuning parameter.

In the presence of Kerr nonlinearity, using CMT, the NLCM equations can be written as

$$i\frac{\partial \vec{E}_f}{\partial z} + i\frac{\bar{n}}{c}\frac{\partial \vec{E}_f}{\partial t} + \kappa \vec{E}_b + \left(\Gamma_s\left|\vec{E}_f^2\right| + 2\Gamma_x\left|\vec{E}_b^2\right|\right)\vec{E}_f = 0$$
$$-i\frac{\partial \vec{E}_b}{\partial z} + i\frac{\bar{n}}{c}\frac{\partial \vec{E}_b}{\partial t} + \kappa \vec{E}_f + \left(\Gamma_s\left|\vec{E}_b^2\right| + 2\Gamma_x\left|\vec{E}_f^2\right|\right)\vec{E}_b = 0$$
(46)

where E_+ and E_- are the slowly varying amplitudes of forward and backward propagating waves, $\bar{n}$ is the average refractive index, and Γ_s and Γ_x are SPM and cross-phase modulation terms. In Eq. (46) the material and waveguide dispersive effects are not included, because the dispersion arising from the periodic structure dominates the rest near Bragg resonance condition. We note that the above NLCM equations are valid only for wavelengths close to the Bragg wavelength. Now, we substitute the stationary solution to the above coupled-mode equations by assuming $E_\pm(z,t) = e_\pm(z)e^{-i\hat{\delta}ct/\bar{n}}$. Using the stationary solution in Eq. (46), we obtain

$$i\frac{de_f}{dz} + \hat{\delta}e_f + \kappa e_b + \left(\Gamma_s|e_f|^2 + 2\Gamma_x|e_b|^2\right)e_f = 0$$
$$-i\frac{de_b}{dz} + \hat{\delta}e_b + \kappa e_f + \left(\Gamma_s|e_b|^2 + 2\Gamma_x|e_f|^2\right)e_b = 0$$
(47)

Equation (47) represents the time-independent light transmission through the grating structure, and it has been extensively investigated by many authors. The NLCM equations are nonintegrable in general. However, in a few cases, NLCM equations have exact analytical solutions representing the solitary wave solutions. For the first time, Christodoulides and Joseph [31] obtained the soliton solution to the NLCM equations, called the slow Bragg soliton, under the integrable massive Thirring model where the SPM and detuning parameters equal zero. Even though it is integrable, it does not find any direct application in nonlinear optics. However, using suitable transformation, it is used in nonlinear optics as a simple model to explain the self-induced transparency effect. The most general form of the solitary wave solutions to the NLCM equations was derived by Aceves and Wabnitz [32]. de Sterke and Sipe [33] also extensively investigated the NLCM equations in a fruitful review article on gap solitons. In their review article, for the first time, using the Stokes parameters they derived the relation for energy density (stationary solution) for the NLCM equations in terms of the Jacobi elliptic function. It should be noted that prior to this review, this Stokes parameter formalism was applied to the nonlinear directional coupler by Diano et al. [46].

Apart from these stationary solutions, some interesting soliton-like solutions are also possible. In FBG, these soliton-like solutions for the NLCM equations carry a lot of practical importance and hence form a major discussion in this chapter. For our investigation, we took the well-known and well-studied NLCM equations in FBG. At this juncture, using the Stokes parameters, we analyze the above NLCM equations in terms of two different conditions as follows. First, we analyze NLCM equations under the Bragg resonance condition, where $\omega_B = \omega_0$ and hence the detuning parameter $\hat{\delta} = 0$, and then analyze the second condition, that is, when the system is far away from the Bragg resonance condition (where the carrier frequency lies far away from the Bragg resonance, i.e., $\omega_B \neq \omega_0$). Hence, in the second case, the detuning parameter $\hat{\delta} \neq 0$ and can have any other arbitrary values. Thus, based on the detuning parameter we discuss the formation of bright and dark solitons in FBG.

To investigate these solitons, we choose the Stokes parameters because they provide useful information about the total energy and energy difference between the forward and backward propagating modes. For our further investigation, we use the following Stokes parameters [47]

$$A_0 = |e_+|^2 + |e_-|^2, \quad A_1 = e_+ e_-^* + e_+^* e_-,$$
$$A_2 = i(e_+ e_-^* - e_+^* e_-), \quad A_3 = |e_+|^2 - |e_-|^2$$

with the constraint $A_0^2 = A_1^2 + A_2^2 + A_3^2$. In FBG theory, the above NLCM equations require that the total power $P_0 = A_3 = |e_+|^2 - |e_-|^2$ inside the grating is a constant along the grating structure. Now, rewriting the NLCM equations in terms of the Stokes parameters, we get

$$\frac{dA_0}{dz} = -2\kappa A_2, \quad \frac{dA_1}{dz} = 2\hat{\delta}A_2 + 3\Gamma A_0 A_2$$
$$\frac{dA_2}{dz} = -2\hat{\delta}A_1 - 2\kappa A_0 - 3\Gamma A_0 A_1, \quad \frac{dA_3}{dz} = 0 \quad (48)$$

In the above equation, we drop the distinction between the SPM and cross-phase modulation effects and hence we have $3\Gamma = 2\Gamma_\times + \Gamma_s$. As expected, it is explicitly clear from Eq. (48) that the total power P_0 ($=A_3$) inside the grating is found to be constant or conserved along the grating structure.

To construct the bright and dark gap solitons from the NLCM equations, after constructing the suitable conserved quantity, we derive the anharmonic oscillator type equation from Eq. (48). As already discussed, we analyze the anharmonic oscillator under two different conditions. First, we consider the simple case of the Bragg resonance condition where the detuning parameter is zero. Under this condition, the anharmonic oscillator equation reduces to the well-known unforced and undamped Duffing oscillator equation. By solving this equation, we provide the explicit analytical solutions for both bright and dark gap solitons of the NLCM equations. In the second case, we apply another condition in which the detuning parameter $\hat{\delta}$ is not zero, that is, for arbitrary values of $\hat{\delta}$, we construct another family of soliton solution for NLCM equations. Before turning to this investigation, we discuss the construction of the anharmonic

oscillator type equation. For this, it is necessary to use the conserved quantity, and it is obtained in the form $\hat{\delta}A_0 + \frac{3}{4}\Gamma A_0^2 + \kappa A_1 = C$, where C is the constant of integration. Using this in Eq. (48), we obtain

$$\frac{d^2 A_0}{dz^2} - \alpha A_0 + \beta A_0^2 + \gamma A_0^3 = 4\hat{\delta}C \quad (49)$$

where $\alpha = \left[-4\hat{\delta}^2 + 4\kappa^2 + 6\Gamma C\right]$, $\beta = 9\hat{\delta}\Gamma$, and $\gamma = \frac{9}{2}\Gamma^2$. It should be noted that the reduced system [see Eq. (49)] allows us to find the soliton-type solutions of the NLCM equations in FBG. Equation (49) describes the motion of a particle moving with the classic anharmonic potential,

$$V(A_0) = -\alpha \frac{A_0^2}{2} + \beta \frac{A_0^3}{3} + \gamma \frac{A_0^4}{4} \quad (50)$$

Equation (50) represents the potential energy distribution in the FBG structure. At this juncture, it should be emphasized that the potential function in our system is found to be the same as the one predicted by Conti and Mills [44]. As discussed above, we discuss Eq. (49) under two different conditions as follows. For the first condition, Eq. (49) reduces to the well-known Duffing oscillator type equation

$$\frac{d^2 A_0}{dz^2} - \alpha A_0 + \gamma A_0^3 = 0 \quad (51)$$

The stationary solutions in terms of elliptic function to the above equation are well known [46]. By applying suitable conditions to the physical parameters of Eq. (51), we also get interesting bright and dark optical solitons in the periodic media. To construct the bright soliton envelope within PBG, we choose the positive nonlinearity. It is well known that the periodic structure induces the dispersion and here it is negative, that is, it is anomalous dispersion. Under suitable conditions the grating-induced dispersion gets exactly balanced with the positive nonlinearity. As a result, there is a formation of a bright soliton within PBG called the *bright gap soliton*. To obtain this, we assume $\alpha < 0$, $\gamma > 0$ and the resulting constant of integration to be zero, and hence we obtain

$$\frac{1}{2}\left(\frac{dA_0}{dz}\right)^2 - \alpha \frac{A_0^2}{2} + \gamma \frac{A_0^4}{4} = 0 \quad (52)$$

From the above equation, we get the bright soliton solution as

$$A_0 = \sqrt{\frac{2\alpha}{\gamma}} \operatorname{sech}(\sqrt{\alpha})z$$
$$= \sqrt{\frac{8(2\kappa^2 + 3\Gamma C)}{9\Gamma^2}} \operatorname{sech}\left(\sqrt{(4\kappa^2 + 6\Gamma C)}\right)z \quad (53)$$

Under the above-mentioned conditions, the frequencies within the PBG form an envelope after the exact balancing of grating-induced dispersion with nonlinearity. Thus, for the frequencies within the PBG, the forward wave decays and hence the energy is now completely transferred to the backward wave by the Bragg reflection process.

From the total energy of the system, we find that the potential energy function has two symmetrical minima at $A_0 = \pm 1$ and these minima are separated by a local maximum at $A_0 = 0$ as shown in Fig. 17.9a. To understand the motion of the photons in the double-well potential, we also depict the phase-plane diagram as shown in Fig. 17.9b.

In addition to the bright gap soliton, there is another type of soliton for the same parameters of NLCM equations called the dark gap soliton, that is, a "hole" of a fixed shape in a continuous wave background field of constant intensity, which has gained considerable physical interest in the soliton-based communication system. To generate the dark soliton, we consider the negative nonlinearity, that is, it is self-defocusing. In this case, the grating-induced dispersion is positive. So we have the formation of a dark soliton envelope within the PBG after exact balancing of these two effects and the corresponding envelope is called *dark gap soliton*. Here, assuming $\alpha > 0$, $\gamma < 0$ and the constant of integration is $\alpha^2/4\gamma$, we get

$$\frac{1}{2}\left(\frac{dA_0}{dz}\right)^2 + \alpha \frac{A_0^2}{2} - \gamma \frac{A_0^4}{4} = \frac{\alpha^2}{4\gamma} \quad (54)$$

By solving the above equation, we get the dark soliton solution of the form

$$A_0 = \sqrt{\frac{\alpha}{\gamma}} \tanh\left(\sqrt{\frac{\alpha}{2}}\right) z$$
$$= \sqrt{\frac{4(2\kappa^2 + 3\Gamma C)}{9\Gamma^2}} \tanh\left(\sqrt{(2\kappa^2 + 3\Gamma C)}\right) z \quad (55)$$

We generally call the solutions in Eqs. (53) and (55) gap solitons of bright and dark type, based on the fact that the field envelope function inside the medium resembles the hyperbolic secant and tangent functions of the NLS equation, which are used to describe the propagation of optical soliton in fibers [1–5]. Here also the forward wave decays, and consequently the energy is now transferred to the backward wave through the process of Bragg reflection as discussed in the case of the bright gap soliton. In this case, the potential function has two symmetric maxima at $A_0 = \pm 1$ and has a local minimum at $A_0 = 0$ as shown in Fig. 17.10. Here, it should be noted that the potential well plots for bright and dark gap solitons are found to be the same as those depicted in resonantly absorbing gratings [39, 45].

Like Iizuka and Wadati [48], we also derived the explicit analytical bright and dark gap soliton solutions (from the grating solitons as a special case) to the NLCM equations. Apart from the discussed bright and dark gap solitons, there is another family of bright soliton existing in the periodic structure for the arbitrary values of the detuning parameter. Under this criterion, we have to solve the anharmonic oscillator Eq. (49). As we are interested in constructing a bright gap soliton solution, we choose the constant of integration to be equal to zero and Eq. (49) takes the form similar to the form presented in Kazantseva et al. [49]:

$$\frac{d^2 A_0}{dz^2} - \alpha A_0 + \beta A_0^2 + 2\gamma A_0^3 = 0 \quad (56)$$

On integrating the above equation, we obtain

$$\left(\frac{dA_0}{dz}\right)^2 - \alpha A_0^2 + \frac{2}{3}\beta A_0^3 + \gamma A_0^4 = C_2 \quad (57)$$

where C_2 is an arbitrary constant of integration. As stated above, once again we choose the constant of integration as zero, so that the Eq. (57) yields

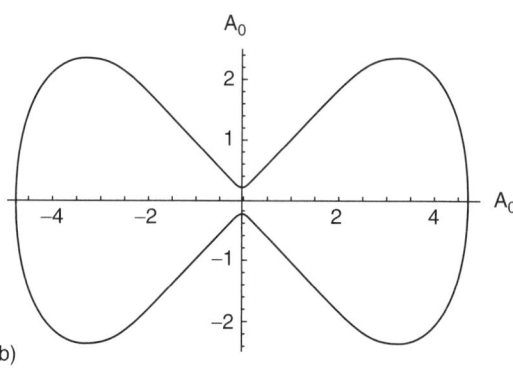

FIGURE 17.9 (a) Under Bragg resonance condition the system possesses double-well potential (for the bright gap soliton) when $\alpha = 0.5$, $\beta = 0$, and $\gamma = 0.23$. (b) Phase-plane diagram for potential function when $\alpha = 0.5$, $\beta = 0$, and $\gamma = 0.23$.

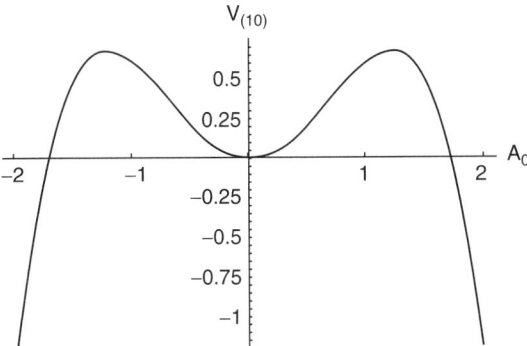

FIGURE 17.10 The system possesses a double hump potential well when $\alpha = 0.52$, $\beta = 0$, and $\gamma = 0.3$.

$$\frac{dA_0}{dz} = \pm A_0 \sqrt{\alpha - \frac{2}{3}\beta A_0 - \gamma A_0^2} \quad (58)$$

When we substitute $A_0 = \frac{1}{y}$, the above equation transforms into the following new form:

$$\frac{dy}{dz} = \sqrt{\alpha \left[\left(y - \frac{\beta}{2\alpha}\right)^2 - \frac{\beta^2 + 9\alpha\gamma}{9\alpha^2} \right]} \quad (59)$$

We introduce a new dependent variable ξ, defined by

$$y - \frac{\beta}{2\alpha} \equiv \sigma \left(\frac{\beta^2 + 9\alpha\gamma}{9\alpha^2}\right)^{1/2} \xi(z) \quad (60)$$

In terms of ξ, Eq. (59) can be written as $\sigma \frac{d\xi}{dz} = \pm\sqrt{\alpha(\xi^2 - 1)}$, from which we find $\xi(z) = \cosh(\sqrt{\alpha})z$. Now substituting the $\xi(z)$ value in Eq. (60) and rearranging the terms we obtain

$$A_0 = \frac{3(4\hat{\delta}^2 - 4\kappa^2 - 6\Gamma C)}{\sqrt{(9\hat{\delta}\Gamma)^2 + 9(4\hat{\delta}^2 - 4\kappa^2 - 6\Gamma C)\left(\frac{9}{2}\Gamma^2\right)}} \quad (61)$$
$$\cdot \cosh\sqrt{(4\hat{\delta}^2 - 4\kappa^2 - 6\Gamma C)}z + 1$$

Similarly, we can also construct another form of bright soliton solution for negative quadratic nonlinearity by performing suitable mathematical manipulations as above

$$A_0 = \frac{3(4\hat{\delta}^2 - 4\kappa^2 - 6\Gamma C)}{\sqrt{(9\hat{\delta}\Gamma)^2 + 9(4\hat{\delta}^2 - 4\kappa^2 - 6\Gamma C)\left(\frac{9}{2}\Gamma^2\right)}} \quad (62)$$
$$\cdot \cosh\sqrt{(4\hat{\delta}^2 - 4\kappa^2 - 6\Gamma C)}z + 1$$

Equations (61) and (62) represent the bright soliton solutions for the arbitrary values of the detuning parameter. Because of the presence of the detuning parameter in the presented bright soliton solutions, some of the spectral components of the carrier wave do not obey Bragg resonance conditions. This means that some of them lie outside the PBG structure and hence they are transmitted. As a result, we have the transmitted wave in addition to the reflected one. Therefore, in this case, because of the violation of the Bragg resonance condition, energy is shared between backward and forward propagating modes. As a result, the symmetry between the counterpropagating waves is broken [44, 50].

7.2. Gap Solitons in Quadratic Media

Apart from the studies on gap solitons due to Kerr nonlinearity, the studies on gap solitons due to quadratic nonlinearity have also gained momentum in recent times. It has been shown that a Bragg grating with quadratic or second-harmonic generation nonlinearity can also give rise to a rich spectrum of solitons [51–60].

In a quadratic medium, a forward propagating beam at ω generates a beam at 2ω in the backward direction. In a quadratic medium where the susceptibility $\chi^{(2)}$ is periodic, the forward propagating field E_f at frequency ω ($=\omega_0 + \Delta\omega$) and backward propagating field E_b at frequency 2ω obey the following CMT [51]:

$$-i\frac{d\vec{E}_f}{dz} = \chi_2(z)\vec{E}_b\vec{E}_f^* e^{i\Delta kz}, \quad i\frac{d\vec{E}_b}{dz} = \chi_2(z)\vec{E}_f^2 e^{-i\Delta kz}$$

where $\Delta k = \Delta k(\omega) = \beta_2(\omega) + 2\beta(\omega)$ is the nonlinear wavevector mismatch, and $\chi_2(z)$ is Λ periodic susceptibility that is normalized in such a way that $|E_{f,b}|^2$ yields the intensity.

This model has been shown to possess a number of remarkable features. That is, in second-harmonic generation nonlinear grating, the gap in which solitons may exist is partly empty, only part of it being filled with actual existing soliton solutions, which is contrary to the Kerr nonlinear Bragg grating [55]. Kivshar [56] has shown that gap solitons of a genuine parametric nature can propagate in quadratic media, even far from the cascading limit. These gap solitons are two-color envelopes bound together through second-harmonic generation. Conti et al. [60] demonstrated the localized energy states of these two-color gap solitons in a periodic structure in the presence of a frequency doubling nonlinearity. The quadratic gap solitons are experimentally observed in homogeneous media [61, 62]. In this review, we also show the existence of quadratic-type soliton solutions in FBG. This can be achieved in our model only when $\gamma \to 0$ in Eq. (49). Subsequently, our model leads to purely quadratic nonlinearity and the corresponding soliton solution can be written as

$$A_0 = \frac{3\alpha}{2\beta \cosh^2(\sqrt{\alpha})z/2} \quad (63)$$

Under the above criterion, another form of bright gap soliton leads to

$$A_0 = \frac{3\alpha}{2\beta \sinh^2(\sqrt{\alpha})z/2} \quad (64)$$

The bright gap soliton solutions (63) and (64) resemble the solutions of two coupled nonlinear equations for the envelopes of the fundamental and second-harmonic field components. The above investigated quadratic-like bright gap solitons can also be compared with the already known results [63–65].

8. BLOCH WAVE ANALYSIS

As we discussed in Section 5.3.1.3, the solutions to the linearized set of equations can be written as a suitable linear combination of the eigenvectors $v_\pm$ each multiplied by the plane-wave factor $e^{i(qz - c\delta t/\bar{n})}$.

Now returning to the fully nonlinear system of Eqs. (38), it is clear that at any given point in space and time, we can write the solution to the general system as a linear combination of the eigenvectors $v_\pm$ with their associated plane-wave factors, but the amplitudes are not constant now. Further, our Bloch wave solutions satisfy the linearized Eq. (9) but do not satisfy the nonlinear Eq. (38). Therefore, we assume that the nonlinearity is weak and these field amplitudes vary on much slower scales than k_0^{-1} and ω_0^{-1}. We can use several mathematical techniques to separate the slowly varying amplitudes from the rapidly varying plane-wave functions. This leads to the idea of searching for solutions that vary on different length and time scales. To implement this idea we use a technique called the method of multiple scales. As already shown [32,33], the electric field envelope does satisfy the NLS equation. The soliton solution of this NLS-type equation at the edges of the PBG is termed the Bragg soliton. Using the multiple scale analysis, we discuss the investigation of such a Bragg grating soliton at both the upper and lower branches of the dispersion curves.

8.1. Bragg Grating Solitons

Here, we consider pulse propagation under the low intensity limit and hence we reduce the NLCM equations to a well-known NLS-type equation that describes the nonlinear pulse propagation at the edges of the PBG structure. In addition, while study-

ing the impact of nonlinearity on the pulse propagation through FBG, the higher order linear and nonlinear effects such as TOD and self-steepening gain importance. Subsequently, the NLS equation becomes the perturbed NLS (PNLS) equation. The solitons existing near the PBG edge are called Bragg grating solitons. Now we turn to investigate these solitons by solving the NLS and PNLS equations through the coupled amplitude-phase method and then discuss the qualitative difference between them.

8.1.1. Theoretical Model

The nonlinear pulse propagation in a periodic nonlinear structure consisting of N alternating layers with different linear refractive indices and different Kerr nonlinearities governed by the NLCM equations of the form [20, 21]

$$i\frac{\partial \vec{E}_f}{\partial z} + i\frac{\partial \vec{E}_f}{\partial t} + \kappa \vec{E}_b + \alpha\left(|\vec{E}_f|^2 + 2|\vec{E}_b|^2\right)\vec{E}_f$$
$$+ \beta\left[\left(2|\vec{E}_f|^2 + |\vec{E}_b|^2\right)\vec{E}_b + \vec{E}_f^2 \vec{E}_b^*\right] = 0$$
$$-i\frac{\partial \vec{E}_b}{\partial z} + i\frac{\partial \vec{E}_b}{\partial t} + \kappa \vec{E}_f + \alpha\left(|\vec{E}_b|^2 + 2|\vec{E}_f|^2\right)\vec{E}_b$$
$$+ \beta\left[\left(2|\vec{E}_b|^2 + |\vec{E}_f|^2\right)\vec{E}_f + \vec{E}_b^2 \vec{E}_f^*\right] = 0 \quad (65)$$

where $\vec{E}_f$ and $\vec{E}_b$ are the slowly varying amplitudes of forward and backward propagating waves. The term $\kappa\ (=n_{0k})$ is the variance of the linear refractive index or strength of the linear grating, $\alpha\ (=n_{nl})$ is the average Kerr nonlinearity across the structure, and $\beta\ (=n_{2k})$ is the variance of the Kerr nonlinearity between the layers. Pelinovsky et al. [20] and Brzozowski and Sargent [21] extensively investigated the above equations to study the all-optical limiting in the nonlinear periodic structures.

Before embarking into the discussion of NLCM equations, first we briefly discuss pulse propagation in the linear regime under which the nonlinear effects α and β become zero. Now, we discuss the dispersion relation associated with the LCM equations as follows. In the linear case, the solutions to the LCM equations are given by

$$A_\pm = U_\pm e^{i(k_z z - \Omega t)} + c.c \quad (66)$$

where k_z and Ω satisfy the following dispersion relation $\Omega^2 = \kappa^2 + k_z^2$. In this relation, when we impose the condition $k_z = 0$, we get $\Omega = \pm \kappa$. Under this condition, the linear problem has a solution of the form

---

c.c: complex conjugate.

$$\begin{pmatrix} A_+ \\ A_- \end{pmatrix} = \begin{pmatrix} 1 \\ -1 \end{pmatrix} e^{-i\kappa t} + c.c \tag{67}$$

It is seen that the above solution satisfies the relation

$$L \begin{pmatrix} A_+ \\ A_- \end{pmatrix} = 0$$

Here L represents the operator and is given by

$$L = \begin{pmatrix} i\frac{\partial}{\partial t} & \kappa \\ \kappa & i\frac{\partial}{\partial t} \end{pmatrix}$$

When we introduce the nonlinearity into the system, LCM equations change to nonlinear coupled-mode equations. The construction of bright and dark soliton solutions in the upper and lower branches of the dispersion curve follows from the multiple scale analysis, which is discussed next.

8.1.2. Multiple Scale Analysis

As discussed earlier, to investigate the formation of solitons near the band edge, the carrier frequency of the laser pulse is detuned out of PBG. The process of detuning the laser pulse outside the band edge can be achieved in several ways, such as the reductive perturbation method and method of multiple scale analysis. Recently, under the low intensity limit, Aceves [66] derived a perturbed two-dimensional NLS equation to describe the gap soliton bullets in the Kerr-type planar waveguides by the method of multiple scale analysis. Therefore, in this chapter, to describe the pulse propagation in FBG with deep index modulation, we also adopt the same analysis. This would also mean that the center frequency of the pulse is being detuned out of the PBG structure. Now the pulse propagation in FBG is governed by the NLS-type equation, which in turn can be easily integrable by the standard inverse scattering transform.

First we find the solution of the governing equation in the linear case and then introduce the nonlinearity into the coupled-mode equations. To introduce the multiple scale analysis, we extend the linear solution to the following form:

$$\begin{pmatrix} \vec{E}_f \\ \vec{E}_b \end{pmatrix} = \varepsilon^{1/2} \vec{E}(\tau_1, \tau_2, X, Z) \begin{pmatrix} 1 \\ -1 \end{pmatrix} e^{-i\kappa t} \tag{68}$$
$$+ \varepsilon U_1 + \varepsilon^{3/2} U_2 + \varepsilon^2 U_3 + \cdots$$

where $\tau_1 = \varepsilon t$, $\tau_2 = \varepsilon^2 t$, $X = \varepsilon^{1/2} x$, and $Z = \varepsilon^{1/2} z$. Now we proceed to solve for $(\vec{E}_f, \vec{E}_b)$ for successive orders in ε. Balancing the $O(\varepsilon)$ terms gives

$$LU_1 = -i \frac{\partial \vec{E}}{\partial z} \begin{pmatrix} 1 \\ 1 \end{pmatrix} e^{-i\kappa t} \tag{69}$$

The solution of the above equation is found to be

$$U_1 = -\frac{i}{2\kappa} \frac{\partial \vec{E}}{\partial z} \begin{pmatrix} 1 \\ 1 \end{pmatrix} e^{-i\kappa t} \tag{70}$$

Next, we turn to compute the higher order corrections to $(\vec{E}_f, \vec{E}_b)$. Balancing the $O(\varepsilon^{3/2})$ terms gives

$$LU_2 = \left(-i \frac{\partial \vec{E}}{\partial \tau_1} - \frac{1}{2\kappa} \frac{\partial^2 \vec{E}}{\partial z^2} - (3\alpha - 4\beta)|\vec{E}|^2 \vec{E} \right) \tag{71}$$
$$\times \begin{pmatrix} 1 \\ -1 \end{pmatrix} e^{-i\kappa t} + c.c$$

To solve the above equation, the secular terms should be equated to zero and therefore

$$i \frac{\partial \vec{E}}{\partial \tau_1} + \frac{1}{2\kappa} \frac{\partial^2 \vec{E}}{\partial z^2} + (3\alpha - 4\beta)|\vec{E}|^2 \vec{E} = 0 \tag{72}$$

The above equation represents the well-known NLS equation, which is used to describe the picosecond pulse propagation in the fiber. In FBG, Eq. (72), represents the pulse propagation outside the PBG structure. In Eq. (72), the variable E represents the amplitude of the envelope associated with the Bloch wave formed by a superposition of E_f and E_b. For the first time, Sipe and Winful [30] derived this kind of NLS equation from the NLCM equations. de Sterke and Sipe [33, 67] derived the NLS equation based on the envelope function approach and also presented the soliton solution outside the PBG but within FBG. Without deriving the NLS equation from NLCM equations, Feng and Kneubuhl [43] investigated the formation of new types of solitary wave solutions called out-gap solitary wave solutions (bright and dark) in the periodic structure. Iizuka and Wadati [48] used the reductive perturbation method and derived a similar NLS equation in FBG. They obtained the explicit form of both the bright and dark analytical soliton solutions in FBG, which are presented below:

$$\vec{E} = \sqrt{\frac{1}{\kappa r}} C_1 \operatorname{sech}\left(C_1 z - \frac{C_1 C_2}{\kappa} \tau_1 \right) \tag{73}$$
$$\times \exp i \left\{ C_2 z + \frac{1}{2\kappa}(C_1^2 - C_2^2)\tau_1 \right\}$$

$$\vec{E} = \sqrt{-\frac{1}{\kappa r}} C_1 \frac{1 + \exp(C_3 z - C_4 \tau_1 + i\phi)}{1 + \exp(C_3 z - C_4 \tau_1)} \tag{74}$$
$$\times \exp i \left\{ C_2 z - \frac{1}{2\kappa}(2C_1^2 + C_2^2)\tau_1 \right\}$$

where $r = (3\alpha - 4\beta)$.

The factors C_1, C_2, C_3, and C_4 are constants. The constant C_4, for $\hat{\delta} = 0$, is given by

$$C_4 = (C_3/2\kappa)\left[2C_2 + \sqrt{4C_1^2 - C_3^2}\right],$$

$$\exp(i\phi) = \frac{C_3 + i\sqrt{4C_1^2 - C_3^2}}{C_3 - i\sqrt{4C_1^2 - C_3^2}}$$

Equations (73) and (74) represent the solutions of the bright and dark Bragg solitons. These solitons were simply referred to as grating solitons by Iizuka and Wadati [48]. Further, they showed that these grating solitons become gap solitons when the group velocity becomes zero. Thus, Eq. (72) has been extensively investigated by many authors.

8.1.3. Grating Solitons due to GVD

The solitons discussed below arise as a result of interplay between anomalous grating-induced quadratic dispersion (normal grating-induced dispersion) and focusing (defocusing) nonlinearities. By using the coupled amplitude-phase method, we solve the NLS equation and end up with the formation of either bright or dark Bragg grating solitons depending on whether the laser carrier frequency is detuned to the upper branch or the lower branch of the PBG structure. For this purpose, we rewrite Eq. (72) in terms of the normalized form

$$i\frac{\partial \vec{E}}{\partial t} + \beta_2 \frac{\partial^2 \vec{E}}{\partial z^2} + \Gamma |\vec{E}|^2 \vec{E} = 0 \quad (75)$$

where $\beta_2 = \frac{1}{2\kappa}$ and $\Gamma = (3\alpha - 4\beta)$. The coefficient β_2 represents the GVD and Γ represents Kerr nonlinearity due to nonlinear polarization. Now we solve Eq. (75) by the coupled amplitude-phase method, which has been introduced in [68,69]. To start with, we consider the solution of the form

$$\vec{E}(z,t) = Q(t + \beta_1 z) \exp\left[i(kz - \omega t)\right] \quad (76)$$

where the function Q is a real one. The unknown parameters k and ω are directly related to the shifts in the wavenumber and frequency respectively, whereas the factor β_1 is the group velocity of the wave. Using Eq. (76) in Eq. (75) and removing the exponential term, we get

$$i\beta_1 Q_x - kQ + \beta_2[Q_{xx} - 2i\omega Q_x - Q\omega^2] + \Gamma Q^3 = 0$$

Now, separating the real and imaginary parts, we have

$$kQ - \beta_2(Q_{xx} - \omega^2 Q) - \Gamma Q^3 = 0 \quad (77)$$

$$(\beta_1 - 2\beta_2\omega)Q_x = 0$$

Integrating and rearranging Eq. (12), we get

$$\left(\frac{dQ}{d\chi}\right)^2 = \left(\frac{k - \omega^2 \beta_2}{\beta_2}\right)Q^2 - \left(\frac{\Gamma}{2\beta_2}\right)Q^4 + C \quad (78)$$

where C is an arbitrary constant of integration. From the above equation, it is possible to get the different analytical solutions for different values of the constant of integration C. Among these solutions, our aim is to investigate only the bright and dark soliton solutions. Now, we discuss the formation of bright solitons outside the PBG but inside the FBG. Thereafter, we apply the same condition to the physical parameters in Eq. (78) and finally we obtain the bright soliton solution analytically. To discuss the bright soliton formation, we choose the positive nonlinearity and hence we have the self-focusing effect in the FBG structure. Because of self-focusing effect, the central frequency of the carrier wave is tuned close to but outside the PBG of the periodic structure. This physically means that the central frequency is shifted to the upper branch of the dispersion curve, where the grating-induced GVD is anomalous. This anomalous GVD gets exactly balanced with the positive nonlinearity (self-focusing effect), and as a result, we have the bright soliton formation outside the PBG but inside the periodic (FBG) structure, which is termed the bright Bragg soliton. As we consider positive nonlinearity in the formation of bright soliton, we choose the cubic nonlinear term as positive and $C = 0$ in the above equation and obtain the following bright soliton solution:

$$\vec{E} = \sqrt{\frac{2(k - \beta_2\omega^2)}{\Gamma}} \operatorname{sech}\left(\sqrt{\frac{(k - \beta_2\omega^2)}{\beta_2}}\right) \chi e^{i(kz - \omega t)} \quad (79)$$

In the above case, the envelope satisfies the bright soliton whose carrier frequencies lie close to but outside the band edge of the PBG and hence we call it the bright Bragg soliton. It should be noted that the existence of solitary waves in the upper branch of the dispersion curve has already been experimentally demonstrated [40, 41].

Similarly, there is another interesting class of solitons called dark solitons, and now we discuss the formation of the same in the FBG. Instead of positive nonlinearity, we consider the negative nonlinearity, which gives rise to the self-defocusing effect in the FBG. This self-defocusing effect shifts the central frequency of the carrier wave to the lower branch of the dispersion curve where we have normal GVD. This normal GVD gets exactly balanced with the negative nonlinearity (self-defocusing effect) and as a result we get the dark soliton formation outside the PBG but inside the FBG structure. This soliton is referred to as the dark Bragg soliton. For analytical purposes, considering the negative nonlinearity, the constant in Eq. (78) is chosen in such a way that the value of the expression inside the square root is a perfect square and hence we obtain the dark solitary wave solution as follows:

$$\vec{E} = \sqrt{\frac{(k-\beta_2\omega^2)}{\Gamma}} \tanh\left(\sqrt{\frac{(k-\beta_2\omega^2)}{2\beta_2}}\right) \chi e^{i(kz-\omega t)} \tag{80}$$

As stated above, here also the field satisfies the dark soliton solution and their carrier frequencies are close to the band edge, and hence we call the above solution the dark Bragg soliton solution.

8.1.4. Grating Solitons due to TOD

In the previous section, we discussed the formation of Bragg grating solitons on the upper and lower branches of the dispersion curve by making use of second-order dispersion and nonlinearity. As pointed out in Section 2, the grating not only exhibits second dispersion but also cubic dispersions and HOD. We know that cubic dispersion in the grating is 10 orders greater than the dispersion in the conventional fiber. Hence, it cannot be neglected and should be considered for further investigation. It was shown that TOD (alone) makes asymmetry and affects the formation of Bragg soliton compression. Therefore, in the following section, for the first time, we discuss the generation of Bragg grating solitons by making use of higher linear (TOD) and nonlinear effects (self-steepening).

Aceves [66], in his recent work, considered higher order effects in FBG and hence derived perturbed NLS equations to describe gap soliton bullets in planar waveguides. Based on this fact, to analyze the impact of nonlinearity on the pulse propagation, we also consider higher order effects in FBG and derive the perturbed NLS equation. Then, we solve it for studying the formation of bright and dark Bragg solitons on both upper and lower branches of the dispersion curve in FBG depending on the sign of nonlinearity.

These solitons are mainly due to interplay between TOD of the grating and Kerr nonlinearity (self-steepening). To see the impact of higher order effects, we continue to balance $O(\varepsilon^2)$ terms, and this gives rise to

$$LU_3 = \left(i\frac{\alpha}{2\kappa}\left(2|\vec{E}|^2\frac{\partial \vec{E}}{\partial z} + A^2\frac{\partial \vec{E}^*}{\partial z}\right) - \frac{i}{2\kappa}\left(-i\frac{\partial \vec{E}}{\partial \tau_1}\right)\right)$$
$$\times \binom{1}{1} e^{-i\kappa t} + c.c \tag{81}$$

From Eq. (70), we have

$$\frac{\partial}{\partial z}\left(-i\frac{\partial}{\partial \tau_1}\right)\vec{E} = \frac{1}{2\kappa}\frac{\partial^3 \vec{E}}{\partial z^3} + (3\alpha - 4\beta)\left(2|\vec{E}|^2\frac{\partial \vec{E}}{\partial z} + E^2\frac{\partial \vec{E}^*}{\partial z}\right) \tag{82}$$

Therefore, the expression for U_3 can be written as

$$U_3 = -\frac{i}{4\kappa^2}\left[(3\alpha - 4\beta)\left(2|\vec{E}|^2\frac{\partial \vec{E}}{\partial z} + \vec{E}^2\frac{\partial \vec{E}^*}{\partial z}\right) + \frac{1}{2\kappa}\frac{\partial^3 \vec{E}}{\partial z^3}\right]\binom{1}{1} e^{-i\kappa t} + c.c \tag{83}$$

Equation (83) represents the perturbation terms that must be added to the NLS equation when we consider the higher order effects in the FBG structure. With this result, the NLS equation changes into the PNLS equation, which is presented below:

$$i\frac{\partial \vec{E}}{\partial \tau_1} + \frac{1}{2\kappa}\frac{\partial^2 \vec{E}}{\partial z^2} + (3\alpha - 4\beta)|\vec{E}|^2\vec{E} + \frac{1}{8\kappa^3}\frac{\partial^3 \vec{E}}{\partial z^3}$$
$$+ \frac{i}{4\kappa^2}(3\alpha - 4\beta)\left(2|\vec{E}|^2\frac{\partial \vec{E}}{\partial z} + \vec{E}^2\frac{\partial \vec{E}^*}{\partial z}\right) = 0 \tag{84}$$

The above PNLS equation represents the nonlinear pulse propagation in a periodic medium with higher order effects outside the PBG structure in FBG.

As stated above, in this section, we construct both bright and dark Bragg solitons for the PNLS equation. Using the coupled amplitude-phase method, we solve the PNLS equation and discuss formation of bright and dark Bragg solitons due to TOD and Kerr nonlinearity. For this purpose, we rewrite Eq. (84) in the form

$$i\frac{\partial \vec{E}}{\partial \tau_1} + a\frac{\partial^2 \vec{E}}{\partial z^2} + ib\frac{\partial^3 \vec{E}}{\partial z^3} + c|\vec{E}|^2\vec{E} + 2id\left(|\vec{E}|^2\vec{E}\right) = 0 \tag{85}$$

where $a = \frac{1}{2\kappa}$, $b = \frac{1}{8\kappa^3}$, $c = (3\alpha - 4\beta)$, and $d = \frac{(3\alpha-4\beta)}{4\kappa^2} = ca^2$. The coefficients a and b represent second- and third-order dispersions. The last term in the above equation accounts for self-steepening, which results from including the first derivative of the slowly varying part of the nonlinear polarization. Using Eq. (76) in Eq. (85) and removing the exponential term, we get

$$i\beta_1 Q_x - kQ + a[Q_{xx} - 2i\omega Q_x - Q\omega^2]$$
$$+ b[iQ_{xxx} + 3\omega Q_{xx} - 3i\omega^2 Q_x - Q\omega^3] + cQ^3$$
$$+ d[3iQ^2 Q_x + \omega Q^3] = 0$$

Now, separating the real and imaginary parts, we have

$$\beta_1 Q_x + bQ_{xxx} + (-2a\omega - 3b\omega^2 + 3dQ^2)Q_x = 0 \quad (86)$$

$$-kQ + (a + 3b\omega)Q_{xx} + (c + d\omega)Q^3$$
$$- (a\omega^2 + b\omega^3)Q = 0 \quad (87)$$

Because Eq. (86) possesses only third-order and first-order derivatives, it can be written in the following form:

$$bQ_{xxx} = (-\beta_1 + 2a\omega + 3b\omega^2 - 3dQ^2)Q_x$$

Integrating this, we get

$$Q_{xx} = \left(\frac{2a\omega + 3b\omega^2 - \beta_1}{b}\right)Q - \left(\frac{d}{b}\right)Q^3 \quad (88)$$

Writing Eq. (87) in the following form

$$Q_{xx} = \left(\frac{k + a\omega^2 + b\omega^3}{a + 3b\omega}\right)Q - \left(\frac{c + d\omega}{a + 3b\omega}\right)Q^3 \quad (89)$$

It is clear that the Eqs. (88) and (89) can be equivalent only under the following conditions:

$$\left(\frac{2a\omega + 3b\omega^2 - \beta_1}{b}\right) = \left(\frac{k + a\omega^2 + b\omega^3}{a + 3b\omega}\right) \quad \text{and}$$
$$\left(\frac{d}{b}\right) = \left(\frac{c + d\omega}{a + 3b\omega}\right)$$

From the above relations, we find ω and k as

$$\omega = \frac{cb - da}{2bd} \quad \text{and}$$
$$k = \frac{(2a\omega + 3b\omega^2 - \beta_1)(a + 3b\omega) - ab\omega^2 - b^2\omega^3}{b}$$
(90)

Equation (88) can also be written as

$$\left(\frac{dQ}{d\chi}\right)^2 = \left(\frac{2a\omega + 3b\omega^2 - \beta_1}{b}\right)Q^2 - \frac{1}{2}\left(\frac{d}{b}\right)Q^4 + C$$
(91)

where C is an arbitrary constant of integration. Here, it can be noted that the above equation has the same form as that of Eq. (78) for which the construction of soliton type solutions has been discussed in the previous section. As we consider positive nonlinearity in the formation of bright soliton, we choose the cubic nonlinear term as positive, and $C = 0$, in the above equation, and obtain the following bright soliton solution

$$\vec{E} = \sqrt{\frac{2(2a\omega + 3b\omega^2 - \beta_1)}{d}} \operatorname{sech}\left(\sqrt{\frac{2a\omega + 3b\omega^2 - \beta_1}{b}}\right)$$
$$\times \chi e^{i(kz - \omega\tau_1)} \quad (92)$$

Note that the soliton found in Eq. (92) has been formed near the frequency band edge due to interplay between grating induced cubic dispersion and the positive nonlinearity (self-steepening effect).

From the experimental point of view, it is necessary to know the magnitude of the peak power P_0 to excite the bright Bragg soliton. Similarly, the soliton period T_0 turns out to be another important physical parameter that is involved in the formation of Bragg soliton. Based on [69], from the bright Bragg soliton solution, we calculate the important and interesting physical parameters such as soliton power and pulse width in the form

$$T_0 = \sqrt{\frac{1}{(2a\omega + 3b\omega^2 - \beta_1)}}, \quad P_0 = \frac{2(2a\omega + 3b\omega^2 - \beta_1)}{d}$$
(93)

With the known values of the parameters a, b, c, and d in an FBG, we can calculate ω using Eq. (90). After calculating the value of ω from Eq. (90), for a given T_0 we can easily calculate the value of β_1 from Eq. (93). By computing all the values of parameters, we can calculate the power required for generating the bright Bragg soliton.

The bright soliton is depicted in Fig. 17.11a. Further, to have an idea about the variation of pulse width and amplitude with respect to input power P_0, we also made two-dimensional plots for various values of the input power P_0. From the plot, it is clear that the pulse width reduces and hence amplitude increases as the value of input power P_0 increases. This is clearly depicted in Fig. 17.11b. In addition, we also found the relationship between the input power P_0 and pulse width T_0 as $T_0 = \sqrt{\frac{2}{dP_0}}$, which is obtained from Eq. (93) [70]. This relationship is depicted in Fig. 17.11c.

Similarly, for the dark soliton formation we choose the self-defocusing effect, which shifts the central frequency of the carrier wave to the lower branch of the dispersion curve. From Eq. (91), we get the dark solitary wave solution as follows:

$$\vec{E} = \sqrt{\frac{(2a\omega + 3b\omega^2 - \beta_1)}{d}} \tanh\left(\sqrt{\frac{2a\omega + 3b\omega^2 - \beta_1}{2b}}\right)$$
$$\times \chi e^{i(kz - \omega\tau_1)} \quad (94)$$

The dark soliton in Eq. (94) results from the cubic dispersion with the higher nonlinearity in the system

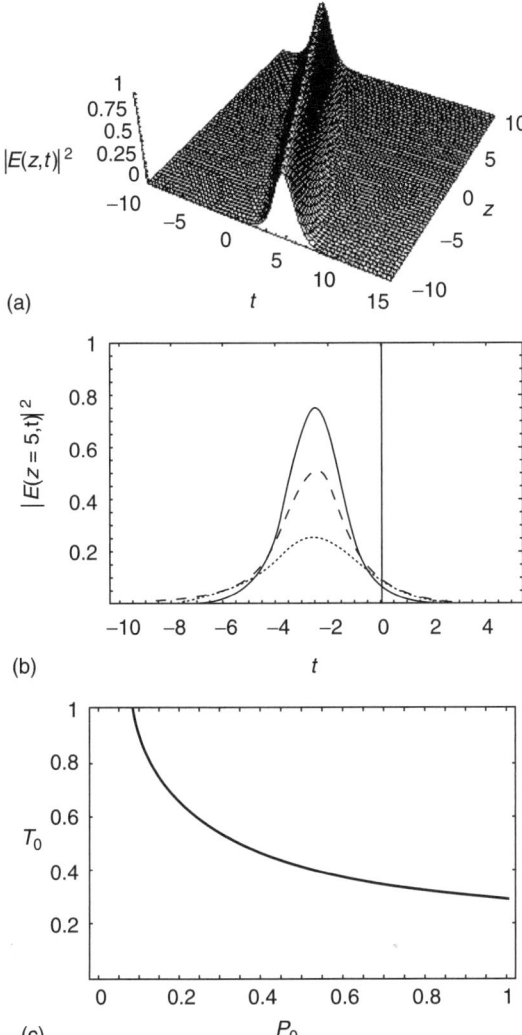

As discussed in the bright soliton case, by knowing all the parameter values, one can calculate the power required to generate the dark Bragg soliton.

The dark soliton in the lower branch is depicted in Fig. 17.12a. In addition, we also made the two-dimensional plots for various values of the critical power. From the plot, it is clear that the pulse width reduces and hence amplitude increases as the value of input power P_0 increases as in the case of bright soliton. This is clearly depicted in Fig. 17.12b.

Note that the PNLS Eq. (85) possesses relatively a large amount of grating-induced dispersion compared with the NLS Eq. (75), since the former also includes cubic dispersion in addition to quadratic dispersion. So, a large amount of nonlinearity is required to compensate for this huge amount of dispersion provided by the grating (PNLS equation). Based on this fact, we can say that the solitons generated due to higher order effects (solitons due to PNLS) will definitely be more intense than the Bragg grating solitons formed due to quadratic dispersion alone (NLS equation).

Thus, in this section we showed that for the range of frequencies discussed above, the description of

FIGURE 17.11 (a) Bright soliton solution in the upper branch of the dispersion curve for the physical parameters $P_0 = 0.5$, $\alpha = 1.4$, $\beta = 0.1$, $\beta_1 = 0.5$, and $\kappa = 0.2$. (b) Intensity plot for various values of input power: $P_0 = 0.25$ (dotted curve), $P_0 = 0.5$ (dashed curve), and $P_0 = 0.75$ (solid curve). (c) Variation of pulse width T_0 with respect to the input power P_0.

with their carrier frequencies close to the band edge, and hence we call the above solution the dark Bragg grating soliton solution.

As has been discussed in the bright soliton case, it is also possible to calculate the power and pulse width for dark Bragg soliton case and the same are given below:

$$T_0 = \sqrt{\frac{2}{(2a\omega + 3b\omega^2 - \beta_1)}}, \quad P_0 = \frac{(2a\omega + 3b\omega^2 - \beta_1)}{d} \tag{95}$$

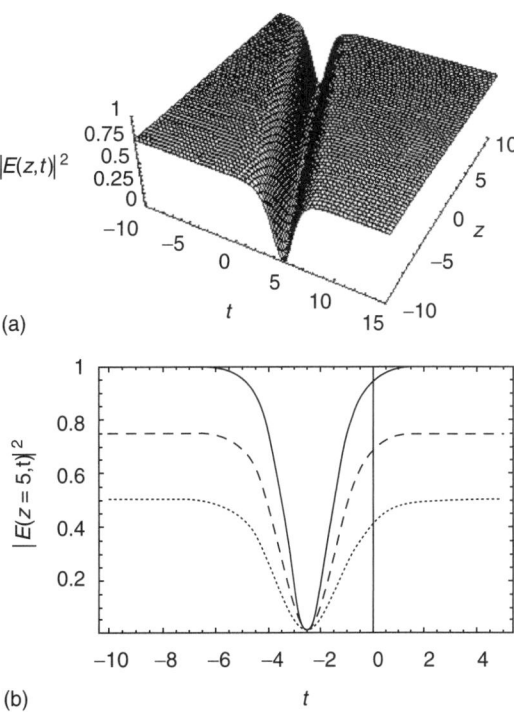

FIGURE 17.12 (a) Dark soliton solution in the lower branch of the dispersion curve for the same physical parameter values as in the bright soliton case. (b) Intensity plot for various values of input power $P_0 = 0.5$ (dotted curve), $P_0 = 0.75$ (dashed curve), and $P_0 = 1.0$ (solid curve).

the electric field in terms of forward and backward propagating waves cannot be adequate. Instead, it is more appropriate to write the electric field in terms of Bloch functions. This has the advantage of reducing the NLCM equations into a single NLS equation. So, in this case, one would not use the coupled mode equations at all but would start directly from the wave equation for the electric field.

After discussing the gap solitons using the CMT and Bragg grating solitons by using Bloch wave analysis, we come to the conclusion that one can use either method to describe the nonlinear pulse propagation in the periodic media. The first of these is the usage of the coupled mode equations. These are valid anywhere in the PBG but require that the index modulation depth be weak ($\alpha P_0 \gg \kappa$). The second approach is Bloch wave analysis, where one can apply the method of multiple scale analysis directly to the wave equation. In this case, one can deal with the systems with arbitrary index modulation depth ($\alpha P_0 \ll \kappa$) but one is restricted to electric fields with frequency content close to the edges of the PBG. Based on the above facts, it can be inferred that the NLCM equations describe the high intensity pulse propagation in FBG, whereas the NLS equation considers the low intensity pulse propagation.

9. RESULTS AND DISCUSSION

After reviewing the fundamentals of FBG, we considered the nonlinear pulse propagation through the same area where the pulse propagation is governed by nonlinear coupled equations. Solitons existing in FBG have been investigated in detail. Based on the pulse spectrum with respect to PBG, solitons in FBG were classified into gap solitons and Bragg grating solitons.

A simple method to obtain the explicit analytical gap soliton solutions to the NLCM equations has been developed. Using the Stokes parameters, we constructed Duffing oscillator equations. Under the Bragg resonance condition, we derived the explicit analytical solutions for both bright and dark gap solitons. In addition to these studies, we also showed the existence of another family of gap soliton for arbitrary values of the detuning parameter. The experimental investigations of bright and dark gap solitons in the Kerr media have gained momentum in recent times. We believe that the experimental verification of the dark gap soliton will be taken up in the near future, and there is no doubt it will receive much attention, like bright gap solitons in the case of temporal solitons in optical fibers. To realize the gap soliton experimentally within the PBG, we need to have very high intensity pulses. Though such intense laser sources are not available at present, we strongly believe that these solitons will be demonstrated experimentally once such high power lasers are engineered.

We also discussed the formation of Bragg grating solitons near the band edge when the carrier frequency of the laser pulse is detuned to either the upper or lower branch of the dispersion curve depending on the sign of nonlinearity. Hence, the governing NLCM equations were reduced to NLS and PNLS equations using multiple scale analysis. The PNLS equation, which incorporates both the HOD effects and self-steepening effects, has been derived to analyze the impact of nonlinearity on the pulse propagation in FBG. From both the NLS and PNLS equations, bright and dark Bragg solitary wave solutions have been constructed by the coupled phase-amplitude method. From our results, we could conclude that the Bragg grating solitons generated due to the PNLS equation would be more intense than the solitons formed due to the NLS equation. Further, we also calculated important and interesting physical parameters such as power and pulse width for both bright and dark Bragg solitons. By knowing all the physical parameters, one can calculate the minimum power required to generate the solitons in the FBG structure.

We thus believe that these solitons found in FBG could find applications and play a vital role in the field of OFCs and sensors as mentioned in Section 4.1.

9.1. Experimental Considerations

We point out that to observe these solitons experimentally, very high power laser sources are required to compensate for the huge amount of grating-induced dispersion. Such intense high power laser sources are yet to be commercially available. To overcome this problem, one can choose the materials having large nonlinear index n_2. One choice is to use chalcogenide glass such as As_2S_3 with $n_2 = 2 \times 10^{-14}$ cm^2/W, the value being roughly 100 times larger than the case of silica fibers [71]. Another choice of material is an AlGaAs integrated Bragg waveguide whose nonlinear index $n_2 = 1.5 \times 10^{-13}$ cm^2/W [72]. Chalcogenide fibers with a nonlinear index 500 times larger than that of silica fibers have been reported [73]. The use of such fibers with a larger nonlinear index will require peak power on the order of 10 kW. Similarly, the corresponding peak power requirement for the As_2S_3 glass is $\sim$100 KW. These are obviously the typical power levels used in recent experiments with nonlinear propagation effects in FBGs [74].

10. ACKNOWLEDGMENTS

K. P. expresses his thanks to DST, UGC (Research Award), and CSIR, Government of India, for financial support through their research grant. K. S. N. acknowledges CSIR for the award of an SRF.

11. REFERENCES

1. A. Hasegawa and Y. Kodama, *Solitons in Optical Communications* (Oxford University, New York, 1995).
2. G. P. Agrawal, *Nonlinear Fiber Optics* (Academic Press, San Diego, 2001).
3. G. P. Agrawal, *Applications of Nonlinear Fiber Optics*, second edition (Academic Press, New York, 2001).
4. Y. Kivshar and G. P. Agrawal, *Optical Solitons: From Fibers to Photonic Crystals* (Academic Press, New York, 2003).
5. A. Ghatak and K. Thyagarajan, *Introduction to Fiber Optics* (Cambridge University Press, 1998).
6. S. Trillo and W. Torruellas, Eds., *Spatial Solitons* (Springer, New York, 2001).
7. G. Keiser, *Optical Fiber Communications* (New York, McGraw-Hill, 1991).
8. G. Drazin and R. J. Johnson, *Solitons: an Introduction* (Cambridge University Press, Cambridge, 1989).
9. N. J. Zabusky and M. D. Kruskal, Phys. Rev. Lett. **15**, 240 (1965).
10. A. Hasegawa and F. Tappert, Appl. Phys. Lett., **23**, 142 (1973).
11. L. F. Mollenauer, R. H. Stolen and J. P. Gordon, Phys. Rev. Lett., **45**, 1095 (1980).
12. K. O. Hill, Y. Fujii and D. C. Johnson, Appl. Phys. Lett. **32**, 647 (1978).
13. B. S. Kawasaki, K. O. Hill and D. C. Johnson, Opt. Lett. **3**, 66 (1978).
14. R. Kashyap, *Fiber Bragg Gratings* (Academic Press, San Diego, CA, 1999).
15. A. Othonos and K. Kalli, Fiber Bragg Grating: Fundamentals and Applications, in *Telecommunications and Sensing*, (Artech House, Boston, London, 1999).
16. A. Yariv, *Quantum Electro Dynamics*, third edition (Wiley, New York, 1989).
17. E. Yablonovitch, Phys. Rev. Lett., **58**, 2059 (1987).
18. N. W. Ashcroft and N. David Mermin, *Solid State Physics* (Holt, Rinehart and Winston, 1976).
19. C. M. de Sterke, B. J. Eggleton and J. E. Sipe, in *Spatial Solitons*, edited by S. Trillo and W. Torruellas (Springer, New York, 170, 2001).
20. D. E. Pelinovsky, L. Brzozowski and E. H. Sargent, Phys. Rev. E. **62**, R4536 (2000).
21. L. Brzozowski and E. H. Sargent, IEEE J. Quantum Electron. **36**, 550 (2000).
22. D. Taverner, N. G. R. Broderick, D. J. Richardson, M. Ibsen and R. I. Laming, Opt. Lett. **23**, 259 (1998); 328 (1998).
23. N. G. R. Broderick, D. Taverner, D. J. Richardson, M. Ibsen and R. I. Laming, Opt. Lett, **22**, 1837 (1997).
24. K. Senthilnathan and K. Porsezian, Opt. Comm. **227**, 275 (2003).
25. A. Melloni, M. Chinello and M. Martinelli, IEEE Photonics Tech. Lett. **12**, 42 (2000).
26. D. Pelinovsky, J. Sears, L. Brzozowski and E. H. Sargent, J. Opt. Soc. Am. B **19**, 43 (2002).
27. H. G. Winful, J. H. Marburger and E. Gamire, Appl. Phys. Lett. **35**, 379 (1979).
28. W. Chen and D. L. Mills, Phys. Rev. Lett. **58**, 160 (1987).
29. D. L. Mills and S. E. Trullinger, Phys. Rev. B **36**, 947 (1987).
30. J. E. Sipe and H. G. Winful, Opt. Lett. **13**, 132 (1988).
31. D. N. Christodoulides and R. I. Joseph, Phys. Rev. Lett. **62**, 1746 (1989).
32. A. B. Aceves and S. Wabnitz, Phys. Lett. A. **141**, 37 (1989).
33. C. M. de Sterke and J.E. Sipe, in *Progress in Optics* XXXIII, edited by E. Wolf (Elsevier, Amsterdam, 1994).
34. K. Senthilnathan, K. Porsezian and P. Malathi, J. Opt. Soc. Am. B **20**, 366 (2003).
35. K. Senthilnathan, K. Porsezian P. Ramesh Babu and V. Santhanam, IEEE J. Quantum Electron **39**, 1192 (2003).
36. K. Porsezian and K. Senthilnathan (Submitted to J. Opt. Soc. Am. B).
37. C. M. de Sterke, N. G. R. Broderick, B. J. Eggleton and M. J. Steel, Opt. Fiber Technol. **2**, 253 (1996).
38. A. B. Aceves, Chaos, **10**, 584 (2000).
39. G. Kurizki, A. E. Kozhenkin, T. Opatrny and B. A. Malomed, in *Progress in Optics* XXXXII, edited by E. Wolf (Elsevier, Amsterdam, 2001).
40. B. J. Eggleton, R. E. Slusher, C. M. de Sterke, P. A. Krug and J. E. Sipe, Phys. Rev. Lett. **76**, 1627 (1996).
41. B. J. Eggleton, C. M. de Sterke, and R. E. Slusher, J. Opt. Soc. Am. B. **14**, 2980 (1997).
42. D. Taverner, N. G. R. Broderick, D. J. Richardson, M. Ibsen and R. I. Laming, Opt. Lett. **23**, 259 (1998); 328 (1998).
43. J. Feng and F. K. Kneubuhl, IEEE QE, **29**, 590 (1993).
44. C. Conti and S. Trillo, Phys. Rev. E **64**, 036617 (2001).
45. T. Opatrny, B. A. Malomed and G. Kurizki, Phys. Rev. E **60**, 6137 (1999).
46. B. Diano, G. Gregori and S. Wabnitz, Appl. Phys. **58**, 4512 (1985).
47. E. Collett, *Polarized Light—Concepts and Application* (Marcel-Decker, New York, 1989).

48. T. Iizuka and M. Wadati, J. Phy. Soc. Jpn, **66**, 2308–2313 (1997).
49. E. V. Kazantseva, A. I. Maimistov and B. A. Malomed, (www.arxiv.org), nlin. PS/00100050 (2000).
50. K. Senthilnathan and K. Porsezian, Opt. Comm. **227**, 295 (2003).
51. S. Trillo and C. Conti, Optical Solitons in: *Theoretical and Experimental challenges*, edited by K. Porsezian and V. C. Kuriakose, Lecture Notes in Physics, vol. 613 (Springer-Verlag, Berlin, 2003).
52. S. Trillo, C. Conti, G. Assanto and A. V. Buryak, Chaos, **10**, 590 (2000).
53. C. Conti, S. Trillo and G. Assanto, Phys. Rev. Lett. **78**, 2341 (1997).
54. H. He and P. D. Drummond, Phys. Rev. Lett. **78**, 4311 (1997).
55. T. Peschel, U. Peschel, F. Lederer and B. Malomed, Phys. Rev. E. **55**, 4730 (1997).
56. Y. Kivshar, Phys. Rev. E. **51**, 1613 (1995).
57. C. Conti, S. Trillo and G. Assanto, Phys. Rev. E. **57**, R1251 (1998).
58. C. Conti, S. Trillo and G. Assanto, Opt. Lett. **23**, 334 (1998).
59. C. Conti, S. Trillo and G. Assanto, Phys. Rev. Lett. **85**, 2502 (1998).
60. C. Conti, G. Assanto and S. Trillo, Opt. Express. **3**, 389 (1998).
61. W. E. Torruellas, Z. Wang, D. J. Hagan, E. W. Van Stryland, G. I. Stegeman, L. Torner and C. R. Menyuk, Phys. Rev. Lett. **74**, 5036 (1995).
62. R. Schiek, Y. Baek and G. I. Stegeman, Phys. Rev. E **53**, 1138 (1996).
63. K. Hayata and M. Koshiba, Phys. Rev. Lett. **71**, 3275, (1993).
64. M. J. Werner and P. D. Drummond, J. Opt. Am. B, **10**, 2390 (1993).
65. A. Buryak and Y. S. Kivshar, Opt. Lett., **19**, 1612 (1994)
66. A. B. Aceves, Optical Solitons in: *Theoretical and Experimental Challenges*, edited by K. Porsezian and V. C. Kuriakose, Lecture Notes in Physics, vol. 613 (Springer-Verlag, Berlin, 2003).
67. C. M. de Sterke and J. E. Sipe, *Phys. Rev. A*, **38**, 5149 (1988).
68. M. Du, K. Chan and K. Chui, *IEEE J. Quantum Electron.*, **31**, 177 (1995).
69. S. L. Palacios, A. Guinea, J. M. Fernandez-Daz and R. D. Crespo, Phys. Rev. E., **60**, R45 (1999).
70. R. Ganapathy, K. Senthilnathan and K. Porsezian, J. Opt. B Quant. SemiClassical Opt., **6**, S436 (2004).
71. M. Asobe, Opt. Fiber Technol., **3**, 142 (1997).
72. P. Millar, R. M. De La Rue, T. F. Krauss, J. S. Aitchison, N. G. R. Broderick and D. J. Richardson, Opt. Lett., **24**, 685 (1999).
73. G. Lenz, J. Zimmermann, T. Katsufuji, M. E. Lines, H. Y. Hwang, S. Spalter, R. E. Slusher, S. W. Cheong, J. S. Sanghera and I. D. Agrawal, Opt. Lett., **25**, 254 (2000).
74. B. J. Eggleton, C. M. de Sterke and R.E. Slusher, J. Opt. Soc. Am. B, **16**, 587 (1999).

18
CHAPTER

Advances in Dense Wavelength Division Multiplexing/Demultiplexing Technologies

Vikram Bhatia*
Avanex Corporation
Erwin Park, New York, USA

1. INTRODUCTION

Optical communication systems are evolving to accommodate the larger information content that needs to be transmitted over point-to-point or meshed networks. Although use of the L-band is being advocated to supplement the bandwidth in the C-band, most system providers prefer to use denser channel spacing in the C-band to increase the information-carrying capacity of their networks. The components for applications such as dispersion compensation, amplification, and multiplexing/demultiplexing (mux/demux) are more easily available for the C-band (than for the L-band) and offer better performance and lower cost. The transition to tighter channel spacing in dense wavelength division multiplexing (DWDM) systems is forcing the component vendors to make constant improvements in technology to keep pace with performance requirements. The products used for multiplexing (combining) and demultiplexing (separating) optical channels, in particular, have undergone big improvements in performance over the past few years.

Additionally, worldwide over the last 5 years the telecommunications market has slowed down, putting pressure on the component suppliers from the point of view of cost. System integrators are asking for better performing products that are comparable or even lower in price than the previous generation of technology. This problem of pricing is exacerbated with the migration of networks closer to the consumer. Metro and access networks have cost as the key factor that regulates their widespread deployment. Some performance metrics can be sacrificed for lower cost, but these trade-offs vary with the type of network—long haul, metro, or access—which the engineer needs to consider during system design.

Although many different technologies are available for mux/demux DWDM channels, the three most common commercial ones are thin film filters (TFFs), arrayed waveguide gratings (AWGs; see Chapter 21), and fiber Bragg gratings (FBGs; discussed in Chapter 15). We discuss the application space for each of these products based on the advantages and limitations of the corresponding products. Over the past few years, optical interleavers have been commonly used to assist in mux/demux applications. We review the role of interleavers and how they can be used with the base technologies to design improved mux/demux products.

2. KEY PERFORMANCE CHARACTERISTICS

Mux/demux products need to have a few critical performance attributes that depend on the type of application they are being used for. The important optical parameters are low insertion loss, flat passband, and high adjacent channel isolation. In addition the impairments such as chromatic dispersion, polarization dependent loss (PDL) and polarization mode dispersion need to be as low as possible. For typical multiplexing applications the insertion loss is a key optical attribute, wherease for demux modules the cross-talk from adjacent and non-adjacent channels needs to be minimal. One way to measure the bandwidth efficiency of a filtering technology is by evaluating the figure of merit (FOM), which is typically the ratio of the 0.5-dB bandwidth to the 25-dB bandwidth. The objective is to design the filter for as high an FOM as possible.

Moreover, because of the premium placed on rack space, the size of the mux/demux modules has to be as small as possible. System integrators are also pushing their vendors to integrate additional components, such as taps, photodiodes, and variable optical attenuators, into the mux/demux modules to enhance functionality. Hence, integrated optics

*E-mail: Vikram_Bhatia@avanex.com

platforms are favored in such cases because the optical monitoring and control functions can be provided without using too much extra shelf space. Mux/demux modules typically need to operate over a 70°C temperature range in the central office environment. Products that can function over this range without temperature control are preferred over those that require active compensation because this adds cost and complexity to the system.

The overall cost of the mux/demux modules is a key factor in determining the product platform. Although the initial cost is critical, the expense and the ease to upgrade the system to larger channel counts are also important. The recent approach by system houses to reduce cost has been to migrate to modular solutions. This requires the slicing of the C-band into multiple subbands, such as five bands of eight channels each at 100-GHz channel spacing. The first installed cost is reduced by initially installing the transmitter and receivers (and hence mux and demux) modules for only one or two bands. As the required capacity increases, the other bands are gradually added over time. Hence, modularity of mux/demux modules at a low cost is fast becoming a key parameter of interest to system designers.

Although good optical performance, low cost, and small size are critical selection criteria for mux/demux modules, the manufacturability and reliability of these modules is also important in determining the appropriate technology for a particular application. TFFs, arrayed waveguide devices, and fiber gratings have been around for a number of years and a lot of field data exist to prove their optical reliability and mechanical robustness. Other technology platforms such as diffraction gratings have also been used for specific mux/demux applications but have lacked widespread deployment because of certain technical limitations and the uncertainty associated with moving away from proven solutions. Hence, this chapter focuses on analyzing TFFs, AWGs, and FBGs as mux/demux products.

2.1. Thin Film Filters

TFF mux/demux modules are typically assembled using multiple three-port devices concatenated in a certain sequence. The building block for each device is a dielectric component that consists of multiple cavities, each of which is bounded by alternating thin film layers of high and low index materials [1]. The filter (at a certain angle of incident light) is designed to pass a specific band of wavelengths while the remaining input wavelengths are reflected. Hence, a three-port TFF device consists of an input port, a transmitted port, and a reflected port. The use of dual and single-fiber collimators allows the light to be coupled from the fiber to the free space and vice versa. The insertion loss of this device is a function of the alignment of the micro-optic component, whereas the spectral properties, such as bandwidth and isolation, are determined by the construction of the filter element. Values of insertion loss in transmission (with respect to the input) are in the 0.4- to 0.5-dB region, whereas the loss in reflection is in the 0.8- to 1.0-dB range. Increasing the number of cavities in the filter typically increases the FOM of the device.

Figure 18.1 illustrates the layout for an eight-channel 100-GHz demux module that is implemented by concatenating eight three-port devices. The first device removes the first 100-GHz channel in transmission and reflects all other channels. The reflected port from this device is spliced to the input port of the second device that transmits only the second channel. This demux chain continues until all channels are separated. The insertion loss obviously increases from the first to the last channel, but this can be overcome by intentionally introducing excess loss in the output port of the first few channels through offset splicing. In metro systems that do not have

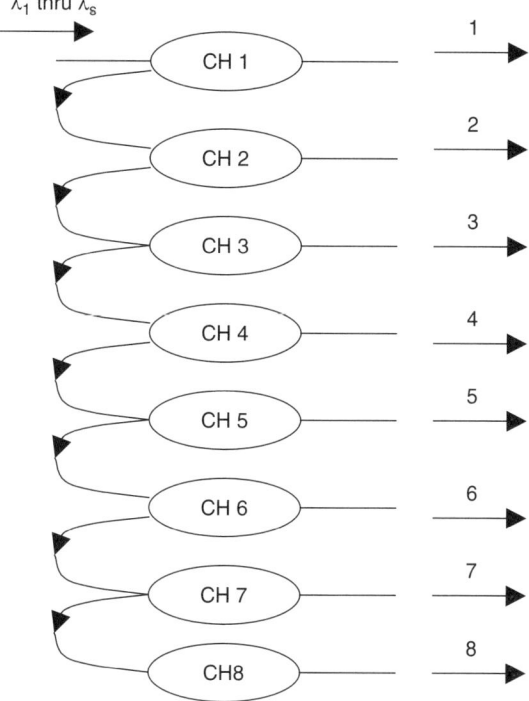

FIGURE 18.1 The configuration of an eight-channel demux using 100-GHz TFF devices. Each three-port device removes a particular channel while reflecting all others.

any optical amplifiers, the equalization of the insertion loss is not as critical. In such cases the total insertion loss of a particular channel through the mux and demux needs to be minimized. This can be done by arranging the filters in the mux in an order that offsets the channel-to-channel insertion loss variation introduced by the demux. For the example considered in Fig. 18.1, this can be done by concatenating the TFF devices in the corresponding mux module such that the insertion loss increases from channel 8 to channel 1.

The spectrum of the three-port device is largely a function of the TFF element. The filters can be fabricated to pass either one channel or a band of adjacent channels. Figure 18.2 depicts the transmission spectrum of a so-called 100-GHz 4-skip-0 filter, one that passes a band of four consecutive 100-GHz channels while at the same time introducing minimal loss for the out-of-band channels on shorter and longer wavelength edges of the band. Such filters have a very high FOM due to the wide bandwidth required to pass four channels while having a steep spectral slope to maximize isolation at the adjacent channels. These skip-0 filters can effectively be used in DWDM systems to eliminate dead channels in banded architectures. The 4-skip-0 filter can also be used in the demux illustrated in Fig. 18.1 to reduce the worst-case insertion loss. By placing this filter at the eight-channel input, the first four channels can be transmitted (and demuxed using corresponding 100-GHz devices) while the other four can be reflected (and again separated using 100-GHz devices). In this modified configuration, the last channel in the chain has only four reflections as compared with seven for the architecture in Fig. 18.1. Although this increases the cost of the demux by about $200–300 (due to the additional 4-skip-0 device), the worst-case insertion loss is lowered by 0.5–1.5 dB from the case in Fig. 18.1. This is a classic example of the price/performance trade-off that the TFF module vendor and the system integrator have to consider during the design phase.

The bit rate and reach of the DWDM system are limited by the chromatic dispersion of the mux/demux and other optical components in the optical path. The dispersion of a three-port device depends on the construction of the filtering element. The phase and amplitude are related through the Hilbert transform, and one would expect the 50-GHz filter would have approximately four times the dispersion magnitude from that of a 100-GHz filter. The reflected dispersion of a device becomes critical in applications where several devices are concatenated in a module. For the example, in Fig. 18.1 the first channel sees only the dispersion in transmission, but the second and subsequent channels have the reflected dispersion added on top of the transmitted dispersion. The dispersion in reflection is much stronger at the adjacent channel and fades rapidly thereafter, and practically only adjacent channels on

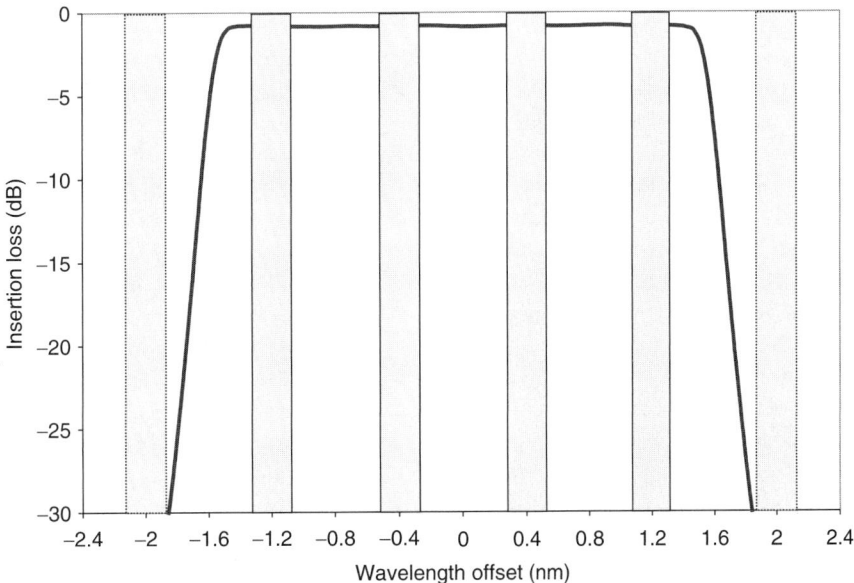

FIGURE 18.2 The transmission spectrum of a 100-GHz 4-skip-0 device. The filter center wavelength has been subtracted to obtain the wavelength offset. The rectangular boxes represent the 100-GHz channels (± 15-GHz passband) within (solid lines) and outside (dashed lines) the filter band.

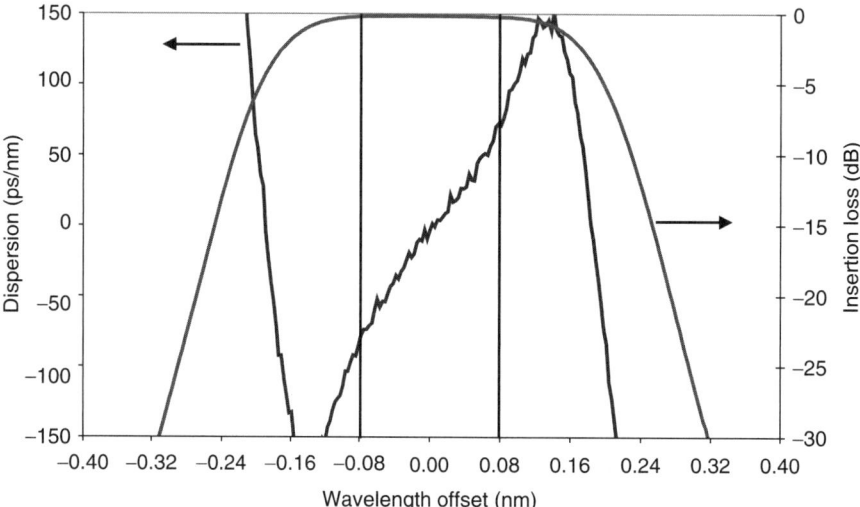

FIGURE 18.3 Transmission spectrum (right axis) and dispersion of a 50-GHz three-port device. Also shown is the ±10-GHz passband for a standard 50-GHz channel.

either side are affected by the dispersion. Figure 18.3 shows the spectrum and dispersion of a 50-GHz three-port device.

The insertion loss of TFF modules increases with channel count, which limits the number of channels that can be muxed/demuxed. With the advent of band filters and clever arrangement of devices, 40-channel 100-GHz mux/demux solutions are commercially available with insertion loss of around 6–7 dB. The passive nature of these devices is achieved by using substrates for filters with the appropriate coefficient of thermal expansion. Typical drift in the center wavelength of such passive devices is about 0.5 pm/°C. The athermalization of TFF mux/demux is a very powerful value proposition for networks where availability of electrical power is limited. The TFF elements are fabricated at the plate level where hundreds of filters might be available per plate. The low fabrication and assembly cost make this an attractive technology for low- to medium-channel count systems where cost is a critical factor.

TFFs have also been commonly used in coarse WDM systems because of their passive nature and low cost. Their advantages over other technologies are more pronounced for low-cost access networks operating in an uncontrolled environment (−40° to 75°C) where rack space is also premium.

2.2. Arrayed Waveguide Gratings

An AWG typically consists of one input and one output multimode interference coupler connected by a series of waveguides [2]. This array of waveguides along with the multimode interference couplers is responsible for the mux/demux functionality of the AWGs, as shown in Fig. 18.4. A set of n input waveguides brings in the optical signals. The input $n \times m$ multimode interference coupler splits the signal into m identical (in amplitude) parts, whereas the m arrayed waveguides act to modify the phase of these parts before they reach the output coupler. The interference of these parts in the $m \times n$ output coupler is a function of the lengths of the arrayed waveguides, and by proper design the n signals at the output waveguides can be made wavelength specific. For example, if the AWG is to be used as a demux, all the channels are present on one input waveguide while the demuxed channels are mapped individually on the output waveguides. In that case, only one of the input waveguides is coupled to an input fiber and a fiber array at the output carries the demuxed channels. Silicon is usually the material of choice for the common substrate of the AWG, whereas the waveguides themselves are made of silica.

The advantage of the planar technology is that other components such as taps and the variable optical attenuator can also be integrated on the same substrate. These planar lightwave circuits (PLCs) are becoming increasingly popular with component vendors looking to provide increased functionality in a limited amount of space. Although the commercial availability of such integrated optic products is still very limited, the system providers are pushing their design and development. The synergy between AWG manufacturing and the semiconductor industry is being used to further drive down the cost of this technology.

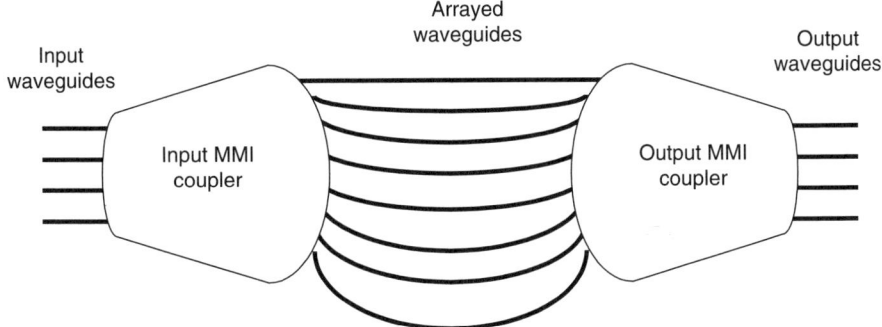

FIGURE 18.4 Layout of a typical AWG device. The input and output MMI (multimode interference) couplers and the waveguides have a common substrate.

Unlike TFF-based mux/demux devices, the insertion loss of the AWGs does not go up with channel counts. Hence, for systems with a large number of channels (32 or larger), AWGs are becoming the solution of choice. The AWGs are available in Gaussian as well as flattop passband profiles. Figure 18.5 shows the spectrum of a 40-channel, 100-GHz, flattop AWG demux. For a small channel count (eight or smaller), the larger insertion loss as well as higher cost of the AWGs gives the TFF solution an advantage. One of the major drawbacks of the AWGs is the requirement of active temperature control. This typically involves heating the AWG to a temperature higher than the maximum operating temperature. If the ambient temperature is very low, the required electrical power can be large. Although passive AWGs are increasingly being marketed, their performance is not as good as that of their active counterparts. Also, albeit the previous generations of AWGs had issues with PDL and cumulative cross-talk, these problems have been addressed in the current version of these devices. Another big advantage of AWGs over TFFs is their inherent low chromatic dispersion. For example, 50-GHz AWGs have been shown with dispersion less than ± 20 ps/nm across a ± 10-GHz passband.

One of the most powerful value propositions of the AWG technology is that of "colorless" AWGs [3]. These devices are fabricated in a way such that the output is cyclic in the frequency domain with a specific free spectral range (FSR). For example, Fig. 18.6 illustrates the concept of a 2×8 demux cyclic AWG. In this case the FSR of the device is set to be eight times the channel spacing. This implies that the output has a cyclic spectrum that repeats every second band of eight channels. One of the two

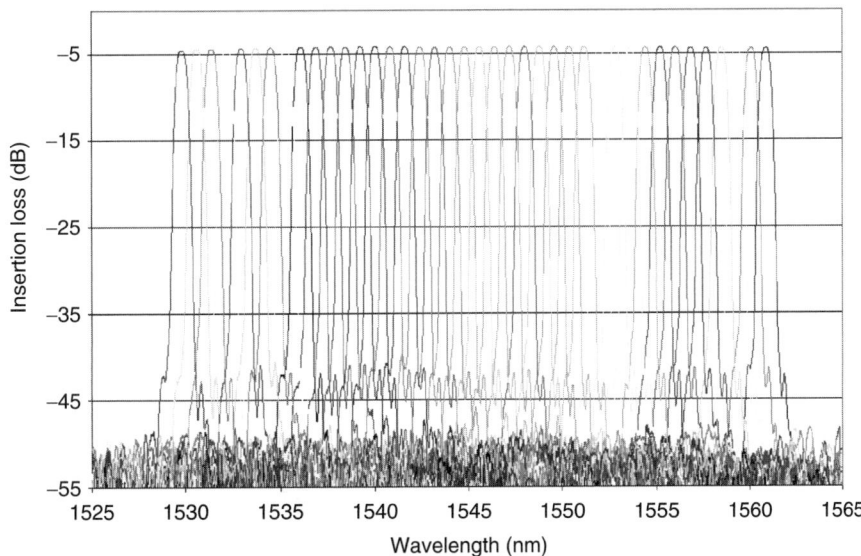

FIGURE 18.5 Spectrum of a 40-channel 100-GHz AWG-based demux. This AWG has flattop passbands; the Gaussian version is also available commercially.

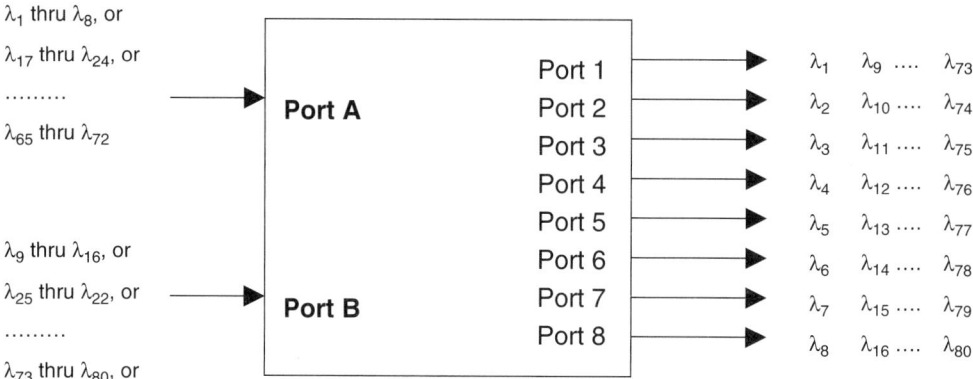

FIGURE 18.6 The basic colorless AWG [3]. If an odd-numbered band needs to be demultiplexed, it will be input at port A, whereas if an even-numbered band needs to be demultiplexed, it will be input at port B. The output is received from port 1 through port 8 and then cycles back to port 1 for the next band.

input ports is connected to the fiber to be demuxed depending on which band is being received. Hence, the same module can be used to demux any band, drastically reducing the different part numbers that would otherwise be needed (as in the case of TFF mux/demux). This provides a powerful incentive to the system integrators because it does not require them to hold inventory for all the different bands.

2.3. Fiber Bragg Gratings

FBGs are formed by exposing a hydrogen-loaded germanosilicate fiber to an ultraviolet interference pattern [4] (see Chapter 15). The ultraviolet pattern modifies the refractive index of the fiber core to produce a grating that serves to reflect a portion of the input optical spectrum while transmitting the rest. The strength of the grating determines the amount of reflectivity, whereas wavelengths that are reflected are a function of the grating pitch. Hence, by tailoring the grating parameters, channels that need to be removed can be reflected and accessed by placing a circulator before the grating. Apodization of the grating can be used to improve the spectral efficiency of the grating. This process involves keeping the average refractive index flat along the grating length while maintaining a certain profile (Gaussian, etc.) for the refractive index modulation of the core.

FBGs are capable of providing optical spectra with very high FOM. Figure 18.7 illustrates the spectrum of a 50-GHz FBG in transmission and reflection. As can be seen, spectral slopes approaching several hundreds of dB/nm can be obtained using this technology. Bragg gratings have been demonstrated for channel spacing as narrow as 12.5 GHz.

An FBG is made passive by mounting it on a substrate with a negative coefficient of thermal expansion. Using this approach, the average wavelength shift over temperature has been reduced to less than 0.3 pm/°C [5]. The limitation of the FBG technology is that, unlike TFFs and AWGs, each grating has to be fabricated separately. This lengthens the cycle time and increases the overall cost. Moreover, the use of a circulator further increases the cost of the systems using this technology. Although the use of Mach-Zehnder–based configurations has been proposed to eliminate the need for circulators, such devices are complex to fabricate and have low yields.

FBGs have traditionally been limited in the application base due to the presence of cladding mode loss in transmission on the shorter wavelength side of the Bragg wavelength (Fig. 18.7). By choosing adequate dopants, the photosensitivity of the cladding can be increased to enhance the overlap between the grating and the fundamental mode supported by the fiber [6]. This drastically reduces the magnitude of the cladding modes, as shown in Fig. 18.8. Such cladding mode–suppressed fibers are commercially available from several vendors in the industry.

FBGs have much lower through loss and hence are ideal for use in add/drop multiplexers where one or a few channels need to be added/dropped while most channels pass through. To build large channel count mux/demux modules, prolific use of circulators is required, which limits the FBGs to low channel count (four or less) systems. Also, these devices are typically favored over TFFs and AWGs in DWDM systems with narrower channel spacing, such as 25 GHz and below.

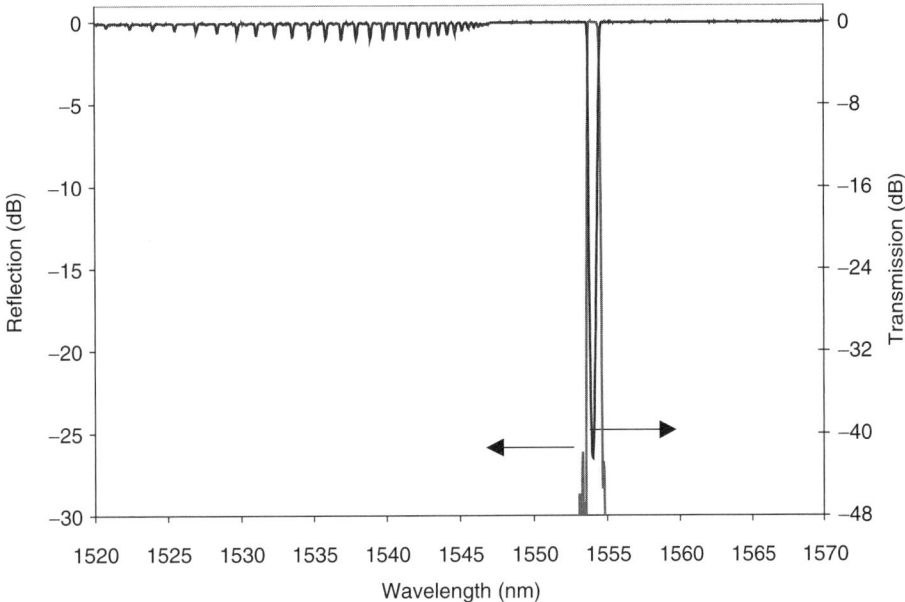

FIGURE 18.7 The transmission and reflection spectrum of a 50-GHz FBG. The loss due to cladding mode coupling is evident on the shorter wavelength edge of the Bragg peak in transmission.

3. OPTICAL INTERLEAVERS

Optical interleavers, or interleaving filters (ILFs), are typically 1 × 2 devices that assist in mux/demux applications. An ILF used for demux with an input channel spacing of f GHz has two outputs, each with $2f$ GHz channel spacing but offset from each other by f GHz. For example, a 50/100-GHz ILF takes the 50-GHz input and maps the odd and even channels on the two output ports. This results in the output ports having 100-GHz spacing, and hence 100-GHz demux solutions can be used. A mux ILF is a 2 × 1 device that combines the odd and even channels into a single output of half the frequency spacing as the input. Interleavers are useful in DWDM systems

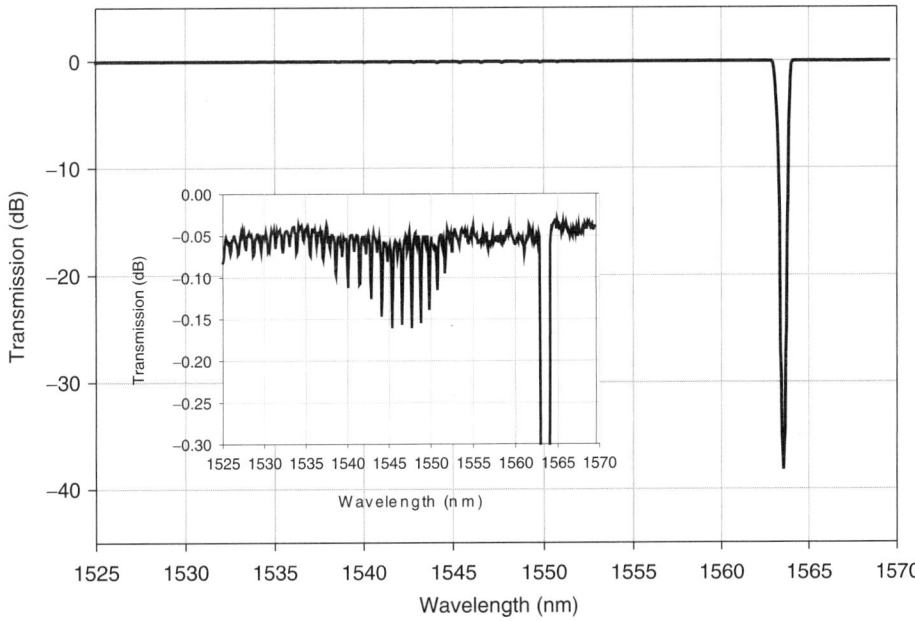

FIGURE 18.8 Cladding mode coupling for a Bragg grating in a specialty fiber. The inset shows a magnified view of the cladding mode loss [6].

where the designer wants to use cheaper mux/demux solutions at twice the system channel spacing.

Interleavers have been proposed using many different technologies, including Mach-Zehnder interferometers (fiber and planar), birefringent crystals, and TFF Fabry-Perot filters. All these involve the splitting of the input light (by polarization or amplitude) and propagating the two halves through a series of interferometers to obtain the required FSR at the output. The FSR needs very tight control over the operating temperature range and the device lifetime because even small deviations from the ideal value can lead to channels walking off from the ITU (International Telecommunication Union) grid.

Interleavers can be designed to have very low chromatic dispersion (around ± 15 ps/nm or less). We have already seen that 50-GHz TFFs have dispersion higher than ± 100 ps/nm (in transmission and in reflection), virtually eliminating this technology from most 50-GHz mux/demux applications. A 100-GHz TFF demux can though be used with 50/100-GHz ILF to obtain very low dispersion. Figure 18.9 shows the spectrum of an eight-channel 50-GHz demux implemented using 50/100-GHz ILF and 100-GHz TFF devices. Adjacent channel isolation higher than 27 dB was obtained while insertion loss was about 6.0 dB (this module also had taps at the input and the output that increase the insertion loss). The worst-case dispersion for the eight channels was ± 25 ps/nm over the ± 10-GHz pass band. The pass band shape and dispersion for one of the eight channels is shown in Fig. 18.10. This shows the value of combining the interleaver with a standard mux/demux technology. AWGs can similarly replace the TFF demux in this example.

4. DISCUSSION

We discussed the three most common commercially available technologies for multiplexing and demultiplexing. The advantages and limitations of TFFs, AWGs, and FBGs were analyzed. It was observed that although all these technologies are suited for mux/demux functionality, the cost and performance trade-offs typically determine their niche applications. For example, TFF technology is strongly suited for low to medium channel count (4–16 channels) applications for systems with channel spacing 100 GHz or larger. On the other hand, because the AWG price per channel reduces considerably with increasing channel counts, this technology is attractive for medium to higher channel counts (8–40 channels) for channel spacing as small as 50 GHz. The modularity of the system might also sway the decision between the different technologies. For a system with modularity of eight channels or below, TFFs might be preferred, whereas for modularity of eight and higher, AWGs might have a cost advantage.

We also observed that other considerations such as physical size and power consumption play an important role in technology choice. For example, a system provider that has limited rack space might pick the AWG technology for a 16-channel 100-GHz demux due its smaller footprint and ability to integrate other optical functions on the same substrate. Another system integrator might select TFFs if passive operation is more important that size. The TFFs are also limited by high chromatic dispersion, especially for 50-GHz and smaller channel spacing. The AWGs and FBGs can be designed to lower the chromatic dispersion. Because of the requirement of

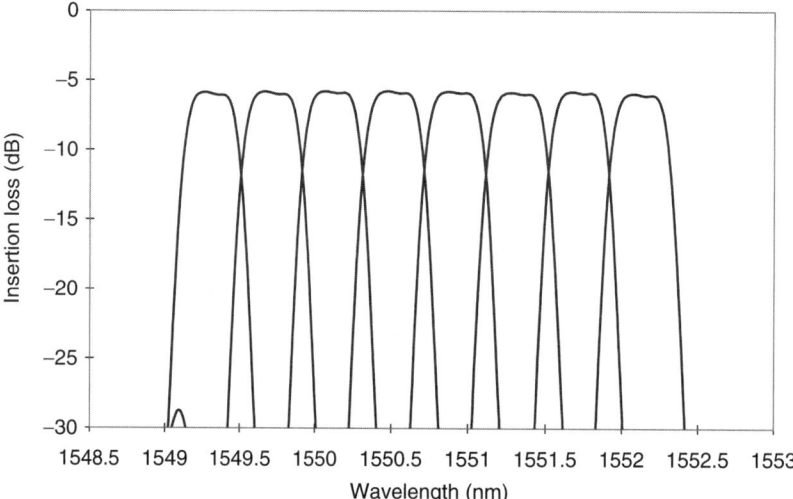

FIGURE 18.9 Spectrum of an eight-channel 50-GHz demux implemented using a 50/100-GHz ILF and 100-GHz TFF devices.

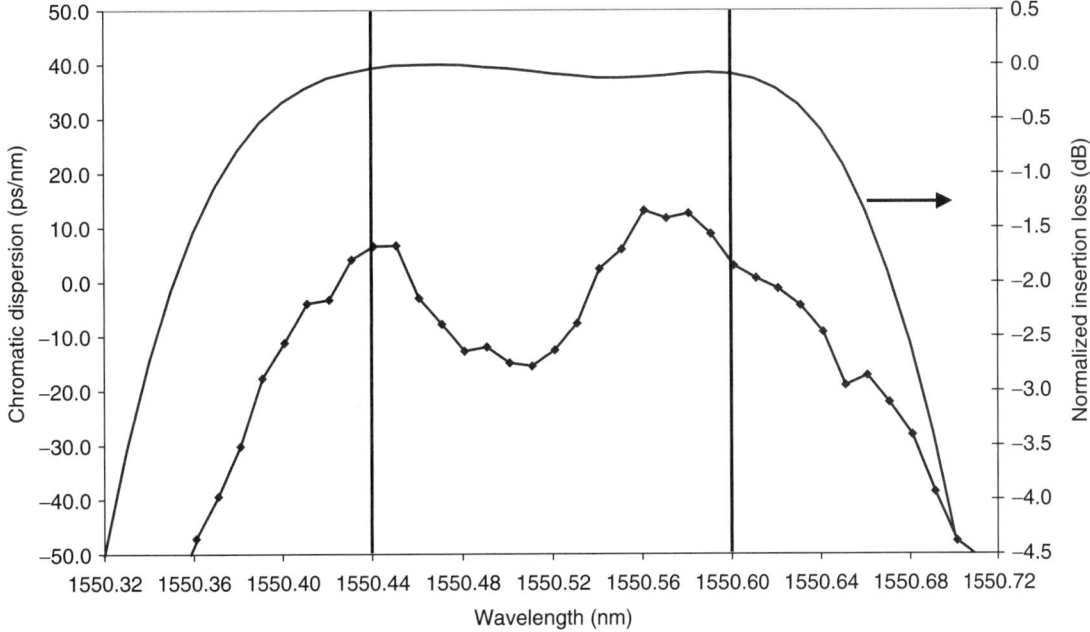

FIGURE 18.10 The pass band profile (right axis) and dispersion of the fourth channel of the eight-channel demux depicted in Fig. 18.9. Note the low insertion loss ripple and dispersion within the ±10-GHz pass band.

optical circulators, the deployment of FBGs has not been as widespread as that of TFFs and AWGs. The advent of cladding mode–suppressed fibers has helped the FBG's cause by making them suitable for adding/dropping a few channels in a large channel count system. FBGs have also been shown to work effectively for low to medium channel count mux/demux applications in channel spacings of 50 GHz or smaller.

We also analyzed the functionality of an optical interleaver. It was shown that optical interleavers help lower the cost of large channel count DWDM systems. The use of the interleaver is appropriate when the channel count is high because the cost of the interleaver can then be spread over a larger number of channels. We showed experimental results where 100-GHz TFFs in conjunction with a 50/100-GHz ILF were shown to produce dispersion better than ±25 ps/nm in a 50-GHz demux system.

5. CONCLUSION

System houses have many choices for the mux/demux functionality. We discussed the principle of operation and applications of the three most important technologies: TFFs, AWGs, and fiber Bragg gratings. It is obvious that other than performance and cost there are several other important considerations that a designer has to keep in mind while selecting the appropriate technology. Some of the trade-offs were listed, and it was shown that each of these technologies has a certain niche area of operation in the mux/demux space.

6. REFERENCES

1. Z. Knittl, *Optics of Thin Films*, John Wiley, New York (1976).
2. M. K. Smit, *Electronics Letters*, **24**, 385 (1988).
3. From the website http://www.neophotonics.com.
4. K. O. Hill, Y. Fujii, D. C. Johnson, and B. S. Kawasaki, *Applied Physics Letters*, **32**, 647 (1978).
5. V. Bhatia, K. P. Reddy, M. A. Marro, and D. L. Weidman, *Trends in Optics and Photonics (TOPS) series*, WDM Components 12 (1999).
6. L. Dong, G. Qi, M. Marro, V. Bhatia, L. L. Hepburn, M. R. Swan, A. Collier, and D. L. Weidman, *Journal of Lightwave Technology*, **18**, 1583 (2000).

19
CHAPTER

Dispersion-Tailored Higher Order Mode Fibers for In-Fiber Photonic Devices

Siddharth Ramachandran*
OFS Laboratories
Murray Hill, New Jersey, USA

1. INTRODUCTION

In-fiber devices have attracted a lot of attention since the invention of the erbium-doped fiber amplifier. These devices have become mainstays in fiber optic communication systems, because they provide the advantages of low loss, polarization insensitivity, high reliability, and compatibility with the transmission line. They have several applications, ranging from signal conditioning (amplification, dispersion control, etc.) to network management (add/drop multiplexing, optical monitoring, etc.). Fiber waveguide engineering has enabled dispersion control of optical signals, and dispersion compensating fibers (DCFs) are the preferred commercially deployed dispersion management devices in transmission systems [1]. Fiber-based resonant couplers, fabricated by either fusing fibers or by inducing periodic perturbations (gratings within the core of a fiber), have become the device schematic of choice for a variety of functions, such as signal or pump multiplexing/demultiplexing (mux/demux) [2], optical monitoring, tunable dispersion compensation [3], and static [4, 5] as well as dynamic spectral shaping [6, 7].

Common to several fiber devices is the single-mode fiber (SMF), which serves as a platform to propagate the signal. Light propagates in the fundamental mode of the fiber, and the device effect itself is due to some extraneously introduced material or structure (dopants for amplification, gratings for phase matching, etc.). In this respect, they are similar to bulk optic devices.

There exists another relatively less explored degree of freedom afforded by fibers—the ability to copropagate more than one mode in so-called few-mode fibers. These fibers can be manufactured by standard processing techniques used to realize commercial transmission fibers and thus offer several of the economies of scale offered by fiber devices. Each mode may be designed to have a uniquely defined modal dispersion and propagation characteristic. Because most fiber devices, such as in-fiber gratings, fused couplers, and DCFs, primarily exploit changes in the phase of light as it propagates through the fiber device, a few-mode fiber guiding several modes with different propagation constants both expands the design space accessible to existing devices and facilitates novel device effects.

This chapter reviews the variety of fiber devices enabled by few-mode fibers—fibers that typically support two to four modes with suitably tailored dispersive properties. Table 19.1 summarizes the variety of application spaces in the field of photonics that these devices have impacted. Few-mode fiber devices demonstrated to date fall in three general classes: (1) novel device effects that are unique to few-mode fiber devices, such as dispersion-less filters and polarization-dependent loss (PDL) controllers; (2) applications that provide enhanced performance in comparison with conventional fiber devices, thus providing strong competition to existing devices, such as higher order mode (HOM) dispersion compensating modules (DCM) and very low loss and low cost variable optical attenuators (VOA); and (3) devices offering novel functionalities that can be potentially disruptive to current optical networking architectures, such as wavelength-continuous, broadband, tunable dispersion compensators.

The fundamental building block for few-mode fiber devices is a static or tunable mode converter that allows shuffling the optical signal between various modes of the fiber. The discussion in this chapter focuses on the use of long period fiber gratings (LPGs) as mode converters, but it must be noted that several other device schematics, such as fused fiber couplers (see Chapter 13) and bulk holographic devices, exist for providing this functionality and

*E-mail: sidr@ofsoptics.com

TABLE 19.1 Application space covered by photonic devices in few-mode fibers.*

Dispersion control	Signal power management	Switching/routing	Polarization control	Sensing
HOD Dispersion Compensation	*Dispersion-less bandpass filters*	*2×2 Optical Switches*	*Tunable PDL controllers*	*Broadband amplitude sensors*
• Lower nonlinearities	• Arbitrarily sharp	• Low loss all-fiber device	• Broadband	• Novel sensing mechanism
• Longer transmission distance	• ASE filtering	• Spatial mode transformation	• Enables PMD compensation	• Simpler diagnostics
	• λ definition in laser cavities	• Exploits modal diversity		• Low-cost assembly
Tunable dispersion control	*Variable optical attenuators*			
• Wavelength continuous	• Simple fiber assembly			
• Broadband	• Potentially very low cost			
• No pass band/bandwidth trade-offs.	• Use in EDFA control			
• Use in-line as well as at R_x				

could be used with few-mode fibers to achieve many of the device functionalities introduced here.

This chapter is organized as follows. Section 2 introduces the properties of few-mode fibers and provides the physical intuition that enables designing them with desired propagation characteristics for various modes. This is followed by Section 3, which introduces the theory and practice of LPGs (see Chapter 15), with an emphasis on showing how the device spectra and other characteristics are intimately related to the dispersive properties of the modes of a fiber. These principles are applied to obtain unique LPG spectra and properties in specially designed few-mode fibers, as described in Section 4. Following that, Section 5 elucidates demonstrations of novel device applications using the LPGs introduced in Section 4. After discussions on static applications as mentioned above, the rest of the chapter concentrates on tunable devices afforded by the few-mode fiber design space. Section 6 introduces the theory and practice of novel tuning effects achieved with LPGs written in these fibers, and Section 7 discusses the various device applications that have been afforded by them.

2. DISPERSIVE PROPERTIES OF FEW-MODE FIBERS

Figure 19.1a shows the canonical refractive index profile of a few-mode fiber, along with the mode profiles of two modes, in this case the LP_{01} and LP_{02} modes, at various wavelengths. A signature of HOMs is the existence of one or more "sidelobes" of power away from the central core region. Because the wave equations and boundary conditions governing modal properties in optics are analogous to the equations of motion governing the wavefunction

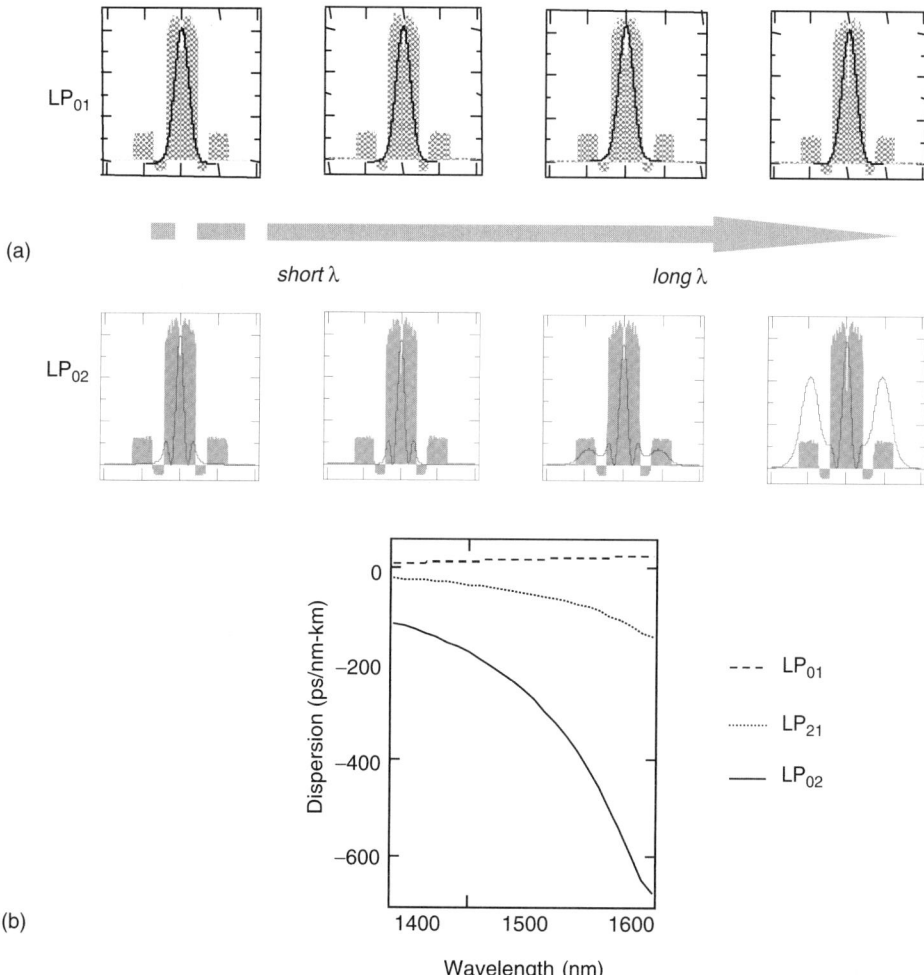

FIGURE 19.1 (a) Mode evolution with wavelength. Gray background is the refractive index profile. Different modes expand at different rates. Modal expansion intimately related to propagation properties. (b) Different modes exhibit significantly different dispersive behavior.

of a particle in a box in quantum mechanics, important physical insight can be gained from quantum mechanical analogies. The canonical refractive index profile of a few-mode fiber is akin to a potential distribution with three discrete segments—a core and a ring that have index greater than silica, which act as attractive wells for light, and a down-doped trench, which acts as a barrier. The indices and thicknesses of these three regions govern the "rate," as a function of wavelength, with which the mode transitions from residing primarily in the core, to "escaping" to the ring. Thus, a fiber can be designed with a given refractive index profile to yield a desired modal rate of expansion, as a function of wavelength. Moreover, the propagation constant, β, of a mode can be expressed in terms of the mode field and the refractive index profile as

$$\beta^2(\lambda) \approx \left(\frac{2\pi}{\lambda}\right)^2 \int\int n^2(r, \lambda) \cdot |E(r, \lambda)|^2 \cdot dA \quad (1)$$

where λ is the free-space wavelength of light, $E(r, \lambda)$ is the mode field distribution, $n(r, \lambda)$ is the refractive index profile of the fiber, and dA signifies an integration across the cross-sectional area of the fiber. Equation (1) shows that the propagation constant of a mode is approximately proportional to the average refractive index of the region in which light exists, weighted by the local intensity profile. Combining this concept with the intuition for designing fibers with specific mode profiles at various wavelengths yields the general recipe for tailoring the refractive index profiles of a fiber to obtain desired values for $\beta(\lambda)$, and all its wavelength-dependent derivates for any spatial mode guided by the fiber. Because $\beta(\lambda)$ and its wavelength-dependent derivatives, such as group velocity and dispersion, are essentially the properties of interest, the intuition elucidated above provides a powerful tool for designing fibers. For example, a mode expanding rapidly with wavelength would have a larger dispersion than a spatially stable mode, and this concept has been exploited to realize low loss high figure of merit DCFs [8, 9].

HOMs provide greater design flexibility in this respect. Note from Fig. 19.1a that the rate of expansion is not only different for different modes, it is also different for the various sidelobes, in the case of HOMs. In addition, a significant proportion of the power remains in the central core, indicating that rapidly expanding modes can be constructed with low bend losses. Previous reports have shown that HOMs can be highly dispersive, with dispersion as high as $-700\,\text{ps/nm}\cdot\text{km}$, while being low loss, which may be useful to realize low-loss compact dispersion compensators [10]. Figure 19.1b shows the dispersion of various modes in a fiber and shows that the modes of different orders are significantly different in dispersive behavior. This design flexibility in achieving a variety of desired propagation properties, in conjunction with the fact that light can be shuffled between various modes, makes feasible a variety of novel devices and effects, as described in the following sections.

3. MODE CONVERSION WITH LPGs: DEVICE PRINCIPLES

LPGs offer coupling between copropagating modes of a fiber and are attractive as spectral shaping elements and mode conversion devices. Coupling between two copropagating modes in a fiber occurs when the grating period is adjusted to match their beat frequency. This behavior can be characterized by a phase matching relationship given by

$$\delta(\lambda) = \frac{1}{2}\left(\Delta\beta(\lambda) - \frac{2\pi}{\Lambda}\right) \quad (2)$$

where δ is a detuning parameter, Λ is the grating period, and $\Delta\beta$ is the difference in propagation constants between the two modes (which is a function of wavelength λ). As a general rule, maximum mode coupling occurs at the resonant wavelength λ_{res}, where the $\delta = 0$ condition (the resonance condition) is satisfied. An intuitive picture of the spectral dependence of LPG couplers can be obtained by considering the phase-matching curve (PMC) of a fiber, which is a plot of the grating period with respect to wavelength at resonance [i.e., a plot of Λ vs. λ_{res} when $\delta = 0$, in Eq. (2)], shown in Fig. 19.2a. Figure 19.2b shows the corresponding grating spectrum obtained. The coupling magnitude decreases monotonically at wavelengths away from λ_{res}, that is, as δ departs from 0, in the PMC. Moreover, the bandwidth of the spectrum in Fig. 19.2b is strongly coupled to the gradient of the PMC (Fig. 19.2a). The slope of the PMC ($d\Lambda/d\lambda$) and the bandwidth of the corresponding spectrum $\Delta\lambda_{bandwidth}$, respectively, are given by

$$\frac{d\Lambda}{d\lambda_{res}} = -2\pi\frac{\Delta\beta'}{\Delta\beta^2} \quad (3)$$

$$\Delta\lambda_{bandwidth} \propto \frac{1}{\Delta\beta'} \quad (4)$$

where $\Delta\beta'$ is the slope with respect to wavelength of the difference in the propagation constants between the two modes, and all other terms have been

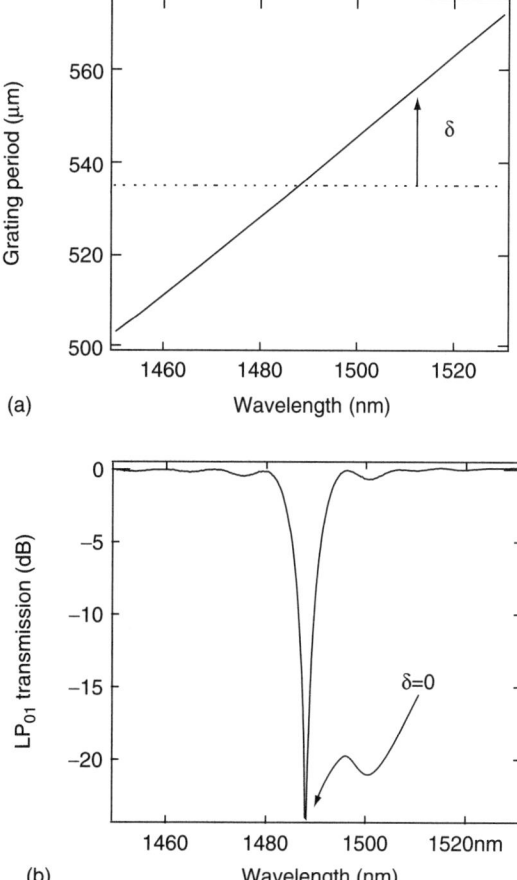

FIGURE 19.2 (a) Phase-matching curve (plot of Λ vs. λ when $\delta = 0$). Dashed line represents grating period. (b) Maximum coupling at $\delta = 0$, decreases as δ departs from 0.

coupled to the dispersive nature of the fiber itself. Equation (2) may be rewritten as a Taylor expansion about the resonant wavelength, λ_{res}

$$\delta(\lambda) = \frac{1}{2}\left(\Delta\beta'(\lambda_{\text{res}}) \cdot \Delta\lambda + \frac{\Delta\beta''(\lambda_{\text{res}})}{2!} \cdot \Delta\lambda^2 \right. \\ \left. + \frac{\Delta\beta'''(\lambda_{\text{res}})}{3!} \cdot \Delta\lambda^3 + \cdots \right) \quad (5)$$

where the primes represent derivatives with respect to wavelength and $\Delta\lambda$ is the difference between the wavelength of interest and the resonant wavelength. These derivatives are related to physically relevant parameters: $\Delta\beta' \approx \Delta\tau$, the differential group delay, $\Delta\beta'' \approx \Delta D$, the differential dispersion, and $\Delta\beta''' \approx \Delta D'$, the differential dispersion slope, and so on. Thus, the spectrum of an LPG may be precisely tailored by adjusting well-known fiber design parameters whose control was demonstrated in Section 2.

This provides an additional degree of freedom in the ability to obtain phase matching that is not available in conventional bulk optic diffraction gratings. LPGs in SMFs have exploited this design space to make sensitive strain and temperature sensors [11, 12], spectrally narrow acoustooptic gratings [13], variable optical attenuators [14], and thermally tunable LPGs [15]. Although some device design flexibility is available with the numerous cladding modes in SMF, fully exploiting the intimate connection between fiber design parameters and LPG spectra is feasible only with few-mode fibers, where both the fundamental mode as well as the HOM can be dispersion engineered, as demonstrated in the following sections.

defined earlier. Thus, the bandwidth of the resonance decreases as the quantity $\Delta\beta'$, and correspondingly the slope of the PMC, increases. The quantity $\Delta\beta'$ is proportional to the difference in group delays between the two modes, and Eqs. (3) and (4) can be understood in the following physically intuitive manner: At the resonant wavelength, the phases of the two modes are matched by the grating vector, yielding the condition $\delta = 0$ in Eq. (2), and maximal coupling. The strength of the resonance away from this wavelength is governed by the rate of dephasing between the two modes, which is governed by the next higher order Taylor term in the expansion of the propagation constant, namely the group delay.

More generally, the spectral response of an LPG is governed by the spectral evolution of the detuning parameter δ. Because the wavelength-dependent quantities in δ are the propagation constants of modes in a fiber, an LPG spectrum is strongly

4. LPGs IN DISPERSION-TAILORED FEW-MODE FIBERS

As illustrated in the previous section, LPG spectral engineering can be achieved by tailoring few-mode fibers with desired dispersive properties. This has proven especially attractive to construct broadband mode–conversion devices, which is described in detail in this section.

4.1. Broadband Mode Converters

The concept of broadband mode conversion was first demonstrated by Poole et al. [10] in a fiber designed to support the LP_{01} and LP_{11} modes with identical group delays at a predetermined wavelength. Note from Eq. (4) that the bandwidth $\Delta\lambda$ of an LPG is inversely proportional to the difference in group delays $\Delta\beta'$, and thus a fiber with identical group delays would be expected to yield very large bandwidths. Periodic microbends, used to couple between the fundamental mode and an LP_{11} mode in this fiber, yielded strong mode conversion ($>10\,dB$) over a bandwidth as large as 74 nm. The few-mode fiber with identical group velocities was designed by operating the HOM (LP_{11} mode, in this case) close to cutoff.

The design principle for such fibers is illustrated in Fig. 19.3, which shows the mode profiles of the fundamental mode as well as the LP_{02} mode at two different wavelengths. At a short wavelength $\lambda = \lambda_1$, when the LP_{01} and LP_{02} modes are well guided and reside predominantly in the core of the fiber (Fig. 19.3a, left), the ray picture (Fig. 19.3b, right) for waveguides accurately predicts that the group delay, $[\beta'(LP_{02})] > [\beta'(LP_{01})]$, because the LP_{02} mode travels at steeper bounce angles compared with the LP_{01} mode. At a longer wavelength $\lambda = \lambda_2$, most of the energy of the LP_{02} mode resides in the low-index ring (Fig. 19.3b), and the mode picture is better suited to gain an intuitive understanding of the mode evolution. As the wavelength of operation is increased, the power fraction of the LP_{02} mode in the low-index ring increases, and its group velocity approaches that of the ring (which is higher than that of the core, because it has a lower index of refraction compared with the core). Now, $[\beta'(LP_{02})] < [\beta'(LP_{01})]$, the LP_{01} mode continues to be well guided and thus no significant change is expected for $n_g(LP_{01})$. Hence, at an intermediate wavelength $\lambda_1 < \lambda_T < \lambda_2$, the group delays of the two modes match, as shown in Fig. 19.3c.

Thus, $\Delta\beta'$ changes signs, and this implies that the slope of the phase-matching relationship [Eq. (3)] changes polarity. The result is a turn-around point (TAP) in the PMC, as shown in Fig. 19.4a. Following the intuitive rule from Section 3, inspection immediately reveals that large bandwidths are attainable if the grating period (shown by the horizontal dashed line) is chosen to couple at the TAP. This is because the dashed line representing the grating period does not depart from the PMC over a large wavelength range. In other words, the resonance condition $\delta = 0$ is satisfied over a larger wavelength range.

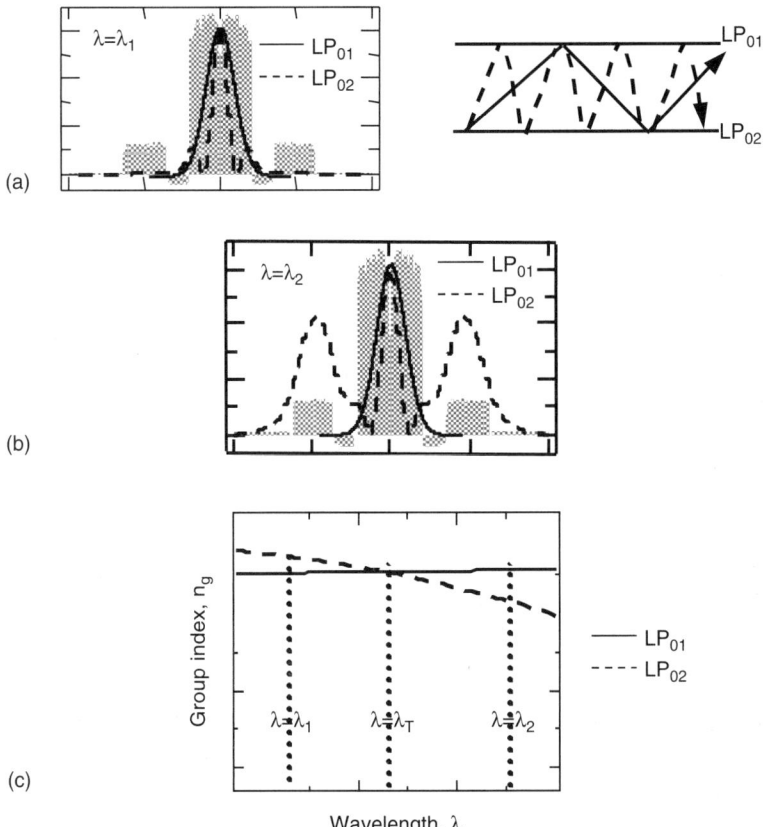

FIGURE 19.3 Dependence of modal delay on mode profile. Gray background is the refractive index profile. (a) Mode profiles at short $\lambda = \lambda_1$ (left), corresponding ray picture (right). (b) Mode profiles at long $\lambda = \lambda_2$; LP_{02} travels in lower index regions. (c) Modal delay for LP_{01} and LP_{02}; equal modal delays at intermediate wavelength $\lambda = \lambda_T$.

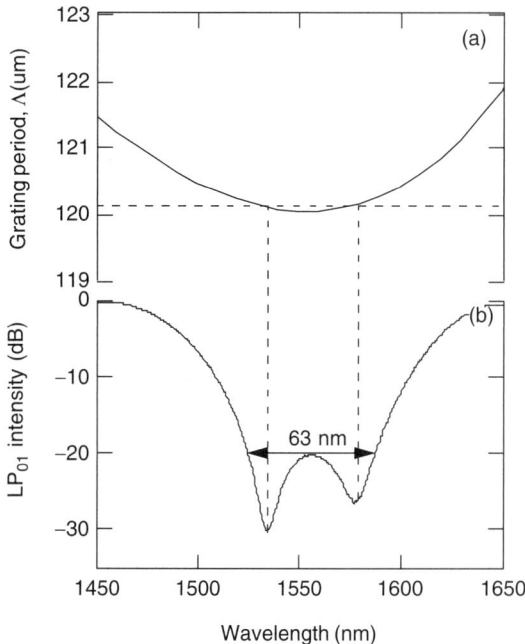

FIGURE 19.4 (a) Simulation: Grating phase matching curve in dispersion-tailored fibers. Large bandwidths obtained at TAP. (b) Experiment: Resultant TAP-LPG spectrum. Peak coupling >30 dB. More than 20 dB (99%) coupling over 63 nm. Horizontal dashed line indicates the grating period and shows the correspondence between the PMC and the grating spectrum obtained.

Fibers with the desired complex index profiles to yield TAPs in the PMC between the LP_{01} and LP_{02} modes were fabricated by modified chemical vapor deposition (MCVD). The LPGs were fabricated in hydrogen-loaded fibers by an ultraviolet laser imaged onto the fiber through an unchirped amplitude mask. The grating period was adjusted to yield a resonance at the TAP.

Figure 19.4b shows the spectrum of light in the LP_{01} mode after the LP_{02} mode has been stripped out for a 1-cm-long LPG in a fiber with a TAP at 1540 nm. The grating period for coupling at the TAP of this fiber is 120.2 μm—at this period the resonance of the PMC is satisfied at two distinct wavelengths because of the existence of a TAP, as shown in Fig. 19.4a. The resultant spectrum yields more than 99% mode conversion over a record wavelength range of 63 nm [16]. In comparison, a 20-dB strong resonance in a conventional LPG typically extends over only ~1 nm. The insertion loss of this LPG is <0.2 dB. In addition, the wavelength range of such coupling may easily be tuned by scaling the fiber dimensions, thus facilitating a mode converter that works over any wavelength range.

The polarization response of these LPGs is obtained by measuring the variation in output intensity of the LP_{02} mode, with respect to variations in the input polarization state. No measurable change is detected within the measurement accuracy of the external cavity laser and photodetector that is used (measurement accuracy ~0.02 dB). Significantly higher accuracy can be achieved by measuring the amount of residual fundamental mode in the fiber after the LP_{02} mode is stripped out—for a sufficiently strong LPG, a minuscule amount of the fundamental mode would remain, and small changes in its intensity would result in large variations on a log (dB) scale. Figure 19.5a shows the grating spectra for one of the strongest TAP LPGs fabricated; the peak coupling efficiency is ~45 dB, with a 40-dB bandwidth of 10 nm and 20-dB bandwidth of 43 nm. We choose the strongest LPG we have written for this investigation, because it requires the largest ultraviolet-induced index change Δn_{UV}, and the polarization dependence of gratings increases with Δn_{UV}. The inset shows the spectra of two orthogonal polarizations corresponding to the largest deviations in

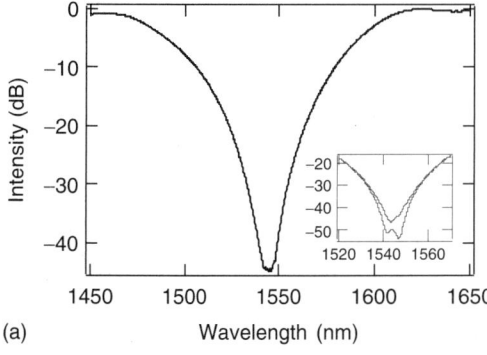
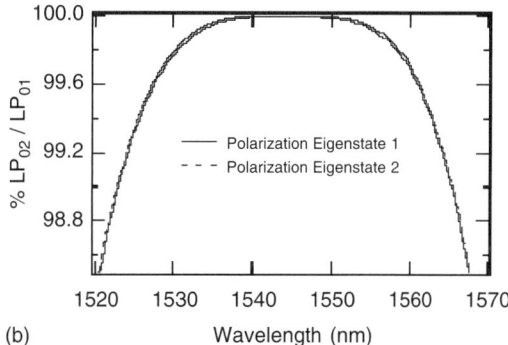

FIGURE 19.5 Polarization dependence of TAP-LPG. (a) Spectrum for a strong mode converter; inset shows polarization-dependent response of the residual fundamental mode. (b) Polarization response of the LP_{02} mode, deduced from data in (a). Spectra barely distinguishable. Peak coupling changes by 0.0002 dB, 20-dB bandwidth changes by 0.2 nm.

grating strength. Figure 19.5b shows the spectral response of the LP$_{02}$ mode calculated from the data in Fig. 19.5a. The spectrum is shown on a linear scale because small deviations from unity are more visible. Even then, the two spectra are barely distinguishable. Variations in input polarization caused the coupling efficiency to range from 99.9960% to 99.9996%, corresponding to a negligible change for applications in optical communications systems. The 20-dB bandwidth also shows a negligible change of ~0.2 nm [17]. The polarization insensitive response of these gratings is expected, because energy for both modes resides in the core of the fiber fabricated by MCVD. Small perturbations in the circularity of the index profile perturb both modes by similar amounts, and because LPG coupling is sensitive only to the difference in the propagation constants (and not the propagation constants themselves) of the two modes, negligible polarization effects are observed.

It is clear that LPGs in dispersion-tailored fibers possessing a TAP provide an excellent platform for building broadband mode converters. The fiber device is naturally adapted for connecting to a fiber-based transmission line and has very low insertion loss and practically no polarization dependence. These properties make it very attractive for numerous applications that use broadband mode converters, as described in Section 5.

4.2. Bandwidth Control of TAP-LPGs

The bandwidth of TAP-LPGs cannot be determined from Eq. (4), because the difference in modal group delays $\Delta\beta'$, is zero at the TAP wavelength. Expanding $\Delta\beta$ as a Taylor expansion [Eq. (5)] and retaining the next higher order term (the next higher order term is the modal dispersion, $D \approx d^2\beta/d\lambda^2$), the 20-dB bandwidth of TAP gratings can be obtained as [16]

$$\Delta\lambda \propto \frac{\lambda_{res}}{\sqrt{\Delta D \cdot L_g}} \qquad (6)$$

where ΔD is the difference in dispersion between the two modes and L_g is the length of the grating. As is expected, when the first-order term in the Taylor expansion of β is zero ($\delta\beta' = 0$, identical group delays), the next order term, differential dispersion $\delta\beta''$, plays a role in determining the bandwidth of the resonance. Just as the gradient of the PMC was related to the difference in group delays $\delta\beta'$, the curvature of the PMC at the TAP is controlled by the differential dispersion. This is shown in Fig. 19.6a, which plots the PMCs for a fiber with a low differential dispersion ΔD at the TAP and one with a high ΔD, respectively. As is evident, the curvature of the PMC at the TAP is lower for the fiber with lower ΔD. Following the insight gained in Section 3, it may be inferred that the PMC with lower curvature at the TAP (lower ΔD) yields a broader LPG bandwidth. This is because the condition $\delta = 0$ is satisfied or nearly satisfied over a larger wavelength range for such a fiber. Indeed, this is also predicted by Eq. (6).

This concept is tested by fabricating several fibers with different differential dispersion values (ΔD) at the TAP. Figure 19.6b is a comparison of the spectra of TAP-LPG written in fibers with ΔDs of ~160, 270, and 520 ps/nm·km, respectively [16]. As expected, fibers with larger differential dispersion show narrower resonances. Note that the values quoted here are the differential dispersion values (i.e., $\Delta D = D_{01} - D_{02}$) and do not correspond to the dispersion of the LP$_{01}$ or LP$_{02}$ mode alone. In fact, TAP grating bandwidths are independent of the dispersive characteristics of individual modes and are a function only of the difference in dispersion values between the two modes.

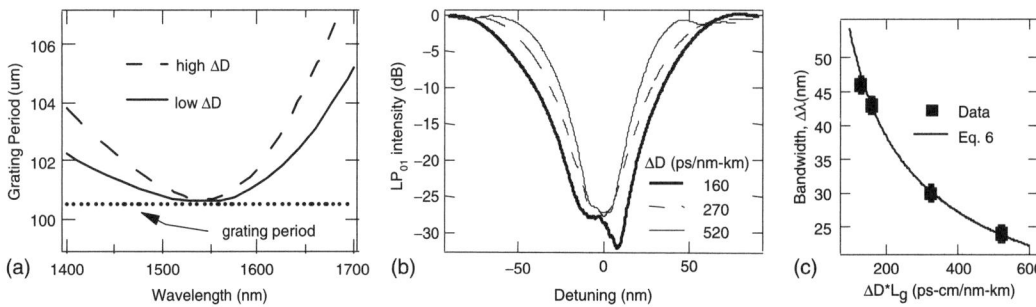

FIGURE 19.6 Control of the LPG spectra by fiber design. (a) PMCs for fibers with low and high ΔD. Higher curvature at TAP for high ΔD fiber. (b) Experimental spectra show that the 20-dB bandwidth increases as ΔD of the fiber decreases. (c) Grating bandwidth vs. $\Delta D \cdot L_g$; L_g = grating length. Experimental data shows an excellent match with theoretical predictions of Eq. (6).

Equation (6) indicates that the bandwidth of TAP gratings is inversely proportional to $\sqrt{\Delta D \cdot L_g}$. Figure 19.6c shows the experimental and theoretical dependence of the 20-dB grating bandwidth with respect to the parameter $\Delta D \cdot L_g$. The excellent match between the prediction and the data indicates that Eq. (6) accurately describes the behavior of LPG spectra in dispersion tailored fibers.

4.3. Spectrally Flat Coupling for VOA

The previous section illustrated how a fiber can be designed to achieve a variety of grating bandwidths by controlling the differential dispersion. This concept can be extended to unique fiber designs in which several orders in the Taylor expansion of $\delta\beta$, as shown in Eq. (5), can be matched. Figure 19.7 shows the dispersive properties of a few-mode fiber that was designed and fabricated to yield identical group delays, dispersions, and dispersion slopes for the LP_{01} and LP_{11} modes. This implies that several Taylor terms of Eq. (5) representing detuning δ, are negligible in value. Thus, δ remains close to zero over a large wavelength range, and phase matching is maintained over this enhanced wavelength range. The corresponding PMC for this fiber is shown in Fig. 19.8. Also shown for comparison are PMCs for conventional LPGs, as well as the TAP-LPGs described in the previous sections. The PMC is essentially flat throughout a spectral range spanning an octave. As is evident, the line representing the grating period is coincident with the PMC over the whole spectral range. Note that the PMC for this fiber is not strictly horizontal (as designed) but possesses a small amount of curvature due to small differences in the dispersive properties for the two modes. Nevertheless, the deviation is negligibly small, and gratings coupling the LP_{01} and LP_{11} modes of this fiber would then be expected to be not only broadband, but also spectrally insensitive to coupling efficiency.

Microbend-induced gratings were obtained by pressing the fiber between two corrugated aluminum blocks that were machined to yield the desired periodicity of corrugations. Figure 19.9a shows the spectrum of a 1-cm-long microbend grating induced in this fiber. The 3-dB spectral width is a record value of 565 nm, and because the dispersive parameters are especially well matched in the 1500-nm range, the coupling is also spectrally flat there. The flat spectral region extends over more than 100 nm, as is evident from Fig. 19.9a, and the insertion loss along with the mode strippers is only 0.2 dB, as is expected of microbend gratings [18].

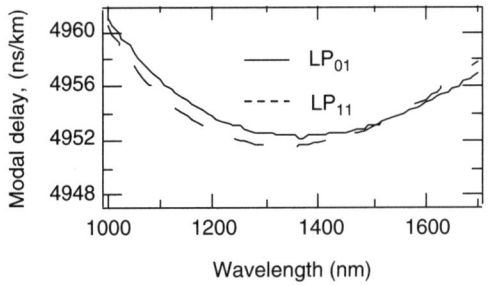

(a)

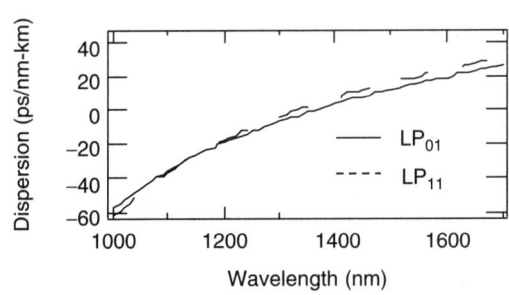

(b)

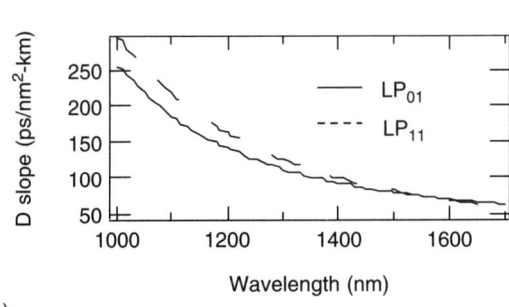

(c)

FIGURE 19.7 Taylor terms [Eq. (5)] for LP_{01} and LP_{11} modes. (a) Group delay; (b) dispersion; (c) dispersion slope. Close match of curves dramatically enhances wavelength range of phase matching.

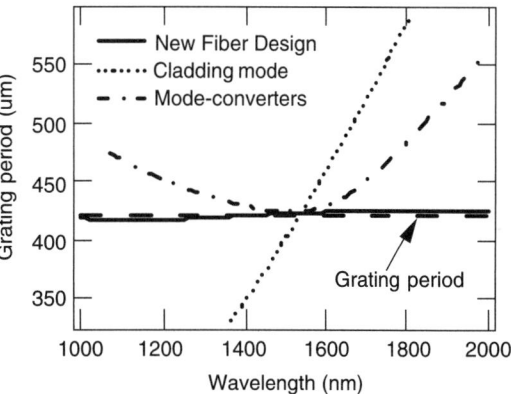

FIGURE 19.8 Grating PMC for various fibers. PMC of new few-mode fiber design is almost invariant with wavelength. Implies phase matching is wavelength insensitive over a large spectral range.

Microbend coupling efficiency is a function of the pressure applied by the corrugated blocks on the fiber. This characteristic can be used to construct a cost-effective wavelength-insensitive VOA, as shown in Fig. 19.9b, because these gratings exhibit flat spectra over bandwidths as large as 110 nm. Continuous tuning between 0 and 10 dB is achieved with a spectral tilt of less than 0.25 dB over the 110-nm spectral range. Because no fiber-coating removal is required to assemble this device and more than a million 1-cm long fiber devices can be obtained from a single preform, this device schematic provides a compact, low-loss, broadband, and potentially cost-effective VOA.

5. STATIC DEVICES USING TAP-LPGs

Section 4 introduced a variety of spectral shaping tools accessible by designing dispersion-engineered fibers. One of the most important devices to arise from these concepts is the broadband mode converter using TAP-LPGs. This section illustrates the device applications it has impacted.

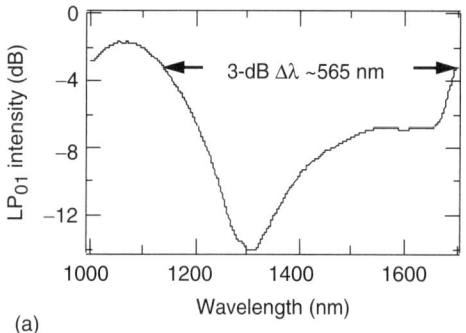

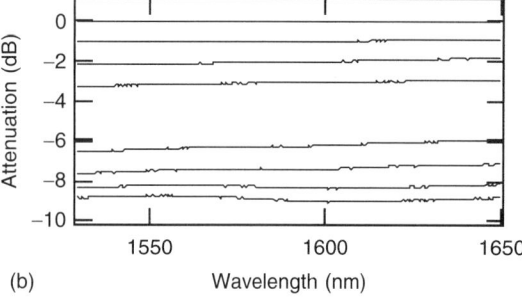

FIGURE 19.9 Microbend grating spectrum in novel few-mode fiber. (a) Grating length = 1 cm. Record 3-dB bandwidth (565 nm) due to flat PMC shown in Fig. 19.8. Spectrally flat coupling in 1550-nm region, where dispersion parameters of fiber are identical (see Fig. 19.7). (b) VOA operation in flat spectral region. Continuous tuning from 0 to 10 dB in C- and L-band, with <0.25-dB spectral tilt.

5.1. HOM-DCM

Long-distance high-speed transmission over broad bandwidths requires careful management of dispersion and nonlinear distortions of the transmitted optical signal. Dispersion, being a linear distortion, can be compensated for by a DCM that provides equal dispersion of the opposite sign. Systems requirements dictate that the DCM of choice must be able to compensate for the variety of dispersion slopes of the transmission fibers used in optical links while being low loss and resistant to nonlinear distortions.

The DCMs widely deployed today are single-mode DCFs, which are relatively low loss and can offer dispersion slope compensation for a wide variety of transmission fibers [19]. However, single-mode DCF solutions result in propagation through small effective area fiber ($A_{\text{eff}} \approx 15–20 \, \mu\text{m}^2$), which causes nonlinear distortions in the signal. Moreover, this technology is also not amenable to providing tunable dispersion in a practical scheme.

An alternate approach is to compensate for dispersion using HOMs, which has received a lot of attention in the recent past due to its ability to offer dispersion slope matching to any transmission fiber currently deployed [20, 21] (Fig. 19.10a). As mentioned in Section 2, the dispersive nature of HOMs results from transitions in the sidelobes of their mode profile, with respect to wavelength, while maintaining significant energy in the central core. Thus, they can be expected to yield low loss and high dispersion. HOM fibers with propagation losses of only 0.4 dB/km and dispersion as high as −220 ps/nm·km have been fabricated, yielding record figures of merit (= dispersion/loss ≈ 550 ps/nm·dB). In addition, the effective area of a HOM is approximately four to five times larger than the fundamental mode in DCFs (Fig. 19.10b), which decreases the impact of nonlinearities.

The schematic of the HOM-DCM is shown in Fig. 19.11. The device consists of a spool of HOM fiber spliced onto two TAP-LPG based mode converters that convert the LP_{01} mode into the LP_{02} mode and vice versa at the input and output of the module, respectively. TAP-LPGs, described in Section 4.1, are ideal choices for mode conversion because they are naturally fiber coupled and offer both broadband and low-loss operation critically required for DCMs. The higher resistance to nonlinear distortions in HOM-DCMs is tested by transmission experiments with 40 Gb/s return to zero (RZ) signals. Links amplified by erbium-doped fiber amplifiers

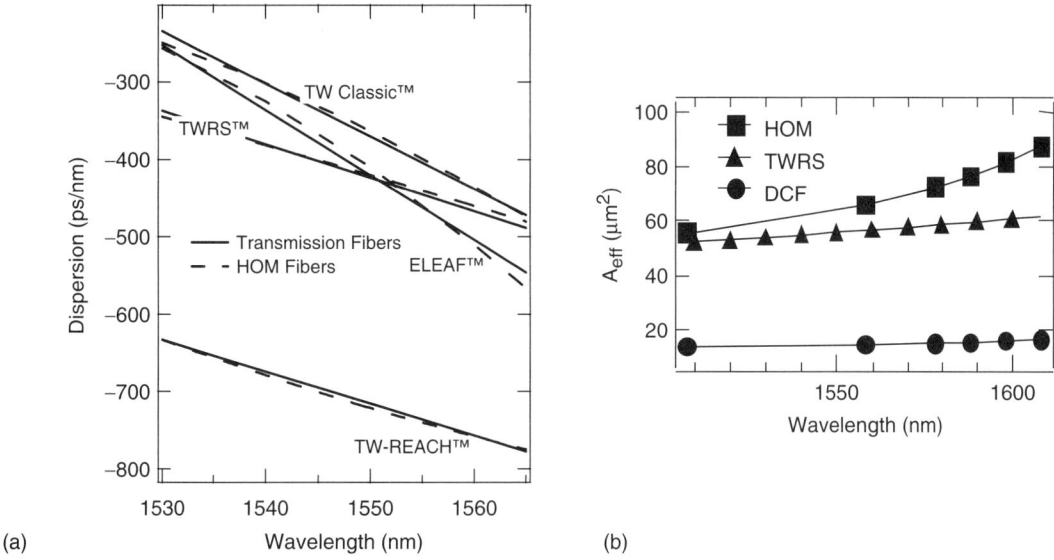

FIGURE 19.10 HOM fibers for dispersion compensation. (a) Solid lines: dispersion compensation requirement for 100-km link with different transmission fibers. Dashed line: corresponding HOM fiber characteristics. Variety of dispersion and dispersion slopes achievable with LP_{02} match dispersion slope of every transmission fiber type. (b) Effective area of HOM ≈ 4-5 × DCF ⇒ lower nonlinearities.

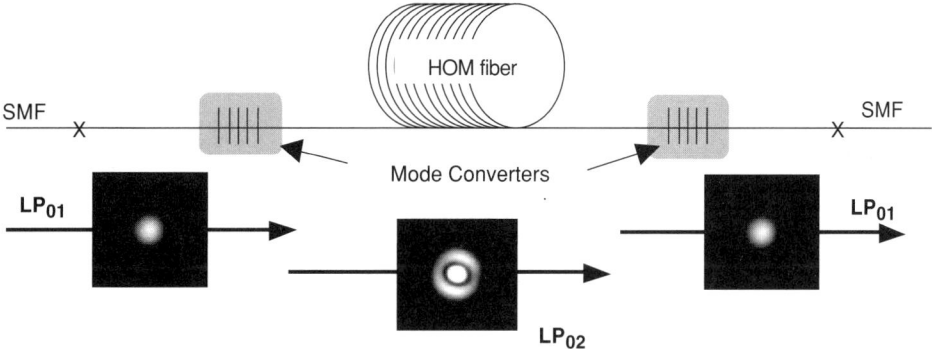

FIGURE 19.11 HOM-DCM schematic. HOM fiber with input and output LP_{01}-LP_{02} TAP-LPG mode converters.

could transmit signals over 12 × 100 km of True-Wave® reduced slope (TWRS) fiber when the HOM-DCM was used and 8 × 100 km when the single-mode DCF was used (Fig. 19.12a) [22]. The 50% increase in transmission distance is attributed to the ability to increase input power into the HOM-DCMs by 10 dB. For hybrid Raman/erbium-amplified systems, 40-Gb/s transmission distances of 1700 km have been enabled by HOM-DCMs, because the optical signal to need ratio was improved by 1 dB in comparison for with an identical link compensated with DCF (Fig. 19.12b) [23]. The transmission distance improvements were primarily enabled by the large effective area for the LP_{02} mode and the use of low-loss polarization-insensitive mode converters fabricated with TAP-LPGs.

5.2. Dispersionless Bandpass Filtering

Broad bandwidth mode converters described in Section 4.1 can also be used as spatial mode transformers to realize band selection filters [24], which have numerous applications such as frequency selectors for fiber-laser cavities and amplified spontaneous emission noise filters in optically preamplified receivers.

The device requires two LPGs that are fabricated in identical fibers that are dimensionally scaled by different amounts—that is, the fiber is drawn from the same preform to different outer diameters (ODs). When a few-mode fiber is dimensionally scaled, the phase-matching relationship shifts such that the TAP now occurs at a different wavelength and grating period, as noted in Section 4.1. This is

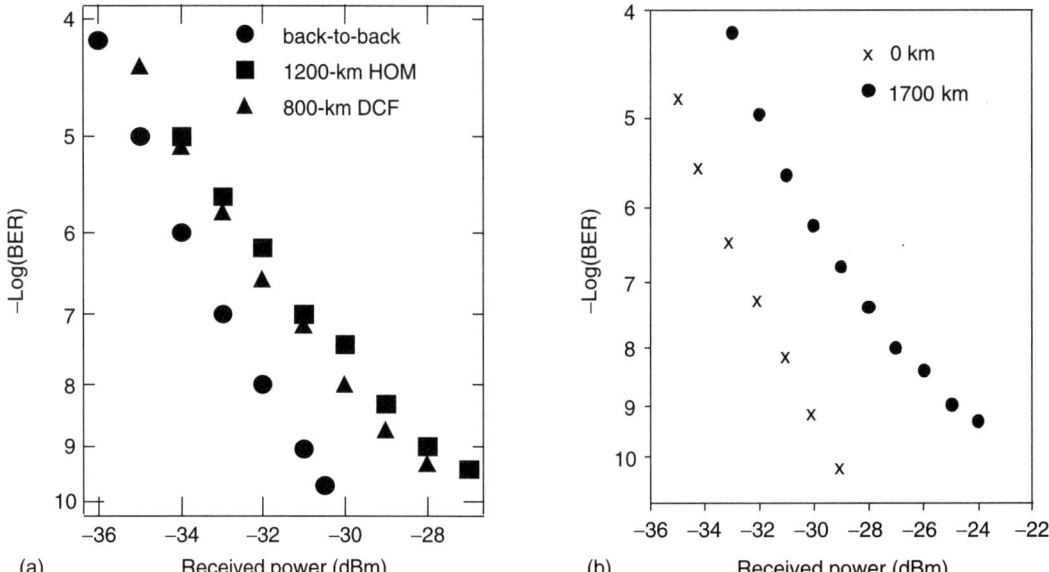

FIGURE 19.12 Transmission experiments with 40 Gb/s RZ signals. 100-km TWRS® + HOM-DCM or DCF. (a) Erbium amplified system—50% transmission distance increase with HOM. (b) Hybrid erbium/Raman system—1-dB optical signal-to-new ratio improvement, enabling 1700-km transmission.

illustrated in Fig. 19.13, which shows the PMC for fibers drawn to ODs ranging from 110 to 121 μm. Figure 19.13 shows that while an LPG written to couple at the TAP of a few-mode fiber yields a broadband mode converter, an identical LPG written in the same fiber drawn to a different OD yields a narrowband resonance expected of conventional LPGs.

The schematic of the bandpass filter is shown in Fig. 19.14a, which illustrates that LPGs in few-mode fibers drawn to two different ODs (denoted I and II) are spliced together to realize the bandpass filter. The spectra obtained by writing LPGs in these two fibers are shown in Fig. 19.14b. The PMC for fiber I (OD = 121 μm) has a TAP at 1540 nm, and an LPG written at the corresponding grating period ($\Lambda = 112.5$ μm) converts the incoming LP_{01} mode

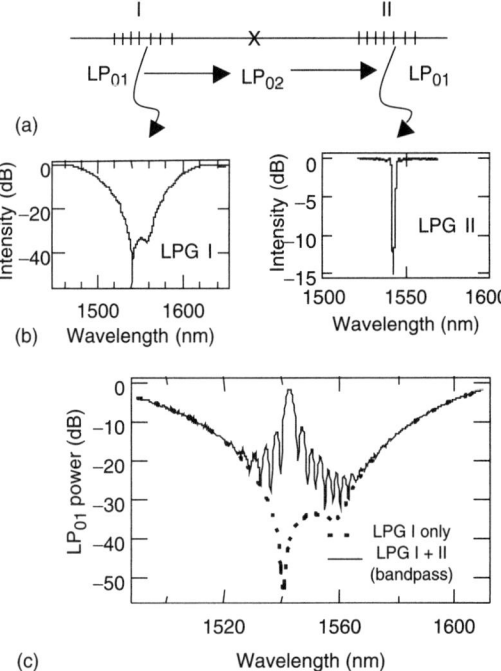

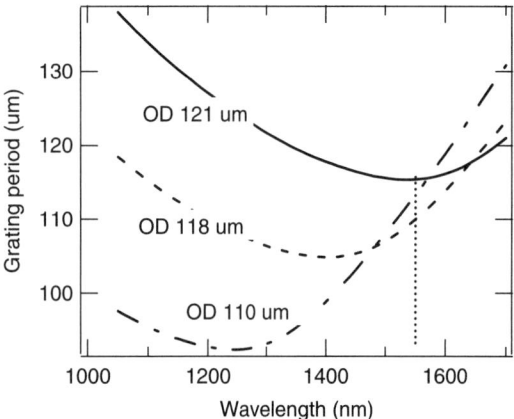

FIGURE 19.13 PMC for LPG in few-mode fibers of different ODs. TAP shifted by dimensionally scaling HOM fiber.

FIGURE 19.14 (a) Bandpass filter schematic. LPG in fiber I and II spliced. Splice ensures adiabatic transition of LP_{02} mode from fiber I to fiber II. Arrows show dominant mode of propagation. (b) LP_{01} transmission spectra of LPG I and II. LPG I, broadband TAP resonance; LPG II, conventional narrowband resonance. (c) Resultant bandpass spectrum. Spectrum defined by characteristics of LPG II alone.

into the LP_{02} mode over the entire C-band. Fiber II is a similar few-mode fiber but is drawn to an OD of 112 μm. In the 1550-nm spectral range, the PMC for this fiber exhibits a monotonic variation, as in conventional LPGs. Figure 19.14b shows the spectrum of a 10-cm-long uniform LPG written in fiber II.

All light in an entire communications band passing through LPG I is converted to the LP_{02} mode. Splicing the two fibers converts the LP_{02} mode of fiber I adiabatically into the LP_{02} mode of fiber II. Then, LPG II selects a desired narrow portion of the spectrum back into the LP_{01} mode. This results in a bandpass filter, as shown in Fig. 19.14c. The 3-dB bandwidth of this filter is ∼2.5 nm, and the peak isolation is as high as −35 dB.

Spectral filters are typically dispersive close to their band edges. For instance, the complex amplitude response of a uniform LPG, as deduced from coupled-mode theory is

$$E_{in} = \left\{ \cos^2\left(L\sqrt{\kappa^2+\delta^2}\right) + \frac{\delta^2}{\kappa^2+\delta^2} \sin^2\left(L\sqrt{\kappa^2+\delta^2}\right) \right\}^{1/2}$$
$$\cdot \exp\left[i\,\mathrm{Tan}^{-1}\left\{\frac{\delta}{\sqrt{\kappa^2+\delta^2}} \tan\left(L\sqrt{\kappa^2+\delta^2}\right)\right\}\right]$$

$$E_{out} = \left\{\frac{\kappa}{\sqrt{\kappa^2+\delta^2}} \sin\left(L\sqrt{\kappa^2+\delta^2}\right)\right\} \cdot \exp[i\pi] \quad (7)$$

where E_{in} is the electric field amplitude of the mode that contains all the power at the input of the LPG, E_{out} is the amplitude of the mode that light is scattered into (which had zero power at the input of the grating), κ is the coupling coefficient, and δ is the detuning. This reveals that while the phase response of E_{out} is not a function of wavelength, the phase response of E_{in} is strongly coupled with its magnitude response. As an LPG becomes spectrally sharp or narrow, the magnitude of E_{in} varies rapidly with wavelength; likewise, the phase response also shows strong wavelength dependence. Thus, it is expected that a signal that passes through a sharp LPG-based band-rejection filter suffers from dispersion-related penalties. On the other hand, the scattered mode E_{out}, has no spectral phase variation and would always be dispersion free.

Because the TAP-LPG–based bandpass filters transmit only the "cross" arm (E_{out}) from LPG II, they are expected to be inherently dispersion free. Figure 19.15a shows the spectra of the gratings in the conventional notch-filter and bandpass configuration, respectively. Figure 19.15b shows their corresponding group delay response. For the notch filter, the group delay rises sharply with wavelength, resulting in dispersion values as high as ±10 ps/nm close to the resonance. This dispersion penalty can potentially limit the performance of 40-Gb/s systems. When the same grating is deployed in the bandpass configuration, the group delay remains constant throughout the resonance, implying that these bandpass filters are dispersion free.

Finally, tuning these bandpass filters simply entails tuning LPG II and thus is amenable to all existing tuning mechanisms for conventional LPGs.

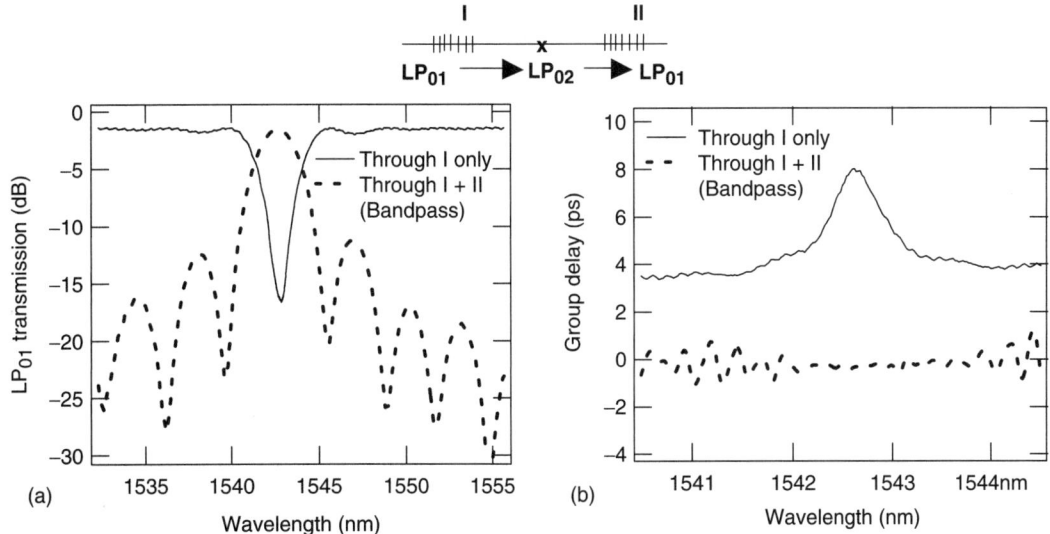

FIGURE 19.15 Transmission (a) and group delay (b) response for conventional and bandpass filters. Conventional filter represented as solid line and bandpass filter shown as dashed line. 3-dB Δλ same for both (2.5 nm). Conventional filter: $D \approx \pm\,10$ ps/nm. Bandpass filter is dispersion free.

Figure 19.16 illustrates the temperature tuning capability of these bandpass filters. Increasing the temperature of the package housing LPG II over a range of 180°C monotonically moves the bandpass resonance from 1538 to 1564 nm. The insertion loss of the filter is only 0.4 dB. The loss varies by approximately 0.2 dB as the filter is tuned over 26 nm.

The device illustrated in Fig. 19.14 provides a general platform for building band selection as opposed to band rejection filters with LPGs. The first mode converter (LPG I) only serves as a device that provides a spatially modified input of light for LPG II and does not define the spectral characteristics of the bandpass filter. The spectrum of the filter is uniquely defined by the inverted spectrum of LPG II alone. This enables the prospect of spectral shaping in the bandpass configuration by varying the spectral properties of LPG II. In addition, this device schematic provides the means to make very narrowband as well as tunable wavelength selectors that are dispersion free regardless of bandwidth.

6. TUNABLE LPGs IN HOM FIBERS

Tuning mechanisms for conventional LPGs comprise either changing the grating period by strain or shifting the PMC by some mechanism that alters the dispersive properties of one or both modes that are coupled through the grating. These mechanisms include the thermo-optic effect (induced by temperature-induced changes of the waveguide properties), stress-optic effect (pressure induced refractive index changes), electro-optic effect, or any other effect that essentially changes the refractive index profile of the fiber. This leads to changes in the propagation constants of one or both modes, as is evident from Eq. (1) in Section 2, which in turn shift the PMC [as is clear from Eq. (2) in Section 3]. Correspondingly, the wavelength at which the line representing the grating period intersects the PMC shifts (Fig. 19.2), moving the entire LPG spectrum to be centered at the shifted resonant wavelength. Strain directly changes the grating period (and thus the position of the grating line in Fig. 19.2a) to shift the resonant wavelength. In either case, the tuning mechanism does not dramatically change the gradient of the PMC, and thus tuning results in reproduction of an identical spectrum at a new resonant wavelength. Such tuning effects have resulted in a variety of tunable devices comprising conventional LPGs and have also been applied to tuning the dispersion-free filters described in Section 5.2. On the other hand, tuning TAP-LPGs results in amplitude changes in the spectrum rather than wavelength shifts due to characteristics unique to their PMCs. This novel effect is described in this section and is applied to novel device functionalities in the following sections.

6.1. Amplitude Modulation: Novel Detuning Effects in TAP-LPGs

Figure 19.17a shows the now-familiar PMC of TAP-LPGs, along with three horizontal dashed lines representing grating periods Λ_T, Λ_A, and Λ_B, where Λ_T corresponds to grating coupling at the TAP and the two other grating periods lie in regions where they do not intersect the PMC at any wavelength. The phase matching condition of Eq. (2) may be recast in terms of this PMC as

$$\delta = \pi \left(\frac{1}{\Lambda_{TAP}(\lambda)} - \frac{1}{\Lambda} \right) \quad (8)$$

where Λ_{TAP} is the grating period required for resonance along the PMC (i.e., the plot of the curve shown in Fig. 19.17a) and Λ is the actual LPG period, taking values Λ_T, Λ_A, and Λ_B, shown in Fig. 19.17a. The corresponding spectra obtained at these periods are shown in Fig. 19.17b. For $\Lambda = \Lambda_T$, coupling occurs at the TAP, and the familiar broadband resonance, described in Sections 4.1 and 4.2, is obtained. However, for a grating with $\Lambda = \Lambda_A$ and $\Lambda = \Lambda_B$, no resonance occurs because the grating period does not intersect the PMC at any wavelength. In terms of Eq. (8), this represents a condition where $\delta \neq 0$ for any wavelength. In this case, the spectrum somewhat resembles that of a TAP resonance but with diminished coupling strength. Moreover, the coupling efficiency for the grating with $\Lambda = \Lambda_B$ is weaker than that for the grating with $\Lambda = \Lambda_A$, because δ is larger in magnitude throughout the spectral range. Thus, the

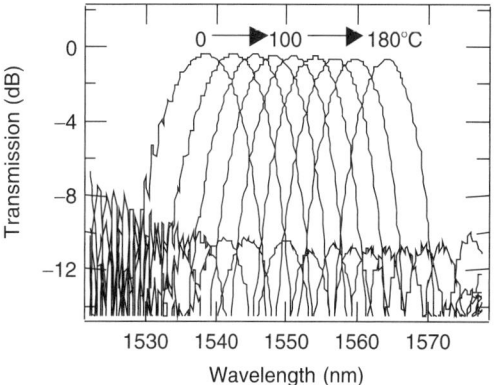

FIGURE 19.16 Tunability and loss characteristics of bandpass LPG. Tuning requires tuning of LPG II only. Thermal tuning of filter −26-nm range with 180°C temperature change. Filter insertion loss ≈ 0.4 dB.

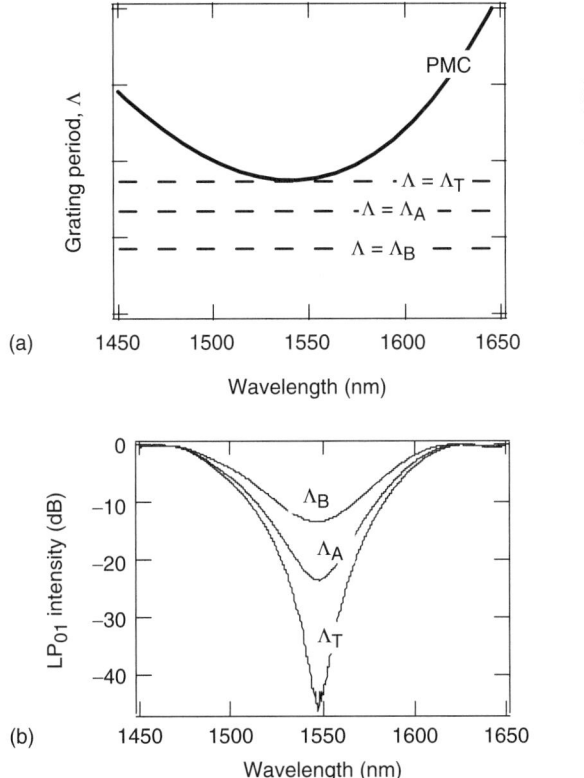

FIGURE 19.17 Novel detuning effects in TAP-LPGs. Coupling efficiency decreases as grating period departs from TAP period Λ_T. Grating periods Λ_A and Λ_B: No phase matching at any λ. Hence diminished resonance with same shape.

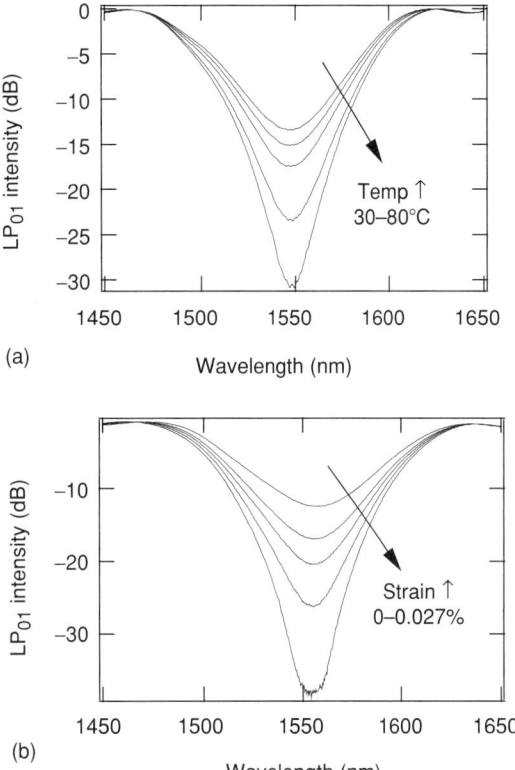

FIGURE 19.18 TAP-LPGs as sensitive temperature (a) and strain (b) sensors. 1-cm-long LPG: 1-dB change in transmission for 4°C temperature shift or 0.002% strain.

coupling efficiency for TAP-LPGs decreases roughly monotonically as the grating period departs from the TAP grating period Λ_T, while maintaining its spectral shape. This is in significant contrast to conventional LPGs, where changing the grating period would merely shift the resonance in wavelength.

This offers a powerful new tuning mechanism that can be applied to the assembly of cost-effective temperature, strain, or ambient index sensors. Changes in the control parameter (strain, temperature, etc.) would serve to change the throughput transmission of the LPG, leading to a sensing system that comprises a cost-effective broadband light-emitting diode (LED) connected to a power detector though a TAP-LPG. Figure 19.18 shows the spectral changes to a TAP-LPG in response to strain and temperature changes. Note that the sensitivity obtained by this mechanism (1-dB change in transmission for 4°C temperature shift or 0.002% strain for a 1-cm-long TAP-LPG) is significantly higher than that for conventional LPG sensors, because small changes in power levels can be measured much more accurately than small wavelength shifts. Moreover, the sensitivity increases linearly with the length of the grating, and because the bandwidth and the length of the grating can be decoupled by designing a fiber with desired dispersive properties, arbitrarily high sensitivities can be achieved.

6.2. Switching and Routing

The novel amplitude tuning mechanism demonstrated in the previous section can be applied for assembling two-state switches instead of continuous modulators. Toggling the LPG between a state that represents coupling at the TAP and another that yields identically zero coupling over a broad bandwidth yields 2×2 switches, as shown in Fig. 19.19. Figure 19.19a shows the spectrum of the switch in the cross- and bar states, respectively. Because LPGs are reciprocal devices, input light in either mode can be switched to an output, again in either mode, simply by switching the LPG between the cross- or bar states, as the experimentally measured mode profiles in Fig. 19.19b show. The switching action was induced by packaging the LPG in a stainless steel package, which when heated applies a strain

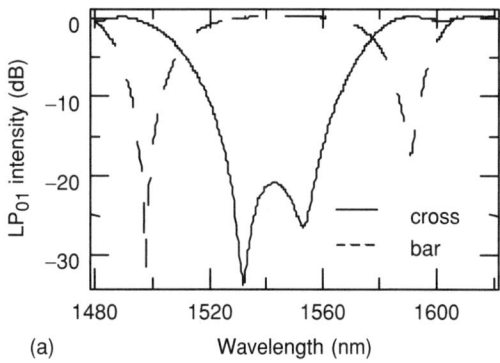

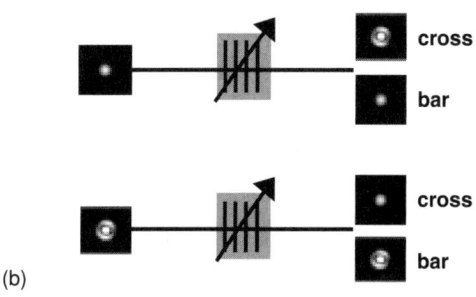

FIGURE 19.19 2 × 2 switches with TAP-LPGs. (a) Spectrum of switch showing strong (cross-state) or no (bar state) mode

as well as a temperature shift in the PMC. Thus, only moderate temperatures are required for the switching action—switching these LPGs from the cross- to the bar states involved changing the temperature on the packages between 50°C and 120°C.

7. TUNABLE TAP-LPG DEVICES

Section 6 introduced some novel tuning capabilities of TAP-LPGs, which allows for both continuous operation and discrete switching. In analogy with the static devices, the tunable LPGs offer a low-loss means to achieve device functionality, and the use of few-mode fibers enables dispersion tailoring of the spectral responses. This section describes examples of devices that exploit the continuous tuning capability as well as the discrete switching functionality of TAP-LPGs.

7.1. Tunable/Adjustable HOM-DCMs

As mentioned in Section 5.1, dispersion management is a critical requirement in optical communications links. Statistical variations in the dispersion of transmission fibers, amplifier-hut spacings, or ambient conditions can lead to significant transmission penalties, because of variations in accumulated dispersion.

One way to address this problem is by introducing tunable dispersion compensators (TDCs) that can provide either dynamic or setable control. Most TDCs demonstrated to date rely on the phase response of optical filters, which have bandwidth versus tunability trade-offs and operate only within well-defined spectrally periodic passbands.

Because few-mode fibers guide more than one mode with different group delays and dispersion, the inherent optical path diversity is naturally suited to realize TDCs that can provide setable control. Figure 19.20a shows the device schematic for an HOM device, which comprises five segments of HOM fibers, arranged in a binary length progression, with switchable TAP-LPGs (see Section 6.2) between each segment. Because this HOM fiber supports the LP_{01} and LP_{02} modes, two optical paths exist in each segment, leading to $2^5 = 32$ distinct optical paths (and thus 32 distinct dispersion values). Figure 19.20b shows the dispersion tuning capability of this device [25]: A tuning range of 435 ps/nm is achieved over a 30-nm bandwidth, with an average insertion loss of only 3.7 dB, which is lower than that of colorless TDCs constructed with any other technology.

To test the performance of these TDCs, carrier-suppressed RZ signals were transmitted through a span comprising TWRS™ and TW-Reach™ fibers (span dispersion about +377 ps/nm at 1545 nm). The device switch state was adjusted to yield −378 ps/nm

conversion. (b) Near field images of the modal output at the cross-state and bar state.

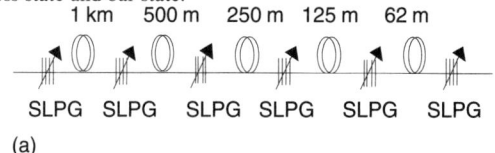

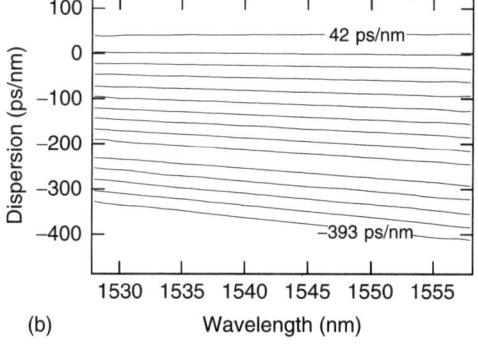

FIGURE 19.20 (a) Adjustable HOM schematic. Binary length progression of HOM fibers with switchable LPGs as

at 1545 nm. The open eye in Fig. 19.21a qualitatively confirms the distortion-free dispersion compensating capability of the device. To quantify this, bit error rate measurements were conducted for the same span at 1535, 1545, and 1555 nm, respectively. Figure 19.21b shows that the bit error rate curves at all three wavelengths essentially overlap with the back-to-back measurement, indicating that the device is broadband and offers penalty-free performance.

The dispersion-tuning characteristic of the device was tested by measuring receiver sensitivities at a bit error rate (BER) of 10^{-9} for 40 Gb/s 1545-nm signals. Several different transmission fiber spans (comprising combinations of TWRS® and TW-REACH™ fibers) with different lengths were assembled, and the adjustable HOM-DCM was tuned to the appropriate values to yield the closest match in dispersion. Figure 19.21c shows the receiver sensitivity for the different spans (with different dispersion values) compensated by the TDC tuned to the corresponding optimal state. The two horizontal dashed lines show back-to-back receiver sensitivities from two separate measurements, indicating 0.4-dB measurement uncertainty. As is evident from Fig. 19.21c, penalty-free transmission (within the measurement error) is achieved for switch states spanning the entire tuning range of the TDC.

The significant distinction from other TDCs is that this device is simultaneously broadband and wavelength continuous. Thus, it may be deployed in-line as well as at the receiver and is not constrained to operation at specific bit rates, bit formats,

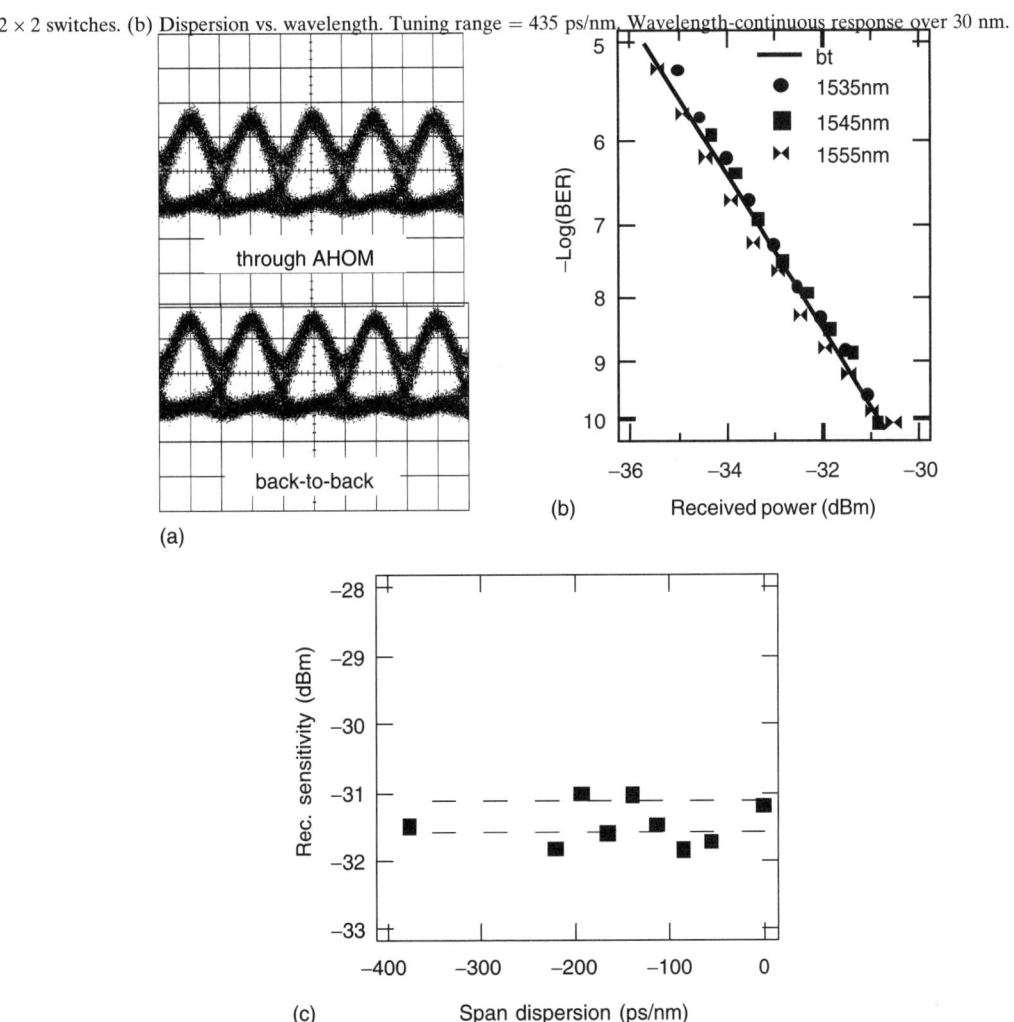

2 × 2 switches. (b) Dispersion vs. wavelength. Tuning range = 435 ps/nm. Wavelength-continuous response over 30 nm.

or channel spacings. Furthermore, there are no inherent trade-offs between tuning range and bandwidth, as is the case for other TDCs.

7.2. Polarizers and PDL Controllers

PDL, which induces random fluctuations in the optical signal-to-new ratio of an optical link, can severely degrade system performance. In-line polarizers and PDL compensation may offer a means to addressing this problem. This section describes the use of HOM fibers and TAP-LPGs to construct such devices.

In Section 5.2, the discussion about the TAP-LPG–based bandpass filters introduced the concept of dimensional scaling to shift the PMC and TAP of the HOM fiber. Similarly, if form birefringence is introduced in the fibers, it can be expected that two distinct PMCs and TAPs would exist for orthogonally polarized light entering the LPG. Figure 19.22a shows the PMCs for an HOM fiber fabricated with 2% ovality in the cross-sectional dimensions of its core. The figure shows that (1) two PMCs are obtained, one for each orthogonal state of polarization (SOP) for light in the fiber, and that (2) both curves show a TAP.

Writing an LPG in this fiber yields a polarizer whose polarizability can be adjusted by varying the index change during fabrication. Figure 19.22b shows the spectra of 20-dB strong 5-cm-long gratings written in these fibers, with a grating period of 90.25 μm, corresponding to the dashed line in Fig. 19.22a. The plots show the grating spectra for the two orthogonal SOPs, + and ||SOP|| yields a broadband spectrum because the grating period is at the TAP for this polarization, but SOP + is not coupled at all, for the same reasons that the mode-converting switch of Section 6.2 does not couple any light in the bar state. This results in a broadband in-fiber polarizer [26].

Furthermore, tuning mechanisms described in Section 6.1 can be used here to modulate the relative coupling efficiencies for the two SOPs of light. The results of strain tuning these gratings (strain range 0–0.05%) are shown in Fig. 19.22c. Varying the strain

FIGURE 19.21 (a) 40-Gb/s eye diagrams for span with 377 ps/nm and TDC at −378 ps/nm at 1545 nm. (b) Bit error rate (BER) curves for 40 Gb/s CSRZ. Penalty-free operation at wavelengths across the C-band. (c) Received power for BER = 10^{-9}, for different fiber spans and TDC states. Penalty-free operation over entire tuning range.

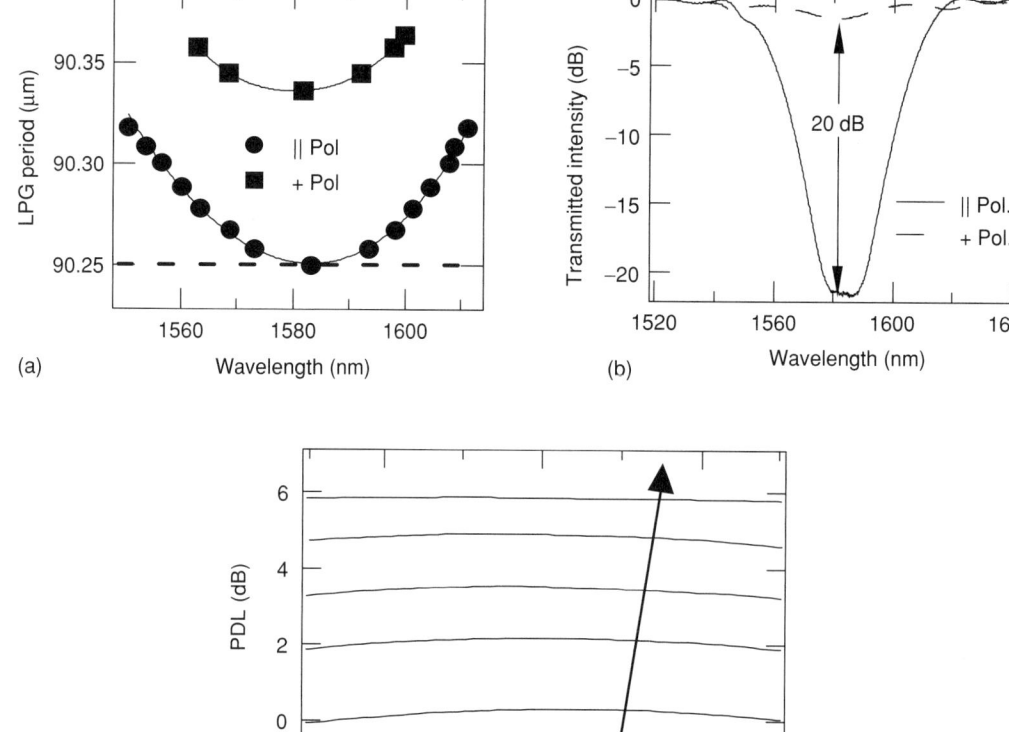

allows for changing the relative amounts of light attenuated from the two orthogonal polarizations, thus yielding a device whose PDL can be continuously tuned between 0 and 6 dB [27]. Although the spectra are only 10-nm wide, fiber dispersion engineering design rules, described in detail in Section 4.2, can be readily applied to achieve any desired bandwidth.

8. CONCLUSION

Few-mode fibers can be designed to possess HOMs with a variety of desired dispersive properties. Coupling between different modes of such fibers with LPGs leads to unique phase matching phenomena. This enables a variety of active and passive devices for signal conditioning applications, such as polarization control, spectral shaping, optical filtering, and dispersion control.

A unique feature of few-mode fiber devices is that it takes the complexity out of the device fabrication process and introduces it into the fiber design process. Dispersion optimized few-mode fibers can be manufactured using techniques used for transmission fibers. Thus, few-mode fiber devices can potentially be highly cost competitive, because the most complex part of the device (the dispersive fibers) can be mass manufactured.

9. ACKNOWLEDGMENTS

The author would like to thank several researchers at Bell Laboratories, Lucent Technologies, and OFS Laboratories, OFS-Fitel, whose experimental and theoretical contributions were essential for the demonstration and development of all the ideas illustrated in this chapter. In particular, contributions of M. F. Yan, L. C. Cowsar, E. Monberg, F. V. Dimarcello, J. W. Fleming, B. Mikkelsen, G. Raybon, S. Chandrsekhar, S. Ghalmi, P. Wisk, and Z. Wang were critical for results presented in this chapter.

10. REFERENCES

1. M.J. Li, "Recent progress in fiber dispersion compensators," *Proc. European Conf. Optical Comm.*, ThM.1.1 (2001).
2. M. Ibsen, P. Petropoulos, M.N. Zervas and R. Feced, "Dispersion-free fibre Bragg gratings," *Proc. Optical Fiber Comm.*, MC1 (2001).
3. B.J. Eggleton A. Ahuja, P.S. Westbrook, J.A. Rogers, P. Kuo, T.N. Nielsen and B. Mikkelsen, "Integrated tunable fiber gratings for dispersion management in high-bit rate systems," *J. Lightwave Tech.*, **18**, pp. 1419–1432 (2000).
4. A.M. Vengsarkar, P.L. Lemaire, J.B. Judkins, V. Bhatia, T. Erdogan and J.E. Sipe, "Long-period fiber gratings as band rejection filters," *J. Lightwave Tech.*, **14**, pp. 58–65 (1996).
5. M.J. Holmes and R. Kashyap, "Side-tap fiber grating filters," *Proc. Bragg Gratings, Photosensitivity and Poling in Glass*, SaC1 (1999).
6. S.H. Yun, B.W. Lee, H.K. Kim and B.Y. Kim, "Dynamic erbium doped fiber amplifier with automatic gain flattening," *Proc. Optical Fiber Comm.*, PD-28 (1999).
7. H. Lebidi, J.-J. Guerin, V. Girardon, X. Bonnet, C. Simonneau, R. Boucenna, C.D. Barros, N. Dely and I. Riant, "Dynamic gain control of optical amplifier using an all-fibre solution," *Proc. European Conf. Optical Comm.*, PD1.8 (2002).
8. L. Gruner-Nielsen and B. Edvold, "Status and future promises for dispersion compensating fibres," *Proc. European Conf. Optical Comm.*, 6.1.1 (2002).
9. J.-L. Auguste, R. Jindal, J.-M. Blondy, M. Claeau, J. Marcou, B. Dussardier, G. Monnom, D.B. Ostrowsky, B.P. Pal and K. Thyagarajan, "−1800 ps/nm·km chromatic dispersion at 1.55 μm in dual concentric core fibre," *Electron. Lett.*, **36**, pp. 1689–1691 (2000).
10. C.D. Poole, J.M. Weisenfeld, D.J. DiGiovanni and A.M. Vengsarkar, "Optical fiber-based dispersion compensation using higher order modes near cutoff," *J. Lightwave Tech.*, **12**, pp. 1746–1758 (1994).
11. X. Shu, L. Zhang and I. Bennion, "Sensitivity characteristics of long-period fiber gratings," *J. Lightwave Technol.*, **20**, pp. 255–266 (2002).
12. H.J. Patrick, A.D. Kersey and F. Bucholtz, "Analysis of the response of long period fiber gratings to the external index of refraction," *J. Lightwave Technol.*, **16**, pp. 1606–1612 (1998).
13. T.E. Dimmick, G. Kakarantzas, T.A. Birks, A. Diez and P.St.J. Russell, "Compact all-fiber acoustooptic tunable filters with small bandwidth-length product," *Photon. Technol. Lett.*, **12**, pp. 1210–1212 (2000).
14. Q. Li, A.A. Au, C.H. Lin, E.R. Lyons and H.P. Lee, "An efficient all-fiber variable optical attenuator via acoustooptic mode coupling," *IEEE Photon. Tech. Lett.*, **14**, pp. 1563–1565 (2002).
15. A. Abramov, A. Hale, R.S. Windeler and T.A. Strasser, "Widely tunable long-period fibre gratings," *Electron. Lett.*, **35**, pp. 81–82 (1999).
16. S. Ramachandran, Z. Wang and M.F. Yan, "Bandwidth control of long-period grating-based mode converters in few-mode fibers," *Optics Lett.*, **27**, pp. 698–700 (2002).
17. S. Ramachandran, M. Yan, L. Cowsar, A. Carra, P. Wisk, R. Huff and D. Peckham, "Large bandwidth, highly efficient mode coupling using long-period

gratings in dispersion tailored fibers," *Proc. Optical Fiber Comm.*, MC1 (2001).

18. S. Ramachandran, M. Yan, E. Monberg, F. Dimarcello, P. Wisk and S. Ghalmi, "Record bandwidth microbend gratings for spectrally flat variable optical attenuators," *IEEE Photon. Tech. Lett.*, **15**, pp. 1561–1563 (2003).

19. M. Wandel, T. Veng, Q. Le and L. Grüner-Nielsen, "Dispersion compensating fibre with a high figure of merit," *Proc. European Conf. Optical Comm.*, PD-A.1.4 (2001).

20. S. Ramachandran, B. Mikkelsen, L.C. Cowsar, M.F. Yan, G. Raybon, L. Boivin, M. Fishteyn, W.A. Reed, P. Wisk, D. Brownlow, R.G. Huff and L. Gruner-Nielsen, "All-fiber, grating-based, higher-order-mode dispersion compensator for broadband compensation and 1000-km transmission at 40 Gb/s," *IEEE Photon. Tech. Lett.*, **13**, pp. 632–634 (2001).

21. A.H. Gnauck, L.D. Garrett, Y. Danziger, U. Levy and M. Tur, "Dispersion and dispersion-slope compensation of NZDSF over the entire C band using higher order mode fibre," *Electron. Lett.*, **36**, pp. 1946–1947 (2000).

22. S. Ramachandran, B. Mikkelsen, L.C. Cowsar, M.F. Yan, G. Raybon, L. Boivin, M. Fishteyn, W.A. Reed, P. Wisk, D. Brownlow, R.G. Huff and L. Gruner-Nielsen, "All-fiber, grating-based, higher-order-mode dispersion compensator for broadband compensation and 1000-km transmission at 40 Gb/s," *Proc. European Conf. Optical Comm.*, PD-2.5 (2000).

23. S. Ramachandran, G. Raybon, B. Mikkelsen, M.F. Yan, L. Cowsar and R-J. Essiambre, "1700-km transmission at 40-Gb/s with 100-km amplifier-spacing enabled by higher-order-mode dispersion-compensation," *Electron. Lett.*, **37**, pp. 1352–1354 (2001).

24. S. Ramachandran, S. Ghalmi, Z. Wang and M. Yan, "Band-selection filters using concatenated long-period gratings in few-mode fibers," *Optics Lett.*, **27**, pp. 1678–1680 (2002).

25. S. Ramachandran, S. Ghalmi, S. Chandrasekhar, I. Ryazansky, M. Yan, F. Dimarcello, W. Reed and P. Wisk, "Wavelength-continuous broadband adjustable dispersion compensator using higher order mode fibers and switchable fiber-gratings," *IEEE Photon. Tech. Lett.*, **15**, pp. 727–729 (2003).

26. S. Ramachandran, M. Das, Z. Wang, J. Fleming and M. Yan, "High extinction, broadband polarisers using long-period fiber-gratings in few-mode fibers," *Electron. Lett.*, **38**, pp. 1327–1328 (2002).

27. M. Das, S. Ramachandran, Z. Wang, J. Fleming and M. Yan, "Broadband, adjustable polarisation-dependent-loss compensators with long-period fiber-gratings," *Proc. European Conf. Optical Comm.*, 10.4.5 (2002).

20
CHAPTER

Acousto-Optic Interaction in Few-Mode Optical Fibers

Helge E. Engan* and Kjell Bløtekjær
Department of Electronics & Telecommunications
Norwegian University of Science and Technology
Trondheim, Norway

1. INTRODUCTION

Optical fibers are usually characterized as single-mode or multimode. Multimode in general means a large number of modes, typically several hundred. There is, however, an interesting regime where only a few modes can propagate; in particular, a two-mode fiber supports only two transverse spatial modes.

Two-mode fibers are primarily used for sensing purposes, using the phase change between the two spatial modes induced by a measurand such as strain, pressure, or temperature. The sensitivity to temperature and strain is comparable with polarimetric sensors, but the pressure sensitivity is poor.

In addition to sensing purposes, few-mode fibers can be incorporated into all-fiber systems to manipulate the propagating beam without having to leave the fiber environment. Thus, examples of frequency shifting [1], optical filtering [2], and add/drop multiplexing [3] functions have been reported. For such purposes, coupling between the LP_{01} mode and the LP_{11} mode or between the LP_{01} polarization modes has been used. At least two such functions have been performed by interaction with an acoustic wave traveling along the fiber. In addition, the LP_{02} mode has been used for dispersion compensation [4] and for sensing purposes [5]. However, acousto-optic coupling into the LP_{02} mode is considered to be difficult and inefficient, and it has not been reported so far. Therefore, it is not discussed further here. This chapter presents a brief introduction to the optical and acoustic properties of few-mode fibers and a more thorough discussion of acousto-optic interaction in such fibers relevant for practical devices.

2. OPTICAL PROPERTIES

The modal properties of optical fibers are determined by the wavelength, core diameter, and numerical aperture. In the weakly guiding approximation, linearly polarized modes (LP modes) give a simplified description of the propagation. In this approximation, it is assumed that the index of refraction is only slightly different in the core and the cladding. For sufficiently long wavelength (or, equivalently, sufficiently small diameter or low numerical aperture), only one transverse spatial mode can propagate. In the LP notation this fundamental mode is referred to as LP_{01}, the two subscripts describing the radial and the circumferential variation, respectively. Assuming the almost linearly polarized field is along the x-axis and the propagation is along the z-direction, the variation of the three perpendicular (x, y, z) components of the electric field E can be most conveniently described by their variation in cylindrical coordinates (r, φ, z) as [6]

$$E = E_0 \begin{cases} \frac{1}{J_0(u)}\left[J_0\left(\frac{ur}{a}\right), \quad 0, \quad -\frac{u}{ink_0 a}J_1\left(\frac{ur}{a}\right)\cos\varphi\right] & r < a \\ \frac{1}{K_0(w)}\left[K_0\left(\frac{wr}{a}\right), \quad 0, \quad -\frac{w}{ink_0 a}K_1\left(\frac{wr}{a}\right)\cos\varphi\right] & r > a \end{cases}$$
(1)

where $u = ak_0(n_{co}^2 - n^2)^{1/2}$, $w = ak_0(n^2 - n_{cl}^2)^{1/2}$ and n, n_{co}, and n_{cl} are the effective, core, and cladding refractive indices respectively, a is the core radius; k_0 is the free space wave number; and J_i and K_i are Bessel and modified Hankel functions. The effective refractive index n determines the wavelength in the fiber, and it is derived from a dispersion relation.

The field distribution of this mode across a circular fiber cross-section is shown schematically in Fig. 20.1. It has full rotational symmetry. The mode is actually twofold degenerate, because the polarization direction is arbitrary.

As the wavelength decreases below cutoff, the next higher order mode, LP_{11}, becomes propagating. Its field distribution is also shown in Fig. 20.1. The mode has a plane of symmetry and a plane of antisymmetry. The orientation of the two lobes is arbitrary; thus, the mode is twofold degenerate. In

---

*E-mail: helge.engan@iet.ntnu.no

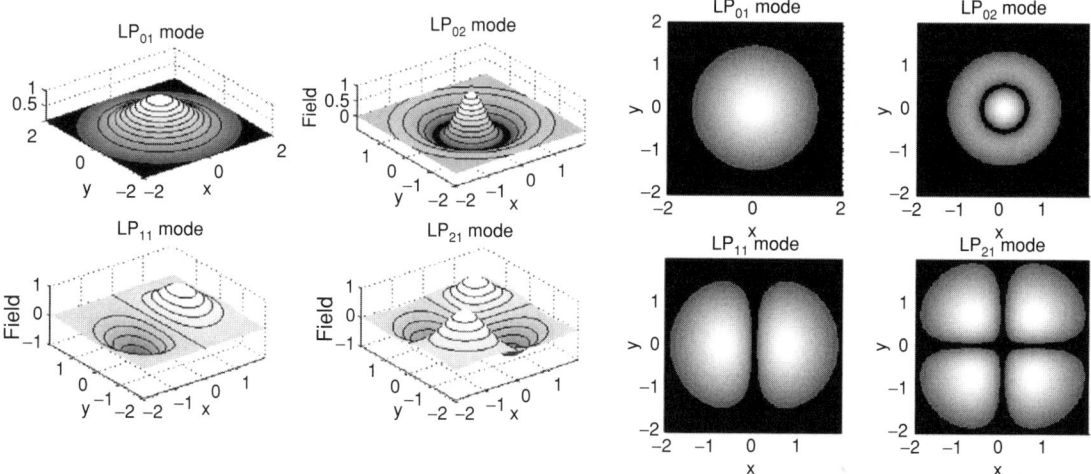

FIGURE 20.1 Transverse electric field (left) and intensity (right) distribution of some lower order LP modes of an optical fiber with circular core.

addition, it may have two orthogonal polarizations. In the LP approximation the polarization modes are also degenerate; hence, the LP_{11} mode is fourfold degenerate. This degeneracy causes problems for some applications. Therefore, elliptical core fibers are used in most experiments and applications involving two-mode fibers.

The field distribution of the lowest order modes of an elliptical core fiber is shown schematically in Fig. 20.2, together with the notation used to characterize them. The even modes have longer cutoff wavelengths than the odd modes. Hence, in a certain wavelength interval only the LP_{01} and LP_{11}^{even} modes can propagate. Even if the odd modes can propagate, experimental setups can usually be arranged such that they can be ignored. Thus, only four modes, including polarization, need to be taken into account.

The propagation constants for the four modes are shown schematically in Fig. 20.3 for an elliptical core fiber and a bow-tie fiber [7]. The bow-tie fiber, shown to the right, has two bow-tie–shaped stress elements that introduce birefringence in the fiber core. The core is not intentionally elliptic, but in practice it has more than sufficient ellipticity to lift the spatial degeneracy. In Fig. 20.3, $k_0 = 2\pi/\lambda$ is the free space propagation constant. The propagation constants β of the modes lie within the interval from $k_0 n_{cl}$ to $k_0 n_{co}$, where n_{co} and n_{cl} are the refractive

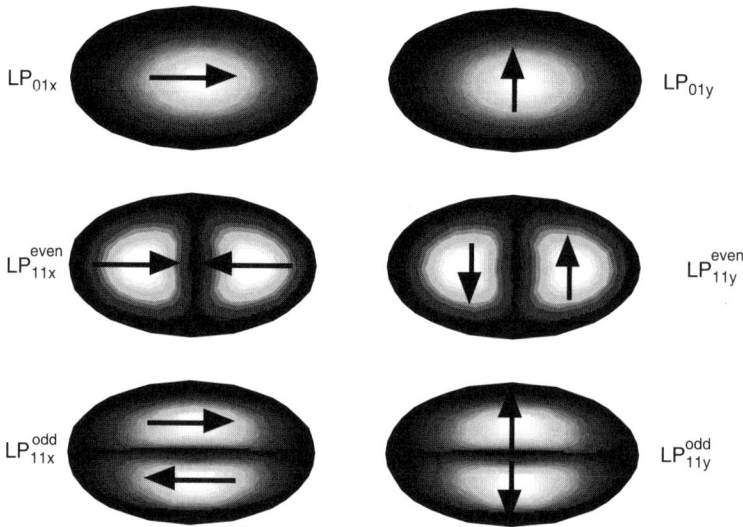

FIGURE 20.2 Sketch of the lowest order spatial and polarization modes of an elliptical core fiber.

Acousto-Optic Interaction in Few-Mode Optical Fibers

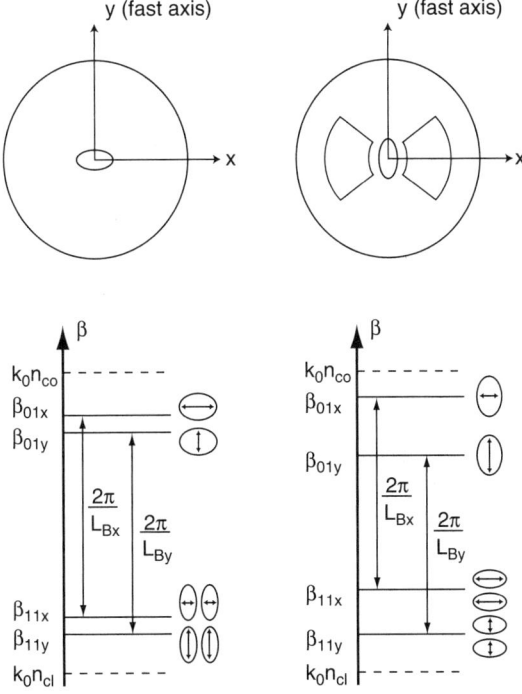

FIGURE 20.3 Schematic drawing of the propagation constants of the lowest order spatial and polarization modes of an elliptical core fiber (left) and a bow-tie fiber (right). (From [7].)

indices of the core and cladding, respectively. Figure 20.3 also introduces the beat lengths $L_B = 2\pi/\Delta\beta$, where $\Delta\beta$ is the difference in propagation constant between two modes. The beat lengths between spatial modes of the same polarization are plotted vs. wavelength in Fig. 20.4a for a bow-tie fiber and in Fig. 20.4b for an elliptical core fiber [7]. The polarization splitting is generally larger in the bow-tie fiber, and it is different for the two spatial modes, resulting in different intermodal beat lengths for the two polarizations. For the elliptical core fiber, these beat lengths are essentially identical. The experimental method used to obtain these curves is described later.

3. ACOUSTIC PROPERTIES

Acoustically, an optical fiber is quite accurately modeled as a homogeneous isotropic beam of circular symmetry. Such beams or rods can support an infinite number of acoustic modes. In the same way as for optical modes, two-mode indices describe the radial (s) and the circumferential (n) variation, respectively, for the acoustic modes. The components of the displacement u of the general mode can be described in cylindrical coordinates (r, ϕ, z) as [8]

$$u_r = \left(AK_d J'_n(K_d r) + BKJ'_n(K_t r) + C\frac{n}{r}J_n(K_t r)\right) \sin(n\phi)e^{i(\Omega t - Kz)}$$

$$u_\phi = \left(A\frac{n}{r}J_n(K_d r) + B\frac{Kn}{K_t r}J_n(K_t r) + CK_t J'_n(K_t r)\right) \cos(n\phi)e^{i(\Omega t - Kz)}$$

$$u_z = -i(AKJ_n(K_d r) - BK_t J_n(K_t r))\sin(n\phi)e^{i(\Omega t - Kz)}$$
(2)

with

$$K_d^2 = \frac{\Omega^2}{V_d^2} - K^2 \quad \text{and} \quad K_t^2 = \frac{\Omega^2}{V_t^2} - K^2$$

where Ω and K are the angular frequency and the wavenumber of the propagating acoustic wave, respectively, and V_d and V_t are the longitudinal and transverse bulk wave velocities, respectively. The stress-free boundary conditions determine the dispersion relations, as well as the constants A, B, and C. The azimuthal order is determined by the integer n, and within each of these mode groups there are several radial orders given by the integer s. As described above for optical modes in a circular

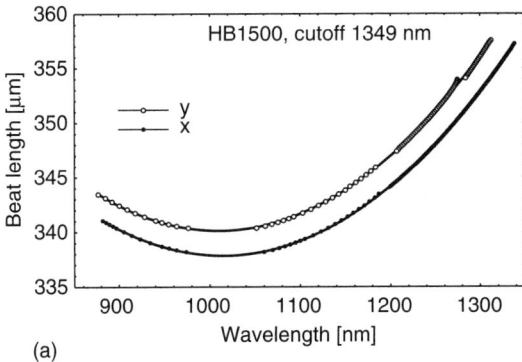

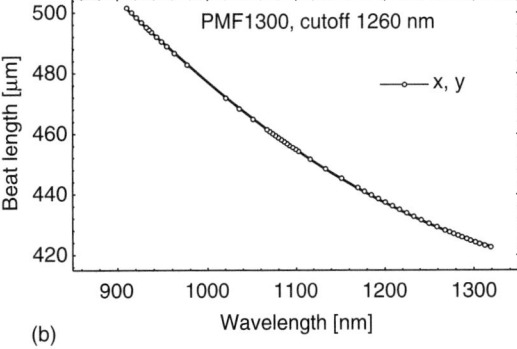

FIGURE 20.4 Beat length between the spatial modes plotted vs. optical wavelength for both polarizations. The fibers are a bow-tie fiber (a) and an elliptical core fiber (b). (From [7].)

symmetry, the acoustic modes are twofold degenerate. To obtain the other solution, in the above equations the sine function can be replaced by cosine, whereas cosine can be replaced by –sine.

In Fig. 20.5 the appearance of the lowest circumferential modes ($n = 0$, 1, and 2, respectively) is indicated. As can be seen, for $n = 0$ there are two solution groups that are denoted the longitudinal and the torsional mode group. The case $n = 1$ defines the flexural mode group. A frequency range of particular importance for acousto-optic interaction is the case in which the acoustic wavelength is large compared with the fiber radius. For $n = 0$ and 1 only the lowest radial modes ($s = 0$) can propagate down to the lowest frequencies. All other modes have cut-off frequencies below which they cannot propagate. Also, higher values of n do not have radial modes that can propagate down to zero frequency. These characteristics are shown in the calculated dispersion relations displayed as frequency–velocity curves in normalized coordinates for fused silica in Figs. 20.6, 20.7, and 20.8.

The mode of most interest is the fundamental flexural mode. It is described in some detail. At low frequencies, the particle motion is purely transverse and uniform over the cross-section [8]. At higher frequencies, when the acoustic wavelength becomes smaller than the diameter, the mode becomes more like a surface wave, with little excitation in the fiber core. This effect contributes to a decrease in efficiency of acousto-optic interaction as the frequency increases. Figure 20.9 shows the measured acoustic wavelength Λ versus frequency for an elliptical core fiber and a bow-tie fiber [7]. The two fibers have the same cladding diameter, and their

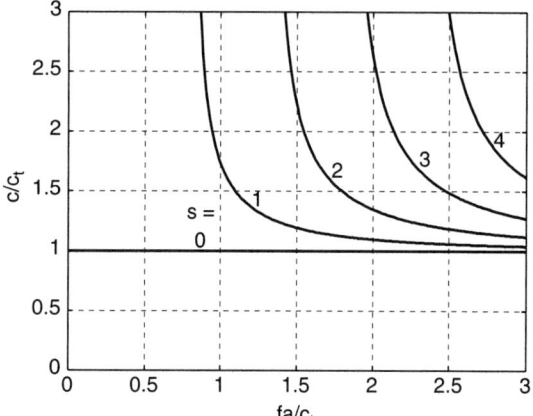

FIGURE 20.6 Phase velocity of torsional modes in quartz fiber normalized to shear wave velocity $c_t = 3764\,\text{m/s}$ as a function of normalized frequency; a is the fiber radius. The five lowest radial modes are shown.

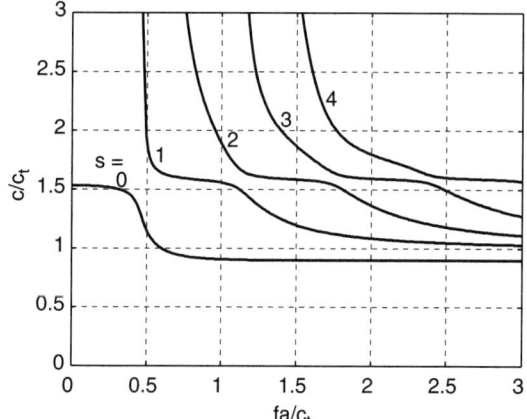

FIGURE 20.7 Longitudinal modes in quartz fiber. Notation as in Fig. 20.6.

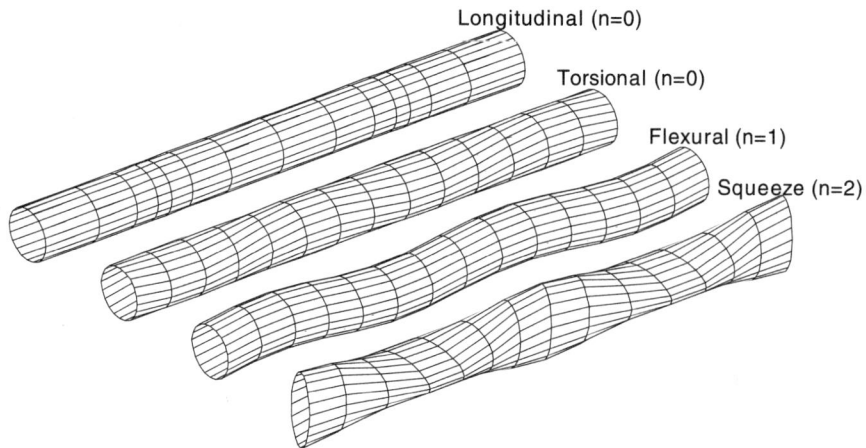

FIGURE 20.5 Appearance of the lowest radial mode ($s = 0$) of each of the longitudinal, torsional, flexural, and squeeze mode groups.

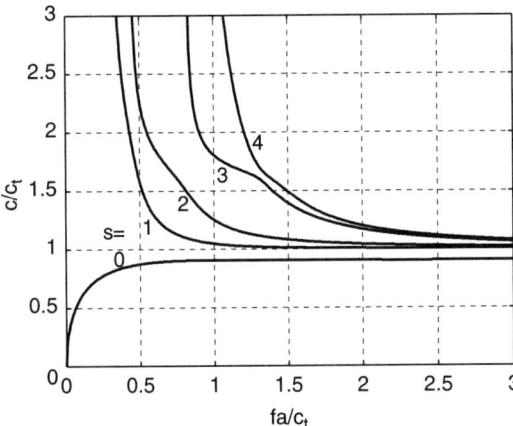

FIGURE 20.8 Flexural modes in quartz fiber. Notation as in Fig. 20.6.

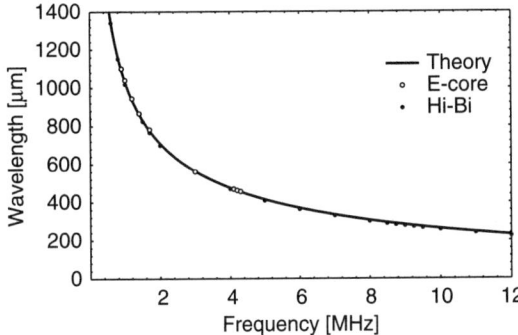

FIGURE 20.9 Acoustic wavelength vs. frequency for an elliptical core fiber and a bow-tie fiber. The two dispersion diagrams are practically coincident. (From [7].)

dispersion diagrams are essentially identical, proving that the nonuniformity caused by the core and the bow-tie has little influence on the acoustic waves. Also note that the agreement between theory [8] and experiment is very good. For acousto-optic interaction, the optical beat length must equal the acoustic wavelength. A comparison of Figs. 20.4 and 20.9 shows that the acoustic frequency required for acousto-optic interaction in the two fibers ranges from 4 to 6 MHz. At these frequencies, there is only one propagating flexural mode.

In a practical fiber there is some acoustic anisotropy. One source is a slight ellipticity of the cladding; another is the presence of the stress elements in a birefringent fiber. The result is that the degeneracy is lifted, so that two flexural modes with slightly different phase velocities can propagate. Their planes of oscillation are orthogonal to each other, but due to unavoidable asymmetry in the fiber cross-section they are not necessarily parallel to the fast and slow optical axes [9]. The acoustic birefringence has implications for acousto-optic interaction. If both acoustic modes are excited, there are acoustic perturbations with two different wavelengths, complicating the interaction. If the acoustic birefringence is weak or absent, the plane of oscillation stays fixed even if the fiber is twisted. The direction of the acoustic perturbation relative to the orientation of the optical modes then changes along the fiber. For stronger birefringence, the plane of oscillation follows the twist and the relation between acoustic and optical polarization remains fixed. As an example, the bow-tie fiber used to obtain the data of Figs. 20.4 and 20.9 has an acoustic birefringence of about 1% and the acoustic and optical principal directions deviated by about 20 degrees [9].

The attenuation of the flexural wave on a bare fiber, that is, a fiber without polymer coating, is on the order of 5 dB/m in the frequency region of interest [10]. Typical components and experimental set-ups use an interaction length of tens of centimeters. The acoustic attenuation may therefore be of some importance for such components.

4. ACOUSTO-OPTIC INTERACTION

4.1. Principles

The coupling of power between two optical modes in a fiber can be described by the coupled-mode theory [11]. The main result of that theory is that coupling is resonant, in the sense that it is strong only if phase match is accomplished, that is, the period of the perturbation (the acoustic wavelength in the present context) is equal to the beat length between the modes. This condition is best illustrated in an ω-β diagram, as shown schematically in Fig. 20.10. Sections of the dispersion diagrams of the two optical modes are shown. An acoustic wave couples the two optical

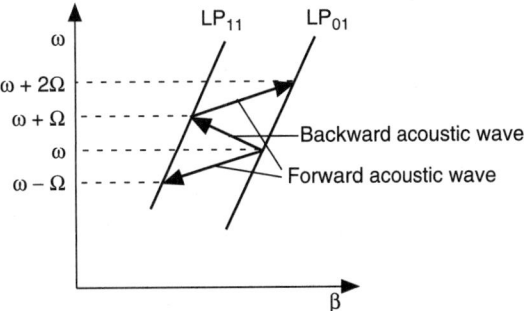

FIGURE 20.10 Dispersion diagram for the LP_{01} and LP_{11} optical modes, illustrating the phase match conditions for acousto-optic coupling.

waves if the vector describing the acoustic ω-β relation spans the gap between the optical waves. Thus, a forward propagating acoustic wave (positive slope) couples an LP_{01} optical wave at frequency ω to an LP_{11} wave at frequency $\omega - \Omega$, Ω being the acoustic frequency. Similarly, a backward propagating acoustic wave (negative slope) couples the LP_{01} wave at frequency ω to an LP_{11} wave at frequency $\omega + \Omega$. The diagram also shows that this LP_{11} wave can be coupled back to an LP_{01} wave at frequency $\omega + 2\Omega$ by a forward propagating acoustic wave. Note that Fig. 20.10 is not to scale. Because of the large difference between acoustic and optical velocities, the slopes of the vectors describing acoustic waves should be much smaller (or the lines describing the optical waves much steeper) than shown. Thus, the acoustic frequencies required for forward and backward interaction are practically the same.

For an interaction region of finite length, the phase match condition does not have to be exactly satisfied. The shorter the interaction region, the more relaxed the phase match requirement.

Coupled-mode theory also establishes requirements for the symmetry of the perturbation to provide coupling between two specific optical modes. Possible coupling between various optical modes by given acoustic modes are indicated in Table 20.1. In principle, an acoustic longitudinal mode should couple between LP_{01} and LP_{02} optical modes. As stated above, this coupling is expected to be inefficient, and reports of this interaction are not known. According to Table 20.1, acoustic torsional waves can couple between optical polarization modes of the same spatial order. For the LP_{01} mode this coupling has been studied both experimentally and theoretically [12, 13]. The demonstrated coupling efficiency for this transition has not so far been sufficient for practical devices. It is believed that the reason for this lies in the difficulty of creating torsional waves in fibers with sufficiently large amplitude. On the other hand, it has been established that flexural waves can be excited with sufficient amplitudes to make a transition between LP_{01} and LP_{11} modes very practical [8]. One condition is that the flexural acoustic wave has a transverse motion in the direction of the lobes of the LP_{11} mode. We concentrate on this particular transition in the following discussion.

4.2. Experimental Setup

A typical experimental setup for acousto-optic interaction studies is shown in Fig. 20.11. Depending on the application, the light source may be a quasi-monochromatic laser source or a wideband incoherent source, such as a super-luminescent diode or a light-emitting diode. The light is coupled into a two-mode fiber. In most cases a pure LP_{01} mode is wanted at the input to the interaction region. This is accomplished by a mode stripper, realized by winding a few turns of the fiber on a small diameter mandrel. Because the LP_{11} mode is less bound to the fiber core, it radiates from the curved fiber, whereas the more tightly bound LP_{01} mode remains.

The rather soft coating of the fiber attenuates the acoustic waves. Hence, the coating is removed in the interaction region. In Fig. 20.11 coated and uncoated parts of the fiber are shown as thick and thin lines, respectively. The acoustic wave is excited by a transducer attached to the fiber. The transducer

TABLE 20.1 Possible acousto-optic coupling between various optical LP modes by acoustic torsional (T), flexural (F), and longitudinal (L) modes as found from symmetry considerations.

LP mode coupling by Torsional, Flexural, and Longitudinal modes:											
			01		11				02		03
					even		odd				
			x	y	x	y	x	y	x	y	x	y
01		x		T	F		F		L		L	
		y	T			F		F		L		L
11	even	x	F			T				F		F
		y		F	T				F		F	
	odd	x	F					T	F		F	
		y		F			T			F		F
02		x	L		F		F			T	L	
		y		L		F		F	T			L
03		x	L		F		F		L			T
		y		L		F		F		L	T	

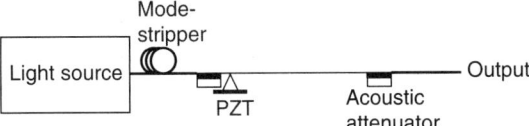

FIGURE 20.11 Schematic diagram of a typical experimental setup for acousto-optic interaction. The light source and the detection and signal processing at the output may vary, depending on the specific application. The part of the fiber drawn as a thin line is stripped of its coating.

is a wafer of a piezoelectric ceramic of the lead-zirconate-titanate (PZT) type resonant at the appropriate frequency. The acoustic amplitude is amplified by an acoustic impedance transformer, consisting of a taper and ending in a tip with a diameter similar to the fiber diameter. The taper may be made of glass or metal. In the first case the tip may be fused to the fiber; otherwise, glue is used. A weak, although for some experimental purposes sufficient, excitation can be obtained by contact between tip and fiber, with a fluid such as a small drop of water or even without any added material so that only a "dry" contact is present [14]. Such methods provide flexibility, because the fiber can be moved and rotated with ease.

The acoustic excitation of the required mode is quite efficient. Complete coupling from LP_{01} to LP_{11} has been obtained with 1 mW of electrical power input to the transducer, with a 72-cm interaction length [10]. The same result is obtained with 80 mW and 9 cm. Some experiments with acousto-optic coupling in so-called null couplers show extremely high efficiency [15]. These components in part rely on technology somewhat different from what is described in this chapter, and they are not discussed further here.

The design of the horn taper is very important for the frequency and impulse response of the transducer. Figure 20.12 illustrates steps in the development of transducers [16]. The top portion of Fig. 20.12 shows the frequency and impulse responses for an early transducer, where the taper was made of fused quartz. The frequency response is very irregular, and the impulse response is wide and spiky. The responses were obtained by means of a laser probe that measures the acoustic amplitude.

A more recent design is shown in the lower portion of Fig. 20.12. The taper is a hollow cone of aluminum. Unwanted reflections are suppressed by epoxy, which acts as an acoustic attenuator. Also,

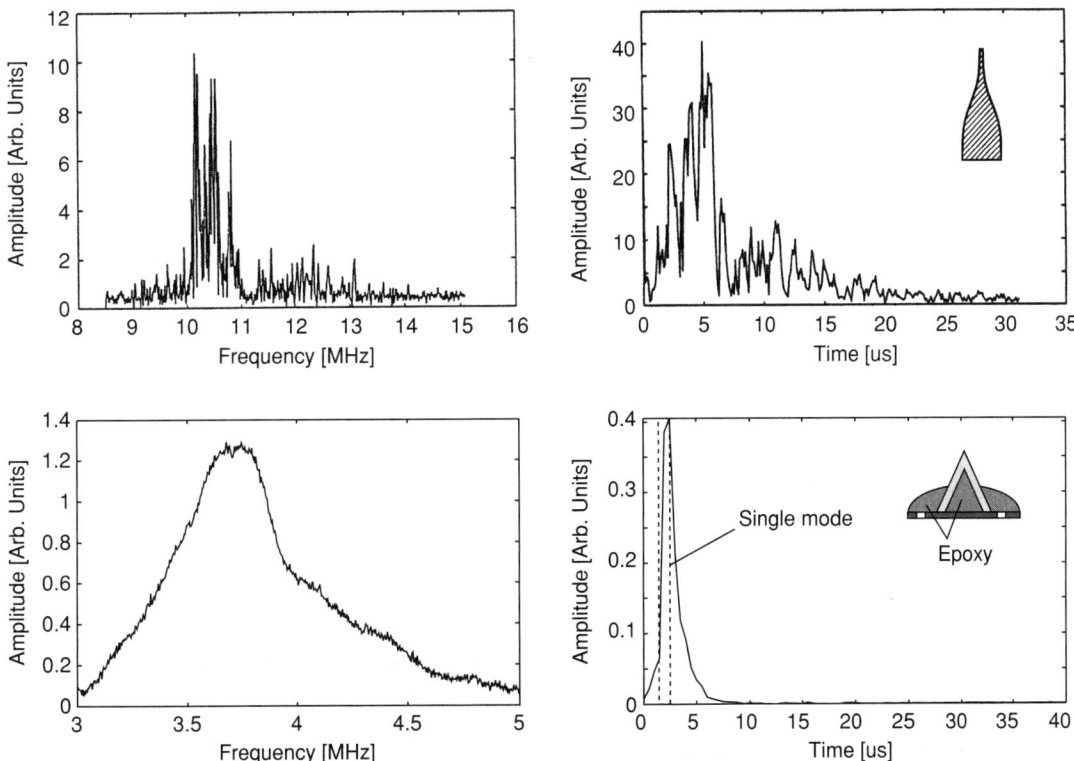

FIGURE 20.12 Frequency and impulse response for a transducer of early design (top) and an improved design (bottom). (From [16].)

it is believed that the newer design preferentially attenuates the higher order radial modes that may be initially excited in the thick part of the horn. This leaves fewer modes and reflected waves to interfere at the opposite end. The resulting frequency response is smooth, and the impulse response has a single narrow peak.

In Fig. 20.11, the acoustic attenuators shown may be a drop of oil or water, or some soft material applied to the bare fiber. The purpose of the attenuators is to reduce reflections of the acoustic wave. The devices and signal processing at the output depend on the application. Examples are shown in the following.

4.3. Frequency Shifter

As demonstrated in Fig. 20.10, the conversion from LP_{01} to LP_{11} occurs with a frequency shift equal to the acoustic frequency. Thus, the setup in Fig. 20.11 acts as an optical frequency shifter. The fact that the frequency-shifted light is in the LP_{11} mode is inconvenient for coupling into a single-mode fiber. However, this problem can be solved by means of a static mode coupler that converts the light back to the LP_{01} mode [2]. The basic principle of two different static couplers is illustrated in Fig. 20.13. The mechanical mode coupler consists of two corrugated plates that impose a periodic bending on the fiber, with a period equal to the beat length at the optical center wavelength [17]. The photorefractive mode coupler is a two-mode fiber where a permanent photorefractive grating has been written [18]. For coupling between the LP_{01} and LP_{11} modes to occur, the grating must be written at a skew angle as indicated in Fig. 20.13. The mechanical mode coupler can easily provide complete coupling between the modes, whereas the coupling in a photorefractive grating is relatively weak.

An improved design of the frequency shifter is obtained by moving the transducer in Fig. 20.11 to the center of the interaction region and doubling the interaction length [19]. The principle of operation can be explained with reference to Fig. 20.10. The input light in the LP_{01} mode first encounters a backward traveling acoustic wave, which couples light to the LP_{11} mode with a positive frequency shift. The acoustic wave amplitude is adjusted to make the conversion complete at the position of the transducer. From there on the light interacts with a forward traveling acoustic wave, causing coupling back to the LP_{01} mode, again with a positive frequency shift. Thus, the output light is shifted by twice the acoustic frequency. This dual frequency shifter has the advantage that the output frequency-shifted light is in the LP_{01} mode. More importantly, the unwanted sidebands are considerably reduced. As shown in the output spectrum of Fig. 20.14, the sidebands are 40 dB down. This is a much better sideband suppression than has been obtained with the simple setup in Fig. 20.11.

4.4. Wavelength Dependence and Tunable Filters

The beat length vs. wavelength dispersion curves of Fig. 20.4 were obtained in an acousto-optic

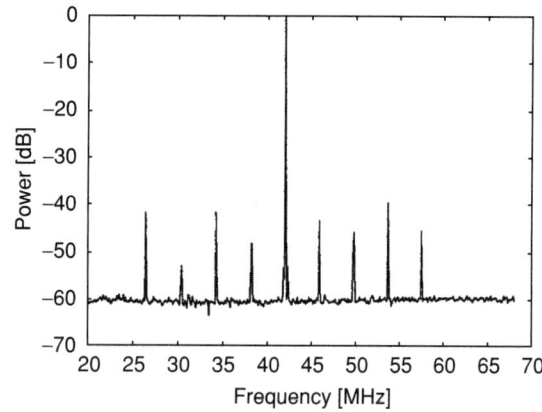

FIGURE 20.14 Output spectrum from an acousto-optic frequency shifter. The spectrum is shifted by heterodyne detection. (From [19].)

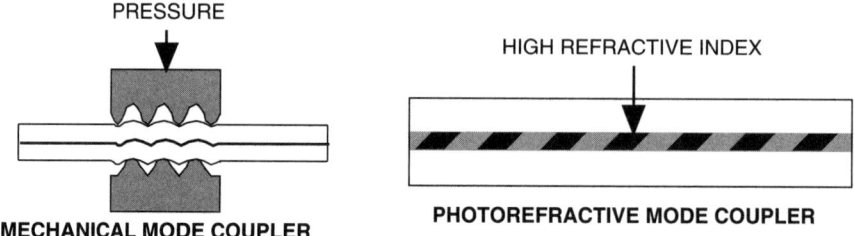

FIGURE 20.13 Principles of static mode couplers. (Left) A mechanical mode coupler consisting of two corrugated plates that impose periodic bends on the fiber. (Right) The photorefractive coupler has a permanent periodic index of refraction written into it.

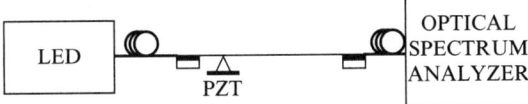

FIGURE 20.15 Schematic diagram of experimental setup for measuring the intermodal beatlength vs. optical wavelength.

interaction experiment [7]. The setup is shown schematically in Fig. 20.15. A wideband source is used. A mode stripper is also applied at the output end of the fiber. Thus, only the LP_{01} mode reaches the optical spectrum analyzer. The wavelengths that are coupled efficiently to LP_{11} show a corresponding attenuation in the LP_{01} mode. The loss versus wavelength as seen on the spectrum analyzer is shown in Fig. 20.16 for various values of the acoustic frequency. By means of polarizers, the polarization associated with the various peaks can be identified. Knowing the acoustic dispersion relation, as in Fig. 20.9, acoustic frequency can be converted to wavelength, which equals the beat length at the attenuation peaks. Thus, the curves in Fig. 20.4 can be derived. Two different sources were used to cover the wavelength interval, a superfluorescent neodymium-doped fiber source for the shorter wavelengths, and a light-emitting diode for the long wavelength region.

The curves of Fig. 20.16 show that the setup of Fig. 20.15 acts as a tunable notch filter. It attenuates the light at certain wavelengths, and these wavelengths can be adjusted by tuning the acoustic frequency. By some modifications, the setup can be transformed into a bandpass filter [2]. The principle is shown in Fig. 20.17. It differs from Fig. 20.15 in that a static mode converter is inserted before the acousto-optic interaction region.

The principle of operation is illustrated by the spectra shown in the lower part of Fig. 20.17. The light is input primarily in the LP_{01} mode, but the LP_{11} mode is also excited to some extent. The mode stripper removes this mode. The mode converter is relatively short; thus, its bandwidth is large. It converts the light to LP_{11} completely at the center wavelength and somewhat less so at other wavelengths. The acousto-optic interaction region is long, resulting in a narrow-band coupling of light from the LP_{11} mode back to the LP_{01} mode. After the second mode stripper, only the narrow-band light in the LP_{01} mode remains.

Figure 20.18 shows improved experimental results for a tunable filter published previously [2]. The center wavelength is tuned by approximately 50 nm by tuning the acoustic frequency from 4.20 to 4.32 MHz. The 3-dB bandwidth is approximately 1 nm.

4.5. Fiber Nonuniformity

The frequency responses shown in Fig. 20.18 are far from ideal. They are asymmetric and show rather

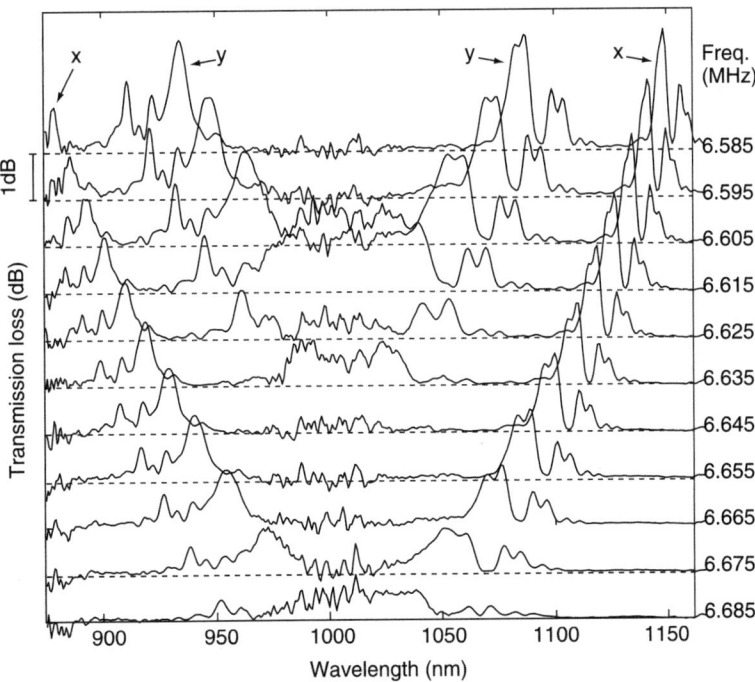

FIGURE 20.16 Attenuation of the LP_{01} mode vs. wavelength in the experiment sketched in Fig. 20.15. The curve parameter is the acoustic frequency. (From [7].)

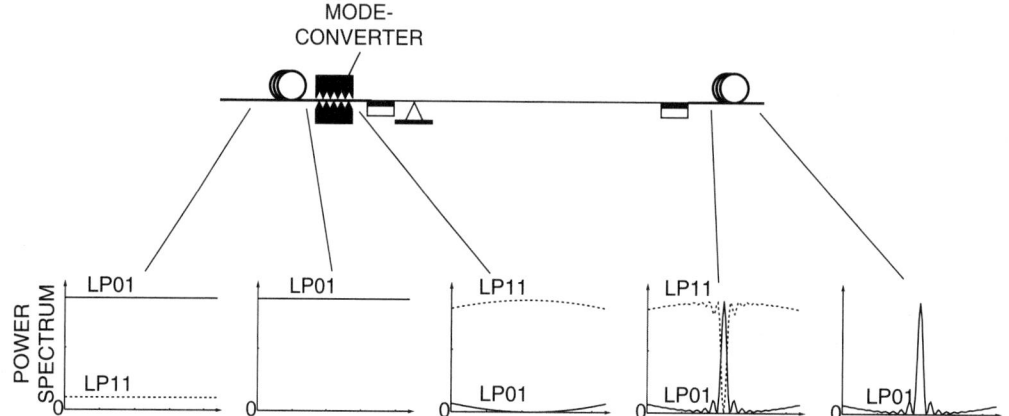

FIGURE 20.17 Schematic drawing of an acousto-optic bandpass filter. The diagrams show spectra of the two optical modes at various points on the fiber.

high side lobes. The reason for this is that the fiber is nonuniform along the interaction region [20, 21]. The optical beat length or the acoustic wavelength varies along the fiber, resulting in phase errors that corrupt the frequency response. The irregularities can be mapped by monitoring the instantaneous frequency shift as a short acoustic pulse propagates along the fiber. The setup is shown schematically in Fig. 20.19 [22]. Because the modes are orthogonal, no intermodal interference is observed if all the light intercepts the detector. Therefore, a knife edge is inserted in front of the detector. The instantaneous interference frequency depends on the phase match condition at the position of the acoustic pulse. The variation in phase mismatch, $\Delta\beta = 2\pi/L_B - 2\pi/\Lambda$, can be derived from the observed interference frequency.

The result for a particular fiber is shown in Fig. 20.20. The nonuniformity corresponds to a variation in frequency shift of about 0.5%. Using the results of Fig. 20.20, the wavelength dependence of loss in a notch filter can be predicted with good accuracy, as shown in Fig. 20.21a. The agreement with experimental results is very good, in contrast to the response predicted when the nonuniformity is neglected, as shown by the dotted line. Fortunately, more uniform fibers exist, as shown by Fig. 20.21b.

4.6. Scanning Heterodyne Interferometer

So-called white light or low coherence interferometry requires a receiving interferometer with a variable delay. Mechanically scanned, bulk, Michelson, and Mach-Zehnder interferometers are standard devices used for this purpose. Acousto-optic interaction offers an attractive alternative [23, 24]. It results in an all-fiber heterodyne interferometer with fast scan and no mechanical motion. The setup is essentially the same as in Fig. 20.19. The input LP_{01} light is partly converted to LP_{11} at the position of the acoustic pulse. Ideally, the coupling should be 50%. The two modes propagate with slightly different group velocity from the position of the acoustic pulse to the output end of the fiber where they are made to interfere on the detector. As the pulse propagates along the fiber, the delay between the two modes at the output is reduced.

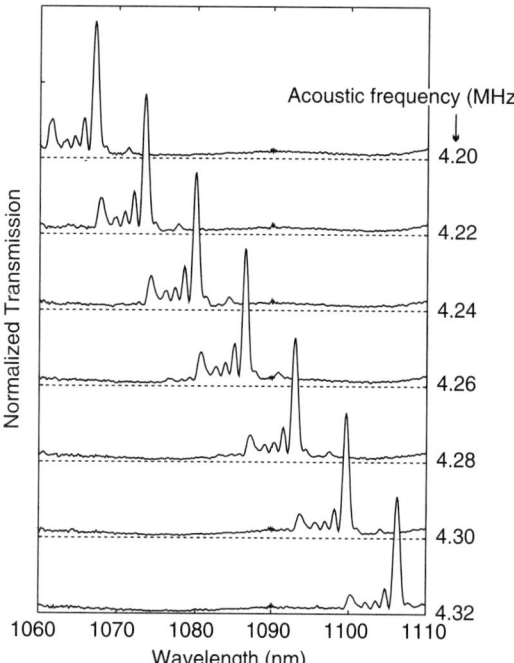

FIGURE 20.18 Bandpass frequency responses for a tunable acousto-optic filter. The curve parameter is the acoustic frequency.

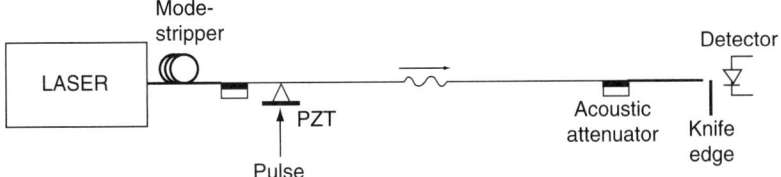

FIGURE 20.19 Schematic drawing of experimental setup for characterizing fiber nonuniformity. The scanning heterodyne interferometer described in Section 4.6 is shown.

As an example of an application of the scanning interferometer, Fig. 20.22 shows a setup to adjust the angular alignment of two birefringent fibers, a and b, before splicing [25]. Linearly polarized light from a broadband source excites both polarizations of fiber a. From the output end of fiber a, the light is coupled into fiber b. The relative orientation of the principal axes of the two fibers can be adjusted. The two fibers have different lengths. The combination of fast and slow axis propagation in the two fibers gives rise to four different delays. Via polarization controllers, the light output from fiber b is sent to a scanning delay interferometer. The four interference peaks appear on the oscilloscope, as shown in Fig. 20.23a. If the angular orientation of the two fibers is adjusted to make the principal axes coincide, there is no cross-coupling between polarizations at the splice, and only one delay remains. This is shown in Fig. 20.23b. Angular alignment accuracy on the order of 0.1 degrees is possible using this method. The method was first realized using a bulk interferometer with mechanical scanning [26]. The fast scan and instantaneous display of the interferogram of the two-mode interferometer is a great advantage for this particular application.

5. PRACTICAL CONSIDERATIONS

For reasons outlined above, circular core fibers are not appropriate for two-mode fiber experiments. Such fibers have been used, but for most purposes an elliptical core fiber is preferred. High birefringence fibers, such as bow-tie fibers, have nominally circular cores, but in practice they are sufficiently elliptic to provide well-defined propagation of the spatial modes.

Two-mode fibers are not commercially available as such. For experimental purposes, fibers that are single mode at one wavelength are used as two-mode fibers at shorter wavelengths. For example, nominally 1550-nm fibers are used at 1300 nm and 800-nm fibers are used at 633 nm. Two-mode fibers at 1550 nm must be specially made.

Some interesting two-mode fibers can be envisioned. For example, a fiber with a high index core,

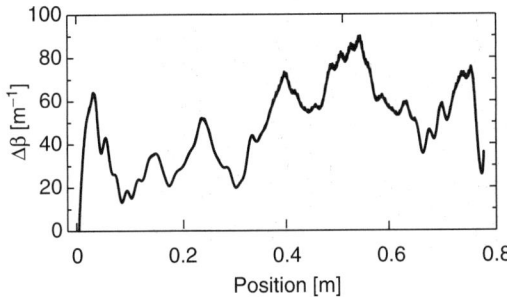

FIGURE 20.20 Phase mismatch vs. position along the fiber, measured by means of the setup of Fig. 20.19. (From [22].)

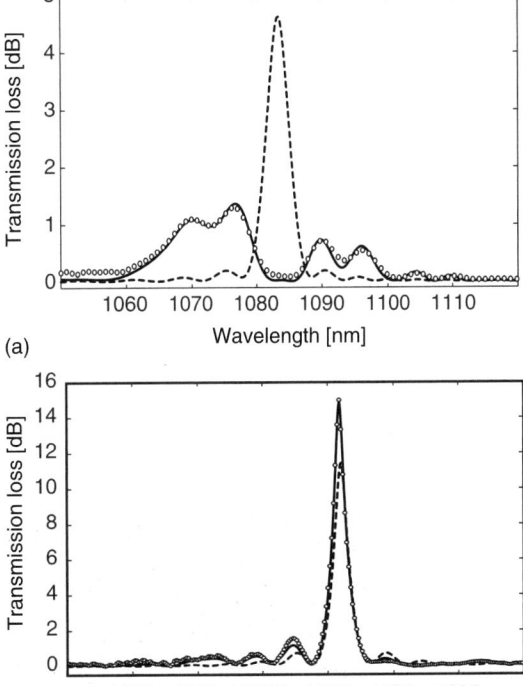

FIGURE 20.21 Acousto-optic notch filter responses. The circles show experimental results. The dashed lines show theoretical predictions assuming a uniform fiber. The solid lines show theoretical results based on the measured nonuniform phase match condition. (a) Bow-tie fiber; (b) elliptical core fiber. (From [22].)

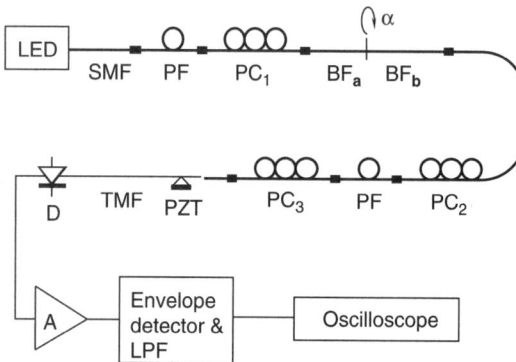

FIGURE 20.22 Acousto-optic scanning interferometer used to align two birefringent fibers. SMF, single-mode fiber; PF, polarizing fiber; PC, polarization controller; TMF, two-mode fiber; D, detector; A, amplifier; LPF, low pass filter. (From [25].)

surrounded by a region of somewhat lower index, could be designed such that the LP_{01} mode would propagate primarily in the central core, whereas the LP_{11} mode occupied primarily the surrounding region. By choosing special glasses in the two regions, beat length and group delay between the modes could be tailored.

The cladding diameter of the fiber influences the acoustic velocity but not the optical properties. The acoustic frequency appropriate for acousto-optic interaction can be adjusted by selecting the cladding diameter, for example, by etching. The coupling efficiency also increases with decreasing fiber diameter. Selective etching could also be used to make the cladding elliptical, with axes coinciding with the optical axes.

A problem with two-mode acousto-optic components is packaging. A bare fiber is not as robust as a coated fiber, and the fiber must be free to execute transverse acoustic oscillations. Metallic coating could be a solution to the first problem, and experiments show that the fiber can be clamped fairly rigidly between two Teflon-coated glass plates parallel to the direction of acoustic oscillation with tolerable increase in attenuation [14]. This could open possibilities for coiling the fiber and packaging the components in reasonably sized packages.

6. INDUSTRIALIZATION

The presentation so far has discussed theoretical and experimental results. We now briefly describe an effort that has given practical devices. Kim et al. [27] described a device that is essentially a notch filter used to flatten the gain of an erbium-doped fiber amplifier. The filter configuration, however, has important differences from the ones that were described above. The configuration is illustrated in Fig. 20.24. The horn structure has a thin center hole so it can be mounted coaxially with the fiber. To obtain acoustic flexural waves in the fiber, an acoustic shear wave transducer is mounted at the thick end of the horn. Second, the LP_{11} mode is slightly cut off (cladding mode), so that the coupled energy can be attenuated by an optical attenuator or some index-matching material at the fiber surface. Third, several horns and frequencies are used together to obtain an overall attenuation curve that has several degrees of freedom, providing a broadband dynamic filtering with great flexibility. Some resulting spectra are shown in Fig. 20.25. A gain flatness of 0.7 dB over a wavelength span of more than 35 nm was reported. This concept has been taken further to become an industrialized product [28].

7. CONCLUSIONS

Acoustic flexural waves on an optical fiber can be made to couple power very efficiently between the LP_{01} and LP_{11} spatial optical modes. Elliptical core fibers are favored for acousto-optic experiments and

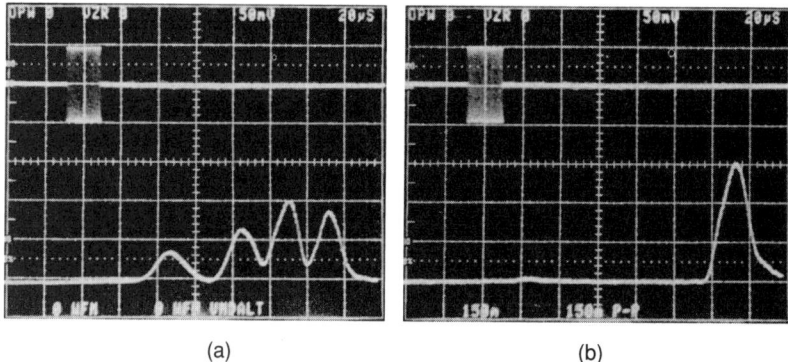

FIGURE 20.23 Interferograms observed with the setup from Fig. 20.22. (a) The fibers are arbitrarily aligned. (b) The fast axes of the two fibers are at 90 degrees to each other. (From [25].)

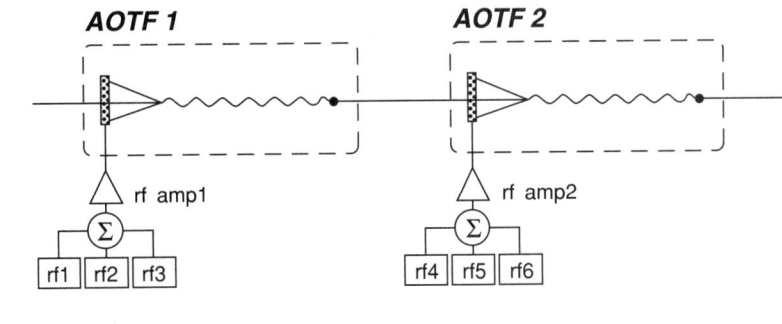

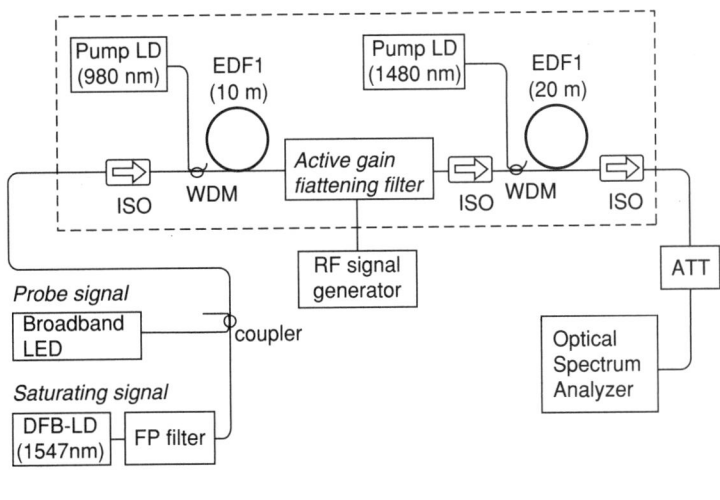

FIGURE 20.24 (a) All-fiber acousto-optic gain flattening filter. (b) Schematic of experimental setup. A dual-stage erbium-doped fiber amplifier (inside the dotted box) and the measurement setup are shown. RF1–RF6, RF signal sources; RF amp1–RF amp2, RF power amplifiers; EDF, erbium-doped fiber; WDM, wavelength division multiplexing coupler; ISO, isolator; ATT, 10-dB attenuator. (From [26].)

components, because the spatial orientation of the modes is well defined. High birefringence fibers, such as bow-tie fibers, have sufficiently elliptical cores, although their cores are not intentionally made elliptic.

Fibers are also more or less acoustically anisotropic, and the principal acoustic and optical axes do not always coincide. This fact must be taken into account in designing acousto-optic components.

A number of acousto-optic two-mode fiber components have been proposed and demonstrated: frequency shifters, tunable filters, and fast scanning heterodyne interferometers. A practical problem with these components is packaging. The fiber must be suspended such that the flexural waves are not quenched. Experiments with Teflon-coated glass plates supporting the fiber show promise for compact packaging.

8. ACKNOWLEDGMENTS

We acknowledge the essential contributions to this chapter made by our graduate students: Jan Ove Askautrud, Erling Kolltveit, Bjørnar Langli, Dan

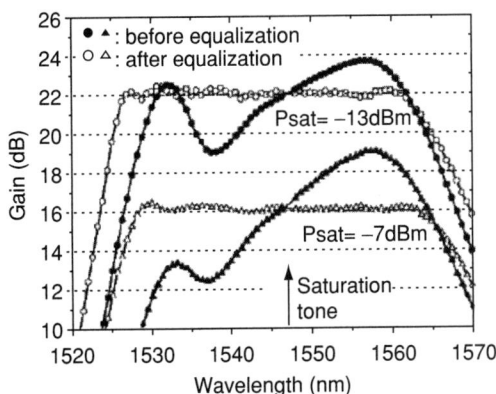

FIGURE 20.25 Gain spectra with and without acoustic power for the setup of Fig. 20.24. (From [26].)

Östling, and Pranay G. Sinha. The Research Council of Norway is acknowledged for financial support.

8. REFERENCES

1. B. Y. Kim, J. N. Blake, H. E. Engan, and H. J. Shaw, "All-fiber acousto-optic frequency shifter," *Opt. Lett.* **11**, 389–391 (1986).
2. D. Östling and H. E. Engan, "Narrow-band acousto-optic tunable filtering in a two-mode fiber," *Opt. Lett.* **20**, 1247–1249 (1995).
3. H. S. Park, S. H. Yun, I. K. Hwang, S. B. Lee, and B. Y. Kim, "All-fiber add–drop wavelength-division multiplexer based on intermodal coupling," *IEEE Photon. Technol. Lett.* **13**, 460–462 (2001).
4. C. D. Poole, J. M. Wiesenfeld, A. R. McCormick, and K. T. Nelson, "Broadband dispersion compensation by using the higher-order spatial mode in a two-mode fiber," *Opt. Lett.* **17**, 985–987 (1992).
5. A. Kumar, N. K. Goel, and R. K. Varshney, "Studies on a few-mode fiber-optic strain sensor based on LP_{01}–LP_{02} mode interference," *J. Lightwave Technol.* **19**, 358–362 (2001).
6. S. Huang, J. N. Blake, and B. Y. Kim, "Perturbation effects on mode propagation in highly elliptical core two-mode fibers," *J. Lightwave Technol.* **8**, 23–33 (1990).
7. D. Östling, B. Langli, and H. E. Engan, "Intermodal beat lengths in birefringent two-mode fibers," *Opt. Lett.* **21**, 1553–1555 (1996).
8. H. E. Engan, B. Y. Kim, J. N. Blake, and H. J. Shaw, "Propagation and optical interaction of guided acoustic waves in two-mode optical fibers," *J. Lightwave Technol.* **6**, 428–436 (1988).
9. B. Langli and K. Bløtekjær, "Effect of acoustic birefringence on acoustooptic interaction in birefringent two-mode optical fibers," *J. Lightwave Technol.* **21**, 528–535 (2003).
10. J. O. Askautrud and H. E. Engan, "Distributed probing of orientation of two-mode birefringent fiber using pulsed acoustic guided waves," *Proc. Fiber Optic Sensors IV (SPIE Proc. 1267)*, 40–49 (1990).
11. A. Yariv and P. Yeh, *Optical Waves in Crystals*, John Wiley & Sons, New York, p. 186 (1984).
12. M. Berwick and D. A. Jackson, "Coaxial optical-fiber frequency shifter," *Opt. Lett.* **17**, 270–272 (1992).
13. H. E. Engan, "Analysis of polarization mode coupling by acoustic torsional waves in optical fibers," *J. Opt. Soc. Am. A* **13**, 112–118 (1996)
14. B. Langli, "Acousto-optic interactions in two-mode birefringent fibers," thesis, Norwegian University of Science and Technology, Trondheim, Norway (1997).
15. T. A. Birks, S. G. Farwell, P. St. J. Russell, and C. N. Pannell, "Four-port fiber frequency shifter with a null taper coupler," *Opt. Lett.* **19**, 1964–1966 (1994).
16. H. E. Engan, D. Östling, P. O. Kval, and J. O. Askautrud, "Wideband operation of horns for excitation of acoustic modes in optical fibers," *Proc. OFS 10, SPIE Proc. 2360*, 568–571 (1994).
17. J. N. Blake, B. Y. Kim, and H. J. Shaw, "Fiber-optic modal coupler using periodic microbending," *Opt. Lett.* **11**, 177–179 (1986).
18. K. O. Hill, B. Malo, K. A. Vineberg, F. Bilodeau, D. C. Johnson, and I. Skinner, "Efficient mode conversion in telecommunication fibre using externally written gratings," *Electron. Lett.* **26**, 1270–1272 (1990).
19. J. O. Askautrud and H. E. Engan, "Fiberoptic frequency shifter with no mode change using cascaded acousto-optic interaction regions," *Opt. Lett.* **15**, 649–651 (1990).
20. W. R. Trutna, Jr., D. W. Dolfi, and C. A. Flory, "Anomalous sidelobes and birefringence apodization in acousto-optic tunable filters," *Opt. Lett.* **18**, 28–30 (1993).
21. D. A. Smith, A. d'Alessandro, and J. E. Baran, "Source of sidelobe asymmetry in integrated acousto-optic filters," *Appl. Phys. Lett.* **62**, 814–816 (1993).
22. B. Langli, D. Östling, and K. Bløtekjær, "Axial variations in the acoustooptic phase-mismatch coefficient of two-mode fibers," *J. Lightwave Technol.* **16**, 2443–2450 (1998).
23. E. Kolltveit and K. Bløtekjær, "Acoustically scanned delay for white-light interferometry," *Proc. OFS 8*, 189–192 (1992).
24. P. G. Sinha, E. Kolltveit, and K. Bløtekjær, "Two-mode fiber-optic time-delay scanner for white-light interferometry," *Opt. Lett.* **20**, 94–96 (1995).
25. P. G. Sinha, L. Bjerkan, and K. Bløtekjær, "Angular alignment of highly birefringent fibers employing acoustically scanned time-delay technique," *IEEE Photon. Technol. Lett.* **7**, 1462–1464 (1995).
26. K. Takada, K. Chida, and J. Noda, "Precise method for angular alignment of birefringent fibers based on an interferometric technique with a broadband source," *Appl. Opt.* **26**, 2979–2987 (1987).
27. H. S. Kim, S. H. Yun, H. K. Kim, N. Park, and B. Y. Kim, "Actively gain-flattened erbium-doped fiber amplifier over 35 nm by using all-fiber acoustooptic tunable filters," *IEEE Photon. Technol. Lett.* **10**, 790–792 (1998).
28. www.noveraoptics.com.

21
CHAPTER

Basic Theory and Design Procedures for Arrayed Waveguide Structures

Christopher R. Doerr*
Lucent Technologies, Bell Laboratories
Holmdel, New Jersey, USA

1. INTRODUCTION

Planar lightwave circuits consist of a planar arrangement of optical waveguides on a substrate. A common waveguide material is silica. Silica waveguides consist of doped glass cores embedded in less-doped glass on a silicon substrate. Optical fibers can often be directly attached to these silica waveguides. Once in the planar lightwave circuit, the light can be filtered, switched, can interfere with other lightwaves, and so forth.

Every waveguide structure has local eigenmodes. A local eigenmode is an optical field distribution at a particular location in the structure that would propagate through the structure unchanged if the structure were to continue unchanged. A waveguide structure can be a single strip of core material, multiple parallel strips, or an essentially infinitely wide strip. If two waveguide structures are very far apart with no intervening core material, one can choose whether to use the eigenmodes of each individual structure or of the whole structure. Provided the strips of core are far enough apart, either approach yields the same conclusions. Ease of understanding the performance may depend on which approach is taken.

Despite the fact that a strip of core is usually a "waveguide," some structures we discuss in this chapter use strips of core that are not waveguides at all. It becomes difficult to tell when a core strip is a waveguide by itself or is part of larger wave-guiding structure. Thus, we often use the term "core strip" rather than "waveguide."

We first discuss the arrayed waveguide lens (AWL). We then discuss the star coupler in detail. We move on to the arrayed waveguide grating (AWG), which is really just a different type of AWL. Finally, we discuss a band demultiplexer that uses two AWGs and one AWL.

---
*E-mail: crdoerr@lucent.com

2. ARRAYED WAVEGUIDE LENS

Figure 21.1 shows an AWL. It consists of two very wide strips of core connected by many approximately parallel narrow strips of core. The wide strips of core plus the transitions between the wide strips and narrow strips are called "star couplers" [1]. All the core strips diverge from a single point in the first star coupler and converge at a single point in the second star coupler. The waveguides are curved in a "w" shape in so that they can all have the same path length.

The AWL performs the operation of a simple lens. Whatever optical field distribution appears at line 1 is imaged onto line 2.

3. THE STAR COUPLER

The purpose of the star coupler is to couple N modes together into N new modes in a special way. Each of the N new modes contains a portion of each of the N original modes.

Each mode starts in a separate strip of core, the separation between strips being very large. There are many ways to couple N core strips together. One way might be to simply bring all the core strips close enough together so that the eigenmodes of the system are no longer confined to a single core strip but never touch the core strips together (using curving waveguides). However, such a coupler is generally difficult to control and is usually sensitive to wavelength, polarization, and fabrication (WPF) changes. Furthermore, logistically it is difficult to see how one can bring N core strips close together and then far apart again in a simple way.

We need to define in what way the star coupler is to combine the modes. We want the star coupler to act like a lens centered between two arrays of far-apart core strips, the distance between the lens and each array being equal to the lens focal length. In this way, the AWL, and other devices, can perform the

FIGURE 21.1 An arrayed waveguide lens.

imaging we want. This design means that the form of the coupling is simply the discrete Fourier transform: The N outputs of the star coupler are approximately the discrete Fourier transform of the N inputs.

The star coupler looks something like a star (Fig. 21.2). It consists of two arrays of core strips radiating outward from two points. The distance between the two points is the star coupler radius, R (actually, there may be extra offsets between the two points due to mutual coupling, which we discuss later). At a distance R from the convergence point, the core strips abruptly transition into one wide strip, which is called the "free space region," "free propagation region," or "slab." The narrow core strips often taper outward before reaching the wide strip. The gap between the narrow core strips must remain finite, however, because of practical fabrication limitations. The taper shape can be anything, as long as it is slow enough to not excite unwanted modes.

We have N eigenmodes in the strips of core far away from the free space region, but we have an infinite (or very large) number of eigenmodes in the free space region. Thus, a big part of the design of a star coupler is trying not to couple to more than N eigenmodes in the free space region. Otherwise, we end up with loss. Because there is an abrupt junction between the free space region and the waveguides, there is generally a loss due to scattering into other modes. This loss can be reduced by making the junction as nonabrupt as possible. To do this, one can make a transition region between the discrete narrow core strips and the continuous wide core strip. However, even if the junction is very gradual, there is a finite loss because of the converging nature of the waveguides (in other words, the waveguides cannot be brought together as slowly as one would like).

A convenient method that requires no extra processing steps is to use "segmentation." Segmentation consists of placing strips of core *parallel* to the free space region edge. The strips have a constant center-to-center spacing, but the strip width gets progressively smaller as the distance from the free space region increases. These strips are not waveguides. A typical number of strips is 15. Segmentation slowly reduces the index step in the gaps between the waveguides as the free space region is approached (Fig. 21.3). The transition happens gradually enough to minimize the net coupling to higher order modes.

The eigenmodes of a periodic array of core strips are called Bloch modes. They are analogous to the eigenmodes of free space, which are plane waves. One example Bloch mode of a periodic array is the same field distribution in every waveguide with the same phase in every waveguide. Because of the periodic nature of the abrupt transition from waveguides to the free space region, the transition is likely to cause coupling from a Bloch mode traveling at angle θ to Bloch modes traveling at angles $\theta \pm \lambda/a$. λ is the wavelength, and a is the center-to-center spacing of the core strips. Let β_f be the propagation constant (recall that propagation constant is the phase velocity – radians per distance) of the fundamental Bloch mode and β_u be the propagation constant of an unwanted mode. The transition is "slow" provided that its length L obeys $(\beta_u - \beta_f)L \gg \pi$. This ensures that the phase of the light added to an unwanted mode is constantly changing, preventing build-up.

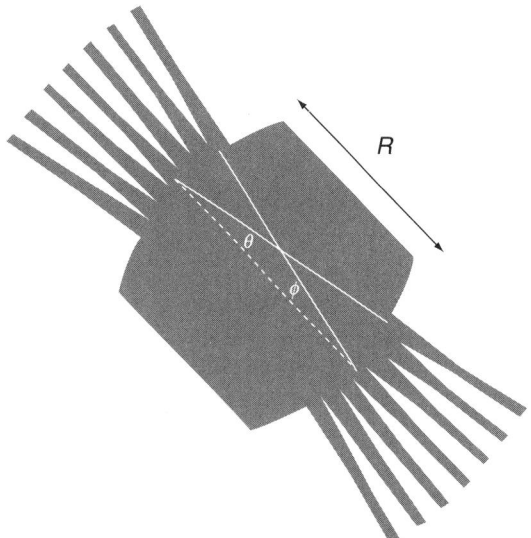

FIGURE 21.2 The star coupler.

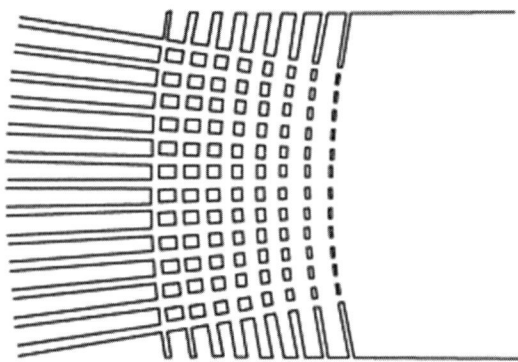

FIGURE 21.3 Segmentation to reduce star coupler loss.

We need to be able to calculate the transmissivity from each waveguide on the left to each waveguide on the right side of the star coupler (far away, where the core strips are uncoupled). Assume we can determine the field distribution from the waveguide along one edge of the free space region (discussed later). An important property that simplifies many optical calculations is that any optical field can be described either in the spatial domain as a function $u(x)$, where u is the complex amplitude and x is distance along the line that the field is defined, or in the angular spectrum domain [2] as $\tilde{u}(k_x)$, where u is the complex amplitude of each plane wave traveling at angle $\sin^{-1}(k_x/k)$ and $k = 2\pi n/\lambda$

$$\tilde{u}(k_x) = \int_{-\infty}^{\infty} u(x) e^{jk_x x} dx \quad (1)$$

The concept of viewing the optical field either in the spatial or angular spectrum domains is often helpful. As long as refractive index jumps are small, which they are in all the structures investigated in this chapter, and we are not concerned about the signal's polarization (whether the electrical field is predominantly vertical or horizontal), we do not need Maxwell's equations.

Once we know the field along one edge of the free space region, we can look at its angular spectrum to determine what the spatial field will be on the other side. This is because each plane wave propagates to a different spatial location on the other side of the free space region. To propagate a plane wave $\tilde{u}(k_x)$, one simply needs to multiply it by

$$\exp(j\sqrt{k^2 - k_x^2}d)$$

where d is the distance traveled along the z-axis. To make the calculation easier, we can define the z-axis separately for each calculation of transmissivity from one port to another port as being a straight line connecting the ports. Then d is conveniently given by

$$R|e^{j\theta} + e^{-j\phi} - 1|$$

where θ is the angular location of one port on one side and ϕ is the angular location of the other port on the other side of the star coupler.

We can calculate the transmissivity from one port to another port by calculating the overlap integral between the field distribution coming from one port propagated to the other side and the field distribution coming from the other port, but not propagated. The overlap integral between u_1 and u_2 is

$$t = \frac{\int u_1(x) u_2^*(x) dx}{\sqrt{\int |u_1(x)|^2 dx \int |u_2(x)|^2 dx}} \quad (2)$$

where t is the transmissivity amplitude and $|t|^2$ is the transmissivity. Equation (2) basically computes the similarity between two field distributions, that is, the transmissivity. If one field is complex conjugated, its propagation direction is reversed.

For example, suppose we take the very simple case of calculating the transmissivity through a structure containing a horizontal waveguide that abruptly tilts by angle θ (Fig. 21.4). Thus the waveguide suddenly accrues a phase tilt of $kx \sin\theta$. If the eigenmode of the waveguide is $u(x)$, then the amplitude transmissivity through the system for the eigenmode is given by

$$t = \frac{\int u(x) u^*(x) \exp(-jkx\sin\theta) dx}{\sqrt{\int |u(x)|^2 dx \int |u(x)|^2 dx}} \quad (3)$$

As mentioned above, to calculate the transmissivities through the star coupler, one needs to know the field distribution $u(x)$ along the free space region boundary when one input is excited. The function $u(x)$ far from the free space region is simply the eigenmode of a single strip of core. However, as $u(x)$ approaches the free space region, it becomes a complicated waveform, being a sum of eigenmodes of the coupled strips of core as they come close to each other. The simplest way to handle this is numerically, via "beam propagation," covered in Section 5.

4. WAVEGUIDE GRATING ROUTER

A device that is a little more complicated than the AWL is the AWG. The only difference is that there is a nonzero path-length difference between adjacent waveguides. This makes the image location on line 2 move with wavelength. If the arm lengths are

FIGURE 21.4 Structure for example transmissivity calculation.

linearly increasing in length by ΔL, which is the usual case, then the angular location of the image obeys the following grating equation:

$$\beta_{\text{waveguide}}\Delta L + \beta_{\text{slab}}(a_1 \sin \theta_1 + a_2 \sin \theta_2) = 2\pi A \quad (4)$$

where a is the center-to-center spacing between core strips at the star coupler free space boundary and A is the grating order. The integer A gives the number of wavelengths in the path-length difference. Waveguide gratings are usually operated at a high grating order; A is often on the order of 100. The above equation can be read as follows: The phase accumulation in the path-length difference between successive arms plus the phase difference caused by radiation direction in the star coupler must equal an integer times 2π.

The AWG, also called the PHASAR (phased array), is called the waveguide grating router [3–5] when combined with the two star couplers (Fig. 21.5). For both the AWL and AWG, the path-length difference between adjacent waveguides must be extremely accurate (within ± 0.01 wavelength, typically) yet the waveguides themselves may be centimeters long. To construct the paths, one uses a combination of straight waveguides and curved waveguides. One issue in achieving an accurate path length is that straight and curved waveguides have different propagation constants.

WPF changes can change the propagation constants for the straight and curved waveguides differently. The best way to counter this is to treat the straight waveguides and curved waveguides independently and make all the straight waveguides have the same width and all the curved waveguides have the same width and radius. For example, for the lens, the total length of the straight waveguides should be the same for all the waveguides, and likewise for the curved waveguides; for the grating, the total length of the straight waveguides should have a constant path-length difference between adjacent arms, and likewise for the curved waveguides. Note that some designers vary the curve radius from arm-to-arm in the array. This can result in a slightly more compact design but requires accurate knowledge of the change in propagation constant with bend radius.

Another good reason to make all bends have the same radius is the transition between a straight waveguide and a bent waveguides. The simplest way to transition between a straight waveguide and a bent waveguide is to use an offset in the waveguide. The bent waveguide is shifted toward the inside of the bend because of the centripetal force in the bend. However, such a transition has some loss. Instead, one can use a short waveguide that is a gradual taper from a straight to a bend. In this transition piece, the radius of curvature varies from infinity to the bend radius. This transition piece is used at every straight-to-bend transition. Because they are identical in every arm, the transitions do not contribute to phase errors. If the designer uses different bend radii in each arm, then the transitions are also different.

After exiting the star couplers, the waveguides must not curve until they are far enough from each other to no longer couple to each other (typically $16\,\mu\text{m}$ for 0.8% index contrast silica waveguides). They also must not get close enough to each other to couple until they are on their way to converging without any more curves in the second star coupler. Mutual coupling in the middle of the array usually leads to terrible performance. Keeping the core strips far apart in the middle of the grating or lens

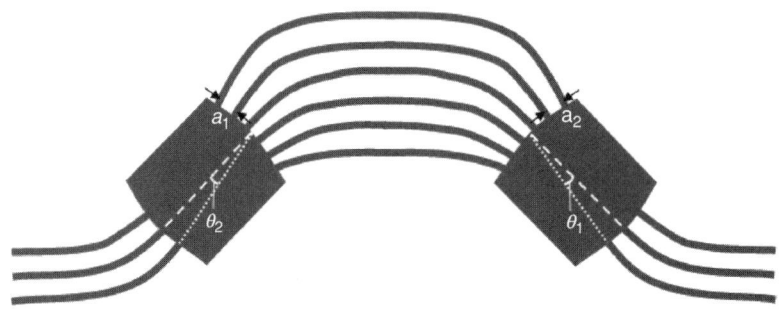

FIGURE 21.5 Waveguide grating router.

Theory and Design of Arrayed Waveguide Structures

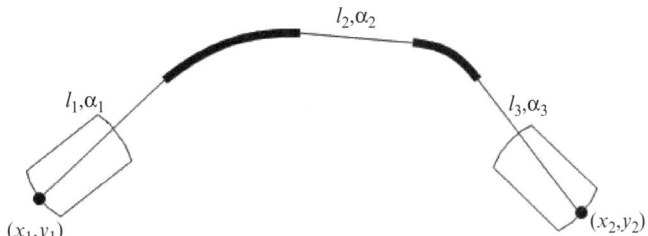

FIGURE 21.6 Structure for a three-straight, two-bend design.

generally means that one uses three straight waveguides and two curved waveguides per arm for arrays where the path-length difference is greater than ~10 wavelengths between adjacent arms (Fig. 21.6) and otherwise five straight waveguides and four curved waveguides per arm (Fig. 21.1). Four straight waveguides and three curved waveguides per arm never seem to work well.

We now describe how to calculate the lengths. For the three-straight/two-curve case, the angle of the center straight length, α_2, can be arbitrarily chosen. However, for convenience, one usually chooses $\alpha_2 = (\alpha_1 + \alpha_3)/2$. The variables α_1 and α_3 are determined by the star coupler design: α_1 and α_3 are chosen such that the waveguides capture the light emanating from the input/output waveguides so that the power in the outermost arms is usually ~13-dB down from the power in the center arm. This gives a good apodization of the grating response (and keeps the side lobes low, but not so low as to make the passband skirts flare undesirably). The straight lengths l_1, l_2, and l_3 are found by solving the following three simultaneous equations:

$$l_1 + l_2 + l_3 = L$$
$$l_1 \cos(\alpha_1) + l_2 \cos(\alpha_2) + l_3 \cos(\alpha_3) = x_2 - x_1 \quad (5)$$
$$l_1 \sin(\alpha_1) + l_2 \sin(\alpha_2) + l_3 \sin(\alpha_3) = y_2 - y_1$$

where L is the total length of the straight lengths of the arm, (x_1, y_1) is the divergence point (in the first star coupler), and (x_2, y_2) is the convergence point (in the second star coupler). One can write these equations in a matrix form and invert the matrix to find l_1, l_2, and l_3. Thus, it is advantageous to use a design program in a language that is good at matrices.

To make the actual grating, one can start with the shortest arm having a very small length and calculate $x_2 - x_1$ and $y_2 - y_1$ from that. One can then calculate the l_1, l_2, and l_3 for all the arms. If any are negative, then the length of the smallest arm must be increased, $x_2 - x_1$ and $y_2 - y_1$ recalculated, and all the l_is calculated again, continuing the cycle until all l_is are non-negative. It may turn out that there is no solution, which usually means the angle difference between the two star couplers is too large or too small. In fact, one usually chooses the angle difference between the two star couplers so as to minimize the length of the shortest arm, thus usually minimizing the accumulation of path-length errors.

The five-straight, four-curve case has more degrees of freedom. Assume we want the total lengths of the curves to be the same for all arms. Then

$$(\alpha_2 - \alpha_1) + (\alpha_2 - \alpha_3) + (\alpha_3 - \alpha_4) \\ + (\alpha_5 - \alpha_4) = \text{constant} \quad (6a)$$

One convenient solution is

$$\alpha_2 = \alpha_1/2 + \text{constant}$$
$$\alpha_4 = \alpha_5/2 + \text{constant} \quad (6b)$$
$$\alpha_3 = \text{constant}$$

We have five unknowns, l_1 through l_5, but only three equations:

$$l_1 + l_2 + l_3 + l_4 + l_5 = L \quad (7a)$$

$$l_1 \cos(\alpha_1) + l_2 \cos(\alpha_2) + l_3 \cos(\alpha_3) + l_4 \cos(\alpha_4) \\ + l_5 \cos(\alpha_5) = x_2 - x_1 = \Delta x \quad (7b)$$

$$l_1 \sin(\alpha_1) + l_2 \sin(\alpha_2) + l_3 \sin(\alpha_3) + l_4 \sin(\alpha_4) \\ + l_5 \sin(\alpha_5) = y_2 - y_1 = \Delta y \quad (7c)$$

We need two more equations. A convenient choice is to make the centers of the third straight fall on a single line that is perpendicular to the third straight with separation d. Thus, the two additional equations are

$$l_1 \cos(\alpha_1) + l_2 \cos(\alpha_2) + l_3 \cos(\alpha_3)/2 + \Delta x_{C1} + \Delta x_{C2} \\ = l_{1p} \cos(\alpha_{1p}) + l_{2p} \cos(\alpha_{2p}) + l_{3p} \cos(\alpha_{3p})/2 \\ + \Delta x_{C1p} + \Delta x_{C2p} + d \cos(\alpha_{3p} + \pi/2) \quad (8a)$$

$$l_1 \sin(\alpha_1) + l_2 \sin(\alpha_2) + l_3 \sin(\alpha_3)/2 + \Delta y_{C1} + \Delta y_{C2} \\ = l_{1p} \sin(\alpha_{1p}) + l_{2p} \sin(\alpha_{2p}) + l_{3p} \sin(\alpha_{3p})/2 \\ + \Delta y_{C1p} + \Delta y_{C2p} + d \sin(\alpha_{3p} + \pi/2) \quad (8b)$$

where subscript p denotes the "previous" adjacent waveguide and Δx_{ci} and Δy_{ci} are the displacements of curve i. We solve these five equations for the five unknowns using linear algebra and place the equations into a loop. If any of the l_is turn out to be negative, that particular l_i for the first arm is incremented. Then the loop runs again. The loop exits when all the l_is are positive and l_1 and l_5 are long enough so that the waveguides are uncoupled at their farthest distance from the star couplers.

5. MUTUAL COUPLING-INDUCED ABERRATIONS

To minimize the loss in the star couplers, the transition from the many narrow core strips to the single wide core strip (free space region) should be as gradual as possible. This implies significant mutual coupling among the waveguides near the free space region. The main effect of this coupling is to blur the boundary between where the free space region ends and the separated waveguides start. In a star coupler with no mutual coupling between waveguides, if one launches light into the center waveguide on one side, one finds that the phase of the light coupled into the waveguides on the other side is the same for all the waveguides. This is simply because the waveguide inlets are placed on a circle centered on the inlet of the central waveguide on the other side. Now let us consider the following thought experiment: Suppose the mutual coupling between waveguides at the star coupler is so strong that there is no longer a gap between the waveguides. In other words, the free space region is made longer by some distance d. Now the inlets on the other side are on a circle whose center is off by d from the inlet on the other side. Now if we launch light into the central waveguide on the mutually coupled side and measure the phase in the central waveguide on the other side, the phases are no longer equal. Instead, the phase has an approximately parabolic distribution.

We can calculate this parabolic distribution. View each point as a point in the complex plane [6]. Then one can use complex math to calculate the length, which is equal to the magnitude, and the angle, which is each to the phase. For example, if the original free space length is R, each inlet is at an angle θ, the mutual coupling offset is d, and the free space propagation constant is k, then the phase in each waveguide is

$$k|d + R\exp(j\theta)|$$

for $\theta \ll 1$, the usual case in star couplers, this is approximately equal to

$$k\left(d + R - \frac{dR}{d+R}\frac{\theta^2}{2}\right)$$

One can see that a nonzero d results in a quadratic dependence in θ.

If the mutual coupling is not too strong, that is, if the light does not spread significantly past the nearest neighboring waveguides to the next nearest neighbors, which is the usual case, the primary deleterious effect is this phase distortion. This causes aberrations, which can, for example, broaden filter passbands and give them shoulders. However, it is simple to correct for the phase aberrations—simply adjust the path lengths of the collecting waveguides to cancel the phase distortion.

The actual phase distribution from mutually coupled waveguides is not purely parabolic, and it must be calculated via a beam propagation program. This is done by launching the fundamental mode of the waveguide into one of the waveguides far enough away from the free space region so that there is no mutual coupling and propagate in until reaching the free space region boundary. (Actually, deciding how far away to start the beam propagation is an interesting problem in itself, but the easiest way to be sure is to start at some distance and propagate all the way in and then start at a little farther distance and again propagate all the way in—except for an overall phase: If one started far enough away, then the final field is the same in both cases.) One can then propagate this field across the free space region using a simple analytic formula. One likewise performs beam propagation for the other side of the star coupler, starting far away and coming toward the free space boundary. One can then calculate the field overlaps [Eq. (2)] to determine the tranmissivities for all the port-to-port combinations, as discussed in Section 3.

There are many commercial beam propagation programs. However, they can be expensive, secretive, and often difficult to incorporate into your own programs. You can easily write your own beam propagation program. The one we show here, the split-step Fourier method, is unfortunately much slower than the commercial ones, which typically use a finite-difference method. However, with a little more effort, you can write one based on the Fourier method that is actually faster than most commercial programs, depending on the structure. This is not presented here, but the C-language code to implement it is given in [7]. Note that Fourier beam propagation methods are accurate only for low index contrast structures (approximately

<1%). Because nearly all silica arrayed waveguide devices are made with an index contrast of ~1%, this is fine.

The split-step Fourier method is as follows (we do just the two-dimensional problem, which is good enough for low-index-contrast waveguide structures). Calculate the starting field $u(x)$. This is usually the eigenmode of the core strip and can be calculated by applying the boundary conditions [$u(x)$ and $du(x)/dx$ are continuous across the boundary]. Break up the transmission into many small equal steps of Δx and Δz in which the core pattern you will be going through is represented by $n(x,z)$ [i.e., $n(x,z)$ is an array of points of value n or $n + \Delta n$]. For each step of Δz,

1. Multiply $u(x)$ by $\exp[j\Delta z k(x,z)]$. Note that $k(x) = 2\pi n(x,z)/\lambda$.
2. Take the discrete Fourier transform to get $\tilde{u}(k_x)$.
3. Multiply $\tilde{u}(k_x)$ by $\exp(j\sqrt{k^2 - k_x^2}\Delta z)$.
4. Take the inverse discrete Fourier transform to get a new $u(x)$.
5. Increase z by Δz and go back to step 1.

Start with relatively large Δx and Δz and make them smaller until the final field is consistent.

You can include segmentation in the beam propagation easily: Every time you reach a segment, widen the waveguides so that there is no longer any gap between them. Note that even though the waveguides terminate on a curved surface, assuming a flat surface for the beam propagation is accurate enough.

To propagate the fields across the free space region and to calculate the transmissivities, we use the information given in Section 3. Calculate the phase of these transmissivities [i.e., $\tan^{-1}(Im/Re)$] and then convert this to a length to calculate how much the waveguide lengths on the side of the star coupler opposite the mutual coupling need to be adjusted to correct for the aberrations. Note that it is best to fit a smooth curve to the calculated lengths. This is because the beam propagation may result in some noise, which is not present in the real aberrations.

6. EXAMPLE DESIGN: DEMULTIPLEXER

Let us design a four-channel demultiplexer. Let it have a grating order >10, so then we can use the simpler three-straight, two-bend design for the waveguide grating. We choose to have simple single-maximum passbands, which gives us passband shapes that are roughly Gaussian. We do this by having the same images for the input and outputs and a linear grating. Suppose we are designing for a dense wavelength division multiplexing system. If we want to match the International Telecommunications Union dense wavelength division multiplexing plan, then we will need the channels to have a constant frequency spacing. We choose the constant frequency spacing of 100 GHz.

First, we need to choose what grating order we want to operate in. The grating order determines the free spectral range of the demultiplexer. The grating order is approximately equal to the central optical frequency divided by the free spectral range. One usually chooses a free spectral range equal to ~1.25 times the span occupied by the channels. Thus, we choose a free spectral range of 500 GHz. We want to operate in the C-band, so the optical frequency is about 194,000 GHz. Thus, we choose a grating order of $A = 388$ (the grating order must be an integer).

Next we calculate ΔL.

$$\Delta L = A\lambda_c/n(\lambda_c) \qquad (9)$$

where λ_c is the center wavelength and $n(\lambda_c)$ is the effective refractive index at λ_c. For example, suppose we want $\lambda_c = 1550$ nm and $n(1550 \text{ nm}) = 1.45$. Then $\Delta L = 414.76$ μm. This is the path-length difference between adjacent arms.

Now we need to determine a, the pitch of the grating waveguides where they connect to the star coupler slabs. We find this by starting with an eigenmode of the grating waveguide far from the slab and propagating it to the slab via beam propagation. We choose a such that approximately half the power has spread to the neighboring waveguides. This rule of thumb gives an amount of mutual coupling that gives a low insertion loss and yet not too much mutual coupling that would result in severe WPF sensitivity. To do the beam propagation, we also need to know the angular difference $\Delta\alpha$ between each grating waveguide at the connection to the slab and the number of grating waveguides M. A good rule of thumb for M is 2.5 times the number of channels in the free spectral range. However, this number needs to be tweaked to get the desired 1-dB passband width and 1- to 20-dB passband width ratio. Let us choose $M = 2.5 \times 5 = 12.5$ and round it up to 13. $\Delta\alpha$ is typically the 13-dB bandwidth of the optical mode from the input/output core strips divided by M. This can be found from calculating the eigenmode of the input/output core strip at its connection to the star coupler slab and then taking the Fourier transform to get the angular spectrum. A typical number is 0.16 rad. Thus, $\Delta\alpha = 0.16/13 = 0.012$ rad. A typical value for a is 10.5 μm.

Finally, we can determine the output port angular locations by using the grating equation, Eq. (4). It is important to realize that the refractive index change with wavelength, even in silica, is not negligible. Thus, one will need to use a slightly different β for each output port.

7. EXAMPLE DESIGN: BAND DEMULTIPLEXER

The demultiplexer we just designed has a round-top passband. Many applications require a flat-top passband. In other words, instead of a passband with a single maximum, we would like to have a passband with multiple maxima. The simplest way to achieve this is to use a structure, such as a *y*-branch coupler, at the input to the grating to achieve a mode shape with multiple peaks [8]. When this mode is imaged onto the output waveguide, more than one transmission peak within the passband can be generated. However, this necessarily adds loss. In fact, one can show that it is impossible to have more than one peak in a passband with zero loss without a coupler in the structure in which multiple inputs *and* multiple outputs are connected to something. It is a necessary but not a sufficient condition.

One way to create a multi–zero-loss peak passband is to connect two waveguide gratings with multiple waveguides. Consider the case shown in Fig. 21.7. In this case, the connecting waveguides are spaced far apart from each other at their connections to the star couplers. The transmissivity from input to output will look like a series of well-separated passbands, shown as "undersampling" in Fig. 21.7. Now suppose one decreases the spacing between the connecting waveguides at their connections to the star coupler. Assume that all the connecting waveguides have the same length. The individual passbands will begin to merge into one wide passband, with multiple zero-loss peaks. At some point, the ripple will become zero and the passband will become perfectly smooth. This is the "perfect sampling" condition. The sampling then meets the Nyquist criterion. This is when the connecting waveguides perfectly sample the spectra from the waveguide gratings. The field along one edge of the star coupler is approximately the spatial Fourier transform of the field along the other edge. Suppose the connecting waveguides have a pitch of b. Then if the grating arms occupy less than the angle λ/nb, the connecting waveguides do not undersample the spectrum. It is analogous to time sampling, such as in a telephone or compact disk player. If the sampling is faster than the inverse

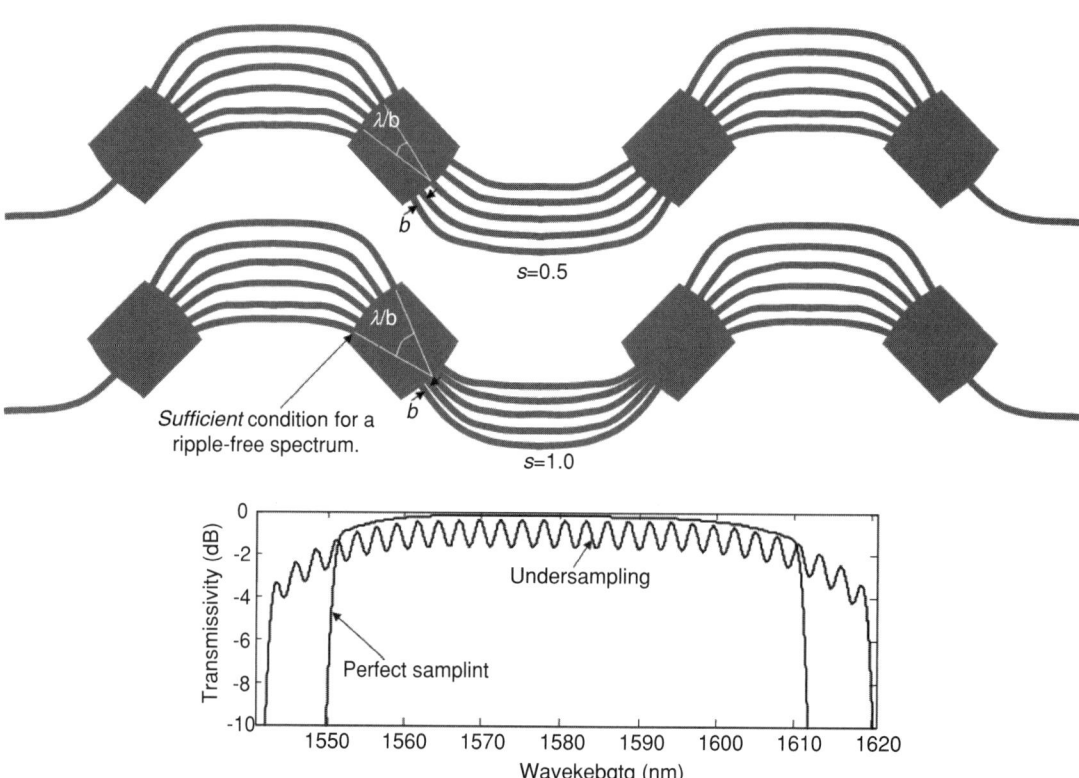

FIGURE 21.7 Spectral sampling in waveguide grating routers.

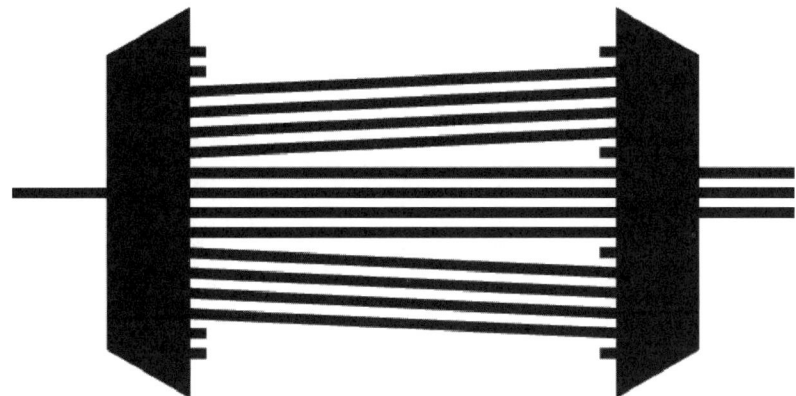

FIGURE 21.8 Band demultiplexer concept.

bandwidth of the signal, no information is lost by the discrete sampling. We can define a sampling coefficient

$$s = \frac{R}{Ma} \frac{\lambda}{n_{\text{slab}} b} \qquad (10)$$

A typical demultiplexer like we designed in the previous section would have $s \approx 0.5$. Perfect sampling is $s = 1.0$.

Two waveguide gratings connected by a set of connecting waveguides act like a simple bandpass filter. If one puts variable optical attenuators that are driven push–pull so as to maintain a constant phase in the connecting waveguides, one can make a dynamic gain equalization filter [9]. If one wishes to make a demultiplexer for a small number of channels and with passbands with very sharp corners, one can use two waveguide gratings connected by *sets* of connecting waveguides, as shown in Fig. 21.8 [10]. The connecting waveguides are approximately all evenly spaced at their connection to the left-hand grating but have increased gaps between sets at the right-hand grating. The increased gaps cause each band to appear at a different port when exiting from the right-hand grating.

Although it has been shown that the connecting waveguides do not have to be exactly the same length, here we make the connecting waveguides within each set the same length. This ensures that the filter has precisely zero chromatic dispersion in the passbands.

Figure 21.9 shows a layout for a 5-band 8-skip-0 band demultiplexer. It consists of two AWGs connected by an AWL. The two waveguide gratings using a three-straight, two-bend design, and the waveguide lens uses a five-straight, four-bend design. The grating orders of the left and right gratings are 30 and 26, respectively. There are 14 waveguides per band. The index step of the waveguides is 0.80%, and they are made of silica. Figure 21.10 shows the measured response. As one can see, the passbands are wide and flat-topped with relatively low loss.

8. CONCLUSION

Arrays of parallel strips of core are a powerful and practical way to make optical devices. I believe there are many devices left to be discovered using such technology. I hope exciting and useful research will continue in this field for many years to come.

9. ACKNOWLEDGMENTS

I thank M. Zirngibl, A. White, and S. Patel for support; C. Dragone for the things he taught me; L. Stulz and L. Buhl for technical assistance; M. Cappuzzo, L. Gomez, A. Wong-Foy, E. Chen, E. Laskowski, R. Pafchek, and many others for

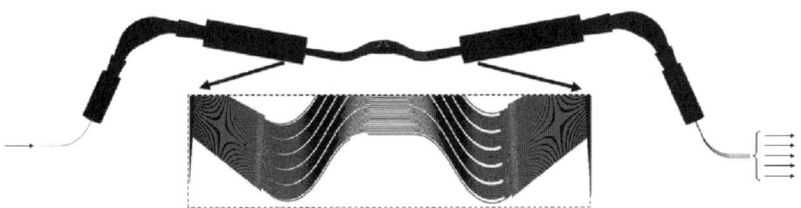

FIGURE 21.9 Waveguide layout of 8-skip-0 band filter.

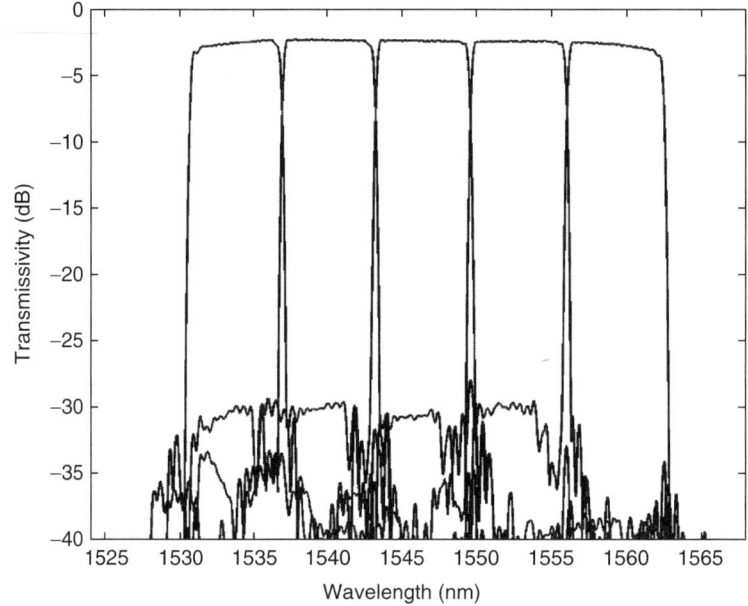

FIGURE 21.10 Measured fiber-to-fiber transmissivities through 8-skip-0 band filter.

device fabrication; P. Bernasconi for many useful discussions; J. Lam, C. Ho, and K. McGreer for discussions and fabrication; and J. Fernandes for assistance.

10. REFERENCES

1. C. Dragone, "Efficient N × N star couplers using Fourier Optics," *J. Lightwave Technol.* **7**, 479–489 (1989).
2. J. W. Goodman, *Introduction to Fourier Optics*, McGraw-Hill, New York (1968).
3. M. K. Smit, "New focusing and dispersive planar component based on an optical phased array," *Electron. Lett.* **24**, 385–386 (1988).
4. H. Takahashi, S. Suzuki, K. Kato, and I. Nishi, "Arrayed-waveguide grating for wavelength division multi/demultiplexer with nanometer resolution," *Electron. Lett.* **26**, 87–88 (1990).
5. C. Dragone, "An N × N optical multiplexer using a planar arrangement of two star couplers," *IEEE Photon. Technol. Lett.* **3**, 812–815 (1991).
6. P. Bernasconi, C. Doerr, C. Dragone, M. Cappuzzo, E. Laskowski, and A. Paunescu, "Large N × N waveguide grating routers," *J. Lightwave Technol.* **18**, 985–991 (2001).
7. C. R. Doerr, "Beam propagation method tailored for step-index waveguides," *IEEE Photon. Technol. Lett.* **13**, 130–132 (2001).
8. M. R. Amersfoort, J. B. D. Soole, H. P. LeBlanc, N. C. Andreadakis, A. Rajhel, and C. Caneau, "Passband broadening of integrated arrayed waveguide filters using multimode interference couplers," *Electron. Lett.* **32**, 449–451 (1991).
9. C. R. Doerr, R. Pafchek, and L. W. Stulz, "16-band integrated dynamic gain equalization filter with less than 2.8-dB insertion loss," *IEEE Photon. Technol. Lett.* **14**, 334–336 (2002).
10. C. R. Doerr, R. Pafchek, and L. W. Stulz, "Integrated band demultiplexer using waveguide grating routers," *IEEE Photon. Technol. Lett.* **15**, 1088–1090 (2003).

CHAPTER 22

Photobleached Gratings in Electro-Optic Waveguide Polymers

Venkata N. P. Sivashankar, Edward M. McKenna, and Alan R. Mickelson*,
Guided Wave Optics Laboratory (GWOL), Department of Electrical & Computer Engineering
University of Colorado at Boulder
Boulder, Colorado, USA

1. INTRODUCTION

In the present work, we review some of our efforts here at the Guided Wave Optics Laboratory (GWOL) related to the writing of holographical gratings in materials that are suited to fabrication of long lifetime electro-optic (EO) waveguide devices. We are motivated by the numerous examples of the profitable use of diffractive structures in waveguides. Fiber Bragg gratings are a prime example of the genre [1, 2]. Quasi-phase matching of $\chi^{(2)}$ waveguides in crystals such as $LiNbO_3$ is also a process that finds application is today's optical marketplace. We discuss how photobleaching of EO polymers can be used as part of the fabrication process. EO polymers are generally EO due to the alignment of the chromophore molecules with which they have been doped. Index changes in the chromophore (dye) doped polymer can be caused by the breaking of electronic bonds in the chromophore molecules. The breaking of dye molecule bonds removes absorption bands, thereby reducing the perception of color in some dye systems. The bond breaking or bleaching of the dye molecules can be induced by illuminating the polymer with light that is resonant with an isomerization reaction in the chromophore molecule. When the illumination is made through a mask or by interference pattern, the resulting index distribution can be quite complex and the feature submicron. Determining an exposure process that results in the polymer system exhibiting the desired mechanical, electrical, and optical characteristics after exposure requires some effort. Experimental results of photobleaching studies at GWOL are presented and discussed.

2. INTEGRATED OPTICS IN EO POLYMERS

2.1. The EO Effect

The EO effect is a second-order nonlinear optical process in which the index of refraction of a material changes linearly due to an applied electric field. The permittivity of a material depends on the the electric polarization within the material [3]. The polarization vector **P** (r,t) of a material is expressible in the form

$$\mathbf{P}(r,t) = \mathbf{P}_{DC}(r,t) + \int_{-\infty}^{t} \chi^{(1)}(r,t-t') \cdot \mathbf{E}^{(1)}(r,t') dt'$$
$$+ \int_{-\infty}^{t} \int_{-\infty}^{t'} \chi^{(2)}(r,t-t'-t'') \mathbf{E}^{(2)}(r,t'') \mathbf{E}^{(1)}(r,t') dt'' dt'$$
$$+ \int_{-\infty}^{t} \int_{-\infty}^{t'} \int_{-\infty}^{t''} \chi^{(3)}(r,t-t'-t''-t''') \cdots \quad (1)$$

where $\mathbf{P}_{DC}$ is a permanent electric dipole moment and $\mathbf{E}^{(i)}$ are applied electric fields. The quantities $\chi^{(n)}$ are the nth order susceptibilities and are tensor quantities. To make this explicit, we can Fourier transform the relation for the **P** expression and write the result using subscript notation

$$P_i = \chi_i^{(0)} + \chi_{ij}^{(1)} E_j + \chi_{ijk}^{(2)} E_j E_k + \chi_{ijkl}^{(3)} E_j E_k E_l + \cdots \quad (2)$$

where the Einstein summation convention that repeated indices are summed over is assumed. In the above, the ith component of the polarization P_i at the frequency ω is now expressed as a power series in the components of the electric fields $\mathbf{E}^{(i)}$ at frequencies $\pm \omega_i$, by where it is assumed that $\omega = \Sigma_i \pm \omega_i$ where either of the $\pm$ may be taken for any of the terms. The term (sequence of pluses and minuses) that dominates the effect is then determined by the macroscopic averaging and phase matching of the the propagating disturbances. Evidently, application of one or more electric fields changes the polarization vector by changing the magnitude and direction of the bound dipoles within our assumed dielectric medium. If the ionic cores (lattice sites) that act as the fixed positive charges about which the negatively charged electron clouds move are randomly located, there is no static electric dipole moment nor any even order susceptibility

---
*E-mail: Mickel@schof.colorado.edu

terms ($\chi^{(2n)}$, n integer) within the medium [3]. When the lattice sites exhibit a regular pattern that possesses no center of symmetry, the internal coulombic repulsion of one electron cloud on another can cause the electron clouds to align themselves relative to the ionic cores such that the material can exhibit an internal static dipole moment. Even when there is no static dipole moment, these noncentrosymmetric crystalline materials still possess all even order susceptibilities $\chi^{(2n)}$. When a material possesses a $\chi^{(2)}$, then a static $\mathbf{E}^{(2)}$ can change the index of refraction seen by a field $\mathbf{E}^{(1)}$. That is, the index of refraction n, that is in general a tensor quantity, can be defined by $\varepsilon = n^2 \varepsilon_0$. The (tensor quantity) ε is related to the polarization through the constitutive relation $\mathbf{D} = \varepsilon \mathbf{E} = \varepsilon_0 \mathbf{E} + \mathbf{P}$. If the $\mathbf{P}$ is dependent on $\mathbf{E}^{(2)}$, then $\mathbf{E}^{(1)}$ sees an index of refraction dependent on the $\mathbf{E}^{(2)}$. A $\chi^{(2)}$ EO effect is referred to as the linear EO effect because the field $\mathbf{E}^{(2)}$ can linearly induce a change in index for a propagating field $\mathbf{E}^{(1)}$ field. One often describes this effect by defining an EO tensor r_{ijk} such that the change in the index tensor $(\frac{1}{n^2})_{ij} = r_{ijk} E_k^{(2)}$.

An amorphous polymeric material does not exhibit the linear EO effect that arises from the $\chi^{(2)}$ susceptibility term in the expansion of the polarization. An amorphous polymer can consist of a single backbone material that will naturally randomly align itself. Chromophores may be mixed with the backbone polymer [4] before material casting. Such materials are referred to as guest–host materials, with the guest being the chromophore and the host the polymer. The chromophore may also be added by chemically attaching the chromophore molecule to the polymer backbone itself. Such a combination is generally referred to as a side-chain polymer, where again the backbone polymer can be referred to as a host or as the main chain. The chromophores that are added to the host material are often dyes. Dyes possess large static electric dipole moments. These intrinsic dipole moments allow the dyes to be aligned in an electric field. Large static dipole moments are always accompanied by even order nonlinearities. Generally, the static dipole moments are aligned along a single direction. If we denote this direction by 3, then the only nonzero elements of the EO tensor r are the elements r_{33} and r_{13}. Evidently, one would like to use whatever techniques one has available to optimize these nonzero coefficients. To access the microscopic parameters, one can apply a treatment similar to that for the macroscopic polarizability P_i to the individual chromophore molecules and expand the molecular dipole moment μ in powers of the local electric fields $E^{(i)}$ as [5]

$$\mu_i = \mu_i^0 + \alpha_{ij} E_j^{(1)} + \beta_{ijk} E_j^{(1)} E_k^{(2)} + \gamma_{ijkl} E_j^{(1)} E_k^{(2)} E_l^{(3)} + \cdots \tag{3}$$

where again the summation convention is assumed and where μ_0 is the permanent dipole moment of the chromophore, α is the linear polarizability, β is the first hyperpolarizability, and γ is the second hyperpolarizability, and so on. For centrosymmetric molecules, all even orders vanish. Chromophores that lack centrosymmetry possess a nonvanishing β. The macroscopic quantity corresponding to the first hyperpolarizability β is the second-order susceptibility $\chi^{(2)}$. The $\chi^{(2)}$ and/or the EO coefficient r_{ijk} can be maximized through adjustment the magnitude of the β as well as macroscopic quantities associated with the sensitivity, distribution, and alignment of the chromophores in the host matrix. It is through this macroscopic averaging [6] that the macroscopic susceptibility depends on the microscopic order parameters.

2.2. The Glass Transition Temperature

Every amorphous polymeric system has a glass transition temperature. We refer to this temperature as T_g. This temperature represents the thermodynamic point at which the number of degrees of freedom of mechanical motion within the polymer system changes discontinuously. Below T_g one can say that rotational motion is frozen out. More precisely, one would say that the specific heat capacity of the material changes discontinuously at T_g, and indeed, a calorimetric measurement is the type generally used to define the glass transition temperature. More heuristically, one says that below T_g, the polymer is in a glassy state. In this glassy state, there is still molecular motion and there is still a relaxation of internal order through molecular motion, but the rate of relaxation is reduced. At another temperature above T_g (between 10% and 70% higher on a Kelvin scale), there is a crossover temperature T_c at which the viscosity of the system reduces significantly and the polymer can flow as a fluid. When the temperature T of a polymer is bracketed between $T_g - 50°C$ and $T_g + 50°C$, one says that there is a time temperature equivalence for the polymer [7–9], such that increasing the temperature causes processes to proceed to completion at an increased rate. The increase in rate is dependent on the process. Here, we consider high temperature polymers to be those whose T_g is more than 50°C above room temperature. This definition is arbitrary but practical as we see in the next section of measured relaxation.

2.3. Poling Lifetime Issues

As was mentioned, to exhibit an EO effect, the chromophores that are doped into a polymer must be aligned. The method by which such alignment is achieved is generally called poling. There are a number of techniques that are used for poling chromophores that are embedded in the polymer matrix. Electric-field poling is one of these methods. In this technique, the polymer is first heated to near its T_g. It is then allowed to come to equilibrium at this elevated temperature. A strong electric field is then applied in the direction into which one would like the static dipole moments to point. The material is subsequently cooled back to room temperature before this field is removed. In corona poling, the doped polymer [10] has a strong field applied across it in such a manner that the field discharges across the polymer, thereby rapidly (rather than gradually as in electric field poling) aligning the chromophores along the poling axis. Optical poling [11] uses the polarization dependence of the chromophore response to apply forces of alignment inside the material. Photobleaching applies such polarization-dependent forces as well and one can strain and/or stress the material if care is not used to remove any internal stresses after the process comes to completion [12].

Strälein et al. [13] measured the relaxation of poled order by monitoring the magnitude of the second-harmonic signals generated by poled polymers under illumination. Their results are schematically depicted in Fig. 22.1. A strong correlation between the characteristic poled order relaxation time τ and the glass transition temperature was measured for all of chromophore–polymer combinations studied. It should be pointed out that all chromophores were "large" molecules in "stable" polymer backbones. The terms large and stable are quite relative ones but here are used to mean that the dye–polymer combinations were such that these materials were candidates for room temperature integrated optical materials. For example, such materials would be competitive for long-lived multi-GHz optical modulators. Composite polymer dye materials whose glass transition is above room temperature satisfy the conditions for large and stable. As will be seen, this condition is not sufficient for the materials to actually be viable candidates for actual devices whose life-times must be minimally on the order of 5 years. Now, at a given temperature, the nonlinear coefficient is shown to relax while exhibiting a stretched exponential dependence on time. Both

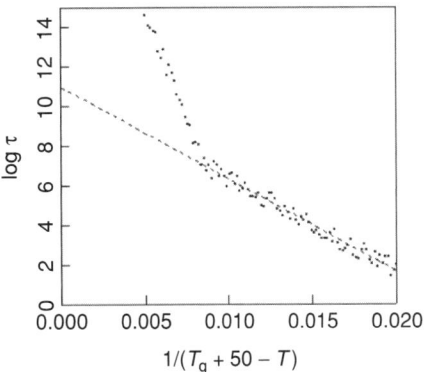

FIGURE 22.1 Presented are some theoretically generated points meant to resemble the experimental points originally measured in [13] for poled order lifetime relaxation in a number of doped polymeric materials. The dotted line represents an empirically derived relation for guest–host polymers. The dark points represent the typical measured behavior for side-chain polymers. The theory of single-strand stress relaxation would yield a smoother version of the side-chain curve located slightly above and to the right of the points depicted on this figure.

guest–host materials and side-chained materials were studied in [14, 15]. For guest–host materials, the relationship between the relaxation time and the temperature is empirically determined to be given by

$$\log[\tau] = \frac{\text{constant}}{T_g + 50 - T} \quad (4)$$

where the temperature is expressed in °Kelvin or °C. For side-chained material, the lifetime decreases rapidly from a value much greater than the guest–host lifetime as the temperature parameter $T_g + 50 - T$ increases up to the limit of validity of the expression that is roughly given by $T = T_g$. This behavior is clearly evident in Fig. 22.1, a typical semilog plot of relaxation time as a function of inverse temperature for a side-chain polymer. The dotted line represents the empirical relation for guest–host materials. It is seen that the side-chain curve deviates strongly from the guest–host results at temperatures much below about $T_g - 50°C$. As is discussed in [15], the side chains relax according to material stress relaxation curves of single-strand polymers. It should be noted that the lifetimes at which the two curves diverge is at a lifetime of several months. For the poling lifetime to exceed several years, then, the guest–hosts need to have higher T_g's than the side chains. This is a reason we study only materials whose T_g exceeds room temperature by at least 50°C.

2.4. High Temperature Polymers

There are a large number of polymers that can be used for waveguiding. Many of these can also be doped with chromophores such that they can exhibit EO coefficients. Most of these polymers are not practical for applications because of the their lifetime. That is, some polymers may degrade rapidly under illumination. Others simply degrade due to deformation or chemical reactions with their environment. A more subtle effect with EO polymers is the tendency for the EO coefficient to relax toward zero after poling, the effect we just discussed in the last paragraph. All poled polymers exhibit this poling lifetime decay even if they are otherwise stable and long-lived. Polymers used in the production of electronics are a class of polymers that have been well tested for a variety of properties. They need to have long-term stability at room temperature. Some of these polmers have good optical characterisitics. There is some discussion of these packaging polymers, for example, in [16]. The glass transition temperature is the most important parameter in determining how long a polymer that is chemically and mechanically stable at room temperature will retain the EO coefficient that it was endowed with during poling.

2.4.1. Some Examples

A workhorse polymer for optical applications is polymethyl-methacrylate (PMMA). It is the material from which most plastic fiber is made. The T_g for *neat*, that is undoped, PMMA is on the order of 100°C, which meets our definition for high temperature. PMMA is therefore rather marginal as a guest–host optical waveguiding material according to the lifetime relation we discussed in the last section. Much of the work we have done is with a side-chained version of PMMA that we call PMMA/DRI, which still has a rather short poling lifetime (less than 1 year) but that is significantly more stable than a guest–host version would be. The low lifetime which still is on the order of a year is sufficient for laboratory experiments but is not usable for applications.

Polyimide is the most common of electronic packaging materials. Polyimides belong to a class of condensation polymers. They are formed from a reaction between an aromatic diamine and an aromatic dianhydride. When both the diamine (i.e., 6FDA) and dianhydride (i.e., 6FDAM) are fluorinated (the F's in the names indicate that some number of the hydrogen groups have been replaced with fluorine atoms), these materials have good optical properties. Without the fluorination, the compounds tend to be yellowish or even black. The only optical polyimides are those that are fluorinated. Guest–host polyimides have been studied in our own work in [20] as well as in work that was carried out at Lockheed [17, 18, 19]. Side-chain polyimides have been studied at IBM.

Benzocyclobutenes (BCB)s are the dielectrics intended for the electronic packaging industry known for their excellent properties like planarization and low dielectric constant. BCB has a backbone that consists of a siloxane-type material. The siloxane strands, however, are interconnected with other siloxane strands with organic (hydrocarbon) bonds. The bonds are arranged such that the material itself forms a network rather than just long serially connected strands as in a usual polymer. The curing temperature can be as high as 450°C, and a T_g is hard to define because of the network that is formed. The BCB family was originally developed as a possible substitute for silicon dioxide as the interconnect layer in the silicon integrated circuits. The original BCB formulation will not serve as this replacement material, but the Dow Chemical Company continues to work in this area as is indicated by the company website [21]. BCB is but one chemical of the product line entitled Cyclotene.

Perfluorinated benzocyclobutane (PFCB) has a name similar to BCB and was produced by Dow for the same target purpose. It is actually quite a different class of chemical compounds. It is a thermocurable material and has propagation loss in waveguide configurations as low as 0.25 dB/cm at 1.3 μm and 0.2 dB/cm at 1.55 μm. The glass transition temperature of this material is as high as 400°C. It is known to exhibit excellent resistance to solvents. Petermann and coworkers [22] fabricated single-mode waveguide in PFCB with losses less than 0.25 dB/cm in the 1300 window as well as in the 1515–1565-nm range.

Polycarbonate is one of an important class of cross-linked polymers. These are materials that are tough, and engineering materials that will withstand a wide range of temperatures. They have been used for fabrication of high-speed optical modulators.

3. PHOTOBLEACHING DYE-DOPED POLYMERS

The dye molecules used to dope polymers to endow them with EO coefficients generally have large dipole moments. These dipole moments are due to a small amount of electrical charge being displaced a large distance from its equilibrium in terms of number of atomic sites. This can be accomplished in conjugated structures where all atoms are double bonded to all others. The σ bonds lie between the bonded cores, whereas the second set of bonds, the π shell of bonds, are delocalized from the atomic centers. The π shell may be quite long, that is, the electron wave cloud of this shell may extend over tens or even hundreds of atomic centers. Many dyes consist of capped aromatic rings at either end connected by a bridge. If a ring is capped with an atom with a large electron affinity (electron acceptor) or one that gives up electrons in the process of bonding (donor), that affinity will be translated onto the π orbital, shifting the electron wave cloud one direction or another along the molecular axis. These donor or acceptor substitutions can then cause the shift of charge of long atomic distance which leads to large dipole moment. Popular dyes for doping can consist of either double-bonded carbons (C = C– group) in the bridge (stilbenes [23]) or double-bonded nitrogens (N=N– group) at the bridge (azos [24]). Evidently, there are (at least) two configurations of such a molecule, one in which the rings can both lie of the same side of the double bond (*cis* state) or one in which the rings lie on either side of the rings (*trans* state). These two states are stereoisomers of each other. Such isomers can have different energies. In the case of the stilbenes and azos used for doping of EO polymers, the *trans* state generally has the lower energy. The transition between these structural states and hence the energy states can be aided by light—a process called photoisomerization [25]. When the chromophore's dipole axis is aligned to the polarization of the incident radiation, it is much more likely to be photoisomerized. Those in the plane perpendicular to the direction of polarization do not interact with the field.

3.1. Early Experiments

Rochfort and coworkers [26] reported the first optical channel formation by photobleaching the chromophores in polydiacetylene. The photobleaching mechanism in polydiacetylene was known to be associated with reduced absorption of an isomerization from an acetylenic to a butatrienic form [27–29]. Waveguide formation in polymers had already been carried out in 1972 [30], along with bleaching studies of guest–host polymers for the purpose of both holographic storage [31] and the study of laser dye lifetimes [32–33]. Garito and coworkers [34, 35] carried out chemical polymerization of polydiacetylene in early 1980s, and the development of first optical guest–host systems by Meredith et al. [36, 37] in liquid-crystal polymers and by Singer and associates [10] in acrylate backbone quickly followed it. This stimulated the studies on the bleaching process in stilbene side chains [38–41], and device fabrication using photobleaching for channel definition [17, 43]. Generally, in azobenzenes [12, 44] and in stilbenes [45], the absorption feature whose absorbance value is reduced most likely corresponds to *cis-trans* isomerization [42, 46, 47], as described above.

3.2. A Theory of Radiation-Induced Chemical Reactions

The process of photobleaching involves chemical reaction within the chromophore molecules in the photosensitive system. The microscopic mechanism of the photochemical reaction is likely to consist of dissociation followed by diffusion of the dissociated chromophore out of the film. The photochemical reactions inside the dye-doped polymers are complicated and have been the subject of research for many years. A semiempirical model based on Beer's law, a local chemical reaction equation, and a stretched exponential time dependence combined with the experimental data on both side-chain and guest–host photoresponsive polymer system is described in [44]. This technique together with the Kramers-Krönig relation [48] can be used to predict the index profile of a photobleached film.

Suppose that there are many species in the system with absorption coefficients α_i. The absorption coefficients are functions of the bleaching wavelength λ_B, the distance into the system that the bleaching light has propagated z, and the time t. If we defined the intensity I as a function of λ_B, z, and t we can write relation between I and α_i's as set of coupled partial differential equations as

$$\frac{\partial}{\partial z}I(\lambda_B, z, t) = -\sum_{i=1}^{n}\alpha_i(\lambda_B, z, t)I(\lambda_B, z, t) \quad (5)$$

for the intensity variation along the incident light path and the rate of absorption as

$$\frac{\partial}{\partial z}\alpha_i(\lambda_B, z, t) =$$
$$-\int_0^\infty \kappa_i(\lambda, \lambda_B, T, t, F)\alpha_i(\lambda_B, z, t)I(\lambda_B, z, t)d\lambda_B$$
$$(6)$$

where we have introduced a function of the quantum yield κ_i for the ith species that is also a function of λ_B, z, and t apart from the temperature T and the flux F of some reactant (e.g., oxygen) that affects the bleaching.

If we were to write the above coupled equations for just one species case, we obtain

$$\frac{\partial}{\partial z} I(\lambda_B, z, t) = \alpha(\lambda_B, z, t) I(\lambda_B, z, t) \quad (7)$$

$$\frac{\partial}{\partial z} \alpha(\lambda_B, z, t) = -\kappa(\lambda, \lambda_B, T, F) \alpha(\lambda_B, z, t) I(\lambda_B, z, t) \quad (8)$$

where we have assumed κ as independent of time. With an initial condition on the absorption coefficient as α_0 and a boundary condition on the intensity at the surface ($z = 0$) as I_0, the analytical solution is written as

$$A(\lambda_0, t) = \int_0^d \alpha(\lambda_0, z', t) dz' \quad (9)$$
$$= \ln\{1 + [\exp(\alpha_0 d) - 1] \exp(-\kappa_0 I_0 t)\}$$

$$I(\lambda_0, z, t) = \frac{I_0}{1 + [\exp(\alpha_0 z) - 1] \exp(-\kappa_0 I_0 t)} \quad (10)$$

where we assume there is no flow of reactants introduced to speed up the bleaching process and the bleaching is done at one wavelength $\lambda = \lambda_B = \lambda_0$. The parameter κ_0 is determined from fitting the experimental data. For the cases of PMMA/DR1 and Ultem/DEDR1, the fitted values of κ can be found in [44].

3.3. A Photobleaching Model for Dye-Doped Polymers

A methodology for determining the bleaching characteristics of dye-doped polymer is presented in [20]. The general phenomenological model predicts the absorption spectrum as well as the change in index of refraction as a function of wavelength for a bleached film without the detailed knowledge of the chemistry or physics of the polymer dye system. In [20], a two-parameter model of the bleaching kinetics is developed by approximating the absorption spectra of the guest–host or side-chained polymer system as being the sum of three Gaussian absorption features, $g_k(\lambda)$, given by the relation

$$g_k(\lambda) = e^{-\frac{(\lambda - \lambda_k)^2}{\Gamma_k}} \quad (11)$$

where λ_k is the center wavelength of the kth Gaussian absorption feature and Γ_k is the width of the kth Gaussian. The three features correspond to the absorption spectra of the polymer backbone and two Gaussians corresponding to the absorption of the chromophore. The situation is illustrated in Fig. 22.2.

The Gaussian form of the absorption feature should hold if thermal broadening dominates inhomogenous broadening. Such a spectrum is referred to by Siegman [49] as "dynamically homogenous." The rationale behind modeling the chromophore with two dye Gaussians has its origin in the electronic structure of the chromophore itself. For molecules that undergo an isomerization reaction about a carbon–carbon double bond, the *trans* state lies at a lower energy level than the *cis* state; thus, the population of the dye at room temperature is largely in the *trans* state. In the *trans* state, there is one absorption peak. The higher energy contributions correspond to the $\pi \rightarrow \pi^*$ singlet and the lower contributions correspond to the $\pi \rightarrow \pi^*$ triplet, which is

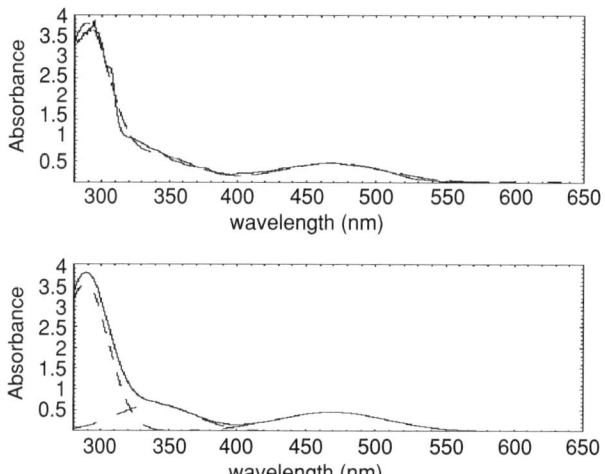

FIGURE 22.2 Shown is how well the absorption spectrum of one specific dye-doped polymer can be approximated as comprising Gaussian shaped absorption features. In the top panel there is an experimental absorption spectrum of the dye/polymer matrix DCM/BCB (solid line) Theoretical gaussian model (dashed line) at a low DCM concentration. In the bottom panel, the theoretical absorption (solid line) used in the top panel is broken out without the experimental curve so that it can be compared with the three Gaussian features that comprise it (dashed line).

nonradiative. These are illustrated in Saltiel's 1967 reaction surface model for stilbene [23]. An energy level diagram illustrating the general features of Saltiel's model is illustrated in Fig. 22.3. Similarly, for molecules that isomerize about a nitrogen–nitrogen double bond or azobenzene-like chromophores, a reaction surface model has been developed by Rau [24]. In both cases isomerization occurs about what Saltiel refers to as a "phantom point" or, as it appears in the figure, a phantom line. Both the singlet and triplet surfaces exhibit a minimum between the *trans* and *cis* branches of these surfaces. This minimum point, the phantom point, may be accessed from either an excited *trans* state or an excited *cis* state. Once at the phantom point, the state may decay to either the *trans* or *cis* ground state. If all molecules were initially in their *trans* ground state, then an incident optical flux will populate the *cis* ground state as some fraction of π^* *trans* states choose this decay mechanism. From the *cis* ground state, absorption to either the triplet or singlet states can occur. These states can then decay back to either the *trans* or *cis* ground states. Another possibility occurs when a ground *cis* state absorbs a photon. An oxygen-assisted photochemical reaction can occur that yields a final state, which for stilbene would yield phenanthrene and in the case of azobenzene would yield anthracene. This is the irreversible bleaching reaction. The model in our case is phenomenological but seems to be a reasonable mathematical model that can be interpreted in terms of the model if one is not too serious about attaching significance to the nonobservable details. An example of an application of the photobleaching model of [20] to the guest-host system of 4-dicyanomethylene-2-methyl-6-(p-dimethylaminostyryl)-4-H-pyran/BCB (DCM/BCB) is illustrated in Fig. 22.4. This photobleaching model has also been applied to describe the bleaching kinetics of other guest-host and side-chained polymer systems such as DCM/PFCB in a more recent work [50].

3.4. Mechanical Effects of Photobleaching

The above described model of photobleaching would be complete if the index contrast of the bleached material were given simply by the Kramers-Krönig relation [20, 54]

$$\Delta n(\omega) = \frac{c}{\pi} \int_0^\infty \frac{\Delta\alpha(\omega)}{\omega^2 - \omega_0^2} d\omega \quad (12)$$

where ω_0 is the resonant angular frequency, c is the velocity of light, and the change in the absorption coefficient $\Delta\alpha$, can be determined from the bleaching models or measurements of the bleached spectra. Unfortunately, the photobleaching process leads volume changes in the material and therefore has mechanical consequences for the material. We presently discuss some of these effects.

3.4.1. Hooke's Law and Tensile Stress

The general relation between the components stress tensor T_{ij} and strain tensor S_{ij} is referred to as Hooke's law. The relation embodying Hooke's law is a linear one relating each component of T_{ij} to each component S_{ij} [53]. There are $9 \times 9 = 81$ elements of the tensor that describe the properties of the material in general. These coefficients form a fourth rank tensor called tensor of elasticity. We can define the components of the C_{ijkl} by the equation

$$T_{ij} = C_{ijkl} S_{kl} \quad (13)$$

where summing on repeated indices is assumed. The strain tensor S_{kl} is related to the vector displacements u(r, t) of material elements inside the material by

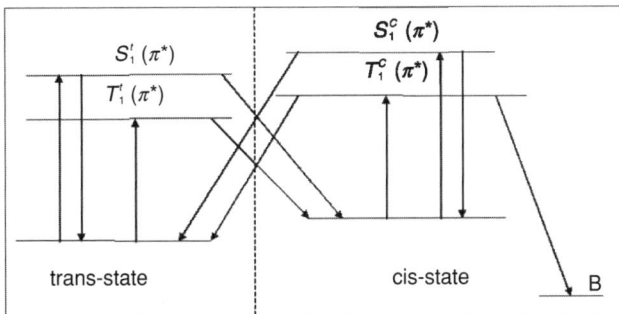

FIGURE 22.3 A possible energy level diagram for a stilbene molecule is shown. A molecule that is initially in its *trans* state can be promoted to a higher level by a photon and subsequently relax to either the *cis* or *trans* ground state. After a time, the *cis* ground state becomes populated. Excitation from the *cis* ground state can result in either a singlet or triplet state or in another state from which decay cannot occur.

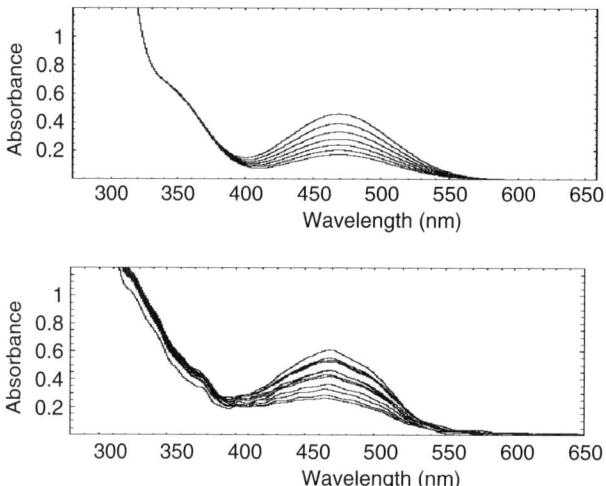

FIGURE 22.4 The efficacy of the use of a Gaussian feature model to yield a set of rate equations that can describe actual bleaching behavior is demonstrated. The experimental set of bleaching curves is illustrated in the lower panel. The fits to these curves based on the rate equation model are illustrated in the upper set of curves. The the rates of the bleaching are well described by the theory with a minimal number of parameters, at least for this regime of bleaching times, the regime of most interest for irreversible bleaching for structure formation.

$$S_{kl} = \frac{1}{2}\left[\frac{\partial u_k(r,t)}{\partial x_l} + \frac{\partial u_l(r,t)}{\partial x_k}\right] \quad (14)$$

where $u_k(r, t)$ are time-dependent deformations of the material element originally located at coordinate $r = \sum_{i=1}^{3} x_i \hat{e}_i$ in the kth direction, where e_i is the unit vector in the x_i direction. The equation of motion for the u_k can be determined from the condition that the divergence of the stress tensor **T**, $\nabla \cdot \mathbf{T}$ is equal to the force, which here is the mass density ρ times the second time derivative of the displacement vector **u**, or in component form

$$\frac{dT_{ij}}{dx_j} = \rho \frac{d^2 u_i}{dt^2} \quad (15)$$

For polymers that are homogeneous and isotropic, Hooke's law simplifies considerably and is expressible as

$$T_{ij} = 2\mu S_{ij} + \lambda \Theta \delta_{ij} \quad (16)$$

where μ and λ are constants and Θ is the trace of the strain tensor $\Theta = S_{xx} + S_{yy} + S_{zz}$. The relation for the linear strain (percent change in length of a material) Θ under a tensile stress T is given by the relation [7]

$$T = E\Theta \quad (17)$$

where $E = 2\mu + \lambda$ is the Young's modulus of the material. Because the bleaching applies a strain to the material, there is also stress and time-varying displacement within the material.

3.4.2. Stress-Modified Indices of Refraction

When static strain is applied or time-varying strain in the form of a propagating acoustic wave is launched into a material, the variation in the index of refraction due to the applied strain can be described by the relation [51]

$$\Delta\left(\frac{1}{n^2}\right)_{ij} = p_{ijkl} S_{kl} \quad (18)$$

where the summation convention (repeated indices are summed over) is assumed and where $\left(\frac{1}{n^2}\right)_{ij}$ is a constant of the index ellipsoid of the material, p_{ijkl} can be called either the strain optic or acousto-optic tensor, and S_{kl} is the strain tensor generated in the material by the static strain or propagating disturbance. This effect is known as the strain-optic or acousto-optic effect. If a unidirectional static strain is assumed in the material, a simpler expression for the change in the index of refraction Δn is obtained:

$$\Delta n = P\Theta \quad (19)$$

where P is the strain-optic coefficient and Θ is the tensile strain. Sometimes this above relation is augmented by the inclusion of Young's modulus (in an isotropic medium) and cast in terms of the material stress ($T = E\Theta$) rather than the strain Θ. The effect is sometimes then referred to as the stress-optic effect. From the previous estimate of the tensile stress, the change in the index of refraction due to the stress–optic effect can be calculated by assuming that the strain–optic coefficient of PMMA/DR1 is approximately the same as that of polycarbonate or $8.843 \times 10^{-12} \frac{\text{cm}^2}{\text{dyne}}$ [7]. Using this value for the strain–optic coefficient, the 2% elongation observed in the atomic force microscope (AFM) should produce an index change on the order of 0.00518. Because the deformation is linearly proportional to the bleaching dose, at the end of the bleaching cycle the change in refractive index in the strained regions will be as high as 0.0236 in the grating. This can completely nullify the index change caused by the bleaching process. It can also be noted that for transient stresses introduced into the film on the order of 2%, the plot of the diffraction efficiency illustrates that the diffraction efficiency can fluctuate quite dramatically from 8% to 4% for an index change of 0.005. This appears to be in line with observations made of gratings at room temperature and may be the cause for the observation of large changes in the first-order diffraction efficiencies of gratings bleached at room temperature in PMMA/DR1 and DCM/PFCB.

3.4.3. Annealing

In earlier investigations at GWOL of waveguides written into PMMA/DR1, AFM measurements showed that waveguides that were irreversibly bleached into PMMA/DR1 contained considerable deformation [12]. It was noted that the deformation and stresses could be relieved in the film by annealing it at 10°C below T_g. It was this observation that convinced us that elevated temperature (from room temperature) bleaching of diffraction gratings was perhaps a good idea. This is the same idea that was discussed in [12] where the original motivation was to speed up processing throughput for channel waveguide structures. The most viable explanation of the measurements of Feng et al. [12] is that the material can be annealed when either processed at higher temperature or when the material is subject to annealing after processing. In general, a material that has been given internal stresses by either deformation or in another manner may relieve the stresses through the process of annealing. Annealing is a process by which thermal energy is added to the material, allowing it to search out local energy minima in the materials structural phase space. To relieve the stresses introduced during the irreversible bleaching of gratings in PMMA/DR1 or DCM/PFCB, it is necessary to anneal the sample. The samples are annealed while bleaching by passing a direct current through a thin film of indium-tin oxide (ITO) coated onto the glass substrate. For samples of PMMA/DR1, the electrodes are mounted onto the sample using either a conductive epoxy or are soldered into place using a low temperature solder containing indium. For DCM/PFCB, the temperatures required to anneal the sample are above the melting point of the solder and only the conductive epoxy can be used. The temperature of the sample is monitored with a thermocouple placed near the grating region. The first-order diffraction efficiency is recorded during the bleaching cycle using a 633-nm HeNe laser. The increased bleaching temperature can dramatically reduce the bleaching time of the sample due to the time temperature equivalence in the temperature regime in which the polymer is bleached [7–9]. Annealing after processing can be an option if the stress induced during processing does not cause irreversible damage. This route can be preferable if the high temperature processing leads to secondary effects, as indicated by the lack of an isobestic point on the photobleaching induced absorption versus time curve [20].

4. DIFFRACTION GRATINGS IN DYE-DOPED WAVEGUIDE POLYMERS

A sinusoidal diffraction grating is developed in a photoactive film by interfering two plane waves in the medium. For an irreversibly bleached polymer, the change in the index of refraction is a function of the intensity of the incident radiation and the period of exposure. For two plane waves of a wavelength λ separated by an angle θ the grating vector $\mathbf{K}_g$ is given by

$$\mathbf{K}_g = \frac{2\pi}{\Lambda_g} = 2k \sin(\theta) \quad (20)$$

4.1. Diffraction Efficiency of Thin Sinusoidal Gratings

For a beam incident at an angle θ_i on a typical sinusoidal diffraction grating of thickness t and index contrast Δn, the diffraction efficiency of the mth order, η_m, is given by [52]

$$\eta_m = J_m^2 \left(\frac{k \Delta n t}{\cos(\theta_i)} \right) \quad (21)$$

where J_m is a Bessel function of the first order, and k is the wavenumber. For thin gratings where $k\Delta n t \ll 1$, the formula may be simplified to the following expression for the first-order diffraction efficiency:

$$\eta = \left(\frac{k\Delta nt}{2\cos(\theta_i)}\right)^2 \qquad (22)$$

The angular dependence of the mth diffracted order is given by the expression

$$\theta_m = \sin^{-1}\left(\sin(\theta_i) + \frac{m\lambda}{\Lambda_g}\right) \qquad (23)$$

where m is the order, Λ_g is the grating period, θ_i is the angle of incidence, and λ is the incident wavelength. The first-order diffraction efficiency for a 633-nm beam incident on a 5-μm sinusoidal grating is illustrated in Fig. 22.5.

4.2. Writing Gratings in Waveguide Polymers

Gratings have been written into samples of PMMA/DR1 and DCM/PFCB using the 514-nm line from an Ar$^+$ ion laser. Other lines have been used, but the 514-nm line of the Ar$^+$ ion laser has the most power, and the bleaching efficiency at the wavelength is sufficient to minimize the bleaching time. The beams are expanded with ×10 objectives and then collimated. The power in each arm of the recording setup is balanced and the diameter of the grating region to be bleached is set using two apertures placed between the collimating lenses and the sample. Typically, the diameter of the apertures is set to approximately 1.1 cm to bleach an area of approximately 1 cm^2. The setup is illustrated in Fig. 22.6.

Initial attempts were made to irreversibly bleach diffraction gratings with approximately 657-nm spacing into films of DCM/PFCB. This material was advantageous because of its high T_g value (>200°C) and the low doses of energy required bleach it to transparency because it is a guest–host and not a side-chained system. For small bleaching doses (<3 kJ) at room temperature, there was visible first-order diffraction, but at the completion of the bleaching cycle the diffraction efficiency of gratings was too small to be measured (≪0.01%). To investigate the process of bleaching gratings into EO polymers further, attempts were made to bleach gratings into PMMA/DR1. This was a system with more-familiar parameters and was used by Ma et al. [44] and Feng et al. [12] to produce waveguides and EO modulators by our group at GWOL. The gratings written into PMMA/DR1 at room temperature exhibited similar behavior to that of the DCM/PFCB. At this point it was observed that the gratings written into DCM/PFCB had a strange discoloration that appeared quite different when observed by transmitted light or by reflection from the surface. This appeared to be the signature of scattering from the film in the grating region. It was hypothesized that stresses induced in the film during the bleaching process may be the cause of the structural failure of the gratings and, ultimately, the observed loss of diffraction efficiency over the bleaching cycle.

To observe the magnitude of the surface deformation caused by room temperature bleaching, an AFM was used to examine the surface of a sinusoidal grating written at room temperature in a sample of PMMA/DR1. Figure 22.7 illustrates an AFM measurement of the surface relief generated by bleaching a diffraction grating in PMMA/DR1 on the 514-nm line of an Ar$^+$ ion laser for approxi-

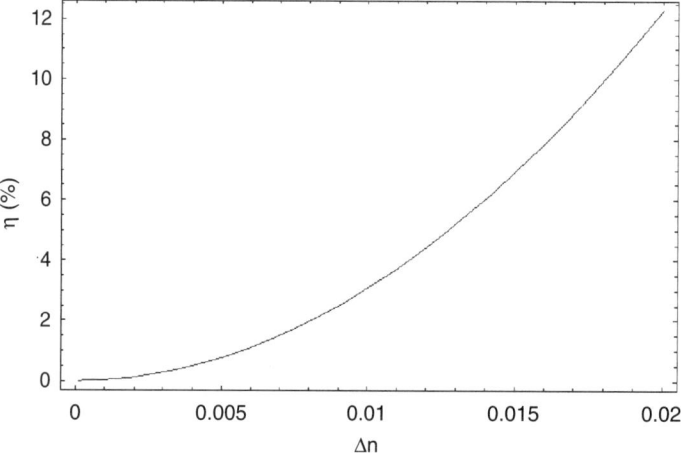

FIGURE 22.5 A theoretical plot of diffraction efficiency vs. index contrast for a 633-nm beam incident on a sinusoidal grating of thickness 5 μm. These numbers were taken to be typical ones for a green bleaching exposure of a film of dye-doped polymer such as PMMA/DR1. This curve can be used in interpreting curves generated from the real-time noninvasive measurement (in red light) of the power in the first diffracted order of a doped polymer grating during exposure by green light.

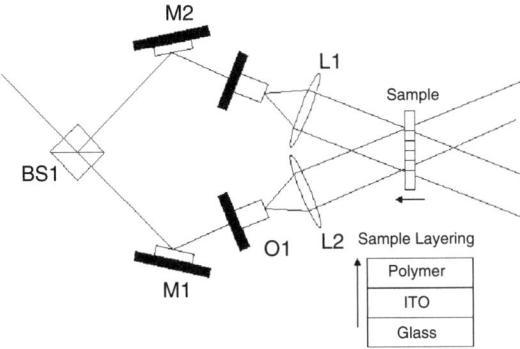

FIGURE 22.6 The experimental apparatus used to photobleach a sinusoidal grating into a doped polymer is depicted. BS1 represents a beam splitter, M1 and M2 mirrors, O1 and O2 are microscope objectives, and L1 and L2 are lenses. The beams forming the Ar laser are first expanded by a factor of 10 before collimation. An aperture that is slightly larger (typically, 1.1 cm^2) than the bleaching area (typically, 1 cm^2) is used before sample illumination.

mately 3 h. The power in each arm of the grating development setup was approximately 66 mW. The total grating area was approximately 0.1 cm^2. Recalling from an earlier section that the strain is related to the tensile stress by the Young's modulus, we can estimate relative magnitudes. The tensile stress between the raised regions of the PMMA/DR1 film may be estimated from the AFM measurements and Hooke's Law by assuming that the film responds linearly and that the linear strain is given by the ratio of the surface relief to half of the grating period. This is 8 nm/375 nm, or approximately 2%. Given that the tensile modulus for acrylic PMMA is approximately 350,000 psi or $2.929 \times 10^{10} \frac{\text{dyne}}{\text{cm}^2}$, the tensile stress between the raised regions is given by Hooke's Law to be approximately $5.858 \times 10^8 \frac{\text{dyne}}{\text{cm}^2}$ [7].

The surface relief of 8 nm over a distance of approximately 0.375 μm generated by the short exposure time produces a deformation in the film on the order of 2%. To completely bleach the sample it requires 24.4 kJ, or an additional bleaching time of 24 hr. Because the deformation increases linearly with the bleaching time, the final deformation introduced into the sample is on the order of 17%. For most materials, a deformation in excess of 15% results in failure of the material. For acrylic PMMA, the tensile elongation at the breaking point is only 2%. Therefore, it is expected that the grating should fail due to stress introduced into the sample upon completion of the bleaching cycle. This expectation has been observed for submicron gratings written in DCM/PFCB and PMMA/DR1 at room temperature.

4.3. Some Study of Transient Gratings

4.3.1. Surface Index Relief Gratings

To characterize the variation of grating efficiencies with the bleaching time and the temperature, the following experiment is set up. Figure 22.8 shows the photobleaching of the polymer thin film. In addition to this setup, a probe beam at 633 nm is incident on the film. A photodetector is used to measure the first-order diffracted beam continuously with time using a personal computer. The film temperature is maintained by passing a current through the ITO layer. The bleaching beam is pulsed with 15-s duration. This causes a stress-induced surface relief that diffracts the probe beam. Because the bleaching beam is shut off after a brief period, the index relief

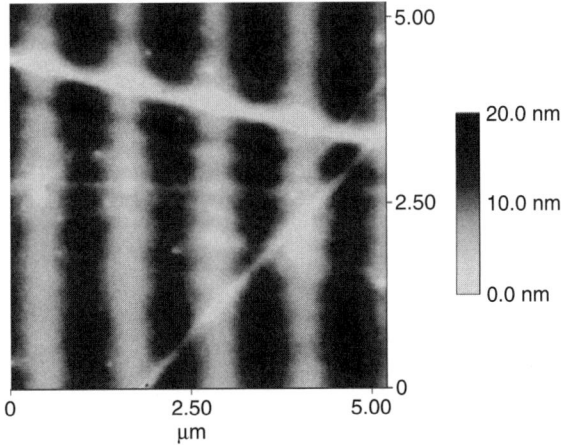

FIGURE 22.7 A picture of the AFM measurement of diffraction grating in DR1/PMMA. The sample was exposed at room temperature for 3 hr at 514 nm. The maximum pitch of the surface relief is 8 nm. The grating period is 0.75 μm. The transverse striations are scratches in the film generated during sample preparation.

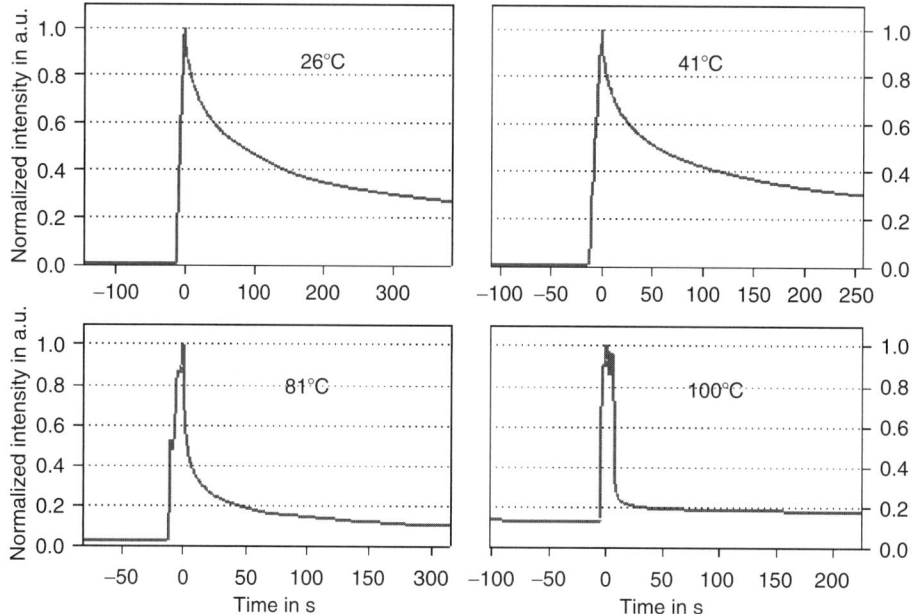

FIGURE 22.8 The results of a monitoring of the first-order diffraction efficiency during and after illumination by a photobleaching beam are shown. The plotted first-order diffraction intensity is that of a 633-nm HeNe laser diffracted from a grating in PMMA/DR1. The bleaching beam is pulsed on for 15 s before being turned off. The transient persists for many pulse periods after the illumination is turned off.

washes off with the relaxation of the stress. Figure 22.8 shows the decay of the first-order diffracted beam at various PMMA/DR1 film temperatures.

The first-order diffraction efficiencies for various film temperatures as a function of time is shown in Figure 22.9. It could be seen that as the temperature gets closer to the glass transition temperature of the film material, the diffraction efficiency gets smaller, clearly an order of magnitude smaller compared with the one at room temperature. Also, it could be noted that the diffraction efficiency falls off rapidly in time as the temperature is closer to the T_g.

4.4. Describing the Grating Formation Process

The equilibrium strain tensor of a material should vanish. The photobleaching process, however, will change the free volume within a polymer either with or without outdiffusion of material. The bleaching process then in and of itself can generate a nonzero strain tensor. The components of the strain tensor are expressible in terms of derivatives of displacements from equilibrium. The photobleaching flux is directed downward into the sample, allowing us to use the natural coordinate system that differentiates the coordinate along the bleaching direction (longitudinal

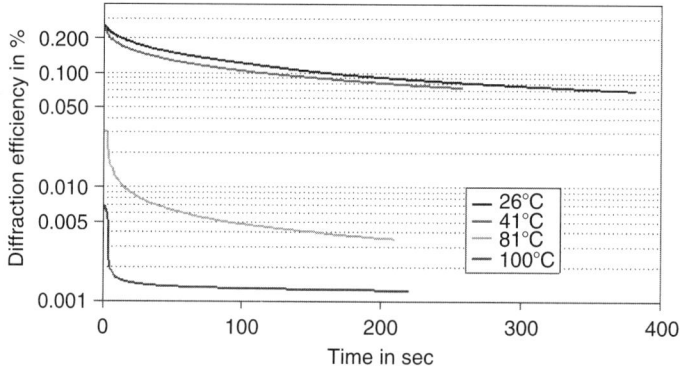

FIGURE 22.9 Experimental curves are depicted that trace the time history of the first-order diffraction efficiencies of 633-mn HeNe beam diffracting from a grating in PMMA/DR1 at various temperatures. It is clearly seen that as the substrate temperature is increased from room temperature to temperatures approaching T_g, the peak diffraction efficiency drops off but that the shape of the decay curve stays approximately constant.

coordinate) from the directions transverse to the bleaching direction. These strain vectors are coupled, but only the longitudinal one is driven. If we write that the overall strain vector is a sum of longitudinal and transverse separable functions of space and time

$$\mathbf{u}(\mathbf{r}, t) = a_l(t)\mathbf{u}_l(\mathbf{r}) + a_t(t)\mathbf{u}_t(\mathbf{r}) \qquad (24)$$

and assume slowly varying time coefficients, we can obtain coupled equations for the time evolution of the material displacement

$$\frac{da_l}{dt} + \gamma_l a_l = \kappa_{lt} a_t \qquad (25)$$

$$\frac{da_t}{dt} + \gamma_t a_t = \kappa_{tl} a_l \qquad (26)$$

The fundamental solutions (which form the matrix of solutions) to these equations are decaying exponentials. Because of coupling of the two degrees of freedom, though, the time evolution of the strain will take on a nonexponential character due to collective energy storage. The solution will depend on the values of the coupling coefficients κ_{tl} and κ_{lt} and damping coefficients γ_l and γ_t. These constants are hard to determine directly either experimentally or theoretically. Taking κ_{lt} equal to κ_{tl} and γ_l and γ_t, we can write a solution set as

$$a_t(t) = C' \exp(-(\gamma + \kappa)t) + (a_t(0) - C') \exp(-(\gamma - \kappa)t) \qquad (27)$$
$$a_l(t) = C\{\exp(-(\gamma + \kappa)t) - \exp(-(\gamma - \kappa)t)\}$$

where the initial condition that the transverse mode is excited before the longitudinal mode is imposed. The C and the C' are the constants of integration that could be determined along with the γ and κ values from the experimental curves presented in Fig. 22.9. Using nonlinear regressional analysis, these constants are determined. A generated plot using these constants superposed on the experimental data is shown in Fig. 22.10. The agreement between theory and measurement validates the assumption that the transient effects are mechanically induced effects.

4.5. Irreversible Gratings in Waveguide Polymers

4.5.1. Experimental Results

A number of diffraction gratings have been made in PMMA/DR1 and a few have been made in DCM/PFCB. The typical intensity in each of the two interfering beams used in the writing process was approximately $50\,\mathrm{mW/cm^2}$. For PMMA/DR1, the typical range of annealing temperatures used was 115 to 130°C. For DCM/PFCB, the typical temperature

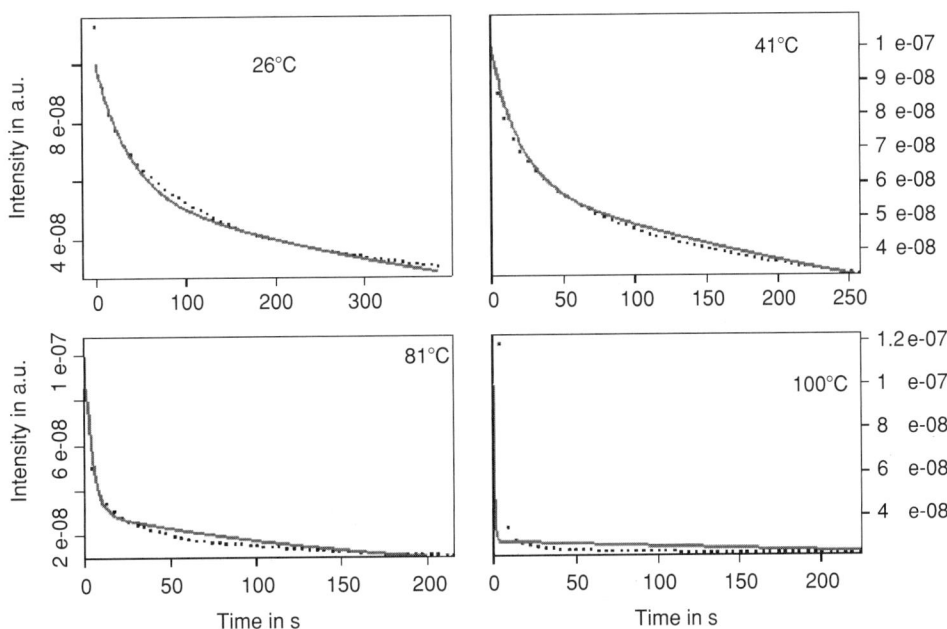

FIGURE 22.10 A comparison of a nonlinear regression analysis of mechanical relaxation with the results for the experimental curves of the decay of diffraction efficiency of a polymeric diffraction grating that was exposed for a 15-s period is displayed. The dots represent the experimental observed data and the solid lines represent the fit using the model. The two-parameter fit to the experimental curves is acceptable. A minimum of two parameters seems to be necessary due to the structure of the curves, which are clearly not purely exponential or stretch exponential.

TABLE 22.1 Experimental grating characteristics.

Material	Thickness	Grating	Diffraction efficiency
DCM/PFCB	3.2 μm	1	0.1%
		2	6.97%
		3	2.02%
		4	0.314%
PMMA/DR1	4.7 μm	5	24.7%
		6	24.04%
		7	2.76%

range used was 178 to 210°C. The diffraction efficiencies of a few gratings are listed in Table 22.1. There is a large variation in efficiencies due to different bleaching times and temperatures as described in a previous section on surface index relief. While the diffraction efficiency of the grating recorded in DCM/PFCB seems low, the value of 0.1% was a large improvement over the unannealed gratings that exhibited diffraction efficiencies that could not be measured. Diffraction from a PMMA/DR1 grating is illustrated in Fig. 22.11.

5. CONCLUSION

In this chapter, we described how we have been able to holographically produce submicron structures in dye-doped polymer materials. The structure that we produced here was a simple diffraction grating, but it is reasonable to believe that we can produce more complex structures, for example, structures that are superpositions of simple gratings. The dye-doped polymer that we used here belongs to a class of polymeric materials that includes the materials presently used for commercial polymeric multigigabit modulators. Polymers doped with conjugated organic dye molecules generally exhibit large third-order nonlinear coefficients and large second-order nonlinear coefficients when they are poled, that is, when the material is prepared in a state where the dye molecules are preferentially aligned along a given axis. The dye molecules used here, and in other nonlinear dye-doped polymer materials, exhibit a *cis-trans* isomerization reaction that can be photoinitiated, generally by visible light. The azobenzene and stilbene dyes that are commonly used to create second-order and/or to enhance third-order nonlinearities are also dyes that exhibit *cis-trans* isomerizations. The dye used here was an azobenzene. Bleaching of a dye molecule after it has been doped into a high temperature polymer, one whose glass transition temperature is higher (possibly significantly) than room temperature, can irreversibly bleach the molecule. Conjugated dyes that exhibit *cis-trans* isomerization reactions generally increase the index of the polymeric medium into which they are introduced. The bleaching of the dye then has the effect of lowering the effective index of the medium. The resulting lowering of the index can be used to write waveguides or even more complex structures in dye-doped polymer thin films. When high temperature polymers are used, then these materials can be used for the production of the types of waveguide devices used in nonlinear optics. The process by which we have altered the index by illumination is quite generally referred to

FIGURE 22.11 A picture of the diffraction of a 633-nm HeNe laser beam incident normally on the surface of a coated slide where a grating has been written holographically in the PMMA/DR1 coating with an Ar+ ion laser lasing on its 514-nm line. Multiple diffractive orders are visible on the viewing screen. The bleached region of the coating on the slide slide is discernible from the unbleached region by the difference in color of the regions. The bleached region appears much less red (significantly less green absorption) than the unbleached region. A corona poling needle is visible to the left of the fixture, which holds the slide. This sample was not poled.

as the photobleaching method. It is in such a high temperature dye-doped material that we have photobleached a diffraction grating in this work.

The material in which we have demonstrated that a diffraction grating can be written is the side-chained material denoted as PMMA/DR1. This material is actually a copolymer comprising 10 methylmethacrylate (MMA) monomers alternated with a single monomer, which is a disperse red 1 (DR1) dye molecule side chain attached to an MMA monomer through a methyl attachment. In this work, we photobleached a grating that we used for free space diffraction of incident light. We were able to show that the structure that we wrote was accurate to submicron detail through observation of the resulting diffraction pattern. However, such detail is only preserved when the bleaching process is carried out with the substrate material stabilized at an elevated temperature from the ambient. We found that we could produce structures with the requisite submicron detail when the temperature was held at roughly 10°C below the glass transition temperature of our material. For our material this glass transition temperature was roughly 121°C. We carried out a study to determine the reason for this heating requirement.

As has been observed in past experiments, the irreversible photobleaching process leads to a reduction of the volume of the material that has been bleached. This change in volume in a bleached region causes a strain between the bleached portion of the material and any unbleached portion as well as strain at interfaces between the bleachable material and materials to which the waveguiding material is attached. Polymeric materials can be strain relieved, either by annealing at temperatures approaching the glass transition temperature of the composite material or by performing the bleaching process at temperatures approaching this glass transition temperature. We have noted that if the strain buildup is large enough, postannealing is not efficacious; in such cases, the material must be bleached at the elevated temperature for the irreversible photobleached material to exhibit the desired submicron structures. To better elucidate the nature of the strain buildup process, we carried out a series of transient photobleaching experiments in this work. From the transient experiments it is clear that strain and surface relief gratings are generated in the material when gratings are written into the material at room temperature. As the temperature of the film is increased it is clear that the lifetime of the strain and surface relief gratings exponentially decreases until the temperature of the film approaches the glass transition temperature, at which point strain and surface relief gratings are no longer supported in the film. The experimental results obtained from the transient experiments clearly show that diffraction from gratings irreversibly bleached into dye-doped polymer films near the glass transition temperature are not due to strain or surface relief gratings in the film.

Although the structure demonstrated in this work is a simple diffraction grating, the diffraction of light by this structure is a rather stringent test of the dimensional integrity and strain relief. That this structure diffracts light in the desired fashion indicates that most any submicron structure can be reliably written in the polymeric material systems that we consider. Many useful structures comprise coherent superpositions of sinusoidal structures [52, 54]. Continuous superpositions of sinusoidal structures are holograms that can store visual or coding information [55, 56]. Double grating structures overlaid on waveguide structures can be used to slow the velocity of the guided light [57]. More complex superpositions of gratings can be used to produce photonic bandgap structures that can be used for exotic applications such as generating vortex solitons [58] or aiding in squeezed light generation [59] or for more practical applications such as reducing spontaneous emission in waveguide structures [60] that are co-doped with lasing dopants as well as the waveguiding/index control dopants as the dyes used here.

6. REFERENCES

1. B. S. Kawasaki, K. O. Hill, D. C. Johnson and Y. Fujii, Narrow-band Bragg reflectors in optical fibers, Opt. Lett. **3**, 66 (1978).
2. U. Österberg and W. Margulis, Dye laser pumped by Nd:YAG laser pulses frequency doubled in a glass optical fiber, Opt. Lett. **11**, 516 (1986).
3. A. Yariv and P. Yeh, *Optical Waves in Crystals*, Wiley Interscience (2003).
4. G. Lindsay and K. Singer (Eds.), *Polymers for Nonlinear Optics*, American Chemical Society, Washington, DC (1995).
5. J. A. Delaire and K. Nakatani, Linear and nonlinear optical properties of photochromic molecules and materials, Chem. Rev. **100**, 1817–1845 (2000).
6. K. D. Singer, M. G. Kuzyk and J. E. Sohn, Second-order nonlinear-optical processes in orientationally ordered materials: relationship between molecular and macroscopic properties. J. Opt. Soc. **4**, 968 (1987).
7. D. Bower, *An Introduction to Polymer Physics*, Cambridge University Press (2002).

8. I. Ward and D. Hadley, *An Introduction to Mechanical Properties of Solid Polymers*. John Wiley and Sons (1993).
9. J. Ferry, *Viscoeleastic Properties of Polymers*, John Wiley and Sons (1980).
10. K. D. Singer, J. E. Sohn and S. J. Lalama, Second harmonic generation in poled polymer films, Appl. Phys. Lett. **49**, 248 (1986).
11. J. M. Nunzi, C. Fiorini, F. Charra, F. Kajar and P. Raimond, All-optical poling of polymers for phase-matched frequency doubling, Am. Chem Soc. Symp. **601** (1995).
12. W. Feng, S. Lin, R. B. Hooker and A. R. Mickelson, Study of UV-bleached channel-waveguide performance in nonlinear optical polymer films, Appl. Opt. **34**, 6885 (1995).
13. M. Strälein, C. A. Walsh, D. M. Burland, R. D. Miller, R. J. Twieg and W. Volksen, Orientational decay in second-order nonlinear optical guest-host polymers: Temperature dependence and effects of polining geometry, J. Appl. Phys. **73**, 8471–8478 (1993).
14. C. A. Walsh, D. M. Burland, V. Y. Lee, R. D. Miller, B. A. Smith, R. J. Twieg and W. Volksen, Orientational relaxation in electric field poled guest–host and side–chain polymers below T_g, Macromolecules **26**, 3720–3722 (1993).
15. D. M. Burland, R. D. Miller and C. A. Walsh, Second-order nonlinearity in poled-polymer systems, Chem. Rev. **94**, 31 (1994).
16. A. R. Mickelson, N. R. Basavanhally and Y. C. Lee, *Optoelectronic Packaging*, Wiley (1997).
17. T. VanEck, A. Ticknor, R. Lytel and G. F. Lipscomb, Complementary optical tap fabricated in an electro-optic polymer waveguide, Appl. Phys. Lett. **58**, 1588–1590 (1991).
18. S. Ermer, J. F. Valley, R. Lytel, G. F. Lipscomb, T. E. VanEck and D. G. Girton, DCM poyimide system for triple-stack poled polymer electro-optic devices, Appl. Phys. Lett. **61**, 2272–2274 (1992).
19. S. Ermer, D. S. Leung, S. M. Lovejoy, J. F. Valley and M. Stiller, Photobleachable donor-acceptor-donor chromophores with enhanced thermal stability, OSA Tech Digest Series, **17**, 50 (1993).
20. D. Tomić and A. R. Mickelson, Photobleaching for optical waveguide formation in a guest-host polyimide, Appl. Opt. **38**, 3893 (1999).
21. Cyclotene product home page, http://www.dow.com/cyclotene/biblio/index.asp
22. G. Fischbeck, R. Moosburger, C. Kostrzewa, A. Achen and K. Petermann, Singlemode optical waveguides using a high temperature stable poylmer with low losses in the 1.55μm range, Electron. Lett. **33**, 518 (1997).
23. J. Saltiel, Perdeuterostilbene. The role of phantom states in the cis-trans photoisomerization of stilbenes, J. Am. Chem. Soc. **89**, 1036 (1967).
24. H. Rau, Photoisomerization of azobenzenes, in *Photochemistry and Photophysics*, ed. F. J. Rabeck, CRC Press, Boca Raton, FL (1990).
25. A. Natansohn and P. Rochon, Photoinduced motions in azo-containing polymers, Chem. Rev. **102**, 4139–4175 (2002).
26. K. B. Rochfort, R. Zanoni, Q. Gong and G. I. Stegeman, Fabrication of integrated optical structures in polydiacetylene films by irreversible photoinduced bleaching, Appl. Phys. Lett. **55**, 1161–1163 (1989).
27. T. Kanetake, Y. Tokura and T. Koda, Photo- and thermo-chromism in vacuum deposited polydiacetylene films, Solid State Comm. **56**, 803–807 (1985).
28. G. Patel, T. Witt and Y. Khanna, Thermochromism in polydiacetylene solution, J. Polymer Sci. **18**, 1383–1391 (1980).
29. R. Chance, G. Patel and J. Witt, Thermal effects on the optical properties of single crystals and solution-cast films of urethane-substituted polydiacetylenes, J. Chem. Phys. **71**, 206–211 (1979).
30. H. W. Weber, R. Ulrich, E. A. Chandross and W. J. Tomlinson, Light-guiding structures of photoresist films, Appl. Phys. Lett. **20**, 143–145 (1972).
31. W. J. Tomlinson, Phase holograms in photochromic materials, Appl. Opt. **11**, 823–831 (1972).
32. I. P. Kaminow, L. W. Stulz, E. A. Chandross and C. A. Pryde, Photobleaching of organic laser dyes in solid matrices, Appl. Opt. **11**, 1563–1567 (1972).
33. F. Higuchi and J. Muto, On the photobleaching quantum yields of heat-treated rhodamine 6G (Rh6G) molecules in the copolymer of methylmethacrylate (MMA) with methacrylic acid (MA), Phys. Lett. **99A**, 121–124 (1983).
34. A. F. Garito, K. D. Singer, K. Hayes, G. F. Lipscomb, S. F. Lalama and K. N. Degai, First single-crystal polymers exhibiting phase-matched second harmonic generation, J. Opt. Soc. Am. **70**, 1399–1400 (1980).
35. A. F. Garito, K. D. Singer and C. C. Teng, Molecular optics: nonlinear optical properties of organic and polymeric crystals, ACS Symposium Series **233**, 1–25 (1983).
36. F. R. Meredith, J. G. VanDusen and D. J. Williams, Optical and nonlinear optical characterization of molecularly doped therotropic liquid crystalline polymers, Macromolecules **15**, 1385–1389 (1982).
37. F. R. Meredith, J. G. VanDusen and D. J. Williams, Characterization of liquid crystalline polymers for electro-optic applications, ACS Symposium Series **233**, 109–133 (1983).

38. M. B. J. Diemeer, F. M. M. Suyten, E. S. Trammel, A. McDonagh, J. M. Copeland, V. W. Jenneskens and W. H. G. Horsthuis, Photo-induced channel waveguide formation in nonlinear optical polymers, Electron. Lett. **26**, 379–380 (1990).
39. R. S. Moshrefzadeh, D. K. Misemer, M. D. Radcliffe, C. V. Francis and S. K. Mohaputra, Nonuniform photobleaching of dyed polymers for optical waveguiding, Appl. Phys. Lett. **62**, 16–18 (1993).
40. T. Zyung, W.-Y. Hang and J.-J. Kim, Accelerated photobleaching of nonlinear optical polymer for the formation of optical waveguides, Appl. Phys. Lett. **64**, 3527–2529 (1994).
41. T. Zyung, J.-J. Kim, W.-Y. Hwang and S.-D. Jung, Effects of photoblaching wavelength on the resulting refractive index profiles in nonlinear optical polymeric thin film, Mol. Cryst. Liq. Cryst. **247**, 49–58 (1994).
42. Y. Shi, W. Steier, L. Lu, M. Chen and L. Dalton, Large stable photoinduced refractive index change in nonlinear optical polyester with disperse red side groupings, Appl. Phys. Lett. **58**, 1131–1133 (1991).
43. K. Chakravorty, Photogeneration of refractive index patterns in doped polyimide films, Appl. Opt. **32**, 2331–2338 (1993).
44. J. Ma, W. Feng, S. Lin, R. J. Feuerstein, R. B. Hooker and A. R. Mickelson, Modeling photobleached optical polymer waveguides, Appl. Opt. **34**, 5352 (1995).
45. S. Lovejoy, S. Ermer, D. Leung, C. Disk, P. Kalameg- ham, L. Zhang and C. Moylan, Design and synthesis of soluble thermally stable nonlinear optical chromophores based on the dicyanomethylenepyran moiety, Organic Thin Films for Photonics Applications, OSA Tech. Dig. Series **212**, 259 (1995).
46. O. Watanabe, M. Tsuchimori and A. Okada, Two-step refractive index change by photoisomerization and photobleaching processes in the films of nonlinear optical polyurethans and a urethane–urea copolymer, J. Mater. Chem. **6**, 1487–1492 (1996).
47. M. Nakanishi, O. Sugihara, O. Okamoto and K. Hiroto, Ultraviolet photobleaching process of azo dye-doped polymer and silica films for fabrication of nonlinear optical waveguides, Appl. Opt. **37**, 1068–1073 (1998).
48. D. C. Hutchings, M. Sheik-bahae, D. J. Hagan and E. W. van Stryland, Kramers-Krönig relation in nonlinear optics, Opt. Quant. Electro. **24**, 1–30 (1992).
49. A. Siegman, *Lasers*, University Science, Mill Valley, CA (1986).
50. E. McKenna, J. Xue, A. Verdoni, M. Yetzbacher, R. Fan and A. R. Mickelson, Kinetic model of irreversible photobleaching of dye-doped polymer waveguide materials, J. Opt. Soc. Am. B **21**, 1294–1301 (2004).
51. J. Xu and R. Stroud, Acousto-Optic Devices: Principles, Design, and Applications, John Wiley and Sons, New York (1992).
52. J. Goodman, *An Introduction to Fourier Optics*, McGraw-Hill (1996).
53. R. P. Feynman, R. Leighton and M. Sands, *The Feynman Lectures on Physics*, Addison-Wesley, Reading, MA (1963).
54. A. Yariv, *Quantum Electronics*, Wiley (1989).
55. P. Hariharan, *Optical Holography: Principles, Techniques and Applications*, Cambridge University Press (1996).
56. F.T.S. Yu, S. Jutamulia and S. Yin, *Introduction to Information Optics*, Academic Press (2001).
57. J. B. Khurgin, Light slowing down in Moiré fiber gratings and its implications for nonlinear optics, Phys. Rev. A **62**, 1–4 (2000).
58. J. W. Fleischer, G. Bartal, O. Cohen, O. Manela, M. Segev, J. Hudock and D. N. Christodoulides, Observation of vortex-ring "discrete" solitons in 2D photonic lattices, Phys. Rev. Lett. **92**, 12, 1–4 (2004).
59. D. Tricca, C. Sibilia, S. Severini, M. Bertolotti, M. Scalora, C. M. Bowden and K. Sakoda, Photonic bandgap structures in planar nonlinear waveguides: application to squeezed-light generation, J. Opt Soc. Am. B **21**, 3, 671–680 (2004).
60. I. Abram and G. Bourdon, Photonic-well microcavities for spontaneous emission control, Phys. Rev. A, **54**, 4, 3476–3479 (1996).

23
CHAPTER

Optical MEMS Using Commercial Foundries

Deepak Uttamchandani*
Department of Electronic & Electrical Engineering
University of Strathclyde
Glasgow, Scotland

1. INTRODUCTION

Microelectromechanical systems (MEMS) describes the technology behind the realization of microsensors, microactuators, and complete microsystems with assemblies of moving parts whose dimensions range from around $1\,\mu m$ to $1000\,\mu m$. The origin of this technology lies in the world of silicon microelectronics [1], and many of the design and microfabrication tools and processes used in MEMS technology are still very similar to those used in silicon microfabrication. The term "micromachining" is used to describe the steps involved in the fabrication of a MEMS device. MEMS technology is very versatile, and today MEMS sensor devices have been successfully engineered into miniature accelerometers and gyroscopes that are deployed in automobile safety and stability systems. From around the mid-1990s, it became apparent that the field of photonics could also benefit from MEMS technology, and in particular the surge of interest and investment in optical fiber communications added an impetus to the development of optical MEMS (O-MEMS[†]). Single-mode fiber optic–based devices such as variable optical attenuators [2], tunable filters [3], optical switches [4], optical cross-connects [5], and add/drop multiplexers [6], so familiar to the photonic engineering community, were realized using MEMS technology. This growth in O-MEMS has given new vigor to the field of free space micro-optics, and various review articles [7, 8] and texts [9, 10] are now available on the subject.

On the face of it, it would appear that becoming involved in research and development activity in the field of O-MEMS would require individual research groups or centers to be equipped with silicon microfabrication facilities, including photo lithography or e-beam lithography; semiconductor doping equipment; deposition equipment for metal, dielectric, and semiconductor layers; etching equipment for metal, dielectric, and semiconductor layers; and wafer sawing and dicing equipment. Such facilities, however, are not available to every research group, and the question arises as to what options exist for researchers who do not have access to such semiconductor processing equipment, that is, researchers who are a part of a "fabless" research environment. One very successful solution is to take advantage of the services being offered by a growing number of commercial foundries specializing in MEMS fabrication.

Commercial foundry–based fabrication of MEMS enables research groups (and small companies) without their own in-house microfabrication facilities to participate in exploiting the advantage of MEMS technology. This can be done without having to incur the very significant costs of installing and maintaining their own facilities to the high standards required to obtain reliable and repeatable fabrication processes. The use of commercial foundries for MEMS fabrication follows the well-established approach in the domain of microelectronics manufacture where microelectronic systems are designed in one part of the world and fabricated wherever an appropriate foundry is available that offers the required processes at competitive costs. This trend is also expected to follow in the field of MEMS, and foundries will be selected by MEMS designers on the basis of the fabrication processes they offer. Academic and commercial MEMS design centers will feed designs to the foundries as appropriate, and the fabricated microsystems will be returned to the design center for experimental evaluation/testing.

When using foundry services, multiuser processes are the most cost-effective way for proof-of-concept and rapid prototyping of MEMS devices. In a multiuser process, the MEMS designer has to adhere strictly to a sequence of processing steps that have

---

*E-mail: du@eee.strath.ac.uk

[†]It should be noted that sometimes the term MOEMS (microoptoelectromechanical systems) is used instead of O-MEMS: Both terms refer to the same technology.

been predetermined by the foundry. These steps, called design rules, will be available directly from the foundry. In this way all designs follow the same rules, and the foundry can sell a fraction of the substrate area to be processed to different customers. Thus, the individual customer does not have to purchase the full area that is processed on the substrate, making it a more economical option. In using multiuser foundry services, MEMS designers have to be both innovative and ingenious with their designs because the design rules issued by the foundry are fixed for all users alike.

This chapter describes three O-MEMS devices that have been fabricated by O-MEMS designers operating in a fabless environment, that is, using a commercial multiuser MEMS fabrication service as described above. The first of these devices is a fiber optic chopper fabricated using the silicon-on-insulator multi user MEMS processes (SOIMUMPs, MEMSCAP, Inc., Durham, North Carolina, USA) and having applications in general optical and fiber optic measurement technology. The second device is a variable optical attenuator using the polysilicon multi user MEMS processes (PolyMUMPs, MEMSCAP, Inc.) where special attention has been paid to achieving self-assembly of a three-dimensional O-MEMS device. The third device is a hybrid tunable optical filter comprising an off-the-shelf dense wavelength division multiplying (DWDM) filter located on a MEMS platform fabricated by SOIMUMPs. Comprehensive technical details and design rules for the SOIMUMPs and PolyMUMPs fabrication processes can be found in [11].

2. OPTICAL CHOPPERS FOR FIBER OPTIC APPLICATIONS

Optical choppers are mainly used in precise optical intensity measurements for both free space and guided-wave optics, particularly where the influence of stray light is perceived to be a problem. When using choppers, light from the optical source is modulated in intensity so that the received light signal is detected as an alternating current at a photoreceiver. The received alternating current (AC) photocurrent is fed into a synchronous detector, such as a lock-in-amplifier, so that its amplitude can be measured without the error induced by direct current (DC) drift of the detector electronics or stray light from the environment.

The most widely available optical choppers are in the form of a motor-driven wheel with regularly spaced apertures through which the light to be "chopped" can pass. Miniaturized optical choppers fabricated by various micromachining technologies and in different material systems have been reported [12–15]. In [12], a micromachined chopper based on a moving diffractive element was reported for infrared measurements. The device was fabricated on LPCVD silicon nitride deposited on a sacrificial LPCVD oxide deposited on a silicon substrate. In [13], an electrostatic actuator was used to drive a single polysilicon shutter integrated with a silicon photodetector. Polysilicon surface micromachining was used in the fabrication process. In [14], the micromachined chopper reported was fabricated on quartz and the device was piezoelectrically driven. In [15], the chopper was also fabricated using polysilicon micromachining but with the inclusion of a back-side etched hole through which the light to be chopped is transmitted.

The chopper reported here was fabricated on silicon-on-insulator (SOI), which is a material that is mechanically superior to polysilicon. The structure is shown schematically in Fig. 23.1. The design is symmetric about its center. Each half of the chopper consists of a blade attached to a suspension and electrostatically driven by a comb microactuator. The two comb drives are electrically connected in parallel. The two blades form a slit that can periodically be closed, resulting in the chopping action if an AC voltage is applied. Alternatively, application of a DC voltage simply results in an optical slit whose width can be controlled by the magnitude of the voltage. The micromachining process includes backside etching of an aperture through the wafer, thereby making it convenient to align and accommodate optical fibers with respect to the chopper blades.

2.1. Fabrication of MEMS Chopper

The SOIMUMPs has been used to produce the device in bonded SOI where the silicon layer is 10 μm

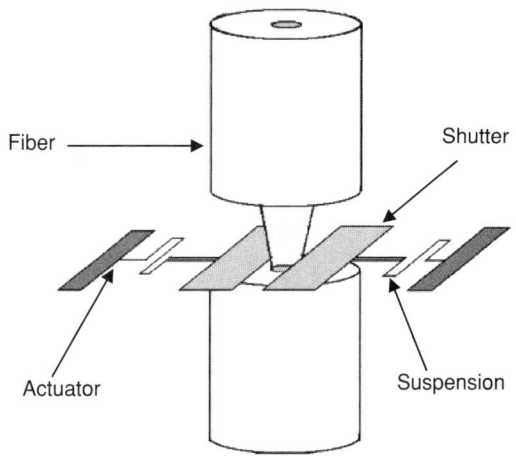

FIGURE 23.1 Schematic diagram of the fiber optical chopper.

thick. A useful feature of the process is the back-etching of through-apertures that enables the positioning of two optical fibers close together to reduce overall loss. Figure 23.2 shows a layout of the device design that was submitted to the foundry, whereas Fig. 23.3 shows a photomicrograph of the microfabricated device after it was returned from the foundry. The size of each shutter blade is 50×100 μm.

2.2. Mechanical Design Considerations

In this section, finite element modeling (FEM) results for static and dynamic characteristics of the chopper are presented.

2.2.1. FEM: Static Analysis

Using ANSYS software (ANSYS Inc., Canonsburg, Pennsylvania, USA) the spring constant K_x for the chopper suspension was obtained as follows. A finite element model of the chopper was constructed and static loading applied to this model. For each value of load applied, the deflection in the x direction was obtained using the software. A plot of the load versus deflection data for the structure yields the spring constant, which was calculated to be 0.51 μN/μm. The relationship between applied voltage and displacement of the movable fingers is likewise obtained through FEM using ANSYS software. During modeling, it is assumed that the travel of the finger is within the constant electrostatic force regime. Figure 23.4 shows the results from the analysis.

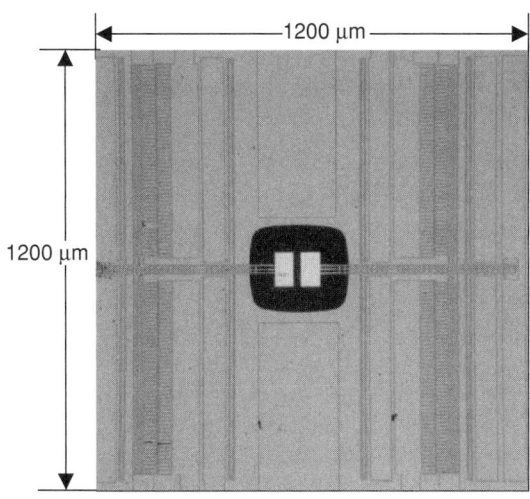

FIGURE 23.3 Photomicrograph of the micro-optical chopper.

2.2.2. FEM: Dynamic Analysis

Modal analysis helps to determine the vibration characteristics (natural frequencies and mode shapes) of a mechanical structure or component, showing the movement of different parts of the structure under dynamic loading conditions, such as those due to the lateral force generated by the electrostatic actuators. The natural frequencies and mode shapes are important parameters in the design of a structure for dynamic loading conditions. Modal analysis of the chopper was performed using ANSYS software. Figure 23.5 shows the mode shape of the chopper at its fundamental frequency, and Table 23.1 shows the natural frequencies of the chopper for its different vibration modes.

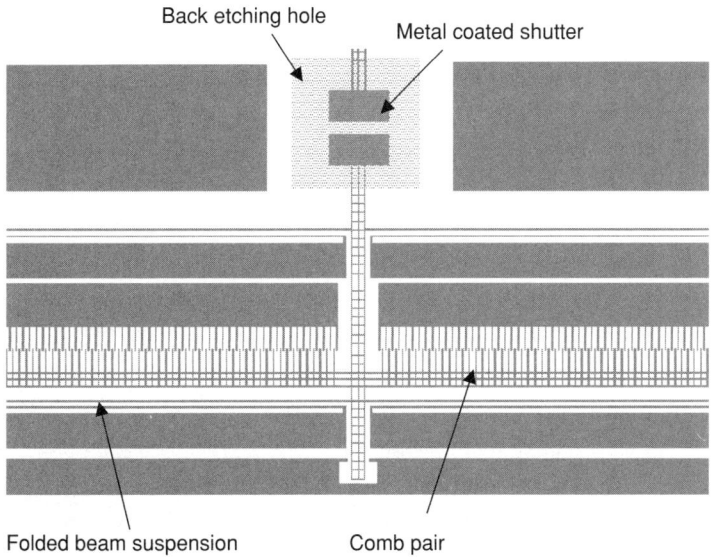

FIGURE 23.2 Layout of comb drive actuator with shutters.

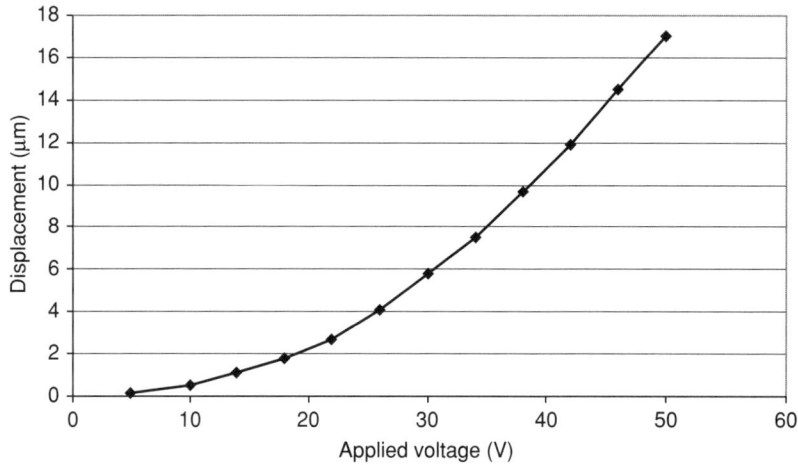

FIGURE 23.4 FEM simulation of relationship between displacement (μm) and applied voltage (V) for comb drive.

2.3. Optical Design Considerations

The optical system consists of the following arrangement. Two identical single-mode fibers are aligned face to face with a separation of around 70 μm. This separation was chosen to minimize the risk of physical damage to the device by a possible collision with the optical fiber ends during experimentation. The input fiber carries an infrared light of 1550-nm wavelength, which is emitted from the fiber end-face toward the output fiber, which collects the light. The chopper is located in the gap between the two fibers. When the shutters of the chopper begin to close, the light received at the output fiber is attenuated. The attenuation is modeled using the notation and parameters shown in Fig. 23.6.

Optical analysis was undertaken in the following steps and with the following assumptions:

1. The light emitted from the input fiber facet is a Gaussian beam with a mode field radius w_0 of 5.1 μm (data obtained from the fiber supplier). The light wavelength λ is 1550 nm.
2. From the input fiber to the aperture, the light retains a Gaussian distribution but the beam waist is broadened: The waist radius at the aperture is w_1 where

TABLE 23.1 Frequencies of different resonant modes.

Mode	Frequency (Hz)
1	3350
2	10,765
3	10,810
4	11,395
5	12,222

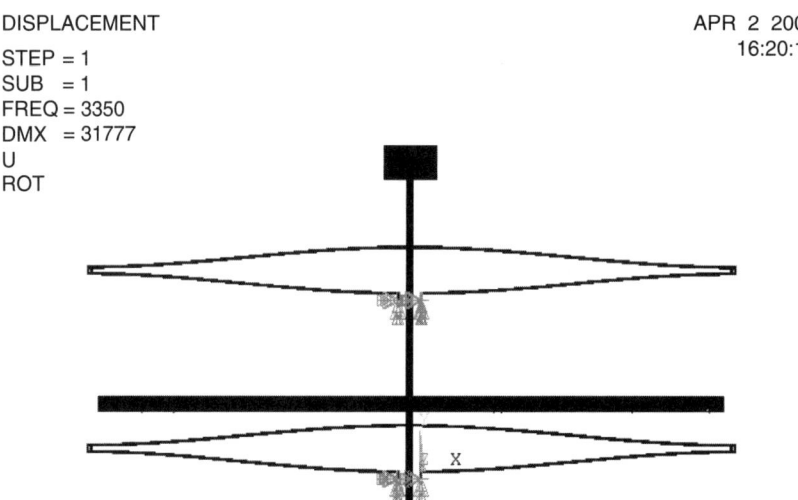

FIGURE 23.5 The fundamental mode shape of the chopper structure.

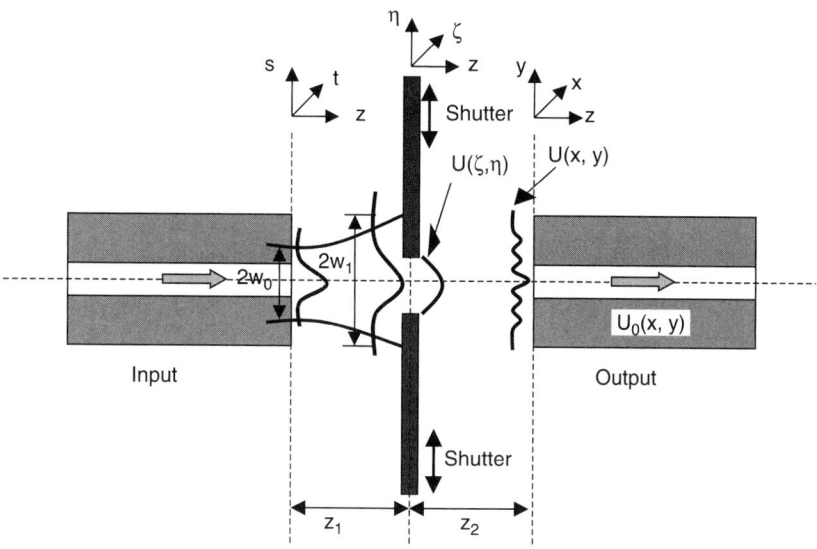

FIGURE 23.6 Shutter system and parameters for optical modeling.

$$w_1 = w_0 \left(1 + \frac{\lambda z_1}{\pi w_0}\right)^{1/2} \quad (1)$$

and z_1 is the distance from input fiber facet to the center position of the aperture and is 35 μm.

3. The light is partially blocked by the blades forming the aperture. The unblocked light diffracts to the output fiber facet. Here the Rayleigh-Sommerfield formula is used [16], which has the form

$$U(x, y, z) = \frac{z_2}{j\lambda} \int_\Sigma \int U(\xi, \eta, 0) \frac{\exp(jkr)}{r^2} \partial\xi\partial\eta \quad (2)$$

where k is $2\pi/\lambda$, r is the distance from the point (ξ, η) in the aperture plane to the point (x, y) in output fiber plane, Σ represents the area of the aperture, and z_2 is the distance the light propagates from the center position of the aperture plane to the output fiber plane and is 35 μm in our device. U is the complex optical wave function. Figure 23.7 is one of the results calculated using the Rayleigh-Sommerfield diffraction theory. The light field in the aperture plane is cut by the symmetric shutter blades that act as a narrow slit and diffracted to the output fiber facet plane.

4. After completing the calculations of step 3, the intensity distribution at the output fiber facet plane is obtained. The proportion of light coupled from the output facet plane into the output fiber is represented as T and is determined by use of the mode-coupling method [17]. The mode coupling can be expressed as

$$T = \left|\int\int_0 U(x, y) \cdot U_0^*(x, y) dx dy\right|^2 \quad (3)$$

where $U_0(x, y)$ is the fundamental mode distribution of the output fiber.

The calculated attenuation resulting from changing the position of the chopper blades is shown in Fig. 23.8. Also in Fig. 23.8, the attenuation of the double chopper blades is compared with that obtained if just a single blade had been used to completely obscure the light field. It is clear that in a double-blade arrangement, a much smaller deflection from the comb actuator is required for each blade than for the single-blade arrangement. This means that a lower operating voltage would be required for the double-blade arrangement because of the smaller deflection.

2.4. Experimental Evaluation

Experimental evaluation consisted of mechanical and optical testing of the device.

2.4.1. Mechanical Testing

Mechanical testing consisted of static and dynamic tests. For static tests, the displacement as a function of the driving voltage was measured while applying a DC voltage between the fixed comb finger array and the moving comb finger array. DC voltage from 0 to 48 V was applied to the device and the blade displacement measured by observation under a high-magnification microscope coupled to a high-resolution camera. Figure 23.9 shows the measured

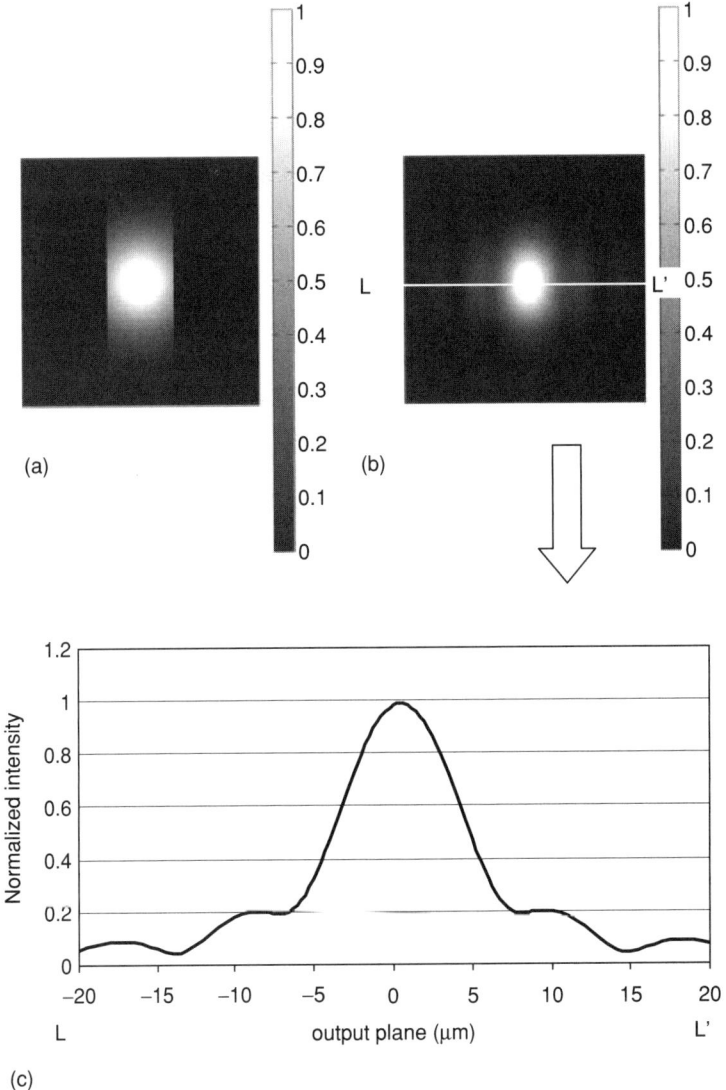

FIGURE 23.7 (a) The light intensity in shutter plane. (b) The light intensity in receiving fiber plane. In (a), the frame dimension is from −15 to 15 μm. In (b) the frame dimension is from −20 to 20 μm. The gray-scale bar represents the normalized intensity value. The two shutters are positioned ±4 μm from the center. (c) The cross-section view of (b).

displacement of the comb drive actuator. The error bars in the experimental data show an uncertainty of ±0.4 μm in the measuring equipment. Figure 23.9 also shows a comparision of the measured data with results of the finite element analysis. The difference between the finite element model and experimental results are partly due to the variation of the fabricated dimensions of the structures. According to the foundry design rules, the thickness of the silicon structure has an uncertainty of ±1 μm. Moreover, the width of the structures can be varied in different chips and processing runs. Bearing in mind the existence of these uncertainties, the match between the model and experimental results in Fig. 23.9 is very good.

The dynamic response of the device was obtained using a high-speed imaging camera (Roper model 4540MX) coupled to an optical microscope. The fundamental mode of the device is in the plane of the chip, and consequently, microscope-coupled laser Doppler systems that are optimized to measure out-of-plane deflections cannot be used. The Roper high-speed camera can capture up to 40,500 frames per second. In this sequence of measurements, it was set to operate at 9000 frames per second, which is above the expected resonant frequency of the

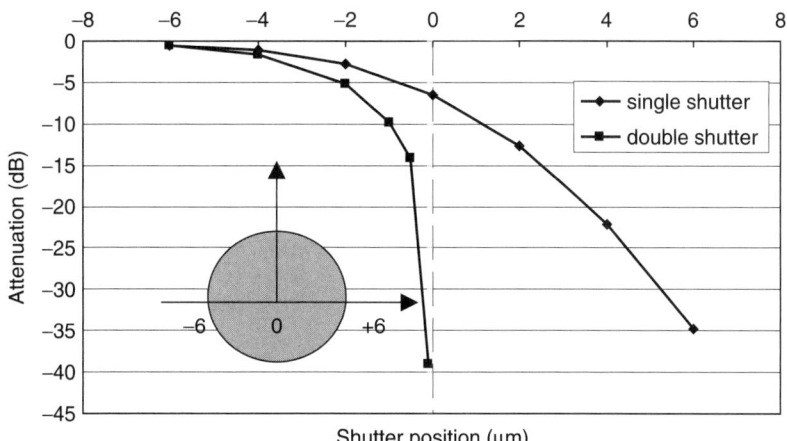

FIGURE 23.8 Calculated attenuation caused by single shutter and double shutter, with inset diagram of mode field at shutter plane.

device. The sequence of frames captured was played back and carefully analyzed to determine the amplitude of deflection at each applied frequency. The measurements obtained are plotted in Fig. 23.10, from where it is evident that the fundamental resonant frequency is around 3 kHz. The FEM result (3.3 kHz) is close to the experimental result (3 kHz), shown of good agreement for a complex structure like the chopper. The Q factor in air is around 0.47. The error bars show that the imaging system has an uncertainty of ±0.4 μm. Although no lifetime tests were conducted, the device was left running for over 10 h at 100 Hz without failure.

2.4.2. Optical Testing

The dies returned from the foundry were of 1-cm sides and, because of the etched hole through the die, were not mounted on conventional circuit boards. Instead, an optical test station was set up that allowed the input and output single-mode fiber to be aligned facing each other and with the chopper placed in the gap between the fiber faces.

Single-mode optical fibers transmitting light of 1550-nm wavelength and with 5.1-μm mode field radius were used to test the performance of the chopper. The aperture formed by the blades of the chopper was initially 20 μm, that is, when the voltage applied was 0 V. When the two shutter blades driven by the comb actuators each travel 10 μm, the aperture was closed entirely. The driving voltage was monitored by a digital oscilloscope. The light coupled to the output fiber was detected by a photodetector and monitored by a different channel of the same oscilloscope.

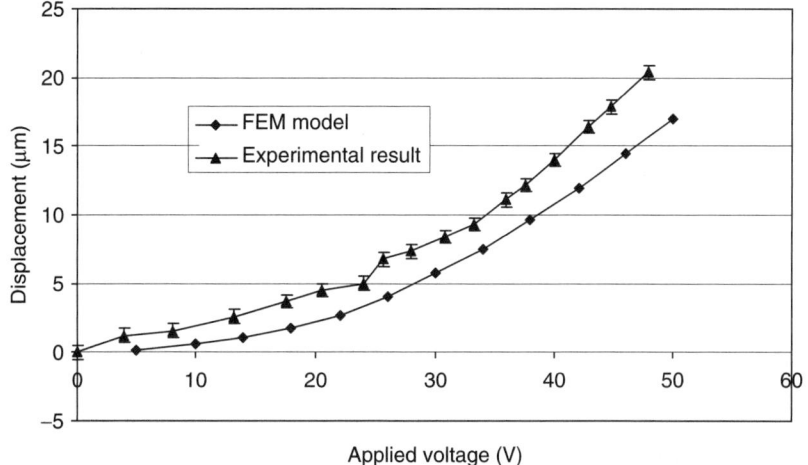

FIGURE 23.9 Voltage vs. displacement (experiment and theory) for the comb drive chopper.

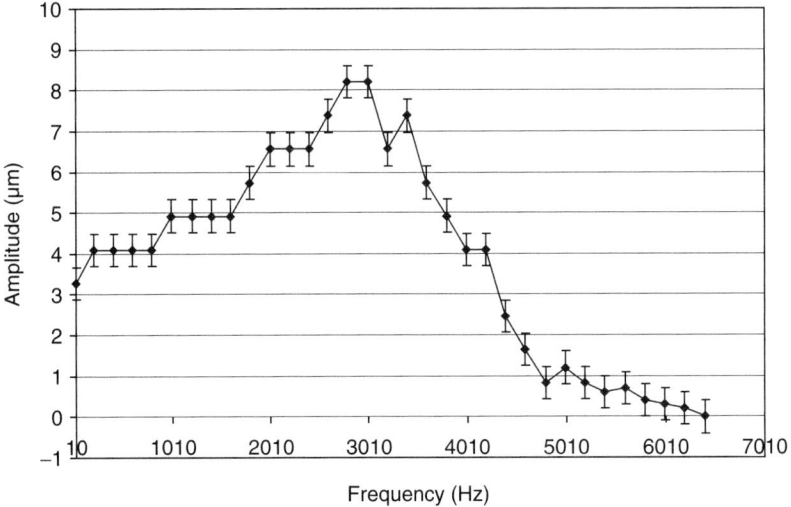

FIGURE 23.10 Mechanical dynamic behavior of the chopper.

The static optical performance was measured by applying a DC voltage to the device and measuring the optical throughput at each voltage. In this mode the chopper functions as a miniaturized optical slit whose width is changed by application of DC voltage. Figure 23.11 shows the results obtained. Silicon is transparent at the wavelength used in this experiment so the blades were gold coated to render them opaque. Thus, when the slit is fully closed, it would be expected that no light would be transmitted. However, slight imperfections in the etching and squareness of the blades led to the closed blades, though contacting, having the possibility of allowing some light through. Moreover, the deposition of gold might not be perfectly to the edge of the blade. Both these effects lead to a measured optical on/off ratio of around 27 dB. The intensity coupled from the input to the output fiber in the on position was measured to be 72%, corresponding to an insertion loss of 1.4 dB. This is caused by the separation of the two fiber faces and any possible misalignments. Between 2 and 20 V, the attenuation changes slowly because the shutters are in the periphery of the light spot where the Gaussian beam distribution of the light spot has low intensity. Previous testing of the comb actuator had established that an applied voltage of 20 V produced a 5-μm displacement of each blade with a further 5 μm to be traversed by each blade. Increasing the electrostatic voltage from 20 to 34 V resulted in a rapid increase in attenuation as expected from this arrangement. In the off position, the power from the output fiber end was measured to be 0.12%, corresponding to a 29-dB attenuation. When the driving voltage was increased to around 40 V, the attenuation remained unchanged, confirming that the chopper blades were fully closed.

The dynamic response of the fiber optic chopper as shown in Fig. 23.12 was obtained from optical measurements. A 0 to 28-V square signal of 1000 Hz frequency (upper trace) was applied to the comb actuator, and the received power was detected by a photoreceiver and displayed on a digital oscilloscope (lower trace). The 0–90% rise time was measured to be 90 μs, and the 100% to 10% fall time was also measured to be 90 μs.

3. THREE-DIMENSIONAL VARIABLE OPTICAL ATTENUATOR

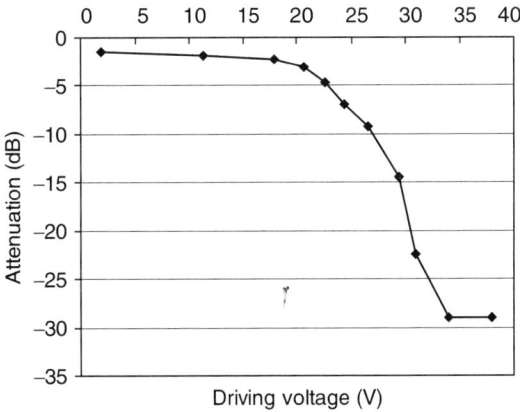

FIGURE 23.11 Static optical performance of MEMS chopper.

This section describes the design and evaluation of a surface micromachined variable optical attenuator fabricated using the PolyMUMPs surface micromachining process (MEMSCAP, Inc.) [11]. Surface

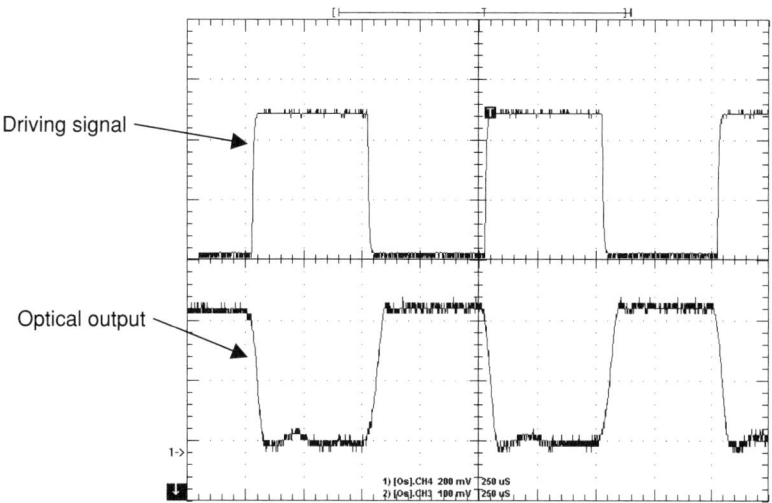

FIGURE 23.12 Dynamic response of the chopper as displayed on an oscilloscope.

micromachining remains a widely used process for the realization of three-dimensional (3D) MEMS devices. The general approach to attaining 3D devices is to initially fabricate planar two-dimensional structures and then subsequently raise the structures out of the plane of the substrate (the x-y plane), thereby orientating them parallel to the z-direction. Considerable research effort has been expended in developing approaches to assembling such 3D MEMS; some of these are described below. Three-dimensional microstructures, including microgratings, microshutters, and micromirrors, are frequently used in free space optical MEMS [18]. More recently, 3D microassembly technology has been applied to radio frequency (RF) MEMS, for example, in achieving out-of-plane micromachined inductors, which are shown to have better electrical characteristics (lower loss, less parasitic capacitance) when compared with in-plane micromachined inductors [19].

Various techniques have been used to assemble such 3D structures. The simplest involves manual assembly using micromanipulators [20]. Although this may be useful for assembling a few discrete devices, it does not lend itself to efficient assembly of large numbers of components or of complex systems. The technique is time consuming, and there is the potential of structures being damaged by the micromanipulators or not being accessible to the micromanipulator probes. Another method is to use dedicated microactuators for 3D lifting and assembly. Approaches using electrically active thermal [21] or electrostatic [22, 23] actuators have been reported. The use of surface tension forces generated from a range of materials is an alternative technique for 3D self-assembly. Meltable pads of thick photoresist have been used to self-assemble micromirrors by utilizing surface tension forces [24]. Molten Pb/Sn solder spheres [25] and thermally shrunk polyimide joints [26] are other materials that have also been used for 3D assembly. Magnetic forces [27], including plastic deformation magnetic assembly [28], have also been used to assemble microstructures, requiring the incorporation of a magnetic material on those sections of the structures that are to be erected. Direct electrostatic forces have also been applied to lift microstructures [29], and potential differences of between 35 and 40 V have been shown to be sufficient to produce lifting forces for hinged micromirrors.

An approach where a stress-induced curved bimorph cantilever is used to self-assemble a movable optical microshutter fabricated by polysilicon surface micromachining is described in the next section. When a bimorph cantilever is formed from two different films having different values of internal stress, the cantilever bends to a concave or convex shape depending on the values and type of stress (i.e., whether tensile or compressive). Polysilicon and metal are materials often used in surface micromachining technology and can be used for forming such self-bending beams because of the different internal stresses present in these materials. Such bent bimorphs have several interesting applications. In micro-optics they have been used in the realization of curled cantilever optical switches [30]. They have also been used for elevating flat micromirrors parallel to the plane of the silicon substrate to obtain a higher angular deflection from the micromirrors [31]. In RF applications, they have been used to form out-of-plane inductors [32].

The following section describes how this principle of a curved bimorph cantilever induced by differential stress has been used to realize self-assembling and holding of movable optical microshutters to form a 3D fiber optic variable optical attenuator (VOA). The theoretical deflection of the bimorph is derived and compared with experimental measurements. Measurements of the angular elevation of a batch of self-assembled microshutters have been made, and these are presented. Thermal measurements on the bimorph beams have also been made, and these are compared with theoretical results. Finally, optical measurements on the fiber optic VOA incorporating the self-assembled microshutters are presented.

3.1. Self-Assembly Applied to a MEMS VOA

A metal/polysilicon bimorph cantilever was applied to the self-assembly of the microshutter in a MEMS fiber optic VOA. A schematic diagram of the MEMS VOA is shown in Fig. 23.13. A movable microshutter is placed between two unlensed single-mode optical fibers whose end-faces are separated by a distance of 60 μm. Optical power coupled from one fiber to the other is controlled by allowing the shutter to partially block the passage of light between the two fibers. The shutter is connected by microhinges to a transport plate that is moved by nanostepping microactuators commonly called scratch drive actuators (SDAs). The shutter that controls the exchange of light between the two optical fibers is initially formed flat and lies on top of the transport plate to which it is connected by microhinges. The transport plate is formed in the poly 1 layer, whereas the shutter is formed in the poly 2 and metal layers of the process (see the PolyMUMPs design rules for a description of poly 1, poly 2, and metal layers [11]). The requirement is to raise and hold the shutter in position so that a 3D microsystem is realized without any post-processing of the foundry fabricated dies.

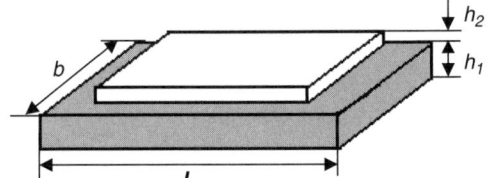

FIGURE 23.14 Dimensions of bimorph beam.

3.2. Design Parameters

The tip displacement of a bimorph cantilever is first calculated. Integration of this structure into the VOA design to achieve self-assembly of the shutter is then illustrated. Figure 23.14 shows the dimensions of a general bimorph beam, which is composed of polysilicon with a metal (gold) layer on its surface. Differential stress in the two layers generates a curvature of the beam and the radius of curvature is denoted as ρ. From [33], the radius of curvature can be obtained as

$$\rho = \frac{1}{6} \frac{4E_p E_m h_1^3 h_2 + 6E_p E_m h_1^2 h_2^2 + 4E_p E_m h_1 h_2^3 + E_m^2 h_2^4 + E_p^2 h_1^4}{h_1 h_2 (h_1 + h_2)(\sigma_p E_m - \sigma_m E_p)} \quad (4)$$

where h_1 and h_2 are the polysilicon and metal thickness; σ_p is the residual stress of polysilicon and σ_m is the residual stress of the gold layer. The value for the residual stress in the polysilicon is 10.31 MPa (compressive) and the residual stress in the gold film is 33.17 MPa (tensile). E_p is the Young's modulus of polysilicon and is taken to be 169 GPa, whereas E_m is the Young's modulus of gold and is taken to be 70 GPa. The thickness of polysilicon h_1 is taken to be 1.508 μm, and the thickness of gold layer h_2, 0.513 μm. The displacement, δ, of the tip is

$$\delta = \rho(1 - \cos(\frac{L}{\rho})) \quad (5)$$

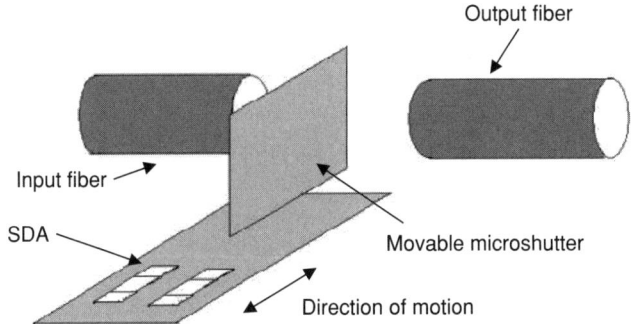

FIGURE 23.13 Schematic diagram of MEMS VOA.

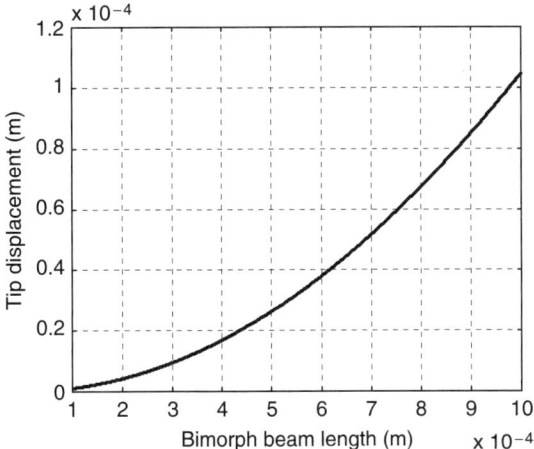

FIGURE 23.15 Calculated result showing the relationship between bimorph beam length and tip displacement.

where L is the beam length. The relationship between tip displacement and beam length has been calculated using Eqs. (4) and (5) and is shown in Fig. 23.15, noting that the opposite signs of stress in the polysilicon and gold films are taken into account when using Eq. (4). In the design described here, the length of the bimorph beam is 400 μm. This value provides enough tip displacement to lift and hold the shutter.

3.3. Device Fabrication

The design layout of the self-assembled shutter of the MEMS VOA is shown in Fig. 23.16. In this layout diagram, the forked holder is made from poly 1 that is partially underneath the poly 2 shutter before release by hydrofluoric acid etching and supercritical CO_2 drying. This holder at the tip of the beam is connected to the stress-induced curved beam formed by poly 2 coated with gold.

After release, the bimorph beam is automatically bent upward due to the large difference of residual stress between polysilicon and metal. The forked holder at the tip of the beam rises and makes the shutter lift up with it, holding it in position. The microhinge is designed to link the shutter to the translation stage, and therefore the shutter is self-assembled and permanently upright on the translation stage. Three-dimensional self-assembly has therefore been achieved relatively simply using the available commercial foundry processes, without resort to postprocessing. Figure 23.17 is an electron micrograph in which the shutter is shown standing on the translation plate and held by the bimorph beam with the holder at its tip. An important point to note is that in this particular application, the inclusion of the bimorph beams has not contributed to an overall increase in size of the device. The angular elevation of the microshutters has been measured following the process described in [28]. The devices were viewed under a microscope to whose stage a dial gauge was fixed. The dial gauge enabled the vertical movement of the stage to be read with an error of 2 μm. The microscope was focused on the bottom edge of the microshutter, and a reading was taken from the dial gauge. The microscope was then focused on the top edge of the shutter, and a second reading was taken from the gauge. The difference in these readings represented the vertical movement of the stage and is the height y shown in Fig. 23.18. The actual length l of the shutter is known a priori, and so the elevation angle θ is calculated using trigonometry. In total, measurements were carried out on a batch of 16 devices from the same PolyMUMPs run. Figure 23.19 shows the range of angular elevations obtained from the measurements.

3.4. VOA Performance

The optical properties of the VOA were measured using a standard InGaAs laser diode with continuous wave output and maximum power of 1 mW at

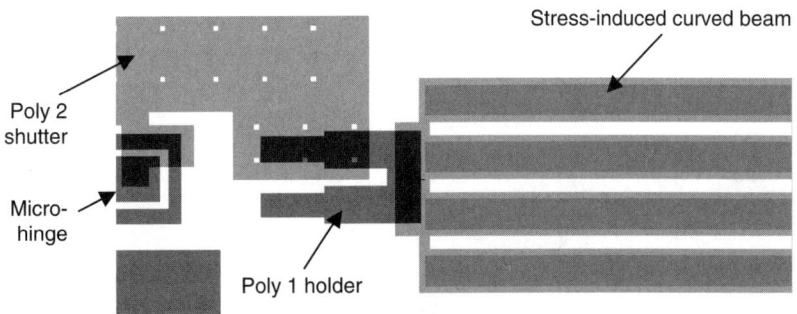

FIGURE 23.16 Layout of self-assembling part of MEMS VOA.

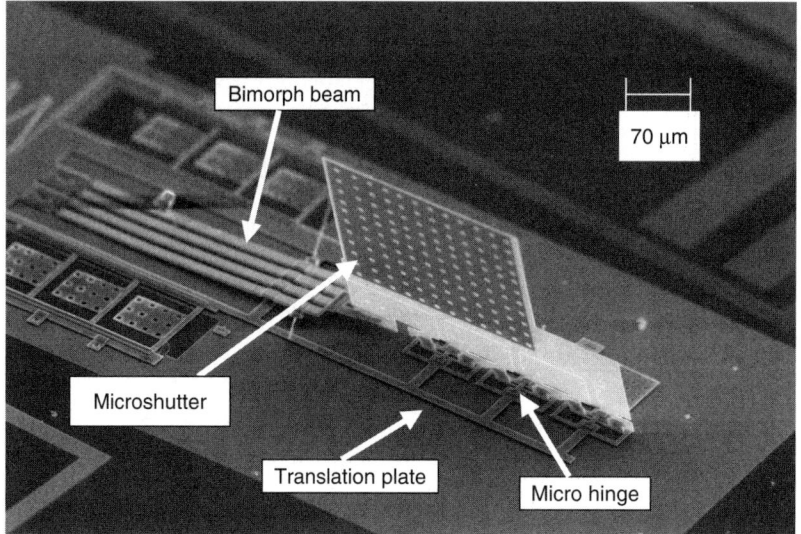

FIGURE 23.17 Electron micrograph of VOA showing self-assembled microshutter.

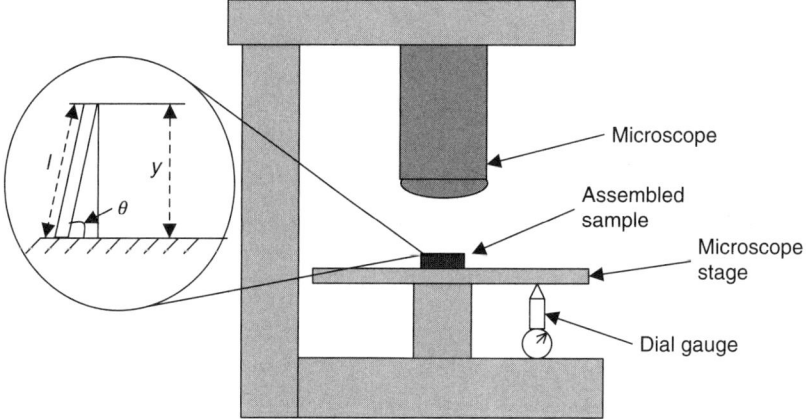

FIGURE 23.18 Experiment setup for measuring the angle of the assembled shutter.

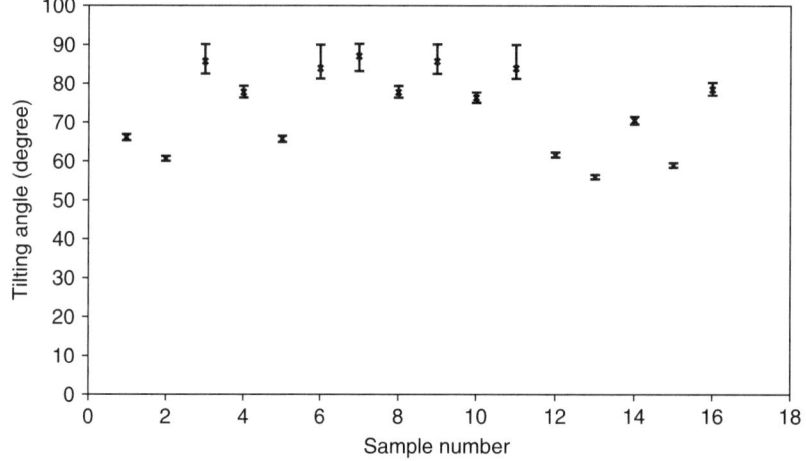

FIGURE 23.19 Self-assembled microshutter angular elevation measurements.

1.55 μm. The optical signal was monitored with a GaAs photodiode. The VOA attached to a custom printed circuit board was mounted under an Olympus BXFM microscope. Standard single-mode optical fiber was used throughout the experiment. One of the fibers was mounted on a rigid holder, whereas the second fiber was aligned accurately using a high precision five-way translation stage. The ends of the optical fibers were placed facing each other at a separation of 60 μm; this resulted in a measured 3-dB insertion loss. The attenuation characteristic of the VOA was tested using a 150-V amplitude square voltage waveform of 50% duty cycle. To precisely measure the attenuation characteristic during one cycle, the device was driven at a relatively low frequency of 80 Hz. The repeatability of the device was measured at 100 Hz or more to observe faster on/off action.

Figure 23.20 shows the attenuation characteristic of the VOA as a function of the travel distance of the shutter when it was placed between two fibers. The total distance that the shutter needs to move from maximum transmission to total attenuation is 18 μm. When the device is electrically grounded at any given point during the VOA motion, a micromachined restoring spring returns the device to its original position and the attenuation cycle can be repeated. The device on/off action was tested for more than 100 cycles. The speed of the SDA stepper can be controlled by adjusting with frequency. The fastest attenuation time observed in the experiments reported here was 36 ms at a driving frequency of 10 kHz with a dynamic attenuation range of 44 dB.

4. HYBRID TUNABLE FILTER

Recent research and development in the field of O-MEMS has been largely concentrated in the area of optical communication applications. The application of O-MEMS in the field of optical sensors is a smaller activity when compared with the communication applications of O-MEMS. There is, however, an increasing interest in applying O-MEMS in the field of optical sensing. In this section, one such sensing application of O-MEMS involving optical fiber sensors is presented.

Fiber optics has been used for many years in a range of sensing applications using fiber Bragg grating (FBG) sensors [34] (also see Chapters 24–26). The sensing mechanism of FBG sensors is the modulation of the reflected wavelength through changes of the effective grating period and refractive index induced by temperature, pressure, and strain. To precisely measure the variation of the reflected wavelength of the FBG, different techniques have been developed [35], including passive detection schemes [36], active detection schemes [37], and other schemes such as wavelength tunable sources and laser frequency modulation [38]. In this section, a novel hybrid O-MEMS tunable filter with the potential for implementing an active detection scheme for FBG sensors is described.

Angular rotation of an interference filter is a well-known principle for tuning its passband [39]. The O-MEMS tunable filter of this section is a "hybrid" optical microsystem, where the combination of a passive optical component—an off-the-shelf DWDM filter—with a MEMS actuated platform,

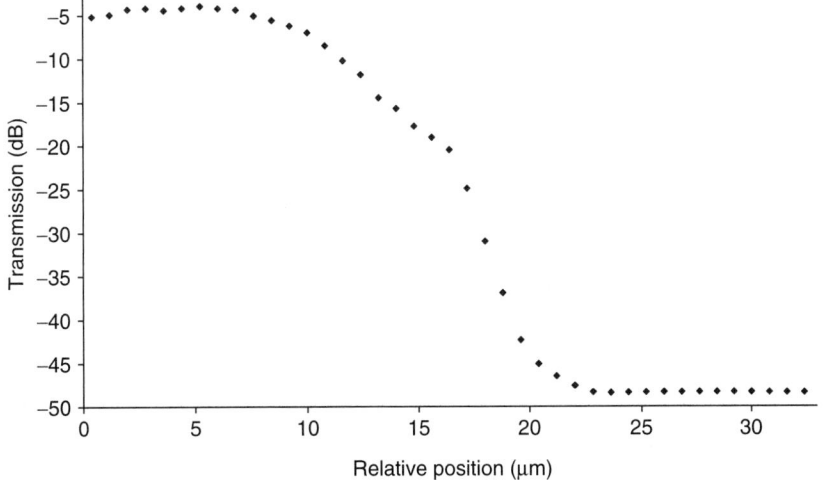

FIGURE 23.20 The attenuation characteristic of the VOA with fibers, with each fiber placed 30 μm from the shutter.

which rotates about its axis, creates an active optical tunable filter.

4.1. Component Characteristics and Fabrication

The tunable filter arrangement consists of an off-the-shelf DWDM filter seated on a silicon MEMS device consisting of a square platform and an actuation system composed of an array of thermal actuators, thereby creating a compact hybrid MEMS device. The filter used for the experiments was an off-the-shelf 200-GHz standard DWDM filter (Koncent Communication, Inc.), with center wavelength of 1560.61 nm (channel 21 of ITU (International Telecommunication Union) DWDM standard) with a -3-dB spectral full width at half-maximum of 1.6 nm. The center wavelength is taken to mean the wavelength of peak measured intensity. The mechanical dimensions are $1.4 \times 1.4 \times 1.4$ mm with the weight of the filter measured to be 6 mg.

The MEMS element consists of a suspended silicon platform supported by a cluster of silicon micromechanical beams through which an electric current can be passed. The beams support the platform and also generate the actuation forces. The platform was fabricated using the SOIMUMPs process described in [11].

Using the process, a square platform ($2000 \times 2000 \times 10 \, \mu m$) held by 20 beams (5 at each corner of the square) was fabricated in SOI, as can be seen in Fig. 23.21. Calculations were made assuming a point load on a beam to select the length and number of beams to support the optical filter. The dimensions of the five supporting beams range from $1000 \, \mu m$ for the shortest to $1100 \, \mu m$ for the longest. All beams are $5 \, \mu m$ wide and $10 \, \mu m$ thick.

The supporting beams as well as the platform are designed to behave as an electrical path constructed of silicon. Hence, electrical connection pads are extended to the fixed ends of the beams. The electrical resistance of the actuators and the platform measured between adjacent corners at room temperature (22°C) was $3.1 \, k\Omega$.

Many publications, for example [40], describe how injecting current through narrow beams generates heat in such beams, increasing the temperature and hence creating an elongation of the beams as an effect of thermal expansion. A characteristic feature of the MEMS structure described here is the conversion of the linear expansion of the beams to the angular rotation of the platform. This mechanism provides a way to achieve the necessary actuation of the platform, thus rotating the filter to obtain the tunability required.

4.2. Experiments and Results

Two different experiments were designed and conducted to analyze the optical performance of the hybrid O-MEMS filter. The first was to analyze the tuning range of the device, and the second was to evaluate the dynamic response of the device.

To achieve accurate readings of the wavelength tunability, a fiber ring laser incorporating the hybrid O-MEMS filter was implemented, as shown in Fig. 23.22. The laser line narrowing means that the ring laser configuration would allow a more precise reading of the wavelength shifts to be measured when the MEMS platform is actuated.

To drive the micromechanical system, the output of a signal generator was connected to a high voltage amplifier (HVA). The HVA had a fixed gain of $\times 20$ and a maximum continuous current output of 100 mA. An erbium-doped fiber amplifier (EDFA) was used as the laser gain medium. The EDFA has a low-signal (less than -25 dBm) amplifier gain of 18 dB for a pump power of 15 mW. An optical spectrum analyzer (OSA) was used to read the values of the wavelength shift due to the incident angle variation of the light beam incident on the filter. The fiber collimators used in the experiment were purchased from Koncent Communications, Inc. and had a nominal beam diameter of $230 \, \mu m$ at a working distance of 1 cm.

4.2.1. Performance of the Hybrid System

To determine the tuning range of the filter, the chip containing the MEMS device was fitted on a customized printed circuit board enabling easier connections to the HVA. This MEMS platform was placed between the two collimators with the filter cube seated on the platform. Multiaxis translation stages were used to achieve maximum alignment

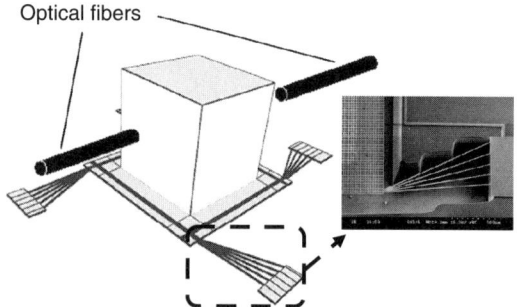

FIGURE 23.21 Schematic of the MEMS platform with optical microfilter seated on it and SEM detail of thermal microactuators acting also as suspension bar.

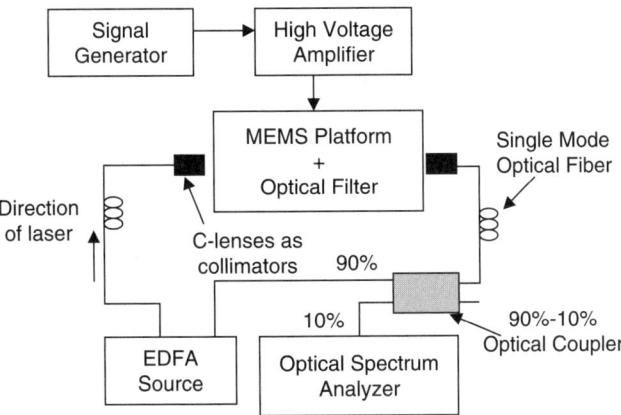

FIGURE 23.22 Ring laser setup used to obtain more accurate readings of the wavelength shift.

between the two collimators. The creation of a fiber gap by the use of two collimating lenses introduced a loss of 3.7 dB measured at 1560.2 nm. This included the loss of the collimators themselves, the diffraction loss through the air gap, and the optical path (excluding the filter cube). The insertion of the filter cube into the fiber gap introduced a further loss of 1.7 dB.

The ring laser setup of Fig. 23.22 was switched on and electric power was applied to the MEMS platform to rotate the filter. Measurements of the center wavelength of the laser were taken for different DC levels of electric power applied to the platform. The experiment was started from 0 V and 0 mA increasing to 36.4 V and 9.3 mA on each electric branch, each of the branches comprising the two adjacent thermal arrays. The total power consumption was 675 mW in the thermal actuators and platform. The maximum power before permanently damaging the device was 1040 mW. Center wavelengths of the laser at intermediate stages were recorded using the OSA while increasing and decreasing the electrical power applied to the actuated platform. The results are shown in Fig. 22.23a. The tuning range of the filter without damaging the thermal actuators was measured to be over 870 pm. A characteristic of the hybrid MEMS filter is the change in transmission loss as the filter is tuned, as seen in Fig. 22.23b. This is caused by deflection of the light path as the filter cube is rotated. However, this characteristic does not affect the measurement of wavelength shift of a Bragg grating sensor because this is only determined by the wavelength at which the power received by the interrogating photodetector is maximum as the filter is tuned.

The wavelength resolution of the tunable filter was measured to be 18 pm when the drive current was increased by 2%. The stability and repeatability of the hybrid MEMS device depends on the performance of the MEMS microactuated platform. To determine the stability, the peak transmission wavelength value was monitored over a 3-h period while keeping the MEMS platform at a fixed position. The peak deviation of the wavelength was measured to be 20 pm. The repeatability was measured by applying the same voltage and current (i.e., power) values to the MEMS actuator over three cycles and in each case monitoring the peak transmission wavelength. The change in peak transmission wavelength was measured to be 11 pm.

4.2.2. Dynamic Analysis of the Hybrid O-MEMS

The dynamic response of the hybrid MOEMS device was also investigated. An AC driving voltage was applied to the hybrid filter. The reflection of a visible laser beam from the filter surface was observed on a white screen located 30 cm from the filter surface while the driving signal frequency was varied. The output of the HVA was set to inject a sine wave of power 280 mW, corresponding to a peak current of 6 mA into each electrical branch. In common with every mechanical structure, this device has a resonance frequency, which was measured to be around 141.2 Hz. At resonance, the line on the screen due to the reflected laser light appeared to double its length.

This hybrid O-MEMS filter has a tuning range of over 870 pm (operating at around 1560.61 nm). The strain and temperature sensitivities for an FBG sensor operating at a 1550-nm wavelength are [41] 1.2 pm/$\mu\varepsilon$ and 13 pm/°C. Therefore, with an 870-pm tuning range it would be possible to measure over 725 $\mu\varepsilon$ and 67°C variations of the FBG.

Various interrogating systems and devices for FBGs have been presented in the literature [42–44],

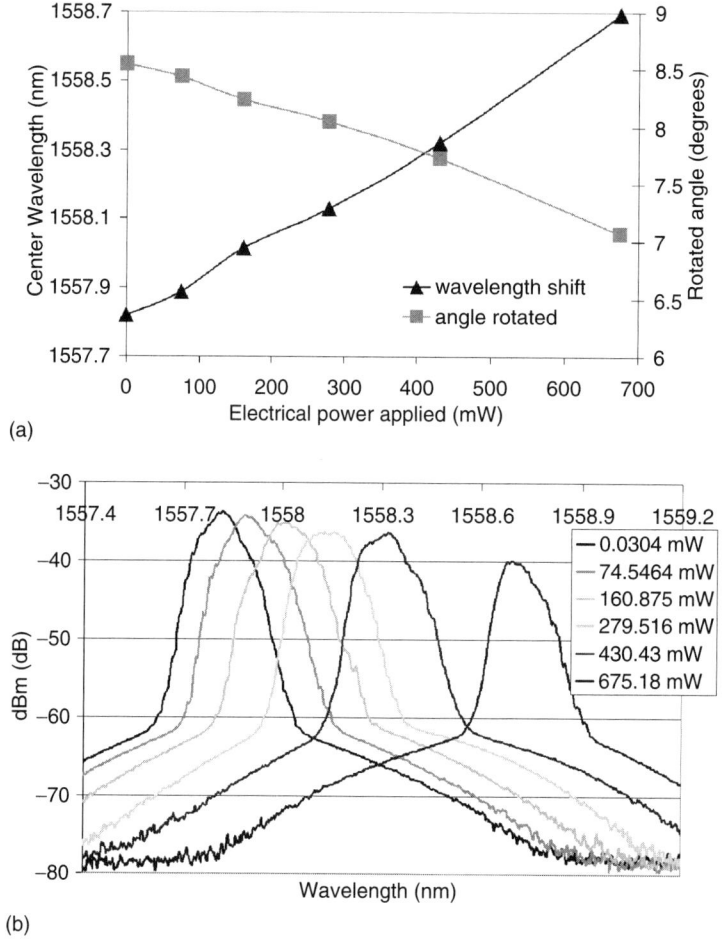

FIGURE 23.23 (a) Wavelength shift and calculated rotated angle of the filter for different power applied to the MEMS platform. (b) Spectral measurement showing the coupling loss.

with a tuning range similar to that obtained from the tunable filter described in this chapter. The tuning range is currently limited to 870 pm, determined by the MEMS rotating platform design. It is not envisaged that the rotation angle (and hence tuning range) can be increased using the microfabrication process that was used here, unless the dimensions of the actuator are increased by 200–300%. This, in turn, means having a device with a larger footprint. However, the narrow tuning range would still be suitable for a Bragg grating sensor array where all the Bragg sensors are operating at the same central wavelength. By using time division multiplexing, it is possible to discriminate Bragg grating sensors having the same central wavelength. This means that the optical interrogation source also does not need to have a wide bandwidth or wavelength excursion, leading to a lower cost system. The features of time division multiplexer Bragg grating sensors stated above are addressed in [41].

5. CONCLUSIONS

This chapter described O-MEMS devices designed at one location (a fabless location) and microfabricated at a commercial MEMS foundry. Three such O-MEMS devices with applications in optical measurement technology and in optical communication systems have been designed and experimentally evaluated. The first of these devices, namely a micromechanical optical chopper for use with optical fiber, has been designed using a double-blade shutter system actuated by an integrated comb drive actuator. Optical modeling of the shutter system has been performed using the appropriate theory suited for the microsystem regime. Through this modeling, it has been shown that a double-blade shutter system has advantages over a single-blade system in terms of the attenuation rate and driving voltage. Electromechanical experiments have shown that 34 V DC applied to the comb drive produced 10-μm displacement, and the resonant frequency of the device is around 3 kHz with a Q factor of 0.47. Optical experiments have been performed using precisely aligned single-mode fibers. It has been shown from the optical measurements that the attenuation of the chopper ranges from 1.4 to 29 dB (1.4 dB is the insertion loss). The dynamic response has also been measured showing rise and fall times of around 90 µs.

A second device, namely a self-assembled 3D VOA, was designed, fabricated at a commercial foundry, and experimentally evaluated. Self-assembly has been achieved with the assistance of metal/polysilicon bimorph beams. The VOAs have been experimentally evaluated, and the dynamic range of attenuation was measured to be 44 dB.

The third device falls into the category of a hybrid O-MEMS device. The hybrid optical MEMS has two parts. The first is the MEMS actuated platform; the second is the optical component, such as the filter, which would normally be an off-the-shelf type component. Hybrid optical MEMS have the advantage that the user can select the optical component with the characteristics that meet their particular requirements. The resolution of the hybrid MEMS device described in this chapter is dependent on the optical bandwidth of the filter selected, and a higher resolution could be achieved by using a narrow-band filter such as a 50-GHz DWDM filter.

The hybrid O-MEMS filter described has a tuning range of over 870 pm (operating around 1560.61 nm). The strain and temperature sensitivities for an FBG sensor operating at a 1550-nm wavelength are [41] 1.2 pm/$\mu\varepsilon$ and 13 pm/°C. Therefore, with an 870-pm tuning range, it would be possible to measure over 725 $\mu\varepsilon$ and 67°C variations of the FBG.

6. ACKNOWLEDGMENTS

Thank all members, past and present, of the Strathclyde Microsystems Group for their contributions to the optical MEMS research described in this chapter. This chapter is dedicated to Arun Uttamchandani.

7. REFERENCES

1. K. Petersen, "Silicon as a mechanical material," *Proc. IEEE*, **70**, pp. 420–457 (1982).
2. A. Bashir, P. Katila, N. Ogier, B. Saadany, and D. A. Khalil, "A MEMS-based VOA with very low PDL," *IEEE Photon. Technol. Lett.*, **16**, pp. 1047–1049 (2004).
3. J. H. Jerman and D. J. Clift, "Miniature Fabry-Perot interferometer in silicon for use in optical fibre WDM systems," *Proc. Int. Conf. on Solid State Sensors and Actuators*, pp. 372–375 (1991).
4. L. A. Field, D. L. Burriesci, P. R. Robrish, and R. C. Ruby, "Micromachined 1 × 2 optical fibre switch," *Sensors Actuators A Phys.*, **53**, pp. 311–315 (1996).
5. M. Kozhevnikov, N. R. Basavanhally, J. D. Weld, Y. L. Low, P. Kolodner, C. A. Bolle, R. Ryf, A. R. Papazian, A. Olkhovets, E. Pordo, D. T. Neilson, V. A. Aksyuk, and J. V. Gates, "Compact 64 × 64 micromechanical optical cross connect," *IEEE Photon. Technol. Lett.*, **15**, pp. 993–995 (2003).
6. C. Pu, L. Y. Lin, E. L. Goldstein, and R. W. Tkach, "Client-configured eight channel add/drop multiplexer using micromachining technology," *IEEE Photon. Technol. Lett.*, **12**, pp. 1665–1667 (2000).
7. M. E. Motamedi, "Micro-opto-electro- mechanical-systems," *Opt. Eng.*, **33**, pp. 3505–3517 (1994).
8. M. C. Wu, L. Y. Lin, S. S. Lee, and K. S. J. Pister, "Micromachined free space integrated micro-optics," *Sensors Actuators A Phys.*, **50**, pp. 127–134 (1995).
9. V. M. Bright (Ed.), "Selected Papers on Optical MEMS," *SPIE Milestone Series*, vol. MS153 SPIE Press (1999).
10. P. Rai-Choudhury (Ed.), "MEMS and MOEMS Technology and Applications," SPIE Press, (2000).
11. D. Koester, A. Cowen, R. Mahadevan, M. Stonefield, and B. Hardy, "PolyMUMPs design rule book," and C. J. Han, A. Cowen, G. Hames and B. Hardy, "SOIMUMPS design rule book," MEMSCAP Inc., http://www.memscap.com/memsrus/crmumps.html
12. M. Ueda, T. Shiono, and T. Ito, "High-efficiency diffractive micromachined chopper for infrared wavelength and its application to a pyroelectric infrared sensor," *Appl. Opt.*, **37**, pp. 1165–1170 (1998).
13. R. Wolffenbuttel and G. Degraaf, "Noise performance and chopper frequency in integrated micromachined chopper-detectors in silicon," *IEEE Tram. Instrum. Meas.*, **44**, pp. 451–453 (1995).
14. H. Toshiyoshi, H. Fujita, and T. Ueda, "A piezoelectrically operated optical chopper by quartz micromachining," *J. Microelectromech. S.*, **4**, pp. 3–9 (1995).
15. M. T. Ching, R. A. Brennen and R. M. White, "Microfabricated optical chopper," *Opt. Eng.*, **33**, pp. 3634–3642 (1994).
16. M. Born and E. Wolf, *Principles of Optics: Electromagnetic Theory of Propagation, Interference and Diffraction of Light*, Cambridge University Press, Cambridge, (1997).
17. H. Kogelnik, "Coupling and conversion coefficients for optical modes," in *Proc. Symp. Quasi-Optics*, New York, pp. 333–347, (1964).
18. R. S. Muller and K. Y. Lau, "Surface-micromachined micro-optical elements and systems," *Proc. IEEE*, **86**, pp. 1705–1720 (1998).

19. J. Zou, C. Liu, D. R. Trainor, J. Chen, J. E. Schutt-Aine, and P. L. Chapman, "Development of three-dimensional inductors using plastic deformation magnetic assembly (PDMA)," *IEEE Transactions on Microwave Theory and Techniques*, **51**, pp. 1067–1075 (2003).

20. M. C. Wu, L. Y. Lin, S. S. Lee, and K. S. J. Pister, "Micromachined free-space integrated micro-optics," *Sensors Actuators A Phys.*, **50**, pp. 127–134 (1995).

21. J. R. Reid, V. M. Bright, and J. T. Butler, "Automated assembly of flip-up micromirrors," *Sensors Actuators A Phys.*, **66**, pp. 292–298 (1998).

22. T. Akiyama, D. Collard, and H. Fujita, "Scratch drive actuator with mechanical links for self-assembly of three-dimensional MEMS," *J. Microelectromech. Syst.*, **6**, pp. 10–17 (1997).

23. N. C. Tien, O. Solgaard, M. H. Kiang, M. Daneman, K. Y. Lau, and R. S. Muller, "Surface micromachined mirrors for laser beam positioning," *Sensors Actuators A Phys.*, **52**, pp. 76–80 (1996).

24. R. R. A. Syms, C. Gormley, and S. Blackstone, "Improved yield, accuracy and complexity in surface tension self-assembled MOEMS," *Sensors Actuators A Phys.*, **88**, pp. 273–283 (2001).

25. K. F. Harsh, V. M. Bright, and Y. C. Lee, "Solder self-assembly for three-dimensional microelectromechanical systems," *Sensors Actuators A Phys.*, **77**, pp. 237–244 (1999).

26. T. Ebefors, E. Kalvesten, and G. Stemme, "New small radius joints based on thermal shrinkage of polyimide in V-grooves for robust self-assembly 3D microstructures," *J. Micromech. Microeng.*, **8**, pp. 188–194 (1998).

27. Y. W. Yi and C. Liu, "Assembly of micro-optical devices using magnetic actuation," *Sensors Actuators A Phys.*, **78**, pp. 205–211 (1999).

28. J. Zou, J. Chen, C. Liu, and J. E. Schutt-Aine, "Plastic deformation magnetic assembly for out-of-plane microstructures: technology and applications," *J. Microelectromech. Syst.*, **10**, pp. 302–309 (2001).

29. R. W. Johnstone and M. Parameswaran, "Self-assembly of surface-micromachined structures using electrostatic attraction," *MOEMS and Miniaturized Systems II, Proceedings of SPIE*, **4561**, pp. 66–76 (2001).

30. R. T. Chen, H. Nguyen, and M. C. Wu, "A high-speed low voltage stress-induced micromachined 2×2 optical switch," *IEEE Photon. Tech. Lett.*, **11**, pp. 1396–1398 (1999).

31. V. A. Aksyuk, F. Pardo, and D. J. Bishop, "Stress-induced curvature engineering in surface-micromachined devices," *Design, Test, and Microfabrication of MEMS and MOEMS, Proceedings of SPIE*, **3680**, pp. 984–993 (1999).

32. V. M. Lubecke, B. Barber, E. Chan, D. Lopez, M. E. Gross, and P. Gammell, "Self-assembling MEMS variable and fixed RF inductors," *IEEE Transactions on MTT*, **49**, pp. 2093–2098 (2001).

33. G. H. Ryder, *Strength of Materials*, Macmillan (1970).

34. Y. J. Rao, "Recent progress in applications of in-fibre Bragg grating sensors," *Opt. Lasers Eng.*, **31**, pp. 297–324 (1999).

35. Y. Zhao and Y Liao, "Discrimination methods and demodulation techniques for fibre Bragg grating sensors," *Opt. Laser Eng.*, **41**, pp. 1–18, (2004).

36. R. W. Fallon, L. Zhang, L. A. Everall, J. A. R. Williams, and I. Bennion, "All-fibre optical sensing system: Bragg grating sensor interrogated by a long-period grating," *Measure. Sci. Technol.*, **9**, pp. 1969–1973 (1998).

37. C. Boulet, D. J. Webb, M. Douay, and P. Niay, "Simultaneous interrogation of fibre Bragg grating sensors using an acoustooptic tunable filter," *IEEE Photon. Technol. Lett.*, **13**, pp. 1215–1217 (2001).

38. G. A. Ball, W. W. Morey, and P. K. Cheo, "Fibre laser source analyzer for Bragg grating sensor array interrogation," *J. Lightwave Technol.*, **12**, pp. 700–703 (1994).

39. J. M. Saurel and J. Roig, "Properties of Fabry-Perot interferential filters illuminated in oblique incidence," *J. Opt.*, **10**, pp. 179–193 (1979).

40. R. Hickey, D. Sameoto, T. Hubbard, and M. Kujath, "Time and frequency response of two-arm micromachined thermal actuators," *J. Micromech. Microeng.*, **13**, pp. 40–46 (2003).

41. Y. J. Rao, "In-fibre Bragg grating sensors," *Measure. Sci. Technol.*, **8**, pp. 355–375 (1997).

42. S. Abad S, F. M. Araujo, L. A. Ferreira, J. L. Santos, and M. Lopez-Amo, "Interrogation of wavelength, multiplexed fibre Bragg gratings using spectral filtering and amplitude-to-phase optical conversion," *J. Lightwave Technol.*, **21**, pp. 127–131 (2003).

43. T. Allsop, K Chisholm, I. Bennion, A. Malvern, and R. Neal, "A strain sensing system using a novel optical fibre Bragg grating sensor and a synthetic heterodyne interrogation technique," *Measure. Sci. Technol.*, **13**, pp. 731–740 (2002).

44. Q. H. Li, C. F. Li, J. Q. Li, and B. S. Liu, "Fibre-grating sensor using dynamic PZT modulation," *Opt. Commun.*, **211**, pp. 129–133 (2002).

24

Principles of Fiber Optic Sensors

Brian Culshaw*
Department of Electronic & Electrical Engineering
University of Strathclyde
Glasgow, Scotland

1. INTRODUCTION

Optical fiber sensors are emerging as an important application of optoelectronic systems. During the 30 or more years since they were first proposed [1], they have matured from a concept into a technology that—typically for sensor systems—is finding important niche applications.

In this chapter, we first introduce the basic concepts that underpin the operation of optical fiber sensors. There are many, and virtually every optical effect has been used in a sensory context. Second, we give a more detailed consideration of specific applications and technologies. Chapter 26 explores physical sensing, so here we spend a little time examining a few areas of the many in which fiber optical chemical sensing appears to be exceptionally promising.

2. FIBER OPTIC SENSORS: THE BASIC PRINCIPLE

The essential features of a fiber optic sensor [2] are shown in Fig. 24.1. Light passing through the sensing region is modulated by the parameter of interest (and frequently by other parameters too) and the modulated light is detected at the optical decoder at the end of the fiber. The modulation process can be linear (sometimes physicists prefer to call these processes elastic). In linear modulation systems, the output contains only the optical frequencies present at the input. In nonlinear (or inelastic) modulators, scattering processes involving the generation of light at frequencies other than those in the input illumination occur, and these scattering processes are related to the quantities to be measured.

There are also two basic architectures available: intrinsic and extrinsic (Fig. 24.2). Intrinsic sensors modulate light while it is still propagating in the fiber. Extrinsic sensors involve optics to extract the light from the fiber and perform the modulation process in another medium. Thereafter, the light is recollected and transmitted for detection.

There are numerous optical processes that can influence the properties of the optical signal:

- Optical attenuation can be introduced through either external modulators in extrinsic sensors or through externally (to the fiber) applied loss-inducing mechanisms, most notably bends and microbends.
- Optical delay—usually phase—is particularly important in intrinsic sensors. The delays within an optical fiber can be modulated by physically changing its length (applying strain), by changing the isotropic pressure around the fiber, or by changing the temperature of the fiber. Typical coefficients for these processes are in the region of $10\,\text{ppm}\cdot\text{K}^{-1}$, $1\,\text{ppm}\cdot\mu\varepsilon^{-1}$, and $1\,\text{ppm}\cdot\text{bar}^{-1}$ (1 bar $\approx 100\,\text{kPa}$).
- Modulation of color/wavelength of the light can be very effective. This can be done extrinsically through moving masks or through absorptive chemical indicators. It can also be implemented intrinsically using variable reflectors and filters (typically Bragg gratings) and through inelastic processes such as Raman or Brillouin scatter for physical measurements. Various forms of luminescence for chemical measurements have also been used in both intrinsic and extrinsic topologies.
- Polarization states are modified when polarized light is passed through a variably birefringent medium. Variations in birefringence usually imply the application of anisotropic physical forces to circular symmetrical fiber or—in some cases—the application of isotropic forces (changes in temperature, strain, etc.) to an inherently birefringent medium such as high birefringence fiber or elliptically cored fiber.

There are also some examples of secondary modulation schemes where the signal representing the

*E-mail: b.culshaw@eee.strath.ac.uk

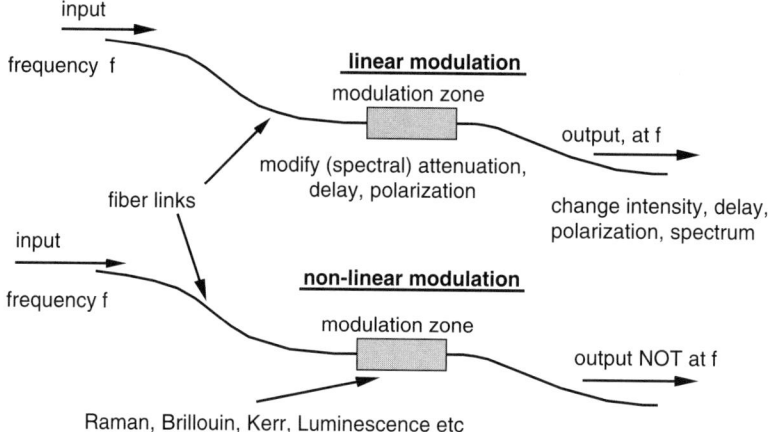

FIGURE 24.1 Basic concepts of optical fiber sensors during linear (elastic) modulation phenomena for which the optical spectral content contains the same frequency distribution at the output as the input. Below nonlinear (inelastic) systems introduce new frequency components.

parameter of interest is put on to some form of subcarrier. Examples of this include decay time measurements for certain types of spectroscopic signals (e.g., looking at the fluorescence lifetime as a function of temperature) and vibrating element structures used to measure physical parameters such as strain and temperature. These mechanisms are presented indicatively in Fig. 24.3.

There is a rich variety of physical and chemical phenomena that influence these optical properties in fiber optic systems. Indeed, most of the phenomena listed above are influenced by more than one measurement parameter. Consequently, cross-talk (particularly from temperature) must always be addressed in sensor design. Indeed, this is common to virtually all other sensor technologies. Arguably, the situation with fiber optics is somewhat more complex because in the final analysis a photodetector simply detects intensity, as appropriate, after a suitable intermediary optical encoding and decoding stage. We modulate in several domains but only detect in one. Consequently, there are obviously additional paths through which the spurious intensity modulated signals may be introduced. These paths often need very careful analysis, though some modulation schemes are ingeniously immune to these influences.

This then encapsulates the basic principles of optical fiber sensors. Another important aspect of the technology is, however, the range of architectures that the sensors can address and in particular the unique features that these architectures give to fiber sensor technology. These architectures are illustrated in Fig. 24.4. Single point sensing is the approach well known and well used in electrical systems. Multiplexed networks also are found frequently in electrical sensing systems. However, these networks do require electrical power supplies at each sensing element. The overwhelming virtue of the fiber optic equivalent is that it is totally electrically passive and therefore readily configured to be intrinsically safe. Furthermore, these optical fiber links in the network assure us that sensors can be interrogated over distances that may be up to several tens of kilometers.

The distributed architecture is probably the most interesting. This is genuinely unique to fiber sensors (with the exception of one or two electrical cable systems with a range of just a few meters). The distributed network enables a single optical fiber through the appropriate measurand interface to measure a parameter of interest as a function of position along the fiber over ranges which can be up to 100 km, though most applications address much shorter interrogation distances.

in the fiber (intrinsic):

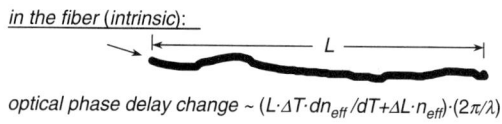

optical phase delay change ~ $(L \cdot \Delta T \cdot dn_{eff}/dT + \Delta L \cdot n_{eff}) \cdot (2\pi/\lambda)$

outside the fiber (extrinsic):

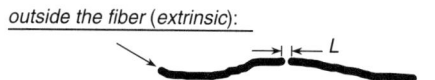

Same relationship, but now applied to material in the gap

FIGURE 24.2 Intrinsic and extrinsic sensing illustrated through length measurement. In the intrinsic sensor, the length measurement concerns the physical length of the fiber. In extrinsic measurement, it is the length of the gap outside the fiber between launch and receive points.

Principles of Fiber Optic Sensors

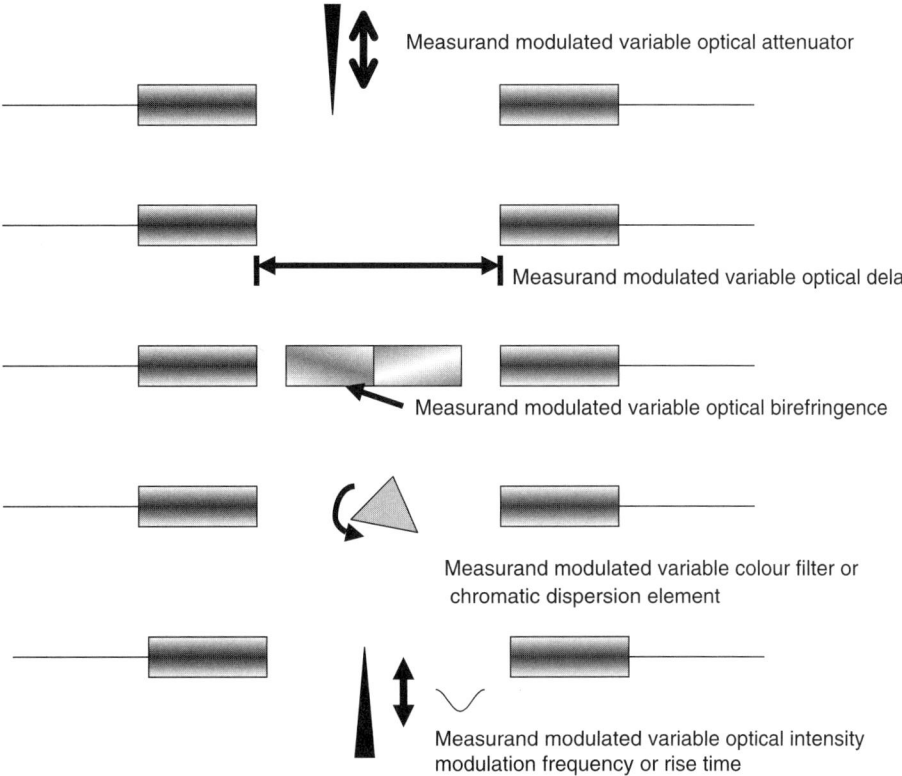

FIGURE 24.3 Indicative linear modulation mechanisms for extrinsic sensors. From the top: direct intensity modulation using a moving mask, length modulation between the faces of the two GRIN lengths optic points, a modulated optical birefringence, a chromatic modulation system (dye, prisms, gratings, etc.) that is measurand dependent, and finally a subcarrier modulation system where, for example, the frequency of intensity modulation carries the information concerning the measurand. The same principle may be applied by producing measurand interfaces that modulate any of the above four phenomena at a frequency dependent on the measurand or with a time decay dependent on the measurand.

Like multiplexed fiber optic systems, distributed networks are entirely passive electrically and so inherently intrinsically safe. However, they are amazingly simple to install—just insert the fiber in the area of interest after protecting it through a suitable measurand interface cable. Additionally, a (usually)

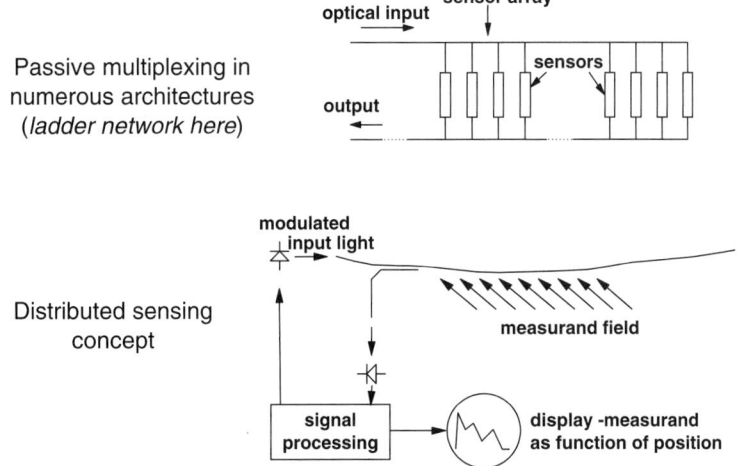

FIGURE 24.4 Optical fiber sensor networks. (Top) A network of discrete sensors coupled passively using optical fiber links. (Bottom) The distributed sensing concept unique to fiber sensor architectures.

simple single interrogation unit can give the value of the measurand of interest as a function of position along the fiber when viewed through a convolution window, which is typically on the order of 1 m in length and can be as small as a few centimeters. Distributed sensor configurations for measuring physical parameters (strain, temperature, etc.) and chemical parameters (particularly liquid spills and gaseous hazardous products) have been realized, and many are well on the way toward commercial production.

A combination of an optical transduction mechanism and an optical transmission medium affords numerous potential benefits to fiber sensor technology when compared with other sensing approaches. These benefits include intrinsic safety and immunity to electromagnetic interference, chemical passivity, long interrogation range, potential light weight and low volume, and tolerance to extremes of temperature and mechanical strain excursions. Silica is one of the most thermally tolerant and mechanically elastic of engineering materials.

These advantages are significant, but there are some inhibitors to real application. Perhaps the most important single one is history. In an environment in which a particular measurement has been made using a specific technology for many years, it is frequently extremely difficult, indeed verging on the impossible, to dislodge this technology. Additionally, fiber sensors are rarely the lowest cost option, and the new architectures, while remarkably flexible and offering entirely novel measurement prospects, are only now beginning to be acknowledged as potentially beneficial in the user community. Therefore, it is inevitable that in common with all other sensor systems, fiber optics must find its niches where its unique properties are potentially extremely beneficial. These are many and include systems such as fiber gyroscopes, fiber optic strain measurement systems using Bragg gratings, distributed architectures where these are of paramount importance, and some chemical and biochemical measurements where safety and chemical passivity (and sometimes long transmission distances or networked systems) are overwhelmingly beneficial. We see that there are many examples where these positive application features come to the fore.

3. FIBER OPTICS IN PHYSICAL SENSING

There is a more detailed account of physical sensors in Chapter 26. Here, for completeness, we briefly discuss some examples of areas in which physical sensing systems have begun to find commercial application or, at very least, to show serious practical promise. At least three system concepts fall into this category exemplifying distributed sensing, multiplexed point sensors, and arrays of individual sensor elements addressed through separate optical fiber links.

The distributed sensing architecture is an area in which fiber optic technology offers unique benefits. For strain sensing, Brillouin scatter provides a frequency encoded signal that can be uniquely related to the acoustic velocity in the fiber provided that the interrogating wavelength is sufficiently accurately known. This in turn can give strain provided that temperature is known or some complementary method for measuring temperature can be used (Fig. 24.5). In some areas, particularly underground and in deep water, temperature is effectively constant so the Brillouin scatter frequency is a direct measure of strain. This approach is finding potential application in the distributed monitoring of strain distribution along high performance polymeric marine ropes as a nondestructive testing technique [3]. It has also (Fig. 24.6) shown some promise in the monitoring of strain distribution along oil transportation pipelines as a means for detecting subsidence or intrusion that could compromise the pipeline's integrity, especially in remote places [4]. In this latter case, temperature compensation is sometimes necessary. One technique is to use spontaneous Brillouin (or spontaneous Raman) as a distributed temperature probe operating in parallel with and in the same fiber as the strain measurement transducer.

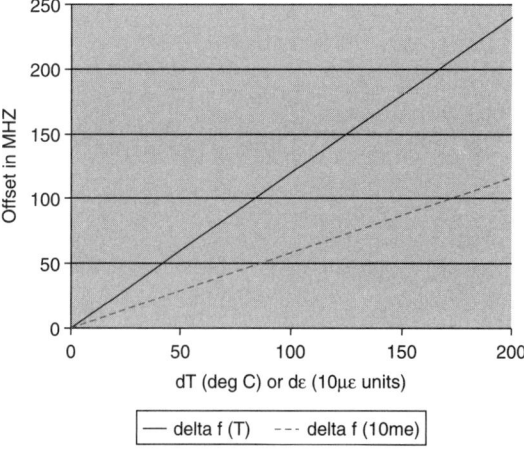

FIGURE 24.5 Distributed sensing using time resolution of Brillouin scatter to give spatial resolution of strain along an optical fiber. The graph gives the changes in offset frequency with strain or temperature.

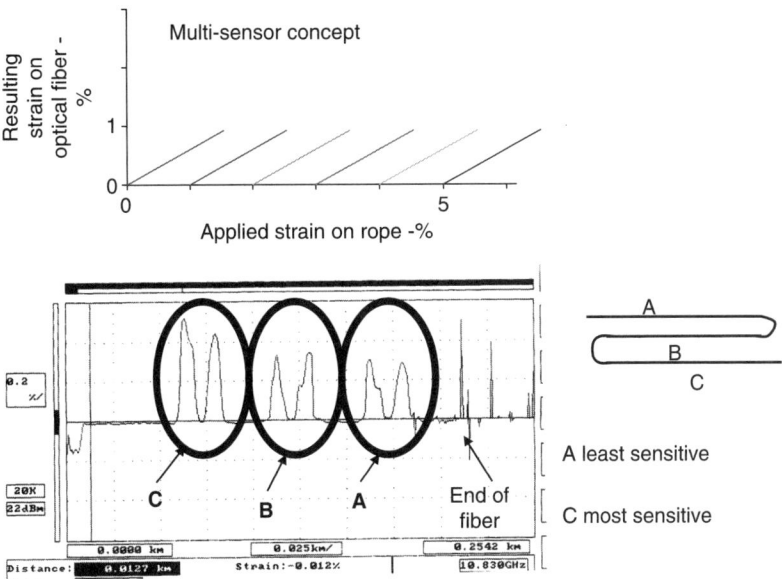

FIGURE 24.6 Some results from the system in Fig. 24.5 illustrating the scaling of strain transfer built into the transducer cable. (The vertical axis is perceived strain on each of the differentially sensitive lengths A, B, and C.)

Fiber Bragg gratings have been extensively evaluated as large arrays of point multiplexed sequentially addressed strain sensors (Fig. 24.7). These arrays have found application in carbon fiber composite parts for aircraft; composite masts in luxury yachts; a wide array of civil structures including bridges, dams, and tunnels; and even monitoring the strain history in thermally stressed works of art [5, 6]. The technical viability of the technology, including temperature compensation through local strain-isolated references, is beyond doubt. The issues that remain are predominantly in the application domain and identifying niches where the current relatively high cost of the system is far outweighed by the benefits that embedded sensors with high degrees of flexibility, electromagnetic interference immunity, and very simple installation processes may offer. Almost certainly, composite material monitoring and

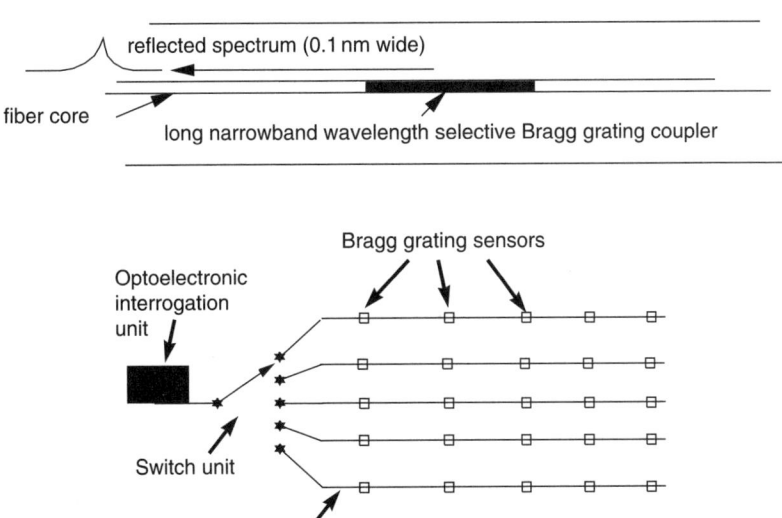

FIGURE 24.7 The fiber Bragg grating in which (top) the peak in the reflected spectrum is varied in response to an applied measurand that changes the period of the grating and (below) a system architecture including strings of gratings (each with separate different center wavelengths) addressed through a single optoelectronic interrogation unit and an array switch. Up to several hundred gratings can be interrogated in this way.

evaluation is one such potential application. Interestingly, Bragg gratings also show some promise for the detection of ultrasound propagating in structures (Fig. 24.8). This prospect offers another application niche—the capability to monitor both dynamic (including ultrasonic) and static strain fields using exactly the same sensors. No other sensor technology offers such flexibility. Bragg gratings are still undoubtedly to evolve into all their niches, but this evolution is well along the way toward happening.

SOFO (surveillance d'ouvrages par senseurs a fibers optiques; Fig. 24.9) is a unique sensor that provides highly stable long-term measurement over integrating gauge lengths from a few tens of centimeters to a few tens of meters, with stabilities on the order of a few microns [7]. This is especially useful in determining average strain: load performance in highly heterogeneous lumpy materials (composites) in which the composite elements are relatively large. Concrete is an example of such a material. In concrete, strain measurement is only meaningful when averaged over a length of many times the dimensions of the aggregate. Consequently, the SOFO sensor has found extensive commercial application in concrete structure evaluation and monitoring. Many thousands of these sensors are now in operation, predominantly in Western Europe, and a comprehensive library of reliability history and application compatibility has been developed. SOFO is interrogated using precision white light interferometry requiring high tolerance mechanics and has also been evaluated in a dynamic microwave subcarrier mode [8]. SOFO is probably the most widely proven of fiber optic strain sensor technologies.

Physical sensing using fiber optics is then an important tool in structural analysis and usage monitoring. As it becomes more widely understood in the user community and as the technology itself expands in application volume (and therefore falls in unit cost), it will undoubtedly become much more widely applied. More details on both principles and applications can be found in Chapter 26.

4. CHEMICAL SENSORS

Fiber optic chemical sensing is a very diverse technology (Fig. 24.10). There are at least five important transduction processes. With one exception, these all involve measuring optical signals at a specific wavelength or alternatively spectral distribution change (color) over a range of wavelengths. It is useful to very briefly examine each of these transduction mechanisms.

Direct spectroscopy predominantly involves measuring the change in optical absorption through a sample species at specific wavelengths corresponding to the optical absorption lines in the species of interest [9]. It is especially useful for measuring gas concentrations because many gases have very specific unique wavelengths at which they produce useful optical absorption. Additionally, it is relatively straightforward to derive ratiometrically corrected absorptions by interrogating around the absorptive region at an immediately adjacent wavelength at which the gas species does not absorb. Direct spectroscopy is less useful in liquids and solids because here the absorption lines are significantly broadened and the unique relationship between spectrum and species is far more difficult to establish. There are a

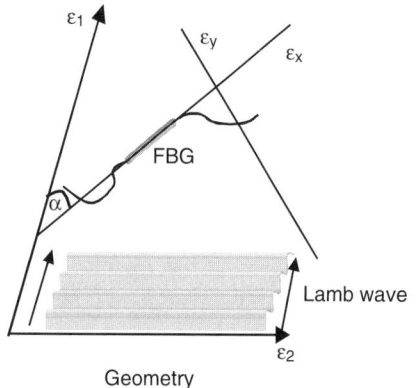
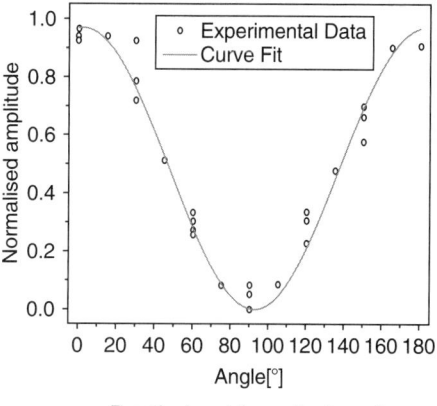

FIGURE 24.8 Fiber Bragg gratings to detect ultrasound—here Lamb waves. At left is indicated the geometry and at right the cosine polar response of the grating with the maximum sensitivity when the direction of propagation of the ultrasound is along the length of the grating, thereby inducing maximum strain.

Principles of Fiber Optic Sensors

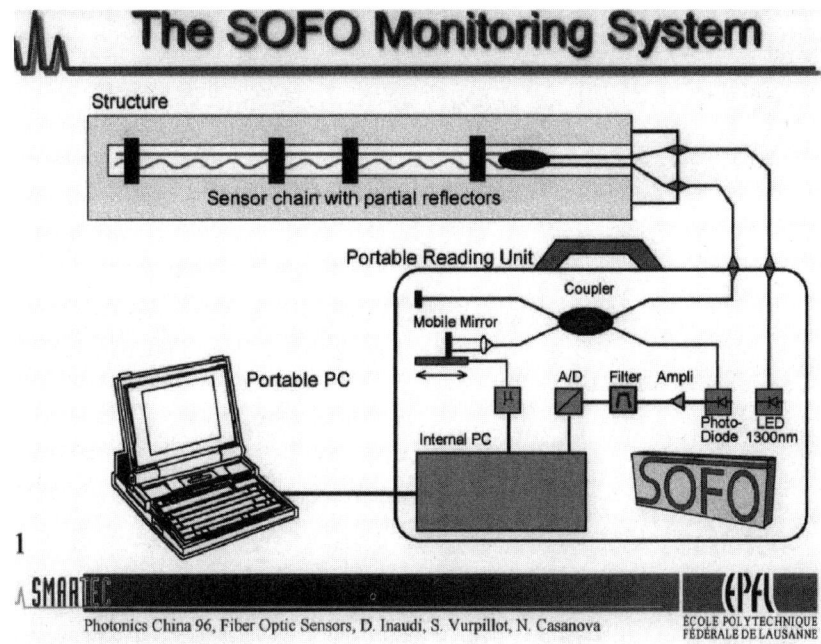

FIGURE 24.9 SOFO schematic (courtesy of Smartec). A white light receiving interferometer is coupled to a remote sensor with one fiber nteracting with the strain measurand and a second (shown wavy) reference fiber. The white light interferometer in the reading unit directly decodes the path difference between the two fibers. The multiple exchanges shown here have different static path lengths in each element, but the system is more frequently used with single sensors.

few—but relatively few—occasions when pattern recognition techniques can be invoked to extract useful information, provided that the choice of the samples is relatively restricted. Some relatively recent work in the classification of olive oils and wines has indicated the potential that this approach offers.

Inelastic spectroscopy, in contrast, is particularly applicable to examining liquids and solid surfaces. There are numerous instruments available commercially for Raman and fluorescence spectroscopy, and some of these have fiber optic coupling probes to reach remote samples. Raman spectroscopy requires

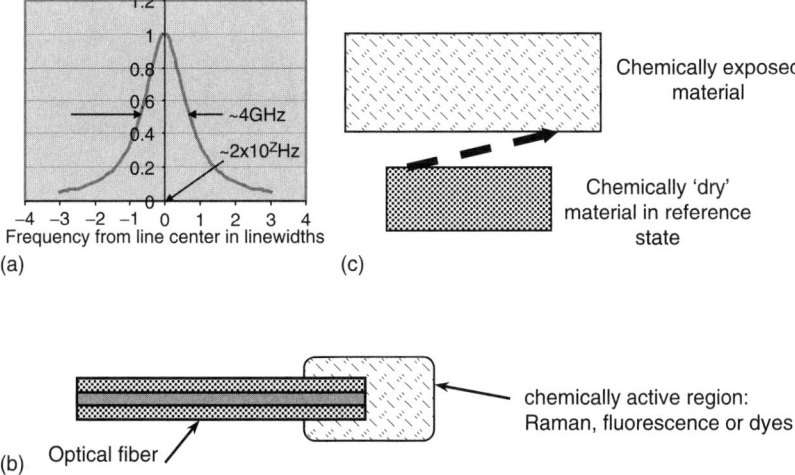

FIGURE 24.10 Indicative techniques for optical fiber–based chemical measurements. (a) Direct spectral absorption, here illustrated using a typical gas line with vertical axis absorption as a function of optical frequency. (b) The optical fiber optrode in which a chemically sensitive region interacts with the measurand of interest and modulates the spectral response of the chemically sensitive material. (c) Chemical to mechanical transduction in which a material (e.g., polymer or catalyst) swells in response to a particular analyte. The volume change is accompanied by density decrease but is also sufficiently robust to introduce optical modulation.

relatively high optical intensities so there is always the prospect of optically induced damage in the sample and additionally the prospect of some interference from the Raman spectrum of the optical fiber (this interference is used to good effect in Raman temperature probes). Consequently, Raman spectroscopy has found relatively little application as a technique focusing on fiber optic systems, though relatively short (a few meters) fiber-linked probes are sometimes available as an accessory for commercial instruments.

Fluorescence spectroscopy [10] sometimes invoking transient measurements has in contrast been quite extensively used. Predominantly, however, its application has been as an intermediate process where the properties of a fluorescent transduction material vary with external influences. In particular, these include oxygen concentration and variations in temperature. Consequently, the fluorescence spectrum is in a known wavelength band, and as is often the case, the fluorescence decay signal (time constant) is largely independent of the intensity of the fluorescent light returned to the interrogating system. Fluorescence from dopants in the optical fiber itself, notably neodymium, has been used as a temperature monitoring system. Fluorescence decay in external transduction elements has found some application in oxygen monitoring. However, in the latter case the stability of the transduction material plays an important part in sensor performance. A frequently encountered difficulty with fluorescent transduction materials is that they often require ultraviolet light to excite them, and ultraviolet transmits very poorly down optical fibers. There are, however, a few infrared fluorophores.

Transducer materials based on the rich variety of indicator chemistry are frequently exploited in fiber optic chemical sensing [11]. The transducer material is immobilized in a suitable host—typically a porous ceramic, a solgel, or through surface chemistry—for interrogation by a specifically tailored broadband optical source. This source may, for example, consist of several light-emitting diodes designed to produce a quasi-white light illumination. This light is absorbed within the host and scattered, and some is collected through the illuminating fiber or occasionally into a local bundle of connecting fibers. The color of the backscatter light is then assessed. The detected color depends on the stability of the detector network, the stability of the source spectrum, and the spectral stability of all the intervening optical components in addition to the color of the chemically sensitive transduction material. Consequently, considerable skill has to be executed in the system design to ensure that the effects of all but the transduction material are minimized. This basic technique has been widely used to measure both obvious color changing inducing phenomena such as pH variations and less obvious ones such as bile concentration in the stomach, blood oxygen, and species-specific gas absorption [12] (e.g., CO_2 or H_2S).

Chemical to mechanical transduction—typically via materials that swell when in contact with the species of interest—has also found some application. Mechanical changes in the material are detected by either ensuring that these changes translate into changes in optical delay through an optical fiber (interferometry) or through changes in optical attenuation, either through moving optical reflectors at the end of the fiber or through introducing microbends. A particularly interesting demonstration of interferometric detection uses palladium as a transducer material and exploits palladium's catalytic properties when in the presence of hydrogen. The hydrogen gas that is adsorbed by the palladium causes it to swell to an extent directly proportional to hydrogen content. This swelling induces length changes in the optical fiber which in turn are monitored interferometrically [13]. The second class of applications uses polymeric materials that selectively swell in the presence of water or solvents dependent on the material. This selected swelling can be translated into a microbend or macrobend response, which in turn induces local attenuation.

Chemical sensing is extremely diverse. Much of it depends on the stability of intermediate chemistry, which itself detects the species of interest. Indeed, this applies to all techniques with the exception of direct spectroscopy, which in turn usually refers to the examination of absorption spectra. Tailoring the chemical transducer is a specific art, and each chemical interface must be designed to operate within the application environment. This can vary immensely. Biological species, especially in vivo measurements, present one set of specific challenges. In contrast, highly unpredictable environments, as experienced in landfill chemistry and water quality monitoring, present yet a different range of needs. Progress in these technologies is best encapsulated in the annual proceedings of the Europtrode meeting (for example the one held in Madrid, April 4–7, 2004) that accurately records international progress in this diverse and challenging technology. The applications examples that appear in the next section exemplify just some of these considerations.

5. CHEMICAL SENSORS: SOME APPLICATION CASE STUDIES

5.1. Multiplexed Fiber Optic Spectroscopy

Responsible waste disposal becomes increasingly problematic as societies become increasingly wealthy. Managed waste disposal sites (i.e., landfills) are now mandatory in many countries and becoming mandatory in very many more. The basic concept (Fig. 24.11) is that a large hole, typically tens of meters in depth and a few kilometers across, is lined with a tough polymeric layer to prevent solvents (leachate) from entering in the local environment and is filled with mixed waste. Much of this waste decomposes anaerobically and in doing so produces predominantly carbon dioxide and methane gas in roughly equal proportions. In some countries the methane gas is simply flared off on site, but increasingly a managed approach to this involves collection and controlled processing. Typically an internal combustion engine for electrical power generation is the preferred, even mandatory, approach. Usually this involves controlled combustion, and consequently it is important to understand the dynamics of the gas production process and the percentage of methane in the gas mixture.

Optical fiber–based methane detection systems have proved to be extremely well suited to this task [14]. They are cost-competitive compared with alternative systems and have better technical performance. The principal competing gas measurement system is based on a device known as a pellistor. This basically burns the gas in a controlled way in the presence of a heated catalyst and measures the temperature of the catalyst as one arm of a resistance bridge. The burn rate depends on the surface state (poisoning) of the catalyst and is therefore subject to change with age. Furthermore, the system does not respond in atmospheres in which there is no oxygen (the anaerobic landfill gas mix should have no oxygen whatsoever in it) and additionally does not respond at high concentrations where the gas–air mixture is no longer inflammable (typically 15% compared with the several tens of percent customarily found in landfill gas). Figure 24.12 shows a sites plan of one such installation, and Fig. 24.13 shows some observed production plots.

Early versions of this system have now accumulated 4 years of site experience, and the technical and economic feasibility have been proven. The records have been found to be extremely useful and have assisted in the landfill management process. Additionally, the applications domain has expanded into examining possibilities for monitoring the performance of peripheral installations and safety monitoring systems. The same gas monitoring principle can be applied to petrochemical site and natural gas storage systems.

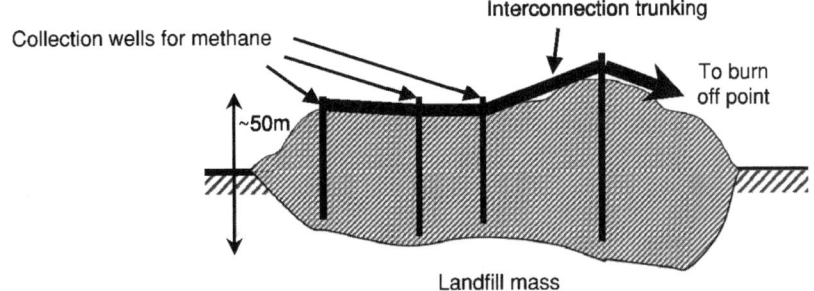

FIGURE 24.11 The application of remote optical fiber spectral measurements to measure methane gas in a landfill site.

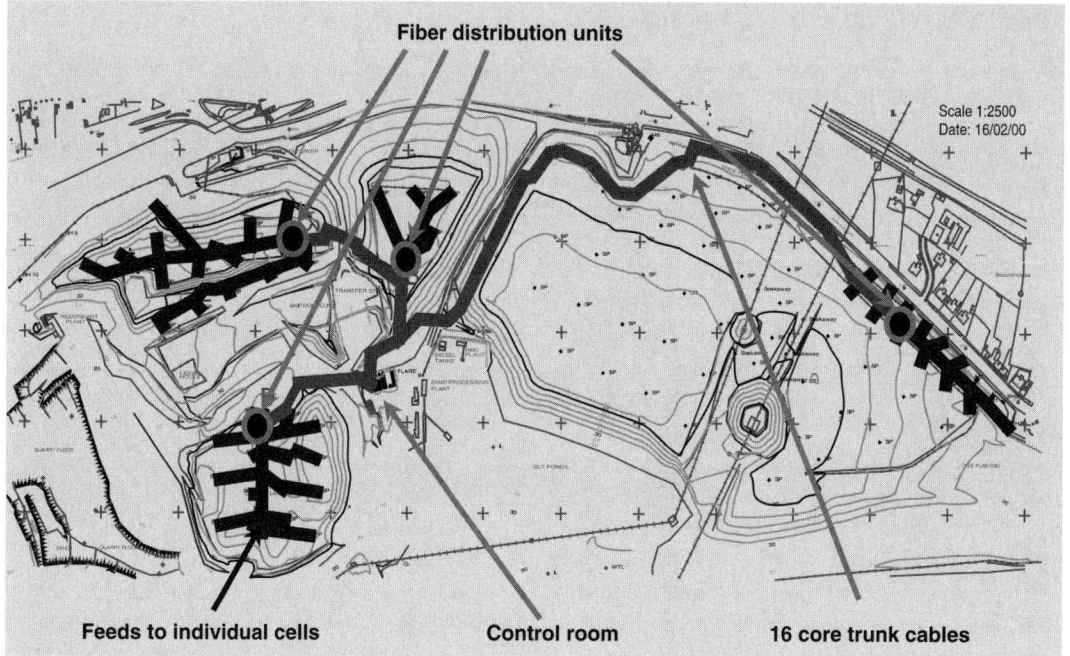

FIGURE 24.12 Site plan of the installation in Fig. 24.11 illustrating the tree network used to address over 60 cells from a single control room with a single laser source.

The basic operating principle is simple (Fig. 24.14). There is a convenient absorption line for methane at the 1.667-μm wavelength and laser diode sources are readily available. These can be current tuned to facilitate detection though lock in systems as illustrated in Fig. 24.14. The laser diode is tuned to a specific methane gas line and so responds uniquely to this particular hydrocarbon. The system can clearly be modified to address other species by changing the source wavelength.

We are still at an early stage in realizing the applications potential of this system, but its distinct technical advantages will in due time make their mark in specialist niches where the high methane density range, long distance monitoring capabilities, and immunity to poisoning are important.

5.2. Olive Oil

Olive oil is the central component of the diet in most Mediterranean countries and is exported from this region internationally. The quality of olive oils can vary enormously, and a recently completed European project (the EU Optimo project) demonstrated an optical assessment technique to ascertain the basic properties of olive oil samples. Both the taste and appearance of olive oil are affected by local climatic conditions and by the techniques used in harvesting. The climate primarily affects the color, and the harvesting and processing techniques affect the turbidity. Additionally, olive oil is divided into broad categories—extra virgin and non–extra virgin depending on the processing techniques. Olive oil is also a highly regional product and varies in taste and appearance depending on the area of production.

Any olive oil classification scheme should therefore be able to discriminate whether a sample is extra virgin or not and the country (preferably region) of origin. There are two basic optical properties to examine, namely color and turbidity. Turbidity (nephelometry) also has different spectral properties dependent on the angle at which it is measured. Consequently, we have a very complex space in which to monitor the properties of olive oils (Fig. 24.15). The through measurement is a direct indicator of optical absorption and the scattering measurements indicate the angular and spectral dependence of turbidity. Figure 24.16 shows photographs of the instrument used to make these measurements.

The data sets produced are complex, and Fig. 24.17 shows two of the four sets of data used to characterize a typical set of samples. Each oil is characterized by four spectra (for four different angles) with 46 points in each spectral measurement, giving a total of 184 values. This data set needs to be reduced to a few characteristic numbers, and standard principal component analysis (PCA) provides a readily available approach to making these calculations. PCA can be tuned to produce as many or as few principal

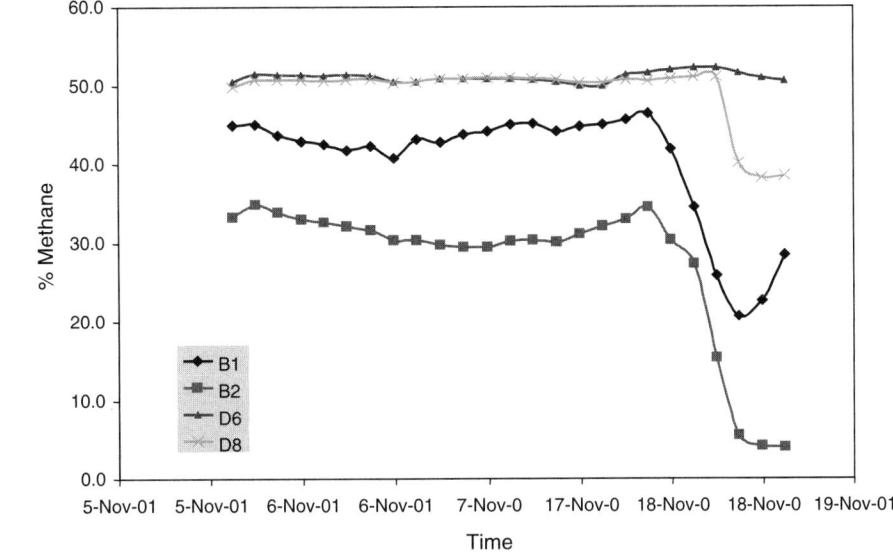

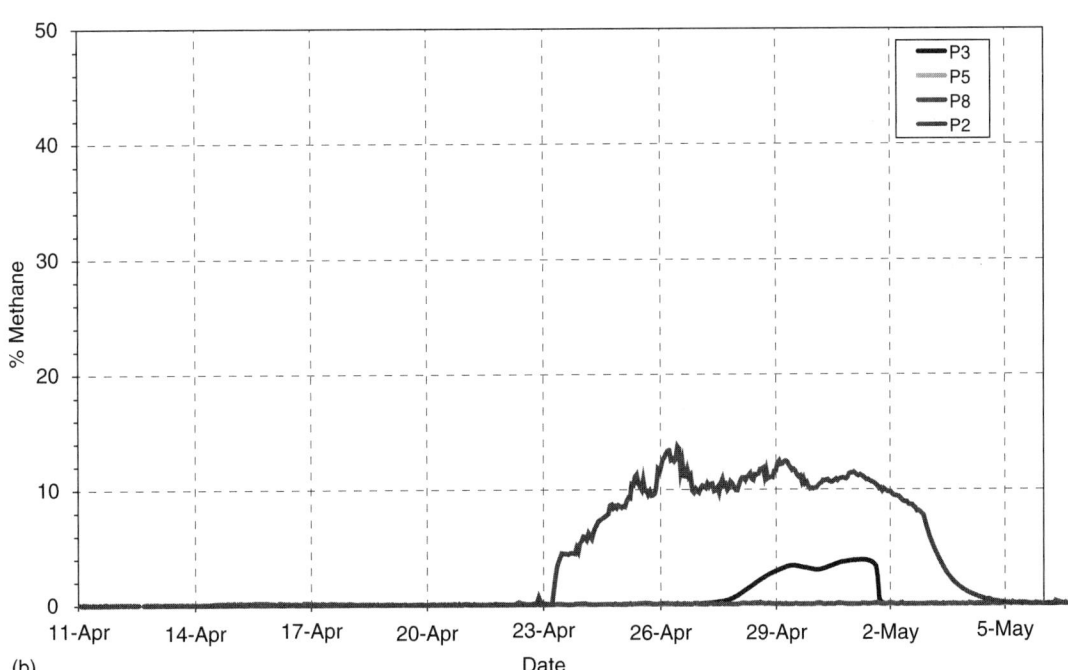

FIGURE 24.13 Some results from the installation in Fig. 24.12. (a) Methane generation at four selected well-head points. (b) Methane gas levels at test points around the site perimeter.

components as required, but selecting three reduces the analysis into a 3D readily visualized space. Figure 24.18 shows the results of one such PCA calculation clustering non–extra virgin oil and extra virgin oil from a particular region (Tuscany) in a total set of extra virgin oils. The extra virgin and non–extra virgin sets are clearly distinguished, and there is only a slight overlap in the regional discrimination.

More complete publications on this technique [15, 16] give extensive additional data verifying the discrimination capabilities that this approach to oil classification is capable of producing. This is an example of an important approach to chemical sensing using optical fibers. The role of the optical fiber is primarily as a collection aperture from the instrument and launch aperture into the spectrometer. It is a particularly simple robust and repeatable system.

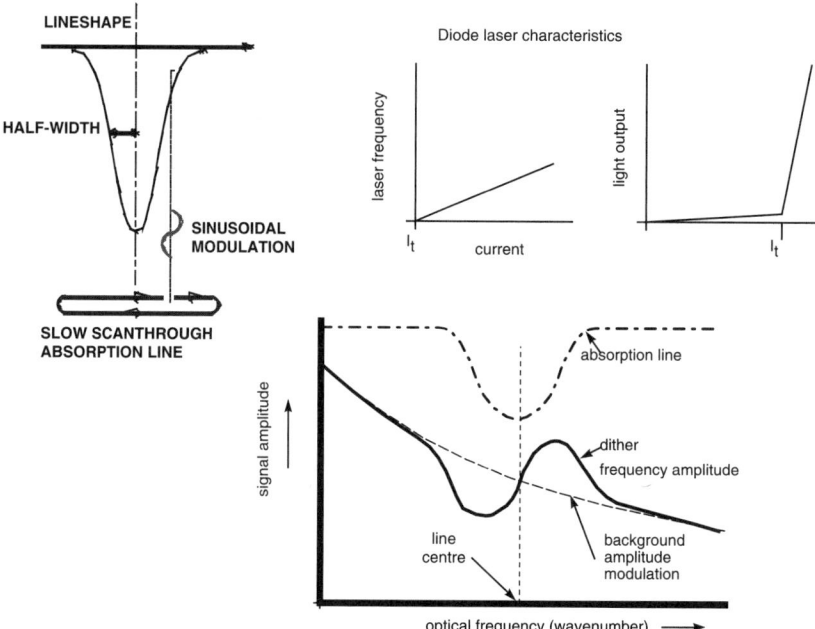

FIGURE 24.14 The basic principle of diode laser spectroscopy. The laser is current or thermally tuned through the absorption line at a relatively slow rate accompanied by high frequency dither. Typical tuning characteristics are shown at top right. The output after absorptions in gas is sketched at the lower right after filtering to select the dither frequency. The background is due to the nonlinearity of the current tuning process where the background amplitude modulation is due to the nonlinear converted to a monotonic dither frequency signal. The deviation from monotonic is used to extract gas concentration.

The fibers also provide fixed launch optics for the four sets of measurements and give simple and robust repeatable conditions. The measurements concerned are in the visible and cover the full visible spectral range. The direct color measurement shows considerable variations in the CIE (International

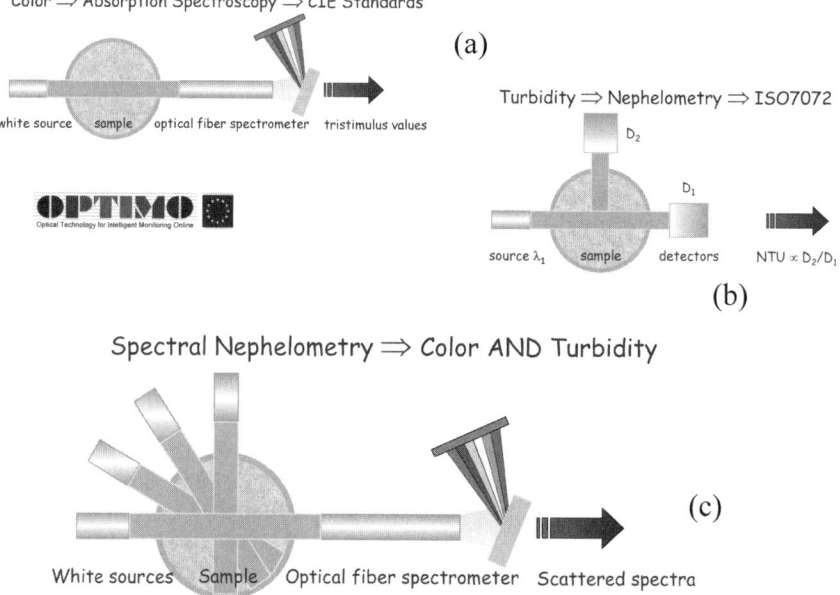

FIGURE 24.15 The concept of liquid sample classification using a color and turbidity measurements. Color (a) is a measure of optical absorption and therefore chemical content. Turbidity is a measure of particulate size and is dependent on the scatter size in wavelengths (b). Combining color and turbidity (c) gives a powerful signature assessment tool. (Courtesy of Anna Mignani, IFAC, CNR, Florence, Italy.)

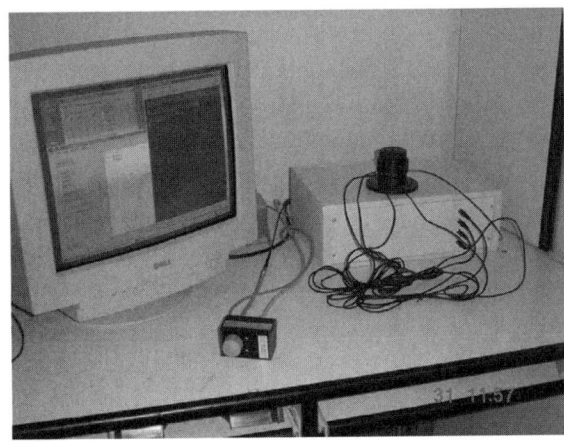
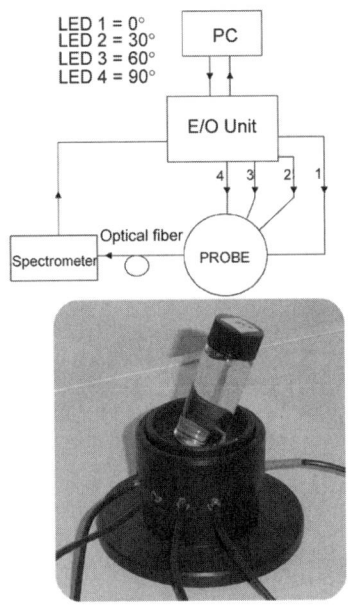

FIGURE 24.16 The apparatus used for olive oil measurements based on the principles in Fig. 24.15. Bottom right shows the scatter cell with a collection fiber (right-hand side) connected to the spectrometer and four white light-emitting diodes connected through input fibers at 0, 30, 60, and 90 degrees to the axis of the collection optics. (Courtesy of Anna Mignani, IFAC, CNR, Florence, Italy.)

Commission on Illumination) (and therefore subjective) color of the various samples. The color is a direct indicator of chemical composition through the absorption spectra and its detailed features are also influenced by scatter (turbidity) from the sample. Although we cannot implement a detailed inversion procedure that extracts individual components, we can take a fingerprint that is characteristic of a particular mix and that discriminates between different mixes, provided the fingerprint has sufficient dimension to it. This is somewhat analogous to the approach that the senses of taste and smell take to determining—in effect—the chemical composition of foods and drinks.

This then is chemical analysis without determining the exact chemistry. However, the chemical analysis does give a repeatable, reliable, and accurate indication of the parameters that are of interest to the final consumer. The PCA software is readily available through the commercial scientific mathematical packages and is relatively simple to use. This very powerful technique deserves greater exploration.

5.3. Distributed Chemical Sensing

Several approaches to chemical sensing using distributed architectures have been demonstrated and some, at least two, are progressing toward useful products. Both involve the use of chemically sensitive coatings, one as a chemical to optical transducer and the other as a chemical to mechanical transducer.

The chemical to optical transduction systems rely on either fluorescence or absorption to detect the presence of the analyte of interest. In both cases—as with all other chemical sensing media—the chemical preparation of the surface coating is critical to the success of the sensor. Typically, the analyte of interest introduces a change in the optical properties of the coating that is detected using time domain reflectometry backscatter. This change does, however, remain in the coating after the analyte has been removed. In many cases the coating chemistry incorporates a rewriting process through which a short high-intensity burst of ultraviolet radiation, which can be launched along the fiber at least over limited distances (many tens of meters), resets the chemistry into its original state. The benefit of this approach is the very wide range of analytes that can be addressed. The principle is finding its niche in security monitoring applications, particularly for alerting personnel to the presence of hazardous chemical or biochemical species [17].

Chemical to mechanical transduction requires a substantially different approach to the detection process [18]. In most of the chemical to mechanical transducers used to date, the transducer material expands in the presence of the analyte of interest. Two basic principles have been used. Gas selective catalysts, notably palladium, expand in the presence of the gas whose reactivity they enhance. This expansion is detected as a change in length of a

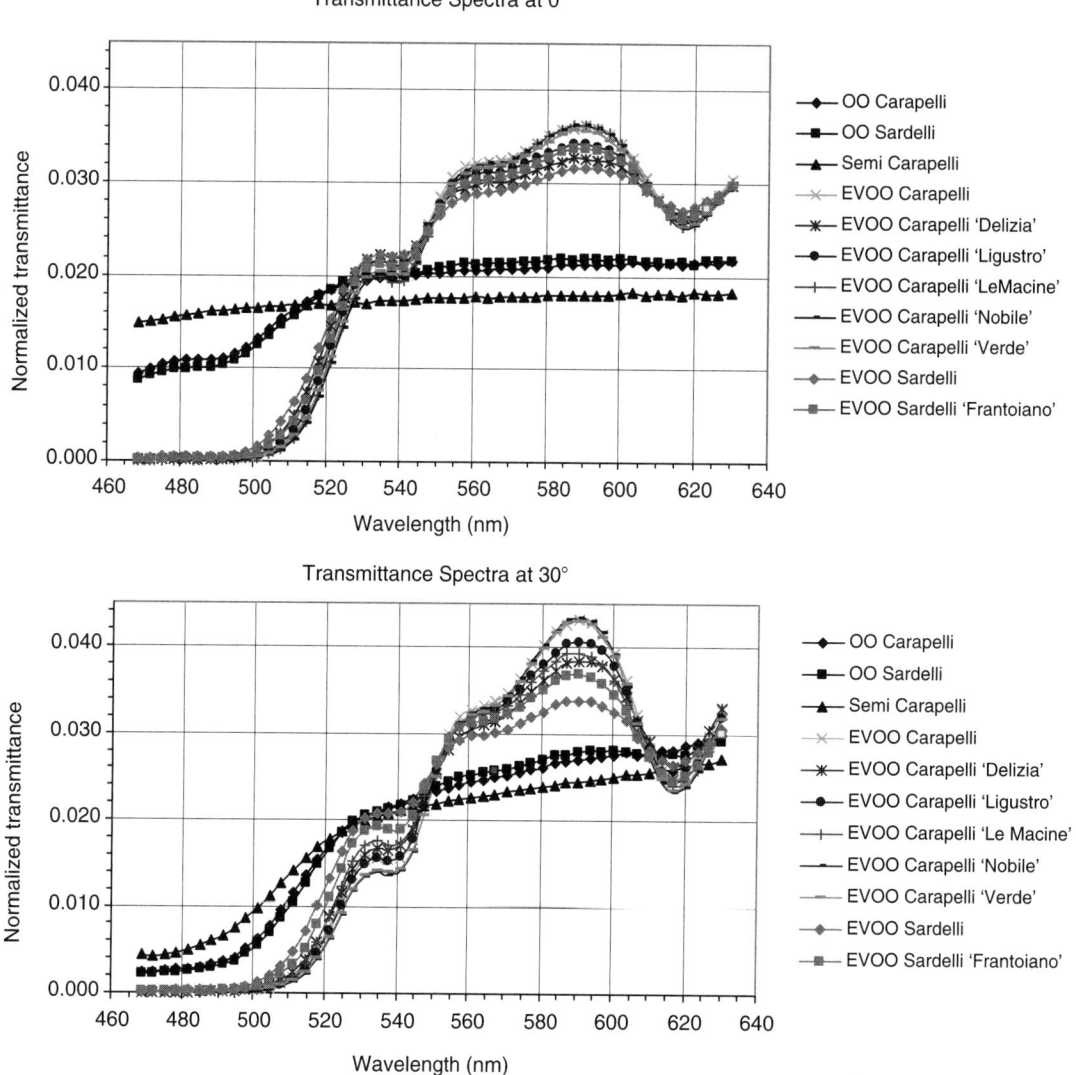

FIGURE 24.17 Two of four angular signatures (0 and 30 degrees) illustrating the spectral variation with wavelength and sample for 11 varieties of olive oil. (Courtesy of Anna Mignani, IFAC, CNR, Florence, Italy.)

coating; this in turn can be detected either interferometrically for small short transducer systems or by using Brillouin scattering in longer distributed networks. The short transducer system has been demonstrated, but to my knowledge the longer version, using Brillouin scattering as an interrogating principle, has not been evaluated. There could be at least two reasons for this. One is the obvious and considerable cost of coating very long lengths of fiber with consistent thicknesses of palladium. The second is the simple expense of the Brillouin scattering readout unit. These two factors together have inevitably discouraged any practical evaluation. Competitive systems based on pellistors, or indeed the point sensitive array described in Section 5.1, offer both greater simplicity and better performance.

The microbending loss mechanism illustrated in Fig. 24.19 is considerably different. It combines straightforward inexpensive polymer chemistry with straightforward inexpensive Rayleigh scattering-based optical time domain reflectometry as the interrogating mechanism. The system cost is then inherently far from daunting. The basic idea is to detect localized liquid spillage using this principle. The sensor cable has been demonstrated for both water and hydrocarbon fuels using different interface polymers [18]. The polymers reverse their state (i.e.,

Principles of Fiber Optic Sensors

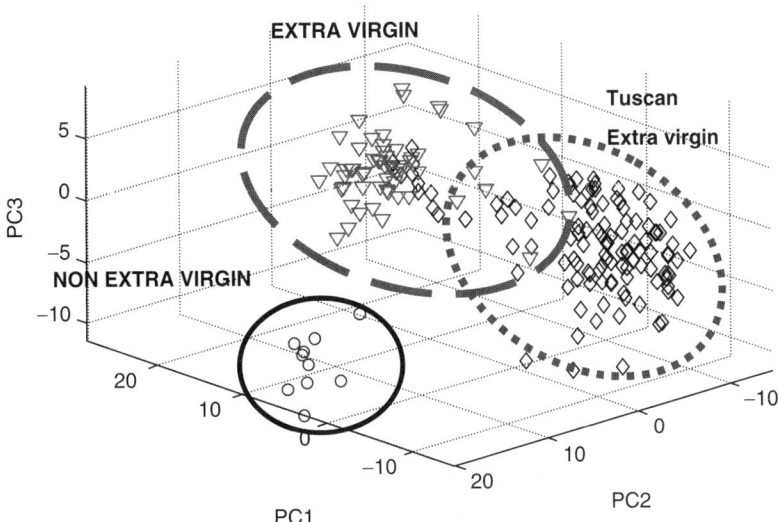

FIGURE 24.18 PCA cluster plots for extra virgin oils (upper left), non–extra virgin oils (bottom center), and Tuscan extra virgin oil (clustered to the right). (Courtesy of Anna Mignani, IFAC, CNR, Florence, Italy.)

shrink again) when the spillage has evaporated or otherwise disappeared so that resetting the sensor is not necessary. This sensor system is in the process of entering field trials for leak detection within the oil industry, and Fig. 24.20 illustrates just one of many possible applications, such as storage tank monitoring, in this sector. As environmental legislation becomes more onerous, the need for sensor systems of this nature will increase.

There are of course competing systems using different technologies. These are predominantly based on electrical interrogation using a dual conductor cable with insulating material designed to change its resistivity in contact with the liquid of interest. In effect, these systems produce a short circuit in the wet part. Additionally, some pipeline monitoring systems rely almost totally on visual inspection, particularly in more remote areas.

There are numerous benefits to using the fiber optic system compared with either of these approaches. The interrogation range of the fiber optic technology does facilitate continuous monitoring in remote areas, thereby both minimizing inspection costs and facilitating early warning to minimize environmental damage. In more local applications, the fiber optic system still has benefits. It is intrinsically safe, it responds much more quickly (seconds rather than many minutes) than the electrical insulator system, it resets after the leakage event has been corrected, and it can detect multiple events while the electrical system can only locate the nearest one. However, the electrical approach is much more

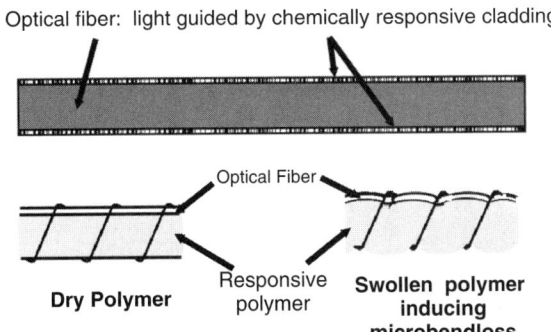

FIGURE 24.19 Distributed architectures for chemical sensing using (upper) chemically sensitive rewritable coatings and (lower) a chemical selective swelling gel.

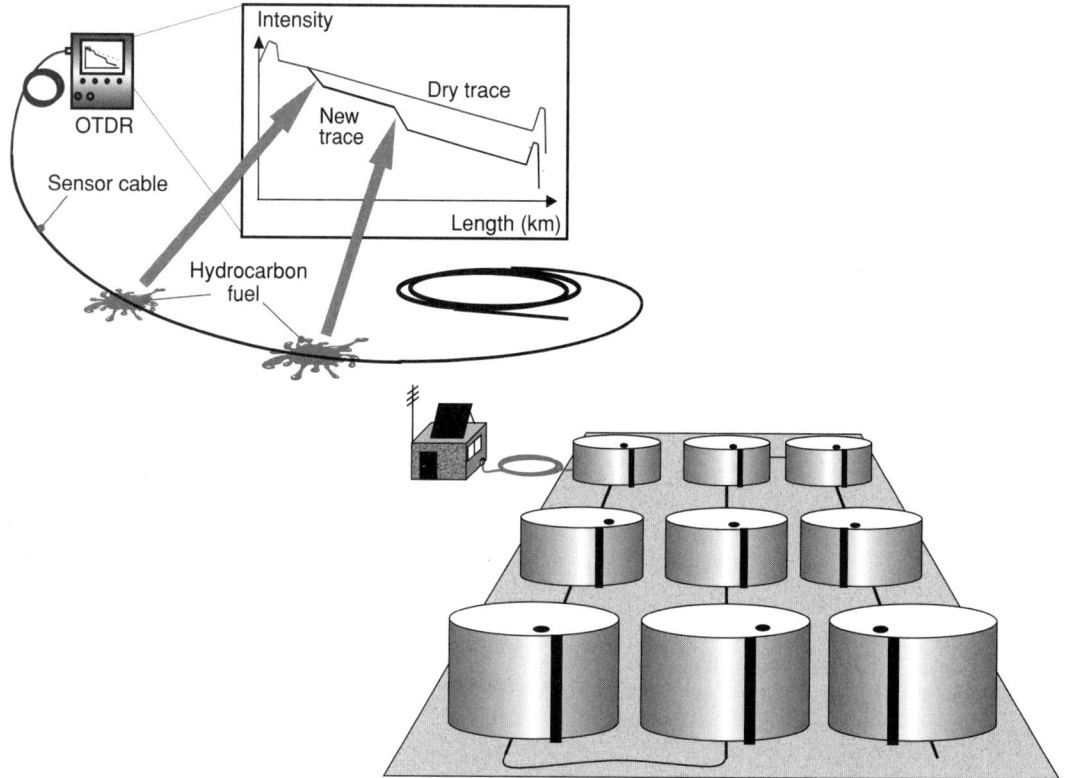

FIGURE 24.20 Leak detection applications for distributed oil or water measurement using a sensor based on a chemically selective gel.

mature and has established itself with its limitations in the marketplace. It is neither cost-competitive with nor offers the application flexibility of the fiber optic approach. The practicalities of the user community will inevitably mean that the initial applications for the fiber optic system will be in areas where the electrical system is either totally unfeasible or has proven to be unsatisfactory. The evolution will take some time.

6. CONCLUSIONS

This chapter has presented a very brief snapshot of optical fiber sensor technologies. Sensing and measurement is a multifaceted diverse science that adapts physical and chemical techniques to very specific measurement issues. Consequently, it is highly fragmented with very different measurement solutions adapted to different measurement environments and applications.

Optics—and fiber optics in particular—is an extremely versatile sensing and measurement technique, and this chapter gives some indication of this versatility. This is a small part of a very considerable body of work that describes many more applications and technologies in reports extending over the past 20 and more years. The OFS[†] and the Europtrode Conference Series are particularly authoritative and are recognized as the technology and science authorities. There are, however, many complementary reports in the applications literature pertinent to a particular industrial commercial or environmental sector.

Sensor technologies gradually find their fit into their application niches. Fiber optic sensing is no exception to this. Fiber sensors offer very substantial benefits. Sometimes these benefits are alien to the user's perception and require careful explanation and promotion. This is particularly true for techniques such as distributed sensing and passive multiplexed arrays. Progress is gradual, but we now see

---

[†]The OFS Conference series started in London in 1983. The first 17 presentations of the conference are compiled on a CD-ROM from SPIE, Bellingham (www.spie.org). OFS (13) (Korea) is available as SPIE vol. 3746; OFS (14) (Venice) as SPIE vol. 4185; and OFS (15) as IEEE Piscataway (ieee.org) catalogue number 02EX533. OFS (16) was in Nara, Japan (www.ee.t.u-tokyo.ac.jp/OFS-16/) in October 2003, and OFS (17) was held in Bruges, Belgium, in May 2005.

significant real application and the rapidly broadening appreciation of the potential that fiber optic sensors offer to the user. The technology will undoubtedly grow in importance.

7. REFERENCES

1. B. Culshaw, "Optical fiber sensor technologies: opportunities and perhaps pitfalls," *J. Lightwave Technol.*, **22**, 39–50 (2004).
2. B. Culshaw and J. P. Dakin (Eds.), *Optical Fiber Sensors*, Artech, Norwood, MA, **1–4**.
3. B. Culshaw, "Fiber optic sensing at Strathclyde: from ropes to eyes, " in *Proc. SPIE*, **4578**, Fiber Optic Technology and Applications, Boston, MA, pp. 74–88, (2001).
4. http://www.sensornet.co.uk
5. E. Udd, S. Calvert, and M. Kunzler, "Usage of fiber grating sensors to perform critical measurements of civil infrastructure," Technical Digest, pp. 496–499, OFS-16, Nara, Japan (2003).
6. R. Willsch, W. Ecke, and H. Bartelt, "Optical fiber grating sensor networks and their application in electric power facilities, aerospace and geotechnical engineering," in *15th Optical Fiber Sensors Conf. Tech. Dig. OFS 2002*, pp. 49–54 (2002).
7. D. Inaudi, "Testing performance and reliability of fiber optic sensing system for long-term monitoring," *Proceedings EWOFS'04*, Santander, Spain (2004).
8. D. Inaudi and D. Posenato, "Dynamic demodulation of long-gauge interferometric strain sensors," *Proceedings of SPIE*, **5384**, San Diego, California, USA (2004).
9. G. Duxbury, *Infrared Vibration-Rotation Spectroscopy: From Free Radicals to the Infrared Sky,* Wiley, New York (1999).
10. J. R. Lakowicz, *Principles of Fluorescent Spectroscopy*, Plenum, New York (1999).
11. A. Grazia Mignani, "Thoughts on the future for chemical sensing," *Proceedings EWOFS'04*, **5502**, p. 59, Santander, Spain (2004).
12. Y. Amao, K. Asai, and I. Okura, "Photofluourescence oxygen sensing using palladium tetrakis porphyrin self assembled membrane in aluminium," *Anal. Comm.*, **36**, 179–180 (1999).
13. X. Bevenot, A. Trouiller, C. Veillas, H. Gagnaire, and M. Clément, "Surface plasmon resonance sensor using an optical fiber," *Measure. Sci. Technol.*, **13**, 118–124 (2002).
14. B. Culshaw, W. Johnstone, A. Maclean, I. Mauchline, D. Moodie, and G. Stewart, "Large scale multiplexing a point sensor for methane gas detection," in *15th Optical Fiber Sensors Conf. Tech. Dig. OFS 2002*, pp. 261–264 (2002).
15. A. G. Mignani, P. R. Smith, L. Ciaccheri, A. Cimato, and G. Sani, "Spectral nephelometry for making extra virgin olive oil fingerprints," *Sensors Actuators B*, **90**, 157–162 (2003).
16. A. G. Mignani, L. Ciaccheri, A. Cimato, G. Sani, and P. R. Smith, "Absorption spectroscopy and multi-angle scattering measurements in the visible spectral range for the geographic classification of Italian extra virgin olive oils," *Proc. SPIE*, **5271B**, *Sensors and Systems for Agriculture and Plant Monitoring*, G. E. Meyer and B. S. Bennedsen (Eds.), pp. 285–288 (2004).
17. L. Cohen, D. Ruiz, W. Huang, and B. Lieberman, "Intrinsic chemical sensor fibers for extended-length chlorine detection," *Proc. SPIE*, **5589**, Optics East, Philadelphia (2004).
18. A. Maclean, C. Moran, W. Johnstone, B. Culshaw, D. Marsh, and P. Parker, "Detection of hydrocarbon fuel spills using a distributed fiber optic sensor," *Sensors Actuators A Phys.*, **109**, 60–67 (2003).

25

Structural Strain and Temperature Measurements Using Fiber Bragg Grating Sensors

Wei Jin,* T. K. Y. Lee, S. L. Ho, and H. L. Ho
Department of Electrical Engineering
The Hong Kong Polytechnic University
Hong Kong

K. T. Lau, L. M. Zhou, and Y. Zhou
Department of Mechanical Engineering
The Hong Kong Polytechnic University
Hong Kong

1. INTRODUCTION

Fibre Bragg grating (FBG) sensors have been the subject of intense research and development since the discovery of photosensitivity in optical fibers by Hill et al. [1]. An FBG is formed within the core of an optical fiber by introducing periodic variation in its refractive index profiles. The resonance (or Bragg) wavelength (λ_B) of the FBG is [2]

$$\lambda_B = 2n_{\text{eff}}\Lambda \quad (1)$$

where Λ is the periodicity (pitch) of the grating and n_{eff} is the effective refractive index of the fiber core. The Bragg wavelength shifts due to applied strain (ε) and temperature changes (T), and the shift is given by [2]

$$\frac{\Delta\lambda_B}{\lambda_B} = K_\varepsilon \varepsilon + K_T T \quad (2)$$

where the first term in Eq. (2) represents the strain effect on the fiber and the coefficient K_ε is determined by the physical elongation of the grating pitch and strain-optic coefficient of the fiber. The second term represents the effect of temperature on the grating and the coefficient K_T is determined by the thermal expansion coefficient and the thermal-optic coefficient of the fiber.

When FBG sensors are attached to the surface of or embedded within a mechanical structure, the values of K_ε and K_T are different from that of the bare fibers. They are affected (and often determined) by the mechanical and thermal properties of the structural materials and the properties of the fiber protective coatings and the materials used to bind the fiber with the structure [3].

Over the past two decades, significant progress has been made on the fabrication, packaging, interrogation, multiplexing, and applications of FBG sensors. Reviews on various aspects of FBG sensor technology can be found in articles by Kersey et al. [4], Othonos and Kalli [2], and Lau [5]. We have been involved in the development of FBG sensor technology and have worked on the improvement of measurement accuracy of FBG sensor arrays by using multiple gas absorption lines as absolute reference [6, 7] and by using various signal processing techniques and algorithms to enhance the signal-to-noise ratio of single [8–10] and multiplexed [11–16] multipoint FBG sensor systems. This chapter reports some of our recent work on applying FBG sensors for structural strain and temperature measurements, including (1) internal strain measurement of composite strengthened concrete bars, (2) dynamic strain measurement of composite materials by using embedded sensors, (3) simultaneous measurement of fluctuating and average temperature of structures in cross-flow, and (4) static and dynamic strain measurement of a train body-shell during static and dynamic load test. These applications make use of the unique features of optical fiber sensors such as small size, and hence these sensors are suitable for embedding, immune to electromagnetic interference, and have remote sensing and multiplexing capability. Some of the results reported here would be difficult to achieve using conventional electrical sensors.

---

*E-mail: eeurgin@foolgee.edu.uk

2. STRAIN MEASUREMENT IN A COMPOSITE-STRENGTHENED CONCRETE BAR

Experimental investigations were carried out to measure the internal strains of composite, concrete, and composite strengthened concrete structures [17–19]. Figure 25.1 shows an example where FBG sensors were used to monitor strains at three particular locations at the composite–concrete interfaces of a composite strengthened concrete beam specimen [18].

The concrete specimen is of a rectangular shape and has dimensions of 45 cm in length, 15 cm in width, and 15 cm in height. To simulate a crack in the concrete specimen, a notch was made in the specimen with a notch-to-width ratio of 0.2 that was filled with epoxy-based resin to avoid any environmental attack on the notch surfaces. The concrete surfaces were pretreated by sanding and cleaning, and three FBG sensors were attached to the surface of the concrete specimen by epoxy resin at three strategic points near the notch (i.e., positions 1, 2, and 3 as shown in Fig. 25.1). Glass fiber mats with epoxy-based (Araldite MY750) resin were used to form composites to strengthen the concrete beam by bonding on the three surfaces, as shown in Fig. 25.1. Six-layer fiber mats were laid up layer by layer directly on the surfaces of the concrete beam.

Figure 25.2 shows a schematic of the FBG sensor system used for the strain measurement. The three sensors (G_1, G_2, and G_3) are connected to the two arms of a fiber directional coupler and are multiplexed using an RF (radio frequency)-band frequency modulated continuous wave (FMCW) technique [16, 18]. Light from a broadband source modulated in intensity by a sawtooth (or triangular) swept frequency carrier generated from a voltage-controlled oscillator (VCO) was launched into the FBG sensor array. The intensity modulation can be performed by using an external modulator or by direct modulation of the driving current if a semiconductor source is used. The reflected signals from FBGs are guided back to a tunable optical filter (TOF) and a photodetector. These signals are subsequently mixed with a reference signal from the VCO. The mixing between the reference signal and the reflected signals produced beat notes with beat frequencies determined by the time delay differences between the reflected FBG signals and the reference signal. The magnitudes of the beat signals are proportional to the convolution of the individual grating reflection spectrum and the transmission spectrum of the TOF. The Bragg wavelength of a particular FBG can be interrogated by scanning the TOF and recording the control voltage of the TOF that corresponds to the peak of that particular beat note signal.

Figure 25.3 shows the strain measured from grating G3 (position 3) when the beam was subjected to

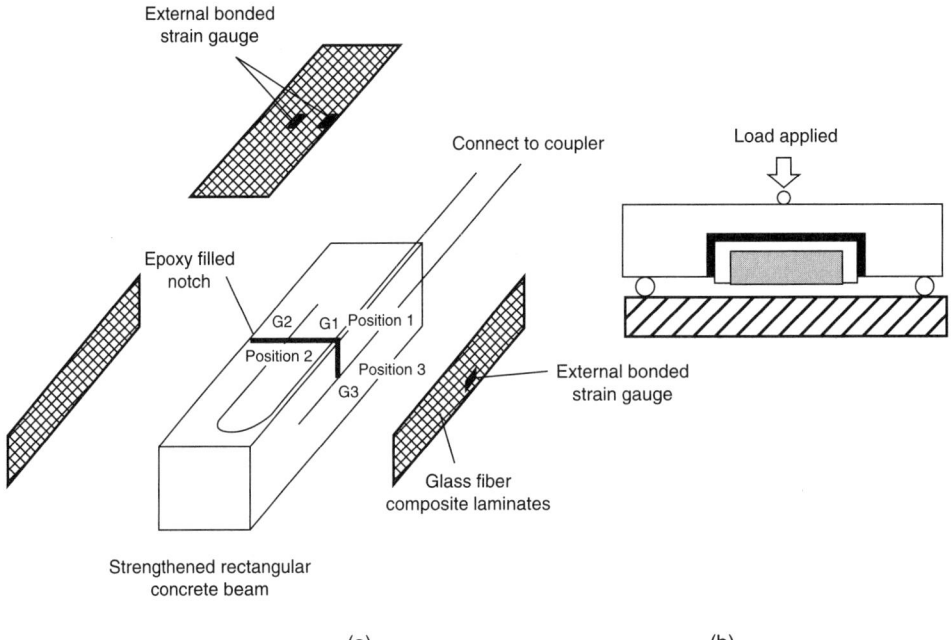

FIGURE 25.1 Schematic of the test specimen showing (a) the location of the FBG sensors and (b) the loading setup.

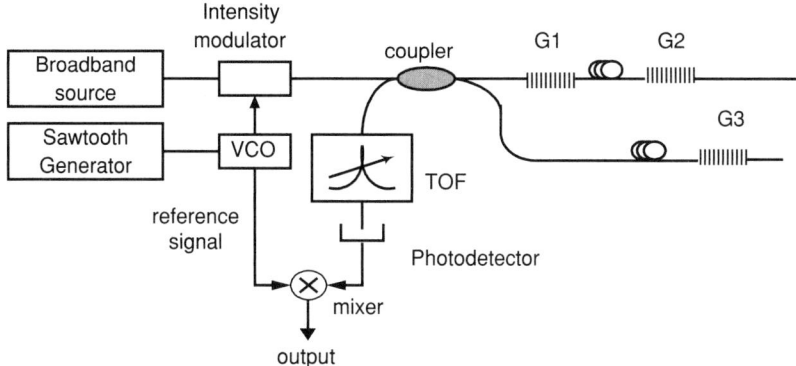

FIGURE 25.2 FMCW multiplexed FBG sensor array.

a three-point bending test (Fig. 25.1b). For comparison, the strain measured from a strain gauge attached on the surface of the reinforcements at the same location of the embedded grating G3 is also showed. The strain values measured from grating G3 and the electric strain gauge are very close to each other when the applied load is below 23 kN. When the load is beyond 23 kN, there is a discrepancy between the readings of G3 and from the electric gauge, indicating debonding has occurred between the composite and the concrete interface. The strain reading of G3 is reduced when the applied load is greater than 30 kN because of the appearance of microcracking at the concrete surface. This is however, not reflected in the reading from the surface mounted strain gauges. FBG sensors, due to their small size and corrosion resistant (compared with conventional electric strain gauge) characteristics, would be ideally suited for embedded applications that measure the internal strain of the composite strengthened concrete structures, which, as indicated from Fig. 25.3, may not be accurately measured by surface-mounted strain gauges.

3. DYNAMIC STRAIN MEASUREMENT OF A COMPOSITE SAMPLE USING AN EMBEDDED FBG SENSOR

Experiments were conducted to test the dynamic response of the embedded FBG sensors within composite materials [20]. The optical setup was based on the use of an edge filter [21] to convert strain-induced Bragg wavelength variation into light intensity change, which was detected by a photoreceiver and analyzed by a Bruel & Kjaer signal analyzer (DK-2850, Naerum, Denmark).

The sample composite beam was fabricated with eight layers of 0/90-degree E-glass fabric and epoxy resin. The dimensions of the beam were $300 \times 25 \times 2.3$ mm. An FBG sensor was embedded between the first and second layers from the top surface of the composite laminate during the lay-up process and was aligned along the beam's longitudinal axis

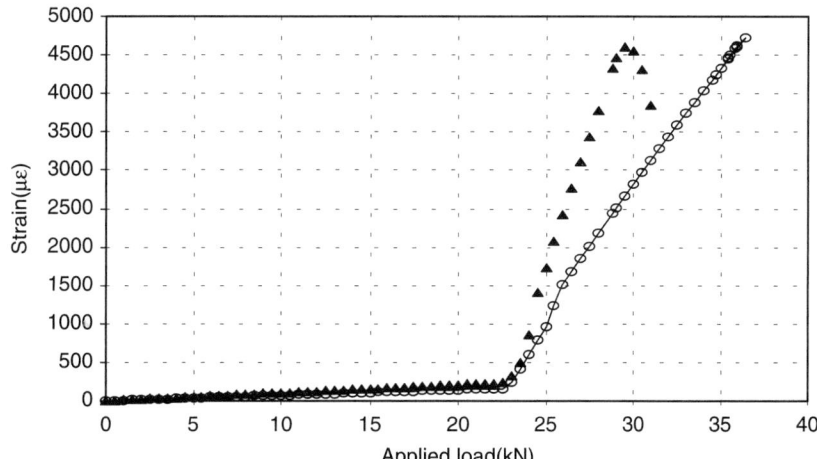

FIGURE 25.3 Strain measured by embedded FBG sensors (▲) and by strain gauges on the surface of the composite (○) at position 3.

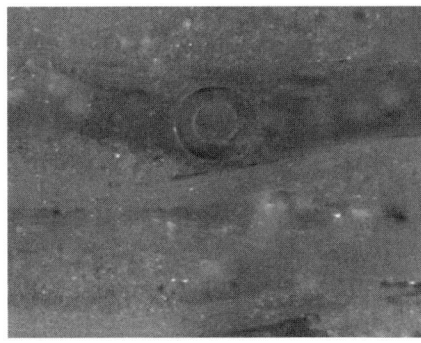
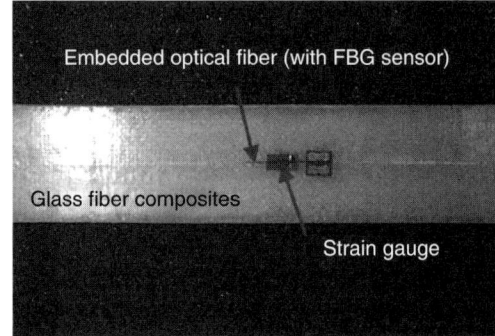

FIGURE 25.4 (a) The cross-sectional and (b) top view of the composite beam.

(Fig. 25.4a). The outer diameter and gauge length of the sensor are 250 μm and 10 mm, respectively. An externally mounted strain gauge was attached on the surface of the beam at the same position as where the FBG sensor was located, as shown in Fig. 25.4b. A previous study showed no significant influence on the mechanical strength of composites with an embedded optical fiber [17].

Both ends of the beam were securely clamped by two rigid mounting devices. An electric-driven shaker was used to produce external excitations to the beam to generate different out-of-plate vibration amplitudes and forces. A schematic illustration of the experiment setup is shown in Fig. 25.5.

To accurately measure the dynamic response of the beam, the location of the FBG sensor and strain gauge was selected to be far away from an expected nodal point, which can be determined using conventional vibration theory [22]. The dynamic responses measured from the strain gauge and from the FBG sensor are plotted in Fig. 25.6. It is obvious that both devices were able to measure the first natural frequency of the beam, $f_1 = 80$ Hz. The strain values measured from the FBG and the strain gauge agree well at frequencies below 100 Hz. Deviations occur when frequencies exceeded 100 Hz. The results from the FBG sensor clearly revealed the second natural frequency of the beam, $f_2 = 230$ Hz, which agrees with the results by Thomson [23]. It is, however, difficult to identify from the strain gauge results.

The discrepancy between the FBG and the strain gauge may be explained as follows. For a clamped-clamped structure, the longitudinal strain of the beam is correlated to its vibration amplitude and excitation frequency. Increasing the excitation frequency would result in a decrease in longitudinal strain in the beam. For frequencies above 100 Hz, the longitudinal strain may be approaching the noise level of the strain gauge, and thus erroneous measurements from the strain gauge resulted. This study provides important information on the feasibility of using embedded FBG sensors as vibration measurement devices to monitor the mechanical performance of composite structures.

4. TEMPERATURE MEASUREMENT ON A HEATED CYLINDER IN A CROSS-FLOW

Over the past few years, we have been applying FBG sensors to the study of flow-induced structural vibrations. FBG sensors, because of their very small dimension coupled with other advantages such as light weight, electrical isolation, and immunity to electromagnetic interference, are ideally suited for applications in wind tunnel environments and would

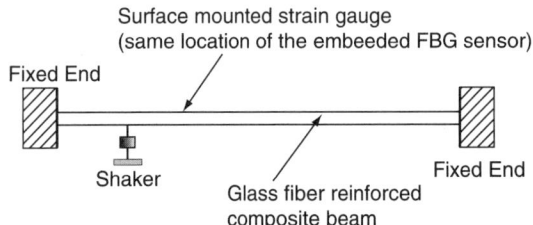

FIGURE 25.5 Experimental setup for dynamic strain measurement.

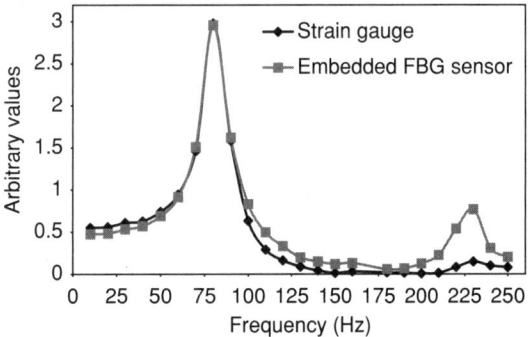

FIGURE 25.6 Dynamic responses measured by the FBG sensor and the strain gauge.

have negligible effects on the flow around test structures and on the vibration characteristics of the structures. We have developed FBG sensor systems for dynamic strain measurements based on either edge filtering [24, 25] or the interferometric demodulation principle [14]. These sensors have been successfully used to measure the dynamic strain of isolated cylinder and coupled cylinder arrays [25, 26] within a cross-flow.

This section reports our recent work on the use of FBG sensors to study the heat transfer characteristics of a heated cylinder in a cross-flow. Figure 25.7 shows the FBG sensor system used in our study. The system can measure static and fluctuating temperatures practically at the same time. The systems consists of a sensing FBG, a reference FBG, a broadband light source, a TOF, a fiber coupler, a photodetector, a low pass electrical filter, a high pass electrical filter, and the data acquisition and signal processing unit. Light from the broadband source first passes through the TOF and is then split into two by the fiber coupler, one to the sensing FBG and the other to the reference FBG. Light reflected from the two FBGs are combined by the same coupler and fed into the photodetector. The photodetector converts the optical signal to an electrical signal that is further processed to give two outputs, one (V_{DC}) for static temperature measurement and the other (V_{AC}) for dynamic measurement. A detailed description of the measurement system can be found in [27].

Experiments were conducted in a suction-type wind tunnel with a 0.5-m-long working section (0.35×0.35 m cross-section) [28]. A brass circular cylinder of diameter $d = 19.0$ mm was vertically mounted in the midplane of the working section, 20 cm from the exit plane of the contraction. The experimental arrangement is shown schematically in Fig. 25.8. The wake fluctuating velocity u was monitored by a single tungsten hot wire of 5 μm in diameter and 2.5 mm in length placed at $x/d = 2$ and $y/d = 1.5$, where x and y are the stream-wise and lateral coordinates, respectively, whose origins are chosen at the cylinder center. The hot wire was operated at an overheat ratio of 1.8 with a constant temperature anemometer.

A groove of about 125 μm deep was made along the cylinder span to lay an optical fiber with a diameter of 125 μm. The fiber, flush with the cylinder surface using heat-conducting silicone, was built with an FBG sensor. The sensor, located at the midspan of the cylinder, measured the static temperature $\overline{\theta_s}$ and the fluctuating temperature θ_s on the cylinder surface. The sensor grating has a finite length of about 10 mm ($\approx 0.5\,d$). The measured temperature represents an average value over this length. Considering that the vortex cell is characterized by a typical

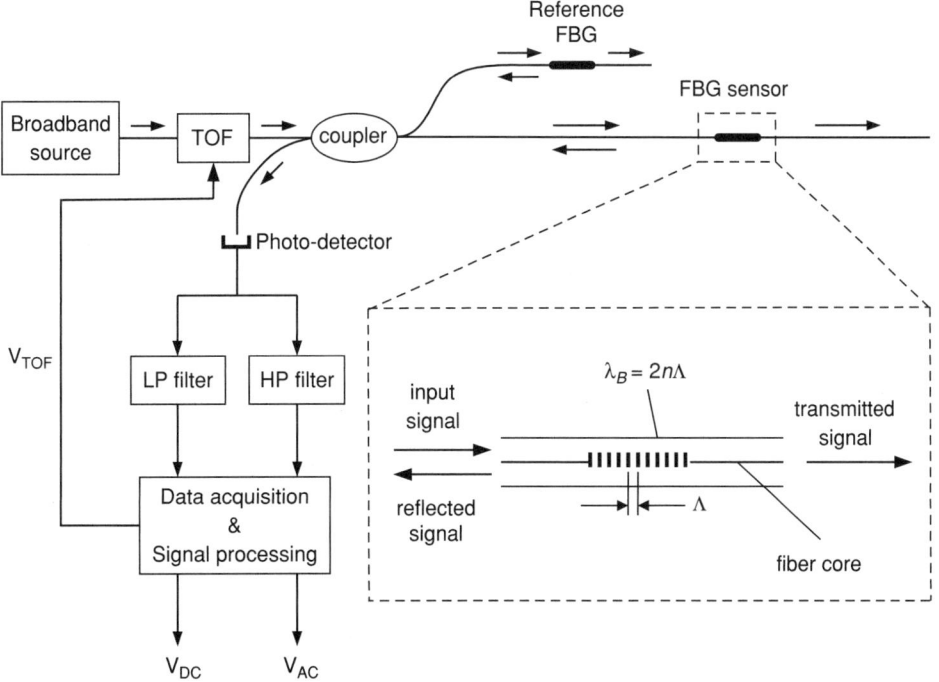

FIGURE 25.7 FBG sensor system for simultaneous measurement of static and fluctuating temperature. LP and HP stand for low pass and high pass, respectively.

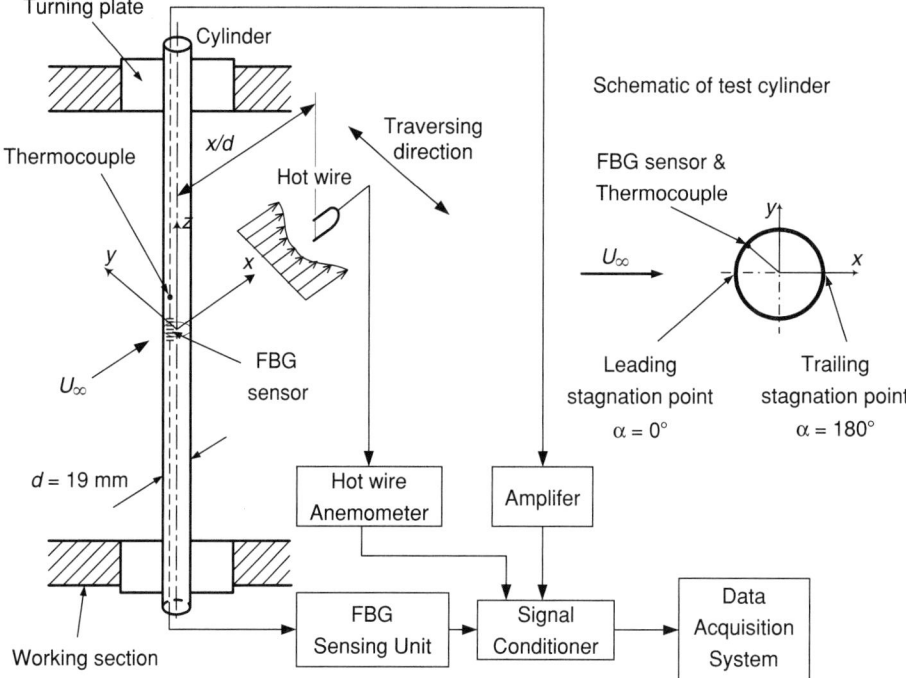

FIGURE 25.8 Experimental arrangement for static and fluctuating temperature measurements.

span-wise extent of $1-3d$ [29, 30], the temperature along the sensor grating should be statistically identical and can be considered to be the local temperature at the midspan of the cylinder. By rotating the cylinder, the circumferential distribution of temperature was measured. Assuming a symmetrical distribution about the x-axis, $\overline{\theta}_s$ and θ_s were measured from $\alpha = 0$ to 180 degrees only, where $\alpha = 0$ and 180 degrees correspond to the leading and trailing stagnation points, respectively. To validate the FBG sensor measurement, $\overline{\theta}_s$ was simultaneously measured using a type-K thermocouple of 0.3-mm diameter placed at the same α as the FBG sensor but 0.02 m($\approx 1d$) away in the span-wise direction. At such small span-wise distance, say $1d$, the static surface temperature captured by the thermocouple should be the same as that by the FBG sensor [29, 30].

Figure 25.9 shows the measured static (average) temperature normalized by the overall mean surface temperature, $\Theta = \frac{1}{n}\sum_{i=1}^{n}\overline{\theta}_s(\alpha_i)$, where $\overline{\theta}_s(\alpha_i)$ represents the measured local static surface temperature

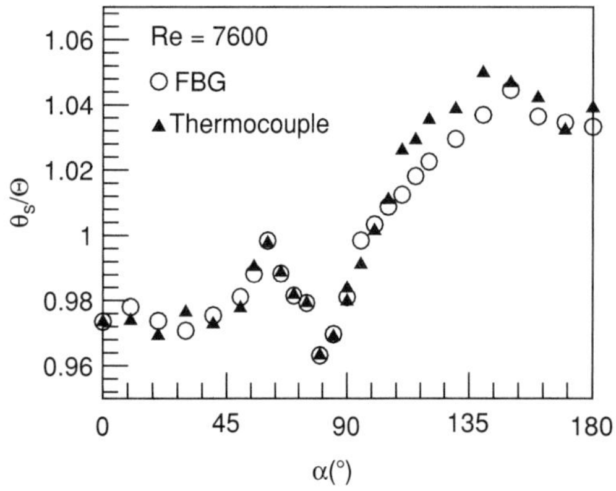

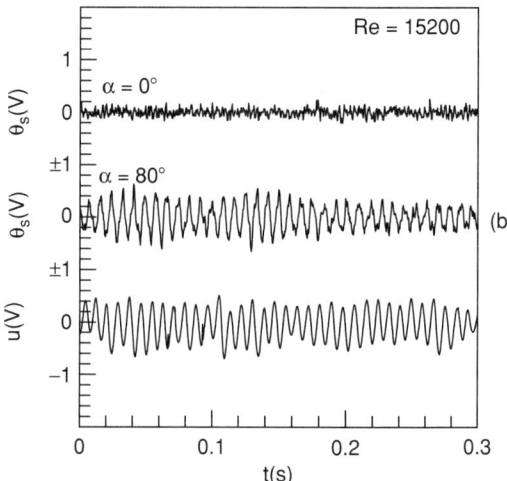

FIGURE 25.9 Circumferential distributions of the local static surface temperature $\overline{\theta}_s/\Theta$ measured using the FBG sensor (○) and a type-K thermocouple (▲).

FIGURE 25.10 Time histories of fluctuating velocity u and fluctuating temperature θ_s at $\alpha = 0$ and 80 degrees. (Time $t = 0$ is arbitrary.)

and n is the total number of $\overline{\theta}_s$ measured around the cylinder surface. The Reynolds number is $Re = 7600$. The results from the thermocouple $\overline{\theta}_s/\Theta$ are also shown in the figure. The two techniques show a good agreement, thus providing a validation for the FBG sensor measurement. The circumferential distribution of $\overline{\theta}_s/\Theta$ is qualitatively consistent with the reported Nusselt number (Nu) data [30, 31].

Figure 25.10 show the time histories of θ_s at $\alpha = 0$ (upper trace) and $\alpha = 80$ degrees (middle trace), where the minimum $\overline{\theta}_s/\Theta$ occurs, along with the simultaneously measured u (lower trace), for a typical $Re = 15{,}200$. The θ_s signal at $\alpha = 80$ degrees exhibits a quasi-periodic fluctuation, which is also evident in the u signal, apparently due to vortex shedding. The θ_s signal at $\alpha = 0$ degrees is quite different from that at $\alpha = 80$ degrees. A pseudoperiodic fluctuation is also evident, but its magnitude is much smaller than that at $\alpha = 80$ degrees.

The test cylinder, fix supported at both ends, may vibrate due to vortex excitation forces, which produces a structural dynamic strain [25, 27]. This would cause a shift $\Delta\lambda_B$ in λ_B and may not be separated from the temperature-induced wavelength shift. Consequently, the measured fluctuating temperature signal might be disturbed. It is therefore important to ensure a negligible contamination due to the residual strain effect. A test was thus conducted at conditions identical to those described for the fluctuating temperature measurements, except that the cylinder was unheated. Figure 25.11 presents the signal (the same scale as that in Fig. 25.10) recorded in the test by the FBG sensor at $Re = 15{,}200$ and $\alpha = 80$ degrees, along with the velocity signal u.

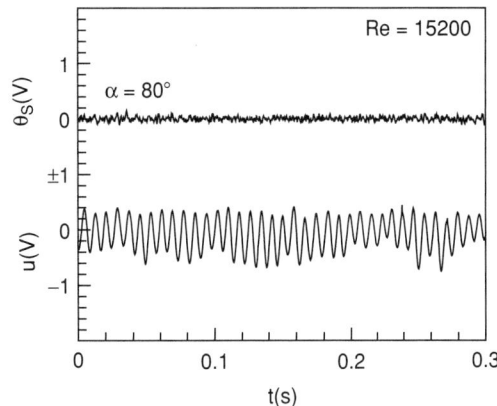

FIGURE 25.11 Time histories of the stream-wise fluctuating velocity u and fluctuating temperature θ_s at $\alpha = 80$ degrees. The cylinder was unheated.

Evidently, the signal is largely due to the structural vibration and background noise. Its amplitude is greatly reduced, compared with that at the same Re and α when the cylinder was heated (Fig. 25.10). These results point to a negligible effect of the structural vibration on the temperature measurement.

The investigation leads to the following conclusions:

1. The local static surface temperature $\overline{\theta}_s/\Theta$ measured using the FBG sensor is in good agreement with that simultaneously obtained by a K-type thermocouple. Furthermore, the circumferential distribution of $\overline{\theta}_s/\Theta$ is qualitatively consistent with the reported Nu data. The results suggest that the FBG sensor can provide reliable measurements for the static temperature on the cylinder surface.
2. The dynamic response of the FBG sensor is excellent. The fluctuating temperature θ_s is closely

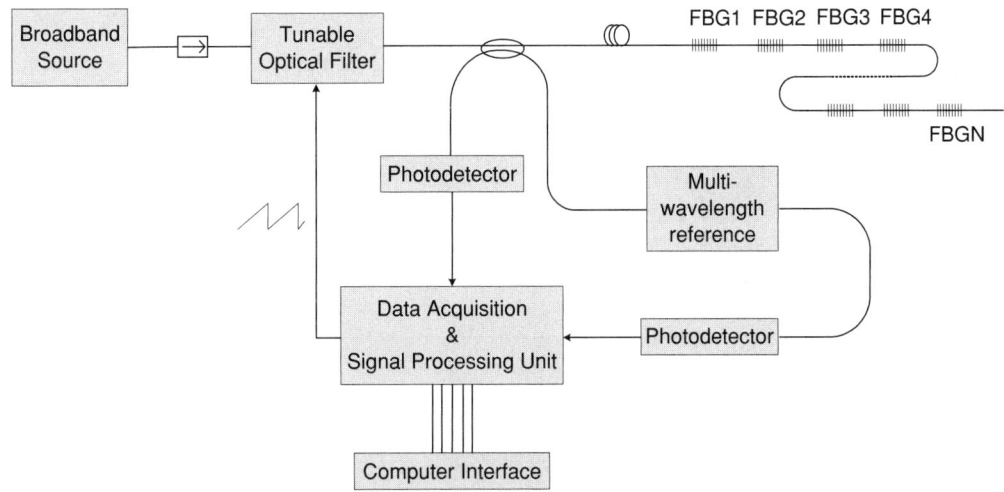

FIGURE 25.12 Multipoint FBG sensing system.

correlated to the hot-wire measurement. These results indicate that the FBG sensor is excellent in resolving the fluctuating surface temperature of a cylinder in a cross-flow.

5. MULTI-POINT STRAIN MEASUREMENT OF KOWLOON CANTON RAILWAY TRAIN BODY SHELL

We have also applied FBG sensing technology to the measurement of strain and temperature of *Kowloon Canton Railway* (KCR) train body shells. A schematic of the FBG system used is shown in Fig. 25.12. The system uses a broadband source and a tunable optical filter to detect the peak wavelength of each FBG within the array [4]. A multiwavelength reference, which can either be a fixed Fabry-Perot comb filter [32] or a gas with multiple absorption lines [7], is used to minimize the nonlinearity and nonrepeatability of the tunable optical filter. The strain measurement resolution of the system is on the order of 1 $\mu\varepsilon$ with a sampling frequency of better than 5 Hz. The system allows about 10 FBG sensors of different wavelengths to be connected serially that are interrogated practically at the same time.

Six FBG sensors were installed on a trailer car and the same number of sensors was installed on the motor car of a KCR train. For the trailer car, three of them were positioned at the corners of the window frame, as shown in Fig. 25.13, with one at the bottom steel bar and the other two at the top surface of the car. For the motor car, four of six FBGs were installed at the four corners of a window frame (see Fig. 25.15). The other two were located

(a)

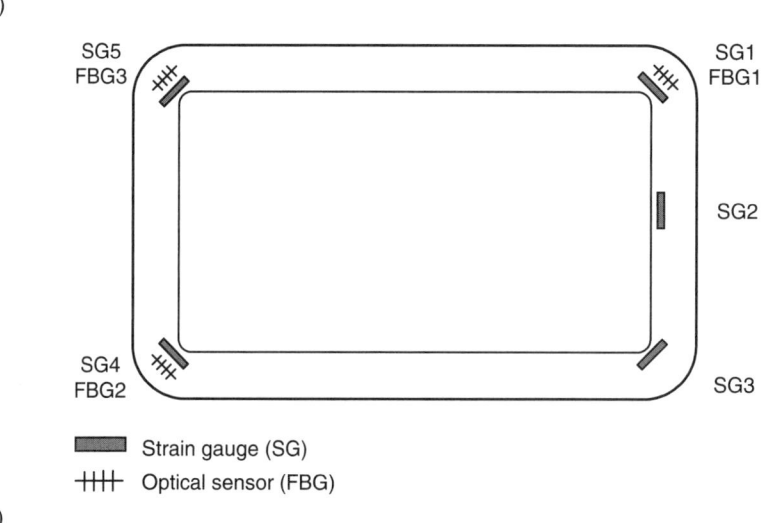

(b)

FIGURE 25.13 (a) Locations of sensors on the trailer car window frame. (b) Schematic showing the details of the sensor installation.

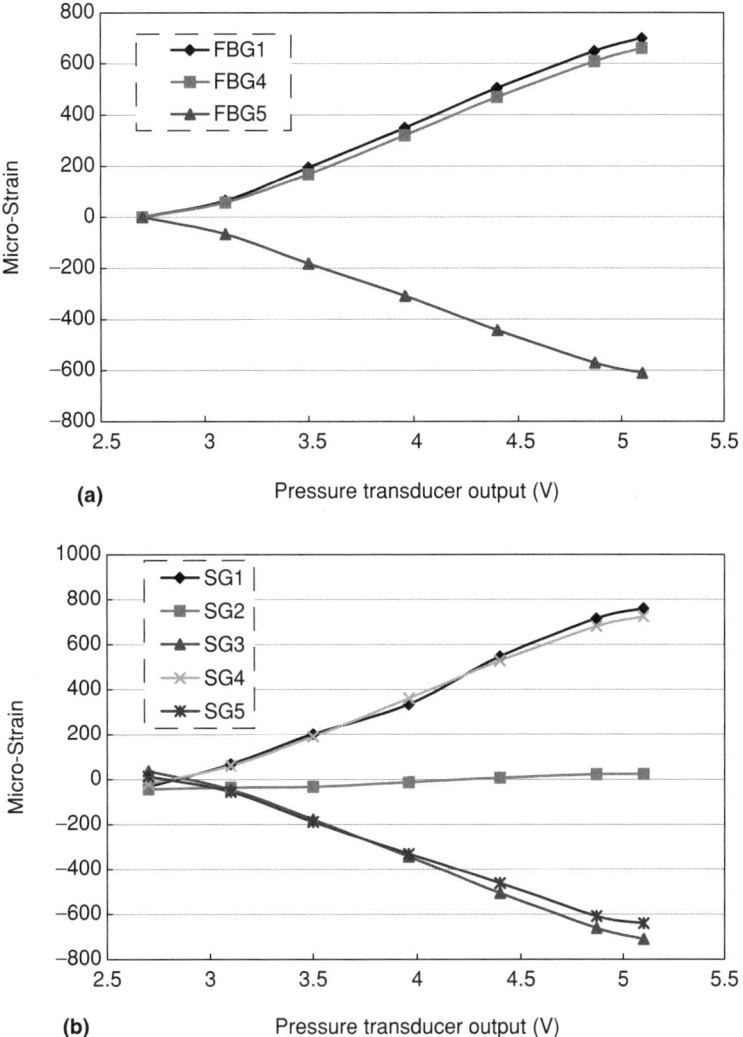

FIGURE 25.14 Results of static load test. (a) FBG sensors; (b) electric strain gauges.

at the top surface and the bottom steel bar, respectively. A similar number of electrical strain gauges were located in close proximity to the FBG sensors, as shown in Figs. 25.13 and 25.15. Figure 25.14 shows the strains measured from the sensors installed on the window frame of the trailer car under the static load test. The load was applied by laying sandbags evenly on the seats and the floor of the trailer car. The horizontal axis in Fig. 25.14 is the output from the pressure transducers located at the bottom of the car and is observed to be proportional to the total load applied to the car. The results from FBG sensors agree well with the electrical strain gauges.

Figure 25.16 shows the dynamic strains measured from FBG and the strain gauges installed on the window frame of the motor car, when the train was running from the *Lo Wu* (time, 0 s) to the *Hung Hom* (time, 2500 s) station. The results from the two types of sensors, for most of the parts, agree with each other with maximum changes of strains occurring at about 2000 s (at *Kowloon Tong* Station

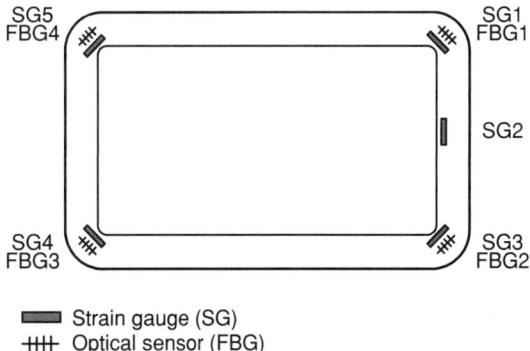

FIGURE 25.15 Schematic showing the installation of the sensors on the motor car window frame.

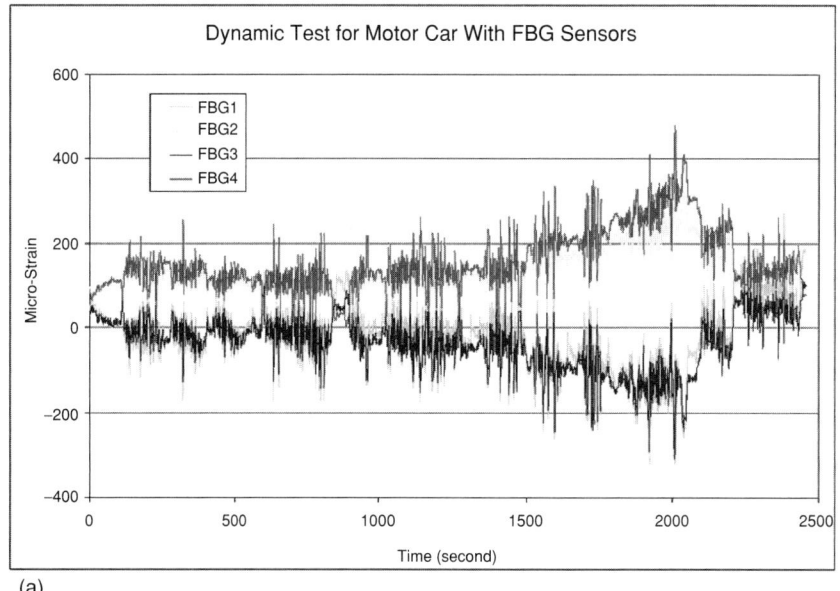

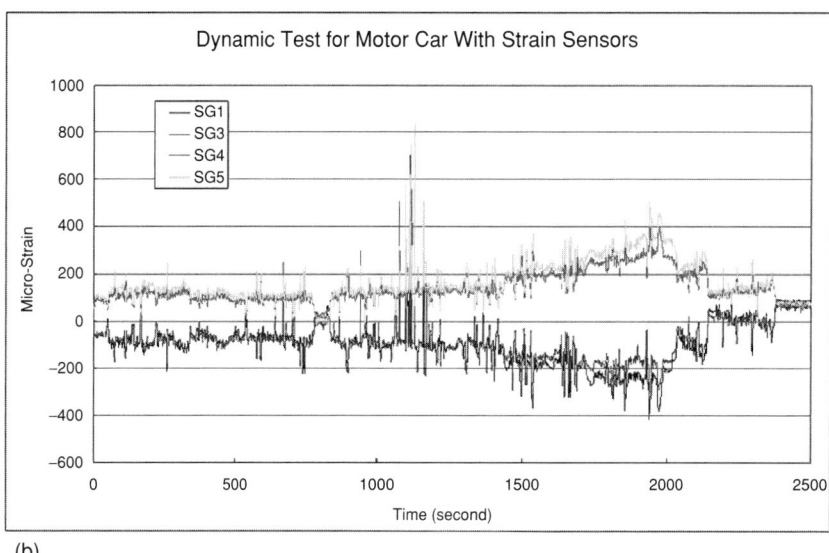

FIGURE 25.16 Results of dynamic load test. (a) FBG sensors; (b) electrical strain gauges.

where many people got off from the car) and the strains were reduced to small values when people alighted from the train at *Lo Wu* and *Hung Hong* terminals. However, there are two noticable differences between the results from the FBG and the electrical strain gauges. The first is that the high frequency dynamic strain due to the movement and vibration of car is more clearly picked up by the FBG sensors than by the strain gauges. This is because that the FBG system has a higher sampling frequency compared with the strain gauge system (the sampling time of the FBG system was about 0.2 s whereas the strain system was about 1 s). The other difference is the large output signals from the strain gauges at around 1100 s (in between *Tai Po Market* and *University* stations). The reason for this is that the train is electrically driven with a 25 kV high voltage overhead line. The cable changes phase at that particular location and this could cause interference to the readings of the electrical strain gauges. These are, however, not reflected at all at the FBG sensor outputs, indicating superior performance of the FBG sensors because of their immunity to electromagnetic interference.

6. SUMMARY

Some of our recent work on the applications of FBG sensors for structural strain and temperature measurements are reviewed. These include the measurement of strain at the strategic locations of composite strengthened concrete structures, internal dynamic strain measurement of composite materials, static and fluctuating temperature measurements of heated cylinders in a cross-flow, and real-time measurement of static and dynamic strains of an electric train body shell. It was shown that for the applications discussed in this chapter, the FBG sensors in general give better results than their electric counterparts, because of their unique features such as small size and ruggedness and are thus suitable for embedding into structures or attaching to small structures. The FBG sensors are of course immune to electromagnetic interference and have multiplexing and remote sensing capability.

7. ACKNOWLEDGMENTS

The work described in this chapter was supported in part by a grant from the Research Grant Council of the Hong Kong Special Administrative Region (Project No. PolyU 5105/99E) and several Hong Kong Polytechnic University Grants. We thank the KCRC for allowing us to publish some of the measurement data on the KCR train body shell.

8. REFERENCES

1. K. O. Hill, Y. Fujii, D. C. Johnson, and B. S. Kawasaki, "Photosensitivity in optical fibre waveguides: application to reflection filter fabrication," *Appl. Phys. Lett.*, **32**, 647–649 (1978).
2. A. Othonos and K. Kalli, "*Fiber Bragg Gratings: Fundamentals and Applications in Telecommunications and Sensing,*" Artech House, Norwood, pp. 301–396, (1999).
3. L. B. Yuan, "Fibre optic white light interferometric sensors for structural monitoring," Ph.D thesis, Department of Mechanical Engineering, the Hong Kong Polytechnic University (2003).
4. A. D. Kersey, M. A. Davis, H. J. Patric, M. LeBlanc, K. P. Koo, C. G. Askins, M. A. Putnam, and E. J. Friebele, "Fiber grating sensors," *J. Lightwave Technol.*, **15**, 1442–1462 (1997).
5. K. T. Lau, "Fibre-optic sensors and smart composites for concrete applications," *Mag. Concrete Res.*, **55**, 19–34 (2003).
6. C. C. Chan, W. Jin, H. L. Ho, D. N. Wang, and Y. Wang, "Improvement of measurement accuracy of fibre Bragg grating sensor systems by use of gas absorption lines as multi-wavelength references," *Electron. Lett.*, **37**, 742–743 (2001).
7. W. Jin, C. C. Chan, and H. L. Ho, "Fibre Bragg grating sensors," Chinese patent no. ZL0120900.X, (2003).
8. J. M. Gong, J. M. K. MacAlpine, C. C. Chan, and W. Jin, "A novel wavelength detection technique for fiber Bragg grating sensors," *IEEE Photon. Tech.*, **14**, 678–680 (2002).
9. J. M. Gong, C. C. Chan, and W. Jin, "Enhancement of wavelength detection accuracy in fiber Bragg grating sensors by using a spectrum correlation technique," *Opt. Commun.*, **212**, 29–33 (2002).
10. C. C. Chan, J. M. Gong, C. Z. Shi, and W. Jin, "Improving measurement accuracy of fiber Bragg grating sensors using digital matched filter," *Sensor Actuat. A. Phys.*, **104**, 19–24 (2003).
11. C. C. Chan, C. Z. Shi, J. M. Gong, and W. Jin, "Enhancement of the measurement range of FBG sensors in a WDM network using a minimum variance shift technique coupled with amplitude-wavelength dual coding," *Opt. Commun.*, **215**, 289–294 (2003).
12. C. Z. Shi, C. C. Chan, and W. Jin, "Improving the performance of a FBG sensor network using a genetic algorithm," *Sensor. Actuat. A Phys.*, **107**, 57–61 (2003).
13. C. C. Chan, C. Z. Shi, and W. Jin, "Improving the wavelength detection accuracy of FBG sensors using an ADALINE network," *IEEE Photon. Tech.*, **15**, 1126–1128 (2003).
14. C.Z. Shi, C. C. Chan, M. Zhang, and W. Jin, "Simultaneous interrogation of multiple fiber Bragg grating sensors for dynamic strain measurements," *J. Optoelectron. Adv.*, **4**, 937–941 (2002).
15. C. C. Chan, W. Jin, H. L. Ho, and M. S. Demokan. "Performance analysis of a time-division-multiplexed fiber Bragg grating sensor array by use of a tunable laser source," *IEEE J. Sel. Top. Quant.*, **6**, 741–749 (2000).
16. P. K. C. Chan, W. Jin, and M. S. Demokan "FMCW multiplexing of fiber Bragg grating sensors," *IEEE J. Sel. Top. Quant.*, **6**, 756–763 (2000).
17. K. T. Lau, C. C. Chan, L. M. Zhou, and W. Jin, "Strain monitoring in composite-strengthened concrete structures using optical fibre sensors," *Compos. Part. B. Eng.*, **32**, 33–45 (2001).
18. P. K. Chan, W. Jin, K. T. Lau, and M. S. Demokan, "Multi-point strain measurement of composite-bonded concrete materials with a RF-band FMCW multiplexed FBG sensor array," *Sensor. Actuat. A Phys.*, **87**, 19–25 (2000).
19. K. T. Lau, L. M. Zhou, C. H. Zhou, C. H. Woo, K. C. Chan, and W. Jin, "Strengthening and monitoring concrete structures using glassfibre composites and

19. fibre-optic Bragg grating sensor," *Materials Science Research International*, **5**, 216–221 (1999).
20. H.-Y. Ling, K. T. Lau, L. Cheng, W. Jin, R. S. Thomson, and M. L. Scott, "Embedded FBG sensor for dyamic strain measurement for a clamped-clamped composite structure," *Key Engineering Materials*, **295**, 21–26 (2005).
21. S. M. Melle, K. Liu, and R.M. Measures, "Practical fiber-optic Bragg grating strain gauge system," *Appl. Opt.*, **32**, 3601–3609 (1993).
22. L. Meirovitch, *Elements of Vibration Analysis*, McGraw-Hill International Edition, Singapore (1986).
23. W. T. Thomson, *Theory of Vibration with Applications*, Prentice Hall, USA (1993).
24. W. Jin, Y. Zhou, P. K. C. Chan, and H. G. Xu, "A fibre-optic grating sensor for the study of flow-induced vibrations," *Sensors Actuators*, **79**, 36–45 (2000).
25. Y. Zhou, R. M. C. So, W. Jin, H. G. Xu, and P. K. C. Chan, "Dynamic strain measurements of a circular cylinder in a cross flow using a fibre Bragg grating sensor," *Exp. Fluid*, **27**, 359–367 (1999).
26. Y. Zhou, Z. J. Wang, R. M. C. So, S. J. Xu, and W. Jin, "Free vibrations of two side-by-side cylinders in a cross flow," *J. Fluid Mech.*, **443**, 197–229 (2001).
27. H. L. Ho, W. Jin, C. C. Chan, Y. Zhou, and X. W. Wang, "A fiber Bragg grating sensor for static and dynamic measurands," *Sens. Actuat. A Phys.*, **96**, 21–24 (2002).
28. Z. J. Wang, Y. Zhou, X. W. Wang, and W. Jin, "A fiber-optic Bragg grating sensor for simultaneous static and dynamic temperature measurement on a heated cylinder in cross-flow," *Int. J. Heat Mass. Tran.*, **46**, 2983–2992 (2003).
29. H. Higuchi, H. J. Kim, and C. Farell, "On flow separation and reattachment around a circular cylinder at critical Reynolds numbers," *J. Fluid Mech.*, **200**, 149–171 (1989).
30. R. King, "A review of vortex shedding research and its application," *Ocean Eng.*, **4**, 141–171 (1977).
31. J. P. Holman, *Heat Transfer*, 8th ed. McGraw-Hill Companies, New York, pp. 248–255 and 303 (1997).
32. C. Miller, T. Li, J. Miller, F. Bao, and K. Hsu, "Multiplexed fiber gratings enhance mechanical sensing," *Laser Focus World*, 119–123 (1998).

26

CHAPTER

Principles and Status of Actively Researched Optical Fiber Sensors

Byoungho Lee,* Yong Wook Lee
School of Electrical Engineering
Seoul National University
Seoul, Korea

Minho Song
Division of Electronics & Information Engineering
Chonbuk National University
Jeonju, Korea

1. INTRODUCTION

The study of optical fiber sensors has a history of some 30 years [1]. Different ideas have been proposed and diverse techniques developed for various measurands and applications. Many of the techniques have found commercial success. Optical fiber sensors have many advantages, such as immunity to electromagnetic interference, light weight, small size, high sensitivity, large bandwidth, ease in signal light transmission, and geometric versatility in that fiber sensors can be configured into arbitrary shapes. However, in many fields of application, optical fiber sensors compete with other more mature technologies, such as electronic measurements. To appeal to users already accustomed to those mature technologies, the superiority of optical fiber sensors over other techniques needs to be clearly demonstrated. In some cases, for items such as optical gyroscopes and optical current sensors, optical fiber sensors should compete with their sibling technologies, for example, optical bulk sensors. Even with these difficulties, considerable effort has been made to advance optical fiber sensors, and many of these are now quite mature.

There have been excellent review books and articles on optical fiber sensors [2–7]. Because Chapter 24 reviews the overall optical fiber sensor technologies, only the current status of some of the actively studied, or well-developed, optical fiber sensors are reviewed in this chapter. Figure 26.1a shows the distribution of papers presented at the 16th International Conference on Optical Fiber Sensors (OFS-16, held in Nara, Japan in October 2003) according to the measurands of interest. The OFS is a major conference in the field of optical fiber sensors. Hence, the statistics of these conferences provide information on recent research trends in the area of optical fiber sensors. The two most highly discussed measurands were temperature and strain, the same as in the OFS-12, held 7 years ago [8]. Figure 26.1b shows the technologies involved in the optical fiber sensors presented at OFS-16. Fiber grating sensors are clearly the most widely studied topic.

In Section 2, fiber grating sensor technology, which is the most popular topic in optical fiber sensors, is reviewed. In Sections 3 and 4, two rather mature topics—fiber optic gyroscopes (FOGs) and fiber optic current sensors, respectively—are reviewed. In Section 5, other sensors are briefly discussed, followed by some concluding remarks.

2. FIBER GRATING SENSORS

The fact that gratings can be written in optical fibers was found accidentally by Hill et al. in 1978 [9]. An interesting history regarding fiber grating can be found in [10]. After Meltz et al. devised a controllable and effective method (the side-writing method) for fiber grating fabrication in 1989 [11], fiber gratings and their applications came to be researched intensively. Fiber gratings have been applied to add/drop filters, dispersion compensators, fiber lasers, amplifier gain flattening filters, and so on for optical communications [5]. Extensive studies have also been performed on fiber grating sensors, some of which have been commercialized.

---

*E-mail: byoungho@snu.ac.kr

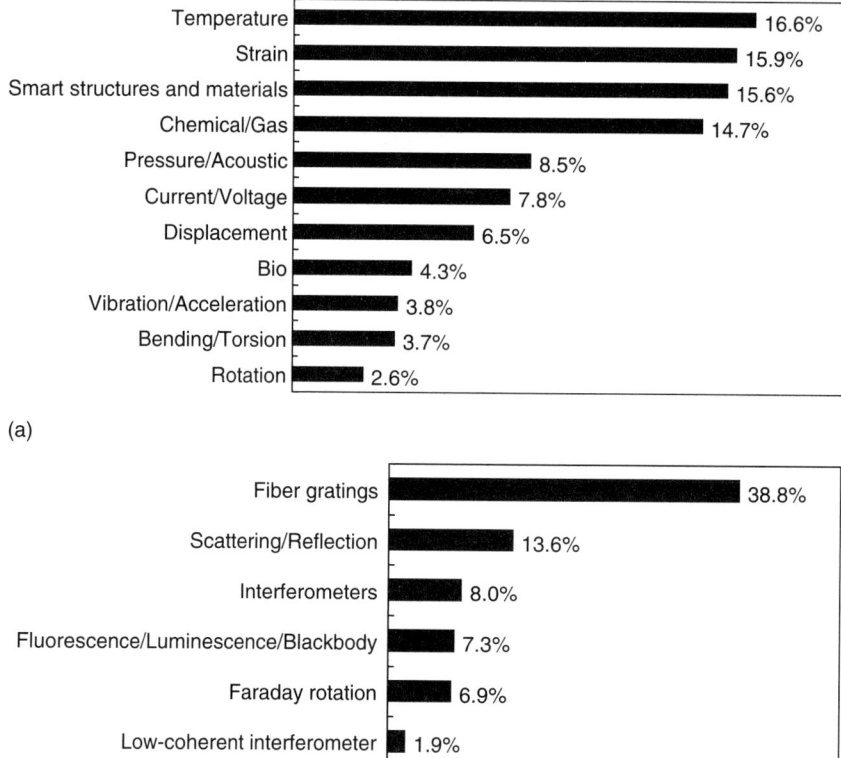

FIGURE 26.1 Distribution of OFS-16 papers according to (a) measurands and (b) technologies. Papers not directly related to measurands, such as those that discuss fiber grating sensor interrogators, multiplexing, and light sources, are not included in these statistics. If one paper deals with more than one measurand (or technology), the paper count is equally divided among the measurands (or technologies). Special session papers on the sensing technologies for smart structures and materials in OFS-16 are included in the statistics and separately notated in (a). Those papers mostly deal with strain or temperature sensors.

Figure 26.2 shows types of fiber gratings and their typical transmission or reflection spectra. Under phase matching conditions, a fiber Bragg grating (FBG) couples the forward propagating core mode to the backward propagating core mode, as can be seen in the propagation constant diagram, Fig. 26.3 [12]. A long period fiber grating (LPG) can couple the forward propagating core mode to one or more of the forward propagating cladding modes, as can also be seen in Fig. 26.3. Here the core mode means a mode of optical field that is confined mostly within the core, and the cladding modes are those that have considerable optical fields within the cladding region of optical fiber. The cladding modes are poorly guided within fiber over distance because they are very sensitive to fiber bending and to the boundary conditions between the cladding and fiber jacket. A chirped fiber grating has a wider reflection spectrum, and each wavelength component is reflected at different positions, which results in a delay time difference for different reflected wavelengths. A tilted fiber grating or blazed grating can couple the forward propagating core mode to the backward propagating core mode and a backward propagating cladding mode or some radiation modes that escape from the optical fiber. In the reflection spectrum of a tilted grating, although the reflected core mode spectrum appears, the backward coupled cladding mode spectrum and the radiation mode spectrum do not appear because the cladding and radiation modes are not guided with distance. In the transmission spectrum of a tilted grating, all dips coming from the backward core mode coupling (main dip), the backward cladding mode coupling (side dip), and the radiation mode coupling (ripple dips) appear as in Fig. 26.2d. A sampled fiber grating can reflect

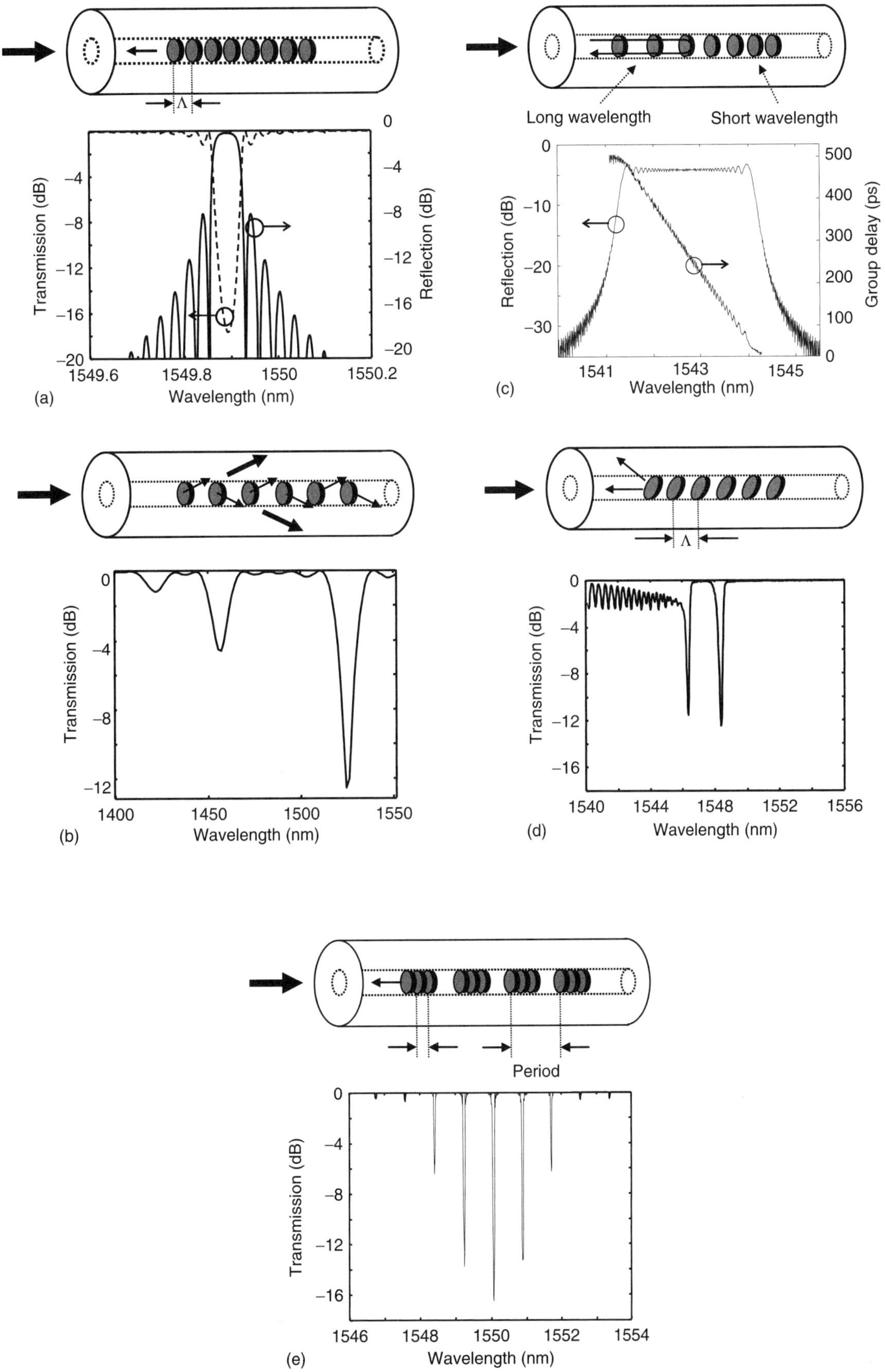

FIGURE 26.2 Types and example spectra of fiber gratings: (a) fiber Bragg grating, (b) long-period fiber grating, (c) chirped fiber grating, (d) tilted fiber grating, and (e) sampled fiber grating.

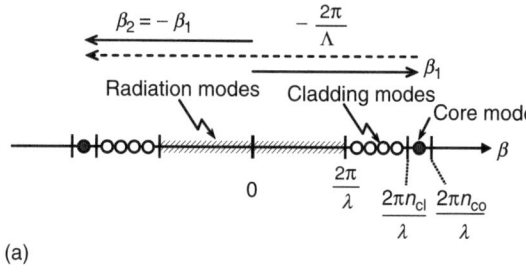

(a)

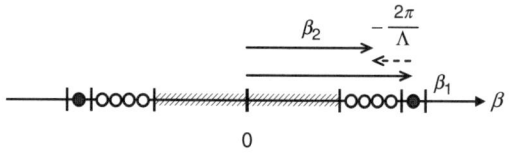

(b)

FIGURE 26.3 The phase matching diagram for (a) a fiber Bragg grating and (b) a long-period grating [12].

several wavelength components with equal wavelength spacing. Their reflectances can be made equal by further applying apodization to the sampled grating. The theory of fiber grating spectra can be found in [5, 12, 13]. All the above types of gratings have been used in various types of fiber grating sensors and wavelength change interrogators. Of these, however, FBGs are the most widely used as sensor heads.

In FBGs, the Bragg wavelength λ_B, or the wavelength of the light that is reflected, is given by

$$\lambda_B = 2n_{\text{eff}}\Lambda \quad (1)$$

where n_{eff} is the effective refractive index of the fiber core and Λ is the grating period. In Eq. (1) it can be seen that the Bragg wavelength is changed with a change in the grating period or the effective refractive index. The former is used for strain and the latter for temperature variation. The grating period can also be changed with temperature variation, but near room temperature the effect of temperature on the refractive index is approximately one order of magnitude larger than that of thermal expansion (or contraction). The typical response of the Bragg wavelength shift to strain is $\sim 0.64\,\text{pm}/\mu\varepsilon$ ($\mu\varepsilon$ = microstrain) near a Bragg wavelength of 830 nm, $\sim 1\,\text{pm}/\mu\varepsilon$ near 1300 nm, and $\sim 1.2\,\text{pm}/\mu\varepsilon$ near 1550 nm [14]. The unit of strain is ε, which is read as strain. It is a relative concept (dimensionless), that is, if a 1-m-long fiber is elongated by 1 μm, the strain is $1\,\mu\text{m}/1\,\text{m} = 1\,\mu\varepsilon$. The typical temperature response of FBGs is $\sim 6.8\,\text{pm}/^\circ\text{C}$ near 830 nm, $\sim 10\,\text{pm}/^\circ\text{C}$ near 1300 nm, and $\sim 13\,\text{pm}/^\circ\text{C}$ near 1550 nm [14], although the values depend on the FBG type [15].

The effect of strain and temperature can be explained by considering the change in Eq. (1) as follows:

$$\Delta\lambda_B \approx 2\left(n_{\text{eff}}\frac{\partial \Lambda}{\partial l} + \Lambda\frac{\partial n_{\text{eff}}}{\partial l}\right)\Delta l + 2\left(\Lambda\frac{\partial n_{\text{eff}}}{\partial T} + n_{\text{eff}}\frac{\partial \Lambda}{\partial T}\right)\Delta T \quad (2)$$

where Δl is the change in FBG length and ΔT is the change in temperature.

The first term in Eq. (2) represents the strain effect on an optical fiber. This corresponds to a change in the grating spacing and the strain-induced change in the refractive index. The above strain effect term may be expressed as (assuming $\Delta T = 0$) [5]:

$$\Delta\lambda_B = \lambda_B(1 - p_e)\varepsilon_z, \quad (3)$$

where $\varepsilon_z = \Delta l/l$ is the strain along the fiber axis and p_e is an effective strain-optic constant defined as

$$p_e = \frac{n_{\text{eff}}^2}{2}[p_{12} - \nu(p_{11} + p_{12})] \quad (4)$$

where p_{11} and p_{12} are components of the strain-optic tensor and ν is Poisson's ratio. For a typical germanosilicate optical fiber, $p_{11} = 0.113$, $p_{12} = 0.252$, $\nu = 0.16$, and $n_{\text{eff}} = 1.482$. Using these parameters in the above equations, the anticipated strain sensitivity at wavelengths near to 1550 nm is a 1.2-pm change as a result of applying $1\,\mu\varepsilon$ to the Bragg grating as described above.

The second term in Eq. (2) represents the effect of temperature change on an optical fiber. The temperature change results in changes in the refractive index and the grating spacing. The wavelength shift for a temperature change ΔT can be written as (assuming $\varepsilon_z = 0$) [5]:

$$\Delta\lambda_B = \lambda_B(\alpha_n + \alpha_\Lambda)\Delta T \quad (5)$$

where $\alpha_n = (1/n_{\text{eff}})(\partial n_{\text{eff}}/\partial T)$ represents the thermo-optic coefficient, which is approximately equal to $8.6 \times 10^{-6}/\text{K}$ for the germania-doped silica-core fiber and $\alpha_\Lambda = (1/\Lambda)(\partial \Lambda/\partial T)$ is the thermal expansion coefficient for the fiber (approximately $0.55 \times 10^{-6}/^\circ\text{K}$ for silica). Clearly, for the temperature change, the index change is by far the more dominant effect compared with the thermal expansion of the grating, as stated above.

The wavelength-encoded measurand information is a unique characteristic of FBGs. In addition to the common advantages of fiber sensors, this wavelength-encoded characteristic provides robustness to noise or power fluctuation and enables wavelength division multiplexing (WDM). Therefore, multipoint sensors can be realized using this technique. Sometimes WDM is combined with spatial division multiplexing, time division multiplexing, and code division multiple access techniques. However, the multipoint FBG sensors are not fully distributed sensors, such as backscattering sensors or optical time-domain reflectometers (OTDRs), because physical parameters can be measured only at the FBG positions.

As described above, FBGs respond to both strain and temperature. Therefore, their effects need to be separated to measure each physical parameter or to measure both simultaneously. Considerable effort has been focused on this topic, and various solutions have been proposed. A straightforward and very practical approach is to use a reference grating [16]. Another FBG (the reference grating) that is isolated from one parameter, for example, strain, is placed near the sensor FBG. The reference grating can be on the same fiber as the sensor FBG [17]. Another method is to use two FBGs with very different Bragg wavelengths (in general, two light sources are needed) that show different responses to the same measurands [18]. FBGs written on different-diameter fibers have also been proposed that give different strain responses, whereas the temperature responses are the same [19, 20]. Conversely, by inscribing FBGs on germanosilicate and boron co-doped germanosilicate fibers [21] or on high birefringence fiber [22], different temperature sensitivities can be obtained, whereras strain sensitivities remain the same. In addition, the simultaneous measurement of strain and temperature based on a superstructure FBG has been reported. This was accomplished by measuring the transmitted intensity and wavelength at one of several narrow-band loss peaks (situated on the slope of a broad-band loss peak) [23]. Recently, an LPG inscribed on a high birefringence fiber was proposed; it exhibits different responses of two adjacent resonant dips to the same measurand (strain or temperature) [24].

Demodulators or interrogators are required for FBG sensors. Their role is to extract measurand information from the light signals affected by the sensor heads. The measurand is typically encoded in the form of a Bragg wavelength change, and therefore the interrogators are typically expected to read the wavelength shift and provide measurand data. Optical spectrum analyzers are not suitable for real sensor systems because they are expensive and their wavelength scanning speed is too slow. Various techniques have been developed for the interrogators [13]. Some are quite simple but are more limited in measurement resolution, dynamic range, or multiplexing, and some are more complicated and provide better resolution but are more expensive or need stabilization. Table 26.1 summarizes interrogator types and Table 26.2 summarizes their performances [25–76]. Many of these are laboratory-scale experimental results. Civic structures such as bridges require the measurement of dynamic strain. When the dynamic signal is measured, the minimum detectable

TABLE 26.1 Fiber grating sensor interrogator types.

Types	Technologies	References
Passive detection schemes	Linearly wavelength-dependent device	25–29
	CCD spectrometer	30–34
	Power detection	35–39
	Arrayed waveguide grating	40
	Identical chirped grating pair	41
	Chirped grating-based Sagnac interferometer	42
Active detection schemes	Fabry-Perot filter	43–45
	Unbalanced Mach-Zehnder interferometer	46–56
	Fiber Fourier transform spectrometer	57, 58
	Acousto-optic tunable filter	59–61
	Matched FBG pair	62–65
	Michelson interferometer	66
	LPG pair interferometer	67
	Modulation phase shift detection	68
Other schemes	Wavelength-tunable source	69–71
	Mode-locked fiber laser with wavelength-time conversion	72
	Optical code division multiple access correlator	73, 74
	Frequency modulation	75
	Intragrating sensing	76

TABLE 26.2 Strain measurement resolution of some fiber grating sensor systems/interrogators.

Static/quasi-static resolution	Dynamic resolution	References (measurement range)
±20 µε		72 (>3500 µε)
~12 µε		57
±8.04 µε		70 (±1000 µε)
4.12 µε		62
3 µε		63
±2 µε		44 (±1300 µε)
±2 µε		39 (1500 µε)
1.9 µε		69
~0.6 µε		34 (160 µε)
~0.4 µε		40
±4.2 µε	0.406 µε/$\sqrt{Hz}$	42 (>500 µε)
Approximately ±3.5 µε	1.5 µε/$\sqrt{Hz}$	26 (~16,000 µε)
3.5 µε	6 nε/$\sqrt{Hz}$	52 (2 kHz)
Approximately ±3 µε	0.5 µε/$\sqrt{Hz}$	28 (>1050 µε)
±0.25 µε	5 nε/$\sqrt{Hz}$	33
40 nε	7 pε/$\sqrt{Hz}$	67
	42 nε/$\sqrt{Hz}$	71
	17 nε/$\sqrt{Hz}$	56
	10 nε/$\sqrt{Hz}$	55
	10 nε/$\sqrt{Hz}$	64 (more than ±100 µε)
	2 nε/$\sqrt{Hz}$	47 (>10 Hz)
	1.5 nε/$\sqrt{Hz}$	54 (10–2000 Hz)
	~7 fε/$\sqrt{Hz}$	48

strain is determined by the background noise level. The magnitude of noise changes with the frequency span because the noise power in the detector and amplifying circuit depends on the frequency span. Generally, the noise magnitude in different bandwidths can be approximated by scaling the power spectrum by the square root of the bandwidth. Therefore, the normalized minimum detectable dynamic strain is displayed in units of $\varepsilon/\sqrt{Hz}$.

Figure 26.4 shows some typical interrogators. Figure 26.4a is an example of an interrogator that uses linearly wavelength-dependent devices (couplers or filters) [28]. The Bragg wavelength shift is monitored by the detected power change. This structure is quite simple and has been commercialized but is not suitable for multiplexed sensors. This type of power detection method does not fully use one of the key advantages of FBG sensors—measurand information is coded as wavelength change and the information is not affected by fluctuations in light power. In Fig. 26.4a, to remove the light power (or loss) fluctuation effect, a power ratio detection technique is used. Figure 26.4b shows the charge-coupled device (CCD) spectrometer interrogator [32], which takes advantage of the fact that the diffraction angle is dependent on the wavelength of the light incident on a diffraction grating.

The structure shown in Fig. 26.4b adopts both WDM and spatial division multiplexing. A one-dimensional CCD can be used instead of the two-dimensional CCD if time division multiplexing is used instead of spatial division multiplexing [33]. Figure 26.4c shows an example of the scanning fiber Fabry-Perot filter interrogator system. The U.S. Naval Research Laboratory tested this for use as a 64-FBG sensor system [44]. The basic principle is to locate the WDM Bragg wavelengths within the free spectral range of the Fabry-Perot filter and monitor their changes by spectrally scanning the filter. In real systems, the dithering of the filter spectrum is adopted for fine measurements and the zero crossing of the time derivative of the received power is monitored.

Figure 26.4d shows an unbalanced Mach-Zehnder interferometer interrogator system adopting a pseudo-heterodyne detection technique [49]. A detailed description of the principle of operation can be found in [13]. The unbalanced Mach-Zehnder interferometer provides very high-resolution interrogation. However, this interferometer is quite sensitive to changes in its environment, such as temperature changes or even air fluctuations. Therefore, a reference grating for the interrogator system (this should not be confused with the reference grating for the sensor head to distinguish

Principles and Status of Optical Fiber Sensors

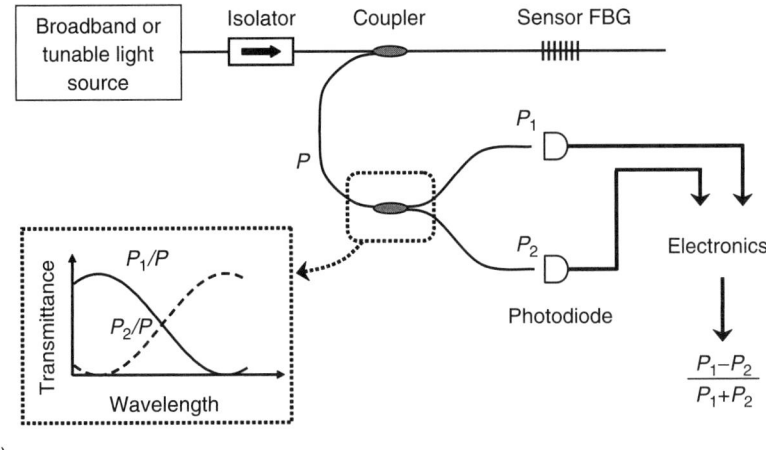

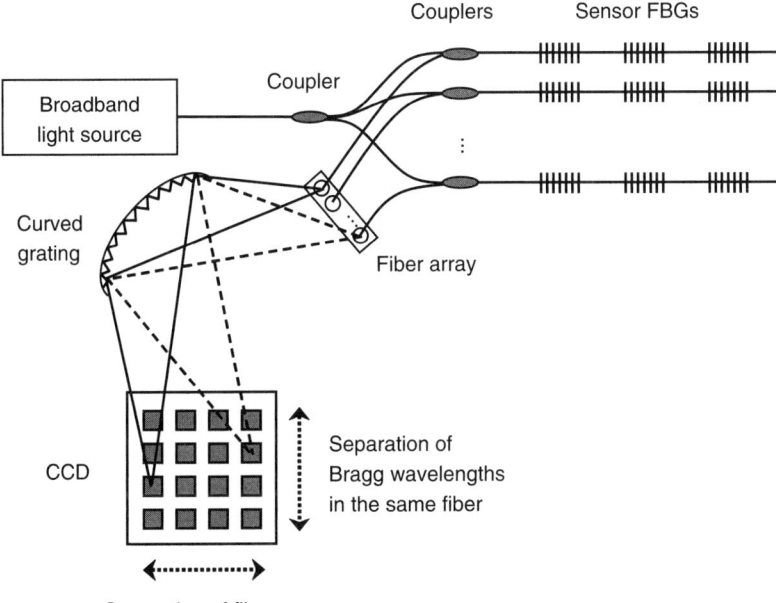

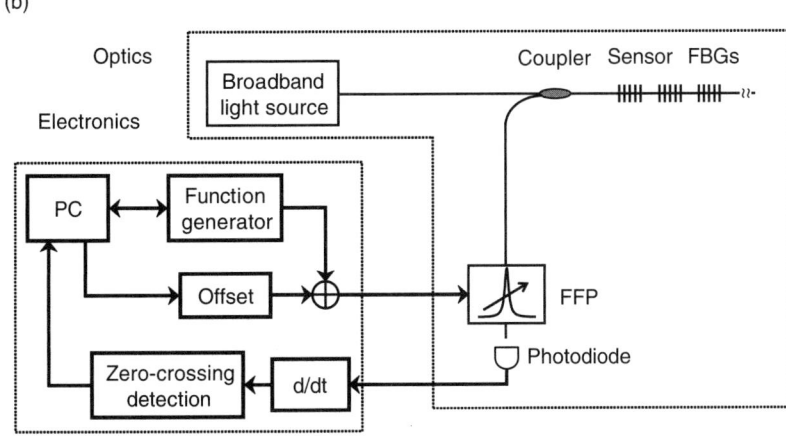

FIGURE 26.4 Some typical interrogators for FBG sensors. (a) Linearly wavelength-dependent device (coupler) method [28]. (b) CCD spectrometer method [32]. (c) Fabry-Perot filter method [44]. FFP, fiber Fabry-Perot filter. *Continued*

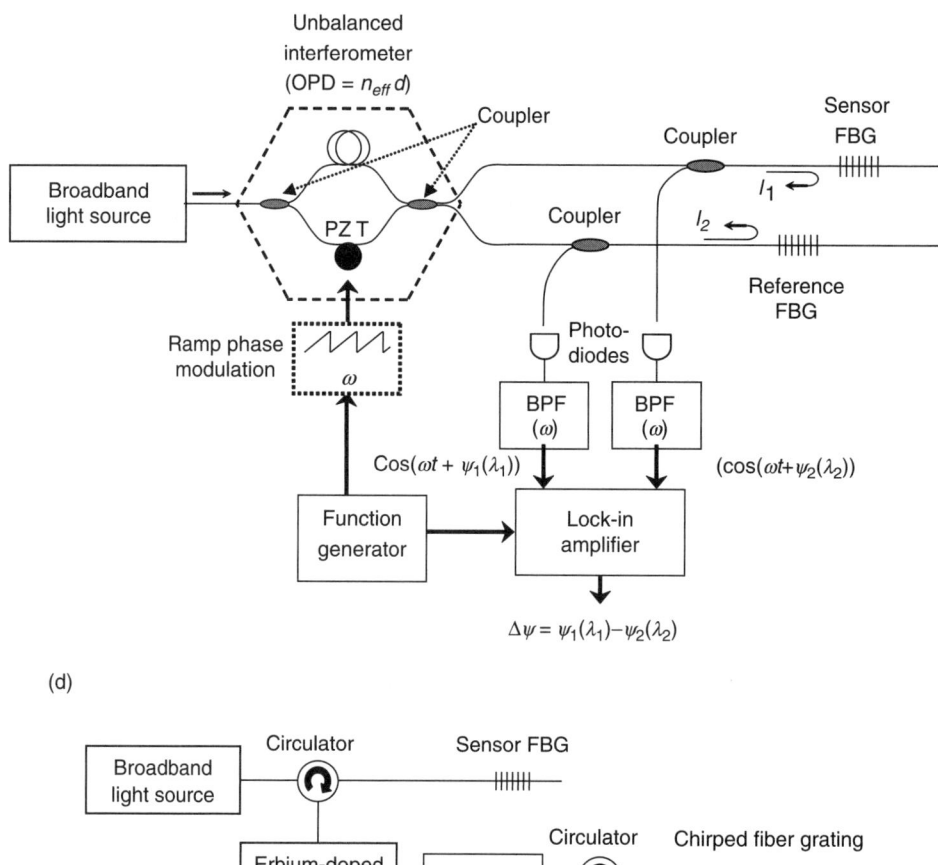

(d)

(e)

FIGURE 26.4, Cont'd (d) Unbalanced Mach-Zehnder interferometer (pseudo-heterodyne detection) method [49]. (e) Modulation phase shift detection method [68].

the temperature and strain effect) is required, as shown in Fig. 26.4d, and the interrogator should be well stabilized.

Figure 26.4e shows a chirped fiber grating interrogator system based on the modulation phase shift detection technique [68]. The interrogator uses the wavelength dependence of the phase group-delay response of a chirped fiber grating to determine the Bragg wavelength shift of the sensing FBG. The sensitivity of the interrogator is determined by the selection of the chirped fiber grating length and the bandwidth. Because of its radio frequency (RF) modulation frequency (several hundred megahertz), the system bandwidth can be increased to several megahertz, which is several orders of magnitude faster than those of Fabry-Perot tunable filters or unbalanced Mach-Zehnder interferometers. In particular, it does not require any referencing to compensate for intensity variations. Multiplexing can also be achieved by using an arrayed waveguide grating that allows the interrogation of more than one FBG sensor with a single chirped fiber grating.

Because of the residual group-delay ripple in chirped fiber gratings (Fig. 26.2c), however, the resolution and the dynamic range of the interrogation system is restricted. Therefore, group-delay ripple of the chirped fiber grating should be controlled to be as small as possible, for example, by apodizing the fiber grating. In addition, any shift in the phase of the chirped fiber grating, due to ambient temperature, is reflected as a measurand change. Therefore, it should be packaged to ensure athermal operation or operated in a temperature-controlled environment.

FBG sensors have been tested for various uses, such as in conjunction with bridges, dams, mines, composite laminates, airplanes, generators, ship water jets, geotechnical fields, and railway systems [33, 55, 77–83]. Some FBG sensors have been commercialized as temperature sensors [84], structural monitoring [85], and oil/gas reservoir monitoring [86]. For example, CiDRA's FBG pressure/temperature sensor was reported to show good functionality at 150°C and 5000 psi for over 2 years with output drift less than 0.015% full scale/year [86]. Such sensors have been deployed in oil/gas wells—for example, in a 15,000-feet down-hole [86].

3. FIBER OPTIC GYROSCOPES

In recent years, FOGs have been replacing mechanical gyroscopes in both newly designed and existing applications. It is generally recognized that the first demonstration of the FOG was achieved in 1976 [87], although there had been a few previous studies. The basic concept is based on the Sagnac effect, which found its first application in the ring laser gyroscope (RLG). The RLG is now used extensively in commercial inertial navigation systems for aircraft, but the high cost is limiting its application because this implementation uses an evacuated mirrored cavity that requires high vacuum and precision mirror technology. However, the interferometric FOG uses the same Sagnac effect in a fiber coil, making reliable, shock-resistant, vibration-resistant, low-cost inertial rotation sensors. Figure 26.5a shows a closed optical fiber loop. Light entering the loop is divided into two counter-propagating waves, that is, a clockwise propagating wave and a counterclockwise propagating wave. When the fiber loop is not rotating, the two waves return perfectly in phase after having traveled along the same path in opposite directions. When the fiber loop rotates, for example, in a clockwise direction as shown in Fig. 26.5b, it is evident that the clockwise wave has to travel a greater distance than the counterclockwise wave, which results in different travel times and

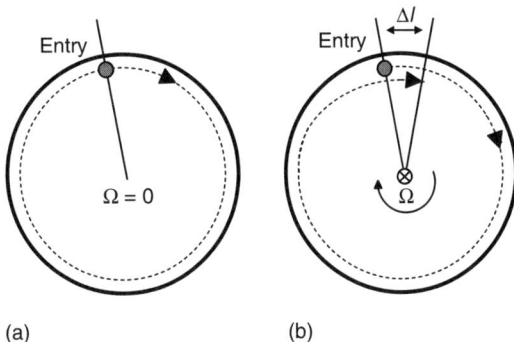

FIGURE 26.5 Principle of the fiber optic gyroscope. The fiber coil is (a) at rest and (b) rotating in a clockwise direction.

a phase difference between the two optical waves. Analyses [88, 89] show that the phase difference $\Delta\phi$ is given by:

$$\Delta\phi = \frac{8\pi N}{\lambda c}\mathbf{A} \cdot \mathbf{\Omega} \quad (6)$$

where N is the number of coil turns, λ is the wavelength in vacuum, c is the speed of light in vacuum, $\mathbf{A}$ is the area vector of the fiber coil (its magnitude is the area enclosed by the single loop and direction is normal to it), and $\mathbf{\Omega}$ is the rotating rate (angular frequency) vector. (In the case of clockwise rotation of Fig. 26.5b, the direction of the vector $\mathbf{\Omega}$ is normally into the paper plane.)

Similar to the output of typical interferometers, the output response of the FOG is a raised cosine function (Fig. 26.6) with intensity I given by:

$$I = \frac{I_0}{2}(1 + \cos\Delta\phi) \quad (7)$$

Here, I_0 is the intensity of the light falling on the detector in the absence of any phase shift. There is an unambiguous operating range of π radians around $n\pi + \pi/2$ (n is an integer) for the phase difference. The unambiguous range and the sensitivity can be scaled up or down by changing the area of the sensing coil or the number of the coil turns, which is one of the major advantages of FOG technology. In the response curve in Fig. 26.6, if $|\Delta\phi|$ is very small, which is usually

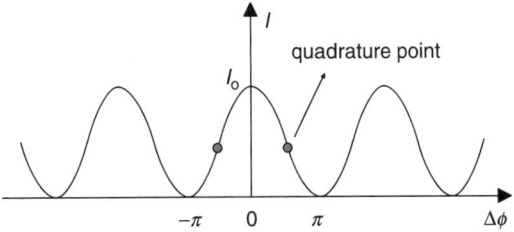

FIGURE 26.6 Raised cosine response of a FOG.

true in most applications, the sensitivity is almost zero and the rotational direction of the fiber coil cannot be distinguished due to the symmetric response of the curve. Therefore, the operating point should be shifted to a quadrature position where the response slope is linear by introducing a phase bias ϕ_b. The use of the nonreciprocal Faraday effect was proposed to generate ϕ_b by controlling the electric current [90]. To achieve high sensitivity, however, ϕ_b must be very stable, which would require a control of the biasing current with an accuracy better than 1 ppm [91]. This phase drift problem can be completely overcome by using a phase modulator placed at the end of the fiber coil.

Figure 26.7a shows the "minimum" configuration of the FOG. A fiber polarizer and a single-mode fiber directional coupler are used to ensure that the counter-propagating light waves have the same polarization and spatial mode and travel in exactly the same path, only in the reverse direction, making the system reciprocal. A dynamic phase modulation is performed at one end of the fiber coil, such that two lights that start at the coupler at the same time and rotate the fiber coil in opposite directions reach the phase modulator (such as a fiber wound around a piezoelectric transducer tube) at different times with the delay time $T = Ln_g/c$, where n_g is the group index of the fiber and L is the length of the fiber coil [92]. When a sinusoidal (or a square-wave) phase modulation $\phi_m(t) = a \sin \omega_m t$ is applied to the phase modulator, a phase difference modulation $\Delta\phi_m(t) = \phi_m(t) - \phi_m(t-T)$ is generated between the two counter-propagating waves, and the overall phase $\Delta\phi_o(t)$ in the interference signal becomes $\Delta\phi_o(t) = \Delta\phi + \Delta\phi_m(t)$. Here, $\Delta\phi$, given by Eq. (6), contains the information regarding the rotation rate of the fiber loop and $\Delta\phi_m(t)$ can be reformed as

$$\Delta\phi_m(t) = 2a \sin \frac{\omega_m T}{2} \cos\left(\omega_m \left(t - \frac{T}{2}\right)\right) = \phi' \cos \omega_m t' \tag{8}$$

where $\phi' = 2a \sin(\omega_m T/2)$ and $t' = t - T/2$. Then, the interference signal at the detector is given by

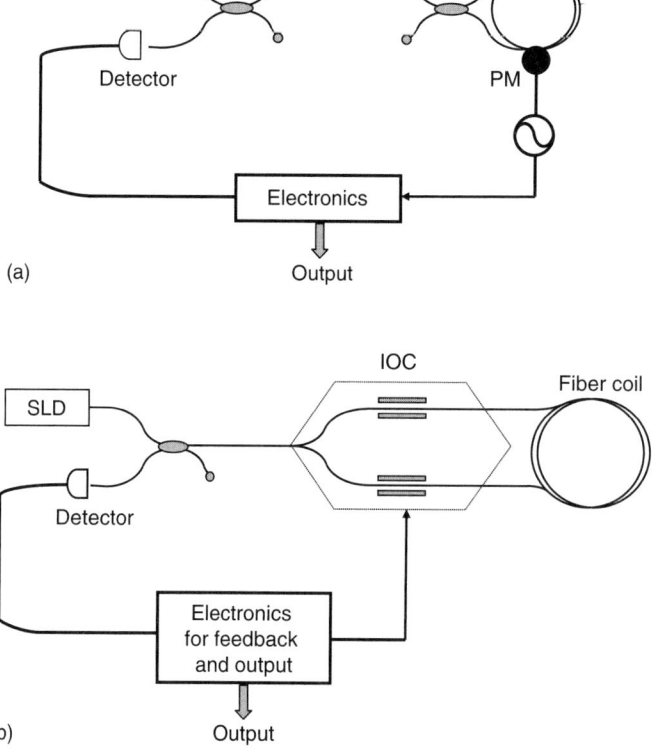

FIGURE 26.7 Basic configurations of FOG: (a) open-loop type (SLD, super-luminescent diode; PM, phase modulator), (b) closed-loop type (IOC, integrated optics chip modulator).

$$I = \frac{I_0}{2}[1+\cos(\Delta\phi_o(t))] = \frac{I_0}{2}[1+\cos(\Delta\phi + \Delta\phi_m(t))]$$
$$= \frac{I_0}{2}[1 + J_0(\phi')\cos\Delta\phi - 2J_1(\phi')\sin\Delta\phi\cos(\omega_m t')$$
$$- 2J_2(\phi')\cos\Delta\phi\cos(2\omega_m t')$$
$$+ 2J_3(\phi')\sin\Delta\phi\cos(3\omega_m t') + \ldots], \quad (9)$$

where $J_n(\ )$ represents the nth-order Bessel function of the first kind. At rest, that is, when $\Delta\phi = 0$, the detector output has a symmetric pattern, as shown in Fig. 26.8a, because the phase modulation is symmetrical. In the frequency domain, only even harmonics of the modulation frequency exist and the ratio of the harmonic amplitudes depends on the modulation depth ϕ' or a. When the coil is rotating, the modulation occurs about the Sagnac phase shifted position of the interferometer response (Fig. 26.8b). This results in an unbalanced response, and the fundamental and odd harmonics are also present.

The intensities of the odd harmonics depend on $\sin\Delta\phi$, as can be seen in Eq. (9). Therefore, if the first harmonic is synchronously demodulated (by using a lock-in amplifier), the maximum sensitivity around the zero rotation rate can be obtained with no directional ambiguity ($\sin\Delta\phi$ is positive or negative if $\Delta\phi$ is positive or negative, respectively, for small values of $|\Delta\phi|$). The choice of the correct modulation frequency is an important consideration in the design of a FOG. At the so-called proper frequency, $f_m = 1/2T$, the intensities of the odd harmonics are at a maximum because the modulation depth ϕ' in Eq. (8) is maximized. By using this frequency, the modulator drive signal can be

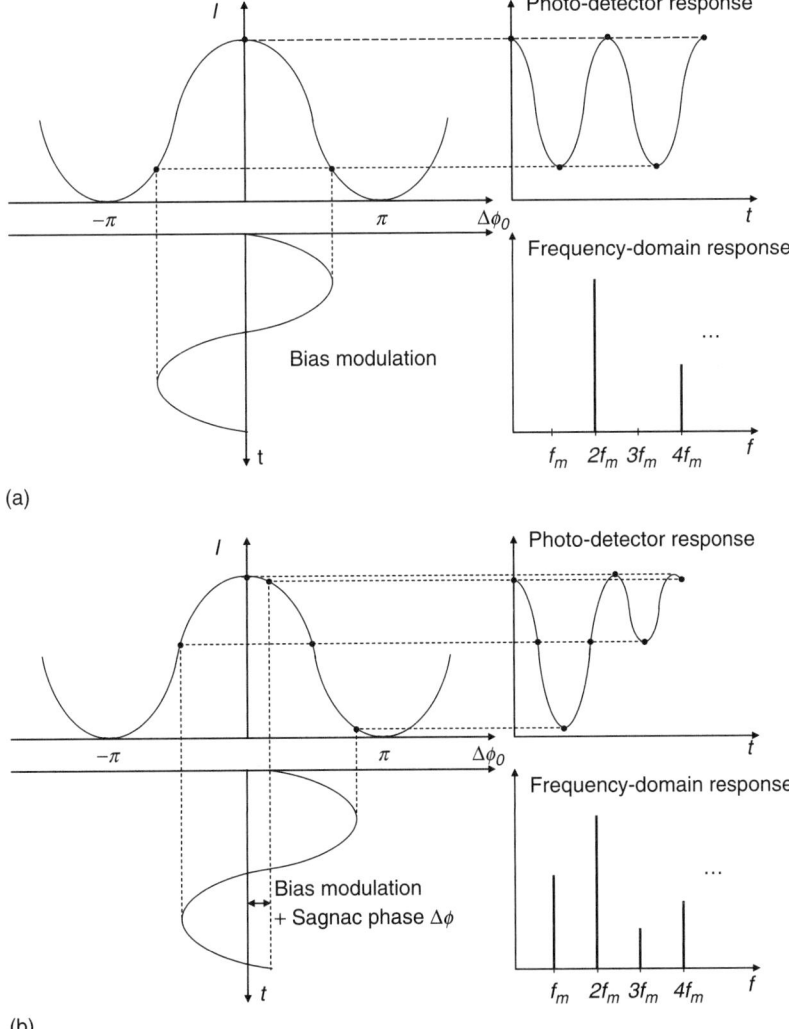

FIGURE 26.8 FOG response to nonreciprocal phase shifts: (a) at rest and (b) rotating.

minimized and parasitic errors associated with defects of the modulation are suppressed [88].

The detected interference signal, I, of this open-loop FOG contains parameters I_o, ϕ', and $\Delta\phi$. Therefore, to accurately measure $\Delta\phi$, it should be well normalized against changes in the light intensity and the modulation depth. A straightforward way to do this is to separately demodulate any three harmonics of the detected intensity and compare them. This type of open-loop signal processing provides a very stable bias, enabling low-cost high accuracy measurements around the zero rotation rate. In gyroscope applications, the measurement of interest is the integrated angle of rotation, and therefore a high-performance gyroscope requires good accuracy over the entire dynamic range of the rotation rate. However, the intrinsic response of this scheme is sinusoidal, making accurate measurements only in a limited range.

This problem is solved by closed-loop signal processing. In this scheme, the demodulated open-loop signal is fed back into the system as an error signal to generate an extra phase shift $\Delta\phi_c$ that has the opposite sign to the Sagnac phase $\Delta\phi$. When the amplitudes of $\Delta\phi_c$ and $\Delta\phi$ become sufficiently close, the odd harmonics are thereby nulled, and the $\Delta\phi_c$ required to do the nulling becomes the output of the closed-loop FOG. This yields an extremely linear response and stable scale factor compared with the raw open-loop output because this feedback is independent of the optical power and the gain of the detection amplifier [93, 94]. The Sagnac effect may be interpreted as a frequency shift (Doppler effect), and therefore a frequency shift in the opposite direction may null out the Sagnac effect. The first proposed closed-loop arrangement used a frequency shift generated using acousto-optic modulators [95, 96]. However, the scheme could not satisfy the full reciprocity requirement of the interferometer, and this is overcome by using a serrodyne phase modulation. A serrodyne signal may be mathematically expressed as a sine wave modulation on top of a frequency shift. That is, the modulation signal $\phi_m(t) = a \sin \omega_m t$ is changed to $\phi_s(t)$, given by

$$\phi_s(t) = \phi_m(t) - 2\pi\Delta f t = a \sin \omega_m t - 2\pi\Delta f t \quad (10)$$

Therefore Eq. (9) should be modified as

$$I = \frac{I_0}{2}[1 + \cos(\Delta\phi_o(t))] = \frac{I_0}{2}[1 + \cos(\Delta\phi + \Delta\phi_s(t))]$$
$$= \frac{I_0}{2}[1 + \cos(\Delta\phi - 2\pi\Delta f T + \Delta\phi_m(t))] \quad (11)$$

When $\Delta\phi = 2\pi\Delta f T$, the Sagnac effect is nulled, suppressing the odd harmonic components, and the frequency shift Δf, which can be measured from the number of serrodyne resets, is used to calculate the rotation angle. A major problem associated with using serrodyne modulation is that the gyroscope with this scheme has a temperature dependence because the refractive index of the glass is related to the serrodyne frequency shift. To suppress this dependence, a digital phase ramp waveform can be used instead of the serrodyne. This digital version allows one to naturally synchronize the ramp resets with the biasing modulation, and the response of the phase modulator is controlled by a second feedback loop checking the error signal at these resets. This all-digital closed-loop processing is intrinsically free of any electronic source of bias drift. Because of these complicated modulation signals, the closed-loop configurations typically use electro-optic modulators that have much wider frequency bandwidth than piezoelectric transducer modulators. Figure 26.7b shows such an example using an integrated optics chip, typically a lithium niobate ($LiNbO_3$) modulator, to modulate the phase.

Another important consideration is the choice of the proper light source for an FOG for more stable and accurate operations. When using FOGs, not only are counter-propagating main waves present, but many spurious waves are also present due to back-reflections, backscattering, or imperfectly filtered cross-polarized waves. With a coherent light source, the spurious interference signals degrade the contrast of the main wave interference. By using a broadband source, these spurious interference effects can be suppressed due to the very short coherence length, while maintaining the main interference signal because the main waves operate about a null path difference. For high-performance applications, the trend is now to use fiber sources in the 1550-nm window using erbium-doped fiber amplifier (usually called EDFA) technology. The high-powered amplified spontaneous emission (usually called ASE) from the Er-doped fiber improves the signal-to-noise ratio, and its unpolarized emission reduces polarization nonreciprocities and allows the use of an ordinary (nonpolarization-preserving) coupler as the source splitter.

Because FOG applications measure very small phase shifts, great care has to be taken to suppress sources of error. The major errors may come from the imperfect polarizer, modulator, or external perturbations, such as nonuniform temperature distribution, vibration, and magnetic field. Time-varying mechanical or thermal gradients in the fiber coil give rise to a nonreciprocal phase shift, which cannot be distinguished from the Sagnac phase shift, because the counter-propagating waves encounter this perturbation at different times [97]. This effect can be reduced by using good housings and winding techniques, such as dipolar winding and quadrupolar winding, to ensure that thermal and mechanical gradients affect the coil in a symmetrical manner. An imbalance in the power levels of the counter-propagating waves can produce a small phase difference, the Kerr effect, because of the nonlinearity of the refractive index of the fiber core. Slow variations in the splitting ratio in the sensing coil may therefore be translated directly into bias drift. However, because a broadband source equalizes the Kerr nonlinearity for both opposite waves, FOGs using broadband sources do not suffer from this error. When a FOG is at rest, the output signal is a random function that is the sum of a white noise and a slowly varying function. The photon shot noise is an important source of white noise and limits the fundamental accuracy of the FOG. The white noise is expressed in terms of the standard deviation of the equivalent rotation rate per square root of bandwidth of detection (similar to the dynamic strain measurement resolution of fiber grating sensors); in other words, the equivalent noise power spectral density in units of deg/hr/$\sqrt{\text{Hz}}$. The white noise causes noise in the measured rotation rate output and results in an angle random walk (ARW) in units of deg/$\sqrt{\text{hr}}$. The slowly varying noise can be induced by several sources and results in long-term drift or bias instability.

The performance of FOGs has been continuously improved and is now quite mature and capable of meeting the most accurate gyroscopic requirements. As can be seen in Fig. 26.9 [7, 89], the FOG has competing technologies such as mechanical gyroscopes and RLGs. The cost of the FOG has been constantly falling in recent years thanks to the decrease in component prices. Currently, FOGs are considered to be the most cost-efficient solution for various inertial navigation applications. One important advantage of the FOG is its ruggedness. It contains no moving parts, unlike its competitors, such as mechanical gyroscopes and RLGs (dithering is involved in RLGs). Because of these unique advantages, it seems likely that the FOG will play a significant role in both military and commercial markets.

Some recent reports indicate that Honeywell Space Systems currently manufactures high-precision FOGs with a flicker noise below 0.0001 deg/hr and an angle random walk of $\sim 80\,\mu\text{deg}/\sqrt{\text{hr}}$ [98], and Ixsea manufactures high-level performance FOGs with

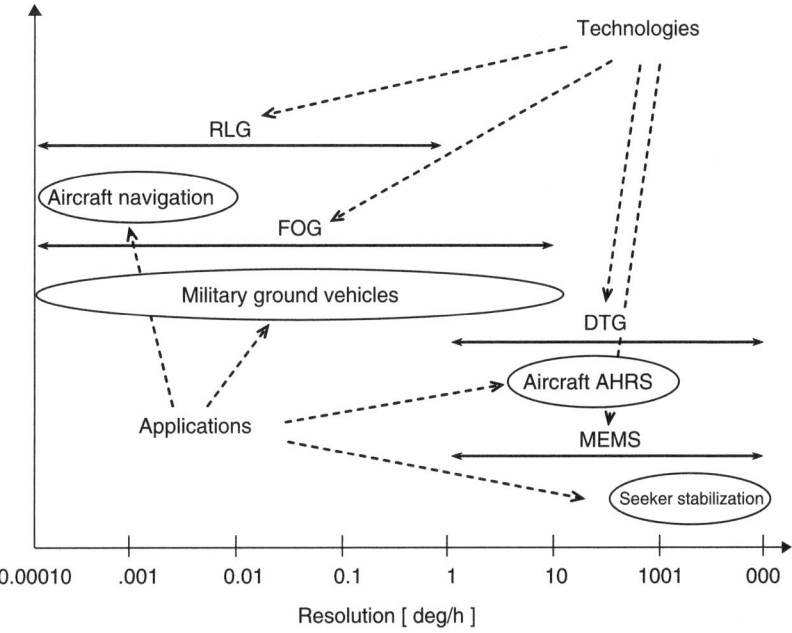

FIGURE 26.9 Comparison of gyroscope technologies and applications. AHRS, attitude heading reference systems; DTG, dynamically tuned gyroscope (mechanical type); FOG, fiber optic gyroscope; MEMS, microelectromechanical system technology; RLG, ring laser gyroscope.

a temperature bias stability better than 0.01 deg/hr (0.003 deg/hr at ambient) for a noise level of 0.001 deg/$\sqrt{\text{hr}}$ [99]. FOGs have been developed by many companies and have been sold or applied in various fields, such as commercial aircraft attitude heading reference systems, military helicopters, missile guidance systems, marine and submarine navigation systems, pipe mapping (gas, power, and communication cables), compasses for tunnel construction, rockets, and automotive navigation systems [89, 98–104].

4. FIBER OPTIC CURRENT SENSORS

With the growth in capacity of electric power systems, the role of protection relaying systems is becoming more important. Such a system should immediately recognize any sudden failure, such as a surge, and separate the failed sections from the power systems to minimize damage and maintain stability of the system. These relaying systems require current sensors, referred to as current transformers or current transducers (CTs), to monitor the current signal to determine whether the monitored part is faulty and should be isolated. Most CTs currently in use are electromagnetic devices that use an iron core and windings to step down the current in the primary line to a smaller current level, mostly 1A or 5A nominal, for secondary devices. When very high currents occur during a fault on the power system, the signals from these CTs may be distorted due to saturation and the residual field in their magnetic cores. Moreover, with the considerable increase in voltages (several hundred kilovolts) in power distribution systems, the insulation of the CTs becomes more difficult and expensive. Therefore, optical current sensors, which do not suffer from electromagnetic interference, with inherent isolation, are good substitutes for conventional CTs.

In addition to protection relaying systems, the deregulation and growth of independent power producers and regional transmission companies have created a need for many new high-voltage revenue-metering points [105]. In the transfer of power from a generation company to a regional transmission company, a 0.5% uncertainty in metering may result in an uncertainty of millions of dollars per year at a high-power metering location [105]. Therefore, the potential use of optical CTs that provide high accuracy metering is promising. Actually, in recent years, optical current sensors are increasingly being installed in high-voltage substations, and, in some areas, they have achieved performance levels exceeding that of conventional magnetic devices.

The concept of optical CTs is quite old [106, 107] and based on the Faraday rotation effect, which states that the polarization of light waves is rotated with the propagation of the light along (or opposite to) a magnetic field inside some material (Fig. 26.10). Optical fiber CTs are not the only optical CTs. Bulk-optic CTs, such as flint glass closed loop–type and crystal-based (Faraday cell) CTs, have also been extensively studied and tested.

The parameter that characterizes the Faraday effect is the Verdet constant that indicates the rotation angle of the polarization per unit magnetic field per unit propagation length. Compared with the bulk devices [7], the Verdet constant of optical fibers is quite small: 1.2×10^{-4} deg/Oe · cm (at 850-nm light wavelength) and 2.13×10^{-4} deg/Oe · cm (at 633 nm), where 1 Oe = 79.6 A/m. However, the optical path length can be increased to compensate for this by winding the fiber around a current-conducting element to create a large number of turns. Although the bulk-optic CTs provide better mechanical stability and smaller sizes, optical fiber CTs provide ease of forming a closed loop (the sensor responds only to the enclosed current), adjustability of the sensitivity and dynamic range, less insertion loss, and a higher signal-to-noise ratio.

In optical fiber CTs, linearly polarized optical waves are input to the optical fiber coil. The linear polarization can be expressed mathematically as a superposition of two circular polarizations (right-handed and left-handed). The magnetic field induced around a current-carrying element induces a circular birefringence inside the optical fiber coil. Therefore, after passing through the coil, a relative phase difference between the two circular polarization components is generated, resulting in rotation of the linear polarization. The Faraday rotation angle is given by

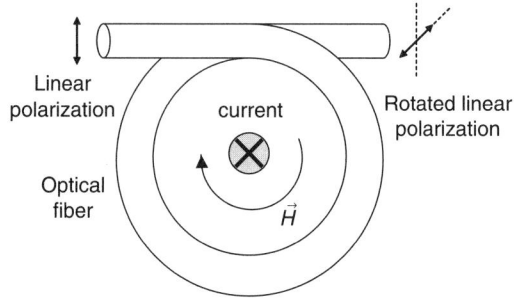

FIGURE 26.10 The Faraday effect.

$$\rho = VN \oint \mathbf{H} \cdot d\mathbf{l} = VNI \qquad (12)$$

where V is the Verdet constant of the optical fiber, N the number of fiber turns, and I is the current enclosed by a single loop, which is equal to the integral of the magnetic field intensity $\mathbf{H}$ over the single loop. This is evidently true (by Ampere's law) regardless of the size or shape of the loop and of the position of the conductor within the loop. The rotation angle, which is proportional to the current, can be measured in a polarimetric scheme by using a 45-degree tilted analyzer that converts the rotation angle into intensity variations at the detector. This process can be represented by the Jones matrix calculations as

$$\begin{bmatrix} E_x \\ E_y \end{bmatrix}_{\text{out}} = P(\theta) F(\rho) \begin{bmatrix} E_x \\ 0 \end{bmatrix}_{\text{in}} \qquad (13)$$

where $P(\theta)$ is the Jones matrix for the analyzer oriented at θ with respect to the x-axis and $F(\rho)$ is the Faraday effect with rotation angle ρ, which are given as follows

$$P(\theta) = \begin{bmatrix} \cos^2\theta & \sin\theta\cos\theta \\ \sin\theta\cos\theta & \sin^2\theta \end{bmatrix} \qquad (14)$$

$$F(\rho) = \begin{bmatrix} \cos\rho & \sin\rho \\ -\sin\rho & \cos\rho \end{bmatrix} \qquad (15)$$

Therefore, the light intensity after the analyzer with $\theta = 45$ degrees becomes

$$I = |E_{x,\text{out}}|^2 + |E_{y,\text{out}}|^2 = \frac{E_{x,\text{in}}^2}{2}(1 - \sin 2\rho)$$
$$= \frac{I_o}{2}(1 - \sin 2\rho) \qquad (16)$$

where a proportional constant that relates the square of the electric field to the light intensity is omitted for simplicity. A Wollaston prism (or a polarization beam splitter) can be used instead of the analyzer to normalize the output against any change in light intensity. It divides the emerging light into two orthogonal linearly polarized components. Each of these polarization components is detected separately, and the difference between the two intensities I_1 and I_2 is normalized to their sum as

$$(\text{Output}) = \frac{I_1 - I_2}{I_1 + I_2} = \sin 2\rho \approx 2\rho \qquad (17)$$

for small ρ. Therefore, the output is independent of the received light power, enabling stable measurement against possible laser power drift and optical loss variations. Another approach to measure the Faraday rotation is to use a Sagnac interferometer. The scheme is nearly identical to that of the open-loop minimum configuration FOG discussed in the previous section, and Fig. 26.11 shows an example of a Sagnac current sensor [108]. The fiber optic polarization controllers, placed in front of the sensing coil, are carefully set so that they transform the incoming linearly polarized waves into circularly polarized waves. Upon leaving the coil, the circularly polarized waves are converted back into linearly polarized waves. The circular birefringence induced by the Faraday effect causes a phase difference between the two counter-propagating circularly polarized waves, and this can be accurately measured using signal processing techniques, such as nonreciprocal phase modulation-demodulation, developed for use in FOG. In this case, the signal is inherently digital, making it more suitable for future digital substation systems.

Although the concept is quite simple, when it is implemented, several difficulties are encountered that limit the resolution of the sensor systems. Optical fibers have some linear birefringence due to

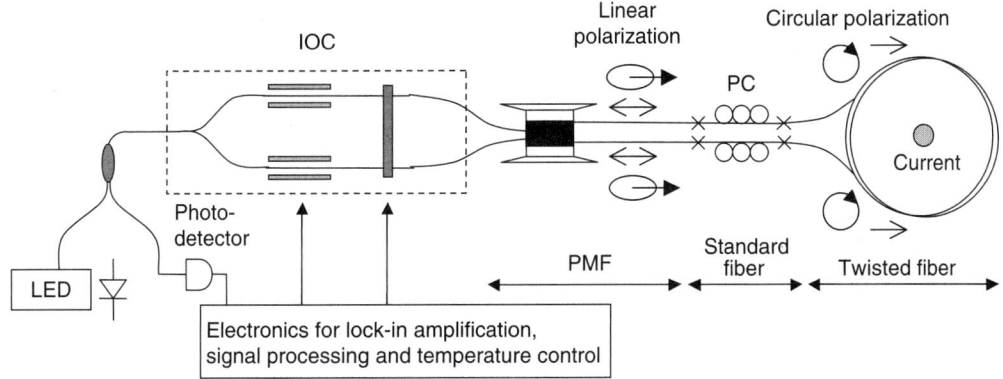

FIGURE 26.11 An example [108] of a high-performance fiber optic current sensor using technology developed for FOG. IOC, integrated optics chip modulator; LED, light-emitting diode; PC, polarization controller; PMF, polarization-maintaining fiber.

imperfect core shape or asymmetrical stress distribution in the core [109]. Linear birefringence implies a velocity difference for two orthogonal polarization components, by which an input linear polarization is converted to an elliptical polarization, which may significantly reduce the output sensitivity of optical fiber CTs. In addition to the intrinsic birefringence of the fiber, birefringence is generated when a fiber is bent into a loop to form a sensor coil, and the effects of vibration, mechanical stress or strain, and temperature variation also add to the total linear birefringence. In some applications, high-current-carrying conductors vibrate. A vibration occurring in a direction that causes the fiber coil to rotate might be critical because the fiber coil will act as an FOG [110].

To overcome the birefringence problem, attempts were first made to produce a low birefringence fiber [111, 112]. However, although the intrinsic birefringence came down to an acceptable level, the bending-induced birefringence could not be avoided. Thermal annealing of the fiber sensor coils considerably reduces linear birefringence [113], but, in practice, because the annealing process weakens the fiber coils, they are difficult to handle. If a birefringence bias corresponding to a 258-degree phase difference is applied, the effect of linear birefringence variation can be minimized [109]. However, the sensitivity is reduced by a factor of 0.217. Tokyo Electric Power Company has been developing current sensors using flint glass fibers that show a low photoelastic coefficient $(4.5 \times 10^{-10}$ cm^2/kg) and a high Verdet constant ($\sim$0.065 min/Oe·cm) [114], but these fibers suffer from higher loss, higher cost, and difficulty in splicing with other fibers. A report indicated that the system was successful in hammering and thermal shock tests for a differential current relaying system [110].

The use of twisted optical fiber has also been proposed in which a high circular birefringence is induced, thereby reducing the effect of linear birefringence [115]. A high birefringent spun fiber was also proposed for this application, which is pulled from a rotating silica preform during the fiber-drawing process [116].

In practice, linear birefringence and intentionally induced circular birefringence may vary with time due to vibration and temperature variation. A simple and efficient way to reduce this effect is to use a reflection-type structure, as shown in Fig. 26.12. The intentionally induced circular birefringence is reciprocal with the reversal of light propagation direction, whereas the Faraday rotation is nonreciprocal.

Therefore, if light travels along the fiber coil and is reflected by a mirror and then travels back along the fiber coil, the Faraday rotation angle is doubled, whereas the effect of the intentionally induced circular birefringence is canceled. However, if a conventional mirror is used, the effect of linear birefringence is also doubled. A Faraday rotator mirror can be used to minimize the linear birefringence effect, in which case the two orthogonal linear polarization components are switched at the Faraday rotator mirror and experience the same phase delay (despite the linear birefringence) after the round trip [116, 117]. Toshiba reported that for its reflection-type optical CT, the influence of mechanical shock from circuit breaker vibration of 8 g (g = gravitational acceleration) was less than the root mean square value of the electronic noise [118].

Sagnac interferometer-type optical fiber CTs are also insensitive to reciprocal effects [119]. With the structure shown in Fig. 26.11, EPFL has achieved $\pm 0.2\%$ accuracy over the ± 500 kA range for a temperature range between 0 and 50°C [108]. The Sagnac structure using a 3×3 fiber optic coupler was also tested [120]. Other reports indicate that Siemens has tested an optical fiber CT with an annealed fiber coil for a generator with a measurement accuracy error of less than 0.25% within a temperature range of 30–90°C [121], and NxtPhase showed a linearity performance better than $\pm 0.2\%$ from 0.5% to 120% of the rated current [105]. ABB Corporate Research has developed a robust fiber optic current sensor with inherent temperature compensation of the Faraday rotation effect [122]. Insensitivity of the sensor to temperatures of less than 0.2% was demonstrated between -35 and 85°C. Recently, a Japanese collaborative group reported an accuracy better than $\pm 0.2\%$ between -40 and 60°C [123].

5. OTHER SENSORS

Other types of optical fiber sensors, based on very simple concepts, have been commercialized. Some examples include a displacement or pressure sensor based on the light coupling of two fibers, a liquid-level sensor based on frustrated total internal reflection, a pressure sensor using a zigzag (periodically bent) fiber, and a temperature sensor based on the detection of radiation from a heated sensor head (blackbody cavity) [124].

One of the most well-developed and commercialized in-line fiber sensors or diagnostic tools is OTDR, which is based on the monitoring of backscattering along the fiber. The concept is also quite old [125], and the OTDR has become a standard

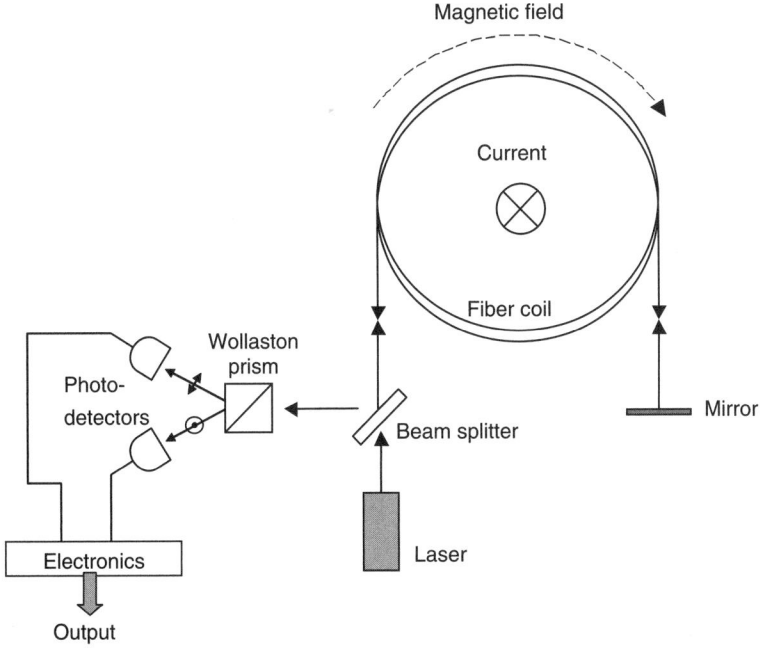

FIGURE 26.12 Reflection-type fiber optic current sensor.

technique for testing optical fiber links [126, 127]. It typically provides submeter spatial resolution, but improved techniques can provide millimeter-order resolution. More complicated methods, such as optical frequency domain reflectometry, have been studied to achieve millimeter or submillimeter ($\sim 10\,\mu$m) spatial resolution [128].

Although OTDRs are generally aimed at monitoring optical fiber communication links, active research on distributed sensors for civil structure monitoring has also occurred. One example is a distributed temperature sensor for monitoring the concrete setting temperatures of a large dam [129]. The principle is based on stimulated Brillouin scattering. An acoustic wave couples two counter-propagating beams that are frequency shifted by an amount that is dependent on temperature or strain. The distributed temperature- and strain-monitoring systems have been commercialized (e.g., see [84]).

Fiber optic low-coherence interferometry has also been commercialized for civil engineering applications [130]. The measurement range of a commercialized deformation sensor using the method is 10 mε for elongation (5 mε for shortening) [84]. Optical coherence tomography that uses fiber optic low-coherence interferometry is becoming an important tool in biomedical applications (see [131] and references therein).

Optical fiber acoustic sensors or optical fiber hydrophone systems have also been studied intensively from the early days of optical fiber sensors [1] and tested for marine or submarine applications [132]. A fiber optic pressure sensor using the fiber Fabry-Perot interferometer method was also commercialized; this has a dynamic range of 0–3000 psi with $\pm 1.0\%$ (of full range) accuracy and $\pm 0.5\%$ (of full range) linearity [133]. Cryogenic fiber optic temperature sensing [134] and fiber optic sensor applications for wide temperature ranges [135] are also attractive fields.

Fiber optic chemical sensors and biosensors have been continuously studied. Even at the first OFS, there was a paper on a fiber optic sensor for medical use [136]. The principles of fiber optic chemical sensors or biosensors are based on the monitoring of absorbance, reflectance, luminescence, refractive index change, or light scattering and are aimed at the measurement of oxygen, pH, carbon dioxide, ammonia, detergents, biochemical oxygen demand, pesticides, and humidity [6, 137]. In many cases, optical fibers are simply used to guide light to the measurement point in the specimen. In some cases, optical fibers are monitoring the response of a material deposited on the end of the fiber [138]. In other cases, well-developed sensor technologies, such as FBG temperature sensors, have been used for biological tests [139]. A good example of optical fiber chemical sensors in which the fiber itself plays a key role in the measurement is the use of LPGs [140, 141]. As discussed in Section 2, an LPG can couple the core mode to a cladding mode. If the coating or jacket is removed from the optical fiber, the evanescent field of the cladding mode that exists outside the cladding experiences the refractive index change of the outside material. The sensitivity can be adjusted by fiber etching [142]. An example of a commercialized biosensor is an in vivo blood pressure sensor that uses white light interferometry [143].

6. CONCLUSION

The current status of optical fiber sensors was briefly reviewed. As soon as optical fibers were developed, it was concluded that they could also be used for sensors. Therefore, the history of research in optical fiber sensors is almost as old as the history of optical fiber communication research. Although optical fiber sensors have not experienced the dramatic commercial success of optical fiber communications, they have been continuously and extensively studied.

Optical fiber sensors have unique advantages, such as high sensitivity, immunity to electromagnetic interference, small size, light weight, robustness, flexibility, and the ability to provide multiplexed or distributed sensing. Fiber grating sensors have been the most widely studied topic of the various optical fiber sensor technologies. Some fiber grating sensors have been commercialized for the health monitoring and oil industries. There are other rather mature optical fiber sensor technologies such as OTDRs, FOGs, and optical fiber current sensors. New ideas are continuously being proposed and tested, not only for various traditional measurands such as strain, temperature, and pressure, but also for new applications such as biosensors or medical imaging, including optical coherence tomography. Articles on optical fiber sensors have already begun to appear in many field-application journals, such as civil engineering journals, sensor journals, power engineering journals, and chemical engineering journals. With continuing effort and enthusiasm, the fields of application for optical fiber sensors will be extended and more successes in commercialization will almost certainly be found.

7. REFERENCES

1. T. G. Giallorenzi, J. A. Bucaro, A. Dandridge, G. H. Sigel, Jr., J. H. Cole, S. C. Rashleigh, and R. G. Priest, "Optical fiber sensor technology," *IEEE Journal of Quantum Electronics*, **18**, no. 4, pp. 626–665, 1982.
2. F. T. S. Yu and S. Yin, Eds., *Fiber Optic Sensors*, Marcel Dekker, New York, 2002.
3. K. T. V. Grattan and B. T. Meggitt, Eds., *Optical Fiber Sensor Technology Volume 2—Devices and Technology*, Chapman & Hall, London, 1998.
4. K. T. V. Grattan and B. T. Meggitt, Eds., *Optical Fiber Sensor Technology Volume 3—Applications and Systems*, Kluwer Academic Publishers, Boston, 1999.
5. A. Othonos and K. Kalli, *Fiber Bragg Gratings—Fundamentals and Applications in Telecommunications and Sensing*, Artech House, Boston, 1999.
6. M. D. Marazuela and M. C. Moreno-Bondi, "Fiber-optic biosensors—an overview," *Analytical and Bioanalytical Chemistry*, **372**, pp. 664–682, 2002.
7. B. Lee, "Review of the present status of optical fiber sensors," *Optical Fiber Technology*, **9**, no. 2, pp. 57–79, 2003.
8. Z. Y. Zhang and K. T. V. Grattan, "Commercial activity in optical fiber sensors," in *Optical Fiber Sensor Technology Volume 3—Applications and Systems* (K. T. V. Grattan and B. T. Meggitt, Eds.), Kluwer Academic Publishers, Boston, pp. 257–306, 1999.
9. K. O. Hill, Y. Fujii, D. C. Johnson, and B. S. Kawasaki, "Photosensitivity in optical fiber waveguides: application to reflection filter fabrication," *Applied Physics Letters*, **32**, pp. 647–649, 1978.
10. K. O. Hill, "Photosensitivity in optical fiber waveguides: from discovery to commercialization," *IEEE Journal on Selected Topics in Quantum Electronics*, **6**, no. 6, pp. 1186–1189, 2000.
11. G. Meltz, W. W. Morey, and W. H. Glenn, "Formation of Bragg gratings in optical fibers by a transverse holographic method," *Optics Letters*, **14**, pp. 823–825, 1989.
12. T. Erdogan, "Fiber grating spectra," *Journal of Lightwave Technology*, **15**, no. 8, pp. 1277–1294, 1997.
13. B. Lee and Y. Jeong, "Interrogation techniques for fiber grating sensors and the theory of fiber gratings," in *Fiber Optic Sensors* (F. T. S. Yu and S. Yin, Eds.), Marcel Dekker, New York, pp. 295–381, 2002.
14. Y.-J. Rao, "Fiber Bragg grating sensors: principles and applications," in *Optical Fiber Sensor Technology*, **2** (K. T. V. Grattan and B. T. Meggitt, Eds.), Chapman & Hall, London, pp. 355–389, 1998.
15. X. Shu, Y. Liu, D. Zhao, B. Gwandu, F. Floreani, L. Zhang, and I. Bennion, "Dependence of temperature and strain coefficients on fiber grating type and its application to simultaneous temperature and strain measurement," *Optics Letters*, **27**, no. 9, pp. 701–703, 2002.

16. M. G. Xu, J. L. Archambault, L. Reekie, and J. P. Dakin, "Thermally-compensated bending gauge using surface-mounted fiber gratings," *Int. J. Optoelectron.*, **9**, pp. 281–283, 1994.
17. M. Song, S. B. Lee, S. S. Choi, and B. Lee, "Simultaneous measurement of temperature and strain using two fiber Bragg gratings embedded in a glass tube," *Optical Fiber Technology*, **3**, pp. 194–196, 1997.
18. M. G. Xu, J. L. Archambault, L. Reekie, and J. P. Dakin, "Discrimination between strain and temperature effects using dual-wavelength fiber grating sensors," *Electronics Letters*, **30**, no. 13, pp. 1085–1087, 1994.
19. S. W. James, M. L. Dockney, and R. P. Tatam, "Simultaneous independent temperature and strain measurement using in-fibre Bragg grating sensors," *Electronics Letters*, **32**, no. 12, pp. 1133–1134, 1996.
20. M. Song, B. Lee, S. B. Lee, and S. S. Choi, "Interferometric temperature-insensitive strain measurement with different diameter fiber Bragg gratings," *Optics Letters*, **22**, no. 11, pp. 790–792, 1997.
21. P. M. Cavaleiro, F. M. Araujo, L. A. Ferreira, J. L. Santos, and F. Farahi, "Simultaneous measurement of strain and temperature using Bragg gratings written in germanosilicate and boron-codoped germanosilicate fibers," *IEEE Photonics Technology Letters*, **11**, no. 12, pp. 1635–1637, 1999.
22. M. Sudo, M. Nakai, K. Himeno, S. Suzaki, A. Wada, and R. Yamaguchi, "Simultaneous measurement of temperature and strain using PANDA fiber grating," *Proceedings of the 12th International Conference on Optical Fiber Sensors (OFS-12)*, Williamsburg, VA, USA, pp. 170–173, 1997.
23. B.-O. Guan, H.-Y. Tam, X.-M. Tao, and X.-Y. Dong, "Simultaneous strain and temperature measurement using a superstructure fiber Bragg grating," *IEEE Photonics Technology Letters*, **12**, no. 6, pp. 675–677, 2000.
24. K. J. Han, Y. W. Lee, J. Kwon, S. Roh, J. Jung, and B. Lee, "Simultaneous measurement of strain and temperature incorporating a long-period fiber grating inscribed on a polarization-maintaining fiber," *IEEE Photonics Technology Letters*, **16**, no. 9, pp. 2114–2116, 2004.
25. S. M. Melle, K. Liu, and R. M. Measures, "A passive wavelength demodulation system for guided-wave Bragg grating sensors," *IEEE Photonics Technology Letters*, **4**, no. 5, pp. 516–518, 1992.
26. A. B. Lobo Ribeiro, L. A. Ferreira, M. Tsvetkov, and J. L. Santos, "All-fibre interrogation technique for fibre Bragg sensors using a biconical fibre filter," *Electronics Letters*, **32**, no. 4, pp. 382–383, 1996.
27. M. Song, S. B. Lee, S. S. Choi, and B. Lee, "Fiber laser strain sensor using an LPG (long period grating) Bragg wavelength demodulation filter," *The 2nd Optoelectronics and Communications Conference (OECC '97) Technical Digest*, Seoul, Korea, pp. 676–677, July 1997.
28. M. A. Davis and A. D. Kersey, "All-fibre Bragg grating strain-sensor demodulation technique using a wavelength division coupler," *Electronics Letters*, **30**, no. 1, pp. 75–77, 1994.
29. Q. Zhang, D. A. Brown, H. Kung, J. E. Townsend, M. Chen, L. J. Reinhart, and T. F. Morse, "Use of highly overcoupled couplers to detect shifts in Bragg wavelength," *Electronics Letters*, **31**, no. 6, pp. 480–482, 1995.
30. A. D. Kersey, M. A. Davis, H. J. Patrick, M. LeBlanc, K. P. Koo, C. G. Askins, M. A. Putnam, and E. J. Friebele, "Fiber grating sensors," *Journal of Lightwave Technology*, **15**, no. 8, pp. 1442–1463, 1997.
31. C. G. Askins, M. A. Putnam, G. M. Williams, and E. J. Friebele, "Stepped-wavelength optical fiber Bragg grating arrays fabricated in line on a draw tower," *Optics Letters*, **19**, pp. 147–149, 1994.
32. Y. Hu, S. Chen, L. Zhang, and I. Bennion, "Multiplexing Bragg gratings combined wavelength and spatial division techniques with digital resolution enhancement," *Electronics Letters*, **33**, no. 23, pp. 1973–1975, 1997.
33. R. Willsch, W. Ecke, and H. Bartelt, "Optical fiber grating sensor networks and their application in electric power facilities, aerospace and geotechnical engineering" *The 15th Optical Fiber Sensors Conference Technical Digest (OFS-15)*, Portland, OR, pp. 49–54, May 2002.
34. K. Zhou, A. G. Simpson, X. Chen, L. Zhang, and I. Bennion, "Fiber Bragg grating sensor interrogation system using a CCD side detection method with superimposed blazed gratings," *IEEE Photonics Technology Letters*, **16**, no. 6, pp. 1549–1551, 2004.
35. S. C. Kang, H. Yoon, S. B. Lee, S. S. Choi, and B. Lee, "Real-time measurement for static and dynamic strain using a fiber Bragg grating and the ASE profile of EDFA," *Proceedings of the 13th International Conference on Optical Fiber Sensors (OFS-13)*, Kyongju, Korea, SPIE **3746**, pp. 530–533, 1999.
36. S. Kim, J. Kwon, S. Kim, and B. Lee, "Temperature-independent strain sensor using a chirped grating partially embedded in a glass tube," *IEEE Photonics Technology Letters*, **12**, no. 6, pp. 678–680, 2000.
37. A. D. Kersey, M. A. Davis, and T. Tsai, "Fiber optic Bragg grating strain sensor with direct reflectometric interrogation," *Proceedings of the Optical Fiber Sensors Conference (OFS-11)*, Sapporo, Japan, pp. 634–637, 1996.
38. V. Grubsky and J. Feinberg, "Long-period fiber gratings with variable coupling for real-time sensing

applications," *Optics Letters*, **25**, no. 4, pp. 203–205, 2000.

39. J. M. Gong, J. M. K. MacAlpine, C. C. Chan, W. Jin, M. Zhang, and Y. B. Liao, "A novel wavelength detection technique for fiber Bragg grating sensors," *IEEE Photonics Technology Letters*, **14**, no. 5, pp. 678–680, 2002.

40. Y. Sano and T. Yoshino, "Fast optical wavelength interrogator employing arrayed waveguide grating for distributed fiber Bragg grating sensors," *Journal of Lightwave Technology*, **21**, no. 1, pp. 132–139, 2003.

41. R. W. Fallon, L. Zhang, A. Gloag, and I. Bennion, "Identical broadband chirped grating interrogation technique for temperature and strain sensing," *Electronics Letters*, **33**, pp. 705–706, 1997.

42. D. Zhao, X. Shu, Y. Lai, L. Zhang, and I. Bennion, "Fiber Bragg grating sensor interrogation using chirped fiber grating-based Sagnac loop," *IEEE Sensors Journal*, **3**, no. 6, pp. 734–738, 2003.

43. A. D. Kersey, T. A. Berkoff, and W. W. Morey, "Multiplexed fiber Bragg grating strain-sensor system with a fiber Fabry-Perot wavelength filter," *Optics Letters*, **18**, pp. 1370–1372, 1993.

44. S. T. Vohra, M. D. Todd, G. A. Johnson, C. C. Chang, and B. A. Danver, "Fiber Bragg grating sensor system for civil structure monitoring: applications and field tests," *Proceedings of the 13th International Conference on Optical Fiber Sensors (OFS-13)*, Kyongju, Korea, SPIE **3746**, pp. 32–37, 1999.

45. P. J. Henderson, D. J. Webb, D. A. Jackson, L. Zhang, and I. Bennion, "Highly-multiplexed grating-sensors for temperature-referenced quasi-static measurements of strain in concrete bridges," *Proceedings of the 13th International Conference on Optical Fiber Sensors (OFS-13)*, Kyongju, Korea, SPIE **3746**, pp. 320–323, 1999.

46. A. D. Kersey, T. A. Berkoff, and W. W. Morey, "High-resolution fibre-grating based strain sensor with interferometric wavelength-shift detection," *Electronics Letters*, **28**, no. 3, pp. 236–238, 1992.

47. R. S. Weis, A. D. Kersey, and T. A. Berkoff, "A four-element fiber grating sensor array with phase-sensitive detection," *IEEE Photonics Technology Letters*, **6**, no. 12, pp. 1469–1472, 1994.

48. K. P. Koo and A. D. Kersey, "Bragg grating based laser sensors systems with interferometric interrogation and wavelength division multiplexing," *Journal of Lightwave Technology*, **13**, pp. 1243–1249, 1995.

49. A. D. Kersey and T. A. Berkoff, "Fiber-optic Bragg grating differential-temperature sensor," *IEEE Photonics Technology Letters*, **4**, no. 10, pp. 1183–2285, 1992.

50. A. Dandridge, A. B. Tveten, and T. G. Giallorenzi, "Homodyne demodulation scheme for fiber optic sensors using phase generated carrier," *IEEE Journal of Quantum Electronics*, **18**, pp. 1647–1653, 1982.

51. Y. L. Lo, J. S. Sirkis, and C. C. Chang, "Passive signal processing of in-line fiber etalon sensors for high strain-rate loading," *Journal of Lightwave Technology*, **15**, pp. 1578–1585, 1997.

52. M. Song, S. Yin, and P. B. Ruffin, "Fiber Bragg grating strain sensor demodulation with quadrature sampling of a Mach-Zehnder interferometer," *Applied Optics*, **39**, no. 7, pp. 1106–1111, 2000.

53. S. C. Kang, S. B. Lee, S. S. Choi, and B. Lee, "A novel demodulation technique for the wavelength shift of fiber Bragg grating sensors using the I/Q signal processing scheme," *Proceedings of the Conference on Lasers and Electro-Optics—Pacific Rim (CLEO/Pacific Rim '99)*, Seoul, Korea, pp. 135–136, Sept. 1999.

54. T. A. Berkoff and A. D. Kersey, "Fiber Bragg grating array sensor system using a bandpass wavelength division multiplexer and interferometric detection," *IEEE Photonics Technology Letters*, **8**, no. 11, pp. 1522–1524, 1996.

55. G. A. Johnson, S. T. Vohra, B. A. Danver, K. Pran, G. B. Havsgard, and G. Wang, "Vibration monitoring of a ship waterjet with fiber Bragg gratings," *Proc. 13th International Conference on Optical Fiber Sensors (OFS-13)*, Kyongju, Korea, Proc. SPIE, **3746**, pp. 616–619, 1999.

56. D. C. C. Norman, D. J. Webb, and R. D. Pechstedt, "Extended range interrogation of wavelength division multiplexed fibre Bragg grating sensors using arrayed waveguide grating," *Electronics Letters*, **39**, no. 24, pp. 1714–1716, 2003.

57. M. A. Davis and A. D. Kersey, "Application of a fiber Fourier transform spectrometer to the detection of wavelength-encoded signals from Bragg grating sensors," *Journal of Lightwave Technology*, **13**, no. 7, pp. 1289–1295, 1995.

58. K. B. Rochford and S. D. Dyer, "Demultiplexing of interferometrically interrogated fiber Bragg grating sensors using Hilbert transform processing," *Journal of Lightwave Technology*, **17**, no. 5, pp. 831–836, 1999.

59. M. Volanthen, H. Geiger, M. G. Xu and J. P. Dakin, "Simultaneous monitoring fibre gratings with a single acousto-optic tunable filter," *Electronics Letters*, **32**, pp. 1228–1229, 1996.

60. H. Geiger, M. G. Xu, N. C. Eaton, and J. P. Dakin, "Electronic tracking system for multiplexed fibre grating sensors," *Electronics Letters*, **32**, pp. 1006–1007, 1995.

61. J. P. Dakin and M. Volanthen, "Distributed and multiplexed fibre grating sensors, including discussion of problem areas," *IEICE Transactions on Electronics*, **E83-C**, no. 3, pp. 391–399, 2000.

62. D. A. Jackson, A. B. Lobo Ribeiro, L. Reeckie, and J. L. Archambault, "Simple multiplexing scheme for a fiber-optic grating sensor network," *Optics Letters*, **18**, pp. 1192–1194, 1993.
63. G. P. Brady, S. Hope, A. B. Lobo Ribeiro, D. J. Webb, L. Reekie, J. L. Archambault, and D. A. Jackson, "Demultiplexing of fibre Bragg grating temperature and strain sensors," *Optics Communications*, **111**, pp. 51–54, 1994.
64. M. A. Davis and A. D. Kersey, "Matched-filter interrogation technique for fibre Bragg grating arrays," *Electronics Letters*, **31**, pp. 822–823, 1995.
65. S. C. Kang, S. Y. Kim, S. B. Lee, S. W. Kwon, S. S. Choi, and B. Lee, "Temperature-independent strain sensor system using a tilted fiber Bragg grating demodulator," *IEEE Photonics Technology Letters*, **10**, no. 10, pp. 1461–1463, 1998.
66. Y. J. Rao, D. A. Jackson, L. Zhang, and I. Bennion, "Dual-cavity interferometric wavelength-shift detection for in-fiber Bragg grating sensors," *Optics Letters*, **21**, no. 19, pp. 1556–1558, 1996.
67. J. Jung, Y. W. Lee, and B. Lee, "High-resolution interrogation technique for fiber Bragg grating strain sensor using long period grating pair and EDF," *The 14th International Conference on Optical Fiber Sensors (OFS-14)*, Venice, Italy, Oct. 2000.
68. A. A. Chtcherbakov and P. L. Swart, "Chirped fiber-optic Bragg grating interrogator in a multiplexed Bragg grating sensor configuration," *Journal of Lightwave Technology*, **22**, no. 6, pp. 1543–1547, 2004.
69. G. A. Ball, W. W. Morey, and P. K. Cheo, "Fiber laser source/analyzer for Bragg grating sensor array interrogation," *Journal of Lightwave Technology*, **12**, pp. 700–703, 1994.
70. T. Coroy and R. M. Measures, "Active wavelength demodulation of a Bragg grating fibre optic strain sensor using a quantum well electroabsorption filtering detector," *Electronics Letters*, **32**, pp. 1811–1812, 1996.
71. S. H. Yun, D. J. Richardson, and B. Y. Kim, "Interrogation of fiber grating sensor arrays with a wavelength-swept fiber laser," *Optics Letters*, **23**, no. 11, pp. 843–845, 1998.
72. M. A. Putnam, M. L. Dennis, J. U. Kang, T.-E. Tsai, I. N. Duling, and I. E. J. Friebele, "Sensor grating array demodulation using a passively mode-locked fiber laser," *Technical Digest of the Optical Fiber Communication Conference*, Dallas, TX, Paper WJ4, pp. 156–157, Feb. 1997.
73. A. D. Kersey, A. Dandridge, and M. A. Davis, "Low-crosstalk code-division multiplexed interferometric array," *Electronics Letters*, **28**, pp. 351–352, 1992.
74. K. P. Koo, A. B. Tveten, and S. T. Vohra, "Dense wavelength division multiplexing of fibre Bragg grating sensors using CDMA," *Electronics Letters*, **35**, pp. 165–167, 1999.
75. P. K. C. Chan, W. Jin, J. M. Gong, and M. S. Demokan, "Multiplexing of fiber Bragg grating sensors using an FMCW technique," *IEEE Photonics Technology Letters*, **11**, no. 11, pp. 1470–1472, 1999.
76. M. LeBlanc, S. Y. Huang, M. Ohn, R. M. Measures, A. Guemes, and A. Othonos, "Distributed strain measurement based on a fiber Bragg grating and its reflection spectrum analysis," *Optics Letters*, **21**, pp. 1405–1407, 1996.
77. Y. J. Rao and S. Huang, "Applications of fiber optic sensors," in *Fiber Optic Sensors* (F. T. S. Yu and S. Yin, eds), Marcel Dekker, New York, pp. 449–490, 2002.
78. K.-T. Lau, L.-M. Zhou, P.-C. Tse, and L.-B. Yuan, "Applications of composites, optical fiber sensors and smart composites for concrete rehabilitation: an overview," *Applied Composite Materials*, **9**, pp. 221–247, 2002.
79. L. Maurin, J. Boussoir, S. Rougeault, M. Bugaud, P. Ferdinand, A. G. Landrot, Y.-H. Grunevald, and T. Chauvin, "FBG-based smart composite bogies for railway applications," *The 15th Optical Fiber Sensors Conference Technical Digest*, Portland, OR, pp. 91–94, May 2002.
80. P. M. Nellen, A. Frank, and A. Kenel, "High strain and high strain gradients measured with fiber Bragg gratings in structural engineering applications," *The 15th Optical Fiber Sensors Conference Technical Digest (OFS-15)*, Portland, OR, pp. 111–114, May 2002.
81. S. Kato and H. Kohashi, "Research on the monitoring system of road slope failures with optical fiber sensors," *Technical Digest of The 16th International Conference on Optical Fiber Sensors (OFS-16)*, Nara, Japan, pp. 488–491, Oct. 2003.
82. K. Shiba, H. Kumagai, K. Watanabe, and H. Iwaki, "Application technologies of OTDR and FBG sensors to civil infrastructures," *Technical Digest of The 16th International Conference on Optical Fiber Sensors (OFS-16)*, Nara, Japan, pp. 492–495, Oct. 2003.
83. E. Udd, S. Calvert, and M. Kunzler, "Usage of fiber grating sensors to perform critical measurements of civil infrastructure," *Technical Digest of The 16th International Conference on Optical Fiber Sensors (OFS-16)*, Nara, Japan, pp. 496–499, Oct. 2003.
84. http://www.smartec.ch.
85. W. Ecke and R. Willsch, "Field trial experience with fiber Bragg grating health monitoring systems and optochemical sensor concepts basing on polychromator signal processing," *Technical Digest of The 16th International Conference on Optical Fiber Sensors (OFS-16)*, Nara, Japan, pp. 500–505, Oct. 2003.

86. D. L. Gysling and F. X. Bostick, III, "Changing paradigms in oil and gas reservoir monitoring—the introduction and commercialization of in-well optical sensing systems," *The 15th Optical Fiber Sensors Conference Technical Digest (OFS-15)*, Portland, OR, pp. 43–46, May 2002.

87. V. Vali and R. W. Shorthill, "Fiber ring interferometer," *Applied Optics*, **15**, pp. 1099–1100, 1976.

88. J. Blake, "Fiber optic gyroscopes," in *Optical Fiber Sensor Technology*, **2** (K. T. V. Grattan and B. T. Meggitt, Eds.), Chapman and Hall, New York, pp. 303–328, 1998.

89. P. B. Ruffin, "Fiber gyroscope sensors," in *Fiber Optic Sensors* (F. T. S. Yu and S. Yin, Eds.), Marcel Dekker, New York, pp. 383–415, 2002.

90. W. C. Davis, W. L. Pondrom, and D. E. Thompson, "Fiber-optic gyro using magneto-optic phase-nulling feedback," in *Fiber-Optic Rotation Sensors* (S. Ezekiel and H. J. Arditty, Eds.), Springer-Verlag, New York, pp. 308–315, 1982.

91. H. Lefevre, *The Fiber-Optic Gyroscope*, Artech House, London, 1992.

92. R. P. Moeller, W. K. Burns, and N. J. Frigo, "Open-loop output and scale factor stability in a fiber-optic gyroscope," *Journal of Lightwave Technology*, **7**, no. 2, pp. 262–269, 1989.

93. J. L. Davis and S. Ezekiel, "Closed-loop, low-noise fiber-optic rotation sensor," *Optics Letters*, **6**, pp. 505–507, 1981.

94. B. Y. Kim and H. J. Shaw, "Gated phase-modulation feedback approach to fiber-optic gyroscopes," *Optics Letters*, **9**, no. 6, pp. 263–265, 1984.

95. J. L. Davis and S. Ezekiel, "Techniques for shot-noise-limited inertial rotation measurement using a multi-turn fiber Sagnac interferometer," *Proceedings of SPIE*, **157**, pp. 131–136, 1978.

96. R. F. Cahill and E. Udd, "Phase-nulling fiber-optic laser gyro," *Optics Letters*, **4**, no. 3, pp. 93–95, 1979.

97. D. M. Schupe, "Thermally induced nonreciprocity in the fiber-optic interferometer," *Applied Optics*, **19**, no. 5, pp. 654–655, 1980.

98. S. J. Sanders, L. K. Strandjord, and D. Mead, "Fiber optic gyro technology trends—a Honeywell perspective," *The 15th Optical Fiber Sensors Conference Technical Digest*, Portland, OR, pp. 5–8, May 2002.

99. T. Gaiffe, "From R&D brassboards to navigation grade FOG-based INS: the experience of Photonetics/Ixsea," *The 15th Optical Fiber Sensors Conference Technical Digest*, Portland, OR, pp. 1–4, May 2002.

100. G. A. Pavlath, "Fiber optic gyro based inertial navigation systems at Northrop Grumman," *The 15th Optical Fiber Sensors Conference Technical Digest*, Portland, OR, p. 9, May 2002.

101. R. Usui and A. Ohno, "Recent progress of fiber optic gyroscopes and applications at JAE," *The 15th Optical Fiber Sensors Conference Technical Digest*, Portland, OR, pp. 11–14, May 2002.

102. J. Nasu, K. Saito, A. Kurokawa, F. Hayashi, and I. Nakatani, "Application of fiber gyros at MPC—inertial navigation and guidance system for M-V rocket," *The 15th Optical Fiber Sensors Conference Technical Digest*, Portland, OR, pp. 15–18, May 2002.

103. R. B. Dyott, S. M. Bennett, D. Allen, and J. Brunner, "Development and commercialization of open loop fiber gyros at KVH Industries (formerly at Andrew)," *The 15th Optical Fiber Sensors Conference Technical Digest*, Portland, OR, pp. 19–22, May 2002.

104. T. Kumagai, W. Ohnuki, S. Yamamoto, A. Hongo, and I. Sone, "Optical fiber sensor applications at Hitachi Cable," *The 15th Optical Fiber Sensors Conference Technical Digest*, Portland, OR, pp. 35–38, May 2002.

105. G. A. Sanders, J. N. Blake, A. H. Rose, F. Rahmatian, and C. Herdman, "Commercialization of fiber-optic current and voltage sensors at NxtPhase," *The 15th Optical Fiber Sensors Conference Technical Digest*, Portland, OR, pp. 31–34, May 2002.

106. E. J. Casey and C. H. Titus, "Magneto-optical electric current sensing arrangement," U.S. Patent no. 3,324,393 (1967).

107. S. Yoshikawa and A. Ueki, "Current-measuring system utilizing Faraday effect element," U.S. Patent no. 3,605,013 (1971).

108. F. Briffod, D. Alasia, L. Thevenaz, G. Cuenoud, and P. Robert, "Extreme current measurements using a fibre optics current sensor," *The 15th Optical Fiber Sensors Conference Technical Digest*, Portland, OR, Postdeadline paper PD3, May 2002.

109. A. J. Rogers, "Optical-fibre current measurement," *Int. J. Optoelectronics*, **3**, pp. 391–407, 1988.

110. K. Kurosawa, Y. Tashiro, and K. Ohkawara, "Differential current relaying system with optical phase detection using flint glass fiber type optical current tansformers," *The 15th Optical Fiber Sensors Conference Technical Digest*, Portland, OR, pp. 549–552, May 2002.

111. A. J. Barlow and D. N. Payne, "Polarisation maintenance in circularly-birefringent fibres," *Electronics Letters*, **17**, no. 11, pp. 388–389, 1981.

112. D. N. Payne, A. J. Barlow, and J. J. Ramskov-Hansen, "Development of low- and high-birefringence optical fibres," *IEEE J. Quantum Electronics*, **18**, no. 4, pp. 477–488, 1982.

113. A. H. Rose, Z. B. Ren, and G. W. Day, "Twisting and annealing optical fiber for current sensors," *Journal of*

Lightwave Technology, **14**, no. 11, pp. 2492–2498, 1996.
114. K. Kurosawa, S. Yoshida, and K. Sakamoto, "Polarization properties of the flint glass fiber," *Journal of Lightwave Technology*, **13**, no. 7, pp. 1378–1384, 1995.
115. S. C. Rashleigh and R. Ulrich, "Magneto-optic current sensing with birefringent fibers," *Applied Physics Letters*, **34**, no. 11, pp. 768–770, 1979.
116. R. I. Laming, and D. N. Payne, "Electric current sensors employing spun highly birefringent optical fibers," *Journal of Lightwave Technology*, **7**, no. 12, pp. 2084–2094, 1989.
117. N. C. Pistoni and M. Martinelli, "Polarization noise suppression in retracing optical fiber circuites," *Optics Letters*, **16**, no, 10, pp. 711–713, 1991.
118. M. Takahashi, H. Noda, K. Terai, S. Ikuta, and Y. Mizutani, "Optical current sensor for gas insulated swtichgear using silica optical fiber," *IEEE Transactions on Power Delivery*, **12**, no. 4, pp. 1422–1428, 1997.
119. J. Blake, P. Tantaswadi, and R. T. de Carvalho, "Inline Sagnac interferometer current sensor," *IEEE Transactions on Power Delivery*, **11**, no. 1, pp. 116–121, 1996.
120. L. R. Veeser and G. W. Day, "Faraday effect current sensing using a Sagnac interferometer with a 3×3 coupler," *Proc. of The 7th Optical Fiber Sensors Conference*, Sydney, Australia, pp. 325–328, Dec. 1990.
121. M. Willsch and T. Bosselmann, "Optical current sensor application in the harsh environment of a 120 MVA power generator," *The 15th Optical Fiber Sensors Conference Technical Digest*, Portland, OR, pp. 407–410, May 2002.
122. K. Bohnert, P. Gabus, J. Nehring, and H. Brändle, "Temperature and vibration insensitive fiber-optic current sensor," *Journal of Lightwave Technology*, **20**, no. 2, pp. 267–276, 2002.
123. M. Takahashi, K. Sasaki, A. Ohno, Y. Hirata, and K. Terai, "Sagnac interferometer-type fiber-optic current sensor using single-mode fiber down leads," *The 16th Optical Fiber Sensors Conference Technical Digest (OFS-16)*, Nara, Japan, pp. 756–759, Oct. 2003.
124. E. Udd, "Overview of fiber optic sensors," in *Fiber Optic Sensors* (F. T. S. Yu and S. Yin, Eds.), Marcel Dekker, New York, pp. 1–39, 2002.
125. M. K. Barnoski and S. M. Jensen, "Fiber waveguides: a novel technique for investigating attenuation characteristics," *Applied Optics*, **15**, pp. 2112–2115, 1976.
126. J. Beller, "OTDRs and backscatter measurements," in *Fiber Optic Test and Measurement* (D. Derickson, Ed.), Prentice Hall, Upper Saddle River, NJ, pp. 434–474, 1998.
127. H. Izumita, "Recent development in fiber optic monitoring system for access networks," *The 16th International Conference on Optical Fiber Sensors*, Nara, Japan, pp. 258–261, Oct. 2003.
128. W. V. Sorin, "Optical reflectometry for component characterization," in *Fiber Optic Test and Measurement* (D. Derickson, Ed.), Prentice Hall, Upper Saddle River, NJ, pp. 383–433, 1998.
129. L. Thevenaz, M. Facchini, A. Fellay, P. Robert, D. Inaudi, and B. Dardel, "Monitoring of large structures using distributed Brillouin fiber sensing," *Proceedings of the 13th International Conference on Optical Fiber Sensors (OFS-13)*, Kyongju, Korea, *Proc. SPIE*, **3746**, pp. 345–348, 1999.
130. D. Inaudi and N. Casanova, "SMARTEC: bringing fiber optic sensors into concrete applications," *The 15th Optical Fiber Sensors Conference Technical Digest*, Portland, OR, pp. 27–30, May 2002.
131. Z. Chen, Y. Zhao, S. M. Srinivas, J. S. Nelson, N. Prakash, and R. D. Frostig, "Optical Doppler tomography," *IEEE Journal of Selected Topics in Quantum Electronics*, **5**, no. 4, pp. 1134–1142, 1999.
132. G. D. Peng and P. L. Chu, "Optical fiber hydrophone systems," in *Fiber Optic Sensors* (F. T. S. Yu and S. Yin, Eds.), Marcel Dekker, New York, pp. 417–447, 2002.
133. http://www.fiberdynamics.com/index.html.
134. Y. W. Lee and B. Lee, "High resolution cryogenic optical fiber sensor system using erbium-doped fiber," *Sensors and Actuators: A Physical*, **96**, no. 1, pp. 25–27, 2002.
135. L. Thevenaz, A. Fellay, W. Scandale, "Brillouin gain spectrum characterization in optical fibres from 1 to 1000 K," *Technical Digest of The 16th International Conference on Optical Fiber Sensors (OFS-16)*, Nara, Japan, pp. 38–41, Oct. 2003.
136. A. M. Scheggi, M. Brenci, G. Conforti, R. Falciai, and G. P. Preti, "Optical fiber thermometer for medical use," *The First International Conference on Optical Fiber Sensors*, IEE, London, pp. 13–16, Apr. 1983.
137. G. Orellana and M. C. Moreno-Bondi, "From molecular engineering of luminescent indicators to environmental analytical chemistry in the field with fiber-optic (bio)sensors," *The 15th Optical Fiber Sensors Conference Technical Digest*, Portland, OR, pp. 115–118, May 2002.
138. F. J. Arregui, D. Galbarra, I. R. Matias, K. L. Cooper, and R. O. Claus, "ZrO_2 thin films deposited by the electrostatic self-assembly method on optical fibers for ammonia detection," *The 15th Optical Fiber Sensors Conference Technical Digest*, Portland, OR, pp. 265–268, May 2002.

139. D. J. Webb, M. W. Hathaway, D. A. Jackson, S. Jones, L. Zhang, and I. Bennion, "First in-vivo trials of a fiber Bragg grating based temperature profiling system," *Journal of Biomedical Optics*, **5**, no. 1, pp. 45–50, 2000.

140. S. W. James, N. D. Rees, R. P. Tatam, and G. J. Ashwell, "Optical fiber long period gratings with thin film overlays," *The 15th Optical Fiber Sensors Conference Technical Digest*, Portland, OR, pp. 119–122, May 2002.

141. T. Allsop, D. J. Webb, and I. Bennion, "A high sensitivity long period grating Mach-Zehnder refractometer," *The 15th Optical Fiber Sensors Conference Technical Digest*, Portland, OR, pp. 123–126, May 2002.

142. S. Kim, Y. Jeong, S. Kim, J. Kwon, N. Park, and B. Lee, "Control of the characteristics of a long-period grating by cladding etching," *Applied Optics*, **39**, no. 13, pp. 2038–2042, 2000.

143. http://www.fiso.com.

Author Biography

CHAPTER 1

Bishnu P. Pal

Bishnu P. Pal obtained his M.Sc. and Ph.D. degrees in Physics from the Jadavpur University (Kolkata) and Indian Institute of Technology (Delhi) in 1970 and 1975, respectively. In late 1977 he joined the academic staff of IIT Delhi as a specialist on Fiber Optics, where he is a Professor of Physics since 1990 and is currently the Head of its Computer Services Center. He has worked as a Visiting Scholar/Visiting Professor in the area of Fiber Optics and Applications at the Norwegian Institute of Technology, Trondheim (Norway), the Fraunhofer Institute für Physikalische Messtechnik, Freiburg (Germany), the National Institute of Standards and Technology, Boulder (USA), University of Strathclyde, Glasgow (UK), Optoelectronics Research Center at City University of Hong Kong, and at the Universities at Nice and Limoges (France) for various periods. He has Guest Edited special issues of Journal of Optics (India) devoted to *Guided Wave Optical Components and Devices* (2004), Proc. IEE (Optoelectronics) devoted to *Guided Wave Optics on Silicon* (1996), International Journal of Optoelectronics devoted to *Optoelectronics in India* (1993) and Joint issue of JIETE (India) and IETE Tech Review (India) devoted to *Optoelectronics and Optical Communication* (1986). Professor Pal has extensive teaching, research, sponsored R&D, and consulting experience (for Indian and US industries) on various aspects of Fiber Optics and related components and he has published and reported over 100 research papers and research reviews (including an independently authored *Guided Wave Optics on Silicon: Physics. Technology, and Status* in the Elsevier series: *Progress in Optics*, vol. XXXII, Ed. Emile Wolf, 1993) in international journals and conferences and has coauthored one each Indian and US patents. He has been a thesis Advisor/Co-Advisor to over 10 Ph.D. students and over 50 M.Tech. and M.Sc. students at IIT Delhi. He is co-author of the book entitled *Fiber Optics and Instrumentation* (in Russian, Mashinostroenie Publishing House, Leningrad, 1987) and has edited the books: *Fundamentals of Fiber Optics in Telecommunication and Sensor Systems* (New Age Publications, New Delhi and John Wiley, New York, 1992, 3rd reprint 2001). He has also contributed 11 chapters in books and monographs. He has been deeply involved with the conception and development of the Fiber Optics Laboratory established at IIT Delhi in late 1970s. Professor Pal is a Fellow of IETE and is a member of the Optical Societies of America and India. Currently he is a member of the Executive Council of the Optical Society of India. He has been an invited speaker at 22 international conferences held in India and abroad and he has been a member of Technical/Advisory Committees of over 15 International Conferences. He is a co-recipient (with K. Thyagarajan) of the *First Fiber Optic Person of the Year award* in 1997 instituted by Lucent Technology in India for his significant contributions in all-fiber components for optical networks and also the *Gowri Memorial Award* for the year 1991 of IETE (India) for his paper (co-author B.D. Gupta) on fiber optic biosensors. He has been recently selected as a Distinguished Lecturer of IEEE/LEOS for 2005–2006 for his significant contributions to Guided Wave Optical Components and Devices. Professor Pal was Chairperson of the National Technical Panel of the Department of Information Technology, Government of India (1998–2001) to oversee growth of All-fiber Components in India and is currently a member of 1) the National Advisory Board on Instrumentation of the Department of Science and Technology, Government of India (2002- to date) and 2) the National Advisory Group on Nanotechnology of the Ministry of Information Technology (MiT), Government of India (2003- to date). His current research interests concern guided wave optical components for DWDM and optical networks, gain flattening in EDFAs, specialty fibers like dispersion compensating fibers and Bragg fibers, and also fiber optic sensors, optrodes, and near field fiber probes.

CHAPTER 2
Pak L. Chu

Professor Pak L. Chu received his B.E. (Hons.), M.E., and Ph.D. degrees from the University of New South Wales, Sydney (Australia), where he served as a faculty of Electrical Engineering and headed the Optical Communications Group for many years.

In July 2001 he took up the positions of Director of Optoelectronics Research Center and Chair Professor of the Department of Electronic Engineering at City University of Hong Kong. His research interests are in optical communication, optical fiber technology, optical waveguide technology, electromagnetic theory, plasma oscillations, and wave propagation in nonlinear media. He has published more than 400 papers in international journals and conferences in these areas. Dr. Chu is a Fellow of the Australian Academy of Technological Sciences and Engineering, a Fellow of the Optical Society of America (OSA), and a Fellow of the Institution of Engineers Australia. The Prime Minister's Department of Australia awarded him in 2003 with the Centenary Medal for his outstanding contributions in optical communications. He is also a consulting professor for four Chinese universities: Shanghai Jiaotong University, Shanghai University, Southeastern University (Nanjing), and the Beijing University of Post and Telecommunication.

CHAPTER 3
Tanya M. Monro

Tanya M. Monro received her Ph.D. degree at the University of Sydney, Australia on theoretical and experimental aspects of self-writing in photosensitive materials. She was awarded the Bragg Gold Medal for the best Ph.D. thesis in Australia in 1998. She joined the Optoelectronics Research Center (ORC) at the University of Southampton, UK in 1998 as a Research Fellow working on the development of holey or microstructured optical fibers. Working at the ORC from 1998 to 2004, she led research in a number of areas related to both silica and soft glass microstructured fibers. In 2000, she was awarded a Royal Society University Research Fellowship (URF) and in 2003 became a Reader at the ORC. Research highlights from this period include the first soft glass microstructured fibers and fibers with record nonlinearity. In 2005, Professor Monro became the Chair of Photonics at the University of Adelaide, Australia.

CHAPTER 4
Sonali Dasgupta

Sonali Dasgupta was born in Kolkata, India in May 1979. She received her B.Sc.(Hons.) degree in Physics from the University of Delhi and her M.Sc. degree also in Physics from the Indian Institute of Technology (Delhi) in 2000 and 2002, respectively. Since 2003, she has been pursuing a Ph.D. in Physics at the IIT Delhi as an awardee of the Shyama Prasad Mukherjee Fellowship of CSIR (India). Her research activities concern modeling and design of photonic bandgap Bragg fibers, Bragg fiber components for DWDM and optical networks, dispersion compensators, and study of nonlinear optical effects in these specialty fibers. Ms. Dasgupta is a student member of the Optical Society of America (OSA) and is also the current president of the IIT Delhi student chapter of OSA. She is also a student member of SPIE.

Bishnu P. Pal (See Chapter 1)

M. R. Shenoy

M. R. Shenoy is currently an Associate Professor of Physics at the Indian Institute of Technology (Delhi), working in the area of Fiber and Integrated Optics. He obtained his Ph.D. degree from Indian Institute of Technology (Delhi) in 1987 and joined the faculty of the Physics Department at IIT Delhi in 1988. He spent about 1 year as a visiting scientist at the Department of Electrical and Electronic Engineering, University of Glasgow during the year 1990, working in the area of Integrated Optical Devices. Dr. Shenoy has also been a visiting scientist at the University of Nice, France during September to December 1992 and May to July 1997 for collaborative research on nonlinear integrated optics. His current R&D interests are in the area of fiber amplifiers and in-line fiber components for optical fiber communication. Dr. Shenoy is an author/coauthor of over 30 research papers and is a co-editor of the book titled *Fiber Optics Through Experiments*, Viva Publishers, New Delhi (1994).

CHAPTER 5
Kin Seng Chiang

Kin Seng Chiang (M'94) received his B.E. (Hon.I) and Ph.D degrees in electrical engineering from the University of New South Wales, Sydney, Australia, in

1982 and 1986, respectively. In 1986, he spent 6 months in the Department of Mathematics of the Australian Defence Force Academy in Canberra. From 1986 to 1993, he was with the Division of Applied Physics of the Commonwealth Scientific and Industrial Research Organization (CSIRO), Sydney. From 1987 to 1988, he received a Japanese Government research award and spent 6 months with the Electrotechnical Laboratory in Tsukuba City, Japan. From 1992 to 1993, he worked concurrently for the Optical Fibre Technology Centre (OFTC) of the University of Sydney. In August 1993, he joined the Department of Electronic Engineering of the City University of Hong Kong, where he is currently Chair Professor. He has published over 280 papers on optical fiber/waveguide theory and characterization, numerical methods, fiber and waveguide devices, optical sensors, and nonlinear guided-wave optics. Dr. Chiang is a Fellow of the Optical Society of America a member of the International Society for Optical Engineering (SPIE), and the Australian Optical Society. He received the Croucher Senior Research Fellowship for 2000–2001.

Vipul Rastogi

Vipul Rastogi received his B.Sc. degree from Rohilkhand University Bareilly, India in 1991, his M.Sc. degree from the Indian Institute of Technology Roorkee (formerly the University of Roorkee), India in 1993, and his Ph.D. degree from the Indian Institute of Technology (Delhi) in 1998. From 1998 to 1999, he was a Postdoctoral Fellow with the Universite de Nice Sophia-Antipolis, Nice, France. During 2000 to 2003 he worked as a Research Fellow in the Optoelectronics Research Center, City University of Hong Kong. In November 2003, he joined the Physics Department of the Indian Institute of Technology Roorkee, where he is currently working as an Assistant Professor His research work has included second-order nonlinear interactions in optical waveguides and periodically segmented waveguides. His current research interests are large mode area optical fibers, microstructured fibers, and long period waveguide gratings. Dr. Rastogi is a member of the Optical Society of America.

CHAPTER 6

Ajoy Ghatak

Ajoy Ghatak is currently Emeritus Professor of Physics at Indian Institute of Technology (Delhi). He obtained his M.Sc. from Delhi University and Ph.D from Cornell University. He has published over 165 research papers and many books. His book *Optics* has been translated to Chinese and Persian. His earlier book *Inhomogeneous Optical Waveguides* (co-authored with Professor M. S. Sodha) has been translated to Russian and Chinese, and his book *Fiber Optics on a PC* (with A. Sharma and R. Tewari) has been translated to Portuguese. Some of his other books are *Optical Electronics, Introduction to Fiber Optics* (both co-authored with Professor K. Thyagrajan and published by Cambridge University Press), and *Quantum Mechanics*. His research interests are in Fiber Optics and in Quantum Mechanics. Professor Ghatak is a recipient of the International Commission for Optics Galileo Galilei award, OSA (Optical Society of America) Esther Hoffman Beller award, the CSIR SS Bhatnagar award, Optical Society of India Amita De Memorial award, Indian Physics Association MM Chugani award, and the UGC Meghnad Saha award. Professor Ghatak is a Fellow of the OSA.

K. Thyagarajan

K. Thyagarajan has been Professor of Physics since 1990 at the Indian Institute of Technology (New Delhi). He received his B.Sc. and M.Sc. degrees from Delhi University and Ph.D. from IIT Delhi in 1971, 1973, and 1976, respectively. During 1977–1978 he worked at the Ecole Normale Superieure, Paris, France and the Central Research Laboratories (LCR), Thomson-CSF, Orsay, France and was on Sabbatical leave at LCR, Thomson-CSF from 1983 to 1984. He was awarded the Fulbright travel fellowship to take up the position of Visiting Professor at the Department of Electrical Engineering, University of Florida, Gainesville, FL from 1993 to 1994. His current research interests are in the fields of optical fiber amplifiers, dispersion compensating systems, fiber gratings, and nonlinear guided wave optics.

In addition to more than 100 research publications to his credit, Professor Thyagarajan has also co-author (with Professor Ajoy Ghatak) five books: *Contemporary Optics* (Plenum Press, New York, 1978), *Lasers: Theory and Applications* (Plenum Press, New York, 1981), *Optical Electronics* (Cambridge University Press, Cambridge, 1989), *Introduction to Fiber Optics* (Cambridge University Press, Cambridge, 1998), and *Lagrangian Optics* (Kluwer Academic Pub., Boston, 2002; co-authors Prof. Lakshminarayanan and Prof Ghatak). The first two books were reprinted in India by Macmillan India Ltd. in 1984 and the next two books by Foundation Books, India in 1991 and 1999, respectively. He has also coauthored (with Prof. Ajoy Ghatak) a review *Graded Index Optical Waveguides: A Review*, *Progress in Optics* (Ed. E. Wolf), North Holland Pub. Co., Amsterdam, Vol. XVIII, 1980, *Nonlinear Optics in Encyclopaedia of Modern Optics* (B. Guenther, Editor) Elsevier, 2004, and *Linear and Nonlinear Propagation Effects in Optical Fibers* in *Optical Solitons: Theoretical and Experimental Challenges*, K. Porsezian and V. C. Kuriakose (Eds.), Lecture Notes in Physics, Vol. 613, Springer Verlag, Heidelberg, 2003.

He was awarded a Research Fellowship of the Indian National Science Academy for carrying out research in the area of Fiber Optic Components during the period 1988 to 1991. He was the joint awardee (with Professor Bishnu P. Pal) of the "*Fiber Optic Person of the Year 1997*" award by Lucent Technologies-Finolex and Voice and Data, India. He was recently awarded the title of "*Officier dans l'ordre des Palmes Académiques*" by the French Government. In 2005 he was elected a Fellow of the Optical Society of America. He has been a consultant to *Tejas Networks India Pvt. Ltd.*, Bangalore specifically looking into advanced issues related to high capacity communication through optical fibers.

CHAPTER 7

Govind P. Agrawal

Govind P. Agrawal has been a Professor at the Institute of Optics at the University of Rochester since 1989. His previous appointments were at the École Polytechnique, France, at City University of New York, New York, and at Bell Laboratories, Murray Hill, NJ. He received his B.S. degree from the University of Lucknow in 1969 and the M.S. and Ph.D. degrees from the Indian Institute of Technology (Delhi) in 1971 and 1974, respectively; IIT Delhi later honored him with Distinguished Alumni award. Agrawal's research interests focus on optical communications, nonlinear fiber optics, and laser physics. He has authored/coauthored more than 300 research papers, several book chapters and review articles, and seven books: *Semiconductor Lasers* (Norwell, MA: Kluwer Academic, 2nd ed., 1993); *Fiber-Optic Communication Systems* (New York: Wiley, 3rd ed., 2002); *Nonlinear Fiber Optics* (San Diego, CA: Academic Press, 3rd ed., 2001); *Applications of Nonlinear Fiber Optics* (San Diego, CA: Academic Press, 2001); *Optical Solitons: From Fibers to Photonic Crystals* (San Diego, CA: Academic Press, 2003); *Lightwave Technology: Components and Devices* (Hoboken, NJ: Wiley, 2004), and *Lightwave Technology: Telecommunication Systems* (Hoboken, NJ: Wiley, 2005). Dr. Agrawal is a Fellow of both the Optical Society of America (OSA) and the Institute of Electrical and Electronics Engineers (IEEE).

Qiang Lin

Qiang Lin received his B.S. degree in Applied Physics and M.S. degree in Optics from Tsinghua University, Beijing, China in 1996 and 1999, respectively. He is currently a graduate student pursuing his Ph.D. degree at the Institute of Optics, University of Rochester. His research interests includes nonlinear fiber optics, optical communications, ultrafast optics, and quantum optics.

Fatih Yaman

Fatih Yaman was born in Diyarbakir, Turkey in 1978. He received his B.S. degree in physics and mathematics from Koc University in Istanbul, Turkey in 2000. He is currently working on his Ph.D. degree at the Institute of Optics at the University of Rochester. His research interests include fiber optic parametric amplifiers, polarization mode dispersion in optical fibers, and soliton transmission.

CHAPTER 8

K. Thyagarajan (See Chapter 6)

CHAPTER 9

Govind P. Agrawal (See Chapter 7)

CHAPTER 10

Namkyoo Park

Namkyoo Park received his B.S. and M.S. degrees in Physics from Seoul National University and Brown University, respectively. After receiving his Ph.D. in Applied Physics from Caltech in 1994, he worked in the specialty optical fiber division of Lucent Bell Laboratories (MH) before his short stay at Samsung Electronics (1996–1997). He joined the School of EECS of Seoul National University in 1997, where he is currently an Associate Professor leading his research group, which was selected by the Korean Ministry of Science and Technology as a National Research Laboratory for next generation optical amplifiers. With over 15 years of research experiences in the area of fiber optics, he has authored/coauthored more than 160 international journals and conference publications and patents. During the past few years, his research efforts have been focused on

Raman amplifier, TDFA, OCDMA, PMD, multi-level transmission, EDWA, and smart applications of OTDR. He is currently an associate editor for IEEE Photonics Technology Letters and Optical Fiber Technology (Elsevier).

Pilhan Kim
Pilhan Kim received his B.S. and M.S. degrees in electrical engineering from the Seoul National University of Korea in 2002. Currently, as a Ph.D. candidate, he works in the optical communication systems laboratory in the department of EECS of Seoul National University. His research interests include general issues in various breeds of fiber optical amplifiers, including EDFA, TDFA, and Raman amplifiers, and also distributed sensor application of OTDR.

Hansuek Lee
Hansuek Lee received his B.S. and M.S. degrees in electrical engineering from the Seoul National University in Korea. Currently, as a Ph.D. candidate, he works in optical communication systems laboratory in the department of EECS of Seoul National University. His research interests include general issues in various kinds of fiber optical amplifiers, such as EDFA, TDFA, Raman, and especially nanostructured optical waveguide amplifiers based on the semiconductor technology.

Jaehyoung Park
Jaehyoung Park received his B.S and M.S. degrees in electrical engineering from the Seoul National University of Korea in 1999 and 2001. Currently, as a Ph.D. candidate, he works in optical communication systems laboratory in the department of EECS of Seoul National University. His research interest is focused in the fiber Raman amplifier.

CHAPTER 11

Niloy K. Dutta
Niloy Dutta is a Professor of Physics at the University of Connecticut. He has published many papers in the area of semiconductor lasers, semiconductor amplifiers, high power lasers, high speed transmission, and photonic logic. He has a Ph.D. in Physics from Cornell University. He is Fellow of IEEE, OSA, and SPIE and is a Member of the Connecticut Academy of Science and Engineering.

Puneit Dua
Puneit Dua is a staff scientist at Technology Research Laboratory, Pennsylvania State University. She has published several papers and has given several presentations in international conferences. She has a Ph.D. in Electrical Engineering from University of Connecticut.

Kunzhong Lu
Kunzhong Lu has a Ph.D. in Physics from the University of Connecticut. At present, he is a staff scientist at Multiplex Inc., in Plainfield, NJ. Dr. Lu has published several papers and has given numerous presentations in international conferences in the area of high power optical amplifiers.

James Jaques
James Jaques is a Member of Technical Staff at Lucent Technologies, Bell Laboratories. He has published many papers in the area of high power lasers, fiber transmission, and photonic logic. He has a Ph.D. in Physics from University of Notre Dame.

CHAPTER 12

Atul Srivastava
Atul Srivastava has over 25 years of research and development experience and is credited with many industry firsts in optics, semiconductor optoelectronics, and high-capacity fiber optic communication systems. His early research work on III-V Semiconductor Materials and Devices was carried out at the prestigious Tata Institute of Fundamental Research in India and at AT&T Bell Labs. From 1995 to 2000, he worked at Lucent Technologies Bell Labs as an Engineer and later as the Director of Optical Amplifiers Department at Agere Inc. While at Bell Labs, Srivastava was credited for demonstrating the first 100-channel long-distance terabit capacity WDM transmission using ultra-wideband amplifiers and several key inventions related to Optical Amplifiers. He co-founded a new start-up company, Onetta Inc., in 2000 and was the VP of Technology at Onetta. He was responsible for leading the research and development in optical amplifiers and WDM subsystems. Recently, Onetta was acquired by Bookham Technology, which is a provider of optical components and subsystems for both telecom and non-telecom applications. Dr. Srivastava joined Bookham technology as Vice President of Product Technology. He is a chair of the Technical Program committee of OSA's topical meeting on "Optical Amplifiers and their Applications." He is also chair of the ITCOM 2005 "Workshop on Control of Optical Components for the Next Generation Dynamic Networks." He is a member of Optical Fiber Communications (OFC) conference technical program committee. Dr. Srivastava is an expert member of the United States National

Committee for International Electro-technical Commission (IEC) for technical committees on Optical Amplifiers and Sub-systems. He has been honored with several awards for his optical networking research, including Bell Labs President's Gold Award and the Trophee du Telephone in Paris. He was elected Fellow of the Optical Society of America in 2003.

Yan Sun
Yan Sun is a vice president at Bookham, Inc. Dr. Sun joined Bookham through its acquisition of Onetta, where he served as the president. Dr. Yan Sun was a co-founder of Onetta, a pioneer in intelligent optical modules. Before Onetta, he worked on optical amplifiers and WDM optical communication systems at Bell Laboratories and optical networking group, Lucent Technologies, as a member of technical staff and then a senior manager. During this period, he and his colleagues discovered, modeled, and studied the control schemes of fast power transients in optically amplified networks and demonstrated the first ultra wideband optical amplifiers and DWDM optical communication systems at terabit level. Between 1985 and 1995, he made numerous contributions in several areas in the fields of microwave electronics, lasers, and optics. Yan is the author/co-author of more than 100 publications and presentations and a co-inventor of 20 patents. He has served on the committees of several international conferences and the editorial board of several technical journals. He was the recipient of several prestigious awards, including the Bell Labs President's Gold Award from Lucent Technologies and the Innovator of the Year Trophy in Paris, France. A fellow of OSA and a senior member of IEEE, Sun has served as a Guest Professor of Tsinghua University, University of Electronic Science and Technology of China, and Southwest Jiaotong University in China

CHAPTER 13

M.R. Shenoy (See Chapter 4)

Bishnu P. Pal (See Chapter 1)

Partha Roy Chaudhuri
Partha Roy Chaudhuri, after obtaining his M.Sc. degree in Physics from the Presidency College, Calcutta in 1991, joined the Time & Frequency Division of the National Physical Laboratory (New Delhi) where he was involved in the interface architecture and different applications of microprocessor-based real time systems for process control, data acquisition, and signal processing. In 1995, he joined the Indian Institute of Technology (Delhi) as a Project Scientist under the Technology Development Mission (TDM) Project on *Photonic Devices* where he also received his Ph.D. degree in 2001. During the TDM project and Ph.D. research, his primary contributions have been the design and fabrication of a computer-controlled electromechanical rig for automated fabrication of fused 2×2 coupler-based various in-line/all-fiber components and also development of a seminumerical model for the coupling process. Later, he pursued postdoctoral research at Kyoto Institute of Technology, Japan, as a Japanese Government Fellow, where he studied propagation in optical waveguides of arbitrary refractive index profiles. In 2002, he joined the Center for Wireless Communication as an associate member in the National University of Singapore (now known as Institute for InfoComm Research) and carried out extensive research in the area of Photonic Crystal waveguides and fibers. Presently he is in the Physics faculty of Indian Institute of Technology (Kharagpur). His current recent research focuses on specialty fiber, e.g. photonic crystal fiber, and biophotonics from the point of view of biomedical application of advanced photonic concepts and instrumentation.

Naveen Kumar
Naveen Kumar received his M.Sc. degree in Physics in 1999 from Kurukshetra University, Kurukshetra (India). Currently he is pursuing his Ph.D. degree from Indian Institute of Technology (Delhi). His current research is focused on design and fabrication of dense wavelength division multiplexed (DWDM) components for optical communication like FBT fiber couplers, wavelength interleaver, and EDFA gain flattening filters, and also on optical fiber sensors.

CHAPTER 14

Walter Johnstone
Walter Johnstone completed his education at the University of Strathclyde, where he graduated with a B.Sc. (Hons) degree in Chemical and Material Sciences in 1977 and a Ph.D. in Laser Physics in 1982. During the years 1980 to 1985 he held various project engineering and management posts involved with laser and optical fiber systems at Pilkington Optronics. In 1985 he was appointed to the post of Development Manager at Logitech Ltd. In this role he was a full member of the company's Management Board with responsibility for all technical functions, including new product development. In 1987 he returned to the University of Strathclyde where he is now a senior member of the academic staff, teaching communications principles, optical communications, and photonics in the Depart-

ment of Electrical and Electronic Engineering. His research interests include fiber optic components and systems for optical communications and optical sensing applications. Work in these fields has led to more than 100 technical publications and several patent applications. In addition to his academic career, Dr. Johnstone is a founding director of OptoSci Ltd. which was established in 1994. Major product lines in OptoSci Ltd. include a range of optical and fiber optic educator kits and a range of optical gas sensors.

CHAPTER 15

K. Thyagarajan (See Chapter 6)

CHAPTER 16

Nirmal K. Viswanathan
Nirmal Viswanathan received his Ph.D. degree in Physics from the University of Hyderabad (India) in 1997. He joined University of Massachusetts at Lowell in 1997 as a postdoctoral researcher, where he carried out multidisciplinary research on all-optical fabrication of surface-relief gratings in azo polymers. In 1999 he joined University of New Mexico, Albuquerque as research associate and worked on fiber lasers and fiber devices. In 2000 he joined 3M Company, Austin, Texas as senior research engineer and was responsible for a number of fiber-based activities, including manipulating the photosensitivity of optical fibers to improving the performance of chirped fiber Bragg grating (CFBG)–based devices for telecom applications. He is currently a scientist at Instruments Research Development Establishment, Dehradun, India and is carrying out research and development activities in fiber and integrated optics. Dr. Viswanathan has published more than 40 papers in international journals and conferences. He also has 6 U.S. patent applications to his credit. His research interests are applied experimental research and device development in optics and photonics technologies that includes fiber optics, polymer optics, and interferometry. Dr. Viswanathan won the Best Thesis award from the Indian Laser Association and the Circle of Technical Excellence and Innovation (CTE&I) award for individual technical achievement from the 3M Company.

CHAPTER 17

K. Porsezian
K. Porsezian received his M.Sc. degree from the University of Madras (India) in 1985 and his Ph.D. degree from Bhrathidasan University, Tiruchirapalli in 1991. His doctoral research focused on nonlinear excitations in ferromagnetic materials. In 1993, he joined the Physics faculty at Anna University (Chennai), where he is currently working as a Reader in the Department of Physics, Pondicherry University, Pondicherry, India. He has published over 75 papers and edited 2 books on optical solitons. His research interests focus on nonlinear fiber optics, SIT solitons, integrability aspects of nonlinear physical systems, modulation instability in nonlinear fiber optics, and beam propagation in photorefractive media. Dr. Porsezian is a recipient of the UGC Research Award (2004–2007), the Sathya Murthy Memorial Award for the Year 1997, given by the Indian Physics Association (1998), the Anil Kumar Bose Memorial Award given by the INSA (1998), the DAAD Post Doctoral Fellow (1995–1997), the Associateship Award from the ICTP, Italy (1997–2004), the AICTE Career Award for Young Teachers (1998–2000), the Indian National Science Academy Young Scientist Medal (1995), and Junior Associateship Award from the ICTP, Italy (1995–1997).

K. Senthilnathan
K. Senthilnathan was born in Kattukkanallur, India on June 12, 1977. He received his M.Sc. and M.Phil. degrees in Physics from the University of Madras (India). Recently he received his Ph.D. degree at the Anna University (India). His primary research interests are based on studying the linear and nonlinear properties of fiber and fiber Bragg gratings, nonlinear wave guides, and beam propagation in photorefractive media.

CHAPTER 18

Vikram Bhatia
Dr. Vikram Bhatia is currently with the Optical Line Modules division of Avanex Corporation, where he is responsible for designing transient-controlled variable-gain amplifiers for next-generation optical networks. He started his professional career with Corning Incorporated in 1996 where he helped commercialize fiber Bragg grating technology. At Corning, Dr. Bhatia held several positions in the Photonics Technology division at various times including those of Senior Development Scientist, Project Manager, and Application Engineering Supervisor. He received his M.S. and Ph.D. in Electrical Engineering from Virginia Tech in 1993 and 1996, respectively. He was awarded his B.E. degree by Birla Institute of Technology (India) in 1992. Dr.

Bhatia holds 6 U.S. patents and has coauthored more than 70 journal publications and conference papers. His technical interests include optical amplifiers, DWDM components, and dispersion compensation.

CHAPTER 19

Siddharth Ramachandran

Dr. Siddharth Ramachandran obtained his Bachelor of Technology (B.Tech.) degree from the Indian Institute of Technology (Kanpur) in 1991, his M.S. from the University of Wisconsin, Madison in 1993, and his Ph.D. in Electrical Engineering from the University of Illinois, Urbana in 1998. His graduate work focused on the optical properties of chalcogenide glasses. Since November 1998, he has worked at Bell Laboratories, Lucent Technologies, and subsequently OFS Laboratories, OFS-Fitel, first as a Member of Technical Staff and since March 2003 as a Distinguished Member of Technical Staff. Dr. Ramachandran's research focuses on investigating fiber and fiber-grating devices in specialty dispersion-tailored fibers. He has authored 58 journal and conference publications, 9 patent applications, and is the editor of an upcoming Springer-Verlag book on fiber-based dispersion compensation. Dr. Ramachandran is a member of *IEEE-LEOS*.

CHAPTER 20

Helge E. Engan

Helge E. Engan received his M.Sc. degree in Physics and his Ph.D. degree in Physical Electronics, both from the Norwegian Institute of Technology, Trondheim, Norway. He worked from 1967 at the Department of Physical Electronics at the Norwegian Institute of Technology. Since 1984 he has been a Professor of Physical Electronics within the Department of Electronics and Telecommunications at the Norwegian University of Science and Technology. He spent one year at Raytheon Research Division, Waltham, Massachusetts and two sabbatical years at Edward L. Ginzton Laboratory, Stanford University, California. He also spent a year with Novera Optics, San Jose, California. He has consulted with numerous institutions and companies internationally and nationally. He has studied electron–phonon interaction in semiconductors as well as surface acoustic wave transduction, propagation, and detection. Also, he has worked on underwater acoustics and acousto-optic interactions in optical fibers. In addition, he has worked with acousto-optic interaction in connection with laser probing of surface acoustic waves and with frequency shifting of light in optical fibers. Professor Engan is a member of the Institute of Electrical and Electronics Engineers, the Optical Society of America, the Acoustical Society of America, the Norwegian Academy of Technical Sciences, and the Norwegian Physical Society.

Kjell Bløtekjær

Kjell Bløtekjær received his diploma in Applied Physics from the Norwegian Institute of Technology, Trondheim (Norway) in 1956, and the degree of Tekn. Dr. from the Royal Institute of Technology, Stockholm, Sweden in 1966. From 1957 to 1966 he was with the Norwegian Defence Research Establishment, and since 1966 he has been with the Norwegian Institute of Technology, which is now the Norwegian University of Science and Technology. He is now Professor Emeritus of Physical Electronics. He has been a visiting scholar at Stanford University, California, IBM Research Laboratories, San Jose, California, and the Royal Institute of Technology, Stockholm, Sweden. He has been engaged in research on statistical noise theory, microwave tubes, acoustic waves in solids, bulk effects and waves in semiconductors, and optical storage in photorefractive and hole burning media. Presently, his research interests are in fiber optics, particularly fiber optic sensors. He is a member and former vice president of the Norwegian Academy of Technological Sciences and a member of the Royal Norwegian Academy of Sciences and Letters, where he is now president of the science class. He is also a member of the Norwegian Physical Society and Optical Society of America.

CHAPTER 21

Christopher R. Doerr

Dr. Christopher R. Doerr earned his B.S. in aeronautical engineering and his B.S., M.S., and Ph.D. (1995) in Electrical Engineering, all from the Massachusetts Institute of Technology (MIT). He attended MIT on an Air Force ROTC scholarship and earned his pilot wings at Chandler AFB, Arizona in 1991. His M.S. thesis was on reducing the noise of a mechanical/optical gyroscope, doing most of the work at the Charles Stark Draper Laboratory. His Ph.D. thesis was supervised by Prof. Hermann Haus and was on constructing a fiber optic gyroscope with noise below the quantum limit. To this end, he worked on high-repetition-rate mode-locked fiber lasers, quadrature squeezing, and a novel fiber optic gyroscope design. He went to Bell Labs in 1995 and has focused on optical communication integrated devices, both in InP and silica. He became a Bell Labs Distinguished Member of Technical Staff in 2000 and received the OSA Engineering Excellence Award in 2002. He is an associate editor for *IEEE Photonics Technology Letters*, is an elected member of the LEOS Board of Governors, is a subcommittee chair for OFC 2005, and has been on the committee of several other conferences.

CHAPTER 22

Alan Mickelson

Alan Mickelson was born in Westport, CT, on May 2, 1950. He received his B.S.E.E. degree from the University of Texas, El Paso, in 1973, and his M.S. and Ph.D. degrees from the California Institute of Technology, Pasadena, in 1974 and 1978, respectively. Following a post-doctoral period at Caltech in 1980, he joined the Electronics Research Laboratory, Norwegian Institute of Technology, Trondheim, Norway, initially as an NTNF Post-Doctoral Fellow, and then as a Staff Scientist. His research in Norway primarily concerned characterization of optical fibers and fiber compatible components and devices. In 1984, he joined the faculty of the Electrical and Computer Engineering Department, University of Colorado, Boulder, where, in 1986, he became an Associate Professor. His current research involves passive and active polymer integrated optic-device fabrication and characterization, and as well as techniques of biosensing. As well as being a university researcher, Professor Mickelson has served as a consultant to various commercial ventures including established companies and start ups.

Venkata Sivashankar

Venkata Sivashankar was born in Madurai, India on April 23, 1972. He received his B.Sc. and M.Sc. degrees in Physics from Madurai Kamaraj University (Madurai) in 1992 and 1994 and his M.Tech. degree in optoelectronics and optical communication from the Indian Institute of Technology, New Delhi, India in 1995, and his M.S. degree in Electrical and Computer Engineering from the University of Colorado, Boulder in 2003. He worked as Senior Software Engineer with IBM GSIL, Bangalore 1996 to 1998 and then as Staff Consultant for network application programming to various industries in USA from 1998 until he started working with Prof. Alan Mickelson as a research assistant in the field of polymer based integrated optics.

Ed M. McKenna

Ed McKenna was born in Santa Rosa, CA, on July 12, 1976. He received his B.S.E.E. and M.S.E.E. from the California Polytechnic State University, San Luis Obispo in 1998 and 2002, respectively. At present he is a research assistant under Dr. Alan Mickelson, pursuing his doctorate in Electrical Engineering from the University of Colorado at Boulder.

CHAPTER 23

Deepak Uttamchandani

Deepak Uttamchandani obtained his Ph.D. from the University College London in 1985. He joined the University of Strathclyde, Glasgow, UK where he is currently Professor of Microsystems Engineering at the Department of Electronic and Electrical Engineering. His research has concentrated on applying photonic and microsystems technology to optoelectronic sensors and systems, such as optically excited micromechanical resonator sensors, and in mechanical characterization of MEMS materials. More recently, his work has expanded to include photonic communication and RF applications of MEMS technology. He has been a part of the UK delegation on the World Micromachine Summit series of meetings and was Local Organiser when the Summit was held in Glasgow, Scotland in 1999. He has also been Co-Chair of a number of SPIE Conferences devoted to MEMS and MOEMS. Professor Uttamchandani is a Fellow of the IEE and a Senior Member of the IEEE.

CHAPTER 24

Brian Culshaw

Brian Culshaw was born in Lancashire, England on September 24, 1945. He graduated from the University College London in 1966 in Physics and thereafter completed (1969) his Ph.D. in Electronic and Electrical Engineering, specialising in microwave semiconductors. After a year at Cornell University, he joined Bell Northern Research (now Nortel) in Ottawa and, whilst continuing to work on microwave semiconductors, developed an interest in fibre optic technology. Late in 1973 he returned to UCL and, after two further years as a post doc working on semiconductor device simulation, developed his interest in fibre optic sensor technologies, their principles and applications. His research has encompassed fibre gyroscopes, hydrophones, spectroscopic analysis systems, and mechanical interferometric sensors. In 1983 he became

Professor of Optoelectronics at Strathclyde University in Glasgow.

He was *de facto* technical chair of the first (1983) International Conference on Optical Fibre Sensors (OFS), now a series regarded as the definitive meeting in the community. He chaired the tenth in Glasgow and was technical co-chair of the 17th in Bruges in 2005. He orchestrated, with SPIE in Bellingham, Washington USA, the CD-ROM of the series proceedings which has recently been reissued. He also initiated European meetings in Smart Structures and the EWOFS workshop series in optical fibre sensor technology. Predominantly with SPIE, he has organised numerous other conferences and workshops in Europe, the US, and Asia. He has recently been confirmed as President Elect of SPIE.

In the mid 1980s he was the founding editor for the *International Journal of Optoelectronics*, and until mid 2004 was a topical editor for *Applied Optics*. He has edited for over a decade with Alan Rogers of Surrey University the *Artech House* series in Optoelectronics, now over 50 titles. He has administered several major research initiatives, particularly multi partner EU programmes in sensing, measurement, fibre optics, and smart structures. He has reviewed research activities and proposals in the UK and elsewhere. He has also acted internationally in Ph.D. and Habilitation examinations.

CHAPTER 25

Wei Jin

Wei Jin received his B.Eng. and M.Sc. degrees from the Beijing University of Aeronautics and Astronautics in 1984 and 1987 respectively. He received his Ph.D. degree in fibre optics from Strathclyde University, Scotland in 1991, and worked as a Postdoctoral Fellow at Strathclyde from 1991 to 1995. He joined the Department of Electrical Engineering at the Hong Kong Polytechnic University in early 1996 and is currently a Professor in the Department, working on various photonics devices and sensors. He authored/co-authored 2 books, 300 journal and conference papers and several patents. He was awarded the 1999 HK PolyU President's Award for Outstanding Performance in Research and Scholarly Activities, and received the 2004 HK PTeC's Outstanding Professional Services and Innovation Awards - the Technology Transfer Award.

Tony Kar Yun Lee

Tony K.Y. Lee obtained his M.Phil. degree in Electrical and Mechanical Engineering and is currently pursuing his doctorate degree to be completed in late 2005 researching on the application of fibre optical sensor on railway application at the Hong Kong Polytechnic University. He has been involved with the research, development, and application of optical Fibre Bragg Gating Sensor (FBG Sensor) in railway engineering, including design and implementation of FBG sensors for disturbance-free axle counter on track, derailment detector on wheel/rail interaction, train identification, train speed detection, and strain/stress monitoring on vehicle structure etc. He also has been actively researching on the harmonic and thermal performance of DC & AC traction systems since 1990. His current interest is the architectural design and implementation of the FBG Sensor to build a high-performance smart railway. He is now the Rolling Stock Design & Systems Engineering Manager in the Kowloon-Canton Railway Corporation Transport Division, responsible for managing all the rolling stock projects, systems engineering & design matters, and train improvement works on reliability, maintainability, availability and safety aspects. He has been a railwayman for almost 20 years since his graduation.

Siu Lau Ho

S.L. Ho received his B.Sc. (with first class honours) and Ph.D. degrees in electrical engineering from the University, U.K. in 1976 and 1979, respectively.

He joined the Department of Electrical Engineering, the then Hong Kong Polytechnic in 1979. He is now a Chair Professor of Electricity Utilisation in Hong Kong Polytechnic University. He has been one of the most active consultants in the University, having won awards as the Most Active Consultant of the University from 1999 to 2003. He has served on many public committees for the Hong Kong Government as Chairman/consultant/members. He is also very active in railway engineering, particularly in the areas of electromagnetic interference, earthing and motor protection. Professor Ho's research interests are in the area of railway engineering, motor protection and design, electric drives, optimization studies and electromagnetic interferences.

He has published around 250 papers in refereed journals and conferences, including around 50 papers in the IEEE Transactions.

Hoi Lut Ho

Hoi Lut Ho received his B.Eng. and Ph.D. Degrees in Electrical Engineering from the Hong Kong Polytechnic University in 1997 and 2002, respectively. He is currently working as a research fellow in the Hong Kong Polytechnic University. He has published over 30 technical papers in international refereed journals as well as international conferences. His research interests include fiber optics sensors, multiplexing techniques in optical sensing system, fiber Bragg grating sensors and photonic crystal fiber sensors and devices.

Kin Tak Lau

Dr. Lau received his Bachelor and Master Degrees in Aerospace Engineering at the Royal Melbourne Institute of Technology (RMIT) University (Melbourne) in 1996 and 1997 respectively. He then received is Ph.D. in Mechanical Engineering at the Hong Kong Polytechnic University in 2001. He has since been working in the same University as Assistant Professor. Dr. Lau received many academic awards including the best paper award (1998), the Sir Edward Youde Memorial Fellowship Award (2000), Young Scientist Award (2002) and Young Engineer of the Year Award 2004. He has been elected as a member of the European Academy of Science. He is a Guest Editor of the Journal of Composites Part B: Engineering, Associate Editor of the International Journal of Structural Health Monitoring, the American Journal of Applied Sciences, and the Hong Kong Institution of Engineers (HKIE) Transactions. He is also a co-chairman of the Symposium of "Nanocomposites for Space and Infrastructure Applications" in European Materials Research Society (E-MRS) Spring Meeting, Strasbourg, 2004.

Li Min Zhou

Dr. L. M. Zhou graduated with a Ph.D. degree in Mechanical Engineering, University of Sydney (Australia) in 1994. He was awarded a prominent Australian Research Fellowship in worldwide open competition in 1996 and subsequently was appointed as Assistant Professor in the Department of Mechanical Engineering of the Hong Kong Polytechnic University in late 1996 and then was promoted to Associate Professor in 2001. Dr Zhou's current research activities have two major directions: one is centred on characterisation, manufacturing and mechanics of advanced engineering materials, including fibre reinforced composites, composites for infrastructure applications, nanocomposites; and the other is on the smart materials/structures and products, including applications of fibre optic sensors and shape memory alloys/polymers. Dr. Zhou has written over 160 technical articles, including 100 refereed journal papers and chapter contributions to refereed books in these fields. He also invented 3 patents in the area of smart products. Dr. Zhou has been honored with several awards, including the Australian Postgraduate Research Award and the President Award for Outstanding Performance/Achievement in Research and Scholarly Activities at the Hong Kong Polytechnic University.

Yu Zhou

Dr. Yu Zhou completed his Ph.D. degree in 1992 at the University of Newcastle (Australia) and joined The Hong Kong Polytechnic University in 1995. His research interests include turbulent flows and control, flow-induced vibrations and control. He has authored or co-authored more than 150 technical papers, including more than 70 international journal and book articles, 5 being published in prestigious *Journal of Fluid Mechanics*. He has acted as a reviewer for many leading international journals in the field of fluid mechanics, and also invited each year as an Expert Assessor of International Standing to assess research proposals by the Expert Advisory Committees of Australian Research Council. Since 1996, he has attracted a research fund exceeding 15 million HK dollars from both internal and external sources and successfully supervised 5 Ph.D. theses. Along with Dr H. Li, he received 2003 VSJ Paper Award from the Visualization Society of Japan.

CHAPTER 26

Byoungho Lee

Byoungho Lee received his Ph.D. degree in electrical engineering from the University of California at Berkeley in 1993. Since 1994, he has been with School of Electrical Engineering, Seoul National University, Korea, where he is now an associate professor. In 2002 he received the Presidential Young Scientist Award of Korea. He became a Fellow of SPIE in 2002 and a Fellow of OSA in 2005. His research interests include optical fiber grating devices and sensors, optical fiber lasers, three-dimensional display and diffractive optical elements.

Yong Wook Lee

Yong Wook Lee received his B.S., M.S., and Ph.D. degrees in electrical engineering from Seoul National University, Seoul, Korea, in 1998, 2000, and 2004, respectively. In 2004, he joined Basic Research Laboratory, Electronics and Telecommunications Research Institute (Korea) where he is a Senior Researcher and he is currently working on high-speed optical semiconductor devices including switches and modulators. His research interests include optical fiber grating devices, optical fiber sensors, optical fiber filters, and optical fiber lasers.

Minho Song

Minho Song studied electrical engineering and received his B.S. degree from the Seoul National University, Seoul, Korea in 1990, and his M.S. and Ph.D. degrees in electrical engineering in 1992 and 1997. In 1997, he joined the electro-optic research center, the Pennsylvania State University, where he was engaged in research and development on fiber-optic sensor systems as a post-doctoral research fellow. In 2000, he joined the faculty of Chonbuk National University, Korea, where he is now an associate professor of Electronics and Information Engineering. His research interests have included fiber gratings, fiber-optic sensors, and optical metrology.

Index

Access network, 18
Acousto-optic interaction
  experimental setup for, 316–318
  fiber nonuniformity in, 319–320
  frequency shifters in, 318
  gain flattening filter, 323f
  principles of, 315–316
  scanning heterodyne interferometer in, 320–321
  tunable filters in, 318–319
  between various LP modes, 316t
  wavelength dependence in, 318–319
Active fibers
  background of, 55–56
  cladding pumped fiber lasers and, 56–57
  tunable fs-solition source, 56
Add/drop multiplexers, FBGs in, 237–238
Adiabatic iteration algorithm, flow diagram of, 161f
AFM. *See* Atomic force microscopes
AIBN (Azobisisobutyronitrile), 28–29
  chemical composition of, 29f
Air-guiding fibers, 62
Air-light overlap, in microstructured optical fibers, 46
Algorithm, adiabatic iteration, 161f
Amplified spontaneous emission (ASE)
  EDFAs and, 126f
  in FRAs, 165f
  in Raman amplifiers, 139, 141
Amplifier gain, in EDFAs, 183–184
Amplifier noise figure, 140
Amplitude modulated with vestigial sideband (AM-VSB), 173
  CNR of, 179
  experimental setup for, 177f
Amplitude modulation, 304–305
AM-VSB. *See* Amplitude modulated with vestigial sideband
Analog transmission, CATV, 173–176
Annealing, 343
Anomalous dispersion, 13
  in Raman amplifiers, 149–150
Anti-Stokes line, 131
Arrayed waveguide grating (AWG), 325, 327–330
  illustration of, 326f

Arrayed waveguide lens (AWL), 325
Arrayed wavelength gratings (AWDs)
  basic colorless, 286f
  in DWDMs, 284–286
  layout of, 285f
ASE. *See* Amplified spontaneous emission
Asymmetric digital subscriber line (ADSL), 35–36
Asymptotic matrix theory, 76
Atomic force microscopes (AFM), 343
  of diffraction gratings, 345f
Attenuation spectrum, of POFs, 28f
Automatic gain control (AGC), 199
Automobile optical networks
  data type and data rate of, 32f
  POFs in, 32
  topologies for, 33f
AWDs. *See* Arrayed wavelength gratings
AWG. *See* Arrayed waveguide grating
AWL. *See* Arrayed waveguide lens

Back plane interconnect
  of POFs, 32–34
  SAN in, 34f
Band filters, waveguide layout of, 333f
Bandpass filtering
  dispersionless, 301–304
  schematic for, 302f, 320f
  transmission and group delay response for, 303f
  tunability and loss characteristics in, 304f
Bandpass overlay filter, 228
  wavelength response of, 228f
Bandstop overlay filter, 228
  wavelength response of, 226f
Bandwidth, in FBGs, 262
Bandwidth control, of TAP-LPGs, 298–299
Beam propagation methods (BPMs), 50
Beam splitter/combiners, 214
Beating, FFC, 211
BERs. *See* Bit error rates
BHT. *See* Butylated hydroxy toluene
Bimorph beams, 362f
  tip displacement and, 363f

Binary phase shift keying (BPSK), 174, 177
Birefringence, 111–114, 416
  gain spectra changes with, 113f
  nonlinear effects of, 255
Bismuth-oxide based gasses, 58–59
Bit error rates (BERs), 15, 34, 173, 185, 198, 307
  in CATV setup, 175f
  measuring, for transmission, 200f
  time-resolved, in EDFAs, 195f
  tolerable mismatch and, 15f
Bit rate transparency, 10
Bloch wave analysis, 272
  in Bragg grating solitons, 272
  multiple scale, 273–274
  theoretical model of, 272–273
Blue-shifts, 13
Bow-tie fiber, acoustic wavelength v. frequency for, 315f
BPMs. *See* Beam propagation methods
BPSK. *See* Binary phase shift keying
Bragg fibers, 71
  air-silica, 81f
  bandgap in one-dimensional periodic medium, 72–75
  dispersion compensating, 78–79
  dispersion of, 78
  electric field components of TE and TM modes in, 77f
  fabrication of, 80–81
  light propagation in, 75–76
  for metro networks, 79–80
  modal characteristics of, 72
  propagation loss, 77–78
  radiation loss of, 80f
Bragg gratings. *See* Fiber Bragg gratings (FBG)
Bragg reflection, in FBGs, 258–259
Bragg resonance condition, 270f
Bragg structures, planar, 73f
Brillouin effects, 53–54
Broadband amplification, 137
Broadband mode converters, in few-mode fibers, 295–296
Butylated hydroxy toluene (BHT), function of, 28f

Carrier-to-noise ratio (CNR), 173, 175
  AM-VSB, 179
  for CATV channels, 176
CATV. *See* Community antenna television
Channel protection, in EDFAs, 195–196
Channel reconfiguration, FRAs and, 166–167
Chemical sensors, 376–379
  distributed, 383–386
  in multiplexed fiber optic spectroscopy, 379–380
  olive oil and, 380–383
Chemically selective gels, 386f
Chirped fiber Bragg gratings, 12
Chromatic dispersion, 3, 5
  in linear effects, 252–253
Cladding mode coupling, for Bragg gratings, 287f
Cladding pumped fiber lasers, 56–57
CMT. *See* Coupled mode theory
CNR. *See* Carrier-to-noise ratio
Coarse wavelength division multiplexing, 20–21
Comb drive actuator, 355f
Comb drive chopper, 359f
Community antenna television (CATV), 173
  analog transmission using, 173–176
  CNR for, 176f
  CSO as function of, 178f
  CTB as function of, 178f
  experimental setup of, 174f
  hybrid digital/analog transmission, 176–177
Composite beams, views of, 392f
Composite second order (CSO), 173
  calculated values of, 180
  of CATV function, 178f
  measurement of, 175
Composite triple beat (CTB), 173
  of CATV function, 178f
  measurement of, 175
Continuous isochronal annealing, for gratings in UV sensitized fiber, 246f
Continuous wave (CW), 105, 134, 147
Coupled mode theory (CMT), 207, 268
  for FBG, 235
  for LPG, 239–240
Coupling region, propagation in, 210–211
Cross-phase modulation (XPM), 91, 147
  in optical fibers, 95–98
Crystal structures, silica/air, 41
CSO. *See* Composite second order
CTB. *See* Composite triple beat
CW. *See* Continuous wave

DCF. *See* Dispersion compensating fibers
DCFA. *See* Double-clad fiber amplifier
Demultiplexers
  band, 332–333
  concept behind, 333f
  example design of, 331–332
Demux
  configuration of, 282
  pass band profile of, 289f
  spectrum of, 288f
Dense wavelength division multiplexing (DWDM), 1, 187, 205, 229–230
  AWGs in, 284–286
  DCFs in transmission of, 12–17
  emergence of, 10
  FBGs in, 286
  fibers for, 10–12
  FWM efficiency in, 11f
  introduction to, 281
  key performance characteristics of, 281–286
  schematic of, 2f
  thin film filters in, 282–284
  in transport, 18
DFB lasers. *See* Distributed feedback lasers
Diamine, 338
Dianhydride, 338
Diffraction efficiency
  decay of, 347f
  first-order, 346
  in thin sinusoidal gratings, 343–344
Diffraction gratings
  AFM measurement of, 345
  in dye-doped waveguide polymers, 343–348
Diffusion lengths, 112–114
Digital/analog transmission
  CATV, 176–177
  results of, 178–180
Diluent gas, 249
Diode laser spectroscopy, 382
Dispersion
  of Bragg fibers, 78
  curve, 260f
  in FBG, 261
  few-mode fibers and, 293–294
  of microstructured optical fibers, 45–46
Dispersion compensating Bragg fibers, 78–79
Dispersion compensating fibers (DCF), 78–79, 291
  coaxial dual-core, 16f
  dispersion spectra of, 15f
  dispersion-induced pulses and, 13f, 14f
  in DWDM transmission, 12–17
  fabricated dual-core, 17f
  first generation, 14
  refractive index profile for, 14f, 16f
  schematic illustration of, 14f
Dispersion compensating modules (DCM), 291
  HOM, 300–301

  tunable HOM, 306–308
Dispersion compensation
  FBGs in, 238–239
  HOM fibers and, 301f
Dispersion curves
  bright solitions in, 277f
  dark solitions in, 277f
  nonlinearity in, 265–266, 267f
Dispersion length, 5
Dispersion power penalty, 7
Dispersion relation, in FBG, 260–261
Dispersion shifted fibers (DSF)
  nonzero, 11–12
  optical transparencies and, 8–9
Dispersion slope compensating fibers (DSCFs), 15–16
  performance parameters of, 19t
Dispersion spectrum
  for DCFs, 15f
  of metro fibers, 20f
  in NZ-DSF, 19f
  in optical transparency, 3–8
  of SMFs, 8f
  TE mode, 79f, 80f
Dispersion-induced pulses, in DCFs, 13f, 14f
Dispersions, for standard bit rates, 8t
Dispersion-tailored few-mode fibers
  broadband mode converters, 295–296
  grating phase matching curve in, 297f
  LPGs in, 295–300
Distributed feedback (DFB) lasers, 8, 10, 174–175
  modulation of, 20
Distribution systems, 19
Doping profile, of C/L EDF, 158
Double hump potentials, 271f
Double Rayleigh backscattering (DRB)
  in FRAs, 165f
  in Raman amplifiers, 143f
Double Rayleigh scattering (DRS), in Raman amplifiers, 142
Double-clad fiber amplifier (DCFA), 173
  gain tilt measurement of, 177–178
Double-pass forward (DPF) EDF super fluorescent source, 230f
DRB. *See* Double Rayleigh backscattering
DRS. *See* Double Rayleigh scattering
DSCFs. *See* Dispersion slope compensating fibers
DSF. *See* Dispersion shifted fibers
Dual pump parametric amplifiers, 107–109
  average gain spectra for, 110f, 113f
  gain spectra for, 108f
  measured and calculated gain spectra for, 109f
  optimized gain spectra for, 106f
DWDM. *See* Dense wavelength division multiplexing
Dye-doped polymer fiber amplifier, 39

Index

Dye-doped polymers
 absorption spectrum, 340f
 photobleaching, 338–343
Dye-doped waveguide polymers,
  diffraction gratings in, 343–348
Dynamic load tests, 398f
Dynamic networks
 channel protection schemes in,
  195–196
 EDFAs for, 187–200
 high capacity networks and, 186–187
 laser control in, 199–200
 link control in, 198–199
 pump control in, 196–198
 transients in, 192–195
Dynamic strain measurement
 of composite samples, 391–392
 experimental setup for, 392

EDF. *See* Erbium-doped fibers
EDFA. *See* Erbium-doped fiber
  amplifiers
Effective index method (EIM), 4, 83
 of HFs, 87f
 of SCF, 86f
Effective noise figure, in Raman
  amplifiers, 140–142
Eigen polarization, 21
Eigenmodes, 325
EIM. *See* Effective index method
Electro-optic (EO) effect, 335–336
Electro-optic (EO) waveguide polymers
 integrated optics in, 335–338
 introduction to, 335
Elliptical core fiber
 modes of, 312f
 propagation constants of, 313f
EO effect. *See* Electro-optic effect
Erbium
 absorption and emission cross section
  spectra, 121f
 energy level diagram of, 9f, 120f, 183f
 signal gain v. length of, 123f
Erbium-doped fiber amplifiers
  (EDFA), 1, 6, 155, 173, 205, 233
 amplifier gain in, 183–184
 ASE spectra for, 126f
 basic characteristics of, 182–183
 channel power in, 188f
 channel protection schemes in,
  195–196
 cross talk between channels in, 188f
 delay in, 191f
 for dynamic networks, 187–200
 emergence of, 9–10
 gain dynamics of, 187–190
 gain flattening of, 124–126, 220, 241
 gain spectra for, 125f, 126f
 for high capacity networks, 181–187
 introduction to, 119
 in laser control, 199–200
 L-band, 155–160
 lifetimes of, 190f

link control in, 198–199
long fiber optic link with, 127f
maximum channel power excursion
  of, 194f
noise in, 126–128, 184–185
optical amplification in, 121–124
optical SNR in, 185–186
output power of, 159f, 189f, 191f
overamplification by, 119–120
parameters, 156t
peaks and numbers of, 193f
power transients for, 189f, 194f
power transients for, chains, 190–192
pump control in, 196–198
pump power in, 197f
refractive index profile for, 128f
regions of power excursion of, 192f
saturation characteristics of, 184f
for S-band, 128
schematic layout of, 9f, 182f
structure of, 124f
system issues in, 185
time resolved BERs in, 195f
Erbium-doped fibers (EDF)
 doping profile of C/L, 156f
 gain and loss coefficient spectra for,
  184f
 inversion distribution of, 156f
 output signal obtained from, 159f
 parameters, 156t
 PCE of, 159
 ring laser, 230f
Experimental grating characteristics,
  348f
External filters, gain flattening using,
  124–125
Extrinsic sensors, indicative linear
  modulation mechanisms for, 373f

Fabry-Perot laser, 8
Faraday effect, 414f
FBG. *See* Fiber Bragg gratings
FDDI, 19
FEM. *See* Finite element modeling
Femtosecond pulse sources, 56
Few-mode fibers. *See also* Dispersion-
  tailored few-mode fibers
 acoustic properties of, 313–315
 acousto-optic interaction, 315–321
 dispersive properties of, 293–294
 industrialization and, 322
 microbend grating spectrum in, 300f
 practical considerations on, 321–322
Few-mode optical fibers
 introduction to, 311
 optical properties of, 311–313
FFC. *See* Fused fiber coupler
Fiber amplifiers
 dye-doped polymer, 39
 emergence of, 9–10
Fiber Bragg gratings (FBG), 233–235,
  256, 375f
 in add/drop multiplexers, 237–238

applications of, 237–239
bandwidth of, 262
Bloch wave analysis of, 272
Bragg reflection in, 258–259
cladding mode coupling for, 287f
CMT for, 235
in dispersion compensation, 238–239
dispersion in, 261
dispersion relation in, 260–261
DWDMs and, 286
experimental considerations of, 278
fabrication of, 257f
fundamentals of, 257–258
gap solitons in, 267–272
grating period in, 238f
introduction to, 233
length variation of reflectivity of, 236f
linear properties of, 258–262
NLCM equations in, 262, 263–264
nonlinear properties of, 262–266
nonphase matched interaction in,
  236–237
PBG in, 260–261
peak reflection in, 234f
phase matched interaction of, 236
in POF, 37f, 38f
pulse propagation in, 259–260
reflection in, 261–262
refractive index during fabrication
  of, 244f
refractive index of, 262–263
simulated transmission spectrum of,
  237f
solitons in, 257, 266–267
spectrum of, 38f
thermal stability of, 245–247
transmission and reflection spectrum
  of, 287f
transmission coefficients in, 261–262
tuning of, 38f
types of, 258
in ultrasound detection, 376f
Fiber Bragg gratings (FBG) sensors
 dynamic strain measurement,
  391–392
 FMCW, 391f
 interrogators for, 407f
 introduction to, 389
 multipoint, 395f
 in temperature measurement, 393f
Fiber design methodologies, 49–51
 confinement loss in, 51
 effective index, 49
 structural, 49–50
 summary of, 51
Fiber grating sensors, 401–409
 interrogator types, 405f
 strain measurement resolution of,
  406t
 types of, 403f
Fiber loop reflector, 218–220
 schematic of, 219f
 wavelength response of, 220f

Fiber material dispersion, 31f
Fiber nonuniformity
 in acousto-optic interaction, 319–320
 schematic of, 321f
Fiber optic current sensors, 414–416
 examples of, 415f
 reflection-type, 417f
Fiber optic gyroscopes (FOG), 401, 409–416
 configurations of, 410f
 cosine response of, 409f
 nonreciprocal phase shifts and, 411f
 principle of, 409f
Fiber optic parametric amplifiers (FOPA), 101, 114
 Dual pump, 107–109
 four wave mixing and, 101–103
 gain spectra for, 104, 108f
 measured and calculated gain spectra for, 109f
 optimized gain spectra for, 106f
 parametric gain for, 103f
 residual fiber birefringence and, 111–114
 single-pump, 103–107
Fiber optic sensors
 basic principle of, 371–374
 intrinsic and extrinsic sensing with, 372f
 introduction to, 371
 networks, 373f
Fiber plane, light intensity in, 358f
Fiber Raman amplifier (FRA), 160
 calculation of higher order terms, 162
 channel reconfiguration and, 166–167
 gain clamping and, 167–168
 in gain prediction, 162–163
 gain profiles and pump evolution of, 166f
 gain spectrum design in, 166f
 initial condition set up of, 162
 inverse scattering and, 164–166
 noise in, 165f
 reiteration in, 162
Fiber structural detuning, in L-band EDFA, 157–160
Fiber to planar waveguide bandstop coupler
 electrode structures for, 229f
 schematic diagram of, 226f, 227f
Fibers
 characteristics of, 12f
 for metro networks, 18–20
Figure of merit (FOM), 78, 281
Finite element modeling (FEM)
 dynamic analysis of, 355–356
 simulation, 356f
 static analysis of, 355
Fluorescence spectroscopy, 378
FMCW. See Frequency modulated continuous wave

FOG. See Fiber optic gyroscopes
FOPA. See Fiber optic parametric amplifiers
Formalism, construction of, in Raman amplifiers, 160
Four wave mixing (FWM), 11, 111
 in DWDM systems, 11f
 frequency generation because of, 99f
 in optical fibers, 98–99
 theory of, 101–103
Fourier method, 331
Fourier transforms, 4–5
Four-wave mixing (FWM), 91
FRA. See Fiber Raman amplifier
Frequency modulated continuous wave (FMCW), 390
 FBG sensor array, 391f
Frequency shifters
 in acousto-optic interaction, 318
 output spectrum from, 318f
Frequency-resolved optical gating (FROG), of signal pulses, 151f
FROG. See Frequency-resolved optical gating
Fused biconical tapered (FBT) fiber coupler, structure of, 206f
Fused fiber coupler (FFC)
 in 3-dB couplers, 214–215
 applications of, 214–221
 in beams splitters/combiners, 214
 beating, 211
 cross-section of, 210f
 discussions regarding, 212–214
 elongation signatures of, 213f
 excess loss during fabrication of, 207f
 fabrication of, 205–207
 in fiber loop reflector, 218–220
 fiber parameters for, 213t
 field distributions of, 211f
 in gain flattening, 220
 introduction to, 205
 modal fields of, 212f
 mode analysis algorithm for, 208–210
 modeling, 207–214
 in MZI-based all-fiber wavelength interleaver, 217–218
 in nonlinear optical loop mirror, 220–221
 polarization of, 212
 propagation in coupling region, 210–211
 pseudowaveguide in, 211f
 schematic of, 215f
 schematic of fabrication of, 206f
 supermodes, 211
 in tap/access couplers, 215
 in tree couplers, 215
 Wavelength interleaver, 216–217
 in WDM couplers, 215–216
Fused over-coupled coupler, recorded pulling signature during fabrication of, 207f
FWM. See Four wave mixing

Gain clamping, FRAs and, 167–168
Gain dynamics, of EDFAs, 187–190
Gain engineering, of Raman amplifiers, 160–169
Gain flattening
 acousto-optic, 323f
 of EDFAs, 220
 of EDFAs using external filters, 124–125
 experimental setup used in, 221f
 intrinsically flat, 125–126
 LPGs in, 241
Gain prediction, FRAs in, 162–163
Gain profiles, of FRAs, 166f
Gain saturation, in fiber optic Raman amplifiers, 135–136, 136f
Gain spectrum
 in FRAs, 166f
 pump adjustment and, 168f
 pump power and, 167f
Gain tilt, of DCFA, 177–178
Gap solitons
 bright, 269
 dark, 270
 in Kerr media, 267–271
 in quadratic media, 271
Gaussian pulses, 7
 broadening of, 5f
 propagation of, at wavelengths, 13f
Ge concentration, UV-sensitized fibers and, 244–245
Gigabit Ethernet, 19
Glass transition temperature, 336–337
Graded-index multimode fibers, bandwidth of, 31f
Grating formation, process of, 346–347
Grating phase matching curve, in dispersion-tailored fibers, 297f
Grating solitions
 due to GVD, 274–275
 due to TOD, 275–276
Group index, 4
Group-velocity dispersion (GVD), 3–8, 13, 148–149, 252
 grating solitions due to, 274–275
Group-velocity mismatch, in Raman amplifiers, 148–149
GVD. See Group-velocity dispersion

H_2 sensitization, diluting, 247–250
HFs. See Holey fibers
High capacity networks
 amplifier gain in, 183–184
 amplifier noise in, 184–185
 dynamic networks and, 186–187
 EDFAs for, 181–187
 optical SNR in, 185–186
 system issues in, 185
High temperature polymers, 338
Higher order dispersion (HOD), 252–253
Higher order mode (HOM), 291, 294
 adjustable, 306f

DCMs, 300–301
  dispersion compensation and, 301f
  tunable, DCMs, 306–307
  tunable LPGs in, 304–306
Highly nonlinear fibers (HNLFs), 105
High-order-mode fibers, 12
HNLFs. See Highly nonlinear fibers
HOD. See Higher order dispersion
Holey fibers (HFs), 42, 44, 45
  cross-section of, 87f
  design methodologies for, 49–51
  effective index profiles of, 87f
  leakage losses and, 88f
  modal intensity of, 55f
  REIM and, 87–88
HOM. See Higher order mode
Home networks, POFs in, 34–36
Honeycomb fibers, 42, 62
Hooke's law, 341–342
Hybrid tunable filter, in O-MEMS, 365–368

Illumination, POFs in, 36, 37f
Industrialization, few-mode fibers and, 322f
Intensity profiles, of solitons, 256
Intermodal beatlength, optical wavelength and, 319f
International Conference on Optical Fiber Sensors (OFS-16), 401, 402f
Internet, 1
Inverse scattering, fiber Raman amplifiers and, 163–166
Inversion distribution, of C/L EDFs, 156f
Irreversible gratings, in waveguide polymers, 347–348
Isothermal annealing, 246f

Kerr coefficients, refractive index profile in oppositely signed, 263
Kerr media, gap solitons in, 267–271
Kerr nonlinearity, 253–254
Kowloon Canton railway (KCR) train body shell, measurement of, 396–398

Large-mode area fibers, 54–55
Large-mode HFs (LMHFs), 54, 56
Laser control, EDFAs in, 199–200
L-band EDFA, 155–160
  channels dropped in, 190f
  fiber structural detuning in, 157–160
  introduction to, 155–156
  NF of, 156f
  output power of, 156f
  pump wavelength detuning in, 156–157
LCM. See Linear coupled mode
Leakage losses
  HFs and, 88f
  of SCFs, 86f

Light lines, 74–75
Light propagation, in Bragg fibers, 75–76
Linear coupled mode (LCM), 260
Linear effects
  chromatic dispersion in, 252–253
  of FBGs, 258–262
  optical loss in, 252
Link control
  EDFAs in, 198–199
  schematic representation of, 199f
Liquid sample classification, 382f
LMHFs. See Large-mode HFs
Long period gratings (LPG), 258, 291
  applications of, 240–241
  bandwidth control of, 298–299
  broadband, 241
  coupled-mode theory for, 239–240
  detuning effects in, 304–305
  in dispersion-tailored few-mode fibers, 295–300
  in gain flattening, 241
  grating couples in, 234f
  introduction to, 233
  mode conversion with, 294–295
  as MZI, 241f
  PMC for, 302f
  polarization dependence of, 297f
  power transmitted by, 240f
  as sensors, 305f
  spectral control of, 298f
  static devices using, 300–304
  switching and routing in, 305–306
  transmission spectrum in, 240f, 241f
  tunable, in HOM fibers, 304–306
  tunable TAP, 306–309
  two by two switches with, 306f
  in WDM filters, 240–241
Loss spectrum
  in optical transparency, 2–3
  sample of, 2f
Low water peak fiber (LWPF), 2
LPG. See Long period gratings
LWPF. See Low water peak fiber

Mach-Zehnder interferometer (MZI), 240
  LPGs as, 241f
  unbalanced, 408f
Mach-Zehnder interferometer (MZI)-based all-fiber wavelength interleaver, 217–218
  schematic of, 218f
  spectral response of, 218f, 219f
Main distribution frame (MDF), 34
MCVD. See Modified chemical vapor deposition
MDF. See Main distribution frame
MEMS. See Microelectromechanical systems
Mercaptan, chemical composition of, 29f
Metro access, 18

Metro core, 18
Metro fibers, dispersion spectra of, 20f
Metro networks
  Bragg fibers for, 79–80
  fibers for, 18–20
Microbend grating spectrum, in few-mode fibers, 300f
Microbending, 3
Microelectromechanical systems (MEMS)
  component characteristics of, 366
  device fabrication of, 363
  hybrid system, 366–367
  introduction to, 353–354
  power applied to, 368f
  schematic of, 366f
  self-assembly applied to, 362
Microelectromechanical systems (MEMS) choppers
  fabrication of, 354–355
  static optical performance of, 360f
Micron-scale structure, fibers with, 41–43
Microshutters, 364f
Microstructured cladding, SEM images of, 61f
Microstructured optical fibers, 41–43, 63–64, 71
  air-light overlap in, 46
  design methodologies for, 49–50
  dispersion of, 45
  fiber drawing of, 48
  introduction to, 43–44
  nonlinearity tailoring of, 44–45
  polarization of, 45–46
  preform fabrication of, 46–48
  representative selection of, 43f
  state-of-the-art manufacturing of, 48
Modal analysis, 83
Modal delay, mode profile and, 296f
Modal power distributions, of different wavelengths, 16, 18f
Mode analysis algorithm, for FFCs, 208–210
Mode conversion, with LPGs, 294–295
Mode profile, modal delay and, 296f
Mode-locked YB^{3+} fiber seed laser, setup of, 57f
Modified chemical vapor deposition (MCVD), 297
Multiplexed fiber optic spectroscopy, chemical sensors in, 379–380
Multiplexing, coarse wavelength division, 20–21
Multipoint strain measurement, of Kowloon Canton railway train body shell, 396–398
Mutual coupling-induced aberrations, 330–331
MZI. See Mach-Zehnder interferometer

NLCM equations. *See* Nonlinear coupled mode equations
Noise figure, of L-band DFA, 156f
NOLM. *See* Nonlinear optical loop mirror
Nonlinear coupled mode (NLCM) equations, 268–269
  in FBGs, 262, 263–264
Nonlinear effects
  birefringence, 255
  Kerr nonlinearity, 253–254
  Self-steepening, 254
  stimulated inelastic scattering, 254–255
Nonlinear optical loop mirror (NOLM), 220–221
Nonlinear polarization, in optical fibers, 91
Nonlinearity
  in dispersion curves, 265–266, 267f
  introduction to, 91
  of microstructured optical fibers, 44–46
  of soft glass fibers, 58
Nonphase matched interaction, of FBGs, 236–237
Nonreciprocal phase shifts, FOG response to, 411f
Nonzero DSF's (NZ-DSFs), 11–12
  dispersion spectra of, 19f
  refractive index profile of, 11f
Normal dispersion, 13
  in Raman amplifiers, 150–151
Normalized fiber length
  pump power/inversion evolution as function of, 157f
  pump propagation/inversion map as function of, 159f
Notch filters, acousto-optic responses to, 321f
NZ-DSFs. *See* Nonzero DSF's

OFC. *See* Optical fiber communications
Office networks, POFs in, 34
Olive oil
  apparatus used for measuring, 383f
  chemical sensors and, 380–383
  PCA cluster plots for, 385f
O-MEMS. *See* Optical microelectromechanical systems
Omnidirectionality, 74
Optical amplification
  in EDFAs, 121–124
  forms of, 120f
  fundamental shape of, 356f
  population inversion and, 120–121
Optical attenuator, three-dimensional variable, 360–365
Optical channels, 8
Optical choppers
  design considerations, 356–357
  dynamic response of, 361f
  for fiber optic applications, 354–360

mechanical dynamic behavior of, 360f
  micro, 355f
  schematic of, 354f
Optical communication systems, block diagram of, 4f
Optical fiber communications (OFC), 251
  soliton-based, 251–252
Optical fibers
  chemical measurements based on, 377f
  FWM in, 98–99
  nonlinear effects in, 91
  nonlinear polarization in, 91
  photosensitivity in, 243
  pulse propagation in, 93–94
  R&D of, 1
  SPM in, 92–93
  third-order nonlinear effects in, 91
  UV sensitization of, 243–247
  XPM in, 95–98
Optical interleavers, 287–288
Optical loss, in linear effects, 252
Optical microelectromechanical systems (O-MEMS)
  design parameters of, 362–363
  experiments with, 366–368
  hybrid system, 367–368
  hybrid tunable filter in, 365–368
  introduction to, 353–354
  mechanical testing of, 357–359
  optical testing, 359–360
  shutters in, 359f
Optical modulation depth (OMD), 175
Optical networks, evolution of, 182f
Optical pulse amplification, in Raman amplifiers, 147–151
Optical pulses, dispersion-induced broadening of, 4f
Optical signal-to-noise ratio (OSNR), 127
  in EDFAs, 185–186
Optical solitons, 251
Optical time domain reflectometer (OTDR), 215, 405, 416–417
Optical transparency
  dispersion shifted fibers and, 8–9
  dispersion spectrum in, 3–8
  loss spectrum, 2–3
Optical wavelength, intermodal beatlength and, 319f
OSNR. *See* Optical signal-to-noise ratio
OTDR. *See* Optical time domain reflectometer
Outer diamters (OD), 301

PBG. *See* Photonic bandgap
PBGFs. *See* Photonic bandgap fibers
PCA. *See* Principal component analysis
PCE. *See* Power conversion efficiency
PDG. *See* Polarization-dependent gain
PDL. *See* Polarization dependent loss
Peak reflection, in FBGs, 234f

Perfluorinated benzocyclobutane (PFCB), 338
PFCB. *See* Perfluorinated benzocyclobutane
Phase matched interaction, of FBGs, 236
Phase matching, 105
  diagram, 404f
Phase velocity, 4
Phase-matching curve (PMC), 294, 296
  for LPGs, 302f
  for various fibers, 299f
Photobleaching
  dye-doped polymers, 338–343
  Gaussian feature models for, 342f
  mechanical effects of, 341
  models for, 340–341
Photonic bandgap (PBG), in FBG, 260–261
Photonic bandgap fibers (PBGFs), 42, 45, 46, 61–63, 71
  boundaries, 63f
  design methodologies for, 49–51
Photonic bandgap structures, 74f
Photonic crystal fibers, 71
Photonic crystals, 71
  schematic representations of, 72f
Photonic devices, application space covered by, 292t
Photosensitivity
  introduction to, 243
  of PMMA fiber, 37f
Physical sensing, fiber optics in, 374–376
PMC. *See* Phase-matching curve
PMD. *See* Polarization mode dispersion
PMMA. *See* Poly-methal-methacrylate
POF. *See* Polymer optical fibers
Polarization, 106–107
  of FFCs, 212
  of microstructured optical fibers, 45–46
Polarization dependence, of TAP-LPGs, 297f
Polarization dependent loss (PDL), 281
  controllers, 308–309
Polarization mode dispersion (PMD), 15, 101
  build-up of, 21f
  combating, 21–22
  in Raman amplifiers, 145–146
Polarization-dependent gain (PDG), 109
  in Raman amplifiers, 146
Poling lifetime issues, 337
Polymer fiber gratings, 36–37
Polymer optical fibers (POF), 27
  applications of, 31–36
  attenutation spectra of, 28f
  in automobile optical networks, 32, 33f
  in back plane interconnect, 32–33
  Bragg gratings in, 37f, 38f
  chemical composition of, 28f

Index

communication applications of, 32
drawing machine, 30f
extrusion method of, 30
fiber drawing of, 29–30
in home networks, 34
in illumination, 36, 37f
manufacture of, 28–30
in office networks, 34
polymerization of, 29
preform, 30f
preform and drawing method of, 28–29
segmented cladding, 37–39
silica fiber bandwidth and, 31
silica fiber minimum bend radius and, 31
silica fiber numerical aperture and, 31
silicon fiber diameter and, 30–31
types of, 27–28
Polymethal-methacrylate (PMMA), 27, 338
chemical composition of, 28f
deuterated, 27
fiber, 27
perfluorinated, 27–28
photosensitivity of, 37f
Population inversion, optical amplification and, 120–121
Power conversion efficiency (PCE), 155, 157
of EDFs, 159
Power splitters, optical characteristics of, 208t
Power transients
in EDFA chains, 190–192
for EDFAs, 189f
Preforms
extruded SF57, 47f
for microstructured optical fibers, 46–47
Principal component analysis (PCA), 380–381, 385f
Pseudowaveguide, in FFCs, 211f
Pulse propagation
in FBG, 259–260
in optical fibers, 93–94
in Raman amplifiers, 147–148
in SPG, 258f
Pump control
in EDFAs, 196–198
surviving channel power transient and, 198f
Pump evolutions, of FRAs, 166f
Pump power
in EDFAs, 197f
gain spectrum and, 167f, 168f
Pump wavelength detuning, in L-band EDFA, 156–157
Pump-noise transfer, in Raman amplifiers, 143–145

Quadratic media, gap solitons in, 271–272

Quartz fiber
flexural mode in, 315f
longitudinal modes, 314
phase velocity of torsional modes, 314

Radial effective index method (REIM)
determination of, 84–85
formulation of, 84
HFs and, 87–88
introduction to, 83–84
SCF and, 85–87
Radiation loss spectrum
of Bragg fibers, 80t
TE mode, 79f
Radiation-induced chemical reactions, theory of, 339–340
Raman amplifiers, 155
anomalous dispersion in, 149–150
ASE in, 139, 141
broadband, 136–137
closed form equation, 160
design of, 137–138
DRB in, 143f
DRS in, 142
effective noise figures in, 140–142
formalism construction in, 160–162
frequency shifts and, 148f
FROG trace of signals from, 151f
fundamental concepts of, 131–132
gain engineering of, 160–169
gain saturation in, 135–136, 136
group-velocity mismatch in, 148–149
introduction to, 160
modern, 136
multiple pump, 137
normal dispersion regime and, 150–151
numerically simulated composite gain, 137f
on-off gain for pumping configurations, 163f
optical pulse amplification in, 147–151
optical SNR in, 140f
parabolic shapes and, 151f
PDG in, 146
performance limiting factors of, 138–146
PMD in, 145–146
pulse-propagation equations in, 147–148
pump-noise transfer in, 143–145
Rayleigh backscattering in, 142–143
RIN in, 144f
schematic illustration of, 132f
signal evolution in, 164f
signal power in, 135f
simple theory behind, 134–135
spectral shift and temporal delay, 150f
spontaneous Raman scattering in, 138–140

spontaneous spectral density and, 140f
total noise figure and, 142f
Raman effects, 53–54, 58, 131
Raman equation, closed form of, 160
Raman fiber amplifier (RFA), 119
Raman gain spectrum, 132–134
for bulk silica, 133f
for different fibers, 133f
numerically simulated composite, 137f
as signal wavelength function, 139f
Rayleigh backscattering
in FRAs, 165f
in Raman amplifiers, 142–143
Rayleigh scattering, 138
Recirculating catalyst, in UV-sensitized fibers, 247
Red-shifts, 13
Reflection, in FBGs, 261–262
Refraction, stress-modified indices of, 342–343
Refractive index
for EDFA, 128f
in FBG fabrication, 244f
in FBGs, 262–263
in H_2 fibers, 248f
layers of, 235f
periodic distribution of, 71
periodic variation of, 85f
REIM. See Radial effective index method
Relative intensity noise (RIN), in Raman amplifiers, 144f
Remote optical fiber spectral measurements, 379f
Research and development (R&D), 1
Residual fiber birefringence, FOPA and, 111–114
Resonant modes, 356t
Reverse/inverse dispersion fibers, 17–18
Ring laser setup, 367f
Root mean square (RMS), 97
variation of, 97f

SAN. See Storage access networks
S-band, EDFAs for, 128
SBS. See Stimulated Brillouin scattering
Scanning electron microscopes (SEMs), 42
Scanning heterodyne
interferometer, in acousto-optic interaction, 320–321
SCF. See Segmented cladding fiber
SCM. See Subcarrier multiplexing
SDH. See Synchronous digital hierarchy
Segmented cladding fiber (SCF)
cross-section of, 85f
effective index profile of, 86f
intensity distribution of, 86f
leakage losses and, 86f
REIM and, 85–87

Self-phase modulation (SPM), 91, 135, 147, 187
  normalized pulse widths in, 96f
  in optical fibers, 92–93
  pulse evolution in, 95f
  spectral broadening due to, 94–95
Self-steepening, 254
Semiconductor optical amplifier (SOA), 119, 221
SEMs. See Scanning electron microscopes
Short period grating (SPG), pulse propagation in, 258f
Shutter plane, light intensity in, 358f
Shutter systems, for optical modeling, 357f
Side-polished evanescently coupled optical fiber overlay devices
  applications of, 229–230
  devices, 227–229
  introduction to, 225–226
  principles of operation, 226–227
Signal attenuation, schematic representations of, 120f
Signal gain, variation of, 123f, 124f
Signal-to-noise ratio (SNR), 119, 140
  optical, in Raman amplifiers, 140
Silbene, energy level diagram for, 341f
Silica fiber
  effective mode area of, 44f
  modal intensity of, 55f
  optical solitons in, 255–257
  polymer fiber bandwidth and, 31
  polymer fiber diameter and, 30–31
  polymer fiber minimum bend radius and, 31
  polymer fiber numerical aperture and, 31
Silica glass, 252
Silica/air microstructured fiber, 42
Silicon oxynitride (SiON), 243
Single pump parametric amplifiers, 103–107
  average gain spectra, 112f
  gain spectra for, 104f
  measured signal gain for, 105f
  optimized gain spectra for, 106f
  parametric gain for, 103f
Single-mode fibers (SMF), 1, 291
  dispersion spectra of, 8f
  high-silica matched clad, 6f
  Raman gain spectrum in, 133
  wavelength dependence of group delay of, 7f
Small core fibers
  background of, 51–52
  design considerations, 52–53
  device demonstrations of, 53–54
SMF. See Single-mode fibers
SNR. See Signal-to-noise ratio
SOA. See Semiconductor optical amplifier
SOFO, 376

schematic of, 377f
Soft glass fibers
  background of, 58
  extreme nonlinearity of, 58–59
  new transmission fibers, 59–60
  solid microstructured fibers and, 60–61
Solid microstructured fibers, soft glass fibers and, 60–61
Soliton pulses, distortionless propagation of, 94f
Solitons. See also Gap solitons
  Bloch wave analysis of FBG, 272
  experimental considerations of, 278
  in FBGs, 266–267
  intensity profiles of, 256f
  linear effects in, 252–253
  nonlinear effects in, 253–255
  in pure silica fiber, 255–257
SOP. See State of polarization
Spectral efficiency, 10
Spectral hole burning, 200
Spectral shifts, Raman-induced, 150f
SPM. See Self-phase modulation
Spontaneous Raman scattering, 138–140
SRS. See Stimulated Raman scattering
Standard single-mode (SSM) telecommunication fiber, 243
Star coupler, 325–327
  illustration of, 326f
  segmentation in, 327f
State of polarization (SOP), 107, 113
  gain vs. signal wavelength for, 114f
Static load tests, 397f
Static mode couplers, principles of, 318f
Stimulated Brillouin scattering (SBS), 106, 254
  distributed sensing using, 374f
Stimulated inelastic scattering, 254–255
Stimulated Raman scattering (SRS), 91, 254
  introduction to, 131
  schematic illustration of, 132f
Stokes line, 131, 136
Storage access networks (SAN), back plane interconnection in, 34f
Strain measurement
  in composite-strengthened concrete bar, 390–391
  of fiber grating sensors, 406t
Strain transfer, in transducers, 375f
Stress-modified indices, of refraction, 342–343
Subcarrier multiplexing (SCM), 173
Super fluorescent sources, output spectra of, 231f
Supermodes, FFC, 211
Surface index relief gratings, 345–346
Synchronous digital hierarchy (SDH), 8

TAP. See Turn-around point
Tap/access couplers, 215
TDFA. See Thulium-doped fiber amplifiers

TE mode. See Transverse electric mode
Tellurite glasses, 59
Temperature measurement, 392–393
  experimental setup for, 394f
Temporal delay, Raman-induced, 150f
Tensile stress, 341–342
Terahertz optical asymmetric demultiplexer (TOAD), schematic configuration of, 221f
Thermal stability, of Bragg gratings, 245
Thin film filters, in DWDMs, 282–284
Thin sinusoidal gratings, diffraction efficiency in, 343–344
Third-order dispersion (TOD), 261
  grating solitons due to, 275–278
Third-order nonlinear effects, in optical fibers, 91
3-dB couplers, 214–215
Thulium, energy levels of, 169f
Thulium-doped fiber amplifiers (TDFA), 155
  introduction to, 169
  inversion analysis of, 169–171
  setup of, 170f
  surviving channels of, 170f, 171f
Tip displacement, bimorph beams and, 363f
TM mode. See Transverse magnetic mode
TOAD. See Terahertz optical asymmetric demultiplexer
TOD. See Third-order dispersion
Trailing edges, 3
Transducers, 378
  frequency and impulse response for, 317f
  strain transfer in, 375f
Transient gratings, 345–346
Translation matrix, 73–74
Transmission coefficients, in FBGs, 261–262
Transmission spectrum
  of 50-GHz three-port device, 284f
  of 100-GHz 4-skip-0 device, 283f
  of FBGs, 237f
  in LPGs, 240f, 241f
Transmissivity calculation, 328f
Transport, DWDM in, 18
Transverse electric field, 312f
Transverse electric (TE) mode, 76, 225
  dispersion spectrum of, 79f, 80f
  electric field components of, 77f
  Hz field variation of, 77f
  radiation loss spectrum of, 79f
Transverse magnetic (TM) mode, 76, 225
  electric field components of, 77f
Tree couplers, 215
True-Wave reduced slope, 301
Tunable filters, in acousto-optic interaction, 318–319
Turn-around point (TAP), 296
  bandwidth control of, LPGs, 298–299
  detuning effects in, LPGs, 304–305

Index

LPGs as sensors, 305f
polarization control devices with, LPGs, 308f
polarization dependence of, LPGs, 297f
static devices using, LPGs, 300–304
switching and routing in, LPGs, 305–306
tunable, LPGs, 306–309
two by two swtiches with, LPGs, 306f

Ultrasound, FBGs and, 376f
Untwisted pairs (UTP), 34
UTP. *See* Untwisted pairs
UV-sensitized fibers, 243–247
absorption spectra of, 244f
continuous isochronal annealing experiments on, 246f
Ge concentration and, 244–245
H_2 sensitization of, 247–250
recirculating catalyst in, 247

Variable optical attenuators (VOAs), 291
attenuation characteristic of, 365f
MEMS, 362
microshutters of, 364f
performance of, 363–364
spectrally flat, 299–300

VCO. *See* Voltage-controlled oscillator
VOAs. *See* Variable optical attenuators
Voltage-controlled oscillator (VCO), 390

Walk-off length, 147
Waveguide grating router, 327–330
designs for, 329f
illustration of, 328
spectral sampling in, 332f
Waveguide polymers
irreversible gratings in, 347–348
writing gratings in, 344–345
Waveguides, 325
Wavelength dependence, in acousto-optic interaction, 318–319
Wavelength division multiplexing (WDM), 34–35, 119, 133–134, 136–137, 155, 181, 187, 189, 192, 205, 405
dropping, channels, 190f, 191f
introduction to, 91
mapping, couplers, 214f
operation of, 216
transients and, 193
for two wavelengths, 216f
wavelength response of, couplers, 214f

Wavelength division multiplexing (WDM) couplers, 215–216
Wavelength division multiplexing (WDM) filters, LPGs in, 240–241
Wavelength interleaver, 216–217
schematic representation of, 217f
Wavelength, polarization, and fabrication (WPF), 325, 328
Wavelength tunable devices, 228–229
Wavelengths
dependence, 7f
Gaussian pulses at, 13f
power modal distributions of, 16, 18f
zero dispersion, 6
WDM. *See* Wavelength division multiplexing
WPF. *See* Wavelength, polarization, and fabrication

XPM. *See* Cross-phase modulation

ZDWL. *See* Zero dispersion wavelengths
Zero dispersion wavelengths (ZDWL), 6, 101, 104, 107, 108, 114
fluctuations of, 109–111
gain spectra and, 111f